ENCYCLOPEDIA OF PHYSICS

EDITED BY

S. FLÜGGE

VOLUME XXXVII/1

ATOMS III – MOLECULES I

WITH 215 FIGURES

SPRINGER-VERLAG BERLIN HEIDELBERG GMBH

1959

HANDBUCH DER PHYSIK

HERAUSGEGEBEN VON

S. FLÜGGE

BAND XXXVII/1

ATOME III – MOLEKÜLE I

MIT 215 FIGUREN

SPRINGER-VERLAG BERLIN HEIDELBERG GMBH
1959

ISBN 978-3-642-45919-1 ISBN 978-3-642-45917-7 (eBook)
DOI 10.1007/978-3-642-45917-7

Ursprünglich erschienen bei Springer-Verlag OHG.
Berlin · Gottingen · Heidelberg 1959
Softcover reprint of the hardcover 1st edition

Contents.

Atomic and Molecular Beam Spectroscopy.

By

P. Kusch and V. W. Hughes.

With 110 Figures.

1. Introduction. Experimental procedures in which the generation and observation of an atomic or molecular beam is an important part have been applied to a large range of problems. The older work [*7*] to [*9*], [*25*] included studies of gas kinetics, molecular scattering cross sections in gases, the interaction of molecules with solid surfaces, and the diffraction of molecular beams. Space quantization of magnetic moments was demonstrated, and the magnetic and electric properties of molecules, atoms, and nuclei were determined by studies of the deflection of neutral atoms or molecules in magnetic and electric fields. High resolution optical spectroscopy utilizes a beam of particles in order to avoid Doppler line broadening[1].

In most of the recent and present applications of molecular and atomic beams in physics, the apparatus is used as a spectrometer of extremely high resolution to measure directly the frequency of transition between energy levels. The range of frequencies over which the spectrometer operates extends substantially from zero as a lower limit to an upper limit determined by the upper frequency limit of available generators of electromagnetic radiation, which corresponds at present to a wavelength of about 1 mm. The frequency range within which the beam spectrometer operates leads to the investigation of a range of phenomena which is not generally open to investigation by the methods of optical spectroscopy, but which does overlap the phenomena observed by the methods of microwave absorption spectroscopy and nuclear resonance spectroscopy.

The purpose of any spectrometer is to detect and measure the emission and absorption lines which occur in the transition of an atom or molecule from one energy level to another. In classical optical spectrometers this is achieved in familiar ways by the use of the light-dispersing properties of optical media or the diffraction properties of light in gratings and interferometers. In all these methods an analysis is made of the light itself. Further, in certain recently developed methods of spectroscopy, as microwave absorption methods or the methods used in the observation of nuclear resonance phenomena, an observation is made of the absorbed or emitted radiation. The method of atomic and molecular beams is unique as a spectroscopic device in that an observation is made of the changed properties of the atoms or molecules consequent to a transition, and not, directly, of the absorbed or emitted radiation.

The emphasis in much of modern beam spectroscopy has been on the study of the properties of nuclei and of the electron. Nuclear magnetic moments and hyperfine structure separations in atoms and molecules due to the nuclear moments—magnetic dipole, electric quadrupole, and magnetic octupole—have been

[1] S. Tolansky: High Resolution Spectroscopy. New York and Chicago: Methuen and Co., Ltd. London 1947.

measured. Nuclear masses have been determined from the study of rotational state transitions of diatomic molecules. The Lamb shift and the anomalous magnetic moment of the electron were discovered by the method of atomic beam spectroscopy, and extensive work has been done on these subjects. In addition, much information on detailed atomic and molecular properties has been obtained.

The present article deals primarily with atomic and molecular beam spectroscopy. The first chapter on methodology includes discussions of the production, characteristics, detection, and deflection of beams of neutral particles, and of the theory and experimental technique associated with the transition process. The second and third chapters deal with atomic and molecular spectroscopy respectively. They include a summary of the theory of the energy levels studied, a discussion of the type of measurements that have been made and of the present limitations to the accuracy of the experiments, and a compilation of the more important data obtained.

A. Methodology.

2. General principles and characteristics of spectrometer operation. The operation of an atomic or molecular beam apparatus as a spectrometer depends essentially on the existence of some observable differential property of the particle in the two states a and b between which a transition is to be observed. Usually this property is the magnetic or the electric dipole moment, and the change in the moment associated with the transition is observed through a change in trajectory of the particle. Trajectories which depend on the magnetic or electric dipole moment of a particle are achieved by allowing the particle to pass through inhomogeneous magnetic or electric fields in which its deflection depends on the value of the moment. In the Lamb shift and related experiments on hydrogen and ionized helium in the $2^2S_{\frac{1}{2}}$, $2^2P_{\frac{1}{2}}$, and $2^2P_{\frac{3}{2}}$ states, detection of transitions is achieved by use of the differential metastability of the magnetic sublevels of the $2^2S_{\frac{1}{2}}$ state in certain magnetic fields.

A schematic diagram of the usual beam spectrometer is shown in Fig. 1. The source of an atomic or molecular beam is an enclosure containing gas or vapor which is allowed to effuse through a slit. An additional collimating slit defines the beam, which ordinarily has a thin ribbon-like cross section. Inhomogeneous magnetic or electric fields occur in the regions A and B in which the particles are deflected in a direction perpendicular to the plane of the beam. The C-region contains the oscillating electromagnetic field, which may cause the transition of the particle from one state to another. Some appropriate detector of neutral particles is placed beyond the end of the B-region. If a particle emerges from the source in a state a and if no transition to another state b occurs in the C-region, the particle may follow a trajectory as shown and finally strike the detector. If, on the other hand, the oscillating field in the C-region is near the Bohr frequency $\nu=(W_b-W_a)/h$ and if the amplitude and polarization of the field are suitable, then the particle may undergo a transition to state b because of the absorption or stimulated emission of radiation. In the state b the particle will experience a different deflection in the B-region, provided that the moment is different than in the state a, and hence will not strike the detector. Thus the transition is observed as a decrease in beam intensity at the detector. A transition from the state b to the state a may in general be observed in a similar manner. In addition to the oscillating electromagnetic field, the C-region usually contains a constant and homogeneous magnetic or electric field, which may be required to develop a splitting of the energy levels of interest and to maintain the particle in a definite quantum state throughout the apparatus in the absence of an induced transition. Many variations of this classic beam spectrometer have been used and will be discussed in this article.

The usual forms of microwave spectroscopy and nuclear resonance spectroscopy, in which the effect of the particles on the radiation is observed, depend for their signal on an excess number of particles in the low energy state at thermal equilibrium, because the probability for spontaneous emission is negligible for the transitions studied and the probabilities for absorption and stimulated emission are equal. In beam spectroscopy, on the other hand, transitions both from state a to state b and from state b to state a result in a decrease in the number of particles reaching the detector. Thus the signal depends on the total number of particles in which the change in state has occurred, regardless of the net absorption or emission of energy. Since the difference in population of the two

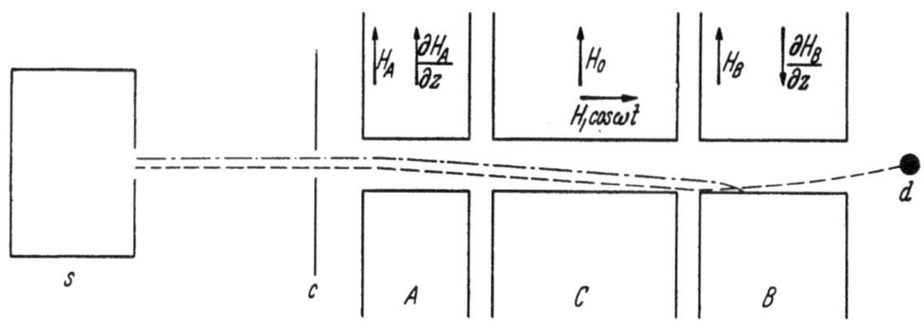

Fig. 1. Schematic diagram of beam spectrometer. s is the source chamber with source slit. c is the collimator slit. The A and B regions are regions with inhomogeneous magnetic or electric fields. The C region contains the oscillating electromagnetic field and, in addition, a constant, homogeneous magnetic or electric field. (Magnetic fields are indicated in the diagram.) d is the detector. The dashed curve (— — —) is the trajectory of a particle which remains in state a throughout the apparatus. The dot-dash curve (— · — · —) is the trajectory of a particle which undergoes a transition from state a to state b in the C-region.

states is ordinarily very small, this characteristic of beam spectroscopy is a distinct advantage.

The line widths in beam spectroscopy are generally determined by the natural width associated with the HEISENBERG uncertainty principle or by inhomogeneity of the constant magnetic or electric field in the C-region. Collision broadening due to other particles or due to the walls is completely negligible, and Doppler broadening can usually be avoided because of the unidirectional character of the beam trajectory. Thus beam spectroscopy has yielded many of the highest precision measurements of transition frequencies.

I. The beam.

a) Sources.

3. Introduction. Gas kinetics. In general the range of problems which have been investigated by the methods of atomic and molecular beams has been much more limited by the restrictions imposed by the detector than by the difficulty of constructing adequate sources of beams of molecules and atoms. Recently, however, particularly in work with radioactive substances where the amount of material available is usually small, the source problem has been severe. Further, the number of sources that are required has been greatly extended by recent developments in detectors.

The source of an atomic or molecular beam is an enclosure filled with gas or vapor and provided with a slit through which a stream of particles effuses. Some of the gas kinetic considerations will be presented in this section. The following sections will deal with evaporation of solids, gas sources, sources of atoms in metastable and in optically excited states, and sources of radioactive atoms.

If a gas is in thermal equilibrium with its surroundings, then the number of particles per unit volume which lie within the velocity interval between v and $v+dv$ is given by the Maxwell distribution law:

$$f(v)\,dv = \frac{4n}{\sqrt{\pi}\,\alpha^3}\,v^2\,e^{-v^2/\alpha^2}\,dv \qquad (3.1)$$

where $f(v)$ is the velocity distribution function, n is the number of particles per unit volume, and $\alpha = \sqrt{2kT/m}$ in which k is BOLTZMANN'S constant ($k = 1.38 \times 10^{-16}$ erg deg^{-1} K), T is the absolute temperature, and m is the mass of the particle. The quantity α is the velocity for which $f(v)$ is a maximum, i.a., the most probable velocity. The average velocity, $\bar{v}$, is given by: $\bar{v} = 2\alpha/\sqrt{\pi} = 1.13\,\alpha$.

Suppose that the gas is contained in an oven, essentially an isothermal cavity, which has a small aperture of area A_s. The aperture is in a wall of infinitesimal thickness and its area is so small that the Maxwellian velocity distribution within the oven is not affected by effusion from the aperture. For this condition of ideal effusion the number of particles, N, which escape from the oven into a vacuum per unit time is:

$$N = \tfrac{1}{4}\,n\,\bar{v}\,A_s. \qquad (3.2)$$

The expression for N may be rewritten:

$$N = \frac{N_0\,p\,A_s}{\sqrt{2\pi R M T}} = \frac{p\,A_s}{\sqrt{2\pi k T m}} \qquad (3.3)$$

where N_0 is AVOGADRO'S number ($N_0 = 6.02 \times 10^{23}$), R is the gas constant per mole ($R = 8.32 \times 10^7$ ergs deg^{-1} K), T is the source temperature in °K, M is the molecular weight, and p is the pressure in dynes/cm². In a typical case the quantities may have the approximate values $p = 10^3$ dynes/cm², $M = 40$ (LiCl, NaF, K), $T = 1000°$ K, $A_s = 10^{-3}$ cm²; hence N will be 1.3×10^{17} particles/sec.

Particles leave the oven according to the cosine law and the number of particles emitted at an angle ϑ from the normal to the aperture and within a solid angle $d\Omega$ is:

$$N(\vartheta)\,d\Omega = \frac{n\,\bar{v}\,A_s}{4\pi}\cos\vartheta\,d\Omega. \qquad (3.4)$$

For the ovens used in atomic and molecular beams, the angle ϑ is approximately zero.

The above formulae are strictly applicable only for ideal effusion. If the source involves evaporation of a solid, then the surface area from which evaporation occurs must be large compared to the aperture area in order that ideal effusion shall occur. If the source is a gas, the effusion rate must be small compared to the rate at which the gas attains thermal equilibrium with its containing vessel. In practice these conditions are ordinarily fulfilled. A not-unrelated requirement for ideal effusion is that the mean free path of the particles in the enclosure be large compared to the slit width. If this were not so, particles emerging from the slit in the direction to constitute the beam would suffer collisions in the neighborhood of the slit and a particle cloud formation exterior to the enclosure would result[1]. The width of the beam would then be increased and not determined simply by the source slit width. It has been found satisfactory in customary practice to have the mean free path of the order of the slit width, and a typical order of magnitude figure for the mean free path—applicable to potassium, for example—is 0.1 mm at a source pressure of 1 mm Hg.

[1] F. KNAUER and O. STERN: Z. Physik 39, 764 (1926).

Source slits with a finite thickness (canals) are unavoidable in practice and have found useful application because of the more directional beams they produce. Provided the canal length is small compared to the mean free path of particles in the source, which is the condition of practical interest, the characteristics of the forward beam remain unchanged by the thickness of the source slit. However, effusion from a canal does not occur according to the cosine law as for the ideal thin aperture, but rather is preferentially in the forward direction. Hence for the same forward intensity the total number of particles emitted from an ideal aperture is greater than the number of particles emitted from a canal. If G is defined as the ratio of these numbers, then G is given by:

$$G = \frac{3l}{4w} \tag{3.5}$$

where l is the length and w is the width of a canal with square cross section. Measurements as indicated in Fig. 2 have shown this expression to be roughly correct provided that the pressure in the oven is sufficiently low (less than 3×10^{-3} mm Hg for the oven in question) so that the mean free path is larger than the length of the canal[1]. The intensity and angular distribution of effusion from a canal are reviewed by FRASER [8] and RAMSEY [25].

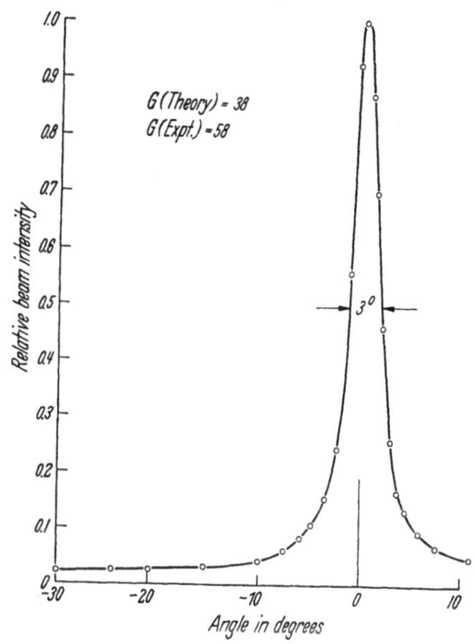

Fig. 2. Relative beam intensity as a function of angle of beam with respect to axis of source canal. The canal dimensions are length = 6.3 mm, width = 0.12 mm, and height = 0.12 mm. The quantity G is a measure of the directivity of the beam.

If the distribution in the source is Maxwellian at a temperature T, then the number of particles effusing from an ideal aperture per unit time within the velocity interval from v to $v + dv$ is:

$$N(v)\,dv = \frac{2N}{\alpha^4} v^3 e^{-v^2/\alpha^2}\,dv. \tag{3.6}$$

Hence the v^2 factor multiplying the exponential in the Maxwellian distribution which describes the velocity distribution among all the particles in the cavity is altered to v^3 for describing the number of particles that effuse per unit time and lie in the velocity interval dv. The most probable velocity for a particle in the beam is $\sqrt{\frac{3}{2}}\,\alpha = 1.22\,\alpha$ and the mean velocity is $\left(\frac{3}{4}\right)\sqrt{\pi}\,\alpha = 1.33\,\alpha$.

4. Evaporation of solids. Ovens to evaporate a material which is a solid of low vapor pressure at room temperature have been the most common source of molecular and atomic beams. The oven must have a well which holds the material from which evaporation is to occur and which is sealed from the surrounding vacuum except for a slit. Heaters must be supplied to maintain the oven at an appropriate temperature, and it is convenient to have a thermocouple or other device for the measurement and control of temperature. The oven material must be such that no significant reaction or alloying between the oven and its load occurs, since a reaction may lead to an excessive rate of loss of material, an excessive rate of evolution of gas, or a destruction of essential geometrical detail such as slits.

[1] DAVIS, NAGLE and ZACHARIAS: Phys. Rev. **76**, 1068 (1949).

Iron and nickel have been found to be satisfactory oven materials for a large fraction of the substances of which molecular beams have been made. In special cases ovens have been made of molybdenum, graphite, silver, copper, and other materials.

A typical oven is shown in Fig. 3. The oven slit jaws are knife edges which are somewhat easier to use than slits made of thin foils. Any distortion of the idealized beam velocity distribution of Eq. (3.6) which is caused by the thickness of the knife edges is not important in most spectroscopic applications. The load is contained in the well and is inserted through a conical hole in the top of the oven which is sealed by a fitted conical plug forced into the hole under considerable pressure. The well communicates with the front of the oven through an approximately horizontal channel whose diameter determines the height of the beam at the slit. The front face of the oven and the bearing surfaces of the slits are lapped. The slits are screwed to the oven and there is a negligible leakage of material through the interface.

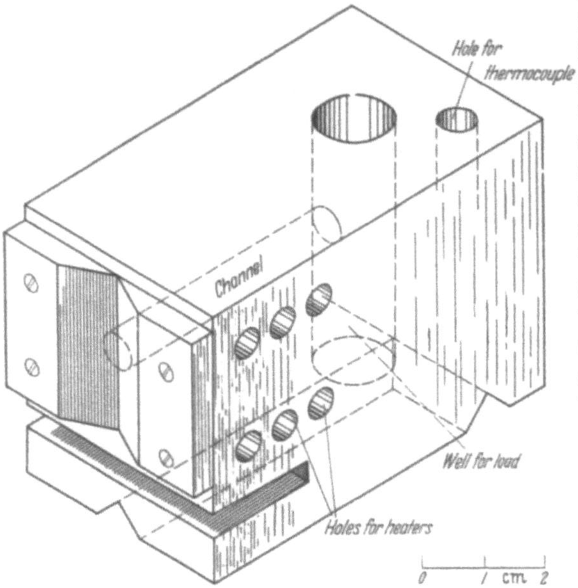

Fig. 3. Typical oven for evaporation of a substance that is solid at room temperature.

In most cases, it is necessary to have only an approximate knowledge of the oven temperature, and it has been found of value to arrange the heating elements in such a way that the slit is raised to a slightly higher temperature than the melt. This procedure has the obvious advantage that condensation on the slit is avoided. It also aids in removing material from the slits which may have been thrown there during the initial heating of the load when the evolution of gas occluded in the load is rapid. In the oven shown in Fig. 3 the preferential heating of the slit is effected by the grouping of the heaters near the slit and by the horizontal cut underneath the front part of the oven. More extreme differential heating can be achieved by a reduction of the cross section connecting the front of the oven which contains the heater and the back of the oven which contains the melt.

The oven is mounted on three tapered tungsten pins which fit into two perpendicular V-shaped notches underneath the oven. The mechanical stability of the oven and the reproducibility of its position are excellent. The heat conduction through the pins can be made quite small. The heaters themselves are usually coils of tungsten wire placed into insulating (ceramic or quartz) tubes.

Ovens of the type shown in Fig. 3 have proved to be satisfactory for the production of beams of all the alkali atoms and alkali halide molecules as well as certain other substances. In the evaporation of LiI it has been found that there is some apparent production of free iodine, which reacts with the iron, especially at the sharp slit edges with a consequent plugging of the slit. A gold plate of moderate thickness on the oven and especially on the slits improves the performance very markedly. It should be noted that Rb and Cs combine with air with explosive violence. These metals may be distilled under vacuum from a mixture of calcium and the alkali chloride into glass capsules which are then sealed off. A glass capsule is placed in the oven in an inert atmosphere. When the plug is inserted the capsule breaks, and there is not enough diffusion of air into the oven through the narrow slit to cause excessive oxidation[1].

Another useful method for producing beams of atomic Cs and Rb is by the so-called reaction technique[2]. A mixture of powdered metallic calcium and the alkali chloride is heated to a temperature of about 600° C and the reaction yielding atomic Cs or Rb then proceeds at a useful rate. An iron oven is entirely satisfactory. There is evidently no hazard of an excessive

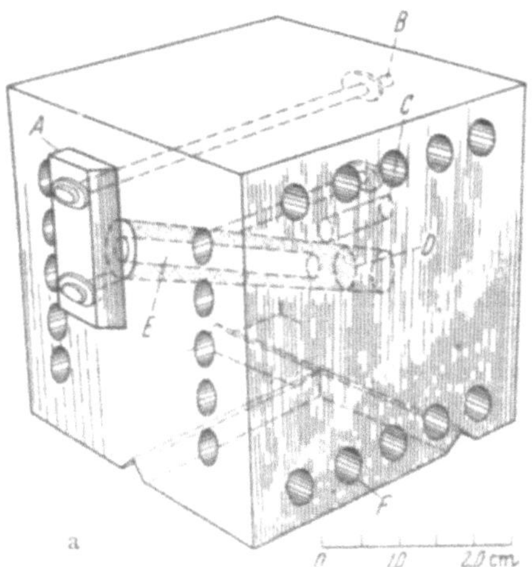

a

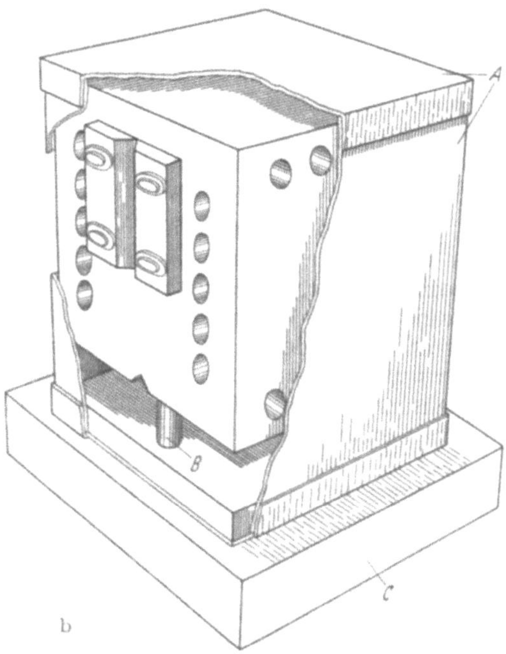

b

Fig. 4a and b. (a) Molybdenum oven for producing gallium beam. *A*, graphite slit jaw; *B*, molybdenum bolts; *C*, thermocouple well; *D*, graphite insert; *E*, well for gallium drilled into graphite insert; *F*, heater hole (total of 20). (b) Oven mount and radiation shield. *A*, molybdenum radiation shield; *B*, front tungsten pin for supporting oven; *C*, lava block.

reaction with air when the oven is exposed to air for reasonable lengths of time. It should be noted, however, that the beam produced by the reaction technique is characterized by a higher velocity than is a beam produced by simple evaporation, and this leads to a reduction of the resolving power of a transition field of fixed length.

Beams of TlCl, the alkali cyanides, the alkali borates, atomic barium, atomic thallium, and a number of salts of the type $NaCl \cdot AlCl_3$ and $KF \cdot BeF_2$ have also been successfully evaporated from these ovens. A typical oven for producing an atomic indium beam is a molybdenum block with the well for the load drilled diagonally down. Molybdenum slit jaws are also used. This well design has been found to be superior to the type shown in Fig. 3 which has a separate opening for loading, because it is easier to make one exit tight to indium vapor than two[3]. An oven for producing an atomic gallium beam (see Fig. 4) differs from an indium oven in that a graphite rod with a hole drilled for the load is inserted into

[1] S. Millman and M. Fox: Phys. Rev. **50**, 220 (1936).
[2] S. Millman and P. Kusch: Phys. Rev. **60**, 91 (1941).
[3] Millman, Rabi and Zacharias: Phys. Rev. **53**, 384 (1938).

5. Gas sources. Extensive work has been done on the radiofrequency spectra of the hydrogen molecules H_2, HD, and D_2. In addition, the radiofrequency spectrum of the molecule O_2 has been investigated. The observations have been made with the gas source near the temperature of liquid nitrogen. This serves the primary purpose of keeping the large fraction of molecules in states of low rotational quantum number, thus making it possible to observe individual lines in the spectrum with a relatively high intensity. It serves the further purposes of reducing the natural widths of the lines, which depend on the transit time of molecules in the transition field, and of increasing the deflecting power of the inhomogeneous magnetic fields. The technique of producing a beam of a gas at low temperature consists simply of attaching to a trap filled with liquid nitrogen a ther-

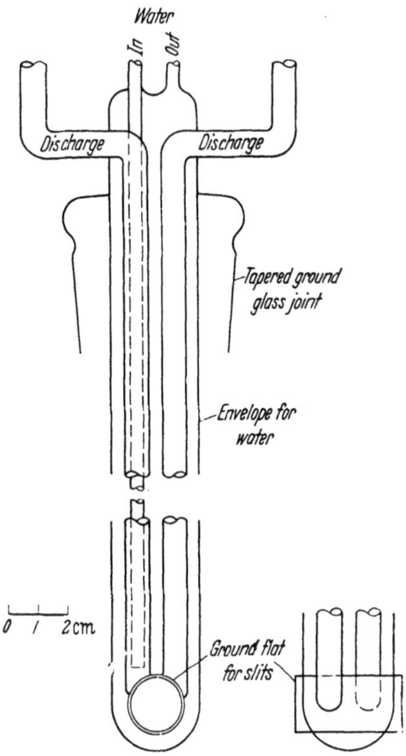

Fig. 7. Wood's discharge tube for producing atomic hydrogen.

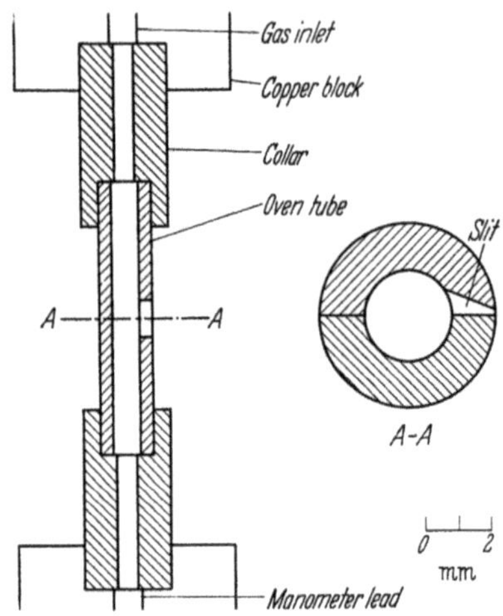

Fig. 8. Oven for producing a beam of atomic hydrogen. The collar and oven tube are of tungsten.

mally conducting cavity into which the gas is admitted and of closing an aperture in the cavity with a slit of suitable dimensions[1].

Numbers of investigations have been made with beams of atomic hydrogen, deuterium, and tritium. The most common method of producing these atoms is by use of a Wood's discharge tube[2]. A long discharge tube has an electrode fitted at both ends, and the molecular gas is admitted to one or both electrode regions from leaks fed from an appropriate gas handling system. The hydrogen reservoir, at approximately atmospheric pressure, is saturated with water vapor since experience has indicated that the presence of water vapor greatly increases the yield of atoms. The central portion of the discharge tube is open and the aperture is closed by a pair of thin glass plates with ground edges to form a slit. The plates may successfully be attached with a wax such as sealing wax. A schematic diagram of a typical tube is shown in Fig. 7. Any foreign material such as metal particles, glass fragments, wax, or dust within the path of the discharge or

[1] Kolsky, Phipps, Ramsey and Silsbee: Phys. Rev. **87**, 395 (1952).
[2] Kellogg, Rabi and Zacharias: Phys. Rev. **50**, 472 (1936).

within a region which is readily reached by the atomic hydrogen greatly reduces the yield of atomic hydrogen through an increased recombination. A glow discharge is maintained by d.c. or a.c. excitation, usually with a voltage of 2000 to 3000 volts and a current of 50 to 100 ma. The slit is located in the plasma region of the discharge. Efficient production of atoms occurs when the spectrum of the discharge consists almost entirely of the Balmer lines with a very low background of the molecular spectrum. It is possible to obtain a beam in which more than 90% of the particles which emerge from the slit are atoms.

HENDRIE[1] has produced a sharply defined beam of hydrogen atoms by a thermal dissociation method[2] (see Fig. 8). Two half cylinders of tungsten rod are each ground on the inside to give semi-cylindrical shells. The wall of one of the shells is then ground away to form a slit of suitable width and height.

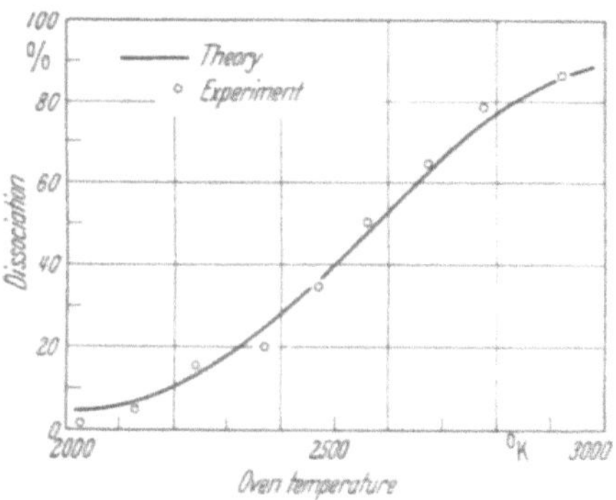

Fig. 9. The dissociation of molecular hydrogen into atomic hydrogen as a function of the source temperature.

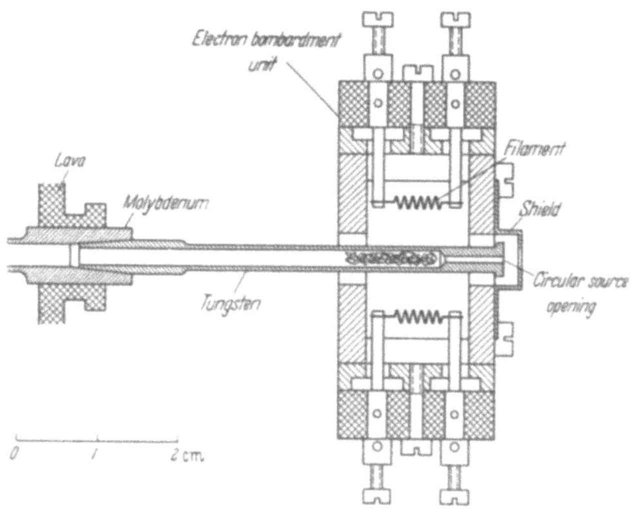

Fig. 10. Electron bombardment source for producing a stream of atomic hydrogen by dissociation of molecular hydrogen.

The two shells are then forced into tungsten tubes at each end and the whole unit may be heated by a high current. A current of about 100 amp, a voltage drop of 2 volts, and a hydrogen pressure of 1 mm Hg represent typical operating conditions to obtain a high fractional dissociation. Hendrie has made an accurate measurement of the fractional dissociation of molecular hydrogen as a function of temperature by making use of the highly differential magnetic moment of atomic and molecular hydrogen. The dissociation curve of hydrogen is shown in Fig. 9. The theoretical curve is computed from the known dissociation energy and entropy of the hydrogen molecule. The same technique applied to molecular nitrogen gave no evidence of thermal dissociation at temperatures up to 3500° K, where the tungsten oven failed. This result enables one to eliminate the lower of two values for the dissociation energy of nitrogen which the interpretation of the molecular spectrum of N_2 would allow.

An electron bombardment oven for dissociating molecular hydrogen is shown in Fig. 10[3]. The tungsten tube through which the hydrogen gas flows is maintained at ground potential, and the remainder of the oven structure including the filament is maintained at a negative potential of about 2000 volts. A current of about 100 ma is used. Pieces of tungsten wire are introduced into the tungsten

[1] J.M. HENDRIE: J. Chem. Phys. **22**, 1503 (1954).
[2] W.E. LAMB and R.C. RETHERFORD: Phys. Rev. **81**, 222 (1951).
[3] HEBERLE, REICH and KUSCH: Phys. Rev. **101**, 612 (1956).

tube to achieve thermal equilibrium between the gas and the metal. High fractional dissociation has been achieved. The oven shown has a small circular opening and is used in an application where a very narrow beam of considerable height is not required.

An electrodeless discharge through a gas in a glass tube in a microwave resonant cavity has been found to be an effective source of atoms[1]. The method is described for producing atomic chlorine (see Fig. 11). A glass or quartz tube is placed at a voltage antinode in the cavity so that a glow discharge occurs in the gas for about a centimeter directly behind the slit. The slit on the end of the tube is directed into the vacuum, but the tube itself is sealed to the walls of the cavity so that the tube may be cooled by a blast of air within the cavity. A power of about 50 watts in a 3000 Mc/sec cavity is required to give a good yield of atomic chlorine. This method has been used successfully to produce hydrogen beams in which the atom content is comparable to that from a Wood's discharge[2] tube. Atomic bromine has also been produced with this microwave discharge source[3]. A special gas handling system for supplying the molecular bromine to the cavity was required to avoid exposure of a metallic needle valve to attack by bromine vapor.

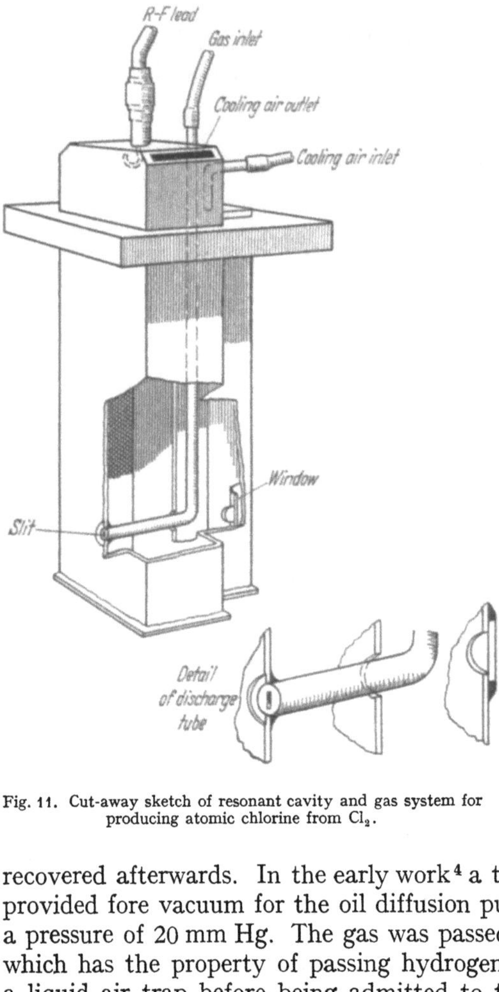

Fig. 11. Cut-away sketch of resonant cavity and gas system for producing atomic chlorine from Cl_2.

A gas recirculation system has been used in the work with tritium, in order that a small sample of the gas would suffice for the experiment and could be recovered afterwards. In the early work[4] a three-stage mercury diffusion pump provided fore vacuum for the oil diffusion pump and could discharge the gas at a pressure of 20 mm Hg. The gas was passed through a hot palladium thimble, which has the property of passing hydrogen gas selectively, and then through a liquid air trap before being admitted to the Wood's discharge tube. A gas sample of about 3 cm³ at STP was recirculated at the rate of about 2 cm³ at STP per minute. A Toepler pump was used to recover the gas at the end of the run. During recirculation of the gas the tritium was gradually lost by chemical exchange and adsorption, and the relative concentration of the tritium was reduced by the evolution of H from the system. The evolution of H was associated with the operation of the discharge tube and was the principal experimental difficulty

[1] Davis, Feld, Zabel and Zacharias: Phys. Rev. **76**, 1076 (1949).
[2] Nagle, Julian and Zacharias: Phys. Rev. **72**, 971 (1947).
[3] J.G. King and V. Jaccarino: Phys. Rev. **94**, 1610 (1954).
[4] E.B. Nelson and J.E. Nafe: Phys. Rev. **75**, 1194 (1949).

with the recirculation system. In a later experiment[1] the Wood's discharge tube was replaced by an electrodeless discharge tube operating at about 4 Mc/sec. The evolution of gases in the electrodes or the absorption of tritium by them was thus avoided.

6. Atoms in metastable and optically excited states. Atomic hydrogen and deuterium in the metastable $2\,^2S_{\frac{1}{2}}$ state have been produced for the experiments on the Lamb shift and on the fine and hyperfine structure of the $n=2$ state. In

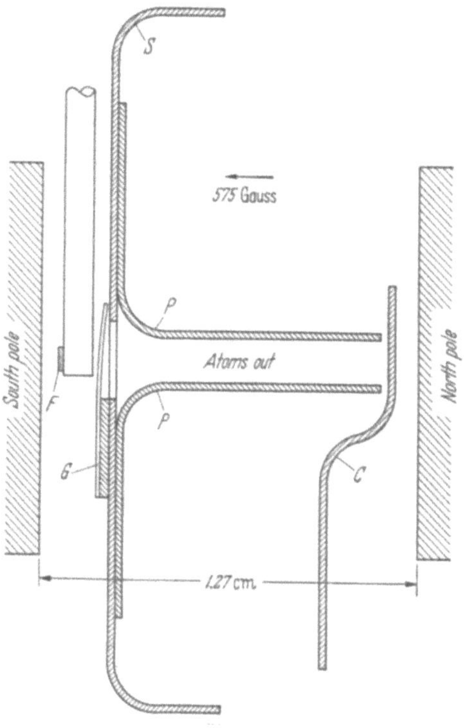

Fig. 12.

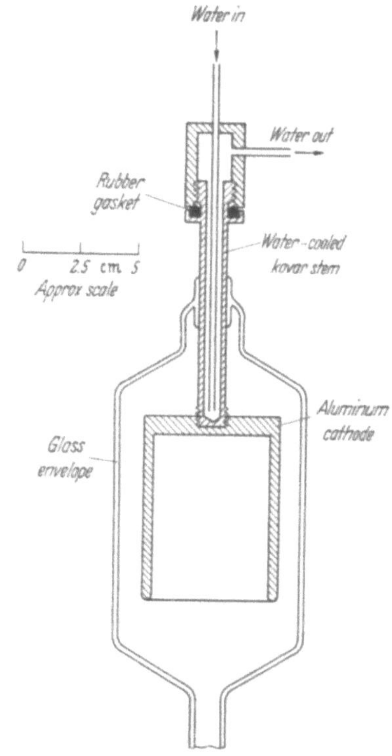

Fig. 13.

Fig. 12. Exciter for producing metastable hydrogen atoms. Cross section perpendicular to the atomic beam. *F*, tungsten filament, 0.01 × 0.12 × 3.8 cm; *G*, molybdenum grid wires, 0.05 mm diameter; *S*, support for grid, molybdenum; *P*, plates for reducing space charge effects; *C*, electron collector.

Fig. 13. Construction of water-cooled electrode.

the original Lamb-Retherford experiment[2] molecular hydrogen was thermally dissociated in an oven of the type shown in Fig. 8 and discussed in Sect. 5. The atoms are in the ground $1\,^2S_{\frac{1}{2}}$ state as they leave the oven. Bombardment by electrons at somewhat above the threshold energy of 10.2 ev for excitation to the $2\,^2S_{\frac{1}{2}}$ state and in a direction perpendicular to the direction of the initial velocity of the beam serves to produce a useful population of atoms in the metastable state. A diagram of an exciter used by Heberle, Reich, and Kusch[3] is shown in Fig. 12, where an excitation voltage of about 13 volts d.c. is applied between the filament and the grid. The magnetic field serves as a polarizer to produce a single magnetic substate and also focuses the electrons. Lamb and Retherford, using the original version of the exciter, have estimated that the

[1] A. G. Prodell and P. Kusch: Phys. Rev. **106**, 87 (1957).
[2] W. E. Lamb and R. C. Retherford: Phys. Rev. **79**, 549 (1950).
[3] Heberle, Reich and Kusch: Phys. Rev. **101**, 612 (1956).

yield of metastable atoms is about one in forty million. HEBERLE, REICH and KUSCH report a substantial increase in the yield with their exciter in which the filament-to-grid separation was increased. The excitation process gives to the atoms an additional component of momentum in a direction perpendicular to the initial direction of the beam. The resulting angular spread in the beam is between 3 and 10° and depends on the velocities of the atoms. In the experiments on the metastable state of hydrogen transitions are detected through a loss in metastability of the atom consequent to the transition, and hence these experiments do not require a well defined beam as in the more usual atomic beam experiment.

Beams of rare gas atoms in the metastable states—He in the $^3S_1$ state[1], Ne, A and Xe in the $^3P_2$ states[2,3]—have been produced from atoms excited in a WOOD's

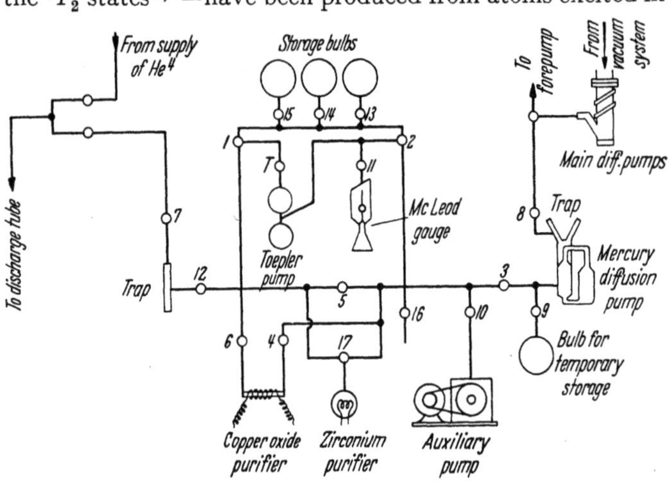

Fig. 14. Gas handling system for helium 3. Numbered circles represent stopcocks.

glow discharge tube. Fig. 13 shows the electrode construction. A current of 100 ma, a voltage of 2500 volts, and a helium pressure of 0.2 mm Hg represent typical operating conditions. The observed metastable atom beam intensity indicated a yield of one metastable He atom in 10^5 normal atoms. This yield agrees in order of magnitude with a theoretical estimate which ascribes the production of metastable atoms to collisions of electrons with ground state He atoms and their loss to wall collisions. Similar yields were observed for Ne and A in their $^3P_2$ states. Light from the discharge tube produced an electron current from the detector wire by photoelectric ejection and gave a light background signal about 7 times the metastable He atom signal. Sputtering of the aluminum cathode on the surrounding glass envelope occurred after many hours of use, and necessitated higher gas pressures to maintain the discharge tube and eventually resulted in decreased metastable atom beam intensity. The gases used were of ordinary commercial grade and required no trapping or purification before being passed into the discharge tube.

A gas recirculation system was required in the work with helium-3 in its metastable $^3S_1$ atomic state[4]. A schematic diagram of the gas handling system is shown in Fig. 14. The three-stage mercury diffusion pump removes gas as it

[1] HUGHES, TUCKER, RHODERICK and WEINREICH: Phys. Rev. **91**, 828 (1953).
[2] WEINREICH, TUCKER and HUGHES: Phys. Rev. **87**, 229 (1952).
[3] H. FRIEDBURG and H. KUIPER: Naturwiss. **44**, 487 (1957).
[4] G. WEINREICH and V.W. HUGHES: Phys. Rev. **95**, 1451 (1954).

comes out of the main oil diffusion pumps and pumps it back into the discharge tube with the pressure in the return line varying from about 1.5 mm Hg at the output of the mercury pump to a few tenths of a mm Hg in the discharge tube. A gas sample of about 1.5 cm^3 at STP was employed. Impurity atoms if present in concentrations greater than about 1% were expected to reduce the metastable He atom density in the discharge tube, and provisions for purification of the recirculated gas were found necessary. Zirconium heated to about 1300° C was used to remove oxygen and nitrogen from the gas sample before a run. During a run cathode clean-up by the aluminum electrodes was capable of removing air leakage into the system, and CuO heated to about 550° C removed hydrogen which was evolved in the system. A major difficulty was experienced with cathode sputtering and the associated absorption of the helium itself, and electrodes had to be changed frequently.

In subsequent work with the rare isotope neon-21 disappearance of the gas into the electrodes was also observed, and an electrodeless discharge was developed[1]. A low frequency of 500 cps is used in order to avoid the shielding problems associated with an a.c. discharge. Two pyrex glass spheres 7 cm in diameter and 1 mm thick serve as electrodes at the ends of a tube 75 cm long and 7 mm in diameter. Electrical connection is made to an aquadag coating on the surface of the spheres. A current of 20 ma, a voltage of 5000 volts rms, and a pressure of 0.1 mm Hg represent typical operating conditions. The loss rate of neon was found to be less than 1 micron-liter per hour, which is negligibly small considering the typical amount of gas of several cm^3 STP.

Certain low-lying metastable states—in particular, the $^2P_{\frac{3}{2}}$ state of gallium[2] with an excitation energy at 800 cm^{-1}, the $^2P_{\frac{3}{2}}$ state of indium[3] at 2200 cm^{-1}, and the $^2P_{\frac{1}{2}}$ state of chlorine[4] at 900 cm^{-1}—are produced in useful amounts in the sources employed for the ground state atoms. Thus at an oven temperature of about 1600 °K at which useful beam intensities are produced, the relative abundance of the metastable state of Ga compared to the ground state is 50% and of the In metastable state, 20%.

The lifetime of a non-metastable excited state of an atom is typically between 10^{-9} and 10^{-7} sec. This lifetime is much too short to allow the propagation of a beam of excited state atoms through an apparatus, but nevertheless the atomic beam magnetic resonance method can be extended to study these excited states as will be discussed in Sects. 31 and 36, and it is of practical interest to discuss the production of a beam of atoms in an excited state. The production is achieved by irradiating a beam of atoms in its ground state or in a metastable state with optical radiation from a light source from the same atomic species. The probability per unit time for an atom in the beam to undergo a transition from its ground or metastable state to an excited state with the absorption of a photon can be derived from standard perturbation theory for radiation processes and is given by the expression[5]:

$$w_{n|0} = \frac{2\pi}{\hbar} |\mathscr{H}'_{n|0}|^2 \varrho \tag{6.1}$$

in which $\mathscr{H}'_{n|0}$ is the matrix element between the initial state (0) and the final state (n) of the Hamiltonian term which gives the interaction energy of the

[1] G.M. Grosof and J.C. Hubbs: Rev. Sci. Instrum. **27**, 171 (1956). — J.C. Hubbs and G.M. Grosof: Phys. Rev. **104**, 715 (1956).

[2] G.E. Becker and P. Kusch: Phys. Rev. **73**, 584 (1948).

[3] A.K. Mann and P. Kusch: Phys. Rev. **77**, 427 (1950).

[4] J.G. King and V. Jaccarino: Phys. Rev. **84**, 852 (1951).

[5] The Quantum Theory of Radiation by W. Heitler, Chap. IV and V. Oxford: Clarendon Press 1954.

atom and the light. The quantity ϱ is the density of initial states per unit energy interval, or the number of radiation oscillators per unit volume and per unit energy interval near resonance. If a plane unpolarized light wave travelling in the x-direction is incident on a beam travelling in the y-direction and if the transition is an electric dipole transition, Eq. (6.1) becomes:

$$w_{n|0} = \frac{2\pi}{3\hbar} \alpha \left(|y_{n|0}|^2 + |z_{n|0}|^2\right) I_0(\nu) \tag{6.2}$$

$x_{n|0}$ is the matrix element of the coordinate x of the electron position relative to the nucleus between the atomic states n and 0. The factor $\frac{1}{3}$ appears from the average taken over all orientations of the atoms (i.e., directions of x) with respect to the direction of polarization of the light. I_0 is the intensity of the light beam in ergs per cm² per sec per unit frequency interval. This intensity is assumed to be constant over the absorption line width for the atom. The Eq. (6.2) was derived for the case in which there is a single excited state that can be excited by the incident light. It will apply also, of course, if there are several excited states which can be independently excited. However, if there are two excited states so closely spaced that their absorption line widths overlap (as can occur when the level separation is due to a hfs interaction), then interference effects must be considered. The interference terms do not in general change the order of magnitude of the transition probability. They may depend critically on the value of an external magnetic field.

The absorption line width for an atom in the beam will be determined by the natural width due to the lifetime of the excited state and perhaps by Doppler broadening. Collision and Stark broadening are negligible for atoms in a beam. This absorption line width will in general be much less than the width of the line from the light source, where all the usual sources of line broadening are present. This implies that only a small fraction of the intensity in the line from the light source will be effective in exciting the beam atoms, and it necessitates that the light source provide high intensity at the center of the spectral line and hence that self-reversal be minimized.

Optically excited states have been produced in beams of alkali atoms[1], of thallium[2], and of helium[3]. As an example to indicate the light intensity required, consider the excitation of a sodium atom from the $3\,^2S_{\frac{1}{2}}$, $m_J = +\frac{1}{2}$ state to the $3\,^2P_{\frac{3}{2}}$, $m_J = +\frac{3}{2}$ state. Use of a hydrogenic wavefunction for the valence electron[4] with an effective nuclear charge $Z_{\text{eff}} = 2.2$ and an effective quantum number $n^* = 3$ gives for the squares of the matrix elements, $|y_{n|0}|^2 = 2.0\, a_0^2$ and $|z_{n|0}|^2 = 0$ in which $a_0 = 0.53 \times 10^{-8}$ cm. If the transition probability for excitation is to be unity for an atom travelling with a velocity of 5×10^4 cm/sec and being irradiated over a path length of 1 cm, then $w_{n|0}$ must equal 5×10^4/sec and use of Eq. (6.2) gives $I_0 = 0.6 \times 10^{-4}$ ergs per cm² per sec per cps. The light flux from a line broadened to 2 parts in 10^5 ($\Delta\lambda/\lambda = 2 \times 10^{-5}$) and not self-reversed at its center must be about 0.6×10^6 ergs/cm² sec.

In practice commercial low voltage arc discharge lamps have been used to excite the $3\,P$ level of sodium. The lamp has been outside of the vacuum system and the light has been focussed onto the beam by a cylindrical lens. An optical

[1] Perl, Rabi and Senitzky: Phys. Rev. **98**, 611 (1955). — B. Senitzky and I. I. Rabi: Phys. Rev. **103**, 315 (1956). — Buck, Rabi and Senitzky: Phys. Rev. **104**, 553 (1956). — P. Buck and I. I. Rabi: Phys. Rev. **107**, 1291 (1957).

[2] G. Gould: Phys. Rev. **101**, 1828 (1956).

[3] White, Hughes, Drake and Lurio: Bull. Amer. Phys. Soc. [II] **2**, 37 (1957).

[4] J. C. Slater: Phys. Rev. **36**, 57 (1930).

illumination in the beam region of about $^1/_{20}$ of the surface brightness of the lamp has been achieved and up to about 15% of the beam atoms have been excited. The lamps have been found to have sufficiently low noise level and long time stability for use in radiofrequency spectroscopy measurements. The $4P$ state of potassium, the $5P$ state of rubidium, and the $6P$ state of cesium have also been produced in this manner.

The production of a beam of ground state sodium atoms with a net orientation of the atomic magnetic moments has been achieved by BROSSEL, KASTLER, and WINTER[1] and by HAWKINS and DICKE[2] by the use of circularly polarized light. Resonance radiation from the sodium D lines excites the atoms to the $3P$ states and the orientation of the atoms is inferred from the polarization of the light re-emitted as the atoms decay to the ground state. A degree of orientation corresponding to an average value of the magnetic quantum number m, associated with the total atomic angular momentum quantum number F, greater than $F/2$ has been obtained[1].

7. Radioactive atoms. Sources of radioactive atoms present many special problems in the production and radiochemical separation of the radioactive species and in the efficient utilization of material. The artificial radioactive isotopes have been produced by pile irradiations, involving usually a single neutron capture and by cyclotron induced reactions of the types (d, α), (p, n) and (α, kn). Some of the longer lived isotopes have been produced as fission products. Often radiochemical separation of the radioactive isotope from the target has been necessary in order to increase the relative abundance of the desired isotope. A carrier material, usually involving a stable isotope of the atom under study, is often added for the purposes of calibration of the magnetic C field or of test and line-up of the apparatus. Further details on source preparation are given in the references included in Table 8 of Sect. 36.

The amount of radioactive isotope used may range typically from 10^{10} to 10^{17} atoms and corresponds to source strengths from a fraction of a millicurie to several hundred millicuries. Such quantities are only a very small fraction (usually less than 10^{-6}) of the material used in an experiment with a stable isotope. They are, however, close to the maximum amounts that are practically available due to limitations in the production process and due to safety requirements.

Several modifications in oven design have been employed in order to use more efficiently the amount of radioactive material available. In one design the usual thin oven slit is replaced by a canal, so that effusion occurs not according to the cosine law [see Eq. (3.4)], but highly preferentially through a small solid angle. This is evidently desirable to make the most efficient use of the limited amount of radioactive material and to reduce radioactive contamination of equipment. An oven of this type used for the study of radioactive alkali isotopes is shown in Fig. 15[3]. For the alkali atoms, the best material which has been found is monel metal. The two blocks are held together by means of a holder which contains all heaters, thermocouples, and positioning adjustments. The two lapped surfaces of the blocks when placed in contact must fit well enough so that the main leak out of the oven is through the exit channel which faces the beam detector. The narrow exit channel is used to diminish the number of atoms diffusing

[1] BROSSEL, KASTLER and WINTER: J. Phys. Radium **13**, 668 (1952). — A. KASTLER: Proc. Phys. Soc. Lond. A **67**, 853 (1954); J. Opt. Soc. Amer. **47**, 460 (1957).

[2] W.B. HAWKINS: Phys. Rev. **98**, 478 (1955).

[3] DAVIS, NAGLE and ZACHARIAS: Phys. Rev. **76**, 1068 (1949).

from the slit for a given intensity toward the detector [see Sect. 3, Eq. (3.5)]. A reduction by about a factor of 50 in the total number of atoms required to produce a given beam intensity in the forward direction was achieved.

Another method of increasing the efficiency of utilization of source material is to use a different type of deflecting field, which permits the acceptance of a larger solid angle of the beam which issues from a canal of circular cross section. With this focussing type of atomic beam apparatus, studies have been made of radioactive Au, Ag, and Cu, and a gain of about 25 in the efficiency of utilization of source material has been reported[1,2] (see Sect. 23).

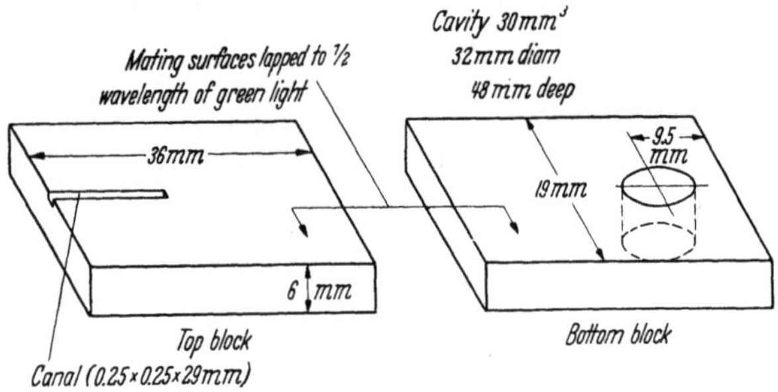

Fig. 15. Channeled oven for producing beams of radioactive alkali atoms. The diameter of the cavity is 3.2 mm and the depth 4.8 mm.

b) Beam characteristics.

8. Beam geometry and intensity. A typical arrangement of slits which determines the geometry of the beam is shown in Fig. 16. The source slit (width w_s) and collimating slit (width w_c) are narrow vertical slits, and a narrow vertical detector is used. The verticality of the elements is not significant except insofar as the vertical direction is the easiest one along which to secure parallelism of the elements. Consider the image of the undeflected beam in the detector plane. The distances between the source slit and collimating slit and between the collimating slit and detector are l_{sc} and l_{cd}, respectively. Both w_s and w_c are very small compared with l_{sc} and l_{cd}. The intensity distribution in the detector plane is shown in Fig. 17. The illumination of the detector plane will drop to zero at the two symmetrical points $\pm z_2$ where

$$z_2 = \frac{1}{2}\left[w_c + (w_s + w_c)\frac{l_{cd}}{l_{sc}}\right]. \tag{8.1}$$

The full aperture of the source slit will illuminate the detector plane within the limits $\pm z_1$ where

$$z_1 = \frac{1}{2}\left[w_c - (w_s - w_c)\frac{l_{cd}}{l_{sc}}\right] \tag{8.2}$$

provided that $w_s/w_c < 1 + (l_{sc}/l_{cd})$. The interval between $-z_1$ and $+z_1$ is the umbra of the image, and the intervals between $-z_2$ and $-z_1$ and between $+z_1$ and $+z_2$ are the penumbra. The intensity drops linearly between z_1 and z_2. Therefore the image has a trapezoidal intensity distribution whose half width at half intensity is:

$$z_3 = \frac{w_c}{2}\left(1 + \frac{l_{cd}}{l_{sc}}\right). \tag{8.3}$$

[1] LEMONICK, PIPKIN and HAMILTON: Rev. Sci. Instrum. **26**, 1112 (1955).

[2] CHRISTENSEN, HAMILTON, LEMONIK, PIPKIN, REYNOLDS and STROKE: Phys. Rev. **101**, 1389 (1956).

If $w_s/w_c > 1 + (l_{sc}/l_{cd})$, no point in the detector plane is illuminated by the full aperture of the source slit. Since no advantage is gained when the full source slit does not illuminate the detector plane, the requirement that $w_s/w_c \leqq 1 + (l_{sc}/l_{cd})$ serves to put an upper limit on w_s for any w_c in the design of an apparatus.

It is seen from Eq. (8.3) that the width of the image is independent of the width of the source slit. As w_s is decreased for constant w_c, the image becomes more nearly rectangular and in the limiting case when $w_s \ll w_c$, $z_1 = z_2$. As w_s is increased for constant w_c, the image becomes more triangular and in the limiting allowed case when $w_s/w_c = 1 + (l_{sc}/l_{cd})$, $z_1 = 0$. The intensity per unit distance within the limits $\pm z_1$ is directly proportional to w_s, if the rate of effusion from unit area of the source is independent of the total aperture of the source. Evidently, all other factors being equal, a nearly rectangular image can be achieved only at the expense of beam intensity. While the specialized requirements of an individual experiment de-

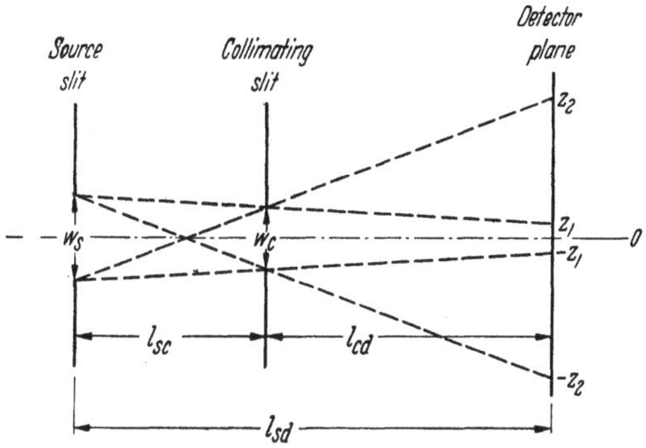

Fig. 16. Schematic diagram illustrating the beam geometry. Dimensions are exaggerated in the direction perpendicular to the direction of propagation of the beam.

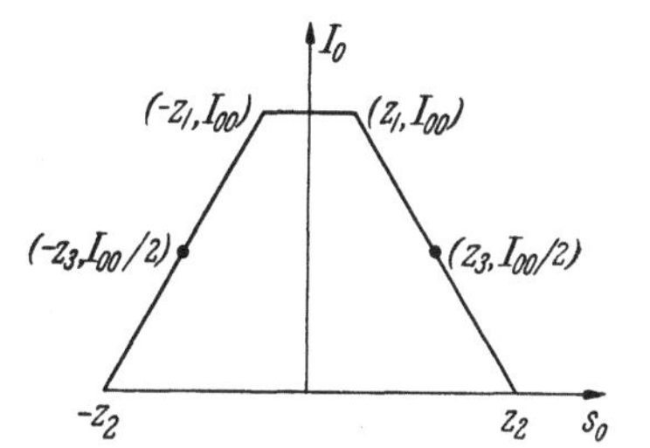

Fig. 17. Intensity distribution in plane of detector as a function of distance s_0 measured in a direction perpendicular to that of beam propagation.

mand a detailed consideration of the effects of slit widths on the observability of lines, on the behavior of the source, and possibly on the efficient utilization of the material of the beam, a very satisfactory arrangement for most purposes is one in which $z_2 = 3z_1$ and $z_3 = 2z_1$. This condition is achieved often with $l_{sc} = l_{cd}$ and $w_s = w_c$.

The detected beam intensity I, or the number of particles impinging on the detector per unit time, is given by use of Eq. (3.4) as:

$$I = \frac{n \bar{v} A_s A_d}{4\pi l_{sd}^2} \tag{8.4}$$

in which A_d is the area of the detector and l_{sd} is the distance from source to detector. This formula applies when the detector area lies wholly within the umbra of the image defined by the source and collimating slits and when the heights of the slits are small compared to l_{sd}. The expression for I may be rewritten using Eq. (3.3):

$$I = \frac{N_0 p A_s A_d}{\pi l_{sd}^2 \sqrt{2\pi M R T}} = \frac{p A_s A_d}{\pi l_{sd}^2 \sqrt{2\pi m k T}}. \tag{8.5}$$

The quantities A_s, A_d, and l_{sd} may vary widely for different experimental conditions. An extreme case occurs in the observation of nuclear resonances by the molecular beam magnetic resonance method for which l_{sd} must be large and the areas A_s and A_d small due to the smallness of the magnetic moment available for deflection. In a typical apparatus designed for the observation of nuclear resonances, the length l_{sd} is 174 cm and both A_s and A_d are rectangular with a height of 0.2 cm and a width of 0.002 cm. If the typical example of Sect. 3 is used, for which $p = 10^3$ dynes/cm^2, $M = 40$, and $T = 1000°$ K, then I is 2.2×10^8 particles per second. This flux of matter to the detector is small compared to that which may often be obtained in atomic or molecular beam experiments. The flux may be increased by a factor of as much as 10^3 when a magnetic moment of a Bohr magneton or an electric dipole moment of several debye is available for deflection, and hence it is possible to use a smaller length between source and detector and larger apertures for source and detector.

9. Velocity distribution. It has been assumed in the preceding sections that the velocity distribution of particles in the oven or other cavity from which the beam effuses is Maxwellian and that effusion occurs through the ideal aperture of kinetic theory. While the detailed validity of this assumption is of the greatest importance in assessing the accuracy of results based on the actual measurement of the intensity in a deflection pattern, it is of much less importance in the spectroscopic applications of atomic and molecular beam techniques.

It is of interest to examine the experimental evidence for the validity of the Maxwellian distribution. An experimental investigation of the velocity distribution in a beam will also yield information about the degree of association of the matter in the beam into complex aggregates.

Of the various devices which have been used to study the velocity distribution in atomic and molecular beams, the most successful[1] has been the spiral velocity selector. It consists of a solid cylinder on whose surface are cut a large number of helical slots and which is capable of rotation about the cylinder axis. This selector has certain inherent advantages over a selector which employs two or more toothed or notched wheels operating on the same shaft at a fixed relative angular position and at a variable angular velocity, or operating on two separate shafts at the same fixed angular velocity but at a variable relative angular position. With the notched wheel selectors the transmission of a beam by the device is possible when the time of transit of the particles over the length of the selector is equal to the time required for the selector to turn through the interval at whose ends the slot in the first disk and that in the second disk transmit the beam. Evidently, if the disks contain a large number of closely spaced teeth, atoms in a particular velocity interval may be transmitted for several angular velocities of the rotor, and a particular rotor velocity will allow the transmission of atoms in several velocity intervals. The elimination of the superposition of several velocity bands, by spacing the slots at an angular interval on each disk which is large compared to the angular interval between slots in the two disks, leads to a great loss in intensity of the particles which lie within a particular velocity interval. However, when a single cylinder is used as a selector, side bands do not occur. The spacing between adjacent slots may be as small as mechanical considerations permit. The width of the slots, a determining factor in the resolution, may in turn be as small as is mechanically feasible and as is consistent with the requirement for adequate transmitted beam intensity.

A rotor for a velocity selector is shown in Fig. 18. The dimensions of the rotor used in the experiments to be described are shown. To permit the adjustment of the beam, which is defined by the oven slit and the detector, parallel to the axis of the rotor, two paraxial slots are provided at opposite sides of a diameter. Slots are placed at 0.5° intervals, and a total of 700 slots can be cut without interference with the paraxial slots.

---

[1] R.C. Miller and P. Kusch: Phys. Rev. **99**, 1314 (1955).

Suppose that the particles in the oven consist of a single atomic or molecular species of mass m and that they are in equilibrium with the oven at a temperature T. If the distribution in velocity is Maxwellian, and if the particles escape from the source through an ideal aperture, then the distribution incident at the entrance to the rotor is given by Eq. (3.6).

The velocity, v_0, of the particle which travels through the slot without changing its position relative to the walls of the slot is:

$$v_0 = \omega L/\varphi_0 \tag{9.1}$$

where both L and φ_0 are fixed properties of the rotor (see Fig. 18) and ω is the angular velocity of the rotor. A parameter γ which describes the resolution properties of the rotor is $\gamma = l/(r_0 \varphi_0)$, where l and r_0 are defined in Fig. 18. For any particular ω and for an incident beam that has a uniform velocity distribution, the width at half intensity of the velocity spectrum transmitted by the rotor is given by $\Delta v = \gamma v_0$. When the velocity distribution of the incident beam is given by Eq. (3.6), the beam intensity, $I(v_0)$, transmitted by the rotor as a function of its angular velocity ω [or of v_0 by Eq. (9.1)] is:

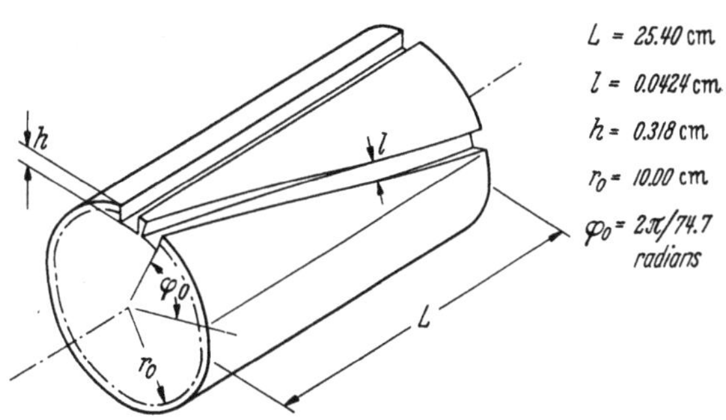

Fig. 18. Schematic diagram of velocity selector.

$$I(v_0) = 2\gamma\, x_0^4\, e^{-x_0^2} \left[1 + 2.5\gamma^2 - \left(\tfrac{13}{6}\right)\gamma^2 x_0^2 + \left(\tfrac{1}{3}\right)\gamma^2 x_0^4\right] \tag{9.2}$$

where $x_0 = v_0/\alpha$. This incident velocity distribution has been normalized to 1 and the range of velocities about v_0 transmitted by the rotor due to the finite width of the slots has been considered in deriving Eq. (9.2). When γ is small, the intensity distribution follows a v^4 distribution law. This modification of the incident v^3 distribution law is imposed by the velocity selector, because the width of the band of velocities transmitted by the rotor is itself proportional to the central velocity in the band. In the apparatus under discussion $\gamma = 0.05033$, and the v^4 distribution is only trivially distorted.

To make a critical test of the reality of the Maxwellian distribution, it is necessary that thermal equilibrium exist within the cavity from which the gas escapes, and that the oven be equipped with a slit which closely approximates an ideal aperture. To achieve the first of these objectives the oven was constructed of pure copper, which, because of its high thermal conductivity, resulted in an oven temperature uniform throughout the oven to within about 0.5%. The ideal aperture was produced by a slit of the design shown in Fig. 19; this slit design is the same as that of Fig. 5 and was described in Sect. 4.

The velocity distributions from beams of thallium and potassium have been studied. No molecules of thallium have been observed in spectroscopic observations, and hence it may be assumed that evaporation of Tl leads to a beam which is purely atomic. The molecule K_2 has frequently been observed[1], but the concentration is low so that to the precision of the results to be quoted the presence of K_2 has no effect on the distribution. In Table 1 are shown the results of

[1] Kusch, Millman and Rabi: Phys. Rev. 55, 1176 (1939).

measurements on the velocity distribution in beams of K and Tl. It is seen that the velocity at which the maximum intensity, I_{max}, is detected after transmission of the beam by the rotor is the same, within experimental error, as that computed on the basis of the assumptions that the distribution in the oven is Maxwellian and that effusion occurs through an ideal aperture. This occurs for a potassium pressure which varies by a factor of about 25 and for a thallium pressure which varies by a factor of about 10. The one exception occurs in the third line of Table 1. For the observations recorded in that line the very thin oven slit was replaced by a pair of copper strips, 0.317 cm thick, and the aperture is far from ideal.

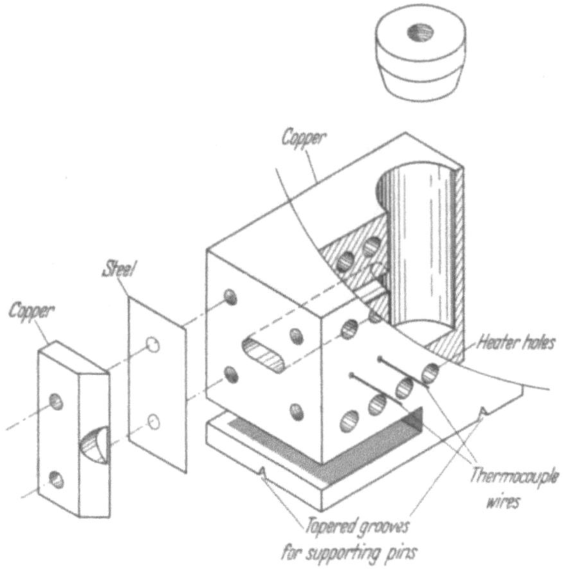

Fig. 19. Copper oven used for studies of velocity distribution showing details of slit construction.

The actual intensity distributions transmitted by the velocity selector are shown in Figs. 20 to 22. In all cases, the reduced velocity is the ratio of the observed velocity, v_0, to the theoretical velocity of the intensity maximum calculated from the measured oven temperature. The theoretical curves are obtained from Eq. (9.2), based on a Maxwellian distribution and an ideal aperture. Fig. 20 corresponds to the first line in Table 1 and the agreement between prediction and observation is excellent, though there is an observable deficiency of low velocity atoms. This deficiency was observed to increase with increasing oven pressure. Fig. 21 corresponds to the third line of Table 1. The effect of the thick slit is to shift the whole distribution to higher velocities. The results of line 4 in Table 1 are shown in Fig. 22. The agreement between observation and prediction is excellent.

Table 1. *Experimental conditions and results for measured velocity distributions.*

Beam	Temperature °K	Oven pressure in mm Hg	Velocity of I_{max} in m/sec: From oven temperature	Velocity of I_{max} in m/sec: From exp. distr.
K	466 ± 2	4.5×10^{-3}	628 ± 2	630 ± 3
K	544 ± 3	1.2×10^{-1}	678 ± 3	679 ± 3
K	489 ± 2	1.4×10^{-2}	644 ± 2	682 ± 3
Tl	870 ± 4	3.2×10^{-3}	376 ± 1	376 ± 2
Tl	944 ± 5	2.1×10^{-2}	392 ± 1	395 ± 2

The experiments just described furnish the best experimental evidence to date of the reality of the Maxwellian distribution. There is, however, a considerable body of other work which has dealt with the velocity distribution in atomic beams. BENNETT and ESTERMANN[1] have used a velocity selector in which two slotted disks are run synchronously on separate shafts and at a variable

[1] A. BENNETT: Phys. Rev. 95, 608 (1954).

phase angle. The results also confirm the reality of the Maxwellian distribution, though the agreement between observed and predicted distributions is less good

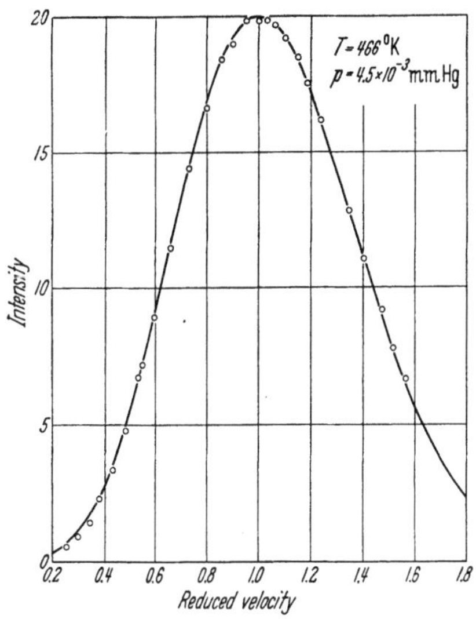

Fig. 20. Typical potassium velocity distribution. Beam intensity is in relative units. The reduced velocity V is the ratio of the observed velocity to the theoretical velocity of the intensity maximum calculated from the observed oven temperature. Solid curve is distribution predicted from Eq. (9.2).

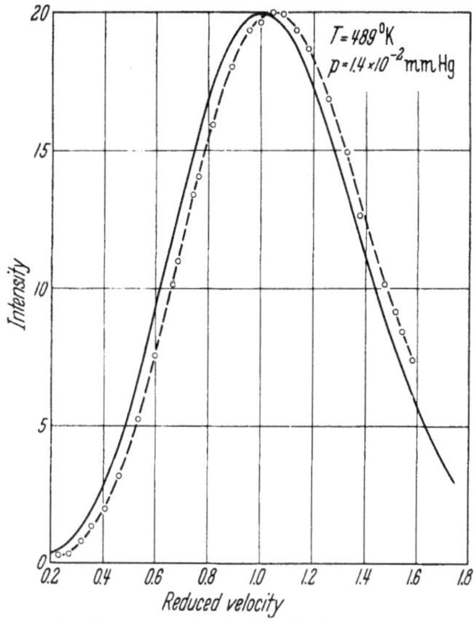

Fig. 21. Potassium velocity distribution from an oven with "thick" slits. Beam intensity and reduced velocity have the same meanings as for Fig. 20. Solid curve is predicted from Eq. (9.2). Dashed curve is empirical curve drawn through the experimental points.

than that in the experiments described in detail above. MILLER and KUSCH[1] discuss other experiments where the interpretation of the results gives less convincing evidence for the reality of the Maxwellian distribution.

10. Composition of molecular beams. When it has been determined that the velocity distribution within the oven is, in fact, Maxwellian and that it is possible to design oven slits which closely approximate the ideal aperture of kinetic theory, a study of the constitution of a molecular beam and of the thermo-chemical properties of certain components of the beam may be made by an analysis of the velocity distribution of the particles in the beam. MILLER and KUSCH[2] have made such a study of ten alkali halides and have found not only molecules of the form MX but also of the form $(MX)_2$ and $(MX)_3$, and have determined the dissociation energies of the polymers by measurement of the equilibrium

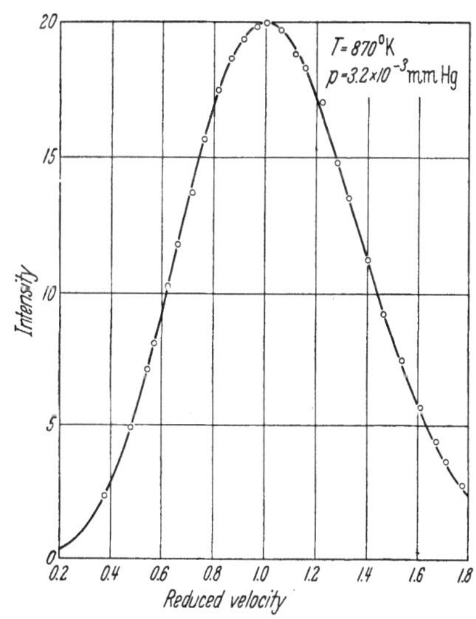

Fig. 22. Typical thallium velocity distribution. Beam intensity and reduced velocity have the same meanings as for Fig. 20. Solid curve is predicted curve using Eq. (9.2).

[1] R. C. MILLER and P. KUSCH: Phys. Rev. **99**, 1314 (1955).

[2] R. C. MILLER and P. KUSCH: J. Chem. Phys. **25**, 860 (1956); **27**, 981 (1957).

an ideal aperture into a collision-free region, the density of particles just outside of the aperture is $n/2$. In the special case when the aperture is rectangular and of dimensions $2a$ by $2b$, the number of particles per cm³, n', on a line normal to the aperture at its midpoint and at a distance y from the aperture is:

$$n' = \frac{n}{\pi} \operatorname{arc\,tan} \left\{ \frac{a b}{y (a^2 + b^2 + y^2)^{\frac{1}{2}}} \right\}. \tag{11.1}$$

For sufficiently large y, this expression becomes:

$$n' = \frac{n a b}{\pi y^2}. \tag{11.2}$$

For typical conditions, $2a = 0.01$ cm, $2b = 0.5$ cm, $y = 50$ cm, and hence $n' = 1.6 \times 10^{-7} n$. If the pressure in the source is 0.1 mm Hg, the effective pressure at the 50 cm distance is 1.6×10^{-8} mm Hg. In many applications, particularly those involving the deflection of a nuclear moment, the pressure at the detector will be even less than this. Furthermore, the change in beam pressure associated with a transition will ordinarily be a small fraction of the total beam pressure.

Since the total number of particles striking the detector per unit time is small, a detector which converts the neutral beam into a current must deal with small currents. Thus for the example given in Sect. 8 of an apparatus designed for the observation of nuclear resonance, 2×10^8 particles per second arrived at the detector. If each particle gives rise to a single electron at the input to a detecting circuit, then the current is 3×10^{-11} amperes. It is evident that the measurement of the small changes of such currents which are involved in the observation of a transition imposes a heavy requirement on the stability and sensitivity of a detecting circuit.

A discussion of the various detection schemes which have been successfully used is given in Sects. 12 through 17, where the Pirani gauge, the surface ionization detector, the universal electron bombardment detector, detectors for metastable atoms and for radioactive atoms, and various less widely used detectors are treated.

12. The Pirani gauge. A Pirani gauge was first applied to the detection of atomic beams by KNAUER and STERN[1]. The gauge has a useful sensitivity for the light gases, and with recent improvements in design it has been used as a detector of beams of molecular nitrogen[2] and oxygen[3]. It has been used extensively in studies of both atomic and molecular hydrogen.

The Pirani gauge consists of two identical cavities which are completely closed except for a deep slit or channel in each cavity through which a beam may enter. The cavities are so arranged that the beam enters one of them without collision with the walls of the slit. Within this cavity the particles of the beam acquire randomly directed velocities, and the rate of escape of the particles from the cavity equals the rate of entry only when the pressure within the cavity exceeds the pressure outside of the cavity by a pressure differential characteristic of the gauge and the gas. The slit of the other cavity is directed so that the beam does not enter this cavity. Fluctuations in the residual pressure in the apparatus in which the gauge is located affect the pressure in each of the two cavities in the same way. Each cavity contains two opposite arms of a Wheatstone bridge which are maintained at a temperature above that of the cavities by the current

[1] F. KNAUER and O. STERN: Z. Physik **53**, 766 (1929).
[2] J. M. HENDRIE: J. Chem. Phys. **22**, 1503 (1954).
[3] J. M. HENDRIE and P. KUSCH: Phys. Rev. **107**, 716 (1957).

flowing through the bridge. With no beam incident on the gauge the bridge is in balance or in a steady state slightly off-balance, and this condition is not affected by fluctuations in residual vacuum except insofar as the vacuum fluctuations are different at the two slits or the two halves of the gauge are themselves different. An incident beam will increase the pressure in one of the cavities; hence the temperature of the two arms of the bridge within that cavity will drop with a consequent drop in the resistance of the arms, and a modified off-balance current will flow in a galvanometer.

Suppose that a gas within a source chamber is at a pressure p_s and a temperature T_s, and that the gas escapes through an ideal aperture of area A_s and falls on the entrance slit, assumed for the present to be an ideal thin slit, of a Pirani gauge at a distance l_{sd} from the source. If the gas within the cavity of the gauge is at a temperature T_d, then when the flow of gas into the gauge equals the outward flow, the equilibrium pressure p_d in the cavity due to the beam is given by use of Eqs. (3.3) and (8.5) as:

$$p_d = \frac{p_s A_s}{\pi l_{s,d}^2} \left(\frac{T_d}{T_s}\right)^{\frac{1}{2}}. \tag{12.1}$$

If $T_d \sim T_s$, $A_s = 10^{-3}$ cm², $l_{sd} = 50$ cm, and $p_s = 1$ mm Hg, then $p_d = 1.3 \times 10^{-7}$ mm Hg. A useful gauge must have a sensitivity (off-balance current for a specified change in the number of molecules or atoms which enter the gauge) adequate for the observation of small changes in the beam intensity.

The sensitivity of the gauge can be greatly increased by using a deep slit or channel on the gauge instead of the ideal slit used in the previous estimate. The sensitivity is directly proportional to a factor $\varkappa$ which is the ratio between the flow resistance of a channel of width b, height a, and length l and that of a very thin slit whose area is the same as the cross section of the channel. $\varkappa$ is the ratio of the pressure of the beam material in the actual cavity and the pressure which would occur in the cavity if the channel were replaced by the thin slit, and $\varkappa$ is given by the expression[1]:

$$\varkappa = \frac{l}{b}\left[\frac{1}{2} + \ln\left(\frac{2a}{b}\right)\right]^{-1}. \tag{12.2}$$

It is possible to achieve values for $\varkappa$ of 200 or 300 without the use of excessively elaborate mechanical design of the gauge. The equilibrium pressure in the gauge may then be of the order of 5×10^{-5} mm Hg.

For bridge arms made of a conductor of fixed cross sectional area, the sensitivity is increased by the use of conductors of maximum surface area, i.e., thin ribbons. It is also desirable to use conductors with a large temperature coefficient of resistance, but in practice this requirement is secondary to that imposed by the desirability of thin ribbons, which implies the need of mechanical properties in the metal which permit fabrication of such ribbons.

The time constant of the gauge is determined by the volume of the cavities in which the ribbons are mounted, by the thermal capacity of the conductors whose temperature is being modified, and by the time constant of the galvanometer used to read the off-balance current in the gauge. In recent designs, the limiting time constant imposed by a high sensitivity galvanometer has been achieved.

The design of several gauges has been described in the literature. A gauge of high sensitivity and low time constant is here described[2]. Two glass plates, about 3 mm thick and flat to within 2 wavelengths of yellow light, are each

[1] P. Clausing: Physica, Haag **9**, 65 (1929).
[2] A. G. Prodell and P. Kusch: Phys. Rev. **88**, 184 (1952).

grooved with two parallel slots 3.5 cm long, 2 mm wide, and 0.25 mm deep. The two glass plates are separated by pieces of smooth gold foil, 0.0025 cm thick, arranged as shown in Fig. 25. The cavities are formed by opposing pairs of grooves in the plates. The gold foil serves to isolate the cavities from each other and from the external vacuum, and to form the channels B, C, and D. Channels B and C are for passage of gas into and out of the cavities. Each cavity contains two opposite arms of a Wheatstone bridge, and the connections to the four arms of the bridge are made through the channels D. The connecting ribbons very nearly fill up these channels, but to minimize leakage from the cavities the external ends of the channels are sealed with a very low vapor pressure wax. The arms of the bridge are made of platinum ribbon, 0.038 cm wide and 1.3×10^{-4} cm thick. The connecting ribbons, also made of platinum, are slightly thinner than the gold foil and about 0.08 cm wide. The whole assembly is tightly clamped in a massive brass frame. Spring pressure is applied over the surface of the glass plates by nine phosphor bronze springs on each of the two outside faces of the glass plates.

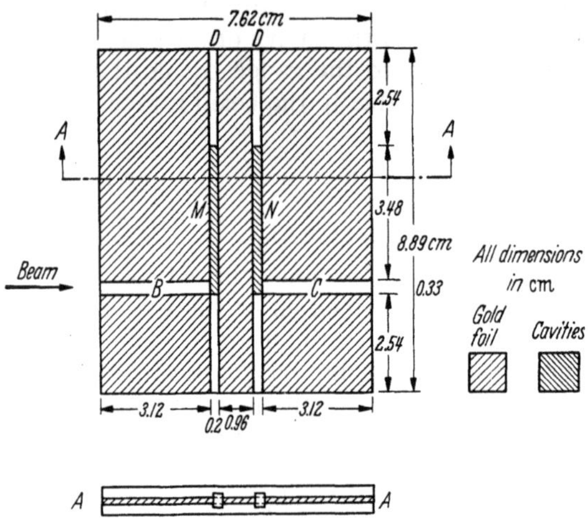

Fig. 25. Schematic diagram of the glass plates of the Pirani gauge showing the arrangement of the gold foil.

A total current of 30 ma in this gauge gave a satisfactory sensitivity and stability. When a high sensitivity galvanometer with a nominal sensitivity of 0.0167 microvolts/mm at three meters from the galvanometer was used to detect the off-balance voltage, the Pirani gauge had a sensitivity of roughly 1 cm deflection on the galvanometer scale for an incident hydrogen atom beam intensity of 10^{10} particles/sec. The stability under optimum conditions corresponded to about 1 mm deflection. The galvanometer showed maximum deflection within 7 or 8 seconds after the beam was turned on, this time being determined by the galvanometer period.

The sensitivity of the gauge may be increased by operating the gauge itself at a low temperature while maintaining a reasonable temperature difference between the ribbons and the cavity. Kolsky *et al.*[1] report the operation of a high sensitivity gauge at a temperature of 195° K while the ribbons are operated at about 300° K.

13. Detection by ionization or electron capture on a surface. The detection of atoms with low ionization potentials and of molecules containing such atoms can be done with the surface ionization detector. If an atom whose ionization potential is less than the work function of a metal strikes the metal surface, it can be re-evaporated as a positive ion. The positive ion current can then be measured with an electrometer tube circuit or with an electron multiplier. Surface ionization detection is, in many cases, extremely efficient and may produce very nearly one ion for each incident atom. It has the remarkable virtue that it is insensitive to all of the usual components of the residual gas in a vacuum system and is, nevertheless, sensitive to a large range of materials.

[1] Kolsky, Phipps, Ramsey and Silsbee: Phys. Rev. **87**, 395 (1952).

If the work function of the metal surface is Φ and if the ionization energy of the atom is I, then the ratio of the number of re-evaporating ions, n^+, to the number of re-evaporating neutral atoms, n, is given by:

$$n^+/n = \exp\left(\frac{\Phi - I}{kT}\right). \tag{13.1}$$

At a temperature of 1000° K, kT is of the order of 0.1 ev, and for a clean tungsten surface Φ is 4.50 ev. The ratio, n^+/n, is large for Cs, Rb, and K where the ionization energies are 3.87, 4.16, and 4.32 ev, respectively. If the detecting surface is maintained at a dull red color, re-evaporation occurs efficiently. For atoms such as Na and Li with ionization energies of 5.12 and 5.36 ev respectively, which are greater than the work function of pure tungsten, the detection efficiency can be brought up to a high level by increasing the work function of the surface by oxidation of the tungsten. The oxidized surface has, in fact, been used for efficient detection of atoms whose ionization energy is as great at 6.07 ev (thallium). Efficient detection has been achieved for all the alkali atoms and for Ga, In, and Tl.

It is not possible to give a very exact prescription for the method of oxidizing a tungsten wire, since the procedure depends markedly on the cleanliness of the tungsten surface, on the excellence of the vacuum, on the nature of the residual gases in the vacuum, and on the density of oxygen in the region of the detector wire which will depend on its rate of admission and on details of mechanical construction. Ordinarily, the wire is cleaned by heating to an elevated temperature of about 2200° K, after which the filament temperature is reduced to about 1500° K and oxygen at a low pressure of about 10^{-2} mm Hg is briefly admitted to the region surrounding the filament. An adequate coat of oxide may persist for long periods of time with no further admittance of oxygen. It is frequently necessary, however, to admit oxygen continuously during the use of the detector, and at such a rate that the overall increase in pressure at the detector is from 10^{-7} to 10^{-8} mm Hg.

The ease with which a particular atomic species may be detected does not depend solely on its ionization potential as compared to the work function of the surface. Another consideration is that the surface must be sufficiently hot to prevent condensation of the beam material. Lew discusses the particularly striking case of the detection of praseodymium[1]. Its ionization potential is 5.76 volts, and it should, from the point of view of ionization potential alone, be detected as readily as indium which has the same ionization potential. However, for Pr unlike for In a wire temperature of about 2000° K is found necessary for re-evaporation. This high temperature precludes the use of an oxidized tungsten surface which is unstable above 1600° K. Successful detection of Pr was achieved on a molybdenum surface. For Mo the time between the arrival of a neutral Pr atom and its re-emission as an ion from a clean surface varies from about one second at 2160° K to about 30 seconds at 1870° K. Optimum detection was obtained with Mo operating at a temperature of 2070° K, which represents a compromise between ionization efficiency and detector life time. The detection efficiency of Pr on Mo is 2×10^{-4}, which is low but still useable.

Aluminum ($I = 5.98$ ev) has been detected on a clean tungsten wire at a temperature of about 1780° K with an efficiency of about 10^{-6}. The low efficiency again relates to the high ionization potential of the atom and to the impossibility of maintaining an oxide coat on a tungsten wire sufficiently hot to re-evaporate aluminum. Furthermore, although the detection efficiency was

[1] H. Lew: Phys. Rev. **91**, 619 (1953).

greater at higher temperatures, the emission of Al impurity atoms from the tungsten wire limited the usefulness of high temperature operation[1].

Barium ($I = 5.19$ ev) can be detected on a clean tungsten surface at about 2000° K, which is adequately high for re-evaporation to occur[2]. Strontium has also been detected by this method.

Tungsten wire contains significant impurities, and when the wire is heated a constant ion current appears. This background current is increased when the wire is oxidized, and it shows a considerable and random fluctuation. The ions arise principally from alkalis embedded in the tungsten. To reduce the background of noise and also to discriminate between the various isotopes of the atomic species under investigation, mass spectrometers have been extensively used in the analysis of the ions produced at the detector[3]. The analyzed current can be amplified through an electron multiplier. The pulses which appear at the output of the multiplier can be counted when the incident ion intensity is small, as, for example, in the case of certain radioactive isotopes[3] and of aluminium[1]. Otherwise the average ion current can be measured.

It has been found that a molecule which contains an atom that is readily detectable on a surface ionization detector may itself be detected with a high efficiency. This fact is of enormous value in extending the range of substances which may be studied by molecular beams methods, thus allowing the determination of further molecular properties and the study of nuclei whose atoms are not readily detectable. The detection process involves both dissociation of the molecule and ionization of the atom and has been extensively discussed for diatomic molecules[4]. It is sufficient to state here that detection of the alkali halides and of molecules which contain an alkali atom occurs with an efficiency near 100%.

The high electron affinity of the halogens indicates that considerable energy is liberated in the process $x + e^- \rightarrow x^-$. Detection of beams of the halogens has been achieved by the use of a hot wire and a mass spectrometer to analyze the negative ions. The use of a mass spectrometer to separate electrons emitted by the hot wire from the ions is essential. Thus the detection of a halogen as compared with the detection of an alkali requires only the reversal of the potential difference which accelerates the ions into a mass spectrometer. All other matters of experimental technique are similar to those employed when the mass spectrometer is used for positive ions. DAVIS *et al.*[5] detected chlorine on a hot tungsten wire with an efficiency of 10^{-4}. This observed efficiency correlated well with the predicted value based on the electron affinity of chlorine, the temperature of the wire, and the work function of tungsten. KING and JACCARINO[6] report a detection efficiency for bromine on tungsten of 2×10^{-5}, but a much higher efficiency with properly activated thoriated tungsten.

The standard electrometer circuit is capable of measuring a current as low as about 10^{-16} amp and is discussed by RAMSEY [*25*]. The most recent designs of mass spectrometers, electron multipliers, and associated electronic circuitry are treated by KING and ZACHARIAS [*18*] and by RAMSEY [*25*].

14. Detection of metastable atoms. If the excitation energy of a metastable atom exceeds the work function of a metal surface, it is energetically possible

[1] H. LEW: Phys. Rev. **76**, 1086 (1949).
[2] R.H. HAY: Phys. Rev. **60**, 75 (1941).
[3] DAVIS, NAGLE and ZACHARIAS: Phys. Rev. **76**, 1068 (1949).
[4] G.E. COGIN and G.E. KIMBALL: J. Chem. Phys. **16**, 1035 (1948).
[5] DAVIS, FELD, ZABEL and ZACHARIAS: Phys. Rev. **76**, 1076 (1949).
[6] J.G. KING and V. JACCARINO: Phys. Rev. **94**, 1610 (1954).

for a collision of the second kind to occur in which the atom returns to the ground state and an electron is liberated from the surface to take up the excess energy. Fig. 26 shows such a process for the $2s$ metastable state of hydrogen which has an excitation energy of 10.2 ev which is substantially higher than the work functions of metals. The electron ejection occurs as an Auger transition, and has been discussed by COBAS and LAMB[1] and by LAMB and RETHERFORD[2] and studied extensively by HAGSTRUM[3] for the similar case of the ejection of electrons from metals by ions. Because the ionization energy for the $2s$ state (3.4 ev) is less than the work function of the metal, an alternative process is also energetically possible in which the atomic electron is captured by the metal. If the proton is neutralized on the surface without being re-emitted, this competing process may seriously reduce the detection efficiency by the electron ejection process. From the experiments of LAMB and RETHERFORD, the detection efficiency for a metastable hydrogen atom by electron ejection from tungsten is estimated to be about $\frac{1}{80}$, which suggests that the competing process is important. The experiments on the fine and hyperfine structure of the $n=2$ state of hydrogen have used this detection technique.

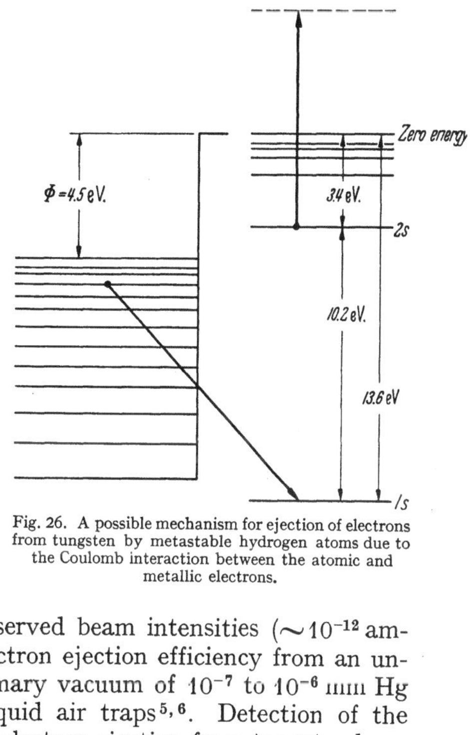

Fig. 26. A possible mechanism for ejection of electrons from tungsten by metastable hydrogen atoms due to the Coulomb interaction between the atomic and metallic electrons.

The metastable $^3S_1$ state of helium has an excitation energy of 19.8 ev, and the efficiency for electron ejection by a $^3S_1$ He atom incident on platinum was measured by DORRESTEIN[4] to be 0.24. The magnetic resonance experiments on the triplet metastable state of helium-4 and helium-3 have used this detection method, and rough estimates from the observed beam intensities ($\sim 10^{-12}$ amperes) give a similar high value for the electron ejection efficiency from an untreated cold tungsten surface in the customary vacuum of 10^{-7} to 10^{-6} mm Hg obtained with oil diffusion pumps and liquid air traps[5,6]. Detection of the metastable $^3P_2$ states of neon and argon by electron ejection from tungsten have been observed with roughly similar efficiencies[7]. Hence the completing process of electron capture by the metal is apparently much less important for the metastable states of He, Ne, and A than it is for the metastable state of H; this is presumably due in part to the fact that the ionization energies of these metastable states (4.8, 5.0 and 4.2 ev, respectively) are higher than the ionization energy of hydrogen (3.4 ev). A considerably smaller beam intensity of krypton in its metastable $^3P_2$ state has been observed[8].

Customarily, the electron current ejected by the metastable atom beam is collected by a plate at a potential of between 2 and 20 volts with respect to the

[1] A. COBAS and W.E. LAMB jr.: Phys. Rev. **65**, 327 (1944).
[2] W.E. LAMB jr. and R.C. RETHERFORD: Phys. Rev. **79**, 549 (1950).
[3] H. HAGSTRUM: Phys. Rev. **96**, 325, 336 (1954); **104**, 672 (1956).
[4] R. DORRESTEIN: Physica, Haag **9**, 433, 447 (1942).
[5] HUGHES, TUCKER, RHODERICK and WEINREICH: Phys. Rev. **91**, 828 (1953).
[6] G. WEINREICH and V.W. HUGHES: Phys. Rev. **95**, 1451 (1954).
[7] WEINREICH, TUCKER and HUGHES: Phys. Rev. **87**, 229 (1952).
[8] DRAKE, HUGHES and LURIO: Bull. Amer. Phys. Soc. [II] **2**, 37 (1957).

Cu, Au, and Ag, and a liquid air cooled copper collecting button[1] and a tungsten button[2] have been used for experiments on Cs^{134m}.

In an extensive series of experiments on radioactive alkali atoms[3] a sulfur button at room temperature has been used as a collector. The collection efficiency is unity and is unaffected by the phase and age of the surface. Further, no measurable loss of radioisotope occurs over a period of several months. In the case of rubidium it is suggested that the success of this sulfur collector is due to the strong chemical binding of the RbS molecule. Also the self-cleaning action of the surface is believed to be important; in air the sulfur forms a coat of sulfurous acid which is rapidly pumped off in the apparatus vacuum to leave a clean sulfur layer. Typical depositions amount to about 10^{-3} of a single molecular layer of RbS. The sulfur collector has also been used in experiments on Cs, Ag, and Tl.

After a typical collection time of 5 to 10 minutes the collector is removed from the vacuum system via some form of airlock and the radioactivity is measured for a time which is usually of the order of the half-life of the isotope. The type of radiation, its energy and half-life, have all been employed to identify the isotope of interest. Often several radioactive isotopes are present due to the production method, and the unwanted isotopes contribute a substantial background counting rate. Changes in counting rates at resonance range typically from 1 to 100 cpm. Isotopes with half-lives as short as 20 minutes have been studied, even in cases where considerable radiochemical work has been required in the source preparation[3].

17. Miscellaneous detectors. The deposition of the material of the beam on a cooled plate was used in the original experiments of STERN and GERLACH with silver atoms [*8*]. It has not found recent application because of the availability of other methods of detection. A considerable discussion of condensation methods of detection is given by ESTERMANN [*7*].

The only common use of detection by chemical interaction of the beam with other matter has been in the detection of atomic hydrogen by its interaction with yellow molybdenum oxide to form blue molybdenum oxide [*7*]. For most experiments involving the use of atomic hydrogen, other detection methods are more convenient; however, the chemical procedure is occasionally used for qualitative observation, as of the effectiveness of a source of atomic hydrogen, where the simplicity of the scheme is very desirable.

The ionization gauge has been used in several investigations. It has the marked disadvantage that it is sensitive not only to the material of the beam but to the residual gas in the apparatus as well. The use of a compensating gauge which detects only the residual gas could, in part, compensate for fluctuations in the residual vacuum. However, the fact that the components of the gauge must operate at high temperatures with the attendant problems of outgassing of components, and the fact that the gauge is limited in application to non-condensable gases suggest that it can hardly compete with other devices as a detector.

A universal-type gauge for the detection of beams has been constructed by ESTERMANN and STERN [*7*]. The operation is based on the modification of the space charge distribution in a diode by the presence of a gas. It has a good sensitivity since the ions of the gas may persist within the gauge for long periods

[1] L. S. GOODMAN and S. WEXLER: Phys. Rev. **99**, 192 (1955).
[2] V.W. COHEN and D.A. GILBERT: Phys. Rev. **95**, 569 (1954).
[3] HOBSON, HUBBS, NIERENBERG, SILSBEE and SUNDERLAND: Phys. Rev. **104**, 101 (1956).

of time. It has, however, many of the limitations of the ionization gauge, in particular that of no differential sensitivity to the beam and to residual gas. The detection of mercury issuing from an oven at $-30°$ C has been reported. In spite of certain useful features of the gauge, it has not often been used in atomic or molecular beam investigations.

II. Beam deflection.

a) General theory.

18. Introduction. Consider a typical apparatus for the production, deflection, and detection of an atomic or molecular beam. All of the essential components of the apparatus are mounted in a vacuum chamber in which the pressure of atoms or of molecules not comprising the beam may be considered to be negligible in its effect on the particles of the beam. The important elements of the apparatus

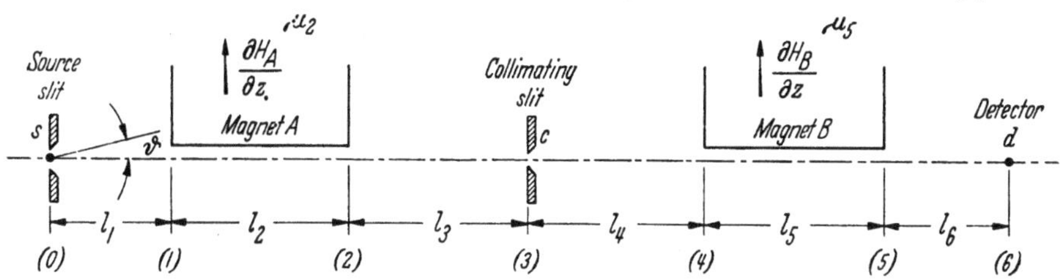

Fig. 28. Schematic diagram of a beam apparatus.

(see Fig. 28) are a source of atoms or molecules with a slit aperture, a collimating slit, various regions which are field-free and whose lengths are at least in part dictated by mechanical considerations, two regions which are occupied by inhomogeneous fields (magnetic or electric), a region in which the transition is to occur and which is ordinarily occupied by a homogeneous field, and a detector. The source slit, the collimating slit, and any other elements which define the aperture of the beam admitted to the detector are all parallel. The slits are perpendicular to the plane of the figure. The regions l_3 and l_4 contain the homogeneous field and the oscillating field which induces the transition. Neither the homogeneous field nor the oscillating field has an effect on the trajectory of the beam, except when a transition occurs and the dipole moment of the particle is changed by the transition. The diagram and the following discussion assume a case where the deflection arises from the force of an inhomogeneous magnetic field on a magnetic dipole moment, but the argument, with a change in notation, is equally applicable when an inhomogeneous electric field acts on an electric dipole moment.

Suppose the energy of a particle in a magnetic field is designated by $W(H)$. Then the force on the particle is:

$$\boldsymbol{F} = -\nabla W = -\frac{\partial W}{\partial H} \nabla H \tag{18.1}$$

where the quantity $-\frac{\partial W}{\partial H}$ is called the magnetic moment, μ, of the particle, and H is the magnitude of the magnetic field. Usually only the force in a single direction

$$F_z = \mu \frac{\partial H}{\partial z} \tag{18.2}$$

is required. The moments of the particles in the two inhomogeneous fields are designated μ_2 and μ_5. They may be different either because the moment is itself a function of field and the field is not the same in the lengths l_2 and l_5, or because a transition occurs within the distances l_3 and l_4.

If a particle leaves the source slit at an angle $\vartheta = 0$ and if no deflecting fields exist, the experimental arrangements are usually such that the particle passes through the collimating slit and onto the detector. It will be assumed that slit widths are negligibly small compared to deflections of particles from the central path *scd*. If a deflecting field is present in the region l_2 (the *A*-magnet), there will be an angle ϑ between the direction of emission of a particle from the oven and the line *scd*, for which the particle will pass through the collimating slit. The angle ϑ depends on the mass, velocity, and magnetic moment of the particle. In the absence of any further deflecting fields, all particles will have a straight line trajectory after leaving the collimating slit, and a distribution of particles will occur in the plane in which the detector lies. For particles of a given mass, the particle density at any position will depend on the velocity and magnetic moment distribution of the particles in the beam.

A deflecting field in the region l_5 (the *B*-field) can be chosen so that the particles which have passed through the collimator after deflection by the *A*-field will strike the detector. This refocussing condition is independent of the mass and velocity of the particles. If the magnetic moment of a particle is independent of magnetic field, as is commonly true, then refocussing can be achieved for a wide range of values of the magnetic fields and their gradients and of the lengths of the various portions of the trajectory. If, however, the magnetic moment depends on the magnetic field, the requirements for refocussing are more restrictive.

Two schemes have been employed for observing transitions of the particles between different quantum states. Suppose the apparatus has been designed so that refocussing occurs if the particles remain in the same state throughout the trajectory. Then refocussing will not occur if the particles undergo a transition in the region l_3 and l_4 to a state with a different magnetic moment, and hence the transition is observed by the decrease in the beam intensity at the detector.

A second method is to arrange the deflecting *B*-magnet in such a way that refocussing occurs at the detector only if a specified moment change occurs in the transition region l_3 and l_4. In this case a transition is detected by an increase in the beam intensity at the detector provided that the transition results in the specified moment change. It is evident that the first method of observing a transition is somewhat more general, since any transition accompanied by a significant moment change is observable; however, the second method has the considerable advantage that it permits observation of transitions of low intensity relative to the total beam intensity without the background of the total beam.

19. Deflection of a beam of single velocity particles. In this section quantitative relations will be derived for the deflection of particles of any particular velocity. Refer to Fig. 28 and consider the slits to be of infinitesimal width. The planes at which deflections are to be calculated are marked by the numbers in parentheses, and the deflections z are taken to be positive if they are above the line *scd*. The accelerations to which the particle is subject in l_2 and in l_5 are given by:

$$a_2 = \frac{\mu_2}{m}\left(\frac{\partial H}{\partial z}\right)_2 \quad \text{and} \quad a_5 = \frac{\mu_5}{m}\left(\frac{\partial H}{\partial z}\right)_5 .$$

Since both H and $\partial H/\partial z$ in the A- and B-regions will actually vary with the position z, the deflecting force also will vary with z. But in most cases the deflection of the beam from the undeflected trajectory is sufficiently small so that this variation of the deflecting force with z can be neglected. It is also adequate usually to disregard the small variations of H and $\partial H/\partial z$ in the direction x perpendicular to the plane of the figure, and the fact that the x coordinate of a beam particle may vary throughout its trajectory. Hence the formulae to be given for a particle travelling in the $y-z$ plane of the figure are applicable for any particle in the beam. A further approximation is that the angle ϑ and the deflections z are sufficiently small so that first order paraxial theory applies. Thus, for example, the velocity of the particle in the direction scd can be taken to have the value v throughout the trajectory.

Suppose a particle travelling with a velocity v in the plane of the figure leaves the oven slit at an angle ϑ with respect to the line scd, where ϑ is positive as shown. With the assumptions mentioned above the deflections of the particle at the planes 1, 2, 3, 4, 5, and 6 are given by:

$$z_1 = l_1 \vartheta, \tag{19.1}$$

$$z_2 = (l_1 + l_2)\vartheta + \frac{a_2 l_2^2}{2v^2}, \tag{19.2}$$

$$z_3 = (l_1 + l_2 + l_3)\vartheta + \frac{a_2 l_2 (l_2 + 2l_3)}{2v^2}, \tag{19.3}$$

$$z_4 = (l_1 + l_2 + l_3 + l_4)\vartheta + \frac{a_2 l_2 (l_2 + 2l_3 + 2l_4)}{2v^2}, \tag{19.4}$$

$$z_5 = (l_1 + l_2 + l_3 + l_4 + l_5)\vartheta + \frac{a_2 l_2 (l_2 + 2l_3 + 2l_4 + 2l_5)}{2v^2} + \frac{a_5 l_5^2}{2v^2}, \tag{19.5}$$

$$z_6 = (l_1 + l_2 + l_3 + l_4 + l_5 + l_6)\vartheta + \frac{a_2 l_2 (l_2 + 2l_3 + 2l_4 + 2l_5 + 2l_6)}{2v^2} + \frac{a_5 l_5 (l_5 + 2l_6)}{2v^2}. \tag{19.6}$$

In order that the particle pass through the collimating slit, which is placed along the line joining the source and the detector, it is necessary that $z_3 = 0$, or

$$\vartheta = -\frac{a_2 l_2 (l_2 + 2l_3)}{2v^2 (l_1 + l_2 + l_3)}. \tag{19.7}$$

Hence Eq. (19.6) becomes:

$$z_6 = \frac{a_2 l_2 (l_2 + 2l_1)(l_4 + l_5 + l_6)}{2v^2 (l_1 + l_2 + l_3)} + \frac{a_5 l_5 (l_5 + 2l_6)}{2v^2}. \tag{19.8}$$

The lengths $l_1, \ldots, l_6$ are usually fixed properties of an apparatus, and it is useful to define Λ by:

$$\Lambda l_5 (l_5 + 2l_6) = \frac{l_2 (l_2 + 2l_1)(l_4 + l_5 + l_6)}{(l_1 + l_2 + l_3)}. \tag{19.9}$$

With this definition Eq. (19.8) reads:

$$z_6 = \frac{l_5 (l_5 + 2l_6)}{2mv^2} \left[\Lambda \mu_2 \left(\frac{\partial H}{\partial z}\right)_2 + \mu_5 \left(\frac{\partial H}{\partial z}\right)_5\right]. \tag{19.10}$$

Refocussing of the particle on an infinitesimal width detector occurs when $z_6 = 0$. This refocussing condition is seen from Eq. (19.10) to be independent of the mass and velocity of the particle, and can evidently be achieved with many combinations of the parameters Λ, μ, and $\partial H/\partial z$. Ordinarily the l's are determined by factors involving the mechanical design of the apparatus, the intensity to be obtained at the detector, the kind of spectrum to be observed, and the width of the image of the undeflected beam which it is desired to obtain at the detector plane. In many cases of interest the moments μ_2 and μ_5 are equal, and then refocussing occurs for $\Lambda\left(\frac{\partial H}{\partial z}\right)_2 = -\left(\frac{\partial H}{\partial z}\right)_5$, which is often obtained with $\Lambda = 1$ and $\left(\frac{\partial H}{\partial z}\right)_2 = -\left(\frac{\partial H}{\partial z}\right)_5$.

The actual finite widths of the slits do not alter the refocussing condition, provided the widths are small enough so that first order deflection theory is applicable. Furthermore, the intensity and shape of the refocussed beam in the detector plane will be the same as that of the undeflected beam. The latter can be understood by noting that for each particle of velocity v which is emitted from the source slit within a small solid angle about the direction $s\,c\,d$ and which arrives at the detector plane in the absence of deflecting fields, there will be a particle of velocity v which is emitted at a small angle ϑ and which will strike the same position in the detector plane in the presence of the deflecting fields, provided the refocussing condition is fulfilled. This theorem will not apply to particles of very low velocity for which ϑ will be large.

In the usual scheme for observing a transition, the refocussing condition ($z_6 = 0$) applies for the particle in the absence of the transition. If the particle undergoes a transition in the region l_3 and l_4 involving a magnetic moment change $\Delta\mu$, it will be deflected to the position

$$z_6 = \frac{l_5(l_5 + 2l_6)}{2mv^2}\,\Delta\mu\left(\frac{\partial H}{\partial z}\right)_5 \tag{19.11}$$

[obtained from Eq. (19.10)], and hence may not strike the detector.

20. Intensity distribution at detector. The intensity distribution in the detector plane of a "Maxwellian" beam which has been deflected as the result of a transition will be considered in this section. The velocity distribution in the beam is given by Eq. (3.6), and the usual case of refocussing in the absence of the transition will be treated. The deflection of a particle with velocity v is given by Eq. (19.11). The number of particles in the undeflected beam with velocities between v and $v + dv$ incident per second on a detector of unit width is given by:

$$I(v)\,dv = \frac{2I_0}{\alpha^4}\,v^3\,e^{-v^2/\alpha^2}\,dv \tag{20.1}$$

where I_0 is the total number of particles incident per second on a detector of unit width. Let s_α be the deflection of a particle with the velocity α. From Eqs. (19.11) and (3.1),

$$s_\alpha = \frac{l_5(l_5 + 2l_6)}{4kT}\,\Delta\mu\left(\frac{\partial H}{\partial z}\right)_5. \tag{20.2}$$

The deflection s_v of a particle with velocity v is:

$$s_v = \frac{\alpha^2}{v^2}\,s_\alpha. \tag{20.3}$$

Since the particles in the beam have on the average higher velocities than α, the deflections of particles in the beam will generally be less than s_α.

Consider an infinitesimal width ds_0 of the refocussed beam in the plane of the detector at the position s_0 where the beam intensity is $I_0(s_0)\,ds_0$. Then the intensity distribution in the detector plane resulting from the deflection of this beam is given from Eqs. (20.1) and (20.3) by:

$$\left.\begin{aligned} dI(s) &= [I_0(s_0)\,ds_0]\,\frac{s_\alpha^2}{(s-s_0)^3}\,e^{-s_\alpha/(s-s_0)} && \text{when } (s-s_0) \text{ has the same sign as } s_\alpha,\\ dI(s) &= 0, && \text{when } (s-s_0) \text{ has the opposite sign from } s_\alpha \end{aligned}\right\} \tag{20.4}$$

in which s is the position coordinate for the deflected beam.

Consider now the total undeflected image in the detector plane shown in Fig. 17. The calculation of the intensity distribution in the deflected beam resulting from this trapezoidal undeflected beam leads to a cumbersome expression[1], and for most purposes it is adequate to consider the deflection of an equivalent rectangular image whose width, $2a$, is equal to the width of the trapezoid at half intensity. Hence the intensity of the undeflected beam is taken as:

$$\left.\begin{aligned} I_0(s_0) &= I_{00} \quad \text{for} \quad -a \leqq s_0 \leqq a, \\ I_0(s_0) &= 0 \ \text{elsewhere}. \end{aligned}\right\} \tag{20.5}$$

The intensity distribution of the deflected beam $I(s)$ is obtained by integrating Eq. (20.4) over the undeflected beam, and is given by:

$$\left.\begin{aligned} I(s) &= I_{00}\left[e^{-s_\alpha/(s+a)}\left(1+\frac{s_\alpha}{s+a}\right) - e^{-s_\alpha/(s-a)}\left(1+\frac{s_\alpha}{s-a}\right)\right] && \text{for} \quad s \geqq a, \\ I(s) &= I_{00}\left[e^{-s_\alpha/(s+a)}\left(1+\frac{s_\alpha}{s+a}\right)\right] && \text{for} \ -a \leqq s \leqq a, \\ I(s) &= 0 && \text{for} \quad s \leqq -a \end{aligned}\right\} \tag{20.6}$$

Eqs. (20.6) apply when s_α is positive; if s_α is negative, obvious modifications of the equation must be made. The total intensity in the deflected beam must equal the total intensity in the undeflected beam and hence

$$\int_{-\infty}^{+\infty} I(s)\, ds = 2a I_{00}. \tag{20.7}$$

If the beam is composed of particles with different magnetic moments, as is often the case, and hence characterized by different values of s_α, then the intensity distribution of the deflected beam is computed as the sum of expressions of the form of Eq. (20.6), weighted in accordance with the relative numbers of particles of the several types.

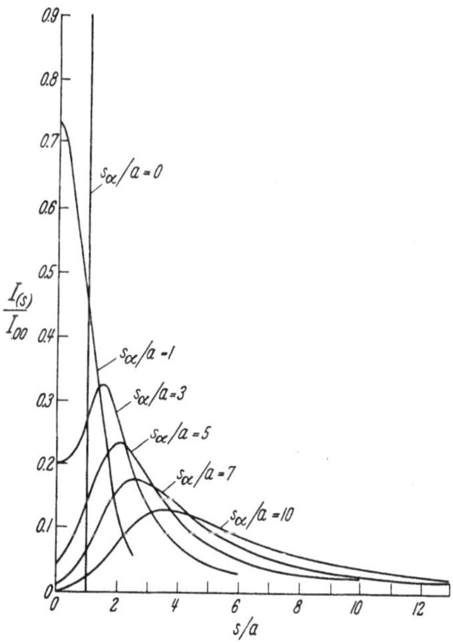

Fig. 29. Intensity distribution in deflected beam for various values of s_α/a, computed from Eq. (20.6). For $s_\alpha/a=0$, $I(s)/I_{00}=1$ for $0 \leqq s/a \leqq 1$.

Table 3. *A comparison of* $(s/a)_{\max}$ *and* s_α/a.

s_α/a	$(s/a)_{\max}$	$s_\alpha/s_{\max}$	s_α/a	$(s/a)_{\max}$	$s_\alpha/s_{\max}$
30	10.067	2.980	6	2.301	2.608
24	8.083	2.969	4	1.745	2.292
18	6.110	2.946	3	1.496	2.005
15	5.131	2.923	2.5	1.382	1.809
12	4.162	2.883	2	1.278	1.565
9	3.212	2.802	1.5	1.186	1.265

Intensity distributions for the deflected beam computed from Eq. (20.6) are shown in Fig. 29 for different values of the dimensionless parameter s_α/a. For large values of s_α/a the position of the maximum in the deflection pattern, $(s/a)_{\max}$, occurs at $\frac{s_\alpha}{3a}$. For moderate values of $\frac{s_\alpha}{a}$, $\left(\frac{s}{a}\right)_{\max} > \frac{s_\alpha}{3a}$. In Table 3 are shown some values of $(s/a)_{\max}$ together with the corresponding values of $s_\alpha/s_{\max}$. It is clear that the deflection of a typical particle in the beam is considerably less than s_α.

[1] N. F. Ramsey [25], Chap. IV.

The change in beam intensity seen by a detector of width w_d due to the transition is the observed signal in an atomic or molecular beam spectrometer. If the beam at the plane of the detector is taken to be a rectangle of width $2a$, then the total refocussed beam intensity seen by the detector, symmetrically placed with respect to the undeflected beam, is

$$\left.\begin{aligned} I &= I_{00} w_d, \quad \text{if} \quad w_d < 2a, \\ I &= I_{00} 2a, \quad \text{if} \quad w_d > 2a. \end{aligned}\right\} \tag{20.8}$$

The beam intensity, I', seen by the detector when a transition occurs is given by integrating Eq. (20.6) over the width of the detector:

$$\left.\begin{aligned} I' &= \int_{-w_d/2}^{w_d/2} I(s)\, ds, \\ I' &= I_{00}\left[\left(a + \frac{w_d}{2}\right) e^{-s_\alpha/\left(a + \frac{w_d}{2}\right)} - \left(a - \frac{w_d}{2}\right) e^{-s_\alpha/\left(a - \frac{w_d}{2}\right)}\right] \quad \text{for} \quad w_d < 2a, \\ I' &= I_{00}\left[\left(a + \frac{w_d}{2}\right) e^{-s_\alpha/\left(a + \frac{w_d}{2}\right)} - \left(\frac{w_d}{2} - a\right) e^{-s_\alpha/\left(\frac{w_d}{2} - a\right)}\right] \quad \text{for} \quad w_d > 2a. \end{aligned}\right\} \tag{20.9}$$

Usually the detector width is less than the width of the undeflected beam ($w_d < 2a$), so that

$$\frac{I'}{I} = \frac{1}{2}\left[\left(\frac{2a}{w_d} + 1\right) e^{-s_\alpha/\left(a + \frac{w_d}{2}\right)} - \left(\frac{2a}{w_d} - 1\right) e^{-s_\alpha/\left(a - \frac{w_d}{2}\right)}\right]. \tag{20.10}$$

In Fig. 30 is plotted the value of I'/I as a function of s_α/a for $2a/w_d = 2$. It is seen that even for large s_α ($s_\alpha = 3w_d$, for example) the decrease in intensity consequent to a transition is considerably less than the number of particles in the beam involved in the transition.

In practice all particles in the beam will not in general undergo the transition, because all particles will not be in the states involved in the transition and because the transition probability is not unity for the particles in the appropriate states. Hence the intensity distribution in the plane of the detector will be a superposition of the intensity distributions of the undeflected beam and of the deflected beam as given by Eq. (20.6). Furthermore, two transitions are usually observed simultaneously—the one corresponding to absorption and the other corresponding to stimulated emission—which may involve moment changes of $+\Delta\mu$ and $-\Delta\mu$. So the intensity distribution is actually the superposition of an undeflected intensity distribution and of the two oppositely deflected intensity distributions.

A second scheme mentioned in Sect. 18 for observing a transition is to arrange the deflecting system so that refocussing occurs only when a specified moment change occurs in a transition, and the transition is then observed as an increase in intensity at the detector. Since the refocussing condition obtained from Eq. (19.10) requires:

$$\Lambda \mu_2 \left(\frac{\partial H}{\partial z}\right)_2 + \mu_5 \left(\frac{\partial H}{\partial z}\right)_5 = 0, \tag{20.11}$$

either the magnetic moment μ or the field gradient $\partial H/\partial z$ must differ in sign in the regions l_2 and l_5 for refocussing to occur. If $(\partial H/\partial z)_2 = (\partial H/\partial z)_5$ and $\Lambda = 1$, as an example, refocussing will not occur if the moment is the same in the two regions but will occur if there is a transition such that $\mu_5 = -\mu_2$. In the absence of a transition s_α can be made large so that the intensity at the detector is low. The intensity may be reduced further by the interposition of a stop wire which serves to block particles with high velocity and therefore with small deflection.

A transition is then observed by an increase in intensity at the detector against a small background. Eq. (20.10) gives the signal intensity at the detector with the altered interpretation that the transition leads to an increase rather than a decrease in the intensity, and hence that I' is the beam intensity without a transition and I is the intensity with a transition. The method is of very great value in the observation of lines arising from an isotope of very low abundance as, for example, from K^{40} in its naturally occurring abundance[1], or from artificially produced radioactive isotopes. It has also been used in several experiments by the electric resonance method[2,3]. The method has the disadvantage that it is not generally possible to observe as many lines as by the method in which refocussing occurs for all states in the absence of transitions, since only a few transitions may result in the appropriate moment change.

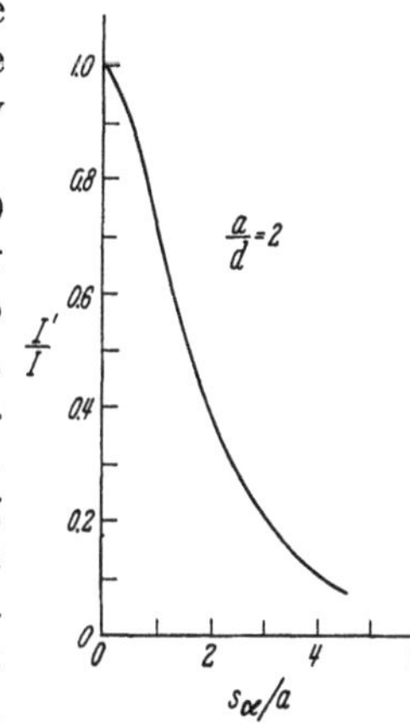

Fig. 30. Relative beam intensity at detector as a function of s_α/a for ratio of undeflected rectangular beam width to detector width of 2, computed from Eq. (20.10). The half width of the undeflected beam is a and of the detector is d where $w_d = 2d$.

In the discussion of the deflection of a beam in Sects. 19 and 20 it has of course been assumed that the atom or molecule retains its state of quantization with respect to the magnetic or electric fields throughout its trajectory, except when an appropriate oscillating field induces a transition in the region between the two deflecting fields. However, even in the absence of an applied oscillating field, a particle travelling through regions with different steady field values may undergo a change in state. If a particle passes through a region of zero magnetic field, it may clearly lose its state of quantization and re-emerge in a different state as it passes into a region where the field is non-zero. A non-adiabatic transition is particularly apt to occur as the particle passes through a region where the field is weak and is changing rapidly in direction[4]. More generally, a particle travelling through a region of non-uniform magnetic field is subjected to a time varying field, and if the amplitude of the effective Fourier component at a possible transition frequency is sufficiently large, a transition can occur. Non-adiabatic interfield transitions have been observed in both magnetic and electric resonance type experiments and have often been found to restrict the lower limit to which the field in the C-region can be taken. Furthermore, the avoidance of non-adiabatic transitions has ordinarily required that the fields in the two deflecting magnets and the homogeneous field be in the same direction. One consequence of this requirement is that the direction of the field gradient in one deflecting magnet cannot be changed relative to the direction of the other simply by changing the direction of its field. Hence an apparatus which has been designed to defocus particles in which no transition has occurred cannot be made to refocus such particles, without some mechanical modifications such as a physical reversal of the direction of the pole tips of one of the inhomogeneous magnets.

b) Deflecting fields.

21. Magnetic dipole moments and magnetic deflecting fields. The magnetic moment of a particle arises from the electron spin and orbital angular momenta,

[1] J. R. Zacharias: Phys. Rev. **61**, 270 (1942).
[2] J. C. Swartz and J. W. Trischka: Phys. Rev. **88**, 1085 (1952).
[3] L. Grabner and V. W. Hughes: Phys. Rev. **79**, 819 (1950).
[4] E. Majorana: Nuovo Cim. **9**, 43 (1932). These non-adiabatic transitions are often called Majorana transitions. See also N. F. Ramsey [25].

from the nuclear angular momentum, and from the molecular rotational angular momentum. The effective moment $\mu (= -\partial W/\partial H)$ depends in general on the state of coupling of the several angular momenta, on the energy level, and on the value of the magnetic field. If the total electronic angular momentum is other than zero, the particle has a magnetic moment of the order of the Bohr magneton $\mu_0 (= 9.27 \times 10^{-21}$ erg gauss^{-1}). If the total electronic angular momentum is zero, as for an atom in the $^1S_0$ state or for a molecule in the $^1\Sigma$ state, the magnetic moment is of the order of the nuclear magneton $\mu_N (= 5.05 \times 10^{-24}$ erg · gauss^{-1}). If the energy is a linear function of the magnetic field, the magnetic moment is constant. This condition is often approximated with the deflecting magnets when the fields are sufficiently large so that the angular momenta are more strongly coupled to the field than to one another. To achieve this condition for atoms, the interaction with the external field must be large compared to the hyperfine structure interaction between the electronic and nuclear angular momenta (usually the magnetic dipole hfs interaction), and for molecules in the $^1\Sigma$ state, the external field interaction must be large compared to the interaction between the rotational and nuclear angular momenta (usually the electric quadrupole interaction). For a single angular momentum the moment is characterized by a g-value, $\mu = -m g \mu_0$, in which m is the magnetic quantum number and g is the negative ratio of the magnetic moment in units of the Bohr magneton to the angular momentum in units of $\hbar$; for two angular momenta separately coupled to the field $\mu = -(m_1 g_1 + m_2 g_2)\,\mu_0$.

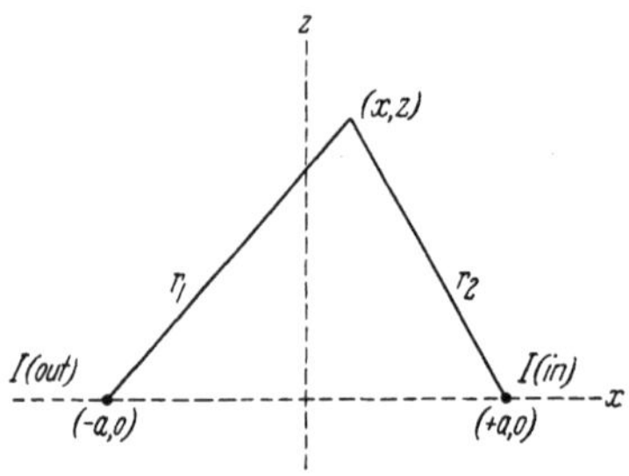

Fig. 31. Geometry of two-wire deflecting field.

It is sometimes convenient to define the g-value as the positive ratio of the magnetic moment in units of the nuclear magneton, μ_N, to the angular momentum in units of $\hbar$; the context will indicate which of the two definitions is appropriate.

The most commonly used deflecting field is one which gives particles in a ribbon-like beam a deflection in a direction perpendicular to the plane of the beam. In this section this classic type of magnetic deflecting field is discussed, and in Sect. 22 the analogous electric deflecting field is treated. Other systems of deflection are considered in Sect. 23.

A system of two straight, parallel wires carrying equal currents in opposite directions (see Fig. 31) gives rise to the magnetic field configuration generally used. The calculation of this magnetic field is a standard problem, and the field and its gradient in the z direction are given by [25]:

$$H = \frac{4Ia}{r_1 r_2}, \tag{21.1}$$

$$\frac{\partial H}{\partial z} = -4Iaz\,\frac{(r_1^2 + r_2^2)}{r_1^3 r_2^3} \tag{21.2}$$

in which H is the magnitude of the total field and the other symbols are defined in the figure. The magnetic equipotentials U and the magnetic field lines V are families of orthogonal circles in a plane perpendicular to the wires whose equations are, respectively:

$$(z - a\cot U)^2 + x^2 = a^2 \operatorname{cosec}^2 U, \tag{21.3}$$

$$z^2 + (x - a\coth V)^2 = a^2 \operatorname{cosech}^2 V. \tag{21.4}$$

Curves of constant U are magnetic equipotentials and pass through the points $x = \pm a$, $z = 0$. Curves of constant V are in the direction of the magnetic field lines and are circles with their centers on the x axis. In Figs. 32 and 33 are shown the field and the gradient of the field in the z direction as a function of x for various values of z. The choice of z/a is dictated by the requirements of a particular experiment, as is the range of x/a within which the beam is propagated. It is often convenient to operate with the beam at a distance of $z = 1.2a$ from the plane defined by the centers of the two wires. The field gradient $\partial H/\partial z$ is reasonably constant with x for this choice of z and is larger than it is for a greater z. The height of the beam is then often taken between the limits $x = -0.7a$ and

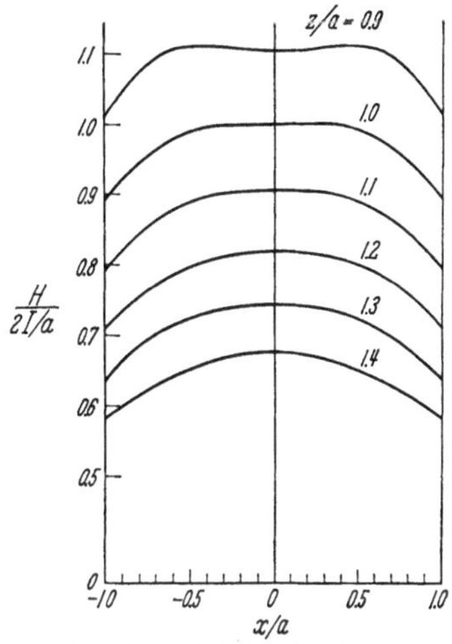

Fig. 32. Magnetic field H in units of $2I/a$ vs. x/a at various values of z/a, for the two-wire field shown in Fig. 31.

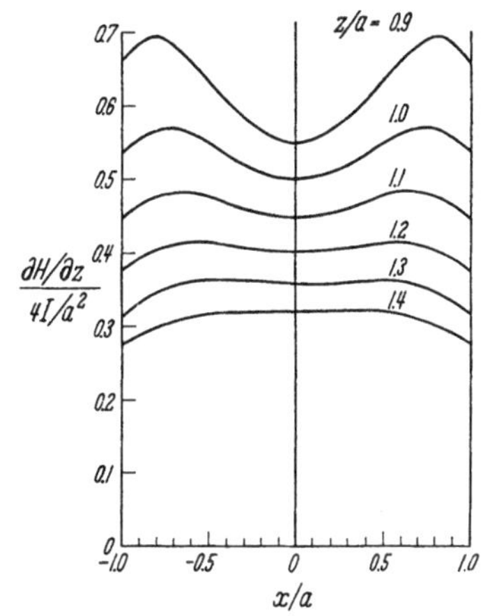

Fig. 33. Magnetic field gradient $\partial H/\partial z$ in units of $4I/a^2$ vs. x/a at various values of z/a, for the two-wire field shown in Fig. 31.

$x = +0.7a$. It has been as large as $2a$, when the field gradient was still sufficiently uniform to permit refocussing of a beam by two deflecting fields. (It will be appreciated that since many particles in the beam traverse the apparatus in such a way that x will vary along the path, the refocussing condition necessitates that $\partial H/\partial z$ shall not vary too markedly with x.) If z is expressed as $z = (1+\delta)a$ and if δ is small, then for $x = 0$ the ratio R of the field gradient to the field is given by:

$$\left.\begin{aligned} R &\equiv \left|\frac{\frac{\partial H}{\partial z}}{H}\right| = \frac{1-\frac{\delta^2}{2}}{a}, \\ R &= \frac{1}{a} \quad \text{for} \quad \delta = 0. \end{aligned}\right\} \tag{21.5}$$

Prodell and Kusch[1] discuss the use of a two wire deflecting field in an investigation of the hyperfine structure and Zeeman effect of hydrogen. The spacing of the wires was 0.5 cm and currents up to 400 amperes were used. In their work it was necessary to place the inhomogeneous deflecting fields within the gap of an electromagnet which produces the uniform field in which the transition frequencies were to be measured. The presence of the electromagnet significantly modifies the properties of the two wire field described above. If the homogeneous field is considerably greater than the field due to the wires,

[1] A. G. Prodell and P. Kusch: Phys. Rev. 88, 184 (1952).

then a region of reasonably constant gradient occurs at $z=0.9a$, which is closer to the wires than the usual value of $z=1.2a$, and hence the ratio of field gradient to field is greater. The presence of the strong homogeneous field also increases the magnetic moment of hydrogen over that obtained in the weaker field due to the wires alone because a Paschen-Back condition is attained.

The maximum fields and field gradients that can be obtained from the two wire system are severely limited by the maximum current which can be sent through the wires consistent with adequate cooling. The considerable length of a two wire deflecting field which would be required in many cases to achieve adequate deflection has the serious disadvantage of reducing the intensity of the beam at the detector.

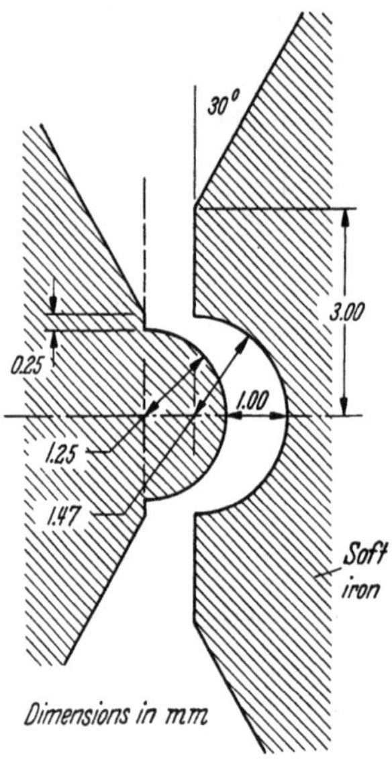

Fig. 34. Cross section of iron deflecting magnet.

The practical solution is to simulate the two wire system with an iron magnet whose pole faces are cylindrical surfaces which coincide with the equipotential surfaces of the two wire system. Evidently the equivalence of the two systems is complete only when the permeability of the iron in the region of the pole faces is so high that the surfaces of the iron are in fact true magnetic equipotentials. The correspondence between the two wire system and the magnet can be obtained by calculating the field at the same point in the two cases. The field for the two wire system is given by Eq. (21.1). The field for the magnet can be computed by use of AMPERE's law

$$\oint \boldsymbol{H}\cdot d\boldsymbol{l} = 0.4\pi N I. \tag{21.6}$$

The integral is taken along a path through the magnet and its gap. NI is the total number of ampere turns through the surface enclosed by the path. It is convenient to choose a path in the median plane of the magnet and gap. If the permeability of the iron is high, the line integral will be appreciable only in the gap and can be evaluated since the dependence of the field on position is the same as for the two wire field given in Eq. (21.1). As an approximation, the variation of the field in the gap may be neglected, and then

$$H = \frac{0.4\pi NI}{L} \tag{21.7}$$

in which L is the length of the air gap in the median plane.

In Fig. 34 is shown a cross section of a magnet with $R=8$. The gap width on the median plane is not a critical dimension, since a change in gap width can be compensated by a change in NI [see Eq. (21.7)]. It is important, however, that the termination of the equipotentials have a small effect on the field in that portion of the gap traversed by the beam. No theoretical solution of the problem of the effect of the discontinuity in the equipotentials has been made. However, for a beam 2 mm in height at $z=1.2a$, the indicated contour is entirely satisfactory. Experimental studies of beam deflections give striking evidence of the close approach of the iron magnet to the two wire system which it approximates[1].

It is useful to consider the deflection properties of a molecular beam apparatus designed for the study of the transitions $\Delta m_I = \pm 1$ in diatomic molecules,

[1] MILLMAN, RABI and ZACHARIAS: Phys. Rev. 53, 384 (1938).

where m_I is the component of a nuclear angular momentum in the field direction[1]. The relevent distances along the beam trajectory (see Fig. 28) are: $l_1 = l_6 = 9.2$ cm, $l_2 = l_5 = 50.8$ cm, $l_3 = l_4 = 27.9$ cm. The long length of $l_3 + l_4$ is provided to allow the use of a long region in which transitions may occur. The resolution of the apparatus is determined by the length of the transition field, and it will be shown in Sect. 25 that $\delta\nu T \simeq 1$ where $\delta\nu$ is the frequency width of a resonance line and T is the time of transit of the molecule through the transition field. The deflecting magnets have a ratio of field gradient to field of about 8 cm^{-1}. This is a property of the magnets which is determined by the contour of the pole faces, provided the magnetic material of which the pole faces are constructed is not magnetically saturated. Permendur is used for the pole faces, and it is estimated to be sufficiently linear up to about 20000 gauss to give a gradient determined by the configuration of the pole faces. The maximum field gradient in the magnet gap is then about 1.6×10^5 gauss/cm. If the moment change occurring in a transition $\Delta m_I = \pm 1$ is measured in terms of the nuclear magneton μ_N, then $\Delta\mu = \pm g_I$. If the source temperature is taken to be 900° K, then for the apparatus under consideration with $\partial H/\partial z = 1.6 \times 10^5$ gauss/cm, use of Eq. (20.2) gives:

$$s_\alpha = 5.7 \times 10^{-3} g_I \text{ (cm)}.$$

The smallest nuclear g value observed by the molecular beam magnetic resonance method is that of K^{39} for which $g_I = 0.260$, and a considerable number of other nuclei have $g_I < 1$.

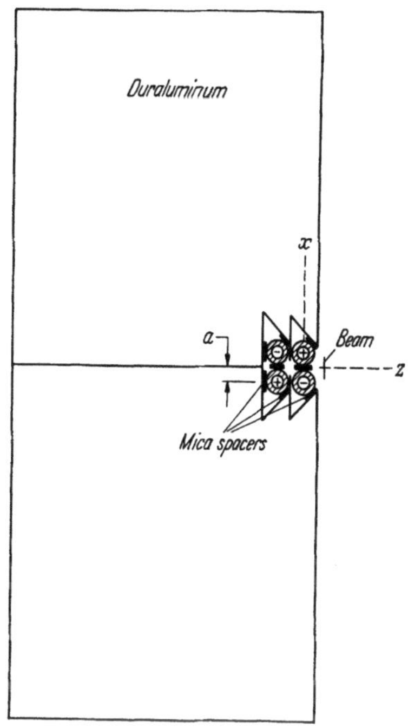

Fig. 35. Cross section of field block showing the arrangement of a four wire deflecting field. + and − indicate opposite directions of current. $a = 0.121$ cm. Beam height = 0.1 cm. All wires are hollow to permit cooling by water. The current in the two wires farthest from the beam is taken as twice that in the two wires near the beam.

The source slit, collimating slit, and detector all have a width of 0.0025 cm for this apparatus. A considerable reduction in the widths of these elements reduces the beam intensity to an excessively low level and increases the already serious problem of maintaining the relative alignment of the elements which define the beam. The ratio of the halfwidth of the undeflected beam to the detector width, $2a/w_d$, is 2, and for K^{39} the value of s_α is 1.5×10^{-3} cm. The parameter s_α/a is then 0.60, and reference to Fig. 30 shows that the change in beam intensity resulting from a transition $\Delta m_I = \pm 1$ in a molecule containing K^{39} is 14% of that which would be observed if all molecules undergoing a transition were to miss the detector completely.

Millman[2] has discussed the properties of a deflecting field which uses a system of four wires arranged so that their centers form a square of side $2a$ (see Fig. 35). A magnetic field and field gradient constant to about 1% is obtained over the interval $-0.8a \leqq x \leqq 0.8a$ at the beam position $z = 1.2a$. This system is of value when quantitative results are to be derived from an exact knowledge of the deflecting field which the beam traverses.

---

[1] Logan, Coté and Kusch: Phys. Rev. **86**, 280 (1952).
[2] S. Millman: Phys. Rev. **47**, 739 (1935).

It is not difficult when observing certain of the transitions between the magnetic components of the hyperfine structure levels of an atom to obtain a decrease in intensity at the detector very nearly as great as the total number of atoms in the beam which have undergone a transition. This follows from the fact that the moment change consequent to a transition may be of the order of a Bohr magneton, and hence an s_α many times the width of the undeflected beam and of the detector may be obtained with relatively short deflecting magnets having moderate field gradients. The discussion above of the deflection properties of a molecular beam apparatus indicates that the observation of nuclear resonances and of internal interactions in diatomic molecules in a $^1\Sigma$ state is much more difficult. The change in magnetic moment in a transition is of the order of a nuclear magneton, and s_α may then be much smaller than required to give a change in intensity equal to the total number of molecules undergoing the transition.

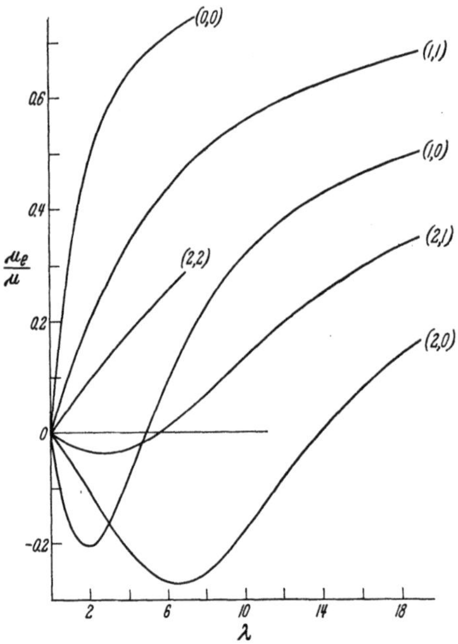

Fig. 36. Effective electric dipole moment, μ_e, in units of the permanent electric dipole moment, μ, vs. the dimensionless parameter $\lambda\left[=\frac{\mu E}{\hbar^2/(2A)}\right]$ for the rotational states $J=0$, $J=1$, and $J=2$. The different curves are designated by their (J, m_J) values.

22. Electric dipole moments and electric deflecting fields. A polar di-

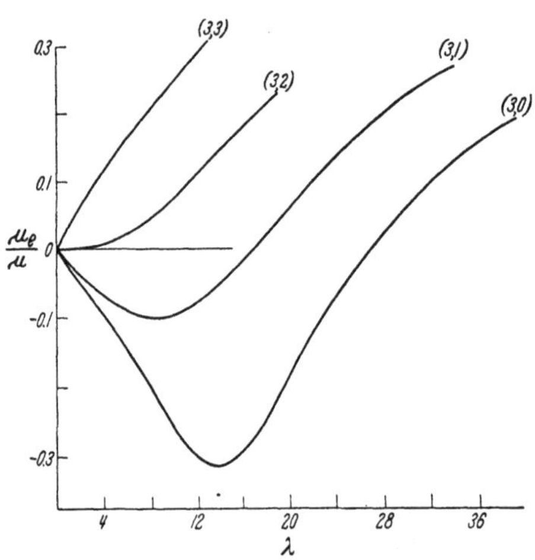

Fig. 37. Effective electric dipole moment, μ_e, in units of the permanent electric dipole moment, μ, vs. the dimensionless parameter $\lambda\left[=\frac{\mu E}{\hbar^2/(2A)}\right]$ for the rotational state $J=3$. The different curves are designated by their (J, m_J) values.

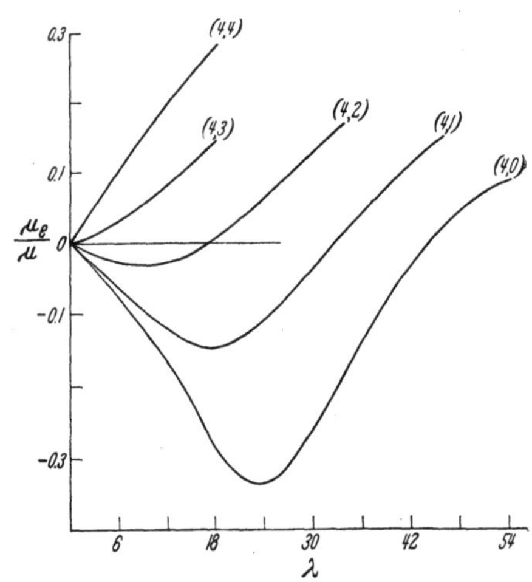

Fig. 38. Effective electric dipole moment, μ_e, in units of the permanent electric dipole moment, μ, vs. the dimensionless parameter $\lambda\left[=\frac{\mu E}{\hbar^2/(2A)}\right]$ for the rotational state $J=4$. The different curves are designated by their (J, m_J) values.

atomic molecule possesses a permanent electric dipole moment, μ. The energy, W, of the molecule is then a function of the electric field E in which the molecule is placed. The electric dipole moment effective for the deflection of a molecular beam by an inhomogeneous electric field, μ_e, is given by:

$$\mu_e = -\frac{\partial W}{\partial E}. \tag{22.1}$$

The effective dipole moment depends on μ, and also on the motion of the nuclei. In particular then, μ_e is a function of the molecular rotational angular momentum (quantum number $= J$) and of its projection in the direction of the applied electric field (quantum number $= m_J$); the variation of μ_e with vibrational state is negligibly small. The theory of the energy levels of a diatomic molecule in an electric field is given in Sect. 51 where the rigid rotor model of a diatomic molecule is used. Figs. 36 to 38 show the effective electric dipole moments as a function of electric field for the rotational states $J = 0, 1, 2, 3$, and 4. The values of μ_e were calculated from the

Table 4. *Permanent electric dipole moments of alkali halide molecules* (in debye or 10^{-18} esu units) (for vibrational state $v = 0$ except for [36 *R 2*]).

	F	Cl	Br	I
Li	6.6 ±0.3 [55 *B 3*]	5.9 ±1.3 [56 *M 1*]	6.19±0.15 [54 *H 3*]	6.64±0.20 [54 *H 3*]
Na		8.5 ±0.2 [54 *T 3*]		
K	7.33±0.24 [50 *G 1*]	10.48±0.05 [53 *L 1*] 10.1 ±0.2 [54 *T 3*]	10.41±0.05 [53 *F 1*]	11.1 [36 *R 2*]
Rb	8.80±0.10 [58 *L 2*]			
Cs	7.88±0.17 [49 *T 4*]	10.5 ±0.25 [51 *L 4*] 10.46±0.14 [53 *L 4*]		12.1 [36 *R 2*]

energy values given in Sect. 51. It will be noted that the effective electric dipole moment depends only on the absolute value of m_J. Actually J is not a good quantum number at high electric field values where $\mu E \gg J \frac{\hbar^2}{2A}$, and it is used simply to designate the energy level. For high rotational states (J large) and for $\mu E \ll J \frac{\hbar^2}{2A}$, the maximum value of μ_e is given approximately from Eq. (51.5) as:

$$\mu_e \simeq \frac{\mu}{2J^2} \frac{\mu E}{(\hbar^2/2A)} \tag{22.2}$$

and hence the effective electric dipole moment will be small. Table 4 gives values of the known permanent electric dipole moments of alkali halide molecules. Most of these values have been determined by molecular beam electric resonance or microwave absorption spectroscopy.

The imhomogeneous electric field which has been used as the deflecting field in almost all of the molecular beam electric resonance experiments is the electric analogue of the two-wire magnetic deflecting field. The basic field configuration is the conjugate of that due to two parallel line charges of opposite sign separated by a distance $2a$. The field and field gradient are given by expressions analogous to Eqs. (21.1) and (21.2):

$$E = \frac{4aq}{r_1 r_2}, \tag{22.3}$$

$$\frac{\partial E}{\partial z} = -4q\,a\,z\,\frac{(r_1^2 + r_2^2)}{r_1^3 r_2^3}. \tag{22.4}$$

Reference should be made to Fig. 31 for the definition of the geometrical symbols. The quantity q is the magnitude of either line charge per unit length. The equipotentials and field lines are given by Eqs. (21.3) and (21.4), respectively. The variation of the field and its gradient with position are given in Figs. 32 and 33, and the related discussion about optimum beam positions and beam height given in the magnetic case is applicable to the electric case as well.

The actual construction of a deflecting field is achieved with two cylindrical metal surfaces which coincide with two of the equipotentials of Eq. (21.3). A typical deflecting field is shown in Fig. 39 for which the equipotentials were constructed of duraluminum and the insulating supports were of lavite. A potential difference of 10 to 15 kv is applied between the electrodes. The field can be computed from Eq. (22.3) with use of the fact that the potential difference between the electrodes equals the line integral of the electric field between the two electrodes taken along the median plane. The ratio, R, of electric field gradient to electric field at the positon of the beam ($z = 1.2a$) is about 1.4 cm^{-1}. No careful study of the effect of the terminations of the equipotentials on the field at the beam position has been made, but the qualitatively satisfactory behaviour of the deflecting field indicates that the effect is small.

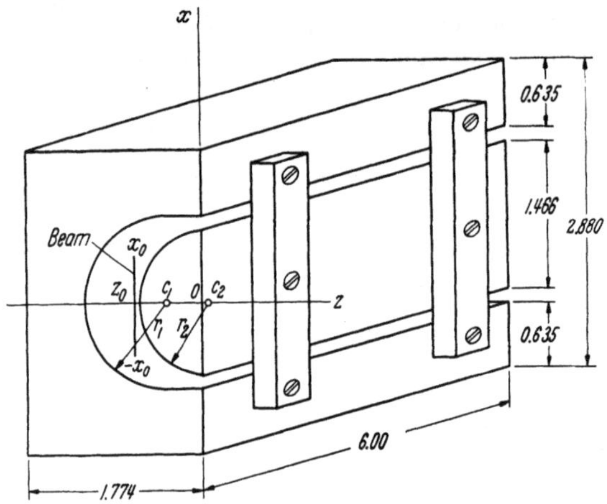

Fig. 39. Electric deflecting field, oblique projection (dimensions in cm), $a = 0.733$, $z_0 = 0.733$, $x_0 = 0.550$, $r_1 = 0.805$, $r_2 = 0.742$, $c_1 = 0.334$, $c_2 = -0.109$.

The deflection obtained in electric experiments may be compared with that obtained in magnetic experiments. The characteristic value for the effective electric dipole moment of a polar molecule in a low rotational state is of the order of 1 debye ($= 10^{-18}$ esu) whereas the characteristic value for the magnetic moment of an atom is of the order of 1 Bohr magneton ($\simeq 10^{-20}$ ergs gauss^{-1}). Hence if the electric and magnetic deflecting fields have the same geometry, then the electric and magnetic forces will be equal if E (esu) $= 0.01\,H$ (gauss). A common strong magnetic deflecting field is 10000 gauss and the corresponding electric field would be 100 esu or 30000 volts/cm, which may readily be achieved. Thus the general requirements for electric deflecting fields for polar molecules closely approximate those for magnetic deflecting fields for atoms. Magnetic deflecting fields as applied to molecules, of course, require much higher values of R and much longer magnets to deflect the molecules usefully, since the effective magnetic moment is of the order of a nuclear magneton rather than a Bohr magneton.

Several refocussing schemes have been employed to select molecules with a particular value of J and $|m_J|$ when J is small. The beam of alkali halide molecules is produced in an oven at a sufficiently high temperature so that many rotational states are populated. Indeed, typically the most probable rotational state for a molecule in the beam is near $J = 50$, and only about 1 in 10^3 of the molecules will be in the rotational state $J = 0$, $J = 1$, or $J = 2$ on which spectroscopic studies are usually made. An obstacle or stop wire is placed in the direct path from source to detector and will prevent molecules of high J from reaching the detector because such molecules have a small effective electric dipole moment and are not appreciably deflected by the inhomogeneous electric fields. On the other hand, a trajectory for molecules of low J can be arranged with a sufficiently large deviation from the path of the undeflected beam so that all except the very high velocity component of the beam is deflected around the obstacle wire. Thus refocussing is achieved for the state of interest and the large background from molecules of higher J is avoided. As an example, a molecule in the $(J, m_J) = (1, 0)$ state can be refocussed if the A- and B-fields are set at $\lambda \left(\lambda = \frac{\mu E}{(\hbar^2/2A)}\right)$ values of about 7 and 2 respectively in an apparatus in which both the field and its gradient have the same direction in both the A- and B-regions[1]. From Fig. 36 it is seen that the effective dipole moment has opposite signs in the two fields so the refocussing condition [see Eq. (20.11)] can be fulfilled for suitable chosen lengths of the deflecting fields. The $(2, \pm 1)$ states may also be refocussed because of the change in sign of the moment between the A- and B-fields, but generally in an apparatus designed to refocus the $(1, 0)$ state the deflection of molecules in the $(2, \pm 1)$ states will be so small that only the low velocity particles will be refocussed around the stop wire. Still, appreciable intensities of molecules in these states are refocussed. Molecules in the other (J, m_J) states shown in Fig. 36 will not be refocussed because the sign of their moment is the same in both the A- and B-fields, and hence their deflection by the two fields will be in the same direction. A transition induced in the region between the deflecting fields from the $(1, 0)$ state to the $(1, \pm 1)$ states, for example, will be observed as a decrease in intensity because the sign of the moments in the A- and B-fields will then be the same and defocussing will occur.

Refocussing of the $(0, 0)$ state has been achieved by reversing the direction of the electrodes in one of the deflecting fields, so that the direction of the field gradient in one field will be opposite to that in the other whereas the direction of the two fields is the same[2]. A background of refocussed molecules in the $(1, \pm 1)$ and $(2, \pm 2)$ states is present. A transition to the $(1, 0)$ state involves a change in sign of the moment and hence leads to defocussing, and a transition to the $(1, \pm 1)$ state involves a sufficiently large change in the magnitude of the moment to lead also to appreciable defocussing.

Another refocussing scheme which relies on non-adiabatic transitions occurring between the regions A and B has been found useful[3]. If the A deflecting field is set near the value $\lambda = 3.6$, then a molecule in the $(1, +1)$ state will suffer the same deflection due to the A-field as did a molecule in the $(1, 0)$ state when the field was set at $\lambda = 7$. Hence if a molecule makes a non-adiabatic (MAJORANA) transition between the A- and B-regions from the $(1, \pm 1)$ to the $(1, 0)$ state and if the B-field is set at $\lambda = 2.8$, then the molecule will be refocussed on the detector. If the non-adiabatic transition occurs between the A- and C-fields and if an oscillating field is present in the C-region which induces a transition from the $(1, 0)$ to the $(1, \pm 1)$ state, then the molecule will be defocussed. The usefulness of this scheme for observing the transition from the $(1, 0)$ to the $(1, \pm 1)$ states, as compared with the more straightforward method discussed above in which quantization of the $(1, 0)$ state is maintained throughout the apparatus in the absence of an induced transition, depends on the probability for occurrence of the non-adiabatic transition between the A- and C-fields. Under low C-field

[1] H. K. HUGHES: Phys. Rev. **72**, 614 (1947).

[2] LEE, FABRICAND, CARLSON and RABI: Phys. Rev. **91**, 1395 (1953).

[3] L. GRABNER and V. W. HUGHES: Phys. Rev. **79**, 819 (1950).

conditions the intensity of the transition observed by this scheme has been found to be greater than that observed by the straightforward method.

To minimize non-adiabatic transitions in the interfield regions it has been found useful to operate the deflecting fields with one electrode at a positive potential and the other at a negative potential so that the beam runs along a plane at very nearly zero potential. A buffer field between the A- and the C-

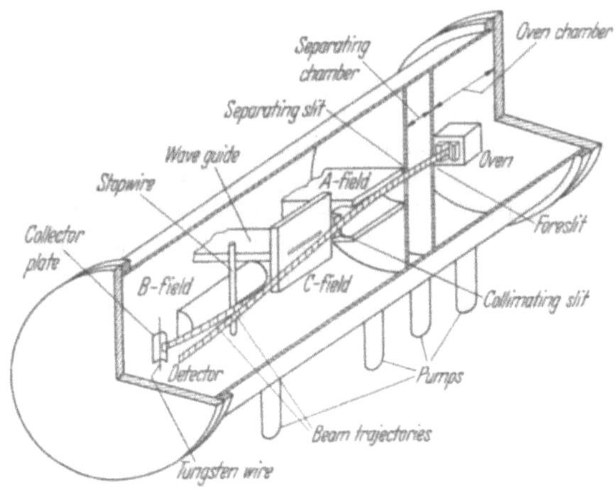

Fig. 40. Schematic diagram of molecular beam electric resonance apparatus.

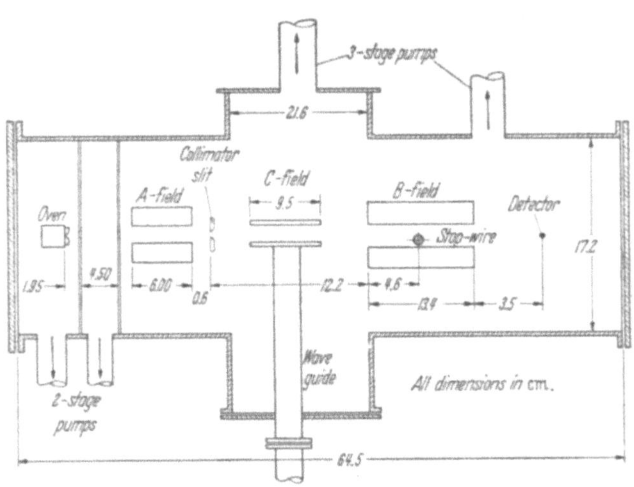

Fig. 41. Top view of molecular beam electric resonance apparatus of Fig. 40, showing dimensions.

regions, which also serves as a collimator, and another buffer field between the B- and C-regions have been found useful for avoiding these non-adiabatic transitions[1].

An apparatus of typical dimensions, which has been used for the study of rotational state transitions, is shown in Figs. 40 and 41[2]. In order to avoid distortion of the electric fields the collimator slits are made of a dielectric such as quartz, and the collimator is placed between the A- and C-fields.

[1] R. BRAUNSTEIN and J. W. TRISCHKA: Phys. Rev. **98**, 1092 (1955).
[2] LEE, FABRICAND, CARLSON and RABI: Phys. Rev. **91**, 1395 (1953).

23. Multipole deflecting fields. Multipole magnetic and electric deflecting fields have been used to focus particles for atomic and molecular beam spectroscopy. If the force on a particle is proportional to its distance from an axis and is directed towards the axis, then the particle will execute simple harmonic motion in the plane perpendicular to the axis. In particular, suppose

$$\boldsymbol{F} = -k\boldsymbol{r} \tag{23.1}$$

in which k is a constant and $\boldsymbol{r}$ is the radial distance from the axis. Then the angular frequency of oscillation is given by:

$$\omega = \sqrt{k/m} \tag{23.2}$$

in which m is the mass of the particle. The particle's initial velocity, v_y, along the axis will be unaffected by the force of Eq. (23.1). Thus all particles emerging from a point source on the axis with an axial velocity v_y, irrespective of their angle of emission, will be focussed on the axis at a distance $(\pi v_y)/\omega$ from the source point. It is seen that the position of the focus point depends on the axial component of velocity of the particle. The action of a deflecting field which exerts a force given by Eq. (23.1) on the particle trajectories is analogous to the action of a converging lens on the path of a light beam. A focus point is achieved for both the particles and the light, and the dependence of the position of the focal point for the particles on the axial velocity (velocity aberration) corresponds to chromatic aberration in the case of light.

Multipole magnetic deflecting fields[1,2] have been applied principally in atomic beam spectroscopy studies of radioactive atoms[3,4]. The principal advantage of these deflecting fields is the increase in beam intensity associated with the larger solid angle subtended by the detector at the source, which can be attained with the cylindrical geometry as compared to the usual geometry determined by narrow vertical slits. If an atom has a net electronic angular momentum and if it is placed in a sufficiently large magnetic field so that the magnetic interaction energy with the external field is large compared to the interaction energy between the nuclear and electronic angular momenta, then the atom will have a magnetic moment $\mu\left(=-\frac{\partial W}{\partial H}\right)$ which is independent of magnetic field and is of the order of a Bohr magneton. In a magnetic field H which is in the radial direction and which varies as:

$$H = \frac{C}{2} r^2 \tag{23.3}$$

where C is a constant, the atom will experience a force:

$$\boldsymbol{F} = \mu C \boldsymbol{r}. \tag{23.4}$$

If μ is negative, the force is directed towards the axis and atoms of a given axial velocity from a point source will all pass through a focus. If μ is positive, the force is directed away from the axis. In order that the magnetic moment of the atom remain constant as the atom traverses a region in which the magnetic field is changing, it is necessary that the changes in magnetic field seen by the atom be sufficiently slow so that non-adiabatic transitions to other quantum

[1] H. FRIEDBURG and W. PAUL: Naturwiss. **38**, 159 (1951).
[2] H. G. BENNEWITZ and W. PAUL: Z. Physik **139**, 489 (1954).
[3] LEMONICK, PIPKIN and HAMILTON: Rev. Sci. Instrum. **26**, 1112 (1955).
[4] CHRISTENSEN, HAMILTON, LEMONICK, PIPKIN, REYNOLDS and STROKE: Phys. Rev. **101**, 1389 (1956).

states, which would have a different projection of μ in the direction of $\boldsymbol{H}$, be avoided. This condition is usually satisfied, provided the atom is not too near the axis $r=0$ where the Larmor frequency is low.

A magnetic field which varies in the manner given by Eq. (23.3) is obtained if the magnetic scalar potential varies as $r^3 \sin 3\varphi$ where φ is the azimuthal polar coordinate in cylindrical coordinates. (More generally, if the magnetic scalar potential is proportional to $r^n \sin n\varphi$, then $\boldsymbol{H}$ is in the radial direction and varies as r^{n-1}.) Field and potential lines are shown in Fig. 42. A magnetic field closely approximating this dependence is produced if six line-poles, alternating in sign, are evenly spaced at a constant distance from the axis of symmetry. A transverse cross section of such a six-pole field is shown in Fig. 43. The distance of the poles from the axis

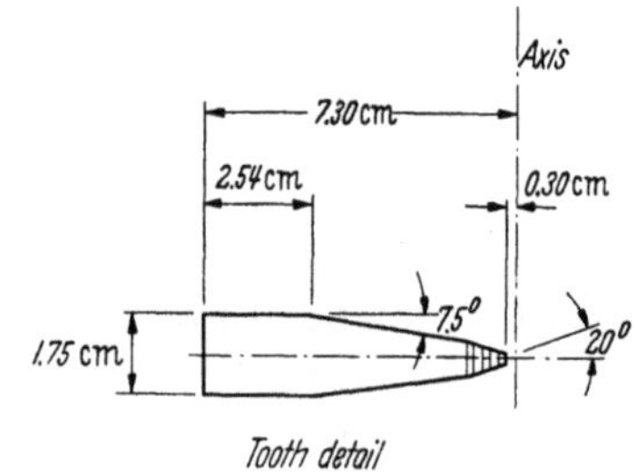

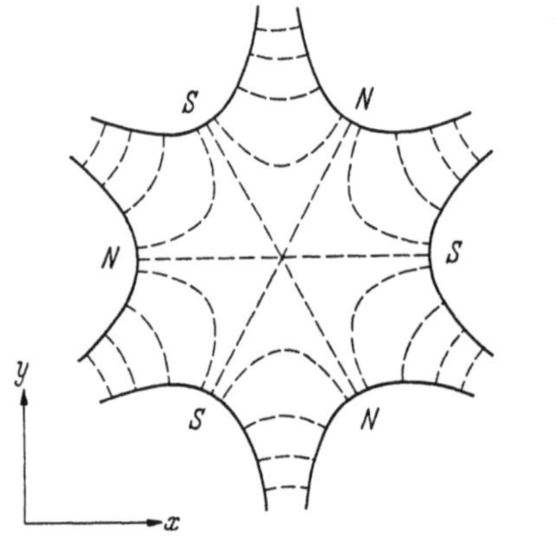

Fig. 42. Field and potential lines for a magnetic field which varies as the square of the distance from the axis. Field lines are dashed (----); potential lines are solid (——); axis is through the geometrical center of diagram perpendicular to figure.

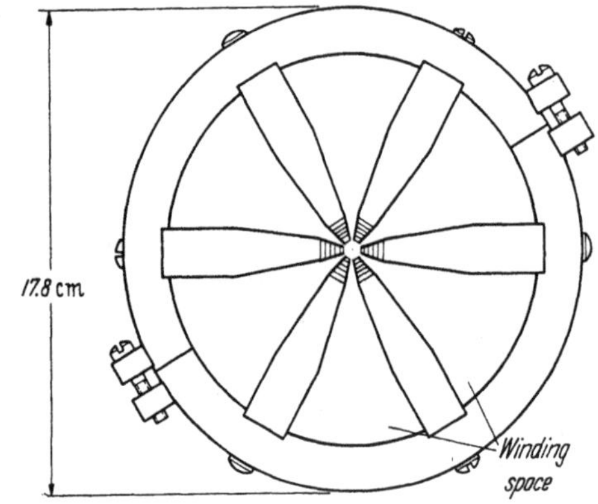

Fig. 43. Six-pole B-magnet, view of the soft-iron structure. Magnet windings (not shown) fill space between teeth. Striations near tips of individual teeth correspond to the longitudinal shaping of the B-magnet which is apparent in Fig. 44.

varies along the length of the magnet. This longitudinal shaping increases the solid angle into which atoms emitted from the source will reach the detector.

A system to detect radiofrequency transitions using these six pole magnetic fields is shown in Fig. 44. The system consists of two deflecting fields corresponding to the conventional A- and B-fields and of a conventional C-field in which radiofrequency transitions are produced. Atoms which have a negative magnetic moment in both deflecting fields (atoms of focussing trajectories) run into the axial stop. Atoms for which the magnetic moment is positive in the A-field never get beyond the entrance to the B-magnet. Only those atoms that make a transition from a state of negative magnetic moment to a state of positive magnetic moment in the C-field, i.e., those atoms which are attracted to the axis in the A-field and are repelled from it in the B-field, reach the detector. The arrangement has the property that none of the atoms which reach the detector have crossed the axis, which is a region where $H \simeq 0$, and thus the danger of non-adiabatic transitions is avoided.

A detector of large size, which consisted of a copper disk 4,5 cm in diameter on which radioactive isotopes were collected, could be used because the scheme for observing a transition does not require the focussing of the beam at a point. Thus the variation of the trajectories with velocity is not a critical matter. An overall improvement in signal intensity of a factor of about 25 as compared with the conventional apparatus has been achieved.

Multipole electric deflecting fields have been used in spectroscopic studies of polar molecules. If a molecule has a quadratic Stark effect so that its electric dipole moment $\mu_e\left(=-\frac{\partial W}{\partial E}\right)$ is proportional to the magnitude of the applied electric field, $(\mu_e/E = \text{constant})$ then in an electric field which is in the radial direction and which varies as:

$$E = C' r \qquad (23.5)$$

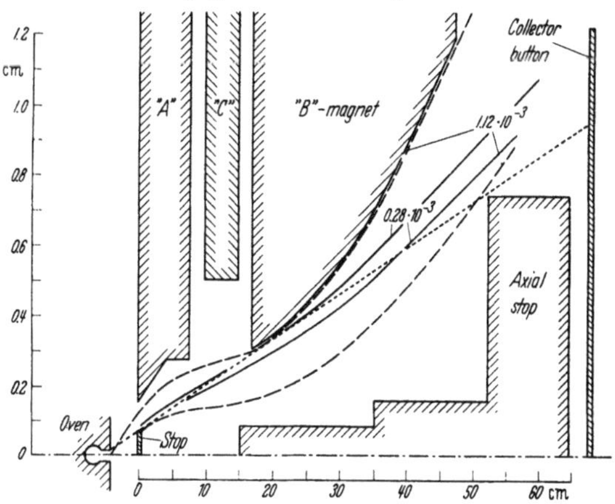

Fig. 44.

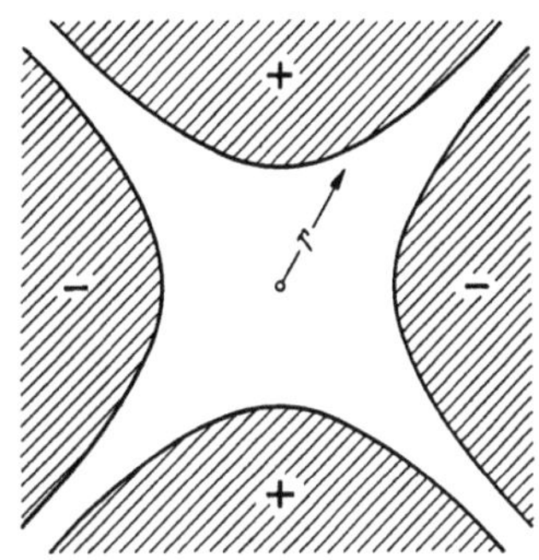

Fig. 45.

Fig. 44. "Focussing" atomic beam apparatus. Scale drawing (note different transverse and longitudinal scales) showing trajectories from center of the oven; stops; magnets; oven. Trajectories are labelled with corresponding values of $\mu H_0/kT$ in which H_0 is the maximum value of the magnetic field at the surface of the A and B magnets, and T is the "kinetic temperature" of the atom, given by $mv^2 = 3kT$. The sign of μ is reversed in the C-region. The two trajectories shown for each value of the parameter correspond to the limiting angles of departure from the source that allow transit of atoms through the magnet. The dotted line corresponds to $\mu H_0/kT = 0$.

Fig. 45. Four pole electric deflecting field.

where C' is a constant, the force on the molecule is:

$$\boldsymbol{F} = 2\left(\frac{\mu_e}{E}\right) C'^2 \boldsymbol{r} = \text{const} \times \boldsymbol{r}. \qquad (23.6)$$

If μ_e is negative, $\boldsymbol{F}$ is directed towards the axis and a focussing is achieved, as discussed previously in this section. On the other hand, if μ_e is positive, the particle trajectory diverges from the axis. The electric field given by Eq. (23.5) is obtained if the electric potential varies as $r^2 \sin 2\varphi$. The system of four cylindrical electrodes shown in Fig. 45, whose inner faces are shaped to form hyperbolae and whose alternate poles are at positive and negative potential, gives rise to this potential.

The four pole electric deflecting field has been used in the maser[1,2] to separate the two inversion levels in NH_3. Molecules in the upper inversion state are focussed into a microwave cavity which serves as the detector, whereas molecules in the lower inversion state are deflected away from the axis and do not enter the detector. This type of deflecting field has also been used for the A and B fields in molecular beam electric resonance experiments on alkali halide molecules[3].

[1] J. P. Gordon: Phys. Rev. **99**, 1253 (1955).
[2] Gordon, Zeiger and Townes: Phys. Rev. **99**, 1264 (1955).
[3] Bennewitz, Paul and Schlier: Z. Physik **141**, 6 (1955).

III. The transition process.

24. Introduction. The determination of the energy level system of an atom or molecule by the method of atomic or molecular beam spectroscopy is made by observing the effect on the atom or molecule of the absorption or emission of a photon of approximately the Bohr frequency for a transition between two levels. Customarily the oscillating field which induces transitions is varied over a range of frequency which will include the resonance line, and an indication is obtained on an appropriate detecting system of the probability of transitions as a function of frequency. All experimental parameters, other than the frequency, of which the most significant are the externally applied steady magnetic or electric field in the C-region and the amplitude of the oscillating field, are held constant. Alternatively, the steady field may be varied while all other experimental parameters are held fixed during the observation of a resonance line. This latter procedure has the faults that the energy level system of an atom or a molecule may often be a complex function of the field, and that it is not easily possible to measure small increments of field with an accuracy comparable to that with which small increments of frequency can be measured.

For a resonance line the transition probability as a function of frequency is of great importance in the design and interpretation of atomic beam experiments. The width of the line in frequency is a basic factor in determining the ultimate accuracy of an experiment, and the determination of the frequency of transition from a resonance line of finite and possibly considerable width requires a detailed theory of the line shape. The radiofrequency power required for an experiment should be such that the probability of transition at the center of a line be a maximum.

The line width, which may be taken as the frequency difference $\delta\nu$ between half intensity points, can be no narrower than the natural width determined from the HEISENBERG uncertainty principle as $\delta\nu T \sim 1$, where T is the time interval over which the atom is subjected to the oscillating transition field. The velocity distribution of the atoms in the beam must be considered to determine the natural width of the resonance line. Usually the lifetimes of the states between which transitions are to be observed are long compared to the time T and hence do not significantly affect the line width, but in experiments involving optically excited states the line width is largely determined by the spontaneous decay rates of the states.

The frequencies of lines are often markedly dependent on magnetic or electric field and the width of a resonance line may be determined principally by the inhomogeneity of the externally applied steady field. The use of the method of separated oscillating fields may reduce the line width to the natural width despite field inhomogeneities. Collision broadening, which is an important factor in optical spectroscopy and in microwave absorption spectroscopy, is negligibly small in beam experiments. Doppler broadening may be eliminated by the design of the oscillating field so that no travelling wave exists in the direction of the particle's velocity.

A variety of transition conditions are encountered in atomic and molecular beam experiments which depend on the states involved and on the type of oscillating field used. Any sufficiently general formulation of the theory of the transition process to cover all the cases would be excessively unwieldy, so various important special cases will be considered separately in the following sections. In general the transition probability for a particle in the beam is large near resonance, so the theory of the transition process involves a complete solution of the time dependent Schrödinger equation rather than a perturbation treatment.

25. Two stable levels; single velocity particles. The Hamiltonian for an atom or molecule in the C-region of a beam spectrometer can be written as:

$$\mathscr{H} = \mathscr{H}_0 + V \tag{25.1}$$

in which $\mathscr{H}_0$ includes all internal interactions of the atomic (or molecular) system and interaction of the atom with an externally applied steady magnetic or electric field. V is the interaction of the atom with the applied oscillating field. Suppose the wave function, ψ, of the atom is expressed as:

$$\psi = \sum_n c_n(t)\,\varphi_n \tag{25.2}$$

in which φ_n are the eigenstates of $\mathscr{H}_0$, and $c_n(t)$ are the time dependent probability amplitudes. The time dependent Schrödinger equation for the probability amplitudes $c_m(t)$ is[1]:

$$i\hbar\,\dot{c}_m = \sum_n (\varphi_m^* \,|\mathscr{H}|\, \varphi_n)\, c_n. \tag{25.3}$$

The special case of two stable (long-lived) energy levels between which a transition may occur is the usual and classic case for atomic and molecular beam experiments. The interaction term V arising from the oscillating field of angular frequency ω is assumed to have the matrix elements

$$V_{pq} = \int \varphi_p^* V \varphi_q\, d\tau = \hbar\, b\, \mathrm{e}^{+i\omega t}; \qquad V_{qp} = \hbar\, b^* \mathrm{e}^{-i\omega t}; \qquad V_{pp} = V_{qq} = 0 \tag{25.4}$$

in which p and q designate the two states. The quantity b is independent of time. For this case Eq. (25.3) is:

$$\left.\begin{aligned} i\hbar\,\dot{c}_p &= W_p c_p + \hbar\, b\, c_q \mathrm{e}^{+i\omega t}, \\ i\hbar\,\dot{c}_q &= W_q c_q + \hbar\, b^* c_p \mathrm{e}^{-i\omega t} \end{aligned}\right\} \tag{25.5}$$

where W_p and W_q are the energy eigenvalues of the states p and q for the Hamiltonian term $\mathscr{H}_0$. It will be assumed that all atoms are in the state p as they enter the oscillating field, i.e.,

$$|c_p| = 1, \quad |c_q| = 0 \quad \text{at} \quad t = 0\,. \tag{25.6}$$

The probability that an atom will be in the state q after spending a time T in the oscillating field is:

$$P_{p,q} = \frac{|2b|^2}{|2b|^2 + (\omega_0 - \omega)^2} \sin^2\left\{\frac{1}{2}\,[|2b|^2 + (\omega_0 - \omega)^2]^{\frac{1}{2}}\, T\right\} \tag{25.7}$$

in which $\omega_0 = (W_q - W_p)/\hbar$.

The function $P_{p,q}$ given by Eq. (25.7) is called the natural line shape. The maximum value of $P_{p,q}$ is 1, of course, and this value is attained at resonance ($\omega = \omega_0$) when $|b|\,T = \frac{\pi}{2}, \frac{3\pi}{2}$, etc. The case $|b|\,T = \frac{\pi}{2}$ is often called the optimum case, since the smallest line width is obtained together with the condition $P_{p,q} = 1$ at resonance. For the optimum case the line width, or frequency difference $\delta\nu$

[1] W. Pauli: Vol. V/1, this Encyclopedia.

between half intensity points, is:

$$\delta\nu = 0.799/T. \tag{25.8}$$

The line shape $P_{p,q}$ is shown in Fig. 46 as a function of the parameter η $\left[\eta = \frac{\pi(\nu-\nu_0)}{|b|},\right.$ where $\left.\nu = \frac{\omega}{2\pi}, \nu_0 = \frac{\omega_0}{2\pi}\right]$. The solid curve is drawn for $|b|\,T = \pi/2$ and the dotted curve for $|b|\,T = \pi$; hence a particular value of η represents twice the frequency difference from the resonance frequency in the dotted curve as in the solid curve. The form of Eq. (25.7) and Fig. 46 show that the line shape is symmetrical about the central frequency ν_0, but that the line intensity need not have its maximum at its center.

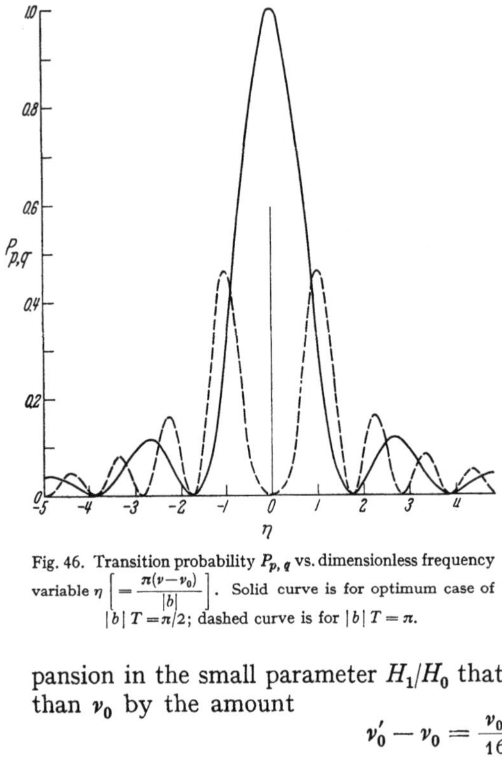

Fig. 46. Transition probability $P_{p,q}$ vs. dimensionless frequency variable η $\left[= \frac{\pi(\nu-\nu_0)}{|b|}\right]$. Solid curve is for optimum case of $|b|\,T = \pi/2$; dashed curve is for $|b|\,T = \pi$.

The actual oscillating field in the C-region will have a time dependence of $\cos\omega t$ rather than of $e^{+i\omega t}$ as assumed for the derivation of Eq. (25.7). The additional term $e^{-i\omega t}$ leads in general to a shift in the resonance frequency of the order of $\frac{|V_{pq}|^2}{(W_q - W_p)^2}\,\nu_0$ and to the addition of a non-resonant term to the line shape $P_{p,q}$ given in Eq. (25.7). Usually both the changes in resonance frequency and in line shape are negligibly small. In particular, for the case of magnetic resonance for a particle of spin $\frac{1}{2}$ in a constant magnetic field H_0 and with a small applied alternating magnetic field $H_1 \cos\omega t$ perpendicular to H_0, it has been shown[1] by an approximate treatment involving an expansion in the small parameter H_1/H_0 that the resonance frequency ν_0' is greater than ν_0 by the amount

$$\nu_0' - \nu_0 = \frac{\nu_0}{16}\left(\frac{H_1}{H_0}\right)^2 \tag{25.9}$$

as the lowest approximation, but that the shape of the resonance curve is unchanged.

The existence of other states besides the states p and q, for which V has nonvanishing matrix elements to the states p and q, will lead to an expression for the line shape $P_{p,q}$ somewhat different from that of Eq. (25.7), even if these states are off resonance for the applied frequency ν. The general theory of magnetic resonance transitions in a rotating magnetic field has been treated and a general formula for the corrections to Eq. (25.7) due to off-resonance states has been given[2]. The corrections increase with increase in the amplitude of the rotating field, but in general they can be neglected for the amplitudes of oscillating fields used in practice.

---

[1] F. BLOCH and A. SIEGERT: Phys. Rev. **57**, 522 (1940); see also N. F. RAMSEY, Phys. Rev. **100**, 1191 (1955), which treats the more general problem of resonance transitions induced by perturbations at two or more different frequencies.

[2] H. SALWEN: Phys. Rev. **99**, 1274 (1955).

A special case of considerable importance is that of magnetic resonance transitions for a spin I with associated magnetic moment μ in a magnetic field. The transition probabilities for transitions between the magnetic substates designated by the magnetic quantum numbers m and m' in a rotating magnetic field of amplitude H_1 and angular frequency ω are given by the Majorana formula:

$$\left.\begin{aligned} P_{m,m'} &= \left(\sin\frac{x}{2}\right)^{4I} (I+m)!\,(I+m')!\,(I-m)!\,(I-m')! \times \\ &\times \left[\sum_{r=0}^{2I} \frac{(-1)^r [\cot(x/2)]^{2r+m+m'}}{r!\,(r+m+m')!\,(I-m-r)!\,(I-m'-r)!}\right]. \end{aligned}\right\} \tag{25.10}$$

where x is defined by

$$\sin^2\frac{x}{2} = \frac{(\gamma H_1)^2}{(\omega-\omega_0)^2+(\gamma H_1)^2}\sin^2\left\{\frac{1}{2}\left[(\omega-\omega_0)^2+(\gamma H_1)^2\right]^{\frac{1}{2}} T\right\} \tag{25.10a}$$

in which ω_0 = Bohr angular frequency for a transition $\Delta m = \pm 1$; γ is the gyromagnetic ratio [see Eq. (28.1)][1].

26. Averages over the velocity distribution. The velocity distribution in a beam produced by thermal effusion from a source is given by Eq. (3.6) as:

$$N(v)\,dv = \frac{2N}{\alpha^4} v^3 e^{-v^2/\alpha^2} dv. \tag{3.6}$$

The velocity range for this distribution is from $v=0$ to $v=\infty$, and the corresponding range for the time T spent by particles in the transition field is from $T=\infty$ to $T=0$. In practice geometrical limitations, the magnitude of deflections, and beam scattering may substantially limit the range of velocities involved in a transition. The shape of the resonance line in a spectroscopic experiment will be the average of the single velocity expression $P_{p,q}$ of Eq. (25.7) taken over the appropriate velocity distribution.

The line shape averaged over the entire velocity distribution of Eq. (3.6) is given by[2, 3]:

$$\langle P_{p,q}\rangle = 2(\beta_0^2/\beta^2)\,K(\beta) \tag{26.1}$$

where

$$\beta_0 = \frac{2L|b|}{\alpha}, \tag{26.1a}$$

$$\beta = \left[(\omega_0-\omega)^2+|2b|^2\right]^{\frac{1}{2}}\frac{L}{\alpha} = \left[(\omega_0-\omega)^2+|2b|^2\right]^{\frac{1}{2}}\frac{\beta_0}{2|b|}, \tag{26.1b}$$

$$K(\beta) = \int_0^\infty x^3 e^{-x^2}\sin^2\frac{\beta}{2x}\,dx, \tag{26.1c}$$

$x = v/\alpha$ and L = length of transition region.

At the resonant frequency $\nu=\nu_0$, $\beta=\beta_0$ so that

$$\langle P_{p,q}\rangle = 2K(\beta_0). \tag{26.2}$$

This expression has its maximum value of 0.766 when $\beta_0 = 1.20\pi = 3.77$. It is usually desirable to obtain this maximum value for the transition probability

[1] E. Majorana: Nuovo Cim. **9**, 43 (1932). — F. Bloch and I.I. Rabi: Rev. Mod. Phys. **17**, 237 (1945); see also general reference [25].
[2] H.C. Torrey: Phys. Rev. **59**, 293 (1941).
[3] H. Salwen: Phys. Rev. **101**, 621 (1956).

Table 5. $K(\beta)$ *and* $(1/\beta^2)\,K(\beta)$.

β	$K(\beta)$	$1/\beta^2 K(\beta)$	β	$K(\beta)$	$1/\beta^2 K(\beta)$	β	$K(\beta)$	$1/\beta^2 K(\beta)$
0.0	0.00000	0.12500	5.5	0.32147	0.010627	12.0	0.23532	0.0016342
0.1	0.00124	0.12427	5.6	0.31611	0.010080	12.2	0.23780	0.0015977
0.2	0.00491	0.12264	5.7	0.31074	0.0095643	12.4	0.24020	0.0015622
0.3	0.01084	0.12044	5.8	0.30539	0.0090781	12.6	0.24251	0.0015275
0.4	0.01885	0.11783	5.9	0.30007	0.0086203	12.8	0.24469	0.0014935
0.5	0.02873	0.11493	6.0	0.29481	0.0081891	13.0	0.24675	0.0014601
0.6	0.04025	0.11180	6.1	0.28962	0.0077834	13.2	0.24868	0.0014272
0.7	0.05318	0.10852	6.2	0.28452	0.0074018	13.4	0.25046	0.0013948
0.8	0.06728	0.10513	6.3	0.27953	0.0070429	13.6	0.25209	0.0013630
0.9	0.08235	0.10167	6.4	0.27466	0.0067056	13.8	0.25358	0.0013315
1.0	0.09817	0.098172	6.5	0.26992	0.0063887	14.0	0.25491	0.0013006
1.1	0.11454	0.094662	6.6	0.26533	0.0060911	14.2	0.25609	0.0012700
1.2	0.13127	0.091161	6.7	0.26089	0.0058117	14.4	0.25712	0.0012400
1.3	0.14819	0.087687	6.8	0.25661	0.0055495	14.6	0.25801	0.0012104
1.4	0.16514	0.084256	6.9	0.25250	0.0053035	14.8	0.25875	0.0011813
1.5	0.18197	0.080877	7.0	0.24857	0.0050728	15.0	0.25936	0.0011527
1.6	0.19856	0.077564	7.1	0.24482	0.0048565	15.2	0.25984	0.0011246
1.7	0.21478	0.074319	7.2	0.24125	0.0046537	15.4	0.26019	0.0010971
1.8	0.23054	0.071154	7.3	0.23787	0.0044637	15.6	0.26043	0.0010701
1.9	0.24574	0.068072	7.4	0.23468	0.0042857	15.8	0.26056	0.0010437
2.0	0.26031	0.065076	7.5	0.23169	0.0041190	16.0	0.26059	0.0010179
2.1	0.27417	0.062170	7.6	0.22890	0.0039629	16.2	0.26053	0.00099272
2.2	0.28728	0.059356	7.7	0.22629	0.0038167	16.4	0.26038	0.00096811
2.3	0.29958	0.056631	7.8	0.22388	0.0036798	16.6	0.26016	0.00094412
2.4	0.31108	0.054007	7.9	0.22166	0.0035516	16.8	0.25987	0.00092076
2.5	0.32171	0.051474	8.0	0.21962	0.0034316	17.0	0.25953	0.00089802
2.6	0.33147	0.049034	8.1	0.21778	0.0033193	17.2	0.25913	0.00087591
2.7	0.34033	0.046685	8.2	0.21611	0.0032141	17.4	0.25869	0.00085443
2.8	0.34831	0.044428	8.3	0.21463	0.0031156	17.6	0.25821	0.00083358
2.9	0.35542	0.042262	8.4	0.21332	0.0030233	17.8	0.25770	0.00081336
3.0	0.36163	0.040181	8.5	0.21219	0.0029368	18.0	0.25717	0.00079375
3.1	0.36702	0.038191	8.6	0.21122	0.0028558	18.2	0.25663	0.00077475
3.2	0.37157	0.036286	8.7	0.21041	0.0027798	18.4	0.25607	0.00075636
3.3	0.37531	0.034464	8.8	0.20975	0.0027086	18.6	0.25551	0.00073856
3.4	0.37826	0.032722	8.9	0.20925	0.0026417	18.8	0.25495	0.00072134
3.5	0.38046	0.031058	9.0	0.20889	0.0025789	19.0	0.25439	0.00070469
3.6	0.38193	0.029470	9.1	0.20867	0.0025198	19.2	0.25384	0.00068859
3.7	0.38272	0.027956	9.2	0.20858	0.0024643	19.4	0.25331	0.00067304
3.8	0.38285	0.026513	9.3	0.20862	0.0024120	19.6	0.25278	0.00065802
3.9	0.38236	0.025139	9.4	0.20877	0.0023627	19.8	0.25228	0.00064351
4.0	0.38130	0.023831	9.5	0.20904	0.0023163	20.0	0.25180	0.00062949
4.1	0.37970	0.022588	9.6	0.20942	0.0022723	21.0	0.24978	0.00056639
4.2	0.37759	0.021405	9.7	0.20990	0.0022308	22.0	0.24841	0.00051324
4.3	0.37502	0.020283	9.8	0.21047	0.0021914	23.0	0.24771	0.00046827
4.4	0.37203	0.019217	9.9	0.21113	0.0021541	24.0	0.24757	0.00042981
4.5	0.36866	0.018205	10.0	0.21187	0.0021187	25.0	0.24782	0.00039651
4.6	0.36493	0.017246	10.2	0.21359	0.0020529	26.0	0.24829	0.00036730
4.7	0.36089	0.016337	10.4	0.21554	0.0019928	27.0	0.24886	0.00034137
4.8	0.35658	0.015477	10.6	0.21770	0.0019376	28.0	0.24940	0.00031811
4.9	0.35203	0.014662	10.8	0.22003	0.0018864	29.0	0.24986	0.00029710
5.0	0.34727	0.013891	11.0	0.22248	0.0018387	30.0	0.25020	0.00027800
5.1	0.34234	0.013162	11.2	0.22501	0.0017938	32.0	0.25051	0.00024464
5.2	0.33726	0.012473	11.4	0.22759	0.0017513	34.0	0.25046	0.00021666
5.3	0.33207	0.011822	11.6	0.23019	0.0017107	36.0	0.25026	0.00019310
5.4	0.32680	0.011207	11.8	0.23278	0.0016718	38.0	0.25006	0.00017317
						40.0	0.24993	0.00015621
						45.0	0.24992	0.00012342
						50.0	0.25001	0.00010000

Table 6. $\beta_{\frac{1}{2}}$ *and* $[\beta_{\frac{1}{2}}^2-\beta_0^2]^{\frac{1}{2}}$ *as functions of* β_0.

β_0	$\beta_{\frac{1}{2}}$	$[\beta_{\frac{1}{2}}^2-\beta_0^2]^{\frac{1}{2}}$	β_0	$\beta_{\frac{1}{2}}$	$[\beta_{\frac{1}{2}}^2-\beta_0^2]^{\frac{1}{2}}$	β_0	$\beta_{\frac{1}{2}}$	$[\beta_{\frac{1}{2}}^2-\beta_0^2]^{\frac{1}{2}}$
0.0	2.088	2.088	5.5	6.896	4.160	12.0	17.762	13.096
0.1	2.101	2.099	5.6	7.015	4.225	12.2	17.947	13.163
0.2	2.130	2.120	5.7	7.136	4.293	12.4	18.132	13.230
0.3	2.169	2.148	5.8	7.260	4.366	12.6	18.319	13.297
0.4	2.216	2.179	5.9	7.386	4.443	12.8	18.507	13.367
0.5	2.269	2.213	6.0	7.515	4.525	13.0	18.698	13.439
0.6	2.327	2.249	6.1	7.648	4.613	13.2	18.892	13.515
0.7	2.390	2.285	6.2	7.784	4.707	13.4	19.090	13.596
0.8	2.456	2.322	6.3	7.925	4.807	13.6	19.290	13.681
0.9	2.526	2.360	6.4	8.070	4.915	13.8	19.496	13.772
1.0	2.598	2.398	6.5	8.220	5.031	14.0	19.705	13.867
1.1	2.672	2.435	6.6	8.375	5.156	14.2	19.921	13.971
1.2	2.748	2.472	6.7	8.538	5.292	14.4	20.140	14.081
1.3	2.826	2.510	6.8	8.707	5.438	14.6	20.366	14.199
1.4	2.906	2.547	6.9	8.885	5.597	14.8	20.596	14.324
1.5	2.987	2.583	7.0	9.072	5.770	15.0	20.831	14.455
1.6	3.070	2.620	7.1	9.269	5.958	15.2	21.072	14.594
1.7	3.153	2.656	7.2	9.477	6.162	15.4	21.319	14.743
1.8	3.238	2.692	7.3	9.698	6.384	15.6	21.571	14.898
1.9	3.324	2.727	7.4	9.932	6.624	15.8	21.828	15.060
2.0	3.411	2.763	7.5	10.180	6.883	16.0	22.090	15.230
2.1	3.498	2.798	7.6	10.440	7.158	16.2	22.359	15.411
2.2	3.586	2.832	7.7	10.713	7.449	16.4	22.633	15.598
2.3	3.676	2.867	7.8	10.995	7.749	16.6	22.910	15.789
2.4	3.765	2.901	7.9	11.284	8.057	16.8	23.195	15.992
2.5	3.855	2.935	8.0	11.574	8.365	17.0	23.485	16.203
2.6	3.947	2.969	8.1	11.864	8.669	17.2	23.777	16.417
2.7	4.038	3.003	8.2	12.148	8.963	17.4	24.074	16.637
2.8	4.131	3.037	8.3	12.425	9.246	17.6	24.377	16.866
2.9	4.224	3.071	8.4	12.693	9.515	17.8	24.682	17.098
3.0	4.317	3.105	8.5	12.950	9.769	18.0	24.989	17.333
3.1	4.412	3.139	8.6	13.196	10.008	18.2	25.301	17.575
3.2	4.506	3.173	8.7	13.431	10.232	18.4	25.614	17.819
3.3	4.601	3.207	8.8	13.654	10.441	18.6	25.928	18.064
3.4	4.697	3.241	8.9	13.868	10.636	18.8	26.245	18.313
3.5	4.794	3.276	9.0	14.072	10.818	19.0	26.564	18.564
3.6	4.891	3.310	9.1	14.267	10.988	19.2	26.882	18.815
3.7	4.988	3.346	9.2	14.452	11.146	19.4	27.200	19.065
3.8	5.087	3.381	9.3	14.630	11.293	19.6	27.518	19.316
3.9	5.185	3.417	9.4	14.800	11.431	19.8	27.837	19.566
4.0	5.285	3.454	9.5	14.961	11.558	20.0	28.154	19.815
4.1	5.385	3.492	9.6	15.117	11.678	21.0	29.718	21.028
4.2	5.487	3.530	9.7	15.266	11.789	22.0	31.237	22.176
4.3	5.588	3.569	9.8	15.410	11.893	23.0	32.709	23.257
4.4	5.691	3.609	9.9	15.548	11.989	24.0	34.138	24.278
4.5	5.795	3.651	10.0	15.681	12.078	25.0	35.533	25.250
4.6	5.899	3.694	10.2	15.933	12.240	26.0	36.908	26.195
4.7	6.005	3.737	10.4	16.170	12.382	27.0	38.275	27.129
4.8	6.111	3.783	10.6	16.394	12.506	28.0	39.643	28.063
4.9	6.219	3.830	10.8	16.608	12.616	29.0	41.018	29.008
5.0	6.328	3.879	11.0	16.812	12.714	30.0	42.404	29.968
5.1	6.438	3.930	11.2	17.010	12.802	32.0	45.203	31.926
5.2	6.550	3.983	11.4	17.202	12.883	34.0	48.037	33.935
5.3	6.664	4.039	11.6	17.391	12.958	36.0	50.885	35.963
5.4	6.779	4.098	11.8	17.577	13.028	38.0	53.735	37.993
						40.0	56.578	40.013
						45.0	63.649	45.013
						50.0	70.711	50.000

at resonance, and the appropriate amplitude of the oscillating field is thus determined from Eq. (26.1a) with $\beta_0 = 3.77$. The corresponding line width $\delta\nu$ between half intensity points is:

$$\delta\nu = \frac{1.073\alpha}{L}. \tag{26.3}$$

In general the line shape can be calculated for an arbitrary amplitude of the oscillating field from Eq. (26.1) with the aid of Table 5, which gives $K(\beta)$ and $(1/\beta^2)K(\beta)$ as functions of β. For a fixed amplitude of the oscillating field, the line intensity varies with the frequency as $(1/\beta^2)K(\beta)$. The half intensity points occur at $\beta = \beta_{\frac{1}{2}}$, where $\beta_{\frac{1}{2}}$ is defined by

$$\frac{1}{\beta_{\frac{1}{2}}^2} K(\beta_{\frac{1}{2}}) = \frac{1}{2\beta_0^2} K(\beta_0). \tag{26.4}$$

Hence

$$\delta\nu = \frac{2|b|}{\beta_0 \pi} (\beta_{\frac{1}{2}}^2 - \beta_0^2)^{\frac{1}{2}}. \tag{26.5}$$

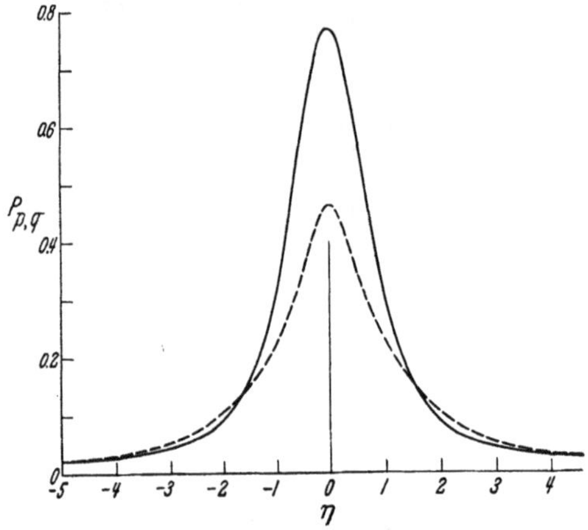

Fig. 47. Transition probability $\langle P_{p,q}\rangle$ averaged over the velocity distribution of the beam vs. dimensionless frequency variable $\eta \left[= \frac{\pi(\nu-\nu_0)}{|b|}\right]$. Solid curve is for optimum amplitude of oscillating field with $\beta_0 = 3.77$; dashed curve is for $\beta_0 = 7.54$.

Eq. (26.5) determines the half width of the line as a function of the oscillating field amplitude. Table 6 gives values of $\beta_{\frac{1}{2}}$ and $(\beta_{\frac{1}{2}}^2 - \beta_0^2)^{\frac{1}{2}}$ as functions of β_0. From Table 5 it is seen that the quantity $(1/\beta^2)K(\beta)$ decreases monotonically with increasing β. Hence the line intensity has its maximum value at the central resonance frequency and decreases monotonically to zero on either side of resonance. The line shape is symmetrical with respect to the resonance frequency.

Fig. 47 shows the line shape $\langle P_{p,q}\rangle$ for two different amplitudes of the oscillating field. The variable η is defined by:

$$\eta = \pi(\nu - \nu_0)/|b|$$

so that

$$\beta = \beta_0 (1 + \eta^2)^{\frac{1}{2}}$$

and

$$\langle P_{p,q}\rangle = \frac{2K(\beta)}{1+\eta^2}. \tag{26.6}$$

The solid curve is for the optimum amplitude when $\beta_0 = 3.77$. The dotted curve is for an amplitude of twice the optimum value when $\beta_0 = 7.54$. In the second case a given η corresponds to twice the frequency differential from the center as in the first case. The line widths of the two curves are, therefore, different by a factor of about 2, which is just the ratio between the values of $(\beta_{\frac{1}{2}}^2 - \beta_0^2)^{\frac{1}{2}}$ for $\beta_0 = 7.54$ and $\beta_0 = 3.77$ in Table 6.

In cases when atoms of velocities less than v_1 and greater than v_2 are eliminated from the beam, the expression (26.1) for the transition probability must be modified by the replacement of $K(\beta)$ by

$$\tfrac{1}{2} K(\beta, x_1, x_2)/N$$

where

$$x_1 = v_1/\alpha; \quad x_2 = v_2/\alpha$$

$$N = \int_{x_1}^{x_2} x^3 \mathrm{e}^{-x^2} dx,$$

$$K(\beta, x_1, x_2) = \int_{x_1}^{x_2} x^3 \mathrm{e}^{-x^2} \sin^2(\beta/2x)\, dx.$$

As an aid in the computation of the theoretical line shape for this case, the integrals $K(\beta, 0, 1)$ and $K(\beta, 1, \infty)$ have been tabulated[1].

27. Comparison with experiment. The measurements of beam spectroscopy are directed towards obtaining the frequency interval between two states. This is usually best found by observation of the line at an amplitude of the oscillating field close to the optimum value corresponding to $\beta_0 = 3.77$ (see Sect. 26), for which the probability of transition at the center of the resonance line is about 0.77 and the natural line width is given by Eq. (26.3). Since the natural line shape is always symmetrical about the resonance frequency, the adjustment of radiofrequency amplitude is not critical in obtaining observations of frequency from which very precise results may be derived, and hence the validity of the theory of the natural line shape given in Sects. 25 and 26 has, in most cases, been subject to only qualitative experimental verification for the region of radiofrequency amplitudes from zero to several times the optimum value.

In certain recent experiments in which multiple quantum transitions between the magnetic components of the hfs levels of K^{39} were observed, the single quantum transitions were observed at approximately the same radiofrequency amplitude as that required to give the multiple transitions[2,3]. This implies that the single transitions were observed at an amplitude many times the optimum amplitude. It was at once apparent that at high amplitude the observable properties of the line, i.e., the probability of transition at the center of the line and the half-width of the line, were in extreme disagreement with the theoretical predictions[4]. Generally the width of the resonance line increases at a much less rapid rate than expected and ultimately, as the amplitude increases, the line width remains constant or suffers a marked decrease. Similarly, the probability of transition at the line center does not fall off as rapidly as predicted from Table 5 after the radiofrequency amplitude is increased greatly beyond the value corresponding to $\beta_0 = 3.77$. Finally, the shift of the central resonance frequency as a function of radiofrequency amplitude does not occur as rapidly as predicted.

The effects are highly variable depending on the details of the configuration of the conductors which produce the radiofrequency fields. It appears, both from the experimental evidence and from theoretical arguments, that the deviations from the idealized theoretical behaviour arise because of a departure of the actual pulse of radiofrequency field applied to the atom from the rectangular pulse assumed in the theory. The experimental evidence is unambiguous that the theoretical behaviour is approximated to higher amplitudes when the circuit design is such that the radiofrequency amplitude is constant over the length of the field and the rate of increase of radiofrequency amplitude is very fast as the atom approaches the transition field. Care was taken to obtain a very homogeneous steady magnetic field, so magnetic field inhomogeneities did not contribute significantly to the line shape.

[1] Document No. 4716, American Documentation Institute Auxiliary Publications Project, Library of Congress, Washington 25, D.C.

[2] P. KUSCH: Phys. Rev. **93**, 1022 (1954).

[3] P. KUSCH: Phys. Rev. **101**, 627 (1956).

[4] H. SALWEN: Phys. Rev. **101**, 623 (1956).

28. Analysis in a rotating coordinate system. It is instructive to consider the rotating coordinate method of describing the transition process. The method can yield quantitative results valid for actual quantum mechanical problems, but it is especially useful in the visualization of the transition process as a classical problem[1,2].

Consider a system of particles each of which has an angular momentum $\boldsymbol{I}\hbar$, a gyromagnetic ratio γ, and hence a magnetic moment $\gamma\hbar\boldsymbol{I}$. In terms of the Landé g-factor $\gamma = -g\mu_0/\hbar$. If a magnetic field $\boldsymbol{H}$ is present, the equation of motion of the system in a stationary coordinate system is

$$\hbar\frac{d\boldsymbol{I}}{dt} = \gamma\hbar\boldsymbol{I}\times\boldsymbol{H}. \tag{28.1}$$

Hence if $\boldsymbol{H}$ is constant, $\boldsymbol{I}$ will precess about $\boldsymbol{H}$ with the angular frequency:

$$\boldsymbol{\omega}_p = -\gamma H\boldsymbol{k} = \omega_p\boldsymbol{k}. \tag{28.2}$$

In a system of coordinates that is rotating with an angular velocity $\boldsymbol{\omega}$

$$\frac{d\boldsymbol{I}}{dt} = \frac{\partial\boldsymbol{I}}{\partial t} + \boldsymbol{\omega}\times\boldsymbol{I} \tag{28.3}$$

where $\partial\boldsymbol{I}/\partial t$ is the rate of change of $\boldsymbol{I}$ in the rotating system. Then

$$\frac{\partial\boldsymbol{I}}{\partial t} = \gamma\boldsymbol{I}\times\left(\boldsymbol{H}+\frac{\boldsymbol{\omega}}{\gamma}\right) = \gamma\boldsymbol{I}\times\boldsymbol{H}_{er}. \tag{28.4}$$

The equation of motion in the rotating system is thus identical to that in the fixed system where the magnetic field $\boldsymbol{H}$ is replaced by an effective field

$$\boldsymbol{H}_{er} = \boldsymbol{H} + \frac{\boldsymbol{\omega}}{\gamma}. \tag{28.5}$$

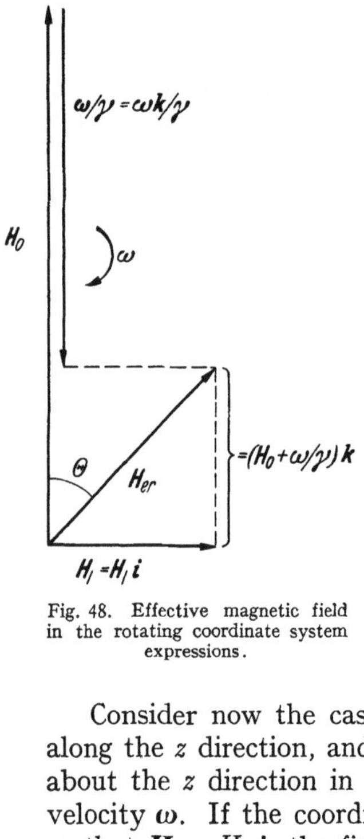

Fig. 48. Effective magnetic field in the rotating coordinate system expressions.

Consider now the case shown in Fig. 48 in which there is a fixed field $\boldsymbol{H}_0$ along the z direction, and another, usually much weaker, field $\boldsymbol{H}_1$ which rotates about the z direction in the plane perpendicular to the z axis with an angular velocity $\boldsymbol{\omega}$. If the coordinate system rotates with the same angular velocity $\boldsymbol{\omega}$ so that $\boldsymbol{H}_1 = H_1\boldsymbol{i}$, the field seen in this system is constant in time and,

$$\boldsymbol{H}_{er} = \left(H_0 + \frac{\omega}{\gamma}\right)\boldsymbol{k} + H_1\boldsymbol{i}. \tag{28.6}$$

Hence

$$H_{er} = \left[\left(H_0 + \frac{\omega}{\gamma}\right)^2 + H_1^2\right]^{\frac{1}{2}} = \frac{a}{\gamma} \tag{28.6a}$$

where

$$\omega_0 = -\gamma H_0, \quad \text{and} \quad a = \left[(\omega-\omega_0)^2 + \left(\frac{\omega_0 H_1}{H_0}\right)^2\right]^{\frac{1}{2}}.$$

The angle Θ of $\boldsymbol{H}_{er}$ relative to $\boldsymbol{H}_0$ is given by

$$\cos\Theta = \frac{\omega-\omega_0}{a}\,; \qquad \sin\Theta = -\frac{\omega_0 H_1}{a H_0}. \tag{28.7}$$

In the rotating coordinate system the magnetic moment μ will precess about the direction of $\boldsymbol{H}_{er}$ with angular frequency $-\gamma\boldsymbol{H}_{er}$. If the moment is positive, ω_0 is negative, and the rotating field must be so chosen that ω is also negative. If

[1] RABI, RAMSEY and SCHWINGER: Rev. Mod. Phys. **26**, 167 (1954).

[2] H. SALWEN: Phys. Rev. **99**, 1274 (1955). SALWEN has used the rotating coordinate method to treat the general quantum mechanical problem of magnetic resonance.

$\omega = \omega_0$, $\Theta = 90°$ and a magnetic moment initially parallel to $\boldsymbol{H}_0$ may precess about $\boldsymbol{H}_{er}$, which is $\boldsymbol{H}_1$, until the moment is antiparallel to $\boldsymbol{H}_0$. This frequency $\omega = \omega_0$ is called the resonance frequency because at this frequency the moment can have its orientation relative to $\boldsymbol{H}_0$ changed most completely. Quantum mechanically, the change in orientation of the moment from parallel to antiparallel with respect to $\boldsymbol{H}_0$ corresponds to a transition from one state of quantization to another.

In atomic and molecular beam magnetic resonance experiments an oscillating rather than a rotating field is commonly applied. The oscillating field may be resolved into two fields rotating in opposite directions, and only the field rotating in a direction to be able to give a resonance in Eq. (28.6a) is important in most cases. Suppose there is a finite length along which the rotating field of frequency ω exists, and that the constant field $\boldsymbol{H}_0$ exists in this length as well as on both sides of it. Before the particle enters the rotating field, $\boldsymbol{H}_{er} = \left(H_0 + \frac{\omega}{\gamma}\right)\boldsymbol{k}$ and hence in the rotating coordinate system the moment precesses about the z axis with the frequency $(\omega_0 - \omega)$. On entering the rotating field, $\boldsymbol{H}_1$ is established, $\boldsymbol{H}_{er}$ is modified to the value given in Eq. (28.6), and the particles precess about $\boldsymbol{H}_{er}$ with angular frequency $-\gamma \boldsymbol{H}_{er}$. Usually the time required for the particle to enter (or to leave) the region with the rotating field is sufficiently short so that the moment does not change its orientation appreciably during this time[1]. At resonance ($\omega = \omega_0$) the precession is about $\boldsymbol{H}_1$ and a large change in the direction of the moment can occur while the particle is in the region of rotating field, corresponding to a high probability of transition. For any given value of H_1 with $\omega = \omega_0$, there are values of the time during which the moment should precess about $\boldsymbol{H}_{er}$ in order to obtain unit probability of transition. At off-resonance frequencies it is possible to attain a high probability of transition by a sufficient increase in H_1 so that $\boldsymbol{H}_{er}$ will be principally determined by H_1.

A quantitative expression for the transition probability in magnetic resonance can be derived for the case of spin $\frac{1}{2}$ particles. Suppose that $\boldsymbol{I}\,(=\frac{1}{2})$ is initially parallel to $\boldsymbol{H}_0$. In the rotating coordinate system $\boldsymbol{I}$ will precess about $\boldsymbol{H}_{er}$ with the angular velocity $-\gamma \boldsymbol{H}_{er}$ at the precession angle Θ. After a time interval T, the angle α between $\boldsymbol{I}$ and $\boldsymbol{H}_0$ is:

$$\cos\alpha = 1 - 2\sin^2\Theta \sin^2 a\,T. \tag{28.8}$$

A correspondence between the classical and quantum mechanical solutions can be established by requiring them to agree on their predicted average z component of angular momentum and on their total probability. If $P_{\pm\frac{1}{2}}$ is the probability that a particle of spin $\frac{1}{2}$ is in the state of magnetic quantum number $m = \pm\frac{1}{2}$, the requirements are:

$$P_{\frac{1}{2}} - P_{-\frac{1}{2}} = \cos\alpha, \qquad P_{\frac{1}{2}} + P_{-\frac{1}{2}} = 1. \tag{28.9}$$

Hence

$$\left.\begin{aligned} P_{-\frac{1}{2}} &= \frac{1-\cos\alpha}{2} = \sin^2\Theta \sin^2\frac{aT}{2} \\ &= \frac{\left(\frac{\omega_0 H_1}{H_0}\right)^2}{(\omega-\omega_0)^2 + \left(\frac{\omega_0 H_1}{H_0}\right)^2} \sin^2\frac{T}{2}\left[(\omega-\omega_0)^2 + \left(\frac{\omega_0 H_1}{H_0}\right)^2\right]^{\frac{1}{2}}. \end{aligned}\right\} \tag{28.10}$$

This result may be found by applying Eq. (25.7) to magnetic transitions for a particle of spin $\frac{1}{2}$.

[1] There are cases, however, both in magnetic and electric resonance experiments in which this is not true (see Sects. 30 and 46), and for magnetic resonance the end region effect has been used to determine the sign of nuclear moments. See also L. GRABNER and V. W. HUGHES: Phys. Rev. 79, 819 (1950).

29. Method of separated oscillating fields. The conventional resonance method utilizes a single oscillating field which is applied uniformly throughout the transition region. The method of separated oscillating fields uses two oscillating fields which are confined to two small regions at the beginning and end of the region within which the energy levels are to be studied[1]. An important advantage of this method is that the resonance line is much less broadened by inhomogeneities of the steady magnetic field than it is when the oscillating field is applied over the entire region.

An analysis of the method of separated oscillating fields with the use of a rotating coordinate system is applicable in the limiting classical case for magnetic resonance of a large angular momentum $\boldsymbol{I}\hbar$ and associated magnetic moment $\mu(=\gamma\hbar\boldsymbol{I})$. Through the first oscillating field the description of the particle's precession in a coordinate system rotating with the angular velocity $\boldsymbol{\omega}$ of the oscillating field is the same as that given in Sect. 28. After leaving the oscillating field, the particle enters an intermediate region where there is only $\boldsymbol{H}_0$ and the magnitude of $\boldsymbol{H}_1$ is zero. Hence at resonance the moment will not precess further in the rotating coordinate system. Upon entering the second oscillating field region where the phase of oscillation is the same as that of the first oscillating field, the precession of μ about $\boldsymbol{H}_1$ continues. The total change in orientation of μ is due to the combined effect of the two oscillating fields. Inhomogeneity of H_0 throughout the intermediate region will not cause any net precession of μ provided the angular velocity, ω_0, of the rotating coordinate system is equal to the space average value of $-\gamma H_0$ over the trajectory of the beam. At off-resonance frequencies the angle of precession in the intermediate field will depend on the particle's velocity, and hence an average over the velocity distribution will yield[2] a maximum transition probability for $\omega=-\gamma H_0$.

A quantum mechanical solution for the transition probability for the method of separated oscillating fields has been obtained for the case of two stable energy levels and an interaction with the oscillating field which has matrix elements only between the two states. Assume that an atom is initially in the state p and has a velocity v. It passes through the first oscillating field of length l where it spends a time τ, then through a region of length L where it spends a time T, and finally through the second oscillating field of length l. The probability for a transition to the state q is given by[3]:

$$P_{p,q}=4\sin^2\Theta\sin^2\frac{a\tau}{2}\left[\cos\frac{1}{2}(\lambda T-\delta)\cos\frac{a\tau}{2}-\cos\Theta\sin\frac{1}{2}(\lambda T-\delta)\sin\frac{a\tau}{2}\right]^2 \tag{29.1}$$

where

$$V_{pq}=\hbar b\,\mathrm{e}^{+i\omega t},\qquad V_{qp}=\hbar b^*\,\mathrm{e}^{-i\omega t},$$

$$\cos\Theta=\frac{\omega_0-\omega}{a};\qquad \sin\Theta=\frac{|2b|}{a},$$

$$a=[(\omega_0-\omega)^2+|2b|^2]^{\frac{1}{2}},$$

$$\omega_0=\frac{W_q-W_p}{\hbar},$$

$$\lambda=\frac{\overline{W}_q-\overline{W}_p}{\hbar}-\omega,$$

$$\tau=\frac{l}{v};\quad T=\frac{L}{v}.$$

---

[1] N. F. RAMSEY: Phys. Rev. **78**, 695 (1950).
[2] RABI, RAMSEY and SCHWINGER: Rev. Mod. Phys. **26**, 167 (1954).
[3] N. F. RAMSEY and H. B. SILSBEE: Phys. Rev. **84**, 506 (1951).

V_{pq} is the matrix element of the interaction of the atom with the oscillating field of angular frequency ω. The quantities W_p and W_q are the energies of the states p and q in the end regions where the steady magnetic field has the value H_0. The quantities $\overline{W}_p$ and $\overline{W}_q$ are the space mean values of the energies of the states p and q over the region L where magnetic field inhomogeneities may exist. The quantity δ is the phase angle by which the second oscillating field leads the

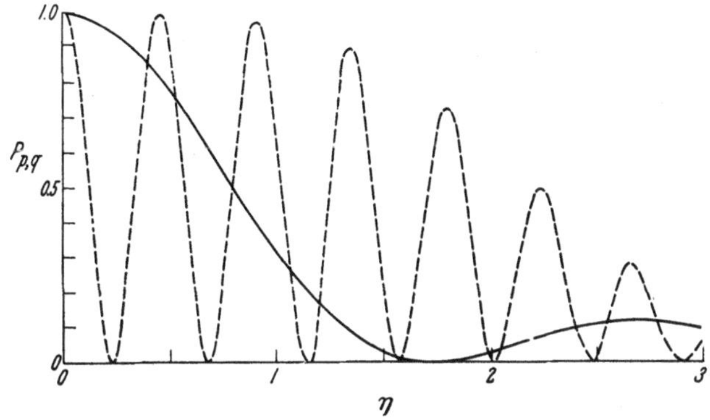

Fig. 49. Theoretical line shapes for the conventional method with a single oscillating field, and for the method of separated oscillating fields. The curve (——) is calculated from Eq. (25.7) with $bT=\pi/2$. The curve (----) is calculated from Eq. (29.1) with $b\tau=\pi/4$, $\delta=0$ and $L/l=10$. $\eta=|(\omega_0-\omega)/2b|$.

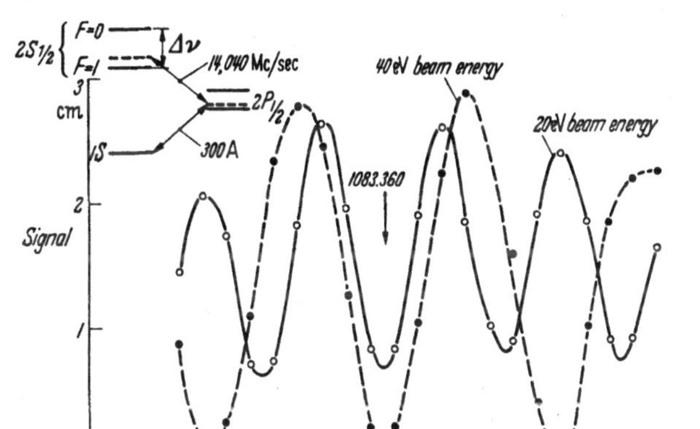

Fig. 50. The energy levels of He^{3+} and the photoelectric current as a function of frequency for two values of the ion beam energy. See also Sect. 36.

first. The theoretical line shape of Eq. (29.1) is shown in Fig. 49 for $\delta=0$, $L/l=10$, and $b\tau=\pi/4$, which gives unit transition probability at resonance. For comparative purposes, the line shape for the conventional resonance method [Eq. (25.7)] is shown for an oscillating field of length $(L+2l)$ and for $\frac{b(L+2l)}{v}=\frac{\pi}{2}$, which also gives unit transition probability at resonance. Fig. 50 shows experimental curves obtained[1] by the separated oscillating fields method for the hyperfine structure transition in the $2S$ state of singly ionized He^3. In the observation of the pattern, δ was made equal to π and a minimum in the transition probability is observed at resonance where $\omega=\omega_0$. The He ions had energies of 20 and 40 ev, and the length L was 20 cm. The line shapes of Figs. 49 and 50,

[1] R. Novick and E. D. Commins: Phys. Rev. **103**, 1897 (1956).

as well as the physical arrangement of the method of separated oscillating fields, indicate a close similarity to the double slit interference phenomena in optics.

The theoretical line shape $\langle P_{p,q}\rangle$ which takes into account the velocity distribution in the beam by the usual atomic beam weighting factor of $v^3 e^{-v^2/\alpha^2}$ is given near resonance $[|\omega_0 - \omega| \ll 2b]$ by the expression:

$$\left.\begin{aligned}\langle P_{p,q}\rangle = \tfrac{1}{4} - \tfrac{1}{2} I(4bl/\alpha) + \\ + \cos\delta\left[\tfrac{1}{2} I(\lambda L/\alpha) - \tfrac{1}{4} I(4bl/\alpha + \lambda L/\alpha) - \tfrac{1}{4} I(4bl/\alpha - \lambda L/\alpha)\right] + \\ + \sin\delta\left[\tfrac{1}{2} K(\lambda L/\alpha) - \tfrac{1}{4} K(4bl/\alpha + \lambda L/\alpha) + \tfrac{1}{4} K(4bl/\alpha - \lambda L/\alpha)\right]\end{aligned}\right\} \quad (29.2)$$

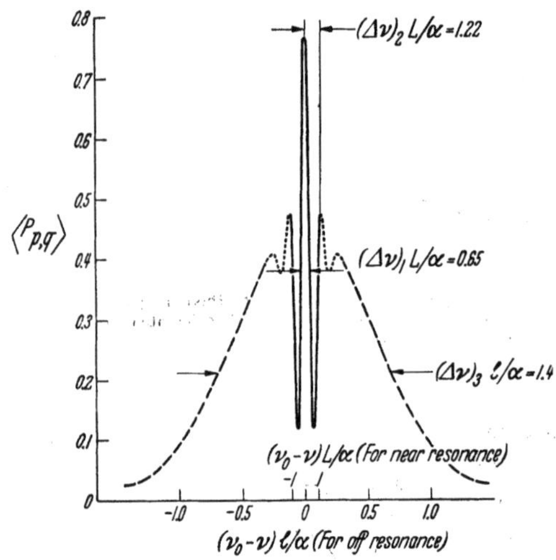

Fig. 51. Transition probability as a function of frequency. The curve (——) is calculated near resonance, the curve (-----) off resonance; the curve (·····) is interpolated. The near resonance curve is a plot of Eq. (29.2) with $\frac{2bl}{\alpha} = 0.600\,\pi$ and $\delta = 0$.

in which the functions $I(x)$ and $K(x)$ are:

$$\left.\begin{aligned} I(x) &= \int_0^\infty e^{-y^2} y^3 \cos(x/y)\, dy, \\ K(x) &= \int_0^\infty e^{-y^2} y^3 \sin(x/y)\, dy. \end{aligned}\right\} \quad (29.3)$$

These functions have been tabulated[1].

The theoretical line shape corresponding to $\delta = 0$ is shown in Fig. 51 for the case in which the experimental parameter $2bl/\alpha$ is chosen as 0.600π to maximize $\langle P_{p,q}\rangle$ at exact resonance ($\lambda = 0$). The half-width of the central peak is $\delta\nu = \frac{0.64\,\alpha}{L}$ whereas the half-width of the line in the conventional resonance method is $\delta\nu = \frac{1.07\,\alpha}{L}$. The smaller off-resonance side peaks are associated with atoms having velocities such that their magnetic moments precess by 360° (or an integral multiple of 360°) in phase relative to the oscillating field in the region between the oscillating fields, and thus the effects of the two oscillating fields can add to produce a complete transition. Since this condition is velocity dependent, only a fraction of the atoms in the beam can contribute to these side peaks, and hence they will be smaller than the central peak to which all the atoms contribute. Note that in the experiments with ionized He^3 (Fig. 50), the particles in the beam are very nearly mono-energetic and the side peaks are well defined. Fig. 52 shows a field-independent hyperfine structure transition in the $2S$ state of deuterium, for which the velocity distribution was somewhat narrower than the usual beam velocity distribution [Eq. (3.6)][2]. The radiofrequency fields in the two regions of length l were in phase ($\delta = 0$) and a maximum of $\langle P_{p,q}\rangle$ occurs at resonance.

Certain advantages are gained by providing a shift between the phases of the first and second oscillating fields. A method of observation in which the line is observed by changing the phase shift between the two oscillating fields by 180° as compared to the method in which the radiofrequency signal is simply

[1] U. E. KRUSE and N. F. RAMSEY: J. Math. Phys. **30**, 40 (1951). — H. SALWEN: Phys. Rev. **101**, 621 (1956); see also Table 5 of present article (p. 58).

[2] REICH, HEBERLE and KUSCH: Phys. Rev. **104**, 1585 (1956).

turned on and off for a fixed phase shift offers certain advantages. The procedure effectively doubles the signal intensity and has been employed for much of the work on the hydrogen molecules[1]. An experimental line observed[2] in this way which agrees well in shape with the theoretically predicted line is shown in Fig. 53. A phase shift of 90° between the first and second oscillating fields produces a dispersion type line and thus affords a sensitive means of determining the resonance frequency.

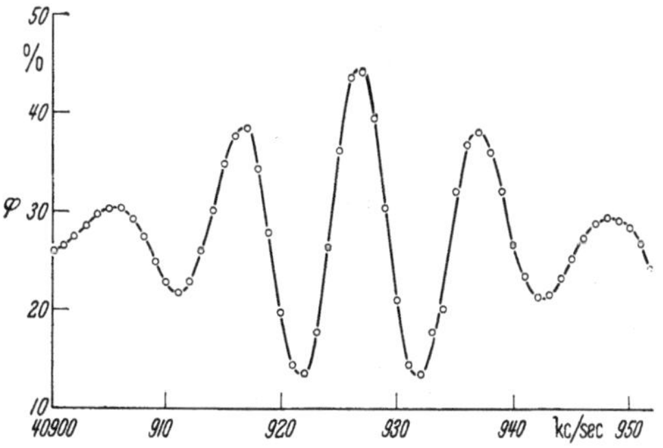

Fig. 52. Magnetic field-independent doublet hfs transition in the 2 S state of deuterium, showing several subsidiary maxima. Since the velocity distribution is considerably sharper than a Maxwellian beam velocity distribution, the subsidiary maxima are not damped as rapidly as in Fig. 51. φ is an observed decrease in intensity of the beam. See also Sect. 36.

In the method of separated oscillating fields the line is much less broadened by inhomogeneities of the steady magnetic field than in the conventional resonance method. Increases in precision by a factor of twenty or more compared to the conventional resonance method have been reported by use of the method of separated oscillating fields in experiments on the hydrogen molecules and on the hyperfine structure of hydrogen and of alkali atoms. This method is used in the cesium atomic clock (see Sect. 39).

The method of separated oscillating fields should also be applicable to electric resonance experiments. For magnetic resonance of a particle of spin greater than $\frac{1}{2}$ the transition probability can be obtained by use of the Majorana formula (see Sect. 25).

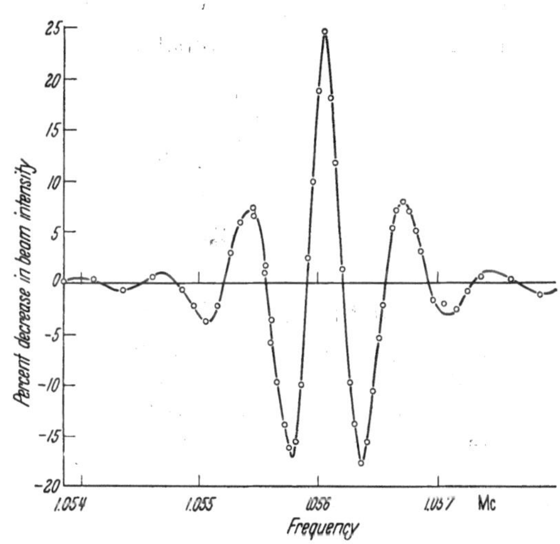

Fig. 53. Experimental change in beam intensity on removing 180° phase shift with ortho-D_2 in zeroth rotational state. The transition is $\Delta m_I = \pm 1$, $\Delta m_J = 0$, where m_I is the magnetic quantum number for nuclear spin and m_J is the magnetic quantum number for rotational angular momentum.

30. Multiple quantum transitions. A multiple quantum transition between two atomic or molecular energy levels is a transition in which the energy is supplied by two or more quanta. The transition is forbidden in a first order perturbation theory which involves the interaction of the atomic system with the radiation field, and occurs through one or more intermediate states. The theory of transitions involving the emission or absorption of two quanta, which are processes closely related to the Raman effect and to dispersion phenomena, was first discussed by M. Goeppert Mayer on the basis of the Dirac radiation theory[3].

[1] See, for example, Kolsky, Phipps, Ramsey and Silsbee: Phys. Rev. **87**, 395 (1952).
[2] N.F. Ramsey and H.B. Silsbee: Phys. Rev. **84**, 506 (1951).
[3] M. Goeppert Mayer: Ann. Physik **9**, 273 (1931); see also G. Breit: Rev. Mod. Phys. **4**, 504 (1932).

A simple theory of induced double quantum transitions will be presented here, in which the applied electromagnetic field is treated as a classical Maxwell field in the manner of the preceding sections[1,2]. Consider a system with three energy levels W_p, W_r, and W_q with the spacing between the adjacent levels (p, r) and (r, q) being unequal. The perturbing interaction, V, with the oscillating field is assumed to have non-vanishing matrix elements connecting only states p and r and states r and q: $V_{rp} = (r|V_0|p)\,\mathrm{e}^{-i\omega t}$ and $V_{qr} = (q|V_0|r)\,\mathrm{e}^{-i\omega t}$. First-order time-dependent perturbation theory gives the probability amplitude that an atom initially in the state p shall be in the state r after it has been subjected to the oscillating field for a time t

$$a_r^{(1)}(t) = -\frac{i}{\hbar}(r|V_0|p)\frac{\{\exp[-i\omega t + i(W_r - W_p)t/\hbar] - 1\}}{[-i\omega + i(W_r - W_p)/\hbar]} \tag{30.1}$$

and

$$a_r^{(1)}(t) \simeq -\frac{it}{\hbar}(r|V_0|p), \quad \text{for} \quad \omega \simeq (W_r - W_p)\,\hbar. \tag{30.1a}$$

The probability amplitude $a_r(t)$ is defined by:

$$a_r(t) = c_r(t)\,\mathrm{e}^{\frac{+iW_r}{\hbar}t} \tag{30.2}$$

together with Eq. (25.2). The usual resonance denominator appears for ω near the Bohr frequency $(W_r - W_p)/\hbar$. Similarly, second-order, time-dependent perturbation theory gives the probability amplitude:

$$\left.\begin{aligned} a_q^{(2)}(t) &= -\frac{1}{\hbar^2}\frac{(q|V_0|r)(r|V_0|p)}{[-i\omega + i(W_r - W_p)/\hbar]} \times \\ &\times \left\{\frac{\exp[-i2\omega t + i(W_q - W_p)t/\hbar] - 1}{[-i2\omega + i(W_q - W_p)/\hbar]} - \frac{\exp[-i\omega t + i(W_q - W_r)/\hbar] - 1}{[-i\omega + i(W_q - W_r)/\hbar]}\right\} \end{aligned}\right\} \tag{30.3}$$

and

$$a_q^{(2)}(t) \simeq \frac{it}{\hbar^2}\frac{(q|V_0|r)(r|V_0|p)}{[-\omega + (W_r - W_p)/\hbar]}, \quad \text{for} \quad 2\omega \simeq (W_q - W_p)/\hbar. \tag{30.3a}$$

Assume that the particle is initially in the state p and that the applied frequency ω is approximately $(W_q - W_p)/2\hbar$. Then there will be negligible transition probability to the state r, which is the only allowed transition in first-order perturbation theory, because ω is far off resonance for this transition. In second-order perturbation theory the transition from p to q is allowed, because of the existence of the intermediate state r. Furthermore, it is seen that one of the terms in the bracket of Eq. (30.3) involves a resonance denominator for $2\omega = (W_q - W_p)/\hbar$. This term is dominant near resonance and Eq. (30.3) reduces to Eq. (30.3a). The occurrence of 2ω instead of ω in the resonance denominator implies that two equi-energetic quanta are involved in the transition. It also implies that the line width for these two quantum transitions is $\frac{1}{2}$ that of the normal single quantum transitions. A comparison of Eqs. (30.1a) and (30.3a) shows that the amplitude for a two quantum transition at its resonance frequency is comparable to the amplitude for a single quantum transition at its (different) resonance frequency, provided that the condition

$$[-\hbar\omega + (W_r - W_p)] \sim (q|V_0|r) \tag{30.4}$$

holds for the two quantum transition. A generalization of this simple theory to higher order multiple quantum transitions is easily made.

[1] V.W. HUGHES and L. GRABNER: Phys. Rev. **79**, **829** (1950). — V.W. HUGHES and J.S. GEIGER: Phys. Rev. **99**, 1842 (1955).

[2] W. PAULI: Vol. V/1, this Encyclopedia.

A double quantum transition was first observed in the electric quadrupole spectrum of $Rb^{85}F$ by use of the molecular beam electric resonance method[1]. The transition observed at half the Bohr frequency was one in which the sum of the angular momentum of the Rb^{85} nucleus ($I=\frac{5}{2}$) and the angular momentum associated with the molecular rotation ($J=1$) changes from $\frac{3}{2}$ to $\frac{7}{2}$. (The sum of these two angular momenta is a constant of the motion to a good approximation, because the coupling to the spin of the F nucleus is small.) Further, this transition was also observed when two different frequencies were applied, provided that the sum of the two frequencies equalled the Bohr frequency for the transition[2]. The perturbing term in the Hamiltonian is the interaction of the electric dipole moment of the molecule with the applied radiofrequency electric field, and the intermediate state is a state with rotational quantum number $J=0$ or $J=2$. (The theory can be generalized to allow for several different intermediate states r.) Although the intermediate state is far removed in energy from either the initial or final states, which have rotational quantum number $J=1$, the double quantum transition is comparable in intensity to the usual single quantum transitions between different electrical quadrupole energy levels of the $J=1$ state, because the single quantum transitions can occur only in the presence of a static electric field which causes an admixture of the rotational states $J=0$ and $J=2$. Hence it is only necessary that the amplitude of the radiofrequency electric field be comparable to that of the static field in order that the intensities of single and double quantum transitions be of the same magnitude. Double quantum transitions within a single rotational state have also been observed for the LiF molecules[3]. The line intensity problem for double quantum transitions seen by the molecular beam electric resonance method has not been considered carefully, and the failure to observe certain other predicted double quantum transitions is not fully understood[1].

Multiple quantum magnetic transitions have been observed[4] by beam spectroscopy in the Zeeman spectra of O_2 and K^{39}. For O_2 the transitions $\Delta J=0$, $\Delta m_J=\pm 2$ and ± 3 were seen, in which J and m_J are the molecular quantum numbers for the total angular momentum and its z component, respectively. For K the transitions $\Delta F=0$ and $\Delta m_F=\pm 2$, ± 3, and ± 4 were seen, in which F and m_F are the quantum numbers for the total angular momentum and its z component. In each case the transition occurred by the absorption or stimulated emission of two or more equi-energetic quanta.

The general theory of magnetic resonance transitions in a rotating magnetic field has been discussed and the following expression is given for the transition probability for single or multiple quantum transitions[5]:

$$P_{p,q}=\frac{|2b|^2}{(\omega_0-\omega)^2+|2b|^2}\sin^2\left\{\frac{|m_q-m_p|}{2}\left[(\omega_0-\omega)^2+|2b|^2\right]^{\frac{1}{2}}T\right\}. \tag{30.5}$$

ω is the frequency of the rotating magnetic field. $\omega_0=\dfrac{W_q-W_p}{\hbar|m_q-m_p|}$, which is the Bohr frequency for the transition divided by the change in magnetic quantum number $|m_p-m_q|$. The amplitude of the matrix element of the interaction with the rotating field which connects the initial and final states is $\hbar b$; it involves

---

[1] V.W. HUGHES and L. GRABNER: Phys. Rev. **79**, 314 (1950).
[2] L. GRABNER and V.W. HUGHES: Phys. Rev. **82**, 561 (1951).
[3] R. BRAUNSTEIN and J.W. TRISCHKA: Phys. Rev. **98**, 1092 (1955). — KASTNER, RUSSELL and TRISCHKA: J. Chem. Phys. **23**, 1730 (1955).
[4] P. KUSCH: Phys. Rev. **93**, 1022 (1954); **101**, 627 (1956).
[5] H. SALWEN: Phys. Rev. **99**, 1274 (1955).

intermediate states in a multiple quantum transition and is proportional to the power $|m_q - m_p|$ of the rotating field. This expression for the transition probability is valid provided only two levels are near resonance, all other levels being off resonance by a frequency separation that is large compared to the natural width of the resonance line. In a higher approximation a shift occurs in ω_0 due to the interaction with the rotating field.

The probability for a multiple quantum transition as given in Eq. (30.5) is seen to differ from the probability for a single quantum transition as given in Eq. (25.7) by the replacement of the usual Bohr condition for resonance by the condition:

$$\nu_0 = \frac{W_q - W_p}{h\,|m_q - m_p|} \qquad (30.6)$$

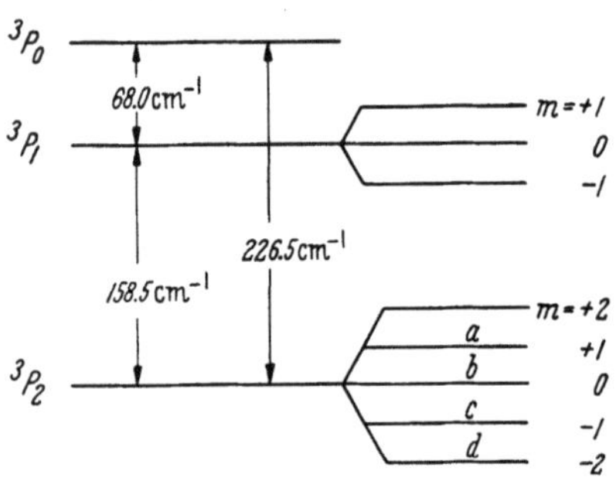

Fig. 54.

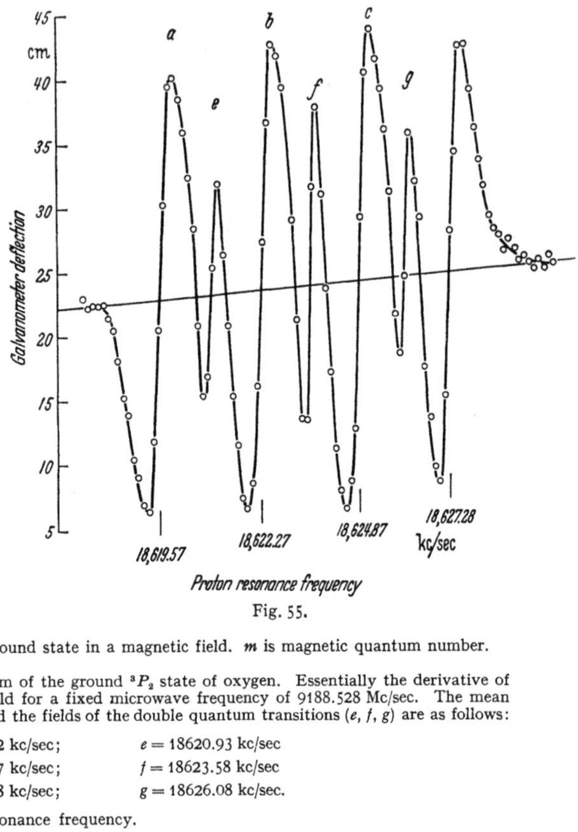

Fig. 55.

Fig. 54. Energy levels of atomic oxygen in the ground state in a magnetic field. m is magnetic quantum number.

Fig. 55. Paramagnetic resonance absorption spectrum of the ground $^3P_2$ state of oxygen. Essentially the derivative of the absorption curve is plotted vs. the magnetic field for a fixed microwave frequency of 9188.528 Mc/sec. The mean fields of the single quantum transitions (a, b, c, d) and the fields of the double quantum transitions (e, f, g) are as follows:

$\frac{1}{2}(a+b) = 18620.92$ kc/sec;	$e = 18620.93$ kc/sec
$\frac{1}{2}(b+c) = 18623.57$ kc/sec;	$f = 18623.58$ kc/sec
$\frac{1}{2}(c+d) = 18626.08$ kc/sec;	$g = 18626.08$ kc/sec.

The magnetic field is measured by the proton resonance frequency.

in which $|m_q - m_p|$ is the number of quanta involved in the transition. The natural line-width for a multiple quantum transition is given by:

$$\delta\nu \sim \frac{1}{|m_q - m_p|\,T} \qquad (30.7)$$

and hence is reduced from that of a single quantum transition by the factor $|m_q - m_p|$. The average of the transition probability Eq. (30.5) can be taken over the velocity distribution in the beam in the manner discussed in Sect. 26[1].

Detailed experimental studies have been made on the line shapes of the multiple and single quantum transitions $\Delta F = 0$, $\Delta m_F = \pm 1, \pm 2, \pm 3, \pm 4$ in the Zeeman spectrum of K^{39} as a function of radiofrequency amplitude[2]. Special care was taken to achieve a uniform magnetic C-field in order that inhomogeneities in the magnetic field would not contribute to the line shape. Calculations

[1] H. SALWEN: Phys. Rev. **101**, 621 (1956).
[2] P. KUSCH: Phys. Rev. **101**, 627 (1956).

of the theoretical line shapes for this spectrum have been made[1]. In general, the behaviour of a resonance line in the neighborhood of the optimum radiofrequency amplitude for the transition is reasonably well described by the theory. The lines are characterized by half-widths which are consistent with theoretical predictions, and the line frequencies agree accurately with those calculated for single and multiple quantum transitions. Within a rather large experimental uncertainty, the radiofrequency amplitude at which the probability of transition is a maximum is consistent with prediction both in absolute and in relative magnitude for single and multiple quantum transitions. However, for higher radiofrequency amplitudes the effects discussed in Sect. 27 for single quantum transitions are observed for multiple quantum transitions as well, in disagreement with the theory. It is believed that the neglect in the theory of the continuous character of the changes in radiofrequency amplitude seen by the atoms at the ends of the transition region is the reason for the disagreement.

Multiple quantum transitions have been rather widely observed in recent years. Multiple quantum Zeeman transitions in the ground state of sodium have been studied by use of optical methods for the production and detection of a polarized beam[2]. Two quantum transitions in the microwave Zeeman spectrum of atomic oxygen in its ground $^3P_2$ state have been observed by the method of microwave magnetic resonance absorption in gases[3]. The energy level diagram and the absorption spectrum for oxygen are shown in Figs. 54 and 55, respectively. The lines e, f, and g are two quantum transitions, and thus line e, for example, occurs at $\frac{1}{2}$ the sum of the resonant field values for transitions a and b. The line width of the two quantum lines is about $\frac{1}{2}$ that of the single quantum lines. Multiple quantum transitions have been used in atomic beam magnetic resonance studies of radioactive gold[4].

31. Transitions involving optically excited states. The transitions considered in the preceding sections have involved only the ground electronic state, and hence the levels could be treated as stable because the spontaneous emission rate for transitions between these levels is negligibly small. However, in a number of important experiments on the $n=2$ state of hydrogen and on optically excited states of alkali atoms, the spontaneous emission rate of the states is an important factor in determining the resonance line shape. In this section transitions involving decaying states will be discussed.

The treatment of the theory of the transition between two stable states given in Sect. 25 can be generalized to the case of two decaying states. The Hamiltonian of Eq. (25.1) is taken without the addition of the interaction with the radiation field which accounts for the spontaneous emission. The wavefunction is expanded as in Eq. (25.2), but for convenience the probability amplitudes $a_n(t)$ will be used, where

$$c_n(t) = a_n(t)\, e^{-\frac{i W_n t}{\hbar}}. \tag{31.1}$$

Consider that there are only two states p and q and that the perturbing interaction with the oscillating field, V, has the matrix elements of Eq. (25.4). Then the time dependent Schrödinger equation for the probability amplitudes

[1] H. Salwen: Phys. Rev. **101**, 623 (1956).

[2] Brossel, Cagnac and Kastler: J. Phys. Radium **15**, 6 (1954). — Besset, Horowitz, Messiah and Winter: J. Phys. Radium **15**, 251 (1954).

[3] V.W. Hughes and J.S. Geiger: Phys. Rev. **99**, 1842 (1955).

[4] Christensen, Hamilton, Lemonick, Pipkin, Reynolds and Stroke: Phys. Rev. **101**, 1389 (1956).

[Eq. (25.3)] can be written with the addition of damping terms to represent the decay of the states:

$$\left.\begin{aligned}\dot{a}_p &= -i\,a_q\,b\,\mathrm{e}^{\left[+i\omega t+\frac{i}{\hbar}(W_p-W_q)t\right]} - \frac{\gamma_p}{2}\,a_p,\\ \dot{a}_q &= -i\,a_p\,b^*\,\mathrm{e}^{\left[-i\omega t+\frac{i}{\hbar}(W_q-W_p)t\right]} - \frac{\gamma_q}{2}\,a_q,\end{aligned}\right\} \tag{31.2}$$

where γ_p and γ_q are the optical decay rates of the states p and q, respectively. (Hence γ_p is the inverse of the mean optical life of an atom in state p.) The justification of this phenomenological treatment of the decay has been considered in connection with hydrogen fine structure measurements[1], and the treatment appears to be valid for all applications made in beam spectroscopy.

The general solutions of Eqs. (31.2) are:

$$a_p = A_1\,\mathrm{e}^{-\left(\delta_1+\frac{\gamma_p}{2}\right)t} + A_2\,\mathrm{e}^{-\left(\delta_2+\frac{\gamma_p}{2}\right)t}, \tag{31.3a}$$

$$a_q = \frac{i}{b}\left\{-\,\delta_1 A_1\,\mathrm{e}^{\left[-\left(\delta_1+\frac{\gamma_p}{2}\right)t-i\omega' t\right]} - \delta_2 A_2\,\mathrm{e}^{\left[-\left(\delta_2+\frac{\gamma_p}{2}\right)t-i\omega' t\right]}\right\} \tag{31.3b}$$

in which $\omega' = \omega + \frac{W_p - W_q}{\hbar}$ and A_1 and A_2 are constants to be determined by the initial conditions.

$$\delta_{1,2} = \left(-\frac{i\omega'}{2} + \frac{\gamma_q-\gamma_p}{4}\right) \pm \frac{1}{2}\left[\left(+i\omega' + \frac{\gamma_p-\gamma_q}{2}\right)^2 - |2b|^2\right]^{\frac{1}{2}}. \tag{31.4}$$

Suppose the initial conditions are:

$$a_p = 1, \quad a_q = 0 \quad \text{at} \quad t = 0. \tag{31.5}$$

Hence,

$$A_1 = \frac{\delta_2}{\delta_2-\delta_1}; \quad A_2 = -\frac{\delta_1}{\delta_2-\delta_1}. \tag{31.6}$$

Then the probability $P_{p,q}$ that the particle will be in the state q at the time T is given by:

$$P_{p,q} = |a_q|^2 = \frac{|\delta_1|^2\,|\delta_2|^2}{|b|^2\,|\delta_2-\delta_1|^2}\,\mathrm{e}^{-\gamma_p T}\left[\left(\mathrm{e}^{-\delta_2 T} - \mathrm{e}^{-\delta_1 T}\right)\left(\mathrm{e}^{-\delta_2^* T} - \mathrm{e}^{-\delta_1^* T}\right)\right]. \tag{31.7}$$

This expression reduces to Eq. (25.7) when $\gamma_p = \gamma_q = 0$.

The time T that a particle spends in the region of the oscillating field is usually long compared to the mean life of an excited state, and the experimental arrangement is usually such as to detect not the probability that a particle is in the state q at time T, but rather the probability that the particle has decayed from the state q in the time interval from $t=0$ to $t=T$. This probability P_q is given by:

$$P_q = \int_0^T \gamma_q\,|a_q|^2\,dt. \tag{31.8}$$

An important application of Eq. (31.8) has been made in the experiments on the Lamb shift[2]. The state p is the $2^2S_{\frac{1}{2}}$ state and its decay rate γ_p can be taken as zero; the state q is the $2^2P_{\frac{1}{2}}$ state (or the $2^2P_{\frac{3}{2}}$ state) which decays to the $1^2S_{\frac{1}{2}}$ state by emission of a photon. When $|b| \ll \gamma_q$ and $|b|\,T \ll 1$, the expression

[1] W.E. LAMB: Phys. Rev. **85**, 259 (1952).
[2] W.E. LAMB and R.C. RETHERFORD: Phys. Rev. **79**, 549 (1950).

for P_q [Eq. (31.8)] reduces to

$$P_q = \frac{\gamma_q |b|^2 T}{\omega'^2 + \left(\frac{\gamma_q}{2}\right)^2}. \tag{31.9}$$

Hence the frequency half-width $\delta\nu$ of the resonance line is $\frac{\gamma_q}{2\pi} \simeq 100$ Mc/sec, as determined by the optical decay rate of the $2P_{\frac{1}{2}}$ state. Another application of Eq. (31.8), which does not involve a beam, has been made in the measurement of the hyperfine structure of positronium where the annihilation decay rates of both states are important[1].

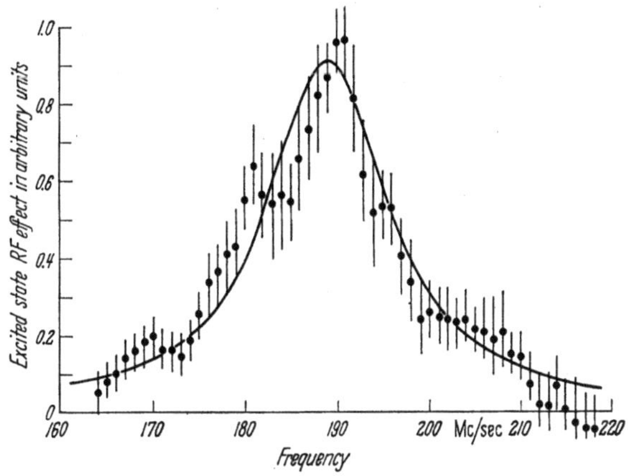

Fig. 56. A typical resonance curve for the $3\,^2P_{\frac{1}{2}}$ state of sodium taken at 2.0 gauss static field. The dots with vertical error lines are the experimental points. The curve is a "best fit" of a Lorentzian curve, Eq. (31.11), to the experimental points.

The theory of transitions between the hyperfine structure levels of the ^2P states of alkali atoms has been discussed principally for the case of two isolated excited levels whose natural widths are small compared to their spacing[2]. If the optical decay rates of the two states are the same, then $P_{p,q}$ as given by Eq. (31.7) reduces to:

$$P_{p,q} = e^{-\gamma T} \frac{|2b|^2}{|2b|^2 + (\omega_0 - \omega)^2} \sin^2 \left\{ \frac{1}{2} \left[|2b|^2 + (\omega_0 - \omega)^2 \right]^{\frac{1}{2}} T \right\} \tag{31.10}$$

in which γ is the common decay rate. The average probability that the atom, which is subjected to the oscillating field during its entire lifetime in an excited state, is in the state q is given by:

$$\langle |a_q|^2 \rangle = \frac{2|b|^2}{|2b|^2 + \gamma^2 + (\omega_0 - \omega)^2}. \tag{31.11}$$

Hence the line width $\delta\nu$ is

$$\delta\nu = \frac{1}{\pi} \left[|2b|^2 + \gamma^2 \right]^{\frac{1}{2}}. \tag{31.12}$$

In practice $\delta\nu$ is of the order of 20 Mc/sec (see Fig. 56).

For the experiments on the excited states of alkali atoms the overlapping of the levels, interference from other nearby states, and perturbations of the levels by the strong radiofrequency fields are appreciable, but in the analysis of the experiments done thus far these effects have not been treated.

32. Transition fields. Electronics. The transition region in a beam spectroscopy apparatus contains the oscillating field which induces the transition, and usually a constant magnetic or electric field as well. In atomic and molecular beam spectroscopy transitions have been studied over essentially the entire available spectrum of electromagnetic radiation. The oscillators employed have

---

[1] V. W. Hughes, S. Marder and C. S. Wu: Phys. Rev. **106**, 934 (1957). Actually in both the Lamb shift and positronium hfs experiments a magnetic field rather than the frequency is varied.

[2] M. L. Perl, I. I. Rabi and B. Senitzky: Phys. Rev. **98**, 611 (1955).

for the most part been of standard design [*10*], [*21*], [*28*][1], although for some of the higher precision experiments special frequency stabilization schemes have been devised. For electric resonance experiments low power sources are used, but for magnetic resonance experiments on nuclear moments and on atoms in shortlived excited states, power sources of up to 100 watts or more are required. The constant magnetic or electric field has usually been produced by simple fields designed with flat, parallel poles or plates.

Typically, frequency measurement is achieved by comparison of the oscillator signal with a quartz crystal secondary frequency standard which itself is calibrated against a signal broadcast from the primary standard in the U.S. Naval Observatory and National Bureau of Standards over station WWV. The laboratory frequency standard usually involves a quartz crystal oscillator at 100 kc/sec, 1 Mc/sec, or 5 Mc/sec, and a chain of frequency multipliers of standard design to produce a series of output signals up to about 300 Mc/sec [*21*]. A difference frequency signal between the oscillator and the frequency standard is generated in a mixer and measured with a heterodyne frequency meter involving its own quartz crystal standard. A typical arrangement for frequency measurement is shown in Fig. 57[2]. The oscillator frequency being measured was about 1500 Mc/sec, and the highest frequency output from the frequency standard was 240 Mc/sec. A harmonic generator involving a coaxial cavity was used to produce a multiple from the frequency signal within 20 Mc/sec of the oscillator frequency. The radio receiver was used simply to amplify the signal from the mixer. Frequency measurement to an accuracy of 1 part in 10^7 can be made by this method.

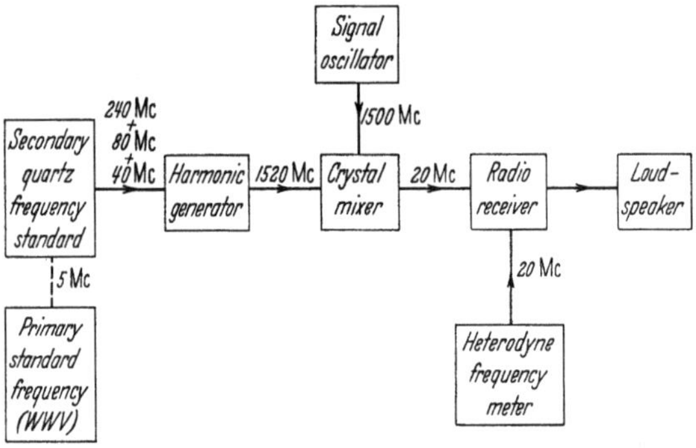

Fig. 57.

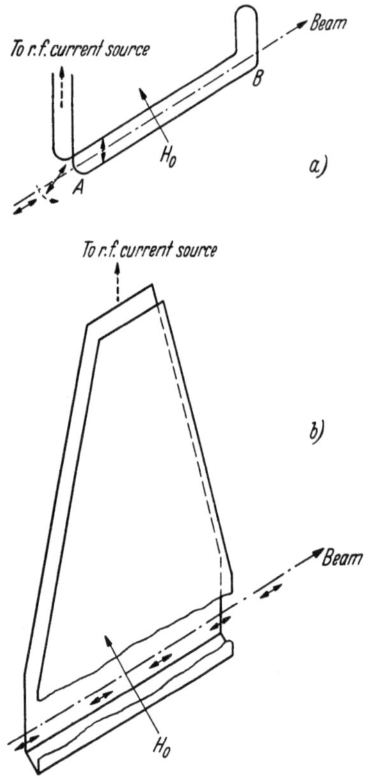

Fig. 58.

Fig. 57. Frequency measurement near 1500 Mc/sec.

Fig. 58. Arrangements for superimposing an oscillating magnetic field on the fixed magnetic field. The orientations of the fixed and alternating fields are indicated by the arrows. (a) The two wire arrangement which exhibits marked "Millman effect". (b) The parallel plane arrangement in which the Millman effect is suppressed.

[1] F. E. TERMAN: Radio Engineer's Handbook. New York-Toronto-London: McGraw-Hill 1943.

[2] HUGHES, TUCKER, RHODERICK and WEINREICH: Phys. Rev. **91**, 828 (1953).

The principal technical feature of the oscillating field peculiar to beam spectroscopy is the electrical and mechanical configuration inside the beam apparatus. For magnetic transitions at frequencies below several thousand megacycles the configurations shown in Fig. 58 have been commonly used[1]. The "hairpin" arrangement in Fig. 58a produces a radiofrequency magnetic field perpendicular to the static field over the major part of the length of the hairpin traversed by the beam, and hence is capable of inducing magnetic transitions in which the z component (z is direction of static magnetic field) of the total angular momentum changes by 1 ($\Delta m = \pm 1$). At the ends of the hairpin a moving particle will effectively experience a time-varying magnetic field with frequencies other than that of the applied frequency because of the change in the direction of the applied radiofrequency field with position. This effect leads to a distortion

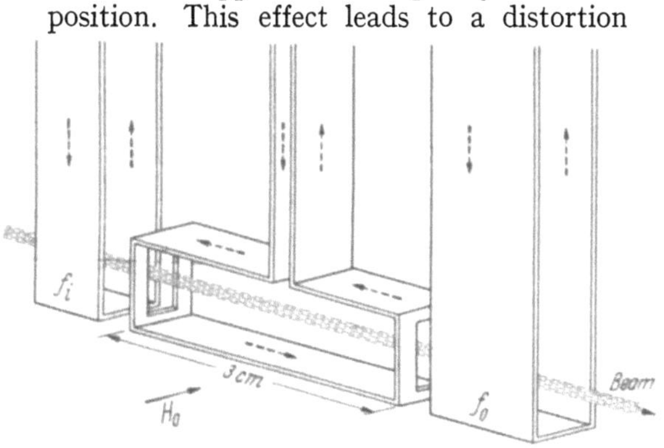

Fig. 59.

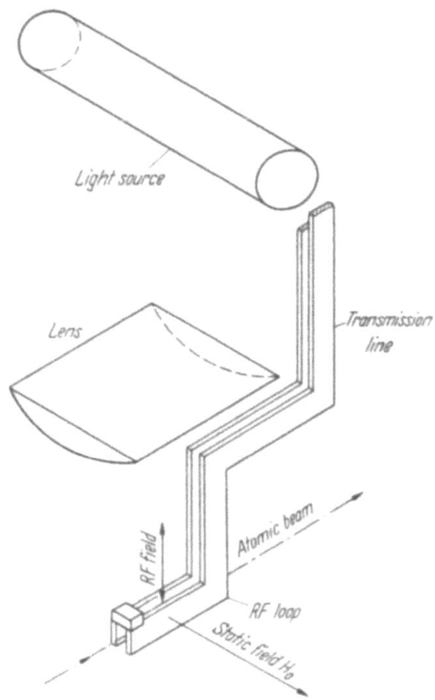

Fig. 60.

Fig. 59. Radiofrequency transition system. The direction of the steady homogeneous field of the C-magnet is indicated by $\boldsymbol{H}_0$, and is perpendicular to the direction of the beam. The wires f_i and f_0 generate oscillating field components in a direction perpendicular to that of $\boldsymbol{H}_0$, whereas the middle wire generates components parallel to $\boldsymbol{H}_0$.

Fig. 60. A diagram of the relative positions of the static and radiofrequency fields, the light source, and the atomic beam. The static magnetic field, the radiofrequency magnetic field, and the atomic beam are mutually perpendicular. The light source lies above the hairpin so that the light beam is propagated parallel to the direction of the radiofrequency field. The lens focuses the light beam into the hairpin.

of a simple resonance curve (called the "Millman" effect) and can be used to determine the sign of the magnetic moment of the particles (see Sect. 46). Doppler broadening of the lines is present for this construction. The radiofrequency arrangement of Fig. 58b with parallel plate "current sheets" also induces transitions with $\Delta m = \pm 1$, but avoids the "Millman" effect. Doppler broadening of the lines is not present because the direction of propagation of the electromagnetic wave is perpendicular to the direction of propagation of the beam. A radiofrequency field of quite uniform amplitude is obtained. A modification of the "hairpin" of Fig. 58a can be made[2] to allow the observation of transitions both with $\Delta m = \pm 1$ and with $\Delta m = 0$. The plane of the two rods parallel to the beam is inclined at an angle of 45° to the direction of the static field, and thus the radiofrequency field has components both parallel and perpendicular

[1] H. Taub and P. Kusch: Phys. Rev. **75**, 1481 (1949).
[2] J.E. Nafe and E.B. Nelson: Phys. Rev. **73**, 718 (1948).

to the static field. Another arrangement for inducing both $\Delta m = \pm 1$ and $\Delta m = 0$ transitions is shown in Fig. 59. In some high precision measurements of Zeeman transition frequencies it has been important to localize the radiofrequency field within a well-defined region, and a copper shield completely enclosing the plates of Fig. 58b except for small entrance and exit slits for the beam has been used[1]. The radiofrequency and optical arrangement used for inducing transitions in the optically excited P states of alkali atoms is shown[2] in Fig. 60. Oscillating magnetic fields of about a gauss are produced.

For magnetic transitions at frequencies above several thousand Mc/sec, the use of a cavity resonator of relatively low Q, with entrance and exit slits for the beam, has been common. Fig. 61 shows the cavity used[3] in the measurement of the hyperfine structure separation in the metastable $^3S_1$ state of helium-3. The cavity operates in the TE_{110} mode (the axial direction is the vertical direction of the figure) and produces an oscillating magnetic field perpendicular to the static magnetic field. The cavity resonant frequency was about 6850 Mc/sec and could easily be varied by about 100 Mc/sec by moving a tightly fitting piece of teflon to different positions in the cavity. The Q of the cavity was approximately 1000. The amplitude of the oscillating field in the cavity varies as $\sin \frac{\pi y}{L}$ along the direction of propagation of the beam, in which L is the cavity dimension in this direction. Oscillating magnetic fields of several tenths of a gauss amplitude were achieved with a 250 milliwatt klystron source.

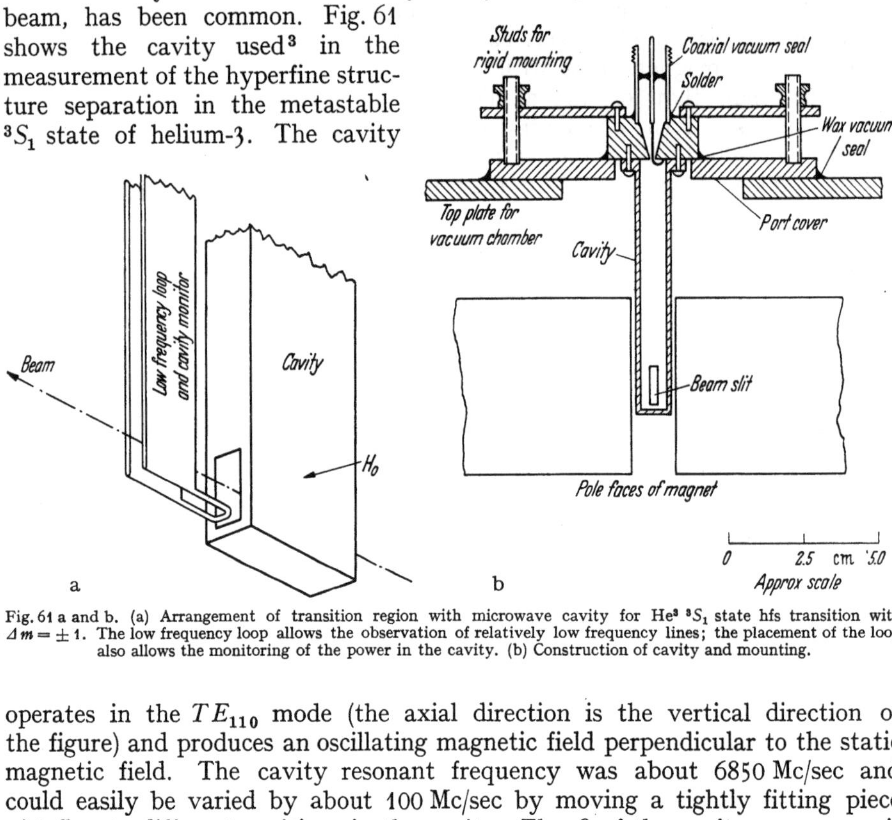

Fig. 61 a and b. (a) Arrangement of transition region with microwave cavity for He^3 $^3S_1$ state hfs transition with $\Delta m = \pm 1$. The low frequency loop allows the observation of relatively low frequency lines; the placement of the loop also allows the monitoring of the power in the cavity. (b) Construction of cavity and mounting.

To induce a thallium hfs transition with $\Delta m = 0$ at about 21000 Mc/sec the microwave circuit shown in Fig. 62 has been used[4]. The wave guide has an oscillating magnetic field parallel to the long dimension of the guide. The wave

[1] HUGHES, TUCKER, RHODERICK and WEINREICH: Phys. Rev. **91**, 828 (1953). — KOENIG, PRODELL and KUSCH: Phys. Rev. **88**, 191 (1952).
[2] PERL, RABI and SENITZKY: Phys. Rev. **98**, 611 (1955).
[3] G. WEINREICH and V. W. HUGHES: Phys. Rev. **95**, 1451 (1954).
[4] A. LURIO and A. PRODELL: Phys. Rev. **101**, 79 (1956).

guide is tapered to allow its insertion between the pole faces of the C-magnet, and the teflon piece (dielectric constant $\simeq 2$) is used to maintain the wave guide cutoff frequency above 21000 Mc/sec.

For the molecular beam electric resonance method transitions have been induced either in the low frequency range up to about 300 Mc/sec or in the centimeter wavelength range. Figs. 63 and 64 show typical arrangements for producing

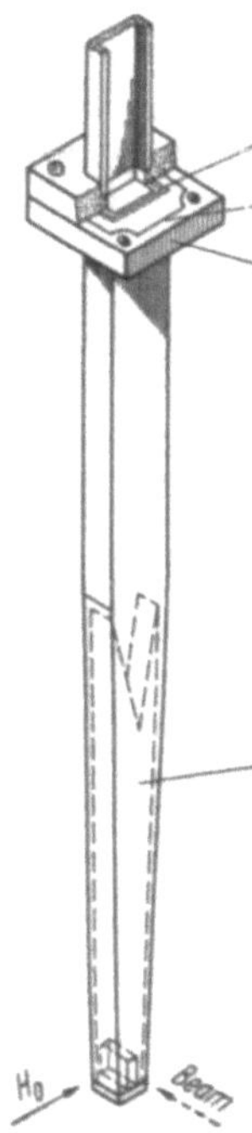

Fig. 62.

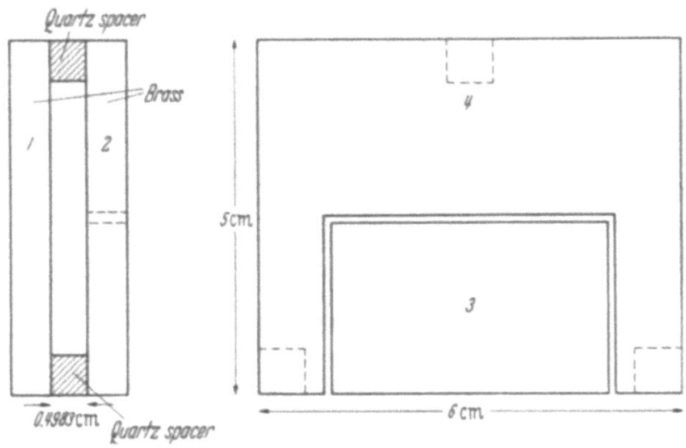

Fig. 63.

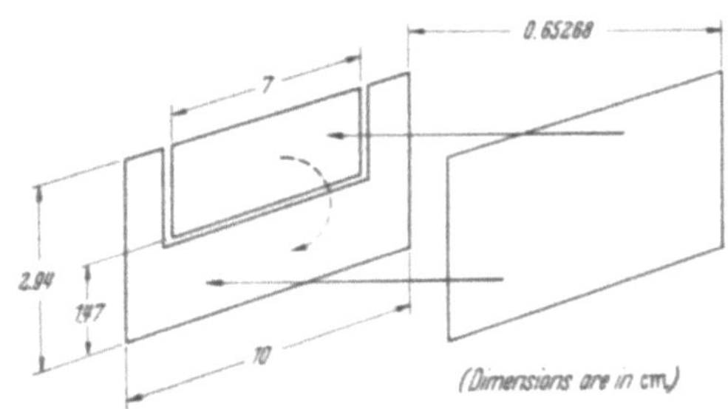

Fig. 64.

Fig. 62. Microwave circuit for inducing $\Delta m = 0$ transitions at 21 kMc. K band wave guide tapers from its standard outside dimensions of 0.63×1.27 cm to 0.625×0.625 cm. A, lead gasket; B, mica window; C, K band wave guide flange; D, teflon piece.

Fig. 63. Transition field for radiofrequency molecular beam electric resonance experiments.

Fig. 64. A three-dimensional view of an electric transition field. The dotted and solid arrows show the directions of the radiofrequency and static fields respectively.

a low frequency oscillating electric field superimposed on a static electric field[1,2]. A potential difference maintained between the parallel plates 1 and 2 produces the static electric field. An alternating voltage is applied between the insulated parts 3 and 4 of plate 2 (3 and 4 are at the same d.c. potential). The oscillating field thus produced will have components both perpendicular and parallel to the static field midway between plates 1 and 2 where the beam travels, and hence transitions in which the total z component of angular momentum, m, changes by ± 1, and 0 can be induced. Usually potential differences with an amplitude of a volt or less are used for the oscillating field. Appreciable oscillating fields near the edges of the parallel plates where the static field is non-uniform are largely avoided in this construction which localizes the field near the opening between 3 and 4. Due to the change in direction of the applied

[1] J. W. Trischka: Phys. Rev. **74**, 718 (1948).
[2] D. T. F. Marple and J. W. Trischka: Phys. Rev. **103**, 597 (1956).

oscillating field along the path of the beam an electric "Millman" effect can be observed[1].

For electric resonance transitions at microwave frequencies the design shown in Fig. 65 has been used[2]. In the center of one of the parallel plates which provide the static electric field there is a narrow slit through which microwave energy radiates into the transition region from the open end of a wave guide. The wave guide gradually tapers to the narrow dimension of the slit in passing through the vacuum envelope to the C-field. Energy is transmitted in the TE_{01} mode whose cutoff wavelength is determined by the long dimension of the guide, so this dimension is maintained the same throughout its length. Around the edges of the parallel plates of the C-field a light coating of aquadag was painted, and across the top and bottom of the plates were clamped sheets of mica with a light

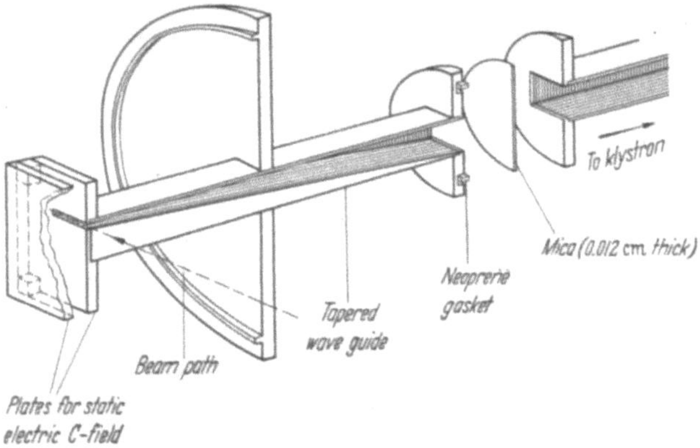

Fig. 65. Microwave circuit, showing wave guide which introduces centimeter wavelength microwave energy into C field region.

coating of aquadag to reduce the amount of energy radiated outside of the C-field region; no direct effect of microwave energy at the detector was ever observed. The termination of the wave guide resulted in considerable reflection, but the system had a broad band-width compared with the width of the spectral lines. The narrow slit in the static field plate did not seriously affect the homogeneity of the static electric field. Only about a microwatt of power is required from the klystron source for optimum transition probability.

The constant magnetic or electric field has usually been produced by simple field designs with flat, parallel poles or plates. Homogeneity over space and constancy in time have been the principal problems. For the attainment of the most homogeneous fields the C-magnet is usually constructed entirely of Armco or high grade Swedish iron[3]. The pole faces are ground and hand-lapped, and uniformity of the gap spacing is achieved with use of quartz or hand-lapped brass spacers accurate to better than 0.001 cm. Substantial variations in field homogeneity are often observed when the magnet is turned off and on again. This effect is believed to be caused by variation in local properties of the iron, and it may necessitate considerable recycling of the field to obtain optimum homogeneity.

[1] L. Grabner and V.W. Hughes: Phys. Rev. **79**, 819 (1950).

[2] Lee, Fabricand, Carlson and Rabi: Phys. Rev. **91**, 1395 (1953).

[3] H. Taub and P. Kusch: Phys. Rev. **75**, 1481 (1949). The use of permendur as pole piece material was found to be less favorable than Armco iron with regard to the homogeneity of the field.

The presence of the iron A- and B-magnets affects the homogeneity in the C-region, and the stray field from the A- and B-magnets may contribute about 50 gauss in the C-region. Slight variations in the A- and B-fields have been used to improve the field homogeneity in the C-region[1]. The homogeneity can be varied by changing the current through small coils of wire held against the pole faces; this adjustment can be made from outside the vacuum envelope. For the precise measurement of the magnetic moment of the electron in hydrogen[2] an unconventional design was employed in which the C-magnet was placed outside the vacuum chamber, and hence could be shimmed in a conventional manner with pieces of nickel, while observing the field homogeneity with a proton resonance probe. A slight tilting of the pole faces could be obtained and was a useful adjustment in attaining field homogeneity. Two-wire deflecting fields were used for the A- and B-fields in order to avoid effects on the C-field of nearby iron magnets.

The C-magnet has customarily been a low voltage, high current, water-cooled electromagnet, powered by submarine batteries. The constancy of the field in time has been obtained by maintaining the current constant. Manual monitoring of the current using a high sensitivity potentiometer connected across a shunt resistor[2], and electronic current monitoring have been employed[3]. Monitoring of the A- and B-field currents has also been found necessary because of their effect on the field in the C-region[1]. Small residual drifts, believed to be due to temperature changes, have remained. Automatic field monitoring with a nuclear resonance probe has also been used, and is more satisfactory.

The method of separated oscillating fields requires two regions of oscillating field, which are separated in space and in which the oscillating fields must have a definite and known phase difference. The radiofrequency or microwave circuit for an individual transition region can be one of the types discussed earlier in this section. If the phase difference between the two fields is to be zero, a feed with symmetrical branches from a common source has been most generally used[4] [*18*]. If the phase difference is to be different from zero, separate feeds, including line stretching elements, from a single oscillator have been employed[3]. Even slight changes in relative phase can cause a change in the resonance line shape [see Eq. (29.1)], which must be considered as a possible systematic error in high precision work, as, for example, in the case of the cesium atomic frequency standard [*18*].

The atomic frequency standard which utilizes the hfs interval in Cs^{133} occurring at about 9193 Mc/sec for a field H_0 of about 60 milligauss involves basically the control of an oscillator to the frequency of the field independent $(F, m_F) = (3, 0) \leftrightarrow (4, 0)$ hfs transition in the ground ${}^2S_{\frac{1}{2}}$ state of cesium. An error signal is derived from the intensity of the line for the frequency control via a servo loop. The basic oscillator operates at 5 Mc/sec, and its frequency is controlled by a quartz crystal and variable air trimmer capacitor. By frequency multiplication and synthesis, which involves a reflex klystron operating at 9180 Mc/sec, a signal at the resonance frequency of 9192.631840 Mc/sec is produced. Long time stability ($\sim$ weeks) and reproducibility of this frequency to about 1 part in 10^{10} have been achieved [*18*].

[1] Hughes, Tucker, Rhoderick and Weinreich: Phys. Rev. **91**, 828 (1953).
[2] Koenig, Prodell and Kusch: Phys. Rev. **88**, 191 (1952).
[3] Kolsky, Phipps, Ramsey and Silsbee: Phys. Rev. **87**, 395 (1952).
[4] Heberle, Reich and Kusch: Phys. Rev. **101**, 612 (1956).

B. Atomic spectroscopy.

33. Introduction. The finer details of atomic spectra, which include hyperfine structure intervals, certain small, fine structure separations—notably those of hydrogen—, and Zeeman splittings, have been studied extensively with high resolution by the method of atomic beam spectroscopy. The interaction energies of the atomic electrons with the higher order nuclear moments—magnetic dipole, electric quadrupole, and magnetic octupole—have been measured with a typical precision of 1 kc/sec and at best with a precision of about 10 cps. The hyperfine structure of the simplest atom, hydrogen, reveals information about the internal structure of the proton itself, and, indeed, the precision to which this hfs separation is known experimentally is well beyond the accuracy of the theoretical interpretation, which is presently limited by the precision to which the fine structure constant α is known. Perhaps the principal application to nuclear physics of the precisely known values of the magnetic dipole hyperfine structure interaction constants occurs through the magnetic dipole hyperfine structure anomaly which arises from the finite size and mass of the nucleus. Its interpretation yields information about the distribution of magnetism in nuclei. The spins and magnetic moments of many radioactive nuclei have been determined by atomic beam spectroscopy.

The electric quadrupole interaction constant, eqQ, is the product of a factor q which is a second derivative of the electric potential due to the electrons at the position of the nucleus and of a factor Q which is the electric quadrupole moment of the nucleus. The value of q can be obtained approximately from a knowledge of the magnetic dipole hyperfine structure interaction constant of the atom, together with the value of the magnetic dipole moment of the nucleus, or, alternatively, from a knowledge of the fine structure splitting; hence the nuclear electric quadrupole moment Q can be determined. This determination of Q is limited at present to an accuracy of roughly 10% because of uncertainties in the atomic theory relating q to magnetic dipole hfs or fine structure separations. Still this method of determining nuclear electric quadrupole moments has yielded most of the better known values of Q. The sizes of many electric quadrupole moments of heavier nuclei, which were larger than could be explained by the independent particle model of nuclear structure, provided the principal early stimulus for the Rainwater-Bohr theory of the collective aspects of nuclear structure.

The nuclear magnetic octupole moment was discovered by atomic beam spectroscopy from the breakdown of the interval rule involving only the magnetic dipole and electric quadrupole hyperfine structure interaction terms. The measured octupole interaction constant is the product of a second derivative of the magnetic field at the position of the nucleus due to the atomic electrons, which can be evaluated approximately from the knowledge of fine structure or magnetic dipole hfs splitting, and of the nuclear magnetic octupole moment. The only information available on nuclear magnetic octupole moments comes from atomic beam spectroscopy.

Most of the measurements of hyperfine structure by atomic beam spectroscopy have been made on atoms in their ground or metastable states. Recently, however, an extension of the method of atomic beams has been made to allow the study of hyperfine structure in optically excited states. This extension involves the combined use of optical and radiofrequency techniques. The large line widths associated with the short lifetimes of the optically excited states result in less accurate determinations of hfs splittings than are obtainable from studies on ground states. The principal application of this method thus far has been the determination of the nuclear electric quadrupole moments of alkali nuclei.

If the dependence of the frequency of a spectral line on external magnetic or electric field is sufficiently small, the natural line width associated with the transit time may be obtained by the conventional beam resonance method, or, often more readily, by the method of separated oscillating fields. Line widths as narrow as 200 cps have been obtained for the Cs^{133} hyperfine structure transition occurring at about 9200 Mc/sec, and have thus allowed the determination of this hfs splitting to about 1 part in 10^9, by use of an astronomical time standard. By use of a servo loop it has been possible to lock an electronic oscillator to this hfs transition with a reproducibility and stability (over time intervals of weeks) of about 1 part in 10^{10}. A similar precision has been attained with the maser which is a molecular beam oscillator utilizing the ammonia inversion frequency. Such a precision is considerably higher than that of any other terrestrial source and over short time intervals is competitive with that of astronomical time standards.

The quantum electrodynamic radiative corrections to the properties of the electron were discovered and precisely measured by the method of atomic beam spectroscopy. Studies of the Zeeman effect, expecially of the alkalis and of gallium and indium, led to the discovery of the anomalous spin magnetic moment of the electron. The present value, which is accurate to about 10 parts per million, is obtained from the measurement of the Zeeman effect of atomic hydrogen. The anomaly in the electron spin magnetic moment stimulated considerable interest in atomic magnetism. The g-values of many atoms have been precisely measured, and the theory of atomic magnetism has been extended to include relativistic and other higher order contributions.

The measurement of the fine structure of the $n=2$ state of atomic hydrogen resulted in the discovery of the Lamb shift, which is the energy separation between the $2\,^2S_{\frac{1}{2}}$ and $2\,^2P_{\frac{1}{2}}$ states. The precisely known values of the Lamb shift—(1057.77 ± 0.10) Mc/sec for H and (14040.2 ± 4.5) Mc/sec for He^+—provide major confirmations for the modern theory of quantum electrodynamics. The measured fine structure separation between the $n=2$ states of deuterium provides the best value of the fine structure constant α.

I. Hyperfine structure.

a) General theory.

34. Hamiltonian and energy levels. The part of the total Hamiltonian of an atom which gives the hyperfine structure and Zeeman interactions of the atom can be written[1]:

$$\mathscr{H} = \sum_k T_e^{(k)} \bullet T_n^{(k)} + \mu_0 g_J \boldsymbol{J} \cdot \boldsymbol{H}_0 + \mu_0 g_I \boldsymbol{I} \cdot \boldsymbol{H}_0. \tag{34.1}$$

The first term is the interaction of the electrons with the non-central electromagnetic field due to the nuclear moments—magnetic dipole, electric quadrupole, magnetic octupole, and higher order multipole moments. The quantities $T_e^{(k)}$ and $T_n^{(k)}$ are tensor operators of rank k and operate in the spaces of the electronic and nuclear coordinates, respectively[2]. The second and third terms give the interaction of the electronic and nuclear magnetic moments, respectively, with the externally applied magnetic field $\boldsymbol{H}_0$; g_J and g_I are the electronic and

[1] C. Schwartz: Phys. Rev. **97**, 380 (1955).

[2] The heavy dot denotes the scalar product of two tensor operators.

$$T_e^{(k)} \bullet T_n^{(k)} = \sum_\mu T_{e,\mu}^{(k)} T_{n,-\mu}^{(k)} (-1)^\mu.$$

nuclear g-factors (the sign convention is taken so that the g-factor for a free electron is positive), $\boldsymbol{J}$ and $\boldsymbol{I}$ are the electronic and nuclear angular momentum operators, and μ_0 is the Bohr magneton.

At zero magnetic field the electronic angular momentum $\boldsymbol{J}\hbar$ and the nuclear angular momentum $\boldsymbol{I}\hbar$ are coupled by the hfs interaction to produce the total atomic angular momentum $\boldsymbol{F}\hbar\,(\boldsymbol{F}=\boldsymbol{I}+\boldsymbol{J})$. The energy eigenvalues W_F are given by first order perturbation theory as:

$$\left.\begin{aligned} W_F &= \langle I J F \,|\, \sum_k T_e^{(k)} \bullet T_n^{(k)} \,|\, I J F\rangle \\ &= \sum_k (-1)^{I+J-F}\, W(I J I J; F k)\, \langle I \| T_n^{(k)} \| I\rangle\, \langle J \| T_e^{(k)} \| J\rangle \end{aligned}\right\} \tag{34.2}$$

where W, the Racah coefficient, is a known algebraic function of its six arguments; the double-barred matrix elements, called the reduced matrix elements, are independent of any magnetic quantum numbers. Eq. (34.2) may be rewritten as:

$$W_F = \sum_k A_k M(I J; F; k) \tag{34.3}$$

in which

$$A_k = \langle T_e^{(k)}\rangle_{J J} \langle T_n^{(k)}\rangle_{I I} \tag{34.3a}$$

and

$$\langle T_e^{(k)}\rangle_{J J} = \frac{(2J)!}{[(2J-k)!\,(2J+k+1)!]^{\frac{1}{2}}} \langle J \| T_e^{(k)} \| J\rangle \tag{34.3b}$$

with a similar expression for $\langle T_n^{(k)}\rangle_{I I}$. The coefficient M is given by the formula

$$\left.\begin{aligned} M(I J; F; k) &= \frac{(2I-k)!}{(2I)!}\,\frac{(2J-k)!}{(2J)!}\,\frac{(I+J-F)!\,(J-I+F)!\,(I-J+F)!}{(I+J+F+1)!}\,[k!]^2 \times \\ &\times \sum_z (-1)^{z+I+J-F} \frac{(2I+2J+1-z)!}{z!\,(2I-k-z)!\,(2J-k-z)!\,[(I+J-F-z)!\,(k+F-I-J+z)!]^2} \end{aligned}\right\} \tag{34.3c}$$

where the sum extends over all positive integral values of z including $z=0$ for which no factorial has a negative argument. M is often expressed in terms of the cosine factor,

$$K = F(F+1) - I(I+1) - J(J+1)$$

and the first four orders of M in terms of K are the following:

$$M(I J; F; 1) = \frac{1}{2 I J} K, \tag{34.4a}$$

$$M(I J; F; 2) = \frac{6}{(2I)(2I-1)(2J)(2J-1)} \left[K(K+1) - \left(\frac{4}{3}\right) I(I+1) J(J+1)\right], \tag{34.4b}$$

$$\left.\begin{aligned} &M(I J; F; 3) \\ &= \frac{20[K^3 + 4K^2 + \frac{4}{5} K\{-3I(I+1)J(J+1) + I(I+1) + J(J+1) + 3\} - 4I(I+1)J(J+1)}{2I(2I-1)(2I-2)(2J)(2J-1)(2J-2)}, \end{aligned}\right\} \tag{34.4c}$$

$$\left.\begin{aligned} M(I J; F; 4) = &\frac{70}{(2I)(2I-1)(2I-2)(2I-3)(2J)(2J-1)(2J-2)(2J-3)} [K^4 + 10K^3 + \\ &+ \tfrac{4}{7} K^2 \{-6I(I+1)J(J+1) + 5I(I+1) + 5J(J+1) + 39\} + \\ &+ \tfrac{4}{7} K \{-34 I(I+1)J(J+1) + 12 I(I+1) + 12 J(J+1) + 18\} - \\ &- \tfrac{24}{35} \times I(I+1)J(J+1) \times \\ &\times \{-2I(I+1)J(J+1) + 4I(I+1) + 4J(J+1) + 27\}]. \end{aligned}\right\} \tag{34.4d}$$

The series (34.3) terminates at the term $k=2J$ or $2I$, whichever is smaller. There are then $2J$ (or $2I$) interaction constants A_k to be solved for from the $2J$ (or $2I$) measured energy intervals. The A_k are given by:

$$\left.\begin{aligned} A_k = (2k+1)\,&\frac{[(2I)!\,(2J)!]^2}{(2I-k)!\,(2I+k+1)!\,(2J-k)!\,(2J+k+1)!}\times\\ &\times\sum_F (2F+1)\,M(IJ;F;k)\,W_F. \end{aligned}\right\} \tag{34.5}$$

The terms involving A_1, A_2, and A_3 are the magnetic dipole, electric quadrupole, and magnetic octupole interactions, respectively, These A's are related to the interaction constants a, b, and c commonly used in the literature for magnetic dipole, electric quadrupole, and magnetic octupole interactions, respectively, by the equations

$$A_1 = IJa, \tag{34.6a}$$

$$A_2 = \frac{b}{4}, \tag{34.6b}$$

$$A_3 = c. \tag{34.6c}$$

The constant A_4 refers to the 2^4-pole electric interaction. Thus far only the terms with A_1, A_2, and A_3 have been observed experimentally, largely because the spectroscopic measurements of a precision that might disclose higher order A's have been made only on atoms in states for which $J \leqq \frac{3}{2}$.

In a weak magnetic field $\boldsymbol{H}_0$ the energy splitting of the magnetic sublevels m of a state with total angular momentum quantum number F is given by:

$$\langle IJFm|\,\mu_0 g_J \boldsymbol{J}\cdot\boldsymbol{H}_0 + \mu_0 g_I \boldsymbol{I}\cdot\boldsymbol{H}_0\,|IJFm\rangle = g_F\,\mu_0 H_0 m \tag{34.7}$$

in which

$$g_F = g_J\frac{F(F+1)+J(J+1)-I(I+1)}{2F(F+1)} + g_I\frac{F(F+1)+I(I+1)-J(J+1)}{2F(F+1)}. \tag{34.7a}$$

The energy levels in any external magnetic field can be analyzed with the Hamiltonian:

$$\mathscr{H} = a\boldsymbol{I}\cdot\boldsymbol{J} + b\,\frac{[3(\boldsymbol{I}\cdot\boldsymbol{J})^2+(\frac{3}{2})\,\boldsymbol{I}\cdot\boldsymbol{J} - \boldsymbol{I}^2\boldsymbol{J}^2]}{2I(2I-1)\,J(2J-1)} + \mu_0 g_J\boldsymbol{J}\cdot\boldsymbol{H}_0 + \mu_0 g_I \boldsymbol{I}\cdot\boldsymbol{H}_0, \tag{34.8}$$

provided that I and J are good quantum numbers for the atom. Only terms through the electric quadrupole interaction are included, since magnetic octupole interactions have been studied only under weak field conditions where Eqs. (34.3) and (34.7) apply.

An important case is for $J=\frac{1}{2}$ for which the Hamiltonian reduces to:

$$\mathscr{H} = a\boldsymbol{I}\cdot\boldsymbol{J} + \mu_0 g_J \boldsymbol{J}\cdot\boldsymbol{H}_0 + \mu_0 g_I \boldsymbol{I}\cdot\boldsymbol{H}_0. \tag{34.9}$$

The energy eigenvalues for any field H_0 are given by:

$$W_{F=I\pm\frac{1}{2},m} = -\frac{\Delta W}{2(2I+1)} + \mu_0 g_I H_0 m \pm \frac{\Delta W}{2}\left[1+\frac{4mx}{(2I+1)}+x^2\right]^{\frac{1}{2}} \tag{34.10}$$

in which ΔW is the zero field hfs separation between the states $F=I+\frac{1}{2}$ and $F=I-\frac{1}{2}$ $[\Delta W \equiv W_{I+\frac{1}{2}}(H_0=0) - W_{I-\frac{1}{2}}(H_0=0)]$ and

$$x \equiv (g_J - g_I)\,\mu_0 H_0/\Delta W. \tag{34.10a}$$

Eq. (34.10) is known as the Breit-Rabi formula[1]. The hfs separation in $\sec^{-1}$, the conventional unit in the measurements here considered, is $\Delta\nu$ with $\Delta\nu = \Delta W/h$. The quantity $[1+4mx/(2I+1)+x^2]^{\frac{1}{2}}$ is to be read as $(1-x)$ for $m = -(I+\frac{1}{2})$. The quantity $\Delta W = a(I+\frac{1}{2})$.

The energy levels given by the Hamiltonian of Eq. (34.8) under strong field conditions, in which the interaction with the external field H_0 is large compared to the hfs interactions, are characterized by the magnetic quantum numbers m_J and m_I associated with J and I. With the inclusion of only the diagonal elements of the hfs terms in this representation, the energy levels are given by:

$$\left.\begin{aligned} W_{m_J, m_I} = \mu_0 g_J m_J H_0 + \mu_0 g_I m_I H_0 + a m_I m_J + \\ + b\frac{[3m_J^2 - J(J+1)]\,[3m_I^2 - I(I+1)]}{4J(2J-1)\,I(2I-1)}. \end{aligned}\right\} \tag{34.11}$$

At intermediate fields it may be necessary to diagonalize the complete secular determinant associated with the Hamiltonian of Eq. (34.8). Second order perturbation theory starting from either the weak field representation (F, m) or the strong field representation (m_J, m_I) can be employed to give corrections to Eqs. (34.3), (34.7), and (34.11)[2]. A useful formula for the Hamiltonian of Eq. (34. 9) is the second order perturbation result starting from the (F, m) representation

$$W_{F=I\pm\frac{1}{2}, m} = W_{F=I\pm\frac{1}{2}}(H_0 = 0) + \mu_0 g_F H_0 m \pm \Delta W \left[\frac{x^2}{4} - \frac{m^2 x^2}{(2I+1)^2}\right] \tag{34.12}$$

in which g_F is defined by Eq. (34.7a)

35. Interpretation of interaction constants[3]. From the measured hfs intervals the hfs interaction constants A_k (or a, b, and c) can be determined, see Eq. (34.5). The general theory of the use of these interaction constants to obtain the magnetic dipole moments, electric quadrupole moments, and magnetic octupole moments of nuclei will be treated in this section. Further and more specific discussions of the nuclear moments and of the magnetic dipole hfs anomalies for particular atoms are given in Sects. 36 to 38.

The electrostatic potential $V(\boldsymbol{r})$ due to the charges in the nucleus at a point exterior to the nucleus can be expressed as:

$$V(\boldsymbol{r}) = \sum_{k=0}^{\infty} r^{-k-1} C^{(k)}(\vartheta, \varphi) \bullet Q_k \tag{35.1}$$

in which

$$C_\mu^{(k)}(\vartheta, \varphi) = \left(\frac{4\pi}{2k+1}\right)^{\frac{1}{2}} Y_{k\mu}(\vartheta, \varphi) \tag{35.1a}$$

where $Y_{k\mu}$ are normalized spherical harmonics of the polar and azimuthal angles ϑ and φ.

Q_k^μ are the general nuclear electric multipole moments:

$$Q_k^\mu = e \int \Psi_N^* \, g_l r'^k C_\mu^{(k)}(\vartheta', \varphi') \, \Psi_N \, dv' \tag{35.2}$$

where $\Psi_N(\boldsymbol{r}')$ is the nuclear wavefunction; e = proton charge; g_l is the orbital g-value of the nucleus:

$$g_l \mu_N \boldsymbol{L}_N = \sum_i g_{l_i} \frac{e\hbar}{2Mc} \boldsymbol{L}_{N_i}; \tag{35.2a}$$

---

[1] G. BREIT and I.I. RABI: Phys. Rev. **38**, 2082 (1931). The total angular momentum is not a good quantum number, and the designation of the energy eigenvalues by a particular F value simply specifies the corresponding zero field energy level.

[2] B.T. FELD and W.E. LAMB jr.: Phys. Rev. **67**, 15 (1945).

[3] C. SCHWARTZ: Phys. Rev. **97**, 380 (1955).

the sum is taken over all the nucleons; $g_{l_i} = +1$ for a proton and 0 for a neutron; M = mass of nucleon; $\boldsymbol{L}_{N_i}$ = orbital angular momentum of nucleon i in units of $\hbar$; μ_N = nuclear magneton; $\boldsymbol{L}_N$ = total orbital angular momentum of nucleus in units of $\hbar$.

The vector potential $\boldsymbol{A}(\boldsymbol{r})$ for the magnetic field due to the nuclear currents and spins at a point exterior to the nucleus can be expressed with the aid of vector spherical harmonics as:

$$\boldsymbol{A}(\boldsymbol{r}) = \sum_{k=1}^{\infty} (-i/k)\, r^{-k-1} (\boldsymbol{L}\, C^{(k)}(\vartheta, \varphi) \bullet M_k) \tag{35.3}$$

in which

$$\boldsymbol{L} = -i\,\boldsymbol{r} \times \nabla. \tag{35.3a}$$

M_k^μ are the general nuclear magnetic multipole moments

$$M_k^\mu = \mu_N \int \Psi_N^* \left[\nabla(r'^{\,k} C_\mu^{(k)}(\vartheta', \varphi'))\right] \cdot \left[g_l \{2/(k+1)\}\, \boldsymbol{L} + g_s \boldsymbol{S}\right] \Psi_N \, dv' \tag{35.4}$$

where g_s is the spin g-value of the nucleus

$$g_s \mu_N \boldsymbol{S}_N = \sum_i g_{s_i} \frac{e\hbar}{2Mc} \boldsymbol{S}_{N_i}. \tag{35.4a}$$

g_{s_i} = spin g-value for nucleon i (proton, $+5.6$; neutron, -3.8); $\boldsymbol{S}_{N_i}$ = spin angular momentum of nucleon i in units of $\hbar$; $\boldsymbol{S}_N$ = total spin angular momentum of nucleus in units of $\hbar$.

Only electric moments of even order and magnetic moments of odd order occur if the nuclear wavefunction has a well-defined parity.

The interaction energy of a single electron (charge $= -e$) with the nuclear fields given by Eqs. (35.1) and (35.3) can be obtained by insertion of the potentials V and $\boldsymbol{A}$ in the Dirac equation for the electron

$$[\boldsymbol{\alpha} \cdot (c\,\boldsymbol{p} + e\boldsymbol{A}) + \beta m c^2 - eV]\, \Psi = W\Psi, \tag{35.5}$$

in which $\Psi(\boldsymbol{r})$ is the electron 4-component spinor wavefunction, $\boldsymbol{p} = -i\hbar\nabla$, $\boldsymbol{\alpha}, \beta$ are Dirac matrices, m is the electron mass, c the velocity of light, W the energy eigenvalue. A perturbation theory treatment is employed. The zeroth order solution $\Psi^{(0)}(\boldsymbol{r})$ is obtained with $\boldsymbol{A} = 0$ and with $V = V_c$, the central part of the electrostatic potential. $\Psi^{(0)}$ will be of the form $\Psi^{(0)} = \binom{\psi}{\varphi}$ where ψ is the large component and φ is the small component of the four-spinor $\Psi^{(0)}$. ψ and φ are of thc form:

$$\psi_{l j m} = \frac{g(r)}{r} \mathscr{Y}_{l j m} \tag{35.6a}$$

$$\varphi_{\bar{l} j m} = \frac{f(r)}{i r} \mathscr{Y}_{\bar{l} j m}. \tag{35.6b}$$

j is the total angular momentum quantum number and m the associated magnetic quantum number; l is the orbital angular momentum quantum number, $\bar{l} = l \pm 1$, according as $j = l \pm \frac{1}{2}$,

$$\mathscr{Y}_{l j m} = (\tfrac{1}{2} m_s\, l\, m_l \,|\, \tfrac{1}{2} l\, j\, m)\; Y_{l m_l}(\vartheta, \varphi)\, \chi_{\frac{1}{2}, m_s}$$

in which χ is a two-component spinor for the electron spin; m_l and m_s are the magnetic quantum numbers for orbital and spin angular momentum, respectively and $m = m_l + m_s$. The perturbation term in the Hamiltonian is then:

$$\mathscr{H}_1 = -e(V - V_c) + e\,\boldsymbol{\alpha} \cdot \boldsymbol{A}. \tag{35.7}$$

In the first order perturbation theory only the diagonal matrix elements of $\mathcal{H}_1$ are needed and the interaction constants A_k of Eq. (34.3) are given by[1]:

$$\left.\begin{aligned} A_k = -Q_k e \int_0^\infty d r\, r^{-k-1} (f^2 + g^2) (-1)^{k/2} \frac{(k-1)!!}{k!!} \frac{(2j)!}{(2j+k)!!\,(2j-k-1)!!}, \\ k \text{ even}, > 0, \end{aligned}\right\} \quad (35.8)^2$$

$$\left.\begin{aligned} A_k = -M_k (e/k)\, 4\varkappa \int_0^\infty d r\, r^{-k-1} f g (-1)^{(k-1)/2} \frac{k!!}{(k-1)!!} \frac{(2j)!}{(2j+k+1)!!\,(2j-k)!!}, \\ k \text{ odd}. \end{aligned}\right\} \quad (35.9)$$

In terms of the specific nuclear moments[3]

$$Q_k = \langle Q_k^0 \rangle_{I, m_I = I}, \qquad M_k = \langle M_k^0 \rangle_{I, m_I = I}\,. \tag{35.10}$$

$\varkappa$ is the Dirac quantum number:

$$\varkappa = \begin{cases} (j+\frac{1}{2}), & j = l + \frac{1}{2}, \\ -(j+\frac{1}{2}), & j = l - \frac{1}{2}. \end{cases}$$

The radial integrals can be evaluated with the assumption that the important contribution to the integrals comes from the region of small values of r. This should be an excellent approximation for the cases $j = l + \frac{1}{2} = k/2$ ($s_{\frac{1}{2}}$ electron, magnetic dipole; $p_{\frac{3}{2}}$ electron, magnetic octupole) where the non-relativistic treatment gives the interaction as due entirely to the electron density at the nucleus ($r=0$). For orbits of larger l, however, the wave function is concentrated farther out and is more slowly rising near the origin, so this approximation is less valid. If the potential V_c is taken as $V_c = Ze/r$ and if the approximation $|mc^2 - E| \ll eV_c$ is made, the values of the radial integrals are:

$$\int_0^\infty r^{-k-1} (f^2+g^2)\, dr = C^2 \left(\frac{2Z}{a_0}\right)^k \frac{(2k-2)!}{k!\,(k-1)!} \frac{[k(2\varkappa+k)(2\varkappa+k-1) - 4\alpha^2 Z^2 (3k-1)]}{(2\varrho+k)(2\varrho+k-1)\dots(2\varrho-k)}, \tag{35.11a}$$

$$\int_0^\infty r^{-k-1} f g\, dr = C^2 \frac{\hbar}{2mc} \left(\frac{2Z}{a_0}\right)^{k+1} \frac{(2k-1)!}{k!\,(k-1)!} \frac{(-k-2\varkappa)}{(2\varrho+k)(2\varrho+k-1)\dots(2\varrho-k)} \tag{35.11b}$$

---

[1] The "double factorial" $k!!$ is defined by:

$$k!! = 2\times 4\times 6\times\cdots\times k \quad (k \text{ even}),$$
$$k!! = 1\times 3\times 5\times\cdots\times k \quad (k \text{ odd}),$$
$$0!! = 1.$$

[2] The quantity $A_2/(2Q_2)$ is often designated as eq_J and

$$eq_J = \left\langle \frac{\partial^2 V}{\partial z^2} \right\rangle_{m_J=J} = \left\langle \frac{\varrho(3\cos^2\vartheta - 1)}{r^3} \right\rangle_{m_J=J}$$

where V is the electrostatic potential produced at the nucleus by all charges outside of the nucleus at which V is to be determined; ϱ is the charge density; r is the distance from the nucleus to the element of charge $\varrho\, d\tau$; ϑ is the angle between $\boldsymbol{r}$ and the z axis, defined through $m_J = J$.

[3] The specific nuclear moments are related to the usually defined moments as follows:

$$\mu = M_1 = \text{magnetic dipole moment}$$
$$eQ = 2Q_2 = \text{electric quadrupole moment.}$$
$$\Omega = -M_3 = \text{magnetic octupole moment.}$$

Q_k and M_k as here used are not to be confused with the tensor operators in Eqs. (35.1) and (35.3).

in which $a_0 = \hbar^2/m e^2$, $\varrho = (\varkappa^2 - \alpha^2 Z^2)^{\frac{1}{2}}$, $\alpha = e^2/(\hbar c)$ and C is a normalization constant. In particular, for the magnetic dipole, electric quadrupole, and magnetic octupole terms:

magnetic dipole:

$$\int_0^\infty r^{-2} f g\, dr = C^2 \frac{\hbar}{2mc}\left(\frac{2Z}{a_0}\right)^2 \frac{F}{(2l+1)\,[\mp(2l+1)-1]}, \quad \mp \text{ as } j = l \pm \tfrac{1}{2}, \tag{35.12a}$$

electric quadrupole:

$$\int_0^\infty r^{-3}(f^2+g^2)\, dr = C^2\left(\frac{2Z}{a_0}\right)^2 \frac{R}{l(2l+1)(2l+2)}, \tag{35.12b}$$

magnetic octupole:

$$\left.\begin{aligned}\int_0^\infty r^{-4} f g\, dr = C^2 \frac{\hbar}{2mc}\left(\frac{2Z}{a_0}\right)^4 \frac{10T}{(2l+3)(2l+2)(2l+1)(2l)(2l-1)\,[\mp(2l+1)-3]},\\ \pm \text{ as } j = l \pm \tfrac{1}{2}.\end{aligned}\right\} \tag{35.12c}$$

The quantities F, R, and T are relativistic correction factors:

$$F = \frac{4j(j+\frac{1}{2})(j+1)}{\varrho(4\varrho^2-1)}, \tag{35.13a}$$

$$R = \frac{l(l+1)(2l+1)}{\varrho(\varrho^2-1)(4\varrho^2-1)}\,[3\varkappa(\varkappa+1) - \varrho^2 + 1], \tag{35.13b}$$

$$T = \frac{(2j+4)!}{(2j-3)!}\,\frac{(2\varrho-4)!}{(2\varrho+3)!}. \tag{35.13c}$$

For an s electron in the non-relativistic limit

$$C^2 = \pi(a_0^2/Z^2)\,\psi^2(0). \tag{35.14}$$

For a non-s electron C can be evaluated in terms of the fine-structure separation between the states $j = l + \frac{1}{2}$ and $j = l - \frac{1}{2}$,

$$C^2 = \frac{\delta}{2.911}\,\frac{l(l+1)}{2HZ^3}\,\frac{1}{a_0} \tag{35.15}$$

where δ is the doublet splitting in cm^{-1} and H is a relativistic correction factor:

$$H = \frac{(\varrho' - \varrho'' - 1)\,2l(l+1)}{Z^2\alpha^2} \tag{35.15a}$$

in which ϱ' and ϱ'' refer to the states $j = l + \frac{1}{2}$ and $j = l - \frac{1}{2}$, respectively. C can also be obtained from measured values of the magnetic dipole interaction constant A_1 and of the nuclear magnetic dipole moment, and then used[1,2] to compute the nuclear electric quadrupole and magnetic octupole moments from the measured interaction constants A_2 and A_3.

The relativistic correction factors and normalization constant have been calculated with wave functions obtained by numerical integration of the Dirac radial equations for an atomic potential which approximates the Thomas-Fermi potential. For the atoms B, Al, Ga, In, Tl, Cl, Br, and I in the doublet P states the directly calculated values[3] differ significantly from the values obtained from Eqs. (35.13) and (35.14).

[1] Davis, Feld, Zabel and Zacharias: Phys. Rev. **76**, 1076 (1949).
[2] Jaccarino, King, Satten and Stroke: Phys. Rev. **94**, 1798 (1954).
[3] C. Schwartz: Phys. Rev. **105**, 173 (1957).

The above treatment is for a single electron bound to the nucleus. It is also correct for describing the interaction of a single valence electron outside closed sub-shells of other electrons, provided that the total wave function for all the electrons can be written in the form of products where the coordinates of the valence electron are separated from those of the core electrons. For configurations such as p^5, d^9 in L-S coupling, or $(\frac{3}{2})^3$, $(\frac{5}{2})^3$ in j-j coupling, where there is just one electron less than the number needed to fill a shell, the even (electric multipole) interactions are just (-1) times the values for a single electron while the odd (magnetic multipole) ones are the same. In the case of more complex electronic configurations the coupling scheme of the several angular momenta involved must be known and the techniques of Racah[1] then show how to calculate the appropriate "projection" factors.

The further approximation has been made of considering the nucleus as a point source and hence of using only that part of the potential corresponding to an observation point outside the nuclear matter. The error made in this way may affect the magnitudes of multipole moments calculated from observed line frequencies. The effects of nuclear size and structure will be considered in Sect. 37.

Second order effects of the hfs interactions involve the product of two matrix elements, each involving a particular multipole interaction, from a state IJF to a different state $I'J'F$. The energy term will involve a part which has the same form as a first order interaction of order k [see Eq. (34.3)] as regards its dependence on F, provided the two multipole interactions k_1 and k_2 satisfy the condition:

$$|k_1 - k_2| \leq k \leq k_1 + k_2. \tag{35.16}$$

The second order effects make difficult the unambiguous separation of the various multipole interactions. Thus, for example, it is seen from Eq. (35.16) that in second order the square of the magnetic dipole interaction can influence the electric quadrupole term, and the cross product of magnetic dipole and electric quadrupole interactions can influence the magnetic octupole term. Usually the principal part of the second order effect involves only the matrix elements to a fine structure state with the same electronic configuration, and a measure of the size of this effect is the ratio of the hfs interaction to the fine structure interaction. The second order effects are particularly important in the determination of the small magnetic octupole interactions since the higher order effects of the lower order multipole interactions may then be relatively large.

The type of atomic wave function used in this section represents a series of closed shells and one valence electron, and only the valence electron contributes to the hfs matrix element. The non-central part of the electrostatic interaction between the electrons leads to the admixture of other electronic configurations and hence to a modification of the hfs matrix element. This configuration mixing occurs primarily due to the electrostatic interactions between a valence electron and the inner core electrons, and can be considered to produce a polarization of the inner core which shields (or anti-shields) the field of the valence electron from the nucleus. The effect on the various multipole interactions has been calculated to be between a few percent and about 30% (see Sect. 38 for further discussions of particular atoms)[2]. The different multipole interactions are affected differently, so that the ratio of different multipole interactions is not independent of configuration interactions.

[1] G. Racah: Phys. Rev. **62**, 438 (1942).
[2] R. Sternheimer: Phys. Rev. **95**, 736 (1954).

b) Methods of measurement and results.

36. Hyperfine structure intervals and nuclear spins. The methods of measuring hyperfine structure intervals can be illustrated by a discussion of the important case for which the electronic angular momentum quantum number J is $\frac{1}{2}$. The hyperfine energy levels of the atom in a magnetic field are given by the Breit-Rabi formula [Eq. (34.10)], and are plotted in Figs. 66 to 68 for the cases of nuclear spin $I=\frac{1}{2}$, 1, and $\frac{3}{2}$, respectively. The abscissa is the dimensionless parameter

$$x = \frac{(g_J - g_I)\,\mu_0 H_0}{\Delta W},$$

which is proportional to the magnetic field H_0. Each level is labeled by its weak field quantum numbers (F, m) and by its strong field quantum numbers (m_J, m_I). The nuclear moment is assumed to be positive for all cases, i.e., the nuclear magnetic dipole moment is in the direction of the nuclear spin. The frequencies of transitions can be computed from Eq. (34.10). The frequencies of transitions which are allowed by the selection rules given by Eqs. (36.5) and (36.7) and which correspond to $\Delta m_J = \pm 1$ at strong fields (and hence are easily observable by the large change in magnetic moment), are given analytically in Table 7 and are plotted in Figs. 69 to 71 for the cases under consideration.

The measurement of hfs intervals is made by the atomic beam magnetic resonance method. The magnetic moment of an atom in a state designated by its low field quantum numbers (F, m) is given by

$$\mu_{F,m} = -\frac{\partial W_{F,m}}{\partial H_0}.$$

Table 7. *Frequencies of transitions for atoms with $J=\frac{1}{2}$ and $I=\frac{1}{2}$, 1 or $\frac{3}{2}$.*

Spin I	Transition $(F,m) \leftrightarrow (F',m')$	Frequency
$\frac{1}{2}$	$(1,1) \leftrightarrow (0,0)$	$\Delta\nu\,[\frac{1}{2}(1+x)+\frac{1}{2}(1+x^2)^{\frac{1}{2}}+y]$
	$(1,0) \leftrightarrow (0,0)$	$\Delta\nu\,(1+x^2)^{\frac{1}{2}}$
	$(1,0) \leftrightarrow (1,-1)$	$\Delta\nu\,[-\frac{1}{2}(1-x)+\frac{1}{2}(1+x^2)^{\frac{1}{2}}+y]$
1	$(\frac{3}{2},\frac{3}{2}) \leftrightarrow (\frac{1}{2},\frac{1}{2})$	$\Delta\nu\,[\frac{1}{2}(1+x)+\frac{1}{2}(1+\frac{2}{3}x+x^2)^{\frac{1}{2}}+y]$
	$(\frac{3}{2},\frac{1}{2}) \leftrightarrow (\frac{1}{2},\frac{1}{2})$	$\Delta\nu\,(1+\frac{2}{3}x+x^2)^{\frac{1}{2}}$
	$(\frac{3}{2},\frac{1}{2}) \leftrightarrow (\frac{1}{2},-\frac{1}{2})$	$\Delta\nu\,[\frac{1}{2}(1+\frac{2}{3}x+x^2)^{\frac{1}{2}}+\frac{1}{2}(1-\frac{2}{3}x+x^2)+y]$
	$(\frac{3}{2},-\frac{1}{2}) \leftrightarrow (\frac{1}{2},+\frac{1}{2})$	$\Delta\nu\,[\frac{1}{2}(1+\frac{2}{3}x+x^2)^{\frac{1}{2}}+\frac{1}{2}(1-\frac{2}{3}x+x^2)-y]$
	$(\frac{3}{2},-\frac{1}{2}) \leftrightarrow (\frac{1}{2},-\frac{1}{2})$	$\Delta\nu\,(1-\frac{2}{3}x+x^2)^{\frac{1}{2}}$
	$(\frac{3}{2},-\frac{1}{2}) \leftrightarrow (\frac{3}{2},-\frac{3}{2})$	$\Delta\nu\,[-\frac{1}{2}(1-x)+\frac{1}{2}(1-\frac{2}{3}x+x^2)^{\frac{1}{2}}+y]$
$\frac{3}{2}$	$(2,2) \leftrightarrow (1,1)$	$\Delta\nu\,[\frac{1}{2}(1+x)+\frac{1}{2}(1+x+x^2)^{\frac{1}{2}}+y]$
	$(2,1) \leftrightarrow (1,1)$	$\Delta\nu\,(1+x+x^2)^{\frac{1}{2}}$
	$(2,1) \leftrightarrow (1,0)$	$\Delta\nu\,[\frac{1}{2}(1+x+x^2)^{\frac{1}{2}}+\frac{1}{2}(1+x^2)^{\frac{1}{2}}+y]$
	$(2,0) \leftrightarrow (1,1)$	$\Delta\nu\,[\frac{1}{2}(1+x+x^2)^{\frac{1}{2}}+\frac{1}{2}(1+x^2)^{\frac{1}{2}}-y]$
	$(2,0) \leftrightarrow (1,0)$	$\Delta\nu\,(1+x^2)^{\frac{1}{2}}$
	$(2,0) \leftrightarrow (1,-1)$	$\Delta\nu\,[\frac{1}{2}(1+x^2)^{\frac{1}{2}}+\frac{1}{2}(1-x+x^2)^{\frac{1}{2}}+y]$
	$(2,-1) \leftrightarrow (1,0)$	$\Delta\nu\,[\frac{1}{2}(1+x^2)^{\frac{1}{2}}+\frac{1}{2}(1-x+x^2)^{\frac{1}{2}}-y]$
	$(2,-1) \leftrightarrow (1,-1)$	$\Delta\nu\,(1-x+x^2)^{\frac{1}{2}}$
	$(2,-1) \leftrightarrow (2,-2)$	$\Delta\nu\,[-\frac{1}{2}(1-x)+\frac{1}{2}(1-x+x^2)^{\frac{1}{2}}+y]$

The frequencies of transitions are computed from the Breit-Rabi formula Eq. (34.10). $x \equiv (g_J - g_I)\mu_0 H_0/\Delta W$; $y \equiv g_I \mu_0 H_0/\Delta W$. Only transitions which are allowed by the magnetic dipole selection rules $\Delta F = 0, \pm 1$; $\Delta m = 0, \pm 1$, and which correspond to $\Delta m_J = \pm 1$ at strong fields are included. See Figs. 69 to 71 for plots of these transition frequencies.

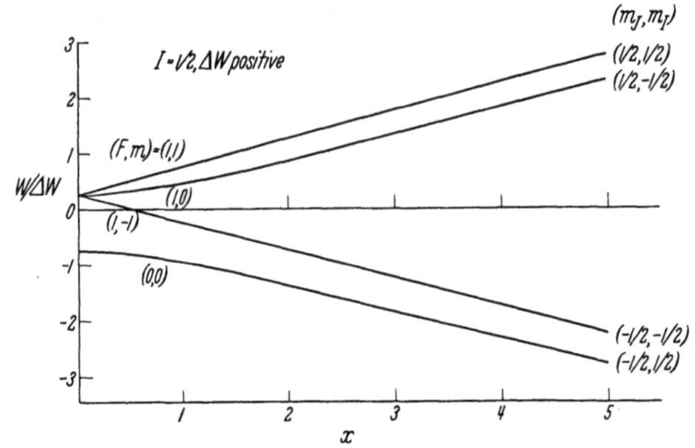

Fig. 66. Energy level diagram for an atom with $J=\frac{1}{2}$ and $I=\frac{1}{2}$ in a magnetic field H_0, obtained from Eq. (34.10). $W=$ energy of atom (the reference zero level is the energy of the atom in the absence of both the hfs interaction and the external magnetic field H_0); $\Delta W \equiv W_{F=I+\frac{1}{2}}(H_0=0) - W_{F=I-\frac{1}{2}}(H_0=0)$; $x=(g_J-g_I)\,\mu_0 H_0/\Delta W$. The levels are designated by both their weak field quantum numbers (F, m) and their strong field quantum numbers (m_J, m_I). The diagram is drawn for a positive nuclear magnetic moment, so the hfs levels are in the normal order and g_I is negative.

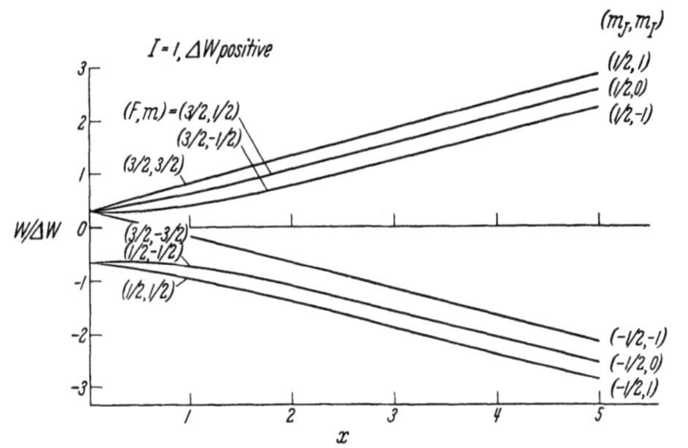

Fig. 67. Energy level diagram for an atom with $J=\frac{1}{2}$ and $I=1$ in a magnetic field H_0. Further description is the same as for Fig. 66.

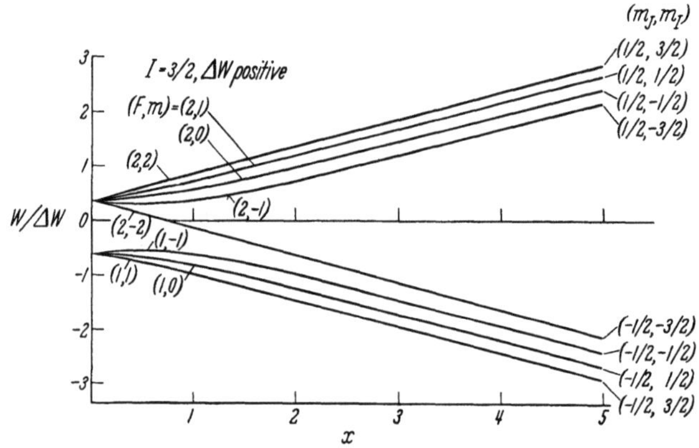

Fig. 68. Energy level diagram for an atom with $J=\frac{1}{2}$ and $I=\frac{3}{2}$ in a magnetic field H_0. Further description is the same as for Fig. 66.

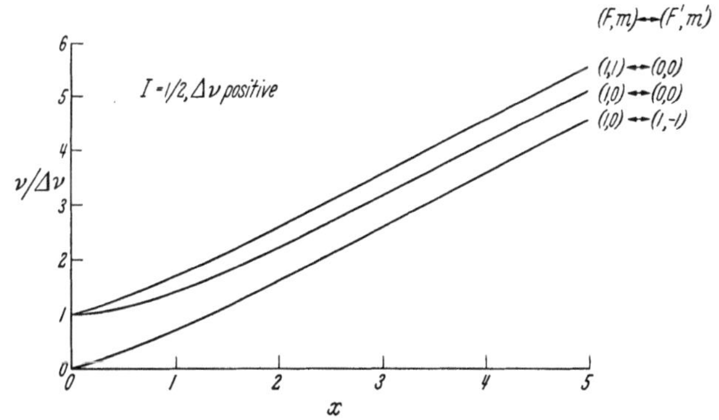

Fig. 69. The frequencies of transitions between magnetic sublevels of the hfs states of an atom with $J=\frac{1}{2}$ and $I=\frac{1}{2}$ as a function of magnetic field H_0 (see Table 7). $\Delta\nu \equiv \Delta W/h$; $x \equiv (g_J - g_I)\,\mu_0 H_0/\Delta W$. The transitions are designated by the weak field quantum numbers of the two levels involved. Only transitions which are allowed by the magnetic dipole selection rules $\Delta m = 0, \pm 1$ and which correspond to transitions $\Delta m_J = \pm 1$ at strong fields, are shown. A positive nuclear magnetic moment is assumed (see caption Fig. 66).

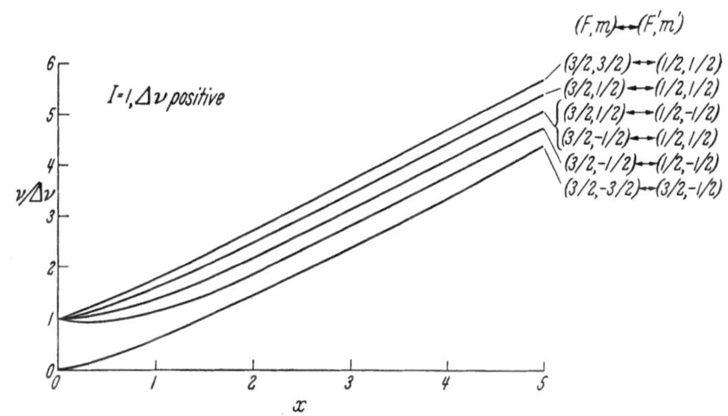

Fig. 70. The frequencies of transitions between magnetic sublevels of the hfs states of an atom with $J=\frac{1}{2}$ and $I=1$ as a function of magnetic field H_0. Further description is the same as for Fig. 69.

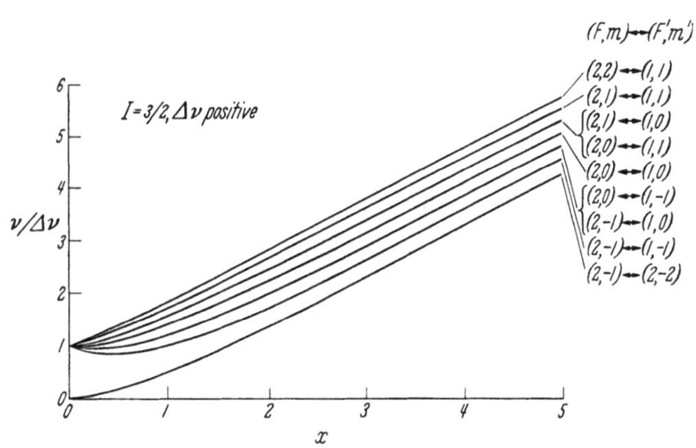

Fig. 71. The frequencies of transitions between magnetic sublevels of the hfs states of an atom with $J=\frac{1}{2}$ and $I=\frac{3}{2}$ as a function of magnetic field H_0. Further description is the same as for Fig. 69.

For an atom with $J=\frac{1}{2}$, differentiation of the Breit-Rabi formula [Eq. (34.10)] gives:

$$\frac{\mu_{F=I\pm\frac{1}{2},m}}{\mu_0} = -g_I m \mp \frac{x/2 + m/(2I+1)}{[1+4mx/(2I+1)+x^2]^{\frac{1}{2}}}(g_J - g_I) \tag{36.1a}$$

for

$$m = I-\tfrac{1}{2},\ I-\tfrac{3}{2},\ \ldots,\ -(I-\tfrac{1}{2}),$$

$$\frac{\mu_{I+\frac{1}{2},m=\pm(I+\frac{1}{2})}}{\mu_0} = \mp(g_J/2+g_I I). \tag{36.1b}$$

The magnetic moments given by Eqs. (36.1) are plotted in Figs. 72 to 74 for $I=\frac{1}{2}$, $I=1$, and $I=\frac{3}{2}$, respectively, for the case of positive nuclear magnetic dipole moments. To the scale of the figures there is no observable effect of the terms in g_I, and hence the figures are applicable to all atoms with the specified nuclear spin I.

Usually the A- and B-deflecting magnets are operated at high values of the magnetic field in order to obtain large deflecting power. At the corresponding large values of $x\,(x\gg 1)$, the magnetic moment is given most simply by use of Eq. (34.11) in terms of the strong field quantum numbers m_J and m_I:

$$\frac{\mu_{m_J,m_I}}{\mu_0} = -g_I m_I - g_J m_J. \tag{36.2}$$

The electronic g-value, g_J, is about 1000 times larger than the nuclear g-value, g_I, and hence significant differential trajectories for an atom due to a transition will occur only if the transition corresponds to a change $\Delta m_J = \pm 1$ at high fields. Only such transitions are observable when the deflecting magnets are operated at high field values. Thus, for example, for the case when $J=I=\frac{1}{2}$ shown in Fig. 66, the only transitions [designated by their (F,m) values], of the five allowed by the magnetic dipole selection rule $\Delta F=0, \pm 1$, $\Delta m = 0, \pm 1$, that are observable are $(0,0)\leftrightarrow(1,1)$, $(0,0)\leftrightarrow(1,0)$ and $(1,0)\leftrightarrow(1,-1)$.

If the deflecting magnets are operated at a suitably chosen low field value, all states may have significantly different magnetic moments and all transitions allowed by the selection rules may be observable. The deflecting power of the A- and B-fields will generally be less than at high fields and all transitions will not be accompanied by the same moment change, but in practice it is generally possible to design deflecting fields so that all the allowed transitions are observable.

The transitions between different hfs levels are induced by an oscillating magnetic field $\boldsymbol{H}_1 \cos\omega t$ in the C-region which contributes the time dependent Hamiltonian term

$$V = \mu_0 g_J \boldsymbol{J}\cdot\boldsymbol{H}_1 \cos\omega t + \mu_0 g_I \boldsymbol{I}\cdot\boldsymbol{H}_1 \cos\omega t. \tag{36.3}$$

In the weak field (F,m) representation (the z direction is the direction of the external magnetic field $\boldsymbol{H}_0$) the non-vanishing matrix elements of the dominant term in V involving interaction of the oscillating field with the electronic magnetic moment are determined from[1]:

$$(F,m|J_z|F,m) = m\frac{F(F+1)+J(J+1)-I(I+1)}{2F(F+1)}, \tag{36.4a}$$

$$\left.\begin{aligned}(F+1,m|J_z|F,m) &= \sqrt{(F+1)^2-m^2}\ \times \\ &\times \sqrt{\frac{(F+1-J+I)(F+1+J-I)(F+2+J+I)(J+I-F)}{4(F+1)^2(2F+1)(2F+3)}},\end{aligned}\right\} \tag{36.4b}$$

[1] E.U. CONDON and G.H. SHORTLEY: The Theory of Atomic Spectra, Chap. III. Cambridge: Cambridge University Press 1953.

$$(F, m+1 \mid J_x + iJ_y \mid F, m) = \sqrt{(F+m+1)(F-m)}\,\frac{F(F+1)+J(J+1)-I(I+1)}{2F(F+1)}, \tag{36.4c}$$

$$\left.\begin{aligned}(F+1, m+1 \mid J_x + iJ_y \mid F, m) &= -\sqrt{(F+m+2)(F+m+1)}\times\\ &\times\sqrt{\frac{(F+1-J+I)(F+1+J-I)(F+2+J+I)(J+I-F)}{4(F+1)^2(2F+1)(2F+3)}},\end{aligned}\right\} \tag{36.4d}$$

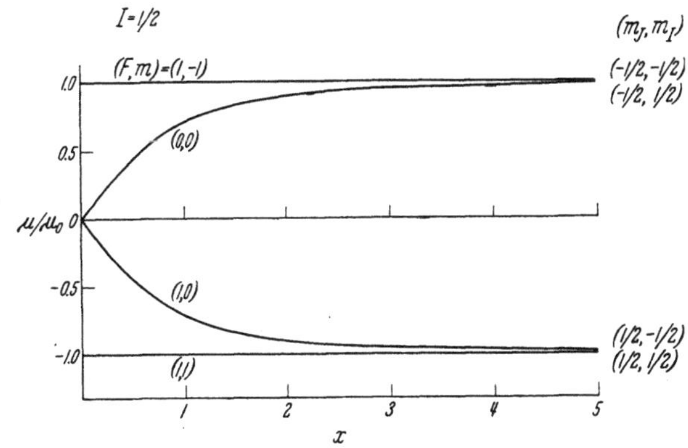

Fig. 72. The magnetic moments of the magnetic sublevels of the hfs states of an atom with $J=\frac{1}{2}$ and $I=\frac{1}{2}$ as a function of magnetic field H_0, obtained from Eqs. (36.1). The levels are designated by both their weak field quantum numbers (F, m) and their strong field quantum numbers (m_J, m_I). A positive nuclear moment is assumed (see figure caption 66).

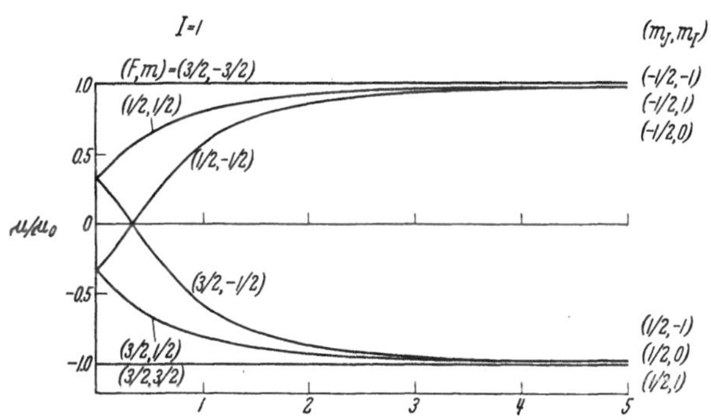

Fig. 73. The magnetic moments of the magnetic sublevels of the hfs states of an atom with $J=\frac{1}{2}$ and $I=1$ as a function of magnetic field H_0. Further description is the same as for Fig. 72.

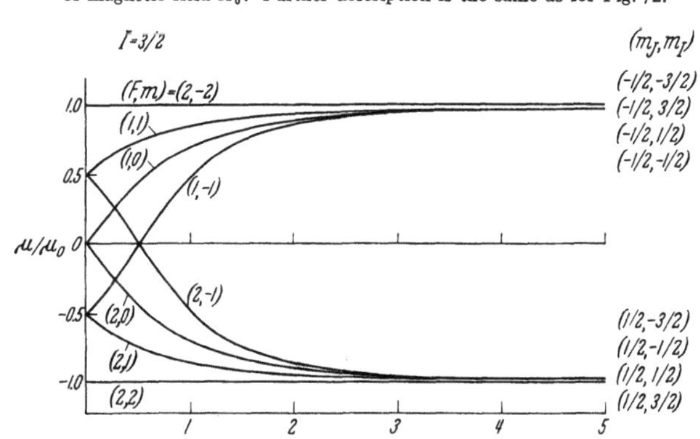

Fig. 74. The magnetic moments of the magnetic sublevels of the hfs states of an atom with $J=\frac{1}{2}$, $I=\frac{3}{2}$ as a function of magnetic field H_0. Further description is the same as for Fig. 72.

$$(F, m-1|J_x - iJ_y|F, m) = \sqrt{(F-m+1)(F+m)}\,\frac{F(F+1)+J(J+1)-I(I+1)}{2F(F+1)}, \quad (36.4\,\mathrm{e})$$

$$\left.\begin{aligned}(F+1, m-1|J_x - iJ_y|F, m) &= \sqrt{(F-m+2)(F-m+1)}\times\\ &\times\sqrt{\frac{(F+1-J+I)(F+1+J-I)(F+2+J+I)(J+I-F)}{4(F+1)^2(2F+1)(2F+3)}}.\end{aligned}\right\} \quad (36.4\,\mathrm{f})$$

These matrix elements determine the selection rules:

$$\left.\begin{aligned}&\Delta F = 0, \pm 1; \quad \Delta m = 0\\ &\qquad (\sigma \text{ transitions}) \text{ for the } z \text{ component of } \boldsymbol{H}_1.\end{aligned}\right\} (36.5\,\mathrm{a})$$

$$\left.\begin{aligned}&\Delta F = 0, \pm 1; \quad \Delta m = \pm 1\\ &\qquad (\pi \text{ transitions}) \text{ for the } x \text{ and } y \text{ components of } \boldsymbol{H}_1.\end{aligned}\right\} (36.5\,\mathrm{b})$$

The same selection rules are obtained from the term in Eq. (36.3) giving the interaction of the oscillating field with the nuclear magnetic moment. The matrix elements for I are obtained from those for J simply by an interchange of I and J for $\Delta F = 0$; for $\Delta F = \pm 1$, in addition to the interchange of I and J, the sign of the matrix element is reversed.

In the strong field (m_J, m_I) representation the non-vanishing matrix elements of the term involving the electronic magnetic moment in Eq. (36.3) are determined from[1]:

$$(m_J, m_I|J_z|m_J, m_I) = m_J, \quad (36.6\mathrm{a})$$

$$(m_J+1, m_I|J_x + iJ_y|m_J, m_I) = \sqrt{(J-m_J)(J+m_J+1)}, \quad (36.6\mathrm{b})$$

$$(m_J-1, m_I|J_x - iJ_y|m_J, m_J) = \sqrt{(J+m_J)(J-m_J+1)}. \quad (36.6\mathrm{c})$$

The selection rules are:

$$\Delta m_J = 0, \quad \Delta m_I = 0 \quad \text{for the } z \text{ component of } \boldsymbol{H}_1, \quad (36.7\mathrm{a})$$

$$\Delta m_J = \pm 1, \quad \Delta m_I = 0 \quad \text{for the } x \text{ and } y \text{ components of } \boldsymbol{H}_1. \quad (36.7\mathrm{b})$$

Corresponding matrix elements and selection rules are obtained for the term in Eq. (36.3) involving the nuclear magnetic moment.

The simplest procedure in principle for determining the hyperfine structure separation between the levels $F = I + \frac{1}{2}$ and $F = I - \frac{1}{2}$ involves the observation of the transition $\Delta F = \pm 1$ at zero field, where the transition frequency is exactly the hfs separation $\Delta\nu$, or at a magnetic field so low that the deviation of the transition frequency from the value $\Delta\nu$ is small compared to the natural width of the line. However, the avoidance of non-adiabatic transitions (see Sect. 20) has necessitated in practice that the C-field be no less than about 0.1 gauss. The transition frequency will then differ in general by a small but significant amount from $\Delta\nu$, and a correction to the observed transition frequency, requiring a knowledge of the field H_0, must be made to obtain $\Delta\nu$. At weak fields $(x \ll 1)$ the line frequencies up to terms of order x^2 are obtained from Eq. (34.10):

$$\left.\begin{aligned}\nu\left(I+\frac{1}{2}, m_1 \leftrightarrow I-\frac{1}{2}, m_2\right) &= (m_1 - m_2)\,\frac{g_I\,\mu_0\,H_0}{h} +\\ &+ \frac{\Delta W}{h}\left[1 + \frac{(m_1+m_2)\,x}{2I+1} + \left\{\frac{1}{2} - \frac{m_1^2+m_2^2}{(2I+1)^2}\right\}x^2\right]\end{aligned}\right\} \quad (36.8)$$

for $\Delta F = 1$ and for both $\Delta m = 0$ (σ transitions) and $m_1 - m_2 = \pm 1$ (π transitions).

[1] E.U. CONDON and G.H. SHORTLEY: The Theory of Atomic Spectra, Chap. III. Cambridge: Cambridge University Press 1953.

Very important measurements of $\Delta\nu$, which illustrate the procedure used for the highest precision determinations, are those of hydrogen and deuterium in their ground $1\,^2S_{\frac{1}{2}}$ state[1,2]. It is seen from Eq. (36.8) that for a σ transition with $m=0$ the transition frequency has no linear dependence on H_0 but only a quadratic dependence. For hydrogen the line $(F, m) = (1, 0) \leftrightarrow (0, 0)$ has a field dependence of only $2.76\, H_0^2$ kc/sec in which H_0 is measured in gauss. This transition was the principal one studied, and at the field of about 0.32 gauss used its frequency is greater than $\Delta\nu$ by about 280 cps. The measurement of the frequency of the π line $(1, 0) \leftrightarrow (1, -1)$ allows the determination of the small correction to be made to the frequency of the σ line to obtain $\Delta\nu$ without explicit knowledge of either H_0 or g_J. Two observations of the σ line are shown in Fig. 75. The half-width of 33 kc/sec for curve (*b*) is the natural width predicted by Eq. (26.3) for the radiofrequency field length of 7.6 cm and a discharge tube source temperature of 300° K. Since this line is substantially independent of the magnetic field H_0, its natural shape is not significantly distorted by small inhomogeneities in the magnetic field. The curve (*a*) is broader than the natural width due to the use of an amplitude of the radiofrequency field greater than optimum (see Sect. 26). The value obtained was: $\Delta\nu = 1420.40573$ Mc/sec. The final stated uncertainty of ± 50 cps corresponds to the choice of the resonance frequency to 1 part in 600 of the line width. Many possible systematic errors—particularly those associated with non-constant amplitude and phase of the radiofrequency field, and with non-uniformity of the steady field H_0, which could distort the natural line shape—were considered in the determination of $\Delta\nu$ to this high precision.

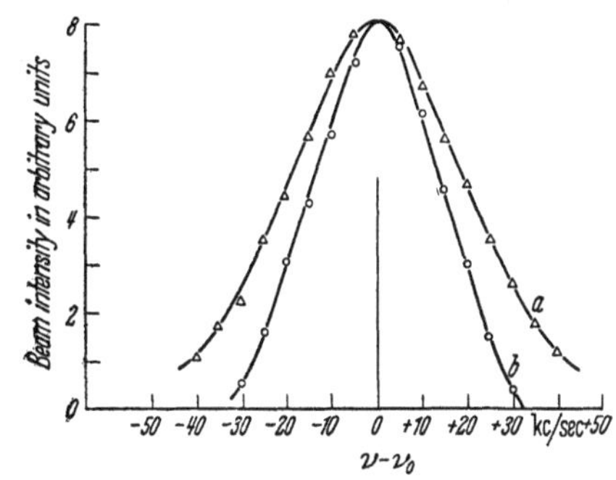

Fig. 75. Observed resonance lines for the transition $(F, m) = (1, 0) \leftrightarrow (0, 0)$ for hydrogen. Change in beam intensity is plotted vs. frequency deviation from the observed line center at about 1420.406 Mc/sec. Curve (*a*) is broadened by the use of excessive radiofrequency power. Curve (*b*) has the natural width associated with the transit time.

For deuterium, the mean of the frequencies of the π lines $(\frac{3}{2}, \frac{1}{2}) \leftrightarrow (\frac{1}{2}, -\frac{1}{2})$ and $(\frac{3}{2}, -\frac{1}{2}) \leftrightarrow (\frac{3}{2}, \frac{1}{2})$ has only a quadratic dependence on H_0. These lines form a close doublet with a frequency separation $2 g_I \mu_0 H_0/h$, and were unresolved at the field of about 0.32 gauss used. The center of this doublet gives $\Delta\nu$ except for the small correction quadratic in H_0. This correction is obtained from observation of the line $(\frac{3}{2}, -\frac{1}{2}) \leftrightarrow (\frac{3}{2}, -\frac{3}{2})$ whose frequency has a linear dependence on H_0. The value obtained was: $\Delta\nu = 327.384302 \pm 0.000030$ Mc/sec.

The hyperfine structure interval $\Delta\nu$ for an atom with $J=\frac{1}{2}$ has also been determined by observing transitions at strong and at intermediate H_0 fields. At strong fields the energy levels for an atom with $J=\frac{1}{2}$ are given by Eq. (34.11) with $b=0$. The fact that the frequencies of transitions $\Delta m_I = \pm 1$, $\Delta m_J = 0$ are approximately equal to $\Delta\nu m_J/(I+\frac{1}{2})$, provided $a \gg g_I \mu_0 H_0$ was used particularly in early magnetic resonance experiments[3] to determine $\Delta\nu$. This method has the technical advantage that $\Delta\nu$ can be obtained from the observation of transition frequencies considerably smaller than $\Delta\nu$. Another method depends on the occurrence of a maximum in the frequencies of the lines $\Delta F = 0$, $\Delta m = \pm 1$

[1] A. G. Prodell and P. Kusch: Phys. Rev. **88**, 184 (1952).
[2] P. Kusch: Phys. Rev. **100**, 1188 (1955).
[3] Kusch, Millman and Rabi: Phys. Rev. **57**, 765 (1940).

at certain values of the field[1]. At the maximum frequencies the lines are quite field insensitive and hence have their natural widths despite moderate field inhomogeneities; a precise determination of $\Delta\nu$ is thus possible. In general, the quantities $\Delta\nu$, x and g_I/g_J may be found from a measurement of the frequencies of three lines by an application of Eq. (34.10). A judicious choice of lines and of H_0 is, however, required to achieve a desired precision in any one of the three quantities.

If the value of $\Delta\nu$ for an isotope under study is not accurately known from optical spectroscopy or from a knowledge of its nuclear magnetic moment, then the observation of a transition $\Delta F = \pm 1$ whose frequency is primarily determined by $\Delta\nu$ may require an extensive search over a wide frequency range. To avoid this problem the common procedure is to observe a transition $\Delta F = 0$, $\Delta m = \pm 1$ at several low values of the magnetic field H_0. For the case $J = \frac{1}{2}$ the frequencies of these lines can be found from the energy levels given in Eq. (34.12) to be:

$$\nu(F = I \pm \tfrac{1}{2}, m \leftrightarrow F = I \pm \tfrac{1}{2}, m - 1) = \frac{\mu_0 g_F H_0}{h} \mp \frac{\Delta\nu(2m-1)\,x^2}{(2I+1)^2} \tag{36.9}$$

in which g_F is defined in Eq. (34.7a). The departure of the transition frequency from a linear dependence on H_0 is given by the term in x^2 and involves $\Delta\nu$, which can thus be determined with a useful accuracy. This method has been used[2] in many recent experiments on radioactive atoms to obtain values of $\Delta\nu$. A high precision value of $\Delta\nu$ can then be obtained by observation of a $\Delta F = 1$ transition. Both $\Delta F = 0$ and $\Delta F = 1$ transitions were used to measure the $\Delta\nu$ of He^3 in the atomic $^3S_1$ state[3].

The determination of the spin of a nucleus is usually achieved most easily by a study of the low field Zeeman effect. For an atom of any J the frequency of a transition $\Delta F = 0$, $\Delta m = \pm 1$ is obtained from Eq. (34.7) as:

$$\nu(F, m \leftrightarrow F, m-1) = g_F \mu_0 H_0/h. \tag{36.10}$$

Since $g_I \ll g_J$, to an adequate approximation:

$$\nu(F, m \leftrightarrow F, m-1) = \frac{g_J \mu_0 H_0}{h} \, \frac{F(F+1) + J(J+1) - I(I+1)}{2F(F+1)}. \tag{36.10a}$$

The g_J (and J) for the atom is usually known from atomic beam spectroscopic studies of an isotope with known nuclear spin or from optical spectroscopy, and the magnetic field can be calibrated, for example, by the observation of a transition for any atom of known properties. Then from a comparison of the observed transition frequencies with Eq. (36.10a) the unknown spin I of the nucleus can be determined. This method has been widely used in recent studies of radioactive nuclei[2,4].

The production of optically excited atomic states in a beam, the technique of producing the optical and radiofrequency radiation needed to produce radiofrequency transitions in excited states, and the resonance line shape for a transition between two excited levels have been discussed in Sects. 6, 32, and 31, respectively. The principal experiments done thus far on atoms in excited states have been measurements of the hyperfine structure intervals in the lowest $^2P_{\frac{3}{2}}$

---

[1] P. KUSCH and H. TAUB: Phys. Rev. **75**, 1477 (1949).
[2] HOBSON, HUBBS, NIERENBERG, SILSBEE and SUNDERLAND: Phys. Rev. **104**, 101 (1956).
[3] G. WEINREICH and V.W. HUGHES: Phys. Rev. **95**, 1451 (1954).
[4] DAVIS, NAGLE and ZACHARIAS: Phys. Rev. **76**, 1068 (1949).

and $^2P_{\frac{3}{2}}$ states of alkali atoms[1]. The method of the experiments can be discussed with reference to the energy level diagram of Fig. 76, which shows the hyperfine levels of the ground $^2S_{\frac{1}{2}}$ and of the excited $^2P_{\frac{1}{2}}$ and $^2P_{\frac{3}{2}}$ states of sodium in a static magnetic field. The A- and B-deflecting magnets are operated at high fields and are arranged so that refocussing of an atom onto the detector will occur only if the m_J value of the atom changes in the C-field. If unpolarized optical resonance radiation from a sodium lamp illuminates an atom in the C-region where a small static magnetic field H_0 is present, the atom may be transferred to an excited P state subject to the electric dipole selection rules $\Delta L = \pm 1$, $\Delta F = 0$, ± 1, $\Delta m = 0$, ± 1. The atom decays from the excited state predominantly by spontaneous electric dipole emission with a mean life of about 10^{-8} sec. In this process of optical excitation and decay the atom may return to a ground state level with a different m_J value than that of its initial level, and it may then be refocussed. If, in addition to the optical radiation, a radiofrequency magnetic field with a frequency equal to the difference between two hyperfine levels in the excited state is applied in the C-region, then an atom excited optically to one of these levels may be transferred to the other by the radiofrequency field subject to the magnetic dipole selecton rules $\Delta F = 0$, ± 1, $\Delta m = 0$, ± 1, and then, decay to a hyperfine level of the ground state. Due to the radiofrequency induced transition the final ground state level to which the atom returns may be changed, and hence the m_J value and the refocussing may also be changed. By this method the radiofrequency spectra of optically excited states can be studied by the atomic beam magnetic resonance method. It may be noted that the observation of a radiofrequency transition depends only on a change in the way in which the atoms from each initial ground state distribute themselves when decaying back to the ground

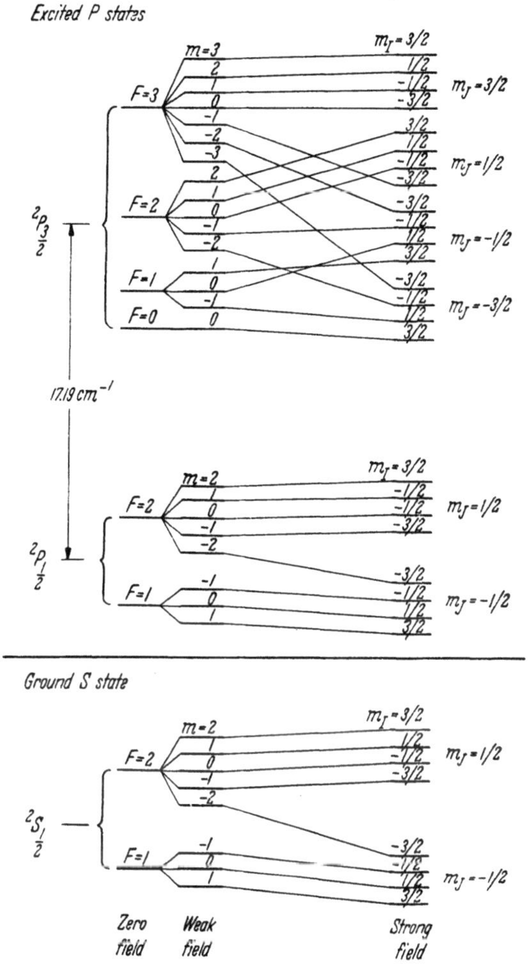

Fig. 76. Energy level diagram of the ground 3 $^2S_{\frac{1}{2}}$ state and excited 3 $^2P_{\frac{1}{2}}$ and 3 $^2P_{\frac{3}{2}}$ states of Na^{23}. The separation of the 3 $^2S_{\frac{1}{2}}$ and 3 $^2P_{\frac{3}{2}}$ states is 5890 Å. The hfs states and their magnetic sublevels at both weak and strong fields are shown. The energy level spacings are not drawn to scale.

[1] Perl, Rabi and Senitzky: Phys. Rev. **98**, 611 (1955). — B. Senitzky and I. I. Rabi: Phys. Rev. **103**, 315 (1956). — Buck, Rabi and Senitzky: Phys. Rev. **104**, 553 (1956). — P. Buck and I. I. Rabi: Phys. Rev. **107**, 1291 (1957).

state from the excited states in which the radiofrequency transitions occur, and it does not depend on a change in the relative populations of the final ground states.

Because of the short lifetime of the excited states, radiofrequency magnetic fields of about a gauss are required to produce a high transition probability and the resonance line widths are broad. Furthermore, the indirectness of the observation of the radiofrequency transition in an excited state, as compared with the usual observation of a transition in the ground state, results in a considerably smaller signal. As a result, the newer techniques of beam detection involving a mass spectrometer, ion multiplier, narrow band amplifiers, and modulation have been required for these experiments[1]. The complete theory of the line shape is complicated by the many levels involved, the interference of overlapping levels, and the presence of the static magnetic field H_0, and no detailed comparisons of observed and calculated line shapes have been made. A typical resonance curve for the $3\,^2P_{\frac{3}{2}}$ state of sodium is given in Fig. 77 and shows two overlapping resonance curves corresponding to the low field transitions, $F=3\leftrightarrow 2$ and $F=2\leftrightarrow 1$. The determination of the hfs intervals can typically be made to only about 1% due to the broadness of the resonance curve and the incompleteness of the theory of the line shape.

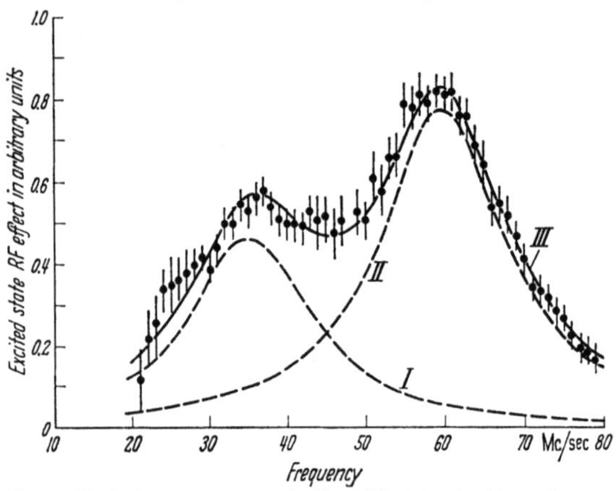

Fig. 77. Typical resonance curve for the $3\,^2P_{\frac{3}{2}}$ state of sodium taken at 1.84 gauss static field, showing two overlapping resonance curves. The dots with the vertical probable error lines are the experimental points. Curves I and II are the individual resonance curves and curve III is their sum. The parameters of the curves are chosen to give a "best fit" of curve III to the experimental points. The two resonance curves correspond to the transitions $F=3\leftrightarrow 2$ and $F=2\leftrightarrow 1$. It is not known with certainty which of these transitions corresponds to *I*.

Measurements of the hyperfine structure intervals in the metastable $2\,^2S_{\frac{1}{2}}$ states of hydrogen and singly ionized helium have used special properties of the states of the $n=2$ level for the functions of polarization and analysis usually accomplished by magnetic deflection in the *A*- and *B*-fields, since in neither case is magnetic deflection useful. For hydrogen[2] the production and detection of the $2\,^2S_{\frac{1}{2}}$ state have been discussed in Sects. 6 and 14, respectively. The energy level diagram showing the Zeeman splitting of the hyperfine structure levels of this state is given in Fig. 78. The upper magnetic sublevel of the $2\,^2P_{\frac{1}{2}}$ state is also shown and is seen to cross the lower two levels of the $2\,^2S_{\frac{1}{2}}$ state at fields near 575 gauss. (The energy difference at zero magnetic field between the $2\,^2S_{\frac{1}{2}}$ and $2\,^2P_{\frac{1}{2}}$ states is the Lamb shift of about 1058 Mc/sec.) The differential lifetimes of sublevels of the $2\,^2S_{\frac{1}{2}}$ hyperfine structure levels are used for polarization and analysis. In the absence of perturbing electric fields the mean lifetime of the $2S$ state lies between 0.11 and 0.16 sec, and is determined by the two-photon decay process[3]. Upon application of an electrostatic field the $2P$ states are mixed into the $2S$ state, and hence the mean life of the $2S$ state is reduced since an electric-dipole single-photon decay is now allowed. A hydrogen atom moving

[1] PERL, RABI and SENITZKY: Phys. Rev. **98**, 611 (1955).
[2] HEBERLE, REICH and KUSCH: Phys. Rev. **101**, 612 (1956).
[3] G. BREIT and E. TELLER: Astrophys. J. **91**, 215 (1940).

through a static magnetic field $\boldsymbol{H}$ experiences an electric field $\boldsymbol{E} = (\boldsymbol{V}/c) \times \boldsymbol{H}$. If the field is about 575 gauss, the admixture of the $m_J = +\frac{1}{2}$ magnetic sublevels of the $2\,{}^2P_{\frac{1}{2}}$ state in the $m_J = -\frac{1}{2}$ magnetic sublevels of the $2\,{}^2S_{\frac{1}{2}}$ state is particularly large because of the near degeneracy of these energy levels, and in practice these magnetic sublevels of the $2\,{}^2S_{\frac{1}{2}}$ state will decay to the $1S$ state in a traversal distance of about 1 cm for a typical H atom velocity of 5×10^5 cm/sec. The upper $m_J = +\frac{1}{2}$ magnetic sublevels of the $2\,{}^2S_{\frac{1}{2}}$ state will not be appreciably quenched to the $1S$ state under these same conditions. This differential property of the $m_J = +\frac{1}{2}$ and $m_J = -\frac{1}{2}$ sublevels is used for polarization and analysis. A similar method has been used in the experiment on the hfs separation for deuterium in the $2\,{}^2S_{\frac{1}{2}}$ state[1]. A resonance line for the deuterium hfs transition is shown in Fig. 52 (Sect. 29).

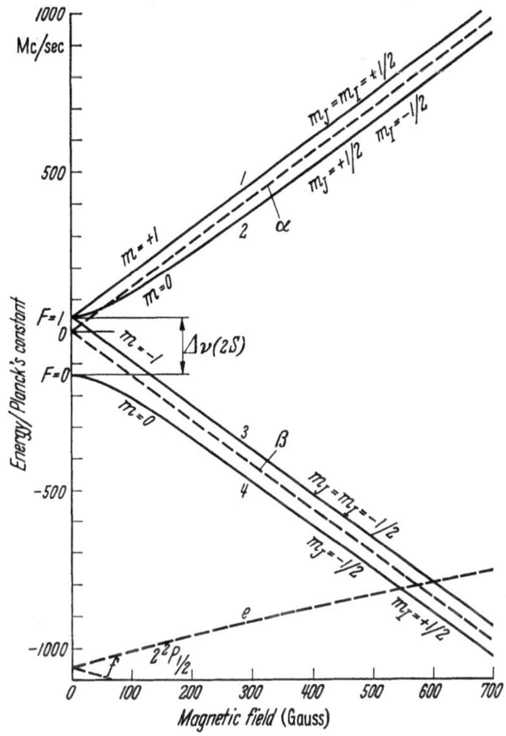

Fig. 78. Zeeman splitting of the hyperfine structure of the $2\,S$ state hydrogen. The dashed lines show the energy levels obtained when hfs is ignored. The upper state e with $m_J = +\frac{1}{2}$ of the $2\,{}^2P_{\frac{1}{2}}$ state is also shown.

In the measurement[2] of the hfs separation in the $2\,{}^2S_{\frac{1}{2}}$ state of ionized He^3 the state was produced by electron bombardment of ground state He atoms. The method of polarization and analysis can be discussed with reference to Fig. 50 (Sect. 29). A microwave electric field of frequency corresponding to the energy difference between the $F = 1$, $2\,{}^2S_{\frac{1}{2}}$ level and the $F = 0$, $2\,{}^2P_{\frac{1}{2}}$ level will cause a preferential depopulation of the $F = 1$, $2\,{}^2S_{\frac{1}{2}}$ level as compared to the $F = 0$, $2\,{}^2S_{\frac{1}{2}}$ level and hence serves as a polarizer. An ion that makes the transition to the $2\,{}^2P_{\frac{1}{2}}$ state decays rapidly to the ground $1\,{}^2S_{\frac{1}{2}}$ state with the emission of a 300 Å photon. Actually a frequency of 13350 Mc/sec, which is somewhat less than the Bohr frequency difference between the $F = 1$, $2\,{}^2S_{\frac{1}{2}}$ level and the $F = 0$, $2\,{}^2P_{\frac{1}{2}}$ level, is used in order to reduce transitions from the $F = 0$, $2\,{}^2S_{\frac{1}{2}}$ level to the $F = 0$, $2\,{}^2P_{\frac{1}{2}}$ level as well; the latter transition can also have appreciable probability because of the great width of the $2\,{}^2P_{\frac{1}{2}}$ level associated with its short lifetime. Transitions are then induced between the $F = 1$ and the $F = 0$ levels of the $2\,{}^2S_{\frac{1}{2}}$ state. The occurrence of transitions is detected by the use of an analyzer similar to the polarizer, which will produce a number of 300 Å photons (observed photoelectrically) proportional to the number of ions occupying the $F = 1$ level. This experiment seems to be the first, and as yet the only, microwave spectroscopy experiment involving an ion beam. It should be noted that the beams used in this experiment and in the experiment previously described on the hfs of H and D in the $2\,{}^2S_{\frac{1}{2}}$ state need not be so well defined

[1] Reich, Heberle and Kusch: Phys. Rev. **104**, 1585 (1956).
[2] R. Novick and E. Commins: Phys. Rev. **103**, 1897 (1956); **111**, 822 (1958).

Table 8. *Nuclear spins and hyperfine structure intervals.*

All measurements were made on atoms in their ground state configuration unless explicit labelling of an excited state configuration is indicated. The symbol (F, F') represents the frequency difference between the hfs levels characterized by the total angular momentum quantum numbers F and F' at zero magnetic field; the level F lies higher in energy than the level F'. With a few exceptions all the hfs intervals tabulated were obtained by correcting measurements made at weak magnetic fields (~ 1 gauss) to zero magnetic field by use of the Breit-Rabi formula Eq. (34.10), or its equivalent for cases in which $J > \frac{1}{2}$. The error estimates are indicated by the numbers in parentheses, which apply to the last figure given for the hfs interval.

Isotope	Nuclear spin	Atomic state	Hyperfine structure intervals at zero field (Mc/sec)	References
$_1H^1$	$\frac{1}{2}$	$^2S_{\frac{1}{2}}$	$(1, 0) = 1420.40573(5)$	55 *K2*; 56 *W3*
$_1H^1$	$\frac{1}{2}$	$2s\,^2S_{\frac{1}{2}}$	$(1, 0) = 177.55686(5)$	56 *H1*
$_1D^2$	1	$^2S_{\frac{1}{2}}$	$(\frac{3}{2}, \frac{1}{2}) = 327.384302(30)$	55 *K2*
$_1D^2$	1	$2s\,^2S_{\frac{1}{2}}$	$(\frac{3}{2}, \frac{1}{2}) = 40.924439(20)$	56 *R1*
$_1H^3$	$\frac{1}{2}$	$^2S_{\frac{1}{2}}$	$(1, 0) = 1516.70170(7)$	57 *P1*
$_2He^3$	$\frac{1}{2}$	$1s\,2s\,^3S_1$	$(\frac{1}{2}, \frac{3}{2}) = 6739.7022(7)$ g	54 *W1*
$_2He^3$	$\frac{1}{2}$	$^2S_{\frac{1}{2}}$ ion	$(0, 1) = 1083.35499(20)$	56 *N2*; 58 *N1*
$_3Li^6$	1	$^2S_{\frac{1}{2}}$	$(\frac{3}{2}, \frac{1}{2}) = 228.208(5)$	49 *K7*
$_3Li^7$	$\frac{3}{2}$	$^2S_{\frac{1}{2}}$	$(2, 1) = 803.512(15)$	49 *K7*
$_5B^{11}$	$\frac{3}{2}$	$^2P_{\frac{1}{2}}$	$(2, 1) = 732.4(1)$	53 *W4*
$_5B^{11}$	$\frac{3}{2}$	$^2P_{\frac{3}{2}}$	$(3, 2) = 222.737(10)$	53 *W4*
			$(2, 1) = 144.00(2)$	
			$(1, 0) = 70.66(20)$	
$_{10}Ne^{21}$	$\frac{3}{2}$	$2p^5\,3s,\,^3P_2$		56 *H3*
$_{11}Na^{22}$	3	$^2S_{\frac{1}{2}}$	$(\frac{7}{2}, \frac{5}{2}) = 1220.64(4)$	49 *D3*
$_{11}Na^{23}$	$\frac{3}{2}$	$^2S_{\frac{1}{2}}$	$(2, 1) = 1771.6262(1)$	51 *L2*; 58 *A1*
$_{11}Na^{23}$	$\frac{3}{2}$	$3p\,^2P_{\frac{1}{2}}$	$(2, 1) = 188.9(10)$	55 *P1*
$_{11}Na^{23}$	$\frac{3}{2}$	$3p\,^2P_{\frac{3}{2}}$	$(3, 2) = 59.8(11)$ a, b	55 *P1*
			$(2, 1) = 35.5(11)$ a, b	
$_{11}Na^{24}$	4	$^2S_{\frac{1}{2}}$	$(\frac{9}{2}, \frac{7}{2}) = 1139.35(10)$	53 *B2*
$_{13}Al^{27}$	$\frac{5}{2}$	$^2P_{\frac{1}{2}}$	$(3, 2) = 1506.14(5)$	53 *L3*
$_{13}Al^{27}$	$\frac{5}{2}$	$^2P_{\frac{3}{2}}$	$(4, 3) = 392.0(2)$	49 *L1*
			$(3, 2) = 274.3(1)$	
$_{17}Cl^{35}$	$\frac{3}{2}$	$^2P_{\frac{3}{2}}$	$(3, 2) = 670.018(6)$ a	51 *J1*
			$(2, 1) = 355.244(6)$ a	
			$(1, 0) = 150.145(6)$ a	
$_{17}Cl^{35}$	$\frac{3}{2}$	$^2P_{\frac{1}{2}}$	$(2, 1) = 2074.383(8)$	51 *K2*
$_{17}Cl^{37}$	$\frac{3}{2}$	$^2P_{\frac{3}{2}}$	$(3, 2) = 555.294(12)$ a	51 *J1*
			$(2, 1) = 298.116(12)$ a	
			$(1, 0) = 127.404(12)$ a	
$_{17}Cl^{37}$	$\frac{3}{2}$	$^2P_{\frac{1}{2}}$	$(2, 1) = 1726.700(15)$	51 *K2*
$_{19}K^{39}$	$\frac{3}{2}$	$^2S_{\frac{1}{2}}$	$(2, 1) = 461.71971(15)$ f	49 *K7*
$_{19}K^{39}$	$\frac{3}{2}$	$4p\,^2P_{\frac{1}{2}}$	$(2, 1) = 57.7(6)$	57 *B1*
		$4p\,^2P_{\frac{3}{2}}$	$(3, 2) = 20.1(4)$	
$_{19}K^{40}$	4	$^2S_{\frac{1}{2}}$	$(\frac{7}{2}, \frac{9}{2}) = 1285.790(7)$	52 *E1*
$_{19}K^{41}$	$\frac{3}{2}$	$^2S_{\frac{1}{2}}$	$(2, 1) = 254.018(6)$	50 *O1*; 49 *K7*
$_{19}K^{42}$	2	$^2S_{\frac{1}{2}}$	$(\frac{3}{2}, \frac{5}{2}) = 1258.9(1)$	53 *B2*
$_{25}Mn^{56}$	3	$^6S_{\frac{5}{2}}$		58 *C1*
$_{29}Cu^{61}$	$\frac{3}{2}$	$^2S_{\frac{1}{2}}$		57 *N1*
$_{29}Cu^{63}$	$\frac{3}{2}$	$^2S_{\frac{1}{2}}$	$(2, 1) = 11733.83(1)$	57 *T1*

Table 8. (Continued.)

Isotope	Nuclear spin	Atomic state	Hyperfine structure intervals at zero field (Mc/sec)	References
$_{29}Cu^{64}$	1	$^2S_{\frac{1}{2}}$	$(\frac{3}{2}, \frac{1}{2}) = 1278\,(20)$ c	54 *L1*
$_{29}Cu^{65}$	$\frac{3}{2}$	$^2S_{\frac{1}{2}}$	$(2, 1) = 12\,568.81\,(1)$	57 *T1*
$_{31}Ga^{66}$	0	$^2P_{\frac{1}{2}}$		57 *H1*
$_{31}Ga^{67}$	$\frac{3}{2}$	$^2P_{\frac{3}{2}}$	$(3, 2) = 597.2\,(26)$	57 *W1*; 57 *H1*
			$(2, 1) = 278.8\,(40)$	
$_{31}Ga^{69}$	$\frac{3}{2}$	$^2P_{\frac{1}{2}}$	$(2, 1) = 2677.9875\,(10)$	56 *L2*
$_{31}Ga^{69}$	$\frac{3}{2}$	$^2P_{\frac{3}{2}}$	$(3, 2) = 634.901\,83\,(20)$	54 *D1*
			$(2, 1) = 319.067\,06\,(20)$	
			$(1, 0) = 128.277\,30\,(20)$	
$_{31}Ga^{71}$	$\frac{3}{2}$	$^2P_{\frac{1}{2}}$	$(2, 1) = 3402.6946\,(13)$	56 *L2*
$_{31}Ga^{71}$	$\frac{3}{2}$	$^2P_{\frac{3}{2}}$	$(3, 2) = 766.695\,80\,(20)$	54 *D1*
			$(2, 1) = 445.469\,60\,(20)$	
			$(1, 0) = 203.043\,40\,(20)$	
$_{31}Ga^{72}$	3	$^2P_{\frac{1}{2}}$	$(\frac{7}{2}, \frac{5}{2}) = 144.8\,(8)$ c	58 *G1*
$_{33}As^{76}$	2	$^4S_{\frac{3}{2}}$		57 *C1*
$_{35}Br^{79}$	$\frac{3}{2}$	$^2P_{\frac{3}{2}}$	$(3, 2) = 2269.552\,(3)$ a	54 *K1*
			$(2, 1) = 2154.499\,(3)$ a	
			$(1, 0) = 1269.702\,(3)$ a	
$_{35}Br^{81}$	$\frac{3}{2}$	$^2P_{\frac{3}{2}}$	$(3, 2) = 2539.794\,(3)$ a	54 *K1*
			$(2, 1) = 2229.056\,(3)$ a	
			$(1, 0) = 1275.291\,(3)$ a	
$_{35}Br^{82}$	5	$^2P_{\frac{3}{2}}$		57 *G1*; 57 *G3*
$_{37}Rb^{81}$	$\frac{3}{2}$	$^2S_{\frac{1}{2}}$	$(2, 1) = 5097\,(13)$	56 *H2*; 54 *H2*; 57 *H7*
$_{37}Rb^{81m}$	$\frac{9}{2}$	$^2S_{\frac{1}{2}}$		56 *H4*
$_{37}Rb^{82}$	5	$^2S_{\frac{1}{2}}$	$(\frac{11}{2}, \frac{9}{2}) = 3094.1\,(24)$	56 *H2*; 57 *H7*
$_{37}Rb^{83}$	$\frac{5}{2}$	$^2S_{\frac{1}{2}}$	$(3, 2) = 3183.3\,(58)$	56 *H2*; 57 *H7*
$_{37}Rb^{84}$	2	$^2S_{\frac{1}{2}}$	$(\frac{3}{2}, \frac{5}{2}) = 3077.5\,(51)$	56 *H2*; 57 *H7*
$_{37}Rb^{85}$	$\frac{5}{2}$	$^2S_{\frac{1}{2}}$	$(3, 2) = 3035.735\,(2)$	49 *K7*; 52 *B7*
$_{37}Rb^{85}$	$\frac{5}{2}$	$5p\,^2P_{\frac{1}{2}}$	$(3, 2) = 362\,(3)$	56 *S3*
$_{37}Rb^{85}$	$\frac{5}{2}$	$5p\,^2P_{\frac{3}{2}}$	$(4, 3) = 120.8\,(10)$ a	56 *S3*
			$(3, 2) = 65.0\,(10)$ a	
$_{37}Rb^{86}$	2	$^2S_{\frac{1}{2}}$	$(\frac{3}{2}, \frac{5}{2}) = 3960\,(20)$	53 *B2*
$_{37}Rb^{87}$	$\frac{3}{2}$	$^2S_{\frac{1}{2}}$	$(2, 1) = 6834.685\,(2)$	53 *D1*
$_{37}Rb^{87}$	$\frac{3}{2}$	$5p\,^2P_{\frac{3}{2}}$	$(3, 2) = 265\,(3)$	56 *S3*
$_{39}Y^{89}$	$\frac{1}{2}$	$^2D_{\frac{3}{2}}$	$(1, 2) = 114.72\,(20)$	57 *E3*
$_{39}Y^{89}$	$\frac{1}{2}$	$^2D_{\frac{5}{2}}$	$(2, 3) = 88.63\,(60)$	57 *E3*
$_{47}Ag^{105}$	$\frac{1}{2}$	$^2S_{\frac{1}{2}}$		58 *E2*
$_{47}Ag^{106}$	6	$^2S_{\frac{1}{2}}$		58 *E2*
$_{47}Ag^{107}$	$\frac{1}{2}$	$^2S_{\frac{1}{2}}$	$(0, 1) = 1712.56\,(4)$	53 *W3*
$_{47}Ag^{109}$	$\frac{1}{2}$	$^2S_{\frac{1}{2}}$	$(0, 1) = 1976.94\,(4)$	53 *W3*
$_{47}Ag^{110m}$	6	$^2S_{\frac{1}{2}}$		58 *E2*
$_{47}Ag^{111}$	$\frac{1}{2}$	$^2S_{\frac{1}{2}}$	$(0, 1) = 2204.54\,(5)$	56 *W1*; 54 *L1*
$_{49}In^{113}$	$\frac{9}{2}$	$^2P_{\frac{1}{2}}$	$(5, 4) = 11385.4300(20)$	57 *E1*
$_{49}In^{113}$	$\frac{9}{2}$	$^2P_{\frac{3}{2}}$	$(6, 5) = 1745.4575(5)$	57 *E2*
			$(5, 4) = 1115.8253(5)$	
			$(4, 3) = 670.9552(5)$	
$_{49}In^{113m}$	$\frac{1}{2}$	$^2P_{\frac{1}{2}}$		56 *C3*
$_{49}In^{114m}$	5	$^2P_{\frac{1}{2}}$	$(\frac{11}{2}, \frac{9}{2}) = 9700(200)$	57 *G4*

Table 8. (Continued.)

Isotope	Nuclear spin	Atomic state	Hyperfine structure intervals at zero field (Mc/sec)	References
$_{49}In^{115}$	$\frac{9}{2}$	$^2P_{\frac{1}{2}}$	(5, 4) = 11409.7506(20)	57 *E1*
$_{49}In^{115}$	$\frac{9}{2}$	$^2P_{\frac{3}{2}}$	(6, 5) = 1752.6865 (2)	57 *E2*
			(5, 4) = 1117.1676 (2)	
			(4, 3) = 668.9631 (2)	
$_{49}In^{116m}$	5	$^2P_{\frac{1}{2}}$	$(\frac{11}{2}, \frac{9}{2})$ = 9000 (200)	57 *G4*
$_{53}I^{123}$	$\frac{5}{2}$	$^2P_{\frac{3}{2}}$		57 *L3*
$_{53}I^{124}$	2	$^2P_{\frac{3}{2}}$		57 *G2*
$_{53}I^{127}$	$\frac{5}{2}$	$^2P_{\frac{3}{2}}$	(4, 3) = 4226.172 (15)	54 *J1*
			(3, 2) = 1965.884 (10)	
			(2, 1) = 737.492 (8)	
$_{53}I^{131}$	$\frac{7}{2}$	$^2P_{\frac{3}{2}}$		57 *L3*
$_{55}Cs^{127}$	$\frac{1}{2}$	$^2S_{\frac{1}{2}}$	(1, 0) = 8790 (250)	56 *N1*; 57 *S6*
$_{55}Cs^{129}$	$\frac{1}{2}$	$^2S_{\frac{1}{2}}$	(1, 0) = 9190 (260)	56 *N1*; 57 *S6*
$_{55}Cs^{130}$	1	$^2S_{\frac{1}{2}}$		56 *N1*
$_{55}Cs^{131}$	$\frac{5}{2}$	$^2S_{\frac{1}{2}}$	(3, 2) = 13200 (110)	53 *B2*
$_{55}Cs^{132}$	2	$^2S_{\frac{1}{2}}$		56 *N4*
$_{55}Cs^{133}$	$\frac{7}{2}$	$^2S_{\frac{1}{2}}$	(4, 3) = 9192.631830 (10)	57 *E4*
$_{55}Cs^{133}$	$\frac{7}{2}$	$6p\,^2P_{\frac{3}{2}}$	(5, 4) = 252.5 (10)	56 *B4*
			(4, 3) = 203.0 (11)	
			(3, 2) = 152.0 (8)	
$_{55}Cs^{134}$	4	$^2S_{\frac{1}{2}}$	$(\frac{9}{2}, \frac{7}{2})$ = 10473.626 (15)	57 *S2*; 53 *B2*
$_{55}Cs^{134m}$	8	$^2S_{\frac{1}{2}}$	$(\frac{17}{2}, \frac{15}{2})$ = 3684.594 (20)	55 *G1*; 55 *G2*
$_{55}Cs^{135}$	$\frac{7}{2}$	$^2S_{\frac{1}{2}}$	(4, 3) = 9724.023 (15)	57 *S2*
$_{55}Cs^{137}$	$\frac{7}{2}$	$^2S_{\frac{1}{2}}$	(4, 3) = 10115.527 (15)	57 *S2*
$_{57}La^{139}$	$\frac{7}{2}$	$^2D_{\frac{3}{2}}$	(5, 4) = 737.967 (15)	57 *T2*
			(4, 3) = 551.987 (5)	
			(3, 2) = 391.603 (10)	
		$^2D_{\frac{5}{2}}$	(6, 5) = 1120.902 (5)	
			(5, 4) = 912.793 (5)	
			(4, 3) = 716.288 (3)	
			(3, 2) = 529.090 (10)	
$_{59}Pr^{141}$	$\frac{5}{2}$	$^4I_{\frac{9}{2}}$	(4, 3) = 3708.05 (5)	53 *L2*
			(3, 2) = 2782.25 (5)	
$_{79}Au^{191}$	$\frac{3}{2}$	$^2S_{\frac{1}{2}}$		57 *E5*
$_{79}Au^{192}$	1	$^2S_{\frac{1}{2}}$	$(\frac{3}{2}, \frac{1}{2})$ = 374.3 (2) c	57 *E5*; 58 *E1*
$_{79}Au^{193}$	$\frac{3}{2}$	$^2S_{\frac{1}{2}}$		57 *E5*
$_{79}Au^{194}$	1	$^2S_{\frac{1}{2}}$	$(\frac{3}{2}, \frac{1}{2})$ = 3600 (120) c	57 *R2*; 57 *H5*
$_{79}Au^{197}$	$\frac{3}{2}$	$^2S_{\frac{1}{2}}$	(2, 1) = 6107.1 (10)	53 *W3*
$_{79}Au^{198}$	2	$^2S_{\frac{1}{2}}$	$(\frac{5}{2}, \frac{3}{2})$ = 21800 (150) d	56 *C1*
$_{79}Au^{199}$	$\frac{3}{2}$	$^2S_{\frac{1}{2}}$	$(\frac{5}{2}, \frac{3}{2})$ = 11110 (130) e	56 *C1*
$_{81}Tl^{197}$	$\frac{1}{2}$	$^2P_{\frac{1}{2}}$		57 *B2*
$_{81}Tl^{198m}$	7	$^2P_{\frac{1}{2}}$		57 *B2*
$_{81}Tl^{199}$	$\frac{1}{2}$	$^2P_{\frac{1}{2}}$		57 *B2*
$_{81}Tl^{200}$	2	$^2P_{\frac{1}{2}}$		58 *M1*
$_{81}Tl^{201}$	$\frac{1}{2}$	$^2P_{\frac{1}{2}}$		58 *M1*
$_{81}Tl^{203}$	$\frac{1}{2}$	$^2P_{\frac{1}{2}}$	(1, 0) = 21105.447 (5)	56 *L2*
$_{81}Tl^{203}$	$\frac{1}{2}$	$^2P_{\frac{3}{2}}$	(2, 1) = 524.0601 (2)	56 *G1*
$_{81}Tl^{204}$	2	$^2P_{\frac{1}{2}}$	$(\frac{5}{2}, \frac{3}{2})$ = 732 (5) c	57 *B2*

Table 8. (Continued.)

Isotope	Nuclear spin	Atomic state	Hyperfine structure intervals at zero field (Mc/sec)	References
$_{81}Tl^{205}$	$\frac{1}{2}$	$^2P_{\frac{1}{2}}$	$(1,0) = 21\,310.835(5)$	56 *L2*
$_{81}Tl^{205}$	$\frac{1}{2}$	$^2P_{\frac{3}{2}}$	$(2,1) = 530.0766(2)$	56 *G1*
$_{83}Bi^{206}$	6	$^4S_{\frac{3}{2}}$		57 *M1*
$_{93}Np^{238}$	2	$(J=\frac{11}{2})$		58 *M2*
$_{93}Np^{239}$	$\frac{5}{2}$	$(J=\frac{11}{2})$		58 *H1*
$_{94}Pu^{239}$	$\frac{1}{2}$	$^7F_1$	$(\frac{3}{2}, \frac{1}{2}) = 7.715(60)$ c	57 *H3*

(a) The errors assigned to the frequency intervals were not given in the original publication; they were estimated from errors assigned in the references to the a's and b's derived from the frequency intervals.

(b) For Na^{23} $3p$ $^2P_{\frac{3}{2}}$ it is possible that the assignments should be $(3, 2) = 35.5(14)$; $(2, 1) = 59.8(14)$.

(c) Sign of μ_I is unknown; hence level with smaller F value may lie above that with larger F value.

(d) Sign of μ_I is uncertain; if negative, $(\frac{3}{2}, \frac{5}{2}) = 22\,500(150)$.

(e) Sign of μ_I is uncertain; if negative $(\frac{3}{2}, \frac{5}{2}) = 11\,180(130)$.

(f) Unpublished data of M. N. McDermott and G. Gould.

(g) Unpublished data of J. White, C. Drake and V. W. Hughes.

geometrically as in the usual atomic beam experiment, because the detection of a transition does not depend upon the observation of small deflections of the beam.

Table 8 presents information on the spins and hyperfine structure intervals of atoms which have been studied by atomic beam spectroscopy. The data given in Table 8 are used to compute the magnetic dipole hfs anomalies and the nuclear moments given in Tables 9 and 10.

37. Nuclear magnetic dipole moments; magnetic dipole hfs and hfs anomalies. The interaction of the nuclear magnetic dipole moment with the applied field $\boldsymbol{H}_0$ $(\mu_0 g_I \boldsymbol{I} \cdot \boldsymbol{H}_0)$ occurs in the general Hamiltonian for an atom in a magnetic field [see Eq. (34.1)], and indeed nuclear magnetic dipole moments or g_I values can be measured directly by the atomic beam magnetic resonance method. Usually however, for the stable and abundant isotopes precise magnetic moment values are more easily obtained by nuclear resonance techniques[1], or by the molecular beam magnetic resonance method (see Sect. 45). If, however, the abundance of an isotope is low or if the total available amount is small, the atomic beam method may be the only one with a sensitivity great enough to determine nuclear g-values.

An important example is the determination[2] of the ratio of the nuclear g values of K^{39} and K^{40}. (The natural abundance of the long-lived radioactive isotope K^{40} is 0.01%.) From the Breit-Rabi formula [Eq. (34.10)] it follows that the spectrum includes doublets whose components, $(F, m) \leftrightarrow (F, m-1)$ and $(F-1, m) \leftrightarrow (F-1, m-1)$, have a frequency separation of $2g_I \mu_0 H_0/h$ at any field H_0.

[1] E. R. Andrew: Nuclear Magnetic Resonance. Cambridge: Cambridge University Press 1955.

[2] Eisinger, Bederson and Feld: Phys. Rev. **86**, 73 (1952).

Hence the ratio of the g-values of two isotopes of the same atom may be obtained directly as the ratio of the corresponding doublet separations. Since both components of the doublets are lines described at high fields by $\Delta m_J = 0$, $\Delta m_I = \pm 1$, they cannot be simply observed on an apparatus designed to observe the transitions $\Delta m_J = \pm 1$, $\Delta m_I = 0$. The ratio $|g(K^{40})/g(K^{39})|$ was measured in an apparatus in which the A- and B-fields were intermediate. By a suitable choice of m it was possible to develop a sufficiently large moment change to allow observation of certain doublets. The measurement was made at a strong field ($H_0 \sim 9000$ gauss) where the doublet separations are relatively large and the mean frequencies of the doublets are quite field insensitive. It was found that $|g(K^{40})/g(K^{39})| = 1.24346 \pm 0.00024$. A similar measurement has been made[1] of the ratio $|g(K^{41})/g(K^{39})|$ on an apparatus in which the A- and B-fields were so low that the atoms were in the (F, m) quantization and all doublets could be observed. A knowledge of $\Delta\nu$, the mean frequency of the doublet (from which x is then obtained), and g_J (see Sect. 41) allows the determination of g_I for an isotope from the observed doublet separation. Measurements of the nuclear moments of gallium[2] and indium[3] from studies on the ground ${}^2P_{\frac{1}{2}}$ states have shown the necessity of considering the perturbation of the state caused[4] by the interaction of the atom with the H_0 field which admixes the ${}^2P_{\frac{3}{2}}$ state.

Nuclear magnetic dipole moments can be determined from the magnetic dipole interaction constants A_1 (or a) together with adequate knowledge about the electronic wave functions [see Sect. 35]. If the hfs interaction is due to an s electron in an atomic state with $J = \frac{1}{2}$, then Eqs. (34.3), (34.4), (35.9), (35.12a) and (35.14) yield the Fermi formula for the hfs interval ΔW in the non-relativistic limit:

$$\Delta W = \frac{(2I+1)}{I} A_1 = \frac{8\pi}{3} \frac{2I+1}{I} \mu_0 \mu \psi^2(0). \tag{37.1}$$

An important modern application[5] of this formula is the determination of the nuclear magnetic dipole moment of a radioactive isotope from a measurement of its $\Delta\nu$ together with a knowledge of the magnetic moment and $\Delta\nu$ for a stable isotope of the same element. Thus the ratio of the hfs splittings for two isotopes is given from Eq. (37.1) as:

$$\frac{\Delta\nu_1}{\Delta\nu_2} = \left(\frac{2I_1+1}{I_1}\right)\left(\frac{I_2}{2I_2+1}\right)\left(\frac{\mu_1}{\mu_2}\right) \tag{37.2}$$

in which μ_1 and μ_2 are the nuclear magnetic dipole moments, and the electronic wavefunction does not appear. Eq. (37.2) is also valid[6] for a non-s electron in an atomic state $J = \frac{1}{2}$.

The hyperfine structure interval $\Delta\nu$ for the simplest atom, hydrogen, has been measured with high precision[7, 8] and its theory has been intensively studied,

---

[1] OCHS, LOGAN and KUSCH: Phys. Rev. **78**, 184 (1950).

[2] P. KUSCH: Phys. Rev. **78**, 615 (1950).

[3] T.G. ECK and P. KUSCH: Phys. Rev. **106**, 958 (1957).

[4] H.M. FOLEY: Phys. Rev. **80**, 288 (1950). — W. W. CLENDENIN: Phys. Rev. **94**, 1590 (1954).

[5] V.W. COHEN and D.A. GILBERT: Phys. Rev. **95**, 569 (1954). — L. S. GOODMAN and S. WEXLER: Phys. Rev. **95**, 570 (1954).

[6] For the limits of validity of Eq. (37.2) see discussion of the hfs anomaly given later in this section.

[7] P. KUSCH: Phys. Rev. **100**, 1188 (1955).

[8] J.P. WITTKE and R.H. DICKE: Phys. Rev. **103**, 620 (1956).

particularly in order to obtain information about the internal structure of the proton[1]. The theoretical expression[2] for the $\Delta\nu$ of hydrogen is:

$$\left.\begin{aligned}\Delta\nu_H=&\left\{\frac{16}{3}\alpha^2 c\, r_{y\infty}\frac{\mu_p}{\mu_0}\right\}\left\{1+\frac{m}{M_p}\right\}^{-3}\left\{1+\frac{3}{2}\alpha^2\right\}\left\{1+\frac{\alpha}{2\pi}-0.328\frac{\alpha^2}{\pi^2}\right\}\left\{1-\alpha^2\left(\frac{5}{2}-\ln 2\right)\right\}\times\\ &\times\left\{1-\frac{\alpha}{\pi}\frac{m}{M_p\mu_p'}\left(\left[3-\frac{3}{4}(\mu_p'-1)^2\right]\ln\frac{M_p}{m}-\frac{1}{8}(\mu_p'-1)^2\right)-\frac{9\alpha}{4\pi}\frac{m}{M_p\,\mu_p'}(\mu_p'-1)^2\ln\frac{2k_0}{M_p}\right\}\times\\ &\times\{1-\delta\}\end{aligned}\right\}\quad(37.3)$$

in which

$r_{y\infty}$ = Rydberg constant for infinite mass in cm^{-1};
μ_p = proton magnetic moment in absolute units;
m = electron mass;
M_p = proton mass;
$\mu_p' = \mu_p/(e\hbar/2M_p c)$;
k_0 = a cut-off factor;
δ = a dimensionless correction associated with the finite size of the proton.

The first bracketed expression is the Fermi formula. The second term is a reduced mass correction. The third term is a relativistic correction. The fourth term arises from the anomalous spin magnetic moment of the electron and the fifth term from the distributed character of quantum electrodynamic effects. The sixth term is a relativistic proton recoil correction calculated with a relativistic two-body equation under the assumption that the proton is a point charge with an anomalous point magnetic moment. The result depends on the proton structure in that an upper frequency cutoff, k_0, is required for the terms in which the proton interacts exclusively via the anomalous part of its moment, but the value of the result is not greatly affected by the particular value chosen for k_0. The important proton structure effect is included in the factor $1-\delta$ with

$$\delta = 2\langle r_{em}\rangle/a_0 \qquad (37.4)$$

in which $\langle r_{em}\rangle$ is the first statistical moment of an electromagnetic distribution which characterizes the proton structure and a_0 is the Bohr radius for hydrogen. The comparison of the theoretical value of $\Delta\nu_H$ with the experimental value is presently limited by the accuracy to which the fine structure constant α is known[3] ($\sim 1/10^5$) and by the absence of a theoretical calculation of the $\alpha^3 \ln\alpha$ quantum electrodynamic correction. Present comparison allows the conclusion that $\langle r_{em}\rangle \leqq 2.5\,\hbar/(\text{Mc}) = 0.5\times 10^{-13}$ cm, which is not inconsistent with the value for the proton size obtained from electron-proton scattering experiments[4].

The theory of the hfs interaction constants A_k given in Sect. 35 treats the nucleus as having a negligible size and an infinite mass. Under these assumptions from Eqs. (34.6a) and (35.9) the ratio of the magnetic dipole interaction constants for two isotopes, a_1/a_2, equals the ratio of the nuclear g-values, g_1/g_2. Actual observations show that the two ratios are not equal and it is useful to define the quantity Δ',

$${}_1\Delta_2' = \frac{a_1}{a_2}\,\frac{g_2}{g_1} - 1\,. \qquad (37.5)$$

[1] A.C. Zemach: Phys. Rev. **104**, 1771 (1956). The references to earlier theoretical work are given in this paper.
[2] See also Sect. 22 of H.A. Bethe and E.E. Salpeter's paper in Vol. XXXV.
[3] E.R. Cohen and J.W.M. DuMond: Vol. XXXV, this Encyclopedia. — C.M. Sommerfield: Phys. Rev. **107**, 328 (1957).
[4] R. Hofstadter: Rev. Mod. Phys. **28**, 214 (1956).

For the common case in which $J=\frac{1}{2}$,

$$ {}_1\Delta_2' = \frac{\Delta\nu_1}{\Delta\nu_2}\left(\frac{I_1}{2I_1+1}\right)\left(\frac{2I_2+1}{I_2}\right)\frac{\mu_2}{\mu_1} - 1. \tag{37.6}$$

The non-zero values of Δ' arise because of the finite size and mass of the nuclei. The effect of nuclear mass is significant only for very light nuclei and is of importance principally in the case of the hydrogens. For a one electron atom in an S state, the inclusion of the principal effect of the finite mass of the nucleus leads[1] to the expression:

$$\left(\frac{\Delta\nu_2}{\Delta\nu_1}\right)_{\text{calc}} = \frac{\mu_2}{\mu_1}\left(\frac{I_1}{2I_1+1}\right)\left(\frac{2I_2+1}{I_2}\right)\left(\frac{m_2}{m_1}\right)^3 \tag{37.7}$$

where m_1 and m_2 are the reduced masses for the two isotopes. The hyperfine structure anomaly, Δ, is defined through,

$$\left(\frac{\Delta\nu_2}{\Delta\nu_1}\right)_{\text{calc}} = \left(\frac{\Delta\nu_2}{\Delta\nu_1}\right)_{\text{obs}}(1-{}_2\Delta_1) \quad \text{and} \quad {}_1\Delta_2 = -{}_2\Delta_1. \tag{37.8}$$

For multi-electron atoms the ratio of the reduced masses for two isotopes is so nearly equal to unity that the principal contribution to Δ arises from nuclear structure effects and $\Delta\approx\Delta'$.

The hyperfine structure anomaly was first observed[2] for hydrogen and deuterium. Its value is now known to be ${}_H\Delta_D = -(1.703\pm0.005)\times10^{-4}$. The anomaly is attributed primarily to the finite size of the deuteron[3]. Because the electron is bound by the charge of the proton, there is an asymmetry between the magnetic interaction of the electron with the proton and with the neutron. Thus when the electron is close to ("inside") the nucleus, it moves rapidly compared to the nucleon motions and therefore will be bound to the apparently stationary proton rather than to the deuteron center of mass. The magnetic interaction of the electron with the proton will be largely unaffected by the deuteron structure, but the neutron moment will appear to the electron as a moment density distributed around the proton and therefore its effect will nearly cancel when the electron is inside the nucleus. Detailed theoretical consideration[4] of the effect of the deuteron structure on the hfs of deuterium predicts an hfs anomaly of $(1.83\pm0.22)\times10^{-4}$, as compared to the observed value of $(1.703\pm0.005)\times10^{-4}$. The theoretical uncertainty is largely due to inadequate knowledge of the deuteron wavefunction. Relativistic and quantum electrodynamic corrections as given in the 3rd, 4th, and 5th terms of Eq. (37.3) cancel in the ratio $(\Delta\nu_H/\Delta\nu_D)_{\text{calc}}$. A complete treatment of nucleon recoil and nucleon structure contributions to the hfs of deuterium has not been given.

Comparison of the measured value of $\Delta\nu$ for singly ionized He^3 in its $2\,{}^2S_{\frac{1}{2}}$ state with a modified Fermi formula including reduced mass, relativistic, and quantum electrodynamic corrections indicates that nuclear structure effects similar to those discussed for the deuteron amount to about 186 ppm[5,6]. No hfs anomaly has been observed[7] for tritium and hydrogen to an accuracy of about $1/10^5$, in agreement with the theoretical expectation[8].

[1] G. BREIT and R.E. MEYEROTT: Phys. Rev. **72**, 1023 (1947).
[2] J.E. NAFE and E.B. NELSON: Phys. Rev. **73**, 718 (1948).
[3] A. BOHR: Phys. Rev. **73**, 1109 (1948).
[4] F. LOW: Phys. Rev. **77**, 361 (1950).
[5] R. NOVICK and E. COMMINS: Phys. Rev. **111**, 822 (1958).
[6] A.M. SESSLER and H.M. FOLEY: Phys. Rev. **98**, 6 (1955).
[7] A.G. PRODELL and P. KUSCH: Phys. Rev. **106**, 87 (1957).
[8] E.N. ADAMS II: Phys. Rev. **81**, 1 (1951).

For heavier nuclei the hfs anomalies may be as large as 0.5%. Two effects associated with the structure of the nucleus predominate. First the finite size of the nuclear charge distribution implies that the electron sees a non-Coulombic field at distances $r<R$, where R is the nuclear radius[1]. Since R may change between two isotopes, the distortion of the electronic wave function gives a contribution to Δ. Secondly, because the nuclear magnetism is distributed over the nuclear volume, a part of the magnetic dipole hfs interaction occurs when the electron penetrates inside the nucleus. For this part of the interaction the character of the nuclear magnetism—whether it arises from the intrinsic spins or the orbital motions of the nucleons—is particularly important, and differences between two isotopes can lead[2] to large values of Δ.

The hfs anomalies are largest if the hfs interaction arises primarily from an s electron, because the electronic wavefunction is then largest close to the nucleus. For $p_{\frac{1}{2}}$ electrons in heavier elements the electronic wavefunction has appreciable values within the nucleus because of relativistic effects, and many hfs anomalies have been found for $P_{\frac{1}{2}}$ atomic states. Due to electronic configuration interaction which can admix[3] excited s-electron states, an hfs anomaly can arise in principle

Table 9. *Magnetic dipole hyperfine structure anomalies.*

For the ${}^2P_{\frac{3}{2}}$ states the ratio of the magnetic interaction constants, a, are given in the column marked $\Delta\nu_l/\Delta\nu_h$. To the accuracy here of interest the ratio of the apparent interaction constants, derived from the intervals in the ${}^2P_{\frac{3}{2}}$ state, is equal to the ratio of the constants corrected for the mixing with the ${}^2P_{\frac{1}{2}}$ state. The spins of all the nuclei in the table are given in Table 8. Each pair of nuclei is described by l (the lighter) and h (the heavier). For nuclei of $Z>3$, the reduced mass correction in Eq. (37.7) becomes trivial and is ignored in the calculation of the anomaly. There are other cases (Br is an example) in which an anomaly might be searched for. Presently available data are not of sufficient accuracy to allow the determination of the anomaly.

Isotopic pair	Atomic state	μ_l/μ_h	$\Delta\nu_l/\Delta\nu_h$	${}_l\Delta_h$	Reference
${}_1H^1 - {}_1H^2$	${}^2S_{\frac{1}{2}}$	3.2571990(10)	4.3386495(4)	$-(1.703\pm0.005)\times10^{-4}$	55 *K2*
${}_1H^1 - {}_1H^2$	$2s\,{}^2S_{\frac{1}{2}}$	3.2571990(10)	4.3386510(23)	$-(1.700\pm0.008)\times10^{-4}$	56 *R1*
${}_1H^1 - {}_1H^3$	${}^2S_{\frac{1}{2}}$	0.937527(10)	0.93650962(6)	0 ± 10^{-5}	57 *P1*
${}_3Li^6 - {}_3Li^7$	${}^2S_{\frac{1}{2}}$	0.252435(6)	0.284013(5)	$+(1.25\pm0.27)\times10^{-4}$	49 *K6*
${}_{17}Cl^{35} - {}_{17}Cl^{37}$	${}^2P_{\frac{1}{2}}$	1.20133(12)	1.201357(13)	$+(2.6\pm10.0)\times10^{-5}$	51 *K2*
${}_{17}Cl^{35} - {}_{17}Cl^{37}$	${}^2P_{\frac{3}{2}}$	1.20133(12)	1.20136(5)	$+(2.5\pm11.0)\times10^{-5}$	51 *J1*
${}_{19}K^{39} - {}_{19}K^{40}$	${}^2S_{\frac{1}{2}}$	−0.30158(6)	−0.359097(7)	$+(4.67\pm0.19)\times10^{-3}$	52 *E1*
${}_{19}K^{39} - {}_{19}K^{41}$	${}^2S_{\frac{1}{2}}$	1.8218(2)	1.81768(1)	$-(2.26\pm0.10)\times10^{-3}$	50 *O1*
${}_{31}Ga^{69} - {}_{31}Ga^{71}$	${}^2P_{\frac{1}{2}}$	0.7870148(12)	0.7870196(6)	$+(6.2\pm2.3)\times10^{-6}$	56 *L2*
${}_{31}Ga^{69} - {}_{31}Ga^{71}$	${}^2P_{\frac{3}{2}}$	0.7870148(12)	0.7869949(13)	$-(25.3\pm0.3)\times10^{-6}$	54 *D1*; 55 *S4*
${}_{37}Rb^{85} - {}_{37}Rb^{87}$	${}^2S_{\frac{1}{2}}$	0.491789(2)	0.4441660(4)	$+(3.515\pm0.004)\times10^{-3}$	52 *O1*; 53 *D1*
${}_{48}Ag^{107} - {}_{47}Ag^{109}$	${}^2S_{\frac{1}{2}}$	0.86985(1)	0.86627(3)	$-(4.12\pm0.04)\times10^{-3}$	53 *W3*
${}_{49}In^{113} - {}_{49}In^{115}$	${}^2P_{\frac{1}{2}}$	0.9978609(12)	0.99786841(28)	$+(7.5\pm1.3)\times10^{-6}$	57 *E1*
${}_{49}In^{113} - {}_{49}In^{115}$	${}^2P_{\frac{3}{2}}$	0.9978609(12)	0.99783716(28)	$-(23.8\pm1.3)\times10^{-6}$	57 *E1*; 57 *E2*
${}_{55}Cs^{133} - {}_{55}Cs^{134}$	${}^2S_{\frac{1}{2}}$	0.86252(25)	0.8776933(13)	$+(1.69\pm0.30)\times10^{-3}$	57 *S2*
${}_{55}Cs^{133} - {}_{55}Cs^{135}$	${}^2S_{\frac{1}{2}}$	0.94500(7)	0.9453527(14)	$+(3.7\pm0.9)\times10^{-4}$	57 *S2*
${}_{55}Cs^{135} - {}_{55}Cs^{137}$	${}^2S_{\frac{1}{2}}$	0.96149(8)	0.9612967(22)	$-(2.0\pm0.9)\times10^{-4}$	57 *S2*
${}_{81}Tl^{203} - {}_{81}Tl^{205}$	${}^2P_{\frac{1}{2}}$	0.990258(1)	0.9903622(5)	$+(1.05\pm0.15)\times10^{-4}$	56 *L2*
${}_{81}Tl^{203} - {}_{81}Tl^{205}$	${}^2P_{\frac{3}{2}}$	0.990258(1)	0.9886498(6)	$-(1.626\pm0.002)\times10^{-3}$	56 *G1*

[1] J. E. Rosenthal and G. Breit: Phys. Rev. **41**, 459 (1932).
[2] A. Bohr and V. F. Weisskopf: Phys. Rev. **77**, 94 (1950).
[3] C. Schwartz: Phys. Rev. **99**, 1035 (1955).

from any atomic state. A very large anomaly has been found[1] in the $^2P_{\frac{3}{2}}$ state of thallium. Second order hfs interactions must be considered in obtaining sufficiently accurate values of the interaction constants to allow the determination of small values of Δ (see Sect. 35).

Table 9 gives the known values of the magnetic dipole hfs anomalies.

38. Nuclear electric quadrupole and magnetic octupole moments. If the first order treatment of the hfs interactions is sufficiently accurate, then the total electronic angular momentum J and the nuclear spin I are good quantum numbers and the interval rule of Eq. (34.3) can be used in the form of Eq. (34.5) to compute the hfs interaction constants A_k (or a, b, and c). When a magnetic octupole interaction can occur ($I \geq \frac{3}{2}$, $J \geq \frac{3}{2}$), it is necessary to consider second order hfs perturbations since the deviations from the interval rule due to such perturbations may be sufficiently great[2,3] to obscure the octupole interaction, which is generally small. In particular, the cross magnetic dipole—electric quadrupole interaction gives terms in the spacing of the hfs levels of a $^2P_{\frac{3}{2}}$ state which are, on experimental grounds alone, indistinguishable from those due to a magnetic octupole interaction. A general expression for this second order effect has been derived[2].

The calculation of the nuclear moments from the hfs interaction constants has been discussed in Sect. 35. If the hfs can be considered to arise from the interaction of a single electron (or hole) with the nucleus, the nuclear moments can be calculated from Eqs. (35.8) to (35.15) or by numerical integration of the Dirac radial equations. Nuclear moments determined in this way are subject to corrections associated with configuration mixing (or polarization) due to the electrostatic interactions between the atomic electrons (see Sect. 35).

The polarization corrections have been studied particularly with reference to the determination of the nuclear electric quadrupole moment Q from the value of the interaction constant b. For a doublet p valence state, which is the most common case, the two most important modes of excitation are the excitation of a core electron in an s-orbit into a higher unoccupied s-orbit and of a core electron in a p-orbit into a higher unoccupied p-orbit. Correction factors, computed for different atoms by use of a Fermi-Thomas potential with the inclusion of exchange effects, have been found[4] to modify the uncorrected result by as much as 25%. Since the measured magnetic dipole interaction constant a together with the nuclear magnetic dipole moment μ is often used for the determination of Q from b (see Sect. 35), it is necessary to consider the effects of polarization on the one-electron formula for a [Eq. (35.9)]. Such corrections have been computed[5], and have also been obtained from the measured a and μ values by assigning the difference between a theoretically calculated hfs, using a single-electron wavefunction together with the measured μ, and the measured hfs, to a polarization correction[3]. These two determinations of the polarization correction differ markedly for a number of atoms. The following equation has been proposed[3] for the determination of Q from measured values of b, a, and μ for $^2P_{\frac{3}{2}}$ states:

$$Q = \frac{0.541\, b\, \mu\, F}{a\, I\, R\, C\,(1-\beta)} \quad \text{barns} \tag{38.1}$$

in which a, b are in Mc/sec and μ is in nuclear magnetons; I = nuclear spin; F and R are obtained from Eqs. (35.13a) and (35.13b), respectively; C is STERNHEIMER'S calculated quadrupole polarization correction[5]; β is the fractional

[1] G. GOULD: Phys. Rev. **101**, 1828 (1956).
[2] C. SCHWARTZ: Phys. Rev. **97**, 380 (1955).
[3] C. SCHWARTZ: Phys. Rev. **105**, 173 (1957).
[4] R. STERNHEIMER: Phys. Rev. **80**, 102 (1950); **84**, 244 (1951); **95**, 736 (1954).
[5] R. STERNHEIMER: Phys. Rev. **86**, 316 (1952).

contribution of s electrons to the hfs. If the $^2P_{\frac{3}{2}}$ state arises from a p hole, Q is obtained from Eq. (38.1) with a change in sign.

For the determination of nuclear magnetic octupole moments Ω, the one-electron octupole integrals have been evaluated by numerical integration of the Dirac radial equation. The following equation is then used to compute Ω for a $^2P_{\frac{3}{2}}$ state arising from a p electron or p hole:

$$\Omega = \frac{c\,\mu\,F\,1.96\times 10^8}{a\,I\,T(1-\beta)\,Z^2} \tag{38.2}$$

in which F and T are obtained from Eqs. (35.13a) and (35.13c), respectively; the unit of Ω is the nuclear magneton-barn:

Table 10 gives values of nuclear electric quadrupole and magnetic octupole moments determined by atomic beam spectroscopy.

Table 10. *Nuclear quadrupole and octupole moments.*

Only those quadrupole moments are given which have been derived from atomic beam hfs data; since the octupole interaction is very small, information about octupole moments has been found only in atomic beams experiments on hfs. Q is calculated from Eq. (38.1). In those cases where the quadrupole polarization correction factor, C, and β have not been calculated, they have been taken to be 1 and 0 respectively. Where C and β have been calculated Q is marked with an asterisk. The octupole moment is calculated from Eq. (38.2). The values of C are taken from [52 *S3*] and of β from [57 *S1*]. Where C and β have not been calculated, the neglect of these factors introduces a large uncertainty in Q; where C and β have been calculated, approximations in Eq. (38.1) and in the calculation of C and β contribute significantly to the uncertainty in Q. Polarization correction factors, applicable to the calculation of Ω from c, have not been investigated.

Isotope	Q (Unit $=10^{-24}$ cm²)	Ω (Unit = nuclear magneton-barn)	References
$_5B^{11}$	0.0318*		53 *W4*
$_{11}Na^{23}$	0.108 (a)		55 *P1*
$_{13}Al^{27}$	0.150*		49 *L1*
$_{17}Cl^{35}$	−0.0750*	−0.0188	51 *J1*; 57 *S1*
$_{17}Cl^{37}$	−0.0591*	−0.0146	51 *J1*; 57 *S1*
$_{19}K^{39}$	0.07		56 *B4*
$_{31}Ga^{67}$	0.206*		57 *W1*
$_{31}Ga^{69}$	0.182*	0.137	54 *D1*; 57 *S1*
$_{35}Ga^{71}$	0.114*	0.180	54 *D1*; 57 *S1*
$_{34}Br^{79}$	0.323		54 *K1*
$_{35}Br^{81}$	0.269		54 *K1*
$_{37}Rb^{85}$	0.276		56 *S3*
$_{37}Rb^{87}$	0.133		56 *S3*
$_{49}In^{113}$	0.816*	0.574	57 *E2*
$_{49}In^{115}$	0.827*	0.565	57 *E2*
$_{53}I^{127}$	0.707*	0.181	54 *J1*; 57 *S1*
$_{55}Cs^{133}$	−0.0033		56 *B4*
$_{57}La^{139}$	0.268 (b)		57 *T2*
$_{59}Pr^{141}$	−0.054 (b)		53 *L2*

(a) Only two of the three hfs transitions have been observed in the $^2P_{\frac{3}{2}}$ state (see Table 8). It is thus possible that Q may be different from that in the table both in magnitude and in sign.

(b) Since the hfs of a state other than the $^2P_{\frac{3}{2}}$ was observed, Eq. (38.1) must be modified.

39. Atomic and molecular frequency standards. Any fundamental frequency standard must provide a frequency which is well defined without reference to any external standards. This requires that the frequency be generated in a system which has a high and constant equivalent Q and for which the physically significant parameters are simply and unambiguously reproducible. The frequencies of certain microwave transitions in atoms and molecules approximate these conditions and hence are suitable for frequency standards. In particular, a transition between the hfs levels in cesium and one between the inversion levels in ammonia have been used as frequency standards. Other transitions, particularly a hfs transition of the more common isotope of thallium, Tl^{205}, have been considered for frequency standard applications.

The ground electronic state of cesium is a $^2S_{\frac{1}{2}}$ state and the nucleus (Cs^{133} with 100% natural abundance) has a spin of $\frac{7}{2}$. Hence there is only a single hfs interval, and at zero magnetic field H_0 there are two energy levels ($F=3$ and $F=4$), separated by a transition frequency of about 9193 Mc/sec. Because of the necessity for maintaining quantization of the states, a small magnetic field H_0 (~ 0.1 gauss) is required (see Sect. 20). In this field each F level is split into $2F+1$ sublevels and the energy levels are given by the Breit-Rabi formula [Eq. (34.10)]. Of the transitions allowed by the magnetic dipole selection rule $\Delta F=0, \pm 1$ and $\Delta m=0, \pm 1$, the σ transition $(F, m)=(4, 0) \leftrightarrow (3, 0)$ is most insensitive to changes in the magnetic field H_0 and hence is the one most useful for a frequency standard. The frequency of this line depends only quadratically on H_0 and is given by:

$$\nu = \Delta\nu + 426 H_0^2. \tag{39.1}$$

Hence if H_0 is 0.1 gauss, the frequency ν is greater than $\Delta\nu$ by 4.3 cps, and if the field is held constant within the limits 0.09 to 0.11 gauss, ν will be constant to about 1 cps or 1 part in 10^{10}. The small field H_0 can be defined in terms of the frequency of another line in the cesium spectrum, for example ,by the requirement that the field sensitive σ line $(3, 1) \leftrightarrow (4, 1)$ whose frequency depends linearly on H_0, be greater in frequency than the $(3, 0) \leftrightarrow (4, 0)$ line by a fixed amount. The dependence of the frequency of the Cs lines on electric field is negligible.

A high equivalent Q is obtained with a spectral line having a large ratio of resonance frequency to line width. The line width is determined by the transit time of the atom through the C-region, with the method of separated oscillating fields (Sect. 29) being used to avoid line broadening due to inhomogeneity of the magnetic field and to allow convenient application of a microwave frequency for a long C-field region. With a C-field length of 100 cm and a low, but useable, oven source temperature of 80° C, a line width of about 200 cps is obtained, and hence the effective $Q(=\nu/\delta\nu)$ is 5×10^7. Fig. 79 shows a resonance line for the $(3, 0) \leftrightarrow (4, 0)$ transition taken at a field H_0 of 0.71 gauss. The $\Delta\nu$ for Cs has been determined to be 9192.63183 (1) Mc/sec in terms of an astronomical time scale[1].

Early work[2] on a Cs frequency standard has been done at the National Bureau of Standards. ZACHARIAS, YATES, HAUN[3] and DALY at MIT led in the development of a commercial Cs frequency standard by DALY and HOLLOWAY of the National Company. A mass spectrometer and electron multiplier following the hot wire surface ionization detector (Sect. 15) yields a high sensitivity of detection and allows the use of a low oven temperature, which conserves source material and gives a narrow line width. A sealed-off vacuum system without pumps, but with a getter, is employed. The transition field and the associated electronics have been discussed in Sect. 32. The important known sources of error for this Cs frequency standard arise from variations of the magnetic field in the C-region, from phase shifts between the two oscillating fields, and from shot noise associated with the random arrival of beam atoms at the detector. The possible overlap of different σ transitions, or even of π transitions induced by components of the microwave field perpendicular to H_0, must also be considered. Reproducibility and both long ($\sim$months) and short time frequency stability of about 1 part in 10^{10} have been established by intercomparison of several Cs standards.

ESSEN and PARRY[1,4] have discussed in detail the design considerations of a cesium frequency and time standard. The effect of variation of a number of

---

[1] L. ESSEN and J.V.L. PARRY: Phil. Trans. Roy. Soc. Lond. A **250**, 45 (1957).
[2] H. LYONS: Ann. New York Acad. Sci. **55**, 831 (1952).
[3] ZACHARIAS, YATES and HAUN: Proc. Inst. Radio Engrs. **43**, No. 3, 364 (1955).
[4] ESSEN, PARRY and BULLARD: Nature, Lond. **176**, 280 (1955).

externally imposed parameters on the measured value of $\Delta\nu(\mathrm{Cs})$ has been investigated. The frequency standard now serves to determine variations in astronomical time.

The ammonia molecule [28] is a symmetric top molecule whose rotational energy levels are characterized by J, the total angular momentum quantum number and K, the quantum number for the component of the total angular momentum along the molecular symmetry axis. Each rotational state is split into two inversion levels. The inversion transition in the $J=3$, $K=3$ rotational state occurs at a frequency of about 23870 Mc/sec, and the main component of this inversion line, which corresponds to no change in the magnetic dipole or electric quadrupole

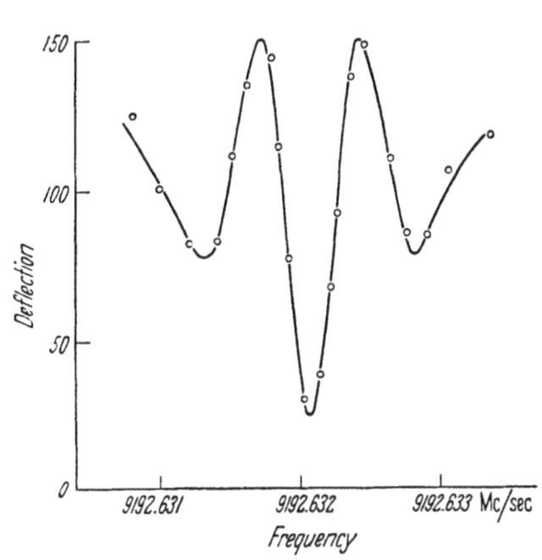

Fig. 79. Cs resonance curve obtained for the transition $(F, m) = (3, 0) \leftrightarrow (4, 0)$ at $H_0 = 0.71$ gauss by the method of separated oscillating fields.

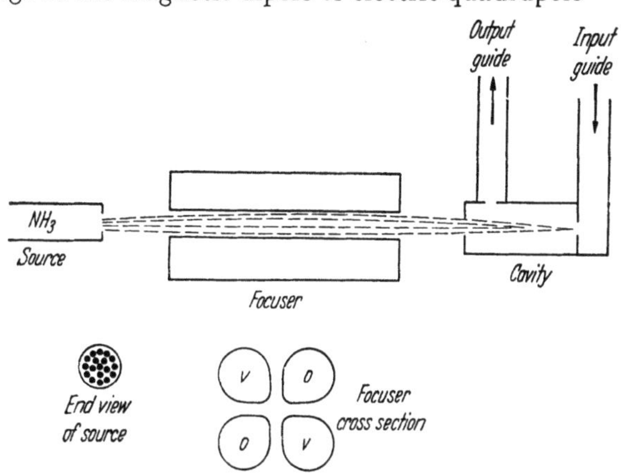

Fig. 80. Simplified diagram of the essential parts of the maser. A directed source of NH_3 molecules is produced by an array of fine tubes. The high Q cavity operates in the TE_{011} mode (z direction is direction of propagation of beam).

interaction energy associated with the hydrogen and nitrogen nuclei, is used as a frequency standard.

Most of the studies of the ammonia inversion lines have utilized the standard methods of microwave absorption spectroscopy, and the minimum line widths achieved have been limited to about 60 kc/sec by Doppler and collision broadening[1]. Spectroscopic lines of considerably smaller width and oscillations in a circuit at the inversion frequency of the $J=3$, $K=3$ rotational state have recently been obtained by use of the maser principle (microwave amplification by stimulated emission of radiation[2]) (see Fig. 80). Molecules in the higher energy inversion level are separated from molecules in the lower energy inversion level by electrostatic deflection in a quadrupole field (see Sect. 23 and Fig. 45), and enter a high Q resonant cavity. If microwave power is fed into the cavity at the inversion frequency, beam molecules may undergo a transition to the lower energy inversion level and thus emit energy at the inversion frequency. This stimulated emission can be detected and hence spectroscopic measurements are possible[3]. Doppler and collision broadening are negligible because of the unidirectional, collision-free nature of the beam, and the line width is primarily the natural width associated with the transit time of the molecules in the cavity. Line widths of about 7 kc/sec have been obtained. Observable emission of power by the beam is achieved with a flux of 10^9 molecules/sec into the cavity.

1 Gunther-Mohr, White, Schawlow, Good and Coles: Phys. Rev. **94**, 1184 (1954).
2 Gordon, Zeiger and Townes: Phys. Rev. **99**, 1264 (1955).
3 J. P. Gordon: Phys. Rev. **99**, 1253 (1955).

Self oscillation of the maser can be achieved if the number of molecules in the beam entering the cavity is sufficiently great. Thermal noise with a white spectrum can be considered present in the cavity, and the molecular beam will act as a highly selective amplifier at the inversion frequency. Consider the diagram of Fig. 81. Thermal noise power is incident on the cavity from guide A. It is selectively amplified by the beam, and power output is taken from guide D and fed back into the input guide A. If the power supplied in emission by the beam is equal to the power losses in the cavity and feedback system, and if the phase of the feedback signal is proper, then oscillations will occur. Analysis of the theory of the oscillations indicates the following characteristics: (1) The center frequency ν_0 of the oscillation is given to a good approximation by the equation:

$$\nu_0 = \nu_B + \frac{\Delta\nu_B}{\Delta\nu_C}(\nu_C - \nu_B) \qquad (39.2)$$

where $\Delta\nu_C$ and $\Delta\nu_B$ are the half-widths of the cavity mode and of the molecular emission line, respectively, and $\nu_C - \nu_B$ is the difference between the cavity resonant frequency ν_C and the inversion line frequency ν_B. (2) The total width at half-power of the spectral distribution of the oscillation is approximately:

$$\delta\nu = 8\pi\, kT\,(\Delta\nu_B)^2/P_B \qquad (39.3)$$

where T is the temperature of the beam source and P_B is the power emitted from the beam.

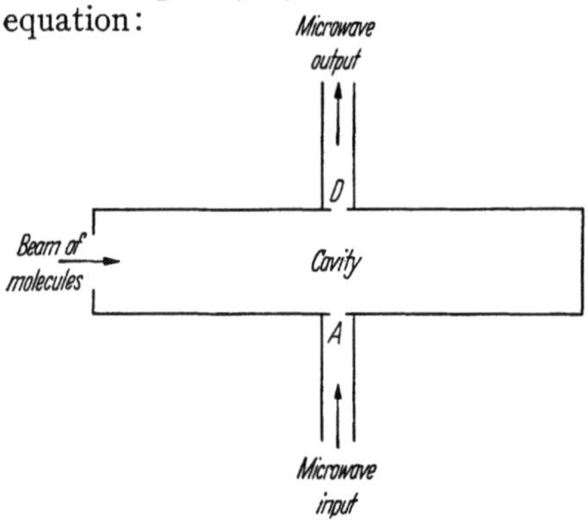

Fig. 81. Schematic diagram of the operation of the maser as an oscillator, showing the resonant cavity and molecular beam.

In practice, a flux into the cavity greater than 10^{13} molecules/sec is required to sustain oscillations. For typical conditions with $T = 300°$ K, $\Delta\nu_B = 3\times 10^3$ cps, and $P_B = 10^{-10}$ watt, Eq. (39.3) gives $\nu_B/\delta\nu = 2\times 10^{12}$. Intercomparison of maser oscillators, which use Invar cavities in contact with ice water to control thermal shifts, has indicated a frequency stability of about 4 parts in 10^{12} for times of the order of a second and a stability of about 1 part in 10^{10} for times of the order of an hour. The dependence of the central frequency ν_0 on the cavity resonance frequency ν_C [Eq. (39.2)] is an important limitation of the maser oscillator as a frequency standard. The frequency has also been observed to depend significantly on the source pressure and focuser voltage, both of which affect the intensity of the molecular beam.

II. Atomic magnetism and fine structure.

40. Magnetic moment of the electron[1]. The history of the anomalous spin magnetic moment of the electron began with the observation of a discrepancy between the first directly measured value of the hyperfine structure separation in hydrogen[2] and the theoretical value obtained from the Fermi formula, Eq. (37.1), with the addition of the $(1 + m/M_p)^{-3}$ reduced mass correction [see Eq. (37.3)]. The value measured by the atomic beam magnetic resonance method was

$$\Delta\nu_H\,(\text{expt}) = (1420.410 \pm 0.006)\ \text{Mc/sec},$$

[1] A short survey of the results has been given by E. R. COHEN and J. W. M. DUMOND in Sect. 24 of their contribution to Vol. XXXV, and theoretical details in Sect. 33 of G. KÄLLÉN's article on quantum electrodynamics, Vol. V/1 of this Encyclopedia.

[2] NAFE, NELSON and RABI: Phys. Rev. 71, 914 (1947); 73, 718 (1948).

whereas the theoretical value was

$$\Delta\nu_H\,(\text{theor}) = (1416.97 \pm 0.54)\ \text{Mc/sec},$$

in which the error quoted in $\Delta\nu_H$(theor) was determined by uncertainties in the values of the fundamental constants appearing in the Fermi formula.

It was suggested[1] that the discrepancy indicates a deviation of the spin magnetic moment of the electron, μ_s, from μ_0 as given by the Dirac theory by an amount of order $\alpha\mu_0$, and, furthermore, that a spin magnetic moment different from μ_0 could be observed by its effect on the Landé g factor of atomic states. Measurements[2] of g_J values in different atomic states by the atomic beam magnetic resonance method yielded values of the spin g value of the electron relative to its orbital g value (g_s/g_l). The deviation of μ_s from μ_0, which is called the anomalous spin magnetic moment of the electron, was shown to arise from virtual radiative corrections to the energy of an electron in an external magnetic field, and μ_s was calculated to be[3]:

$$\mu_s = \mu_0\left(1 + \alpha/(2\pi)\right) = \mu_0(1.00116). \tag{40.1}$$

The experimental and theoretical values of μ_s were found to be in excellent agreement.

The best experimental value for the spin magnetic moment of the electron is obtained by the combination of the results of two types of experiments. One is the measurement of the ratio of the g value of the electron in the ground state of the hydrogen atom, $g_J(H, {}^2S_{\frac{1}{2}})$, to the g value of the proton, g_p. The frequency of the transition $(F, m) = (1, 0) \leftrightarrow (1, -1)$ in hydrogen is observed by the atomic beam method, and then the resonance frequency of the protons in a mineral oil sample is observed in the same magnetic field[4]. Use of the expression for the H atom transition frequency obtained from the Breit-Rabi formula (see Table 7) together with the known value of $\Delta\nu_H$ (see Table 8) and of the relation $\nu_p = 2g_p''\mu_0 H_0/h$ for the proton resonance frequency ($g_p'' = g$ value of proton in the cylindrical mineral oil sample) allows the calculation of g_J/g_p'' from the observed resonance frequencies. By careful shimming of the C-magnet and with the use of deflecting fields of the two wire type (Sect. 21) to avoid the effect of iron deflecting magnets on the field homogeneity in the C-region, the natural line width of 60 kc/sec for the 4 cm radiofrequency region was obtained at the transition frequency of about 3655 Mc/sec. The critical problem for this experiment was to obtain a sufficiently homogeneous and constant field H_0 in the transition region to assure that the same effective field H_0 determines the resonance frequencies for the H atoms and for the protons. A correction of 1.5 ppm is made for the bulk diamagnetism of the cylindrical mineral oil sample. Finally $g_J(H, {}^2S_{\frac{1}{2}})$ differs slightly from g_s for a free electron due to a relativistic correction obtained from the Dirac equation for the H atom in an external magnetic field[5].

$$g_J(H, {}^2S_{\frac{1}{2}}) = g_s\left(1 - \frac{\alpha^2}{3}\right) = g_s(1 - 17.7\times 10^{-6}). \tag{40.2}$$

Hence the final result of the experiment is: $g_s/g_p' = -658.2288 \pm 0.0006$, in which g_p' is the proton g value in a spherical sample of mineral oil.

[1] G. Breit: Phys. Rev. **72**, 984 (1947); **74**, 656 (1948).

[2] P. Kusch and H. M. Foley: Phys. Rev. **72**, 1256 (1947); **73**, 412 (1948); **74**, 250 (1948). — A. K. Mann and P. Kusch: Phys. Rev. **77**, 435 (1950).

[3] J. Schwinger: Phys. Rev. **73**, 416 (1948); **76**, 790 (1949).

[4] Koenig, Prodell and Kusch: Phys. Rev. **88**, 191 (1952).

[5] G. Breit: Nature, Lond. **122**, 649 (1928). — N. F. Mott and H. S. W. Massey: The Theory of Atomic Collisions, 2nd edit., p. 72. Oxford: Clarendon Press 1949.

The quantity g_s/g_p' has also been obtained in a microwave magnetic resonance absorption experiment[1]. The transitions $(F, m) = (1, 1) \leftrightarrow (0, 0)$ and $(1, 0) \leftrightarrow (1, -1)$ were measured at a frequency of about 9000 Mc/sec and in the same magnetic fields the proton resonance absorption frequencies were also measured. The result $g_s/g_p' = -658.2298 \pm 0.0003$ is in good agreement with the atomic beam value. g_s/g_p' has also been determined from a microwave magnetic resonance absorption experiment on deuterium[2]. The value obtained, $g_s/g_p' = -658.2286 \pm 0.0015$, agrees with the hydrogen value and thus indicates the absence of any nuclear effects to within the experimental accuracy.

The second experiment used for the determination of g_s/g_l is the measurement of g_l/g_p'. The cyclotron frequency of the free electron and the proton resonance frequency in a spherical mineral oil sample are measured in the same magnetic field, and their ratio gives $2g_l/g_p'$. Two independent measurements of this ratio have been made. For one measurement the electrons were obtained from a hot filament and the electron cyclotron resonance was observed by its effect on the electron trajectories[3]. The value obtained was $2g_l/g_p' = -657.475 \pm 0.008$. For the other measurement the electrons were obtained by use of the photoelectric effect and the resonance was observed by a microwave absorption method[4]. The value obtained was $2g_l/g_p' = -657.463 \pm 0.007$. The sources of error in these measurements arise from inhomogeneity and nonconstancy of the magnetic field, diamagnetic effects, space charge effects, and the relativistic dependence of the electron mass on velocity. If the average of these two values is taken ($2g_l/g_p' = -657.469$) with an assigned error of ± 0.006, and if the value $g_s/g_p' = -658.2288 \pm 0.0006$ from the atomic beam experiment is used, the value

$$g_s/g_l = 2\mu_s/\mu_0 = 2(1.001156 \pm 0.000010) \tag{40.3}$$

is obtained.

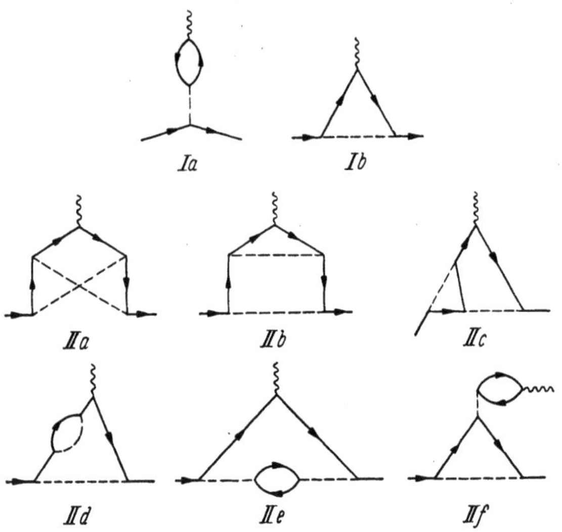

Fig. 82. Feynman diagrams for radiative corrections contributing to the anomalous spin magnetic moment of the electron. Solid lines indicate electrons and dashed lines indicate photons or the external field. Diagrams I give second order radiative corrections of order α; diagrams II give α^2 corrections.

Fourth order virtual radiative corrections of order α^2 to the spin magnetic moment of the electron have been calculated by the modern covariant, renormalized theory of quantum electrodynamics[5]. The Feynman diagrams contributing to the α and α^2 corrections to the electron magnetic moment are shown in

[1] R. BERINGER and M. A. HEALD: Phys. Rev. **95**, 1474 (1954).
[2] GEIGER, HUGHES and RADFORD: Phys. Rev. **105**, 183 (1957).
[3] J. H. GARDNER and E. M. PURCELL: Phys. Rev. **76**, 1262 (1949); **83**, 996 (1951).
[4] P. FRANKEN and S. LIEBES jr.: Phys. Rev. **104**, 1197 (1956).
[5] C. M. SOMMERFIELD: Phys. Rev. **107**, 328 (1957).

Fig. 82. The theoretical value is

$$\left.\begin{aligned} g_s/g_l &= 2\left(1+\frac{\alpha}{2\pi}-0.328\,\frac{\alpha^2}{\pi^2}\right)\\ &= 2\,(1+0.0011614-0.0000018)=2\,(1.0011596). \end{aligned}\right\}\tag{40.4}$$

The experimental value of Eq. (40.3) agrees with the theoretical value within the limit of error, but it does not provide a convincing test of the α^2 correction to the moment.

A direct experimental measurement of the ratio g_s/g_l for free electrons with an energy of 420 kev has been made[1]. Polarization and analysis of the electron spin is achieved by Mott scattering from the Coulomb field of a heavy nucleus. The angle of precession of the electron spin in a magnetic field and the cyclotron motion of the electron in the same field are used to obtain: $g_s/g_l = 2.00 \pm 0.01$. The present low accuracy of this determination makes it unsuitable for the determination of the anomalous magnetic moment of the electron. The transition $\Delta m_s = \pm 1$ for the free electron has recently been observed[2] through polarization of electrons in exchange collisions with Na vapor which has, in turn, been polarized by optical pumping. The method has not yet been developed to the point where it has yielded experimental data of an accuracy comparable to that of the experiments that have been described in this section.

41. Atomic g values. The g_J values of many atoms have been measured by the atomic beam magnetic resonance method. Usually transitions $\Delta m_J = \pm 1$, $\Delta m_I = 0$ are observed at strong H_0 fields; a knowledge of the hfs intervals and of g_I, which may be required in order to determine g_J, is obtained by independent experiments (Sect. 36). In general the ratio of the g_J values for two different atoms or for two different states of the same atom has been obtained. The early motivation for such measurements was the determination of the spin magnetic moment of the electron[3]. However, since the electron's magnetic moment is determined best from a study of hydrogen, the measured g values of atoms are at present used mainly as a test of the theory of atomic magnetism.

The theory of atomic magnetism for a multielectron atom is based on a generalized Dirac-Breit equation, with the addition of a term to represent the interaction of the anomalous spin magnetic moments of the electrons with the external magnetic field[4]. The equation reads:

$$\left\{W+\sum_{h=1}^{n}\left[\beta_k m c^2+\boldsymbol{\alpha}_k\cdot(c\,\boldsymbol{p}_k+e\boldsymbol{A}_k)\right]-V+\sum_{l>k=1}^{n}B_{kl}+\Omega_{H_0}\right\}\Psi=W\Psi \tag{41.1}$$

in which W is the stationary state energy, including the rest energy of the n electrons, V is the sum of the electrostatic interactions,

$$V=-\sum_{k=1}^{n}\frac{Ze^2}{r_k}+\sum_{l>k=1}^{n}\frac{e^2}{r_{kl}} \tag{41.1a}$$

$-e$ is electron charge, Ze is nuclear charge, $\boldsymbol{p}_k$ is the momentum of the k-th electron, $\boldsymbol{A}_k$ is the vector potential of the external field at the k-th electron (for a uniform and constant magnetic field $\boldsymbol{H}_0$, $\boldsymbol{A}_k=\frac{1}{2}\boldsymbol{H}_0\times\boldsymbol{r}_k$), $\boldsymbol{\alpha}_k$, β_k are the usual Dirac matrices, and B_{kl} is the Breit interaction between k-th and l-th electrons,

$$B_{kl}=\frac{-e^2}{2r_{kl}}\left[\boldsymbol{\alpha}_k\cdot\boldsymbol{\alpha}_l+\frac{(\boldsymbol{\alpha}_k\cdot\boldsymbol{r}_{kl})\,(\boldsymbol{\alpha}_l\cdot\boldsymbol{r}_{kl})}{r_{kl}^2}\right] \tag{41.1b}$$

[1] Louisell, Pidd and Crane: Phys. Rev. **94**, 7 (1954).

[2] H. G. Dehmelt: Phys. Rev. **109**, 381 (1958).

[3] P. Kusch and H. M. Foley: Phys. Rev. **74**, 250 (1948).

[4] W. Perl and V. W. Hughes: Phys. Rev. **91**, 842 (1953). — W. Perl: Phys. Rev. **91**, 852 (1953). — A. Abragam and J. H. van Vleck: Phys. Rev. **92**, 1448 (1953).

with $\boldsymbol{r}_{kl}=\boldsymbol{r}_k-\boldsymbol{r}_l$, the vector distance between k-th and l-th electrons.

$$\Omega_{H_0}=\mu_0\left[\alpha/(2\pi)-0.328\,\alpha^2/\pi^2\right]\sum_{k=1}^{n}\boldsymbol{\sigma}'_k\cdot\boldsymbol{H}_0 \tag{41.1c}$$

is the interaction of the anomalous electron spin moments with the external field $\boldsymbol{H}_0$ where $\boldsymbol{\sigma}'$ is the 4×4 Pauli spin matrix. The wave function depends on n space coordinates $\boldsymbol{r}_k$ and n 4-component spinor variables.

The magnetic interaction energy W_{H_0} is computed in first order perturbation theory by evaluating the diagonal matrix element of

$$\sum_{k=1}^{n} e\,\boldsymbol{\alpha}_k\cdot\boldsymbol{A}_k+\Omega_{H_0} \tag{41.2}$$

for the zeroth order wave function computed from Eq. (41.1) without the terms of Eq. (41.2). Another equivalent procedure is to derive a Schrödinger-Pauli equation from Eq. (41.1), and then to evaluate the energy term which depends linearly on H_0 by first order perturbation theory. Higher order terms in H_0 correspond to diamagnetic corrections, which are usually negligible. The result can be expressed:

$$W_{H_0}=\mu_0\,g_J\,H_0\,m_J \tag{41.3}$$

in which m_J is the magnetic quantum number for the total electronic angular momentum. The quantity g_J can then be expressed as a power series in α valid through terms of order α^2.

The lowest order term for g_J is the value obtained from non-relativistic theory, and, if the atom obeys Russell-Saunders coupling so that the total orbital angular momentum $\boldsymbol{L}$ and the total spin angular momentum $\boldsymbol{S}$ are separately constants of the motion,

$$g_J=g_L\frac{J(J+1)+L(L+1)-S(S+1)}{2J(J+1)}+g_S\frac{J(J+1)+S(S+1)-L(L+1)}{2J(J+1)} \tag{41.4}$$

with $g_L=1$, $g_S=2$. The term in α arises entirely from the $\alpha/(2\pi)$ contribution to the anomalous spin magnetic moment of the electron. It may be included by using $g_S=2(1+\alpha/2\pi)$ in Eq. (41.4).

The term in α^2 arises in part from the α^2 contribution to the anomalous spin magnetic moment of the electron and in part from relativistic effects given by the Dirac-Breit equation. The relativistic contributions of order $\alpha^2\mu_0H_0$ can be identified with corresponding $\alpha^2 r_y$ contributions to fine structure, which arise from a relativistic increase of mass, a spin-orbit interaction, spin-other orbit, and orbit-orbit interactions. The part corresponding to spin-other orbit interactions can be interpreted as a diamagnetic effect, which is the reduction of the external field at a given electron due to the alteration of the orbital motion of the other electrons by the external field.

The breakdown of Russell-Saunders coupling for an atom can affect the g_J value. A combination of electrostatic interaction between electrons, which leads to configuration mixing, and of spin-orbit interactions can cause[1] the admixture of states of different L and S, or, indeed, if several fine structure states of the same configuration are present, the spin-orbit interactions alone can lead to an alteration in the coupling[2]. The admixture of other fine structure levels by the external field H_0 may also be important[2]. The motion of the nucleus associated

[1] M. PHILLIPS: Phys. Rev. **88**, 202 (1952).

[2] A. ABRAGAM and J.H. VAN VLECK: Phys. Rev. **92**, 1448 (1953).

Table 11. *Atomic g_J values.*

Atom	State (b)	Experimental value (e): Quantity measured (i)	g_J/g_J (H) (c)	g_J (d)	Theoretical value, g_J (g)	Reference: experimental	Reference: theoretical
Electron (free)		$\frac{g_s}{g_l}$		$g_s = 2.00 \pm 0.01$	$g_s = 2\left(1 + \frac{\alpha}{2\pi} - 0.328\,\frac{\alpha^2}{\pi^2}\right)$	54 *L2*	49 *S1*; 50 *K1*; 57 *S4*
		$\frac{g_s}{g_J(\mathrm{Na})}$	0.999974 (30)	2.00226 (40)	$= 2.0023192$	58 *D1*	
		$\frac{g_l}{g_p}$	$-\frac{2g_l}{g_p} = 657.469\,(9)$ (f)			51 *G4*; 56 *F1*	
H^1	$^2S_{\frac{1}{2}}$	$\frac{g_J(\mathrm{H})}{g_p}$	$-\frac{g_J}{g_p} = 658.2176\,(0.8)$ (f)	2.002277 (10)	2.0022838	52 *K3*; 54 *B5*	28 *B1*; 40 *M1*
	$2^2S_{\frac{1}{2}}$	$\frac{g_J(\mathrm{H})}{g_p}$	1.000016 (12)	2.002309 (16)	2.0023103	55 *W1*	28 *B1*; 40 *M1*
D^2	$^2S_{\frac{1}{2}}$	$\frac{g_J(\mathrm{D})}{g_p}$	0.9999997 (2.3)	2.002276 (10)	2.0022838	49 *N3*; 57 *G6*	
		$\frac{g_J(\mathrm{D})}{g_J(\mathrm{H})}$					
He^4	$1s2s, {}^3S_1$	$\frac{g_J(\mathrm{He})}{g_J(\mathrm{H})}$	0.9999767 (0.8)	2.002230 (10)	2.0022373	53 *H4*; 58 *D2*	53 *P2*
$Li^{6,7}$	$^2S_{\frac{1}{2}}$	$\frac{g_J(\mathrm{Li})}{g_J(\mathrm{Na})}$	1.000016 (25)	2.002309 (27)	2.002297	49 *K7*	52 *P1*; 53 *P1*
N^{14}	$^4S_{\frac{3}{2}}$	$\frac{g_J(\mathrm{N})}{g_p}$	0.9999184 (1.7)	2.002114 (10)	2.002	54 *H4*	
O^{16}	$^3P_1$	$\frac{g_J(\mathrm{O})}{g_p}$	0.7496368 (0.9)	1.500981 (10)	1.500995	52 *R6*; 58 *R1*	53 *A1*; 54 *K5* 58 *I1*
	$^3P_2$		0.7496043 (0.9)	1.500916 (10)	1.500931		
F^{19}	$^2P_{\frac{3}{2}}$	$\frac{g_J(\mathrm{F})}{g_p}$	0.6662 (230)	1.3339 (230)	1.3341	58 *R2*	
Ne^{20}	$2p^5\,3s, {}^3P_2$	$\frac{g_J(\mathrm{He})}{g_J(\mathrm{Ne})}$	0.7495883 (3)	1.500883 (11)	1.50088	52 *W1*; 57 *D1*; 58 *L1*	58 *L1*

Table 11. (Continued.)

Atom	State (b)	Experimental value (e)			Theoretical value g_J (g)	Reference	
		Quantity measured (i)	$g_J/g_J(\mathrm{H})$ (c)	g_J (d)		experimental	theoretical
Na^{23}	$^2S_{\frac{1}{2}}$	$\frac{g_J(\mathrm{Na})}{g_J(\mathrm{K})}$	1.000016(25)	2.002309(27)	2.002293	49 *K 7*	52 *P 1*; 53 *P 1*
P^{31}	$^4S_{\frac{3}{2}}$	$\frac{g_J(\mathrm{P})}{g_J(\mathrm{N})}$	0.99981(200)	2.0019(200)	2.002	55 *D 1*	
A^{40}	$3p^5\,4s\,^3P_2$	$\frac{g_J(\mathrm{A})}{g_J(\mathrm{He})}$	0.7496260(5)	1.500959(12)	1.5011	57 *D 1*; 58 *L 1*	58 *L 1*
K^{39}	$^2S_{\frac{1}{2}}$	$\frac{g_J(\mathrm{K})}{g_J(\mathrm{H})}$	1.000016(4)	2.002309(11)	2.00230	52 *F 2*	52 *P 1*; 53 *P 1*
Cr^{52}	$^7S_3$	$\frac{g_J(\mathrm{Cr})}{g_J(\mathrm{K})}$	0.999536(78)	2.001348(79)	2.002	53 *B 4*	
$Cu^{63,65}$	$^2S_{\frac{1}{2}}$	$\frac{g_J(\mathrm{Cu})}{g_J(\mathrm{Cs})}$	1.00015(500)	2.0026(500)	2.002	57 *T 1*	
$Ga^{69,71}$	$^2P_{\frac{1}{2}}$	$\frac{g_J(\mathrm{Ga},\,^2P_{\frac{1}{2}})}{g_J(\mathrm{Na})}$	0.332534(66)	0.665825(66)	0.66589	48 *K 1*	52 *P 1*; 53 *P 1*
	$^2P_{\frac{3}{2}}$	$\frac{g_J(\mathrm{Ga},\,^2P_{\frac{3}{2}})}{g_J(\mathrm{Ga},\,^2P_{\frac{1}{2}})}$	0.666212(87)	1.333941(88)	1.33411		
Br	$^2P_{\frac{3}{2}}$	$\frac{g_J(\mathrm{Br},\,^2P_{\frac{3}{2}})}{g_J(\mathrm{Cl}^{35},\,^2P_{\frac{3}{2}})}$	0.9997(400)			52 *K 6*; 54 *K 1*	
$Rb^{85,87}$	$^2S_{\frac{1}{2}}$	$\frac{g_J(\mathrm{Rb})}{g_J(\mathrm{Na})}$	1.000066(27)	2.002409(29)	2.00234	49 *K 7*	52 *P 1*; 53 *P 1*
$Ag^{107,109}$	$^2P_{\frac{1}{2}}$	$\frac{g_J(\mathrm{Ag})}{g_J(\mathrm{Cs})}$	1.00002(105)	2.00232(105)	2.002	53 *W 3*	
In^{115}	$^2P_{\frac{1}{2}}$	$\frac{g_J(\mathrm{In},\,^2P_{\frac{1}{2}})}{g_J(\mathrm{Na})}$	0.332531(102)	0.665819(104)	0.6659	50 *M 2*; 48 *K 1*	52 *P 1*; 53 *P 1*

	$^2P_{\frac{3}{2}}$	$\frac{g_J(\text{In}, {}^2P_{\frac{3}{2}})}{g_J(\text{In}, {}^2P_{\frac{1}{2}})}$	0.666392 (120)	1.334301 (12)	1.3341		
Xe (a)	$^3P_2$	$\frac{g_J(\text{Xe}, {}^3P_2)}{g_J(\text{He}, {}^3S_1)}$	0.749593 (60)	1.50089 (60)		57 *F1*	
Cs^{133}	$^2S_{\frac{1}{2}}$	$\frac{g_J(\text{Cs})}{g_J(\text{Na})}$	1.000150 (27)	2.002577 (29)	2.00244	49 *K7*	52 *P1*; 53 *P1*
Pr^{141}	$^4I_{\frac{9}{2}}$	$\frac{g_J(\text{Pr})}{g_J(\text{Cs})}$	0.365 ± 0.002	0.731 ± 0.004	0.727	53 *L2*	
Au^{197}	$^2S_{\frac{1}{2}}$	$\frac{g_J(\text{Au})}{g_J(\text{Cs})}$	1.00095 (57)	2.00420 (58)	2.002	53 *W3*	
Hg (h)	$6s\,6p,\ {}^3P_1$	$\frac{g_J(\text{Hg})}{g_p}$	0.74106 (270)	1.4838 (270)	1.50	52 *B6*	

(a) Experiment was done on xenon with its natural isotopic abundance.

(b) The atomic state is the ground state electronic configuration unless a different electronic configuration is indicated.

(c) The quantity $g_J/g_J(\text{H})$ is obtained by combining the appropriate measured quantities.

(d) The experimental value of g_J for an atom is obtained by combining the directly measured quantity involving the g_J of the atom with other measured quantities given in the table in order to obtain g_J/g_l. (The unit is $g_l = 1$.) For all cases, the measured value of g_l/g_p is required, and for several atoms with the best known g_J values the inaccuracy of about 9 ppm for this quantity provides the principal part of the quoted errors.

(e) Errors are indicated by the numbers in parentheses in parts per million (ppm).

(f) This number is the unweighted average of the two experimental values in the references.

(g) The latest value $\alpha^{-1} = 137.0390 \pm 0.0012$ has been used in calculating theoretical g_J values. See Ref. [57 *S4*].

(h) Result applies to the naturally occurring odd isotopes of mercury.

(i) The quantity g_p is the nuclear g-value of the proton in a spherical sample of mineral oil.

with its finite mass leads to a correction to the contribution of the electron orbital motion to the g_J value[1].

Table 11 gives the measured atomic g_J values (mainly by the atomic beam magnetic resonance method) and the corresponding theoretical values. The most careful comparisons of theoretical and experimental g_J values have been made for He, the alkali atoms, and oxygen. For helium in the metastable $^3S_1$ state the experimental value, accurate to ± 0.8 ppm, agrees with the theoretical value, accurate to ± 1 ppm. For the heavier alkali atoms unexplained deviations of the experimental and theoretical values by as much 70 ppm occur. For oxygen in its $^3P_1$ and $^3P_2$ states experimental and theoretical values of g_J differ by 7 ppm.

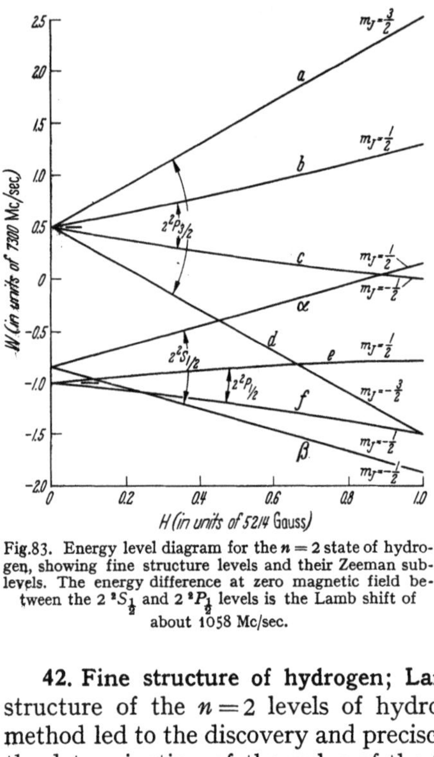

Fig. 83. Energy level diagram for the $n = 2$ state of hydrogen, showing fine structure levels and their Zeeman sublevels. The energy difference at zero magnetic field between the $2\,^2S_{\frac{1}{2}}$ and $2\,^2P_{\frac{1}{2}}$ levels is the Lamb shift of about 1058 Mc/sec.

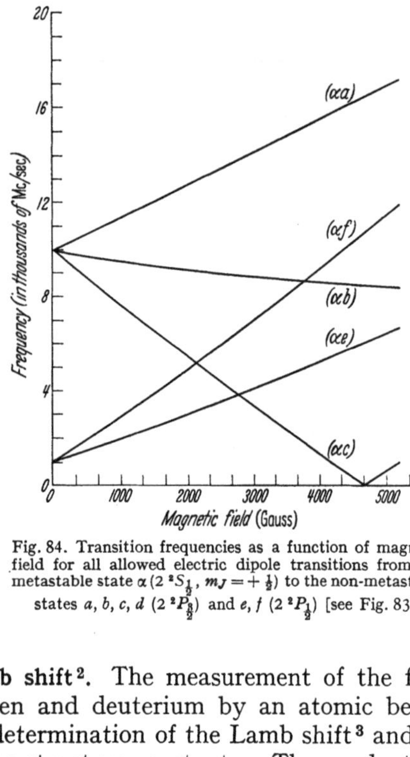

Fig. 84. Transition frequencies as a function of magnetic field for all allowed electric dipole transitions from the metastable state α ($2\,^2S_{\frac{1}{2}}$, $m_J = +\frac{1}{2}$) to the non-metastable states a, b, c, d ($2\,^2P_{\frac{3}{2}}$) and e, f ($2\,^2P_{\frac{1}{2}}$) [see Fig. 83].

42. Fine structure of hydrogen; Lamb shift[2]. The measurement of the fine structure of the $n = 2$ levels of hydrogen and deuterium by an atomic beam method led to the discovery and precise determination of the Lamb shift[3] and to the determination of the value of the fine structure constant α. The production and detection of the metastable $2^2S_{\frac{1}{2}}$ state of hydrogen have been discussed in Sects. 6 and 14, and the basic theory of the resonance line shape is given in Sect. 31.

An energy level diagram showing the Zeeman levels of the $n = 2$ states of hydrogen is given in Fig. 83. The displacement of the $^2S_{\frac{1}{2}}$ from the $^2P_{\frac{1}{2}}$ state at zero magnetic field is the Lamb shift (~ 1058 Mc/sec). The hfs of these states is not included in this diagram. The frequencies of the principal transitions studied are shown in Fig. 84 as a function of the magnetic field H_0. The tran-

[1] M. PHILLIPS: Phys. Rev. **76**, 1803 (1949).

[2] In this Encyclopedia, the Lamb shift has been discussed from the theoretical viewpoint by H. A. BETHE and E. E. SALPETER in Sect. 21 of their article (Vol. XXXV) and by G. KÄLLÉN, Sect. 37, in Vol. V/1.

[3] W. E. LAMB jr. and R. C. RETHERFORD: Part I, Phys. Rev. **79**, 549 (1950); Part II, Phys. Rev. **81**, 222 (1951); W. E. LAMB jr.: Part III, Phys. Rev. **85**, 259 (1952); W. E. LAMB jr. and R. C. RETHERFORD: Part IV, Phys. Rev. **86**, 1014 (1952); TRIEBWASSER, DAYHOFF and LAMB: Part V, Phys. Rev. **89**, 98 (1953); DAYHOFF, TRIEBWASSER and LAMB: Part VI, Phys. Rev. **89**, 106 (1953).

sitions are from the initial level $m_J = +\frac{1}{2}$ of the $^2S_{\frac{1}{2}}$ state to a final level of either the $^2P_{\frac{1}{2}}$ or the $^2P_{\frac{3}{2}}$ state. The selection rules for the electric dipole transitions in the high field region are $\Delta m_J = 0$ if the radiofrequency electric field, $\boldsymbol{E}_{rf}$, is parallel to $\boldsymbol{H}_0$ and $\Delta m_J = \pm 1$ if $\boldsymbol{E}_{rf}$ is perpendicular to $\boldsymbol{H}_0$. A transition from a $2S$ to a $2P$ level leads to a rapid decay to the $1S$ state (decay rate $= 6 \times 10^8$/sec) and is observed as a decrease in the intensity of the metastable atom beam reaching the detector. This method of detecting the transition does not depend on the observation of small deflections of the beam, and hence only a relatively poorly defined beam is required. The method of production of the metastable atom beam described in Sect. 6 is, therefore, adequate. Transitions

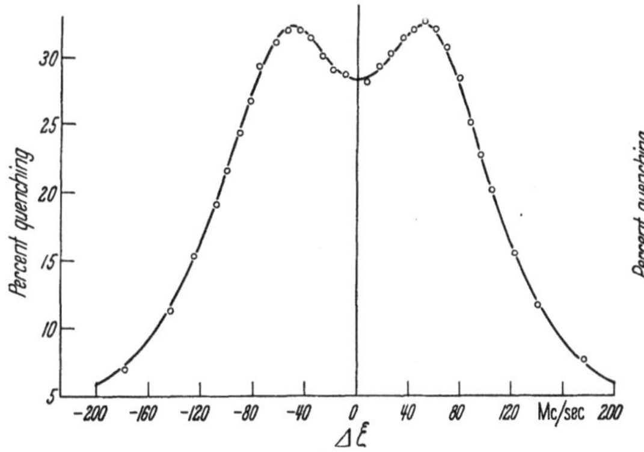

Fig. 85.

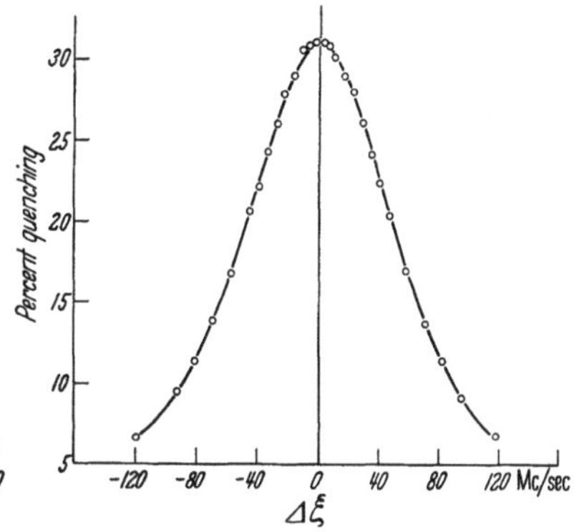

Fig. 86.

Fig. 85. Resonance curve for transition between states α and f of hydrogen (see Figs. 83 and 84) at 2395 Mc/sec. Data are taken as a function of magnetic field but are plotted for convenience against a frequency variable ξ, using the conversion $(d\xi/dH_0) = 1.9216$ Mc sec^{-1} gauss^{-1}. The center of the line occurs at $H_0 = 703.77$ gauss. The double peaks are due to the hfs interaction. The solid curve is the theoretical curve; the dots are experimental points.

Fig. 86. Resonance curve for transition between states α and f of deuterium (see Figs. 83 and 84) at 2395 Mc/sec. Data are taken as a function of magnetic field, but are plotted for convenience against a frequency variable ξ, using the conversion $(d\xi/dH_0) = 1.9216$ Mc sec^{-1} gauss^{-1}. The center of the line occurs at $H_0 = 704.57$ gauss. No splitting due to the hfs interaction is resolved. The solid curve is the theoretical curve; the dots are experimental points.

from the $m_J = -\frac{1}{2}$ level of the $2^2S_{\frac{1}{2}}$ state have not been studied. This level is observed from Fig. 83 to lie close to the $m_J = +\frac{1}{2}$ or the $m_J = -\frac{1}{2}$ level of the $^2P_{\frac{1}{2}}$ state for values of H_0 from 500 to 1500 gauss, which are used in the experiments. Hence due to the motional electric field $\boldsymbol{E} = (\boldsymbol{v} \times \boldsymbol{H}_0)/c$, arising from the velocity $\boldsymbol{v}$ of the atom through $\boldsymbol{H}_0$, a sufficiently large admixture of P state occurs to lead to the decay of this level to the $1S$ state. The complete elimination of this level, which was required in order to avoid the overlap of transitions from this level with transitions from the $m_J = +\frac{1}{2}$, $^2S_{\frac{1}{2}}$ level, can be achieved by the addition of a small static electric field of several volts/cm.

The line widths are determined primarily by the spontaneous radiative decay of the $2P$ states to the $1S$ state and amount to about 100 Mc/sec (Sect. 31). Because of the large breadth of the lines, a resonance curve cannot be taken in practice by varying the microwave frequency, but is obtained rather by varying H_0 with a fixed microwave frequency. A proton resonance probe was used to measure H_0 in the later experiments. Observed resonance curves for H and D are shown in Figs. 85 and 86. The resolved double peak in Fig. 85 arises because of the hfs interaction in hydrogen. The smaller hfs separation for deuterium is not resolved in Fig. 86.

Analysis of the resonance curves together with the theory of the Zeeman effect yields the separation between $2^2S_{\frac{1}{2}}$, $2^2P_{\frac{1}{2}}$, and $2^2P_{\frac{3}{2}}$ levels at zero magnetic field. A detailed theory of the line shape and of the effects of external magnetic

Table 12. *Experimental and theoretical values of the Lamb shift.*

Energy interval	Experimental value (Mc/sec)	Reference (experimental)	Theoretical value (Mc/sec)	Reference (theoretical)
H, $2^2S_{\frac{1}{2}}-2^2P_{\frac{1}{2}}$	1057.77(10)	53 *T1*	1057.99(13)	57 *S4*; 53 *S1*
D, $2^2S_{\frac{1}{2}}-2^2P_{\frac{1}{2}}$	1059.00(10)	53 *T1*	1059.23(13)	57 *S4*; 53 *S1*
He^+, $2^2S_{\frac{1}{2}}-2^2P_{\frac{1}{2}}$	14040.2 (45)	57 *L4*	14056.8 (30)	57 *S4*; 57 *L4*; 53 *S1*

and electric fields and of hyperfine structure on the fine structure levels has provided the basis for the determination of the fine structure splittings to 0.1 Mc/sec. This accuracy corresponds to choosing the resonance frequency to $^1/_{1000}$ of the line width. The experimentally determined fine structure splittings for H, D and He^+ are given in Table 12. The uncertainties which are given in parentheses apply to the last figure of the values.

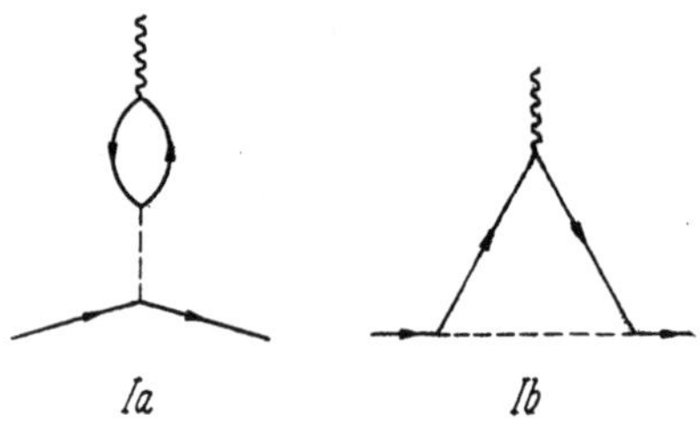

Fig. 87. Feynman diagrams for radiative corrections contributing to the Lamb shift of lowest order $\alpha^3 r_y$. Solid lines indicate electrons and dashed lines indicate photons or the external field. Diagram (Ia) is a vacuum polarization correction and diagram (Ib) is a correction associated with the emission and reabsorption of a virtual photon.

The theory of the Lamb shift is based on the modern renormalized, covariant theory of quantum electrodynamics, and, indeed, the discovery and precise determination of the Lamb shift have provided a principal stimulus and test for this theory. The Lamb shift in hydrogen is due to the contribution of virtual radiative processes to the energy levels of an electron in the external Coulomb field of the proton, and it is of the order $\alpha^3 r_y$. Feynman diagrams for the radiative processes of this order are shown in Fig. 87. The result to lowest order in α is[1]:

$$W(2^2S_{\frac{1}{2}}) - W(2^2P_{\frac{1}{2}}) = \left(\frac{\alpha^3 r_{y\infty}}{3\pi}\right)\left\{\left[\ln\frac{m c^2}{k_0(2,0)} - \ln 2 + \frac{5}{6}\right] - \frac{1}{5} - \left[\ln\frac{r_{y\infty}}{k_0(2,1)} - \frac{3}{8}\right]\right\} \tag{42.1}$$

in which $k_0(2,0)$ and $k_0(2,1)$ are "average excitation energies". The first term in square brackets comes from the shift of the $2^2S_{\frac{1}{2}}$ level associated with the Feynman diagram (Ib) of Fig. 87. The term involving $(-\frac{1}{5})$ is the contribution of vacuum polarization to the shift of the $2^2S_{\frac{1}{2}}$ level and arises from diagram (Ia) of Fig. 87. The last term in square brackets comes from the shift of the $2^2P_{\frac{1}{2}}$ level. Higher order quantum electrodynamic effects and effects associated with the finite mass of the nucleus must also be included in the theory. The theoretical values for H and D are given in Table 12. The agreement between the experimental and theoretical values is very close for both H and D. The experimental and theoretical values for the Lamb shift in singly ionized helium fail to agree by twice the sum of the uncertainties in the two values[2].

The fine structure separation between the $2^2P_{\frac{3}{2}}$ and $2^2P_{\frac{1}{2}}$ states of deuterium is used to determine[3] the value of the fine structure constant α. From the energy

[1] E.E. SALPETER: Phys. Rev. **89**, 92 (1953). References to the theoretical work on the Lamb shift are given in this paper.

[2] E. LIPWORTH and R. NOVICK: Phys. Rev. **108**, 1434 (1957). The measurement of the Lamb shift in singly ionized He did not involve a beam technique and hence is not discussed in this article.

[3] DAYHOFF, TRIEBWASSER and LAMB: Phys. Rev. **89**, 106 (1953). Deuterium is used rather than hydrogen primarily because of its smaller hfs.

separations found between the pairs of states $2^2S_{\frac{1}{2}}$ and $2^2P_{\frac{1}{2}}$ and $2^2S_{\frac{1}{2}}$ and $2^2P_{\frac{3}{2}}$, the separation between the $2^2P_{\frac{3}{2}}$ and $2^2P_{\frac{1}{2}}$ states is found to be 10971.59 ± 0.20 Mc/sec. The frequency separation is given by the theoretical formula:

$$\nu_{\frac{3}{2}\leftrightarrow\frac{1}{2}} = \frac{c\, r_{yD}}{16}\,\alpha^2 \left\{\left[1 + \frac{5}{8}\,\alpha^2\right] + \left[\frac{\alpha}{\pi} - \frac{0.656}{\pi^2}\,\alpha^2\right]\right\}. \tag{42.2}$$

The first bracketed term arises from the Dirac theory and the second bracketed term arises from quantum electrodynamic effects. The constant α can be computed from $\nu_{\frac{3}{2}\leftrightarrow\frac{1}{2}}$ and known values of c and r_{yD} (r_{yD} = Rydberg constant for D in cm^{-1}). The value of α that is obtained[1] is $1/\alpha = (137.0390 \pm 0.0012)$.

C. Molecular spectroscopy.

43. Introduction. The radiofrequency and microwave spectra of many molecules, principally diatomic, have been studied with high resolution by the molecular beam magnetic and electric resonance methods. The observations fall naturally into three groups:

1. Observations of the unresolved spectra of molecules by the magnetic resonance method. The temperatures at which beams of many molecules, notably the alkali halides, are produced are sufficiently high to give a significant population of states with a high rotational quantum number, J, and sometimes with a relatively high vibrational quantum number, v. Since the magnetic deflection of molecules depends primarily on their nuclear magnetic moments, molecules with all the different J and v values are refocussed. The spectrum then consists of a multiplicity of unresolved lines arising from the different J and v states and only the envelope of the lines is observable.

2. Observation of the resolved spectra of diatomic molecules by the electric resonance method. Since the electric deflection of polar diatomic molecules depends on the rotational state J, molecules of a single J value can be selected. The spectra of single rotational states with low J values have been studied for many alkali halide molecules.

3. Observations of the resolved spectra of diatomic molecules by the magnetic resonance method. If the molecular beam can issue from a source at a sufficiently low temperature so that only the lowest few rotational states are populated to a significant extent, as is the case for the hydrogen molecules and for oxygen, then transitions involving a single rotational and vibrational state can be studied.

Those features of molecular spectra which reveal information about nuclei have been studied most intensively. Transitions between low rotational states of alkali halide molecules have been studied, and from the observed resonance frequencies values are obtained for the ratios of atomic masses with an accuracy comparable to that from mass spectroscopic measurements. The study of hyperfine structure in molecular spectra, which arises from the interactions of the nuclear magnetic dipole moment and the nuclear electric quadrupole moment with the molecular fields, yields values for nuclear-molecular interaction constants. From a knowledge of the electric quadrupole interaction in the hydrogen molecules, the electric quadrupole moment of the deuteron has been obtained after calculation of the electric field gradient at the deuteron nucleus. For more

[1] COHEN, DUMOND, LAYTON and ROLLETT: Rev. Mod. Phys. **27**, 363 (1955). From an analysis of all data on the atomic constants they conclude $1/\alpha = 137.0373 \pm 0.0006$. Cf. also Vol. XXXV, this Encyclopedia. This value as well as the one quoted by DAYHOFF *et al.* are in error through an error in the fourth-order radiative correction to the magnetic dipole moment of the electron. See C.M. SOMMERFIELD: Phys. Rev. **107**, 328 (1957).

complicated molecules such a computation is difficult to perform reliably, but the ratio of nuclear electric quadrupole moments for two isotopes is obtained from values of the electric quadrupole interaction constants in two isotopic molecules. The use of an external magnetic field allows a determination of nuclear magnetic moments.

From the viewpoint of molecular physics a great deal of detailed information is available from the molecular beam experiments, particularly on the alkali halide and hydrogen molecules. The rotational and rotational-vibrational constants associated with the model of a vibrating rotator are obtained with high precision from the frequencies of rotational state transitions for several vibrational states observed for alkali halide molecules. The electric quadrupole interaction constant gives important information with regard to molecular binding[1]. This information is most valuable in cases where the nuclear electric quadrupole moment is known from independent studies of atomic spectra, so that the electric field gradient at the nucleus in the molecule can be determined. The measured magnetic hyperfine structure interaction, or the spin-rotation interaction (a term of the form $c\boldsymbol{I}\cdot\boldsymbol{J}$), together with a value of the nuclear magnetic dipole moment, allows calculation of the molecular magnetic field at the nucleus. The value of c for H_2 has been used to evaluate a diamagnetic shielding correction involved in the determination of the proton magnetic moment. The use of an external magnetic field allows a determination of rotational magnetic moments and of related magnetic shielding effects while an external electric field allows the measurement of the electric dipole moments of polar diatomic molecules. Molecular constants for the hydrogen molecules H_2, HD, and D_2 have been determined with unusually high precision, and their theory has been treated in much detail.

44. Hamiltonian for a diatomic molecule. The present section gives the basic Hamiltonian for a diatomic molecule in external electric and magnetic fields. This Hamiltonian has been found adequate for the understanding of all observed radiofrequency and microwave spectra for alkali halide molecules, which have been the principal subject of investigation in both electric resonance experiments and in magnetic resonance experiments with unresolved spectra. The theory of the hydrogen molecules is summarized in a separate section (Sect. 55), because of the greater detail of the experimental studies and the greater completeness of the theory for these simple molecules.

The molecule is assumed to be in a $^1\Sigma$ electronic state, which is the ground state of an alkali halide molecule. The part of the Hamiltonian which does not involve the electronic motion is[2,3] [*25, 28*];

$$\left.\begin{aligned}\mathscr{H} = F_{v,J}^{\mathrm{op}} &- e q_1 Q_1 \frac{3(\boldsymbol{I}_1\cdot\boldsymbol{J})^2 + \frac{3}{2}(\boldsymbol{I}_1\cdot\boldsymbol{J}) - \boldsymbol{I}_1^2\boldsymbol{J}^2}{2I_1(2I_1-1)(2J-1)(2J+3)} - \\ &- e q_2 Q_2 \frac{3(\boldsymbol{I}_2\cdot\boldsymbol{J})^2 + \frac{3}{2}(\boldsymbol{I}_2\cdot\boldsymbol{J}) - \boldsymbol{I}_2^2\boldsymbol{J}^2}{2I_2(2I_2-1)(2J-1)(2J+3)} + c_1\boldsymbol{I}_1\cdot\boldsymbol{J} + c_2\boldsymbol{I}_2\cdot\boldsymbol{J} + \\ &+ \frac{g_{I_1}g_{I_2}\mu_0^2}{R^3}\,\frac{2\left[3(\boldsymbol{I}_1\cdot\boldsymbol{J})(\boldsymbol{I}_2\cdot\boldsymbol{J}) - \boldsymbol{I}_1\cdot\boldsymbol{I}_2\boldsymbol{J}^2\right]}{(2J+3)(2J-1)} - \\ &- \boldsymbol{\mu}\cdot\boldsymbol{E} + \mu_0 g_{I_1}\boldsymbol{I}_1\cdot\boldsymbol{H}_0 + \mu_0 g_{I_2}\boldsymbol{I}_2\cdot\boldsymbol{H}_0 + \mu_0 g_J\boldsymbol{J}\cdot\boldsymbol{H}_0.\end{aligned}\right\}\quad(44.1)$$

The term $F_{v,J}^{\mathrm{op}}$ represents the Hamiltonian of a vibrating rotator [*28*]. Its eigenvalues have been calculated[4] for any potential which can be expanded as

[1] C.H. TOWNES and B.P. DAILEY: J. Chem. Phys. **17**, 782 (1949).
[2] V.W. HUGHES and L. GRABNER: Phys. Rev. **79**, 829 (1950).
[3] LEE, FABRICAND, CARLSON and RABI: Phys. Rev. **91**, 1395 (1953).
[4] J.L. DUNHAM: Phys. Rev. **41**, 721 (1932).

a series of powers of $(R-R_e)$ in the neighborhood of the potential minimum at $R=R_e$, where R is the internuclear distance. The energy eigenvalues can be expressed as:

$$F_{v,J}=\sum_{l,k=0}^{\infty} Y_{lk}(v+\tfrac{1}{2})^l J^k(J+1)^k \tag{44.2}$$

where l and k are summation indices, v and J are the vibrational and rotational[1] quantum numbers, respectively, and Y_{lk} are coefficients which depend on molecular constants. The Y_{lk}'s are related to the constants conventionally used in the analysis of optical band spectra by the approximate expressions [28]:

$$\left.\begin{array}{lll} Y_{10}\simeq\omega_e, & Y_{02}\simeq -D_e, & Y_{11}\simeq-\alpha_e, \\ Y_{01}\simeq B_e, & Y_{20}\simeq\omega_e x_e, & Y_{21}\simeq\gamma_e \end{array}\right\} \tag{44.3}$$

where the symbols on the right side refer to the coefficients in the usual expansion for the vibrational and rotational energies of a molecule:

$$F_{v,J}=\omega_e(v+\tfrac{1}{2})-\omega_e x_e(v+\tfrac{1}{2})^2+\cdots+B_v J(J+1)-D_e J^2(J+1)^2+\cdots \tag{44.4}$$

in which $\omega_e=h\times$(vibration frequency); x_e is the anharmonic coefficient;

$$B_v=B_e-\alpha_e(v+\tfrac{1}{2})+\gamma_e(v+\tfrac{1}{2})^2, \tag{44.4a}$$

$B_v[=\hbar^2/(2M_r R_v^2)]$ is the rotational constant for the vibrational state v; $M_r[=M_1M_2/(M_1+M_2)]$ is the reduced mass of the rotator with masses M_1 and M_2; $B_e[=\hbar^2/(2M_r R_e^2)]$ is the rotational constant corresponding to the equilibrium separation R_e; α_e is the vibrational-rotational interaction constant; γ_e is a higher order anharmonic vibrational-rotational interaction coefficient; $D_e(=4B_e^3/\omega_e^2)$ is the rotational stretching constant. For the present accuracy of the experiments it has been adequate to use a perturbation theory which treats $F_{v,J}^{\text{op}}$ as the dominant term in the Hamiltonian, and hence to consider v and J as good quantum numbers.

The second and third terms in the Hamiltonian of Eq. (44.1) represent the electric quadrupole interactions of the two nuclei with spins I_1 and I_2. Q_1 and Q_2 are the electric quadrupole moments of the nuclei, as defined by footnote 3 on p. 86 referring to Eq. (35.10). The quantities q_1 and q_2, which refer to nucleus 1 and 2 are defined by:

$$q=\left\langle\frac{\partial^2 V}{\partial z^2}\right\rangle-\left\langle\frac{\varrho(3\cos^2\vartheta-1)}{r^3}\right\rangle \tag{44.5}$$

where V is the electrostatic potential produced at the nucleus 1 or 2 by all charges outside of the nucleus 1 or 2; ϱ is the charge density; r is the distance from the nucleus to the element of charge $\varrho\, d\tau$; ϑ is the angle between $\boldsymbol{r}$ and the z axis which is fixed with respect to the molecule (the internuclear axis). The quantity q may vary with v and J. It is related to eq_J in footnote 2, p. 86 through:

$$e\,q_J=-q\,\frac{J}{2J+3}\,. \tag{44.5a}$$

The terms in the Hamiltonian of Eq. (44.1) for the quadrupole interactions give correct results only for matrix elements diagonal in J and I, but for the present experimental accuracies off-diagonal elements can be neglected.

[1] Throughout this discussion of molecular spectroscopy J refers to the rotational angular momentum of the diatomic molecule. In the previous discussion of atomic spectroscopy J referred to the total electronic angular momentum of the atom.

The fourth and fifth terms are the magnetic hyperfine structure interactions or the spin-rotation interactions between the nuclear spins and the rotational angular momentum. The parameters c_1 and c_2 may vary slightly with the rotational quantum number. The sixth term is the magnetic dipole-dipole interaction between the two nuclei, in which R is the internuclear distance, g_{I_1} and g_{I_2} are the g-values for the two nuclei.

The term $-\boldsymbol{\mu}\cdot\boldsymbol{E}$ is the interaction of the electric dipole moment μ of the molecule with an externally applied electric field $\boldsymbol{E}$.

$$\mu \equiv \left\langle \sum_{i=1}^{n} - e\,\boldsymbol{r}_i + \sum_{j=1}^{2} e Z_j \boldsymbol{R}_j \right\rangle \tag{44.6}$$

in which $-e$ is the electron charge; $\boldsymbol{r}_i$ is the position vector of electron with respect to center of mass of the molecule; Z_j is the atomic number of the j-th nucleus; $\boldsymbol{R}_j$ is the position vector of the j-th nucleus with respect to center of mass of the nucleus. The expectation value is taken for the ground electronic state of the molecule with the nuclei fixed at their equilibrium positions[1]. The last three terms in Eq. (44.1) give the interactions of the nuclear magnetic moments with g values g_{I_1}, and g_{I_2} and the molecular rotational magnetic moment with g value g_J with an external magnetic field $\boldsymbol{H}_0$.

The energy levels determined by the Hamiltonian of Eq. (44.1) can be computed by perturbation theory. The principal term is $F_{v,J}^{\text{op}}$, which has energy eigenvalues given by Eq. (44.2) and eigenstates defined by the vibrational and rotational quantum numbers. The contribution to the energy of the internal nuclear-molecular interactions and of the interaction of the molecule with external fields is obtained by first order perturbation theory using a zeroth order eigenstate of $F_{v,J}^{\text{op}}$ defined by the quantum numbers v, J. The coupling scheme of the three vectors $\boldsymbol{J}$, $\boldsymbol{I}_1$ and $\boldsymbol{I}_2$, and the useful approximations for calculating the energy level splittings are determined by the relative magnitudes of the various internal nuclear-molecular interactions and of the interactions with the external fields. Various useful cases are discussed in the following sections.

Higher order effects involve the admixture of other electronic states or other rotational states. A mixing of higher electronic states arises from the interaction of the nuclear spins with the components of molecular electronic angular momentum perpendicular to the internuclear axis, and in second order perturbation theory leads to an energy term with the same dependence on nuclear spin as the electric quadrupole interaction. This "pseudo-quadrupole" effect is contained in the observed interaction constants but is generally negligible[2]. A mixing of other rotational states occurs[3] due to the interaction of the electric dipole moment of the molecule with an external electric field.

I. Unresolved molecular spectra by magnetic resonance method.

a) High field case.

45. Magnitude of nuclear g values. The earliest application of the use of molecular beam techniques to spectroscopy was in the determination of nuclear g values[4]. If a diatomic molecule in a $^1\Sigma$ state is placed in a magnetic field which is sufficiently high so that the interaction energy of the nuclear magnetic dipole

[1] The electric dipole moment $\boldsymbol{\mu}$ reduces to a single component along the internuclear axis.
[2] H.M. FOLEY: Phys. Rev. **72**, 504 (1947).
[3] U. FANO: J. Res. Nat. Bur. Stand. **40**, 215 (1948).
[4] RABI, MILLMAN, KUSCH and ZACHARIAS: Phys. Rev. **55**, 526 (1939).

moments with the applied field is much greater than the interaction energy of the nuclear spin and the molecular rotation, either through the interaction of the nuclear magnetic moment with the magnetic field produced by the molecular rotation or through the interaction of the nuclear electric quadrupole moment with the gradient of the electric field at the nucleus, then the appropriate strong field quantum numbers are the magnetic quantum numbers m_{I_1}, m_{I_2}, and m_J, referred to the direction of the applied field. If all internal nuclear-molecular interactions are neglected, then the energies of the levels are:

$$W = (m_{I_1} g_{I_1} + m_{I_2} g_{I_2} + m_J g_J)\,\mu_0 H_0. \tag{45.1}$$

An oscillating magnetic field $\boldsymbol{H}_1 \cos\omega t$ produces the time-dependent Hamiltonian term:

$$V = \mu_0 g_{I_1} \boldsymbol{I}_1 \cdot \boldsymbol{H}_1 \cos\omega t + \mu_0 g_{I_2} \boldsymbol{I}_2 \cdot \boldsymbol{H}_1 \cos\omega t + \mu_0 g_J \boldsymbol{J} \cdot \boldsymbol{H}_1 \cos\omega t. \tag{45.2}$$

If $\boldsymbol{H}_1$ is perpendicular to $\boldsymbol{H}_0$, selection rules for transitions induced by V are:

$$\Delta m_{I_1} = \pm 1, \quad \text{or} \quad \Delta m_{I_2} = \pm 1, \quad \text{or} \quad \Delta m_J = \pm 1. \tag{45.3}$$

Only one of the quantities m_{I_1}, m_{I_2}, m_J may change while the other two remain fixed. Molecular rotational magnetic moments are small, and the transitions $\Delta m_J = \pm 1$ have not been seen in studies of unresolved spectra. The transitions observed with $\Delta m_{I_1} = \pm 1$ or $\Delta m_{I_2} = \pm 1$ have the characteristic frequencies:

$$\nu_1 = g_{I_1} \mu_0 H_0/h, \quad \nu_2 = g_{I_2} \mu_0 H_0/h. \tag{45.4}$$

For a homonuclear molecule a single magnetic quantum number m_I, associated with the total nuclear spin, together with m_J characterize the system. The energy levels are

$$W = (m_I g_I + m_J g_J)\,\mu_0 H_0, \tag{45.1a}$$

and the frequency of the transition observed is

$$\nu = g_I \mu_0 H_0/h. \tag{45.4a}$$

For a polyatomic molecule under strong field conditions a number of transition frequencies, each of the form $\nu_i = g_{I_i} \mu_0 H_0/h$ are observed.

The internal nuclear-molecular interactions will modify the energy levels given by Eqs. (45.1) and (45.1a). However, at sufficiently high fields, the spectrum which appears for a transition $\Delta m_I = \pm 1$ is symmetrical about the nuclear resonance frequency $g_I \mu_0 H_0/h$, provided a superposition of transitions involving equally positive and negative values of m_I and m_J is obtained (see Sect. 47). Even in the absence of any prior knowledge of the magnitude of the internal interactions, criteria can be established for the existence of strong field conditions. One is that the ratio of the resonance frequency and the magnetic field should be independent of magnetic field. Another is that at a fixed magnetic field the resonance frequency associated with a particular nucleus should be independent of the molecular species in which it is observed.

A number of nuclear g values have been determined by this method. Most of the observations were made by measurements of the central frequency of the resonance line in a magnetic field calibrated by classical procedures[1,2] [*25*]. In all cases the resonance frequency of a particular nucleus was measured in several different molecules and over a range of magnetic field, and in no case was there

[1] S. Millman and P. Kusch: Phys. Rev. **60**, 91 (1941).
[2] J.B.M. Kellogg and S. Millman: Rev. Mod. Phys. **18**, 323 (1946).

any significant deviation of the ratio of the frequency and the magnetic field. Subsequent determinations of nuclear g values by the nuclear resonance absorption and induction methods have given results in complete agreement with those obtained by the molecular beam magnetic resonance method [24]. In certain cases, notably that of iodine, no resonance line has been observed by molecular beam spectroscopy, and it may be inferred that the internal nuclear-molecular interactions are so great in molecules containing iodine that the spectrum in the neighborhood of the frequency $g_I \mu_0 H_0/h$ is very broad and no useful intensity occurs at any one frequency.

Since the detection of a molecular beam is particularly efficiently and simply achieved by ionization on a surface when the molecule contains an alkali atom, such molecules are useful for the determination of nuclear g-values, provided they are stable at temperatures required to produce a reasonable beam intensity. An important example is the use of KOH and NaOH for the determination of the proton magnetic moment[1]. Even though a rather complex polyatomic molecule may be used, such as $NaCl \cdot AlCl_3$ for the measurement of g(Al), for which no adequate theory exists for the details of the spectrum including internal nuclear-molecular interactions, the center of the resonance line at strong fields gives a good value of the nuclear g-factor.

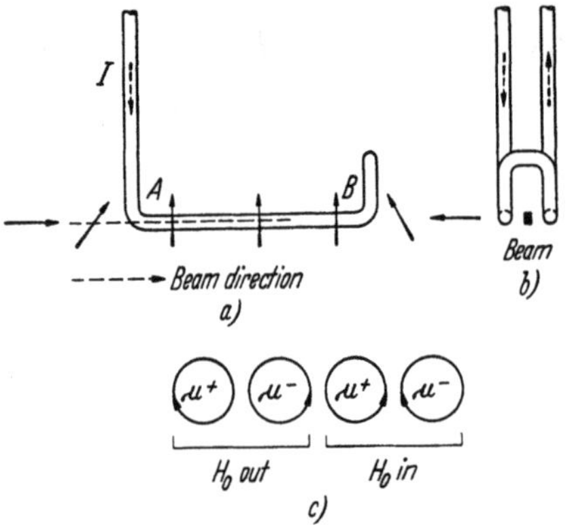

Fig. 88. Schematic diagram of the two-wire oscillating field indicating the changes in direction of the oscillating field along the beam trajectory. The sense of precession of a nuclear moment in the static magnetic field H_0 is indicated.

46. Sign of nuclear g value. An asymmetry in the nuclear resonance curve, which can arise from the change in direction of the radiofrequency field near its end regions and which is dependent on the direction of the applied static magnetic field $\boldsymbol{H}_0$, may be used to determine the sign of the nuclear magnetic moment[2]. Consider the radiofrequency circuit shown in Fig. 88. The beam moves between the two parallel conductors as shown. The radiofrequency magnetic field has a predominantly vertical component over the length AB, but the field direction changes beyond AB as indicated. For the moving molecules the field appears to rotate in a counterclockwise sense. This sense of rotation is unchanged when the direction of the current in the wires is reversed. The vertical oscillating field throughout the region AB can be resolved into two components, one rotating clockwise and the other rotating counterclockwise. The component effective in causing transitions is the one which rotates in the same sense as the precession of the nuclear magnetic moment about the field $\boldsymbol{H}_0$ (Sect. 28), which is determined by the sign of the moment and the direction of the field $\boldsymbol{H}_0$. The directions of precession of the moment μ are shown in Fig. 88 for the different signs of μ and directions of $\boldsymbol{H}_0$.

In the end regions beyond AB where the radiofrequency field direction changes, the rotating field components seen by the moving molecules will have

[1] H. TAUB and P. KUSCH: Phys. Rev. **75**, 1481 (1949). This experiment actually determined the ratio g_p/g_J, in which g_J is the electronic g value of an atom. The zero moment property of an atom at certain magnetic field values is used in order to obtain comparable deflections for both the atoms and the molecules.

[2] S. MILLMAN: Phys. Rev. **55**, 628 (1939).

frequencies different from the applied frequency. For the case shown in Fig. 88 the field component which rotates counterclockwise will have a higher frequency, and the clockwise component a lower frequency. If the nuclear magnetic moment is positive and $\boldsymbol{H}_0$ is directed out from the paper, so that μ precesses clockwise about $\boldsymbol{H}_0$, then to produce a transition in the end regions the applied frequency must be greater than that required to produce the transition in the central region AB. Hence a resonance curve observed at a fixed field $\boldsymbol{H}_0$ and at a sufficiently high radiofrequency amplitude will show an auxiliary maximum, possibly unresolved, at a frequency greater than $\nu_I = g_I \mu_0 H_0/h$. If the direction of $\boldsymbol{H}_0$ is reversed so that it points into the paper in Fig. 88, then the auxiliary maximum will appear on the low frequency side of ν_I. Correspondingly, if the frequency is held fixed, and the resonance curve is observed by varying the field $\boldsymbol{H}_0$, then for $\boldsymbol{H}_0$ out and a positive moment, an auxiliary maximum appears on the low field side of $H_0 = \nu_I h/(g_I \mu_0)$. If the nuclear magnetic moment is negative, the position of the auxiliary maximum with respect to the central resonance is reversed. In Fig. 89 is shown a pair of resonance curves for Li^7 in the molecule Li_2 where the auxiliary maxima clearly appear. The case of Li_2 is particularly favorable for the exhibition of the maxima since the internal interactions in this molecule are small and the line has a small half-width. In Fig. 90 is shown a pair of resonance curves for Li^6 in LiCl, where there is no marked asymmetry in the line but only a shift of the central frequency of the line as the field is reversed.

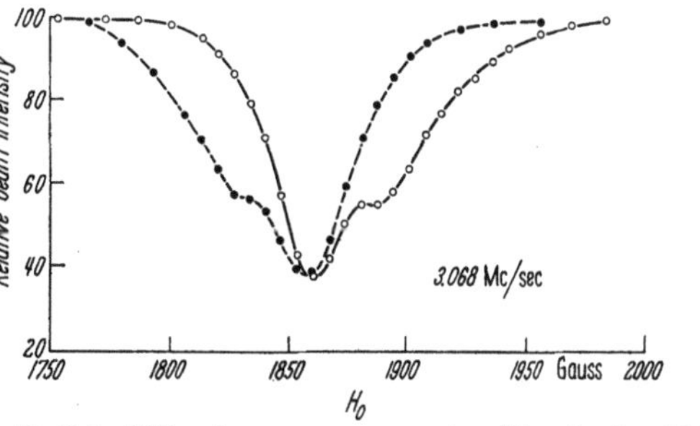

Fig. 89. Pair of Li^7 nuclear resonance curves from Li_2 molecules, which determine the sign of the nuclear magnetic moment of Li^7. The two curves correspond to opposite directions of the static magnetic field $\boldsymbol{H}_0$. For the solid curve if the velocity is in the positive x direction of a right-handed coordinate system, $\boldsymbol{H}_0$ is in the positive z direction. For the dashed curve the direction of $\boldsymbol{H}_0$ is reversed.

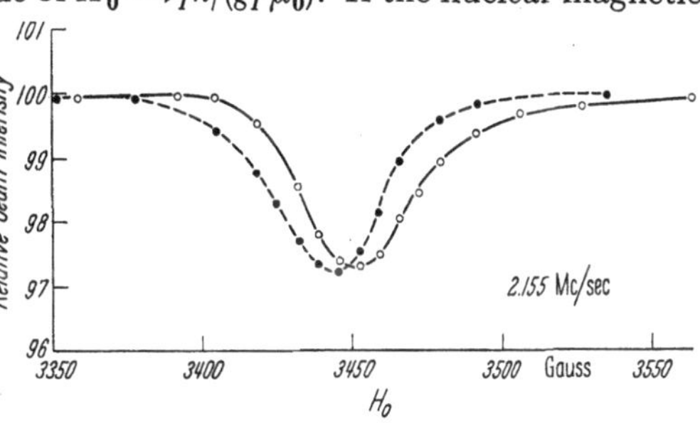

Fig. 90. Pair of Li^6 nuclear resonance curves from LiCl molecules, which determine the sign of the nuclear magnetic moment of Li^6. The solid and dashed curves have the same meaning as in Fig. 89.

For the purpose of determining the sign of a nuclear moment, the line asymmetry can be made large by use of a radiofrequency circuit in which the end regions are large compared to the central region where the oscillating field has a fixed direction. The signs of the considerable number of magnetic moments first measured by the molecular beam magnetic resonance method have been found by the method described in this section[1].

47. Electric quadrupole and spin-rotation interactions. Early observations showed that internal nuclear-molecular interactions play a significant role in determining the intensity distribution within a nuclear resonance curve. The resonance lines always have widths greater than the natural resolution width of the apparatus and sometimes exhibit auxiliary maxima. Thus, for example,

[1] J. B. M. KELLOGG and S. MILLMAN: Rev. Mod. Phys. **18**, 323 (1946).

for Na in Na_2 and for K in K_2 the nuclear resonance line consists of a central maximum accompanied by two or more symmetrically spaced satellites whose frequency separation from the central peak is independent of the magnetic field H_0 at sufficiently high fields[1]. It was tentatively proposed that these satellites arise from the electrostatic interaction of a nuclear electric quadrupole moment with the rest of the molecule.

Theoretical calculations indicated that such satellite peaks are indeed to be expected[2]. A detailed investigation was made of the energy levels and resonance line shapes for a diatomic molecule in which one nucleus has an electric quadrupole moment and the second nucleus has zero spin. The calculations were made for the large molecular rotational quantum numbers which are characteristic of the alkali halides at the temperatures at which they have useful vapour pressures, and an important element in the theory is the effect of the statistical distribution of the molecules among states of various rotational angular momenta on the intensity distribution within the envelope of lines arising from the states of various J and m_J. The assumption that the spin of the second nucleus in the molecule is zero does not detract from the generality of the treatment at high magnetic fields. If both nuclei have electric quadrupole moments, the energy eigenvalue will simply involve the sum of the electric quadrupole energies of the individual nuclei. For a homonuclear diatomic molecule the nuclear magnetic resonance spectrum, which includes the effect of the electric quadrupole interactions, is the superposition of the spectra associated with the individual nuclei, provided that the rotational angular momentum is large[3]. The theory must include the spin-rotation interactions to explain the observed spectra[4].

Consider the special case of the general Hamiltonian of Eq. (44.1) in which one of the two nuclei is assumed to have zero spin.

$$\mathscr{H} = \mu_0 g_I \boldsymbol{I}\cdot\boldsymbol{H}_0 + \mu_0 g_J \boldsymbol{J}\cdot\boldsymbol{H}_0 + c\,\boldsymbol{I}\cdot\boldsymbol{J} - eqQ\,\frac{[3(\boldsymbol{I}\cdot\boldsymbol{J})^2 + \frac{3}{2}\boldsymbol{I}\cdot\boldsymbol{J} - \boldsymbol{I}^2\boldsymbol{J}^2]}{2I(2I-1)(2J+3)(2J-1)}. \tag{47.1}$$

Under strong field conditions the (m_I, m_J) representation is appropriate. In first order perturbation theory only the diagonal matrix elements of the internal nuclear-molecular interactions are used, and the energy eigenvalues are:

$$\left.\begin{aligned} W_{m_I,m_J} &= \mu_0 g_I H_0 m_I + \mu_0 g_J H_0 m_J + c_2 m_I m_J - \\ &\quad - eqQ\,\frac{[3m_I^2 - I(I+1)]\,[3m_J^2 - J(J+1)]}{4I(2I-1)(2J+3)(2J-1)}. \end{aligned}\right\} \tag{47.2}$$

Since the most probable values of J for the molecules are high[5], the approximation is made that $m_J/J (\equiv z)$ is a continuously varying parameter, all of whose values are equally probable within the interval $-1 \leqq z \leqq 1$. Eq. (47.2) can then be written

$$W_{m_I,m_J} = \mu_0 g_I H_0 m_I + \mu_0 g_J H_0 z J + c m_I z J - \frac{eqQ}{16}(3z^2-1)\frac{[3m_I^2 - I(I+1)]}{I(2I-1)}. \tag{47.2a}$$

The frequency of a line for a transition $m_I \leftrightarrow m_I - 1$, $\Delta m_J = 0$ is:

$$\nu = \frac{W_{m_I,m_J} - W_{m_I-1,m_J}}{h} = \frac{\mu_0 g_I H_0}{h} + \frac{czJ}{h} - \frac{3eqQ}{16h}\,\frac{(3z^2-1)(2m_I-1)}{I(2I-1)}. \tag{47.3}$$

[1] Kusch, Millman and Rabi: Phys. Rev. **55**, 1176 (1939).
[2] B.T. Feld and W.E. Lamb jr.: Phys. Rev. **67**, 15 (1945).
[3] H.M. Foley: Phys. Rev. **71**, 747 (1947).
[4] W.A. Nierenberg and N.F. Ramsey: Phys. Rev. **72**, 1075 (1947).
[5] At the temperatures at which beams of useful intensity are obtained, the most probable values of J are between approximately 15 and 100.

The observed line will be the superposition of transitions due to all the states present in the beam. The natural width of the line and the width due to field inhomogeneity are usually negligible compared to the width arising from the internal nuclear-molecular interactions. Furthermore, the transition probabilities for transitions $\Delta m_I = \pm 1$, $\Delta m_J = 0$ are independent of m_J, because the transition matrix element of V, which is the interaction with the radiofrequency field given in Eq. (45.2), does not involve m_J. Hence the line shape for any particular J, $I_J(\nu)$, is given by

$$I_J(\nu) = D_J(z) \tag{47.4}$$

where $D_J(z)$ is the density of transitions and is equal to $|dz/d\nu|$ which is evaluated from Eq. (47.3) and may be expressed as a function of ν through the same equation. In general D_J will be a function of J through the term czJ/h. The total intensity at ν will be proportional to

$$I(\nu) = \int I_J(\nu)\, 2J\, a^2 \exp(-a^2 J^2)\, dJ \tag{47.5}$$

where the limits of integration are taken to cover the range of J that can contribute to the intensity at the frequency ν and $a^2 = \hbar^2/(2AkT)$ where A is the moment of inertia of the molecule. Under our assumptions,

$$N(J)\, dJ = 2J\, a^2 \exp(-a^2 J^2)\, dJ \quad \text{and} \quad \int_0^\infty N(J)\, dJ = 1.$$

When $m_I = \frac{1}{2}$ the term involving eqQ in Eq. (47.3) is zero. If $\nu_I = g_I \mu_0 H_0/h$, then $\nu = \nu_I + czJ/h$ and $I_J(\nu) = h/(cJ)$. At the frequency ν, all J from $\frac{h}{c}|\nu - \nu_I|$ to ∞ contribute to the intensity and:

$$I(\nu) = \frac{h a \sqrt{\pi}}{2c}\left[1 - \Phi\left(\frac{h a |\nu - \nu_I|}{c}\right)\right] \tag{47.6}$$

where Φ is the error function.

Since positive and negative values of z are equally probable, the line is symmetrical about the frequency ν_I. The width of the envelope at half intensity is:

$$\delta\nu = \frac{0.954\, c}{a h}. \tag{47.7}$$

When the spin $I = \frac{1}{2}$, this line comprises the complete nuclear reorientation spectrum. Since the spin-rotation interaction constant c is usually small, strong field conditions are achieved with rather small values of H_0. When I is an odd multiple of $\frac{1}{2}$, the line $m_I = \frac{1}{2} \leftrightarrow -\frac{1}{2}$ also appears, but a larger H_0 may be required to obtain strong field conditions. There is experimental evidence from molecular beam electric resonance experiments (Sect. 54) that the spin-rotation interaction constant c varies somewhat with vibrational and rotational state, but modifications of Eq. (47.6) to account for this effect have not yet been made.

For the case $m_I \neq \frac{1}{2}$ and for a transition $m_I \leftrightarrow m_I - 1$, the density of transitions, $D_J(\nu)$, is:

$$D_J(\nu) = \frac{1}{cJ/h - 6\nu' z} = \frac{1}{[(cJ/h)^2 - 12\nu'(\nu - \nu_I - \nu')]^{\frac{1}{2}}} \tag{47.8}$$

in which $\nu' = \frac{3eqQ}{16h}\,\frac{2m_I - 1}{I(2I-1)}$. The denominator is zero and hence D_J is ∞ when $\nu = \nu_I + \nu' + c^2 J^2/(12\nu' h^2)$ or $z = cJ/(6\nu' h)$. Usually $cJ/h \ll \nu'$, so that the maximum intensity occurs when $\nu = \nu_I + \nu'$ and $z = 0$. For $I = 1$ there are two possible transitions with $\Delta m_I = \pm 1$; one transition is $m_I = 1 \leftrightarrow 0$ and the other is $m_I = 0 \leftrightarrow -1$. The corresponding maxima of D_J are found from Eq. (47.8) to occur at the frequencies $\nu = \nu_I + 3eqQ/(16h)$ and at $\nu = \nu_I - 3eqQ/(16h)$. Hence the spectrum will be a doublet symmetrical about ν_I, and the frequency separation of the two peaks is $3eqQ/(8h)$. Fig. 91 shows the density of transitions, D_J, for $I = 1$ and $I = \frac{3}{2}$ for the case of $c = 0$. The spacings of the maxima are independent of H_0 and determine the electric quadrupole interaction constant. The quantity D_J determines the important features of the spectrum $I(\nu)$. An infinite intensity is predicted at the maxima by Eq. (47.8). Actually, of course, the observed line intensity will be finite, because of the finite resolution of an apparatus associated with the natural line width. Furthermore the term cJ/h in Eq. (47.8) gives a value of ν at which D_J is a maximum that varies slightly with J.

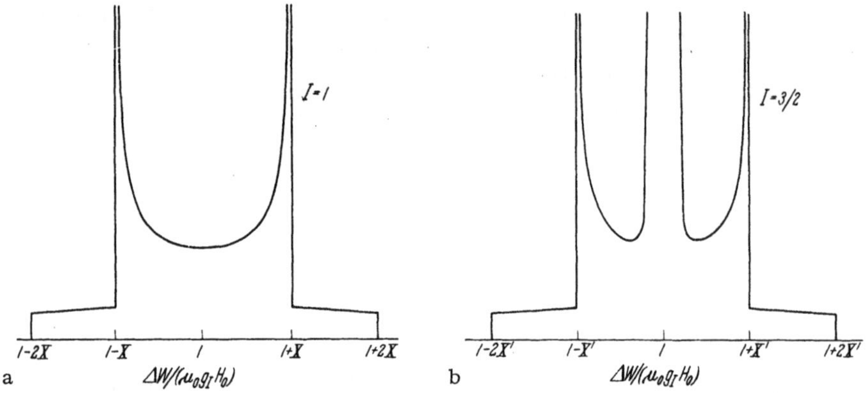

Fig. 91 a and b. (a) Shape of the nuclear resonance line for $I = 1$, corresponding to the transitions $\Delta m_I = \pm 1$, $\Delta m_J = 0$, in strong magnetic fields, as a function of the transition energy. The ordinate for the curve is the value of the density of transitions D computed from Eq. (47.8) with $c = 0$ and shown in an arbitrary unit. The unit for the abscissa is the energy of the transition ΔW divided by $g_I \mu_0 H_0$. The dimensionless parameter X is given by: $X = \frac{3eqQ}{16 g_I \mu_0 H_0}$. (b) Same as for Fig. 91 a except $I = \frac{3}{2}$ and $X' = \frac{eqQ}{8 g_I \mu_0 H_0}$.

Finally the dependence of the interaction constants on v and J further broadens the maxima. The general features of the observed spectrum are described by the theory just presented. However, a unique determination from the experimental data of the coefficients that describe the variation of q with J and v does not appear to be possible.

An important example in which no electric quadrupole satellites have been observed is Li^6 ($I = 1$) in any of the lithium halides[1]. Experimentally, the ratio of the width of the central peak corresponding to the transition $m_I = +\frac{1}{2} \leftrightarrow -\frac{1}{2}$ in the spectra of Li^7 ($I = \frac{3}{2}$) in LiCl, LiBr, and LiI, and the width of the envelope of the spectra of Li^6 in the same molecules is found to be very nearly equal to the ratio $g_I(Li^7)/g_I(Li^6)$. Since the interaction constants c are expected to be proportional to g_I for isotopic nuclei in the same molecule, this result indicates that the spin-rotation interaction is the important factor in determining the intensity distribution in the spectra of Li^6 in the halides. From this evidence it has been possible to conclude that $|Q_6/Q_7| \leq 0.044$. The sign of the ratio Q_6/Q_7 has been found[2] to be positive by observation of the spectra of the lithium halides as described in Sect. 48.

48. Sign of electric quadrupole interaction. The sign of the electric quadrupole interaction constant eqQ in a diatomic molecule is not determined by an observation of the complete high field spectrum, as discussed in Sect. 47. However,

[1] P. KUSCH: Phys. Rev. **75**, 887 (1949).
[2] P. KUSCH: Phys. Rev. **92**, 268 (1953).

if a selection of certain of the magnetic states is made by means of deflections, then the resulting spectrum can yield the sign of eqQ provided the sign of g_I is known[1].

Under strong field conditions, the energy levels of a diatomic molecule for which one nucleus has an electric quadrupole moment and the other nucleus has zero spin are given by Eq. (47.2). The electric quadrupole satellite peaks for the transitions $\Delta m_I = \pm 1$ occur near $z=0$ as discussed in Sect. 47. Hence the part of Eq. (47.2) pertinent for determining the frequencies of these satellite peaks can be written:

$$W = m_I g_I \mu_0 H_0 + \text{const} \times (e\,q\,Q\,m_I^2) \tag{48.1}$$

in which the constant is positive. The four possible energy level diagrams corresponding to the different possible signs of g_I and eqQ are shown in Fig. 92

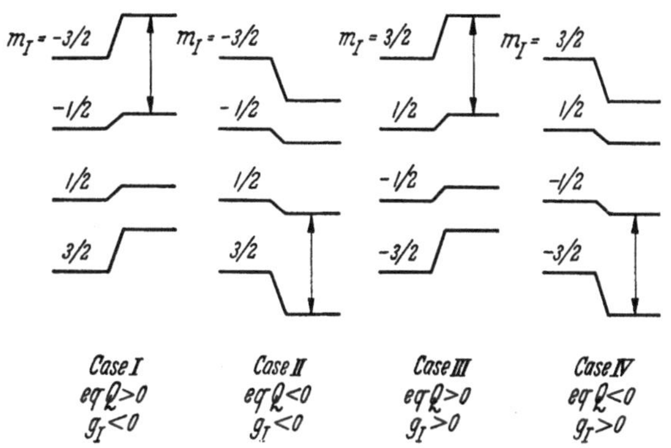

Fig. 92. Energy level diagram for $I = \frac{3}{2}$ showing the shift in energy due to the nuclear electric quadrupole interaction energy for the four cases corresponding to the different signs of eqQ and g_I. In each case the arrow indicates the transition which gives rise to the high frequency satellite in the nuclear resonance spectrum.

for $I = \frac{3}{2}$. The transition $m_I = \frac{1}{2} \leftrightarrow -\frac{1}{2}$ has the frequency $g_I \mu_0 H_0/h$ in all four cases. In the most common case when g_I is negative (μ_I positive) the high frequency electric quadrupole satellite occurs for transitions between states of negative m_I if eqQ is positive and between states of positive m_I if eqQ is negative.

The deviation of the trajectory of the molecule from the direction of the undeflected beam is determined by the sign of m_I for a nuclear moment of a given sign. Hence it is possible to exclude states of either positive or negative m_I by the interposition of an obstacle on one side or the other of the position of the undeflected beam. This removal of molecules with a particular sign of m_I will suppress either the high or low frequency satellite depending on the sign of eqQ. In this way the sign of eqQ relative to the sign of g_I can be determined. If both nuclei in the diatomic molecule have comparable magnetic moments, the selection of molecules with a particular sign of m_I for one of the nuclei is much less favorable. Under strong field conditions the presence of the second nucleus, even if it has an electric quadrupole moment, will not affect the nuclear reorientation spectrum of the first nucleus.

Two examples of interest are shown in Figs. 93 and 94. For Li^7, which has a negative g_I, the electric quadrupole interaction in LiI is positive. For Na, which also has a negative g_I, the electric quadrupole interaction energy in Na_2 is negative. The different behaviour of the resonance curves as the obstacle is interposed is conspicuous.

[1] Logan, Coté and Kusch: Phys. Rev. **86**, 280 (1952).

It is not necessary that the various maxima in the resonance envelope be distinct in order to determine the sign of the quadrupole interaction energy, since the preferential removal of positive or negative m_I states will shift the center of gravity of the observed resonance curve. In this way, for example, it has been found that eqQ for Li^7 in Li_2 is positive. The shift of the center of the curve is only about ± 1.5 kc/sec as the states of positive or negative m_I are removed, but the result is unambiguous.

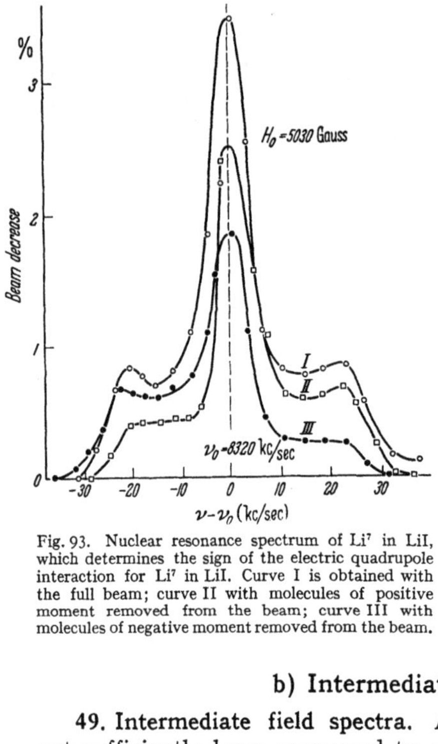

Fig. 93. Nuclear resonance spectrum of Li^7 in LiI, which determines the sign of the electric quadrupole interaction for Li^7 in LiI. Curve I is obtained with the full beam; curve II with molecules of positive moment removed from the beam; curve III with molecules of negative moment removed from the beam.

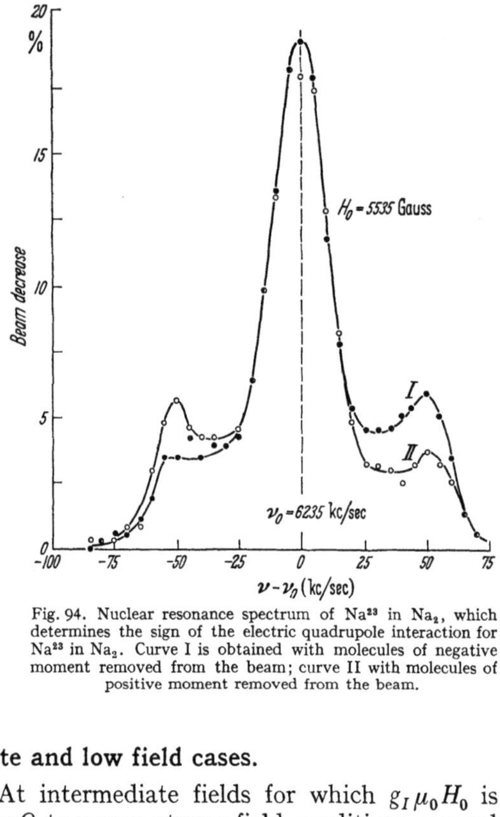

Fig. 94. Nuclear resonance spectrum of Na^{23} in Na_2, which determines the sign of the electric quadrupole interaction for Na^{23} in Na_2. Curve I is obtained with molecules of negative moment removed from the beam; curve II with molecules of positive moment removed from the beam.

b) Intermediate and low field cases.

49. Intermediate field spectra. At intermediate fields for which $g_I \mu_0 H_0$ is not sufficiently large compared to eqQ to assure strong field conditions, second order terms involving $(eqQ)^2$ may contribute significantly to the intensity distribution within the envelope of the lines. Such an effect was first observed as an increase in the width of the central component of the spectrum of Na in NaBr as the field H_0 was decreased[1].

Intermediate field spectra will be exemplified by a discussion of the case $I=\frac{3}{2}$. Only the central line, $m_I=\frac{1}{2}\leftrightarrow -\frac{1}{2}$ needs to be considered, since the two satellite lines $(\frac{3}{2}\leftrightarrow\frac{1}{2})$ and $(-\frac{1}{2}\leftrightarrow -\frac{3}{2})$ have sufficiently broad envelopes so that second order quadrupole effects are negligible. The frequency of the central line is[2]:

$$\nu\left(\frac{1}{2}\leftrightarrow-\frac{1}{2}\right)=\nu_I+\frac{cJz}{h}-\frac{3(eqQ)^2}{256h^2\nu_I}(1-z^2)(9z^2-1). \tag{49.1}$$

The perturbation theory calculation, which is based on the Hamiltonian of Eq. (47.1) and on the strong field representation (m_I, m_J), has been carried to the second order in eqQ, but only to the first order in c, because $c \ll eqQ < g_I \mu_0 H_0$

[1] W.A. NIERENBERG and N.F. RAMSEY: Phys. Rev. **72**, 1075 (1947).
[2] OCHS, COTÉ and KUSCH: J. Chem. Phys. **21**, 459 (1953).

applies. If the small term involving c is neglected, $|d\nu/dz|$ is zero at $z=0$ and at $z=\pm\sqrt{\frac{5}{9}}$. Hence the density of levels, D_J, is a maximum at $\nu_2=\nu_I+3(eqQ)^2/(256h^2\nu_I)$, corresponding to $z=0$, and at $\nu_1=\nu_I-(eqQ)^2/(48h^2\nu_I)$, corresponding to $z=\pm\sqrt{\frac{5}{9}}$. The total interval between the two maxima is $25(eqQ)^2/(768h^2\nu_I)$. The term in c does not significantly modify the high frequency peak ν_2 at $z=0$. However, the low frequency peak is split by the term in c, which gives a contribution of opposite sign at $z=+\sqrt{\frac{5}{9}}$ and at $z=-\sqrt{\frac{5}{9}}$. The frequencies of the two components of ν_1 are $\nu_1'=\nu_1-\sqrt{5}\,cJ/(3h)$ and $\nu_1''=\nu_1+\sqrt{5}\,cJ/(3h)$. The interval between ν_1' and ν_1'' is proportional to J and independent of H_0. The thermal distribution of molecules over different rotational states J broadens these peaks. If c is very small, the peaks will not be resolved; if c is large, the peaks will have a relatively large separation, but they will be of low intensity. In Fig. 95 the calculated intensity patterns for these two cases is shown.

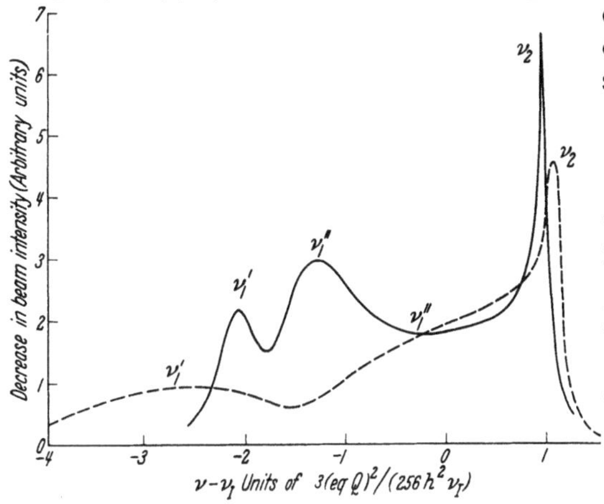

Fig. 95.

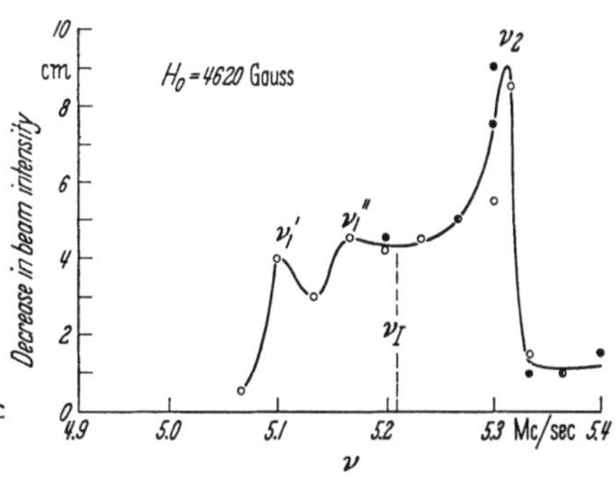

Fig. 96.

Fig. 95. Theoretical line shapes for the nuclear resonance transition $m_I=+\frac{1}{2}\leftrightarrow-\frac{1}{2}$ for $I=\frac{3}{2}$ under intermediate field conditions. The ordinate is the density of transitions D as computed from Eq. (49.1), including the average over the distribution of rotational states in the beam [Eq. (47.5)]. For the solid curve $c\bar{J}=0.18\,(\nu_2-\nu_1)$ with $\bar{J}=50$ and for the dashed curve $c\bar{J}=0.62\,(\nu_2-\nu_1)$ with $\bar{J}=50$.

Fig. 96. Central maximum of nuclear resonance of Na^{23} in NaCl, showing the transition $m_I=+\frac{1}{2}\leftrightarrow-\frac{1}{2}$. Open and closed circles show two different traversals of the spectrum.

The central line of the Na resonance in NaCl is shown in Fig. 96. The electric quadrupole interaction constant eqQ/h for Na in NaCl is -5.6 Mc/sec, and at the field $H_0=4620$ gauss the separation of the two predicted peaks ν_1 and ν_2 is calculated to be 180 kc/sec, which is consistent with the observed line. The similarity of this observed line shape to the theoretically computed line shapes of Fig. 95 is striking.

However, conspicuous quantitative and qualitative deviations from the predictions of the present theory have been noted in the low and intermediate field spectra of many alkali halides. As an example Fig. 97 shows the central resonance line for Cl^{35} in NaCl. The separation of the peaks ν_a and ν_b is inversely proportional to H_0 and has a value consistent with the value of eqQ determined from the satellite peaks. Hence these peaks are identified with the peaks ν_1 and ν_2. No splitting of the peak ν_a into two components corresponding to ν_1' and ν_1'' is observed, presumably because of the small value of c, which is expected from the known small g_I value of Cl. The observed peak at ν_c is unexplained. More generally, for a number of alkali halides the relative intensity of the portion of the envelope which is describable in terms of the theory of diatomic molecules with known molecular constants and of that portion of the envelope which exhibits anomalous details of structure is significantly dependent on both the oven temperature and oven pressure. It has, therefore, been postulated that the gas of the alkali halide consists not only of diatomic molecules but also of polymers of the diatomic molecule. While the theory of the spectra of such complex structures has not been discussed, a peak at ν_I is to be expected for any molecule if the internal interactions are small compared to the external field interaction. Observation of certain details of the unresolved low field spectrum of Na in NaF, has also led to the conclusion that polymers of the alkali halides exist and that their internal nuclear-molecular interactions are small[1]. A velocity analysis of beams of the alkali halides has given a straightforward determination of the dimeric content of such beams, independently of an interpretation of obscure details of a spectrum (Sect. 10). The existence of a substantial concentration of a dimer in beams of the alkali halides implies that quantitative conclusions from the unresolved spectra of these molecules are subject to some uncertainty.

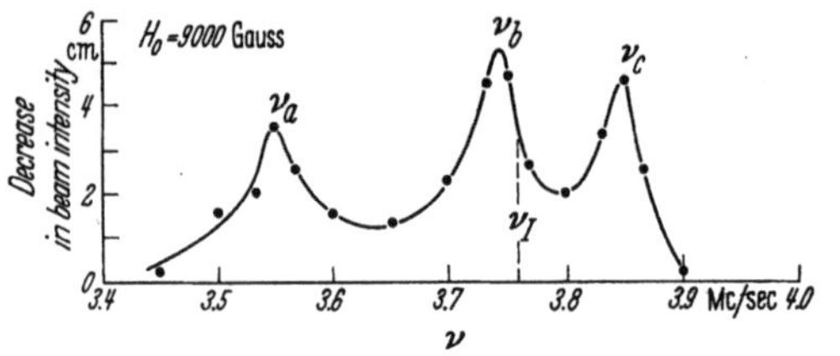

Fig. 97. Central maximum of nuclear resonance of Cl^{35} in NaCl, showing the transition $m_I = +\frac{1}{2} \leftrightarrow -\frac{1}{2}$.

50. Low field spectra. Unresolved nuclear resonance spectra at zero and at weak magnetic fields have been studied[2-5] particularly carefully for the case $I=\frac{3}{2}$. When the rotational quantum number J is large, as is the case for unresolved magnetic resonance spectra, the effect of the second nucleus can be neglected[3]. In zero field the (F, m) representation is useful, in which F is the total angular momentum quantum number ($\boldsymbol{F}=\boldsymbol{I}+\boldsymbol{J}$), and m is the associated magnetic quantum number. The energy eigenvalues of the Hamiltonian of Eq. (47.1) are[6]:

$$W_F = -e\,q\,Q\,\frac{[\frac{3}{4}K(K+1) - I(I+1)\,J(J+1)]}{2I(2I-1)(2J+3)(2J-1)} + \frac{c}{2}K \tag{50.1}$$

in which $K=F(F+1)-I(I+1)-J(J+1)$, and F takes on the values $J+I$, $J+I-1, \ldots, J-I$. The selection rule for transitions is $\Delta F=\pm 1$. Hence for $I=\frac{3}{2}$ the transition frequencies are:

$$\nu = (W_{J+\frac{3}{2}} - W_{J+\frac{1}{2}})/h = -\frac{eqQ}{4h} + \frac{c}{h}\left(J+\frac{3}{2}\right), \tag{50.2a}$$

$$\nu = (W_{J+\frac{1}{2}} - W_{J-\frac{1}{2}})/h = \frac{3eqQ}{4h}\left[\frac{2J+1}{(2J+3)(2J-1)}\right] + \frac{c}{h}\left(J+\frac{1}{2}\right), \tag{50.2b}$$

$$\nu = (W_{J-\frac{1}{2}} - W_{J-\frac{3}{2}})/h = \frac{eqQ}{4h} + \frac{c}{h}\left(J-\frac{1}{2}\right). \tag{50.2c}$$

The transition $F=(J+\frac{1}{2}) \leftrightarrow (J-\frac{1}{2})$ gives rise to a line of relatively low frequency. The transitions $F=(J+\frac{3}{2}) \leftrightarrow (J+\frac{1}{2})$ and $F=(J-\frac{1}{2}) \leftrightarrow (J-\frac{3}{2})$ give rise to lines of frequencies $eqQ/(4h)+cJ/h$ and $eqQ/(4h)-cJ/h$ for high J. The mean of these two frequencies determines eqQ/h.

---

[1] P. Kusch: J. Chem. Phys. **21**, 1424 (1953).
[2] W.A. Nierenberg and N.F. Ramsey: Phys. Rev. **72**, 1075 (1947).
[3] H.J. Zeiger and D.I. Bolef: Phys. Rev. **85**, 788 (1952).
[4] D.I. Bolef and H.J. Zeiger: Phys. Rev. **85**, 799 (1952).
[5] R.E. Coté and P. Kusch: Phys. Rev. **90**, 103 (1953).
[6] B.T. Feld and W.E. Lamb jr.: Phys. Rev. **67**, 15 (1945).

The spectra of Na in the sodium halides have been studied at both high and low magnetic fields, and the values obtained for eqQ from the two types of data

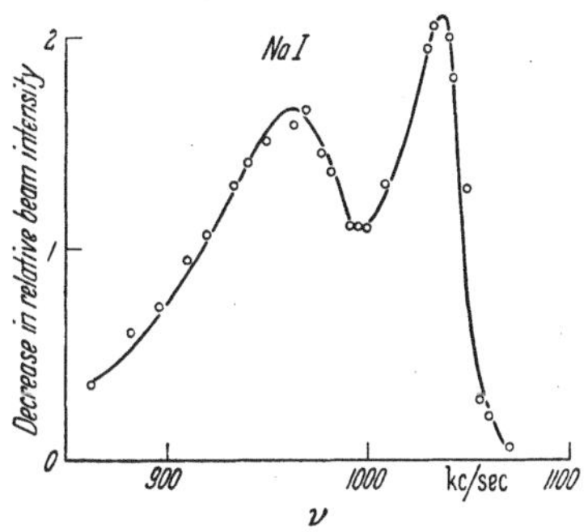

Fig. 98.

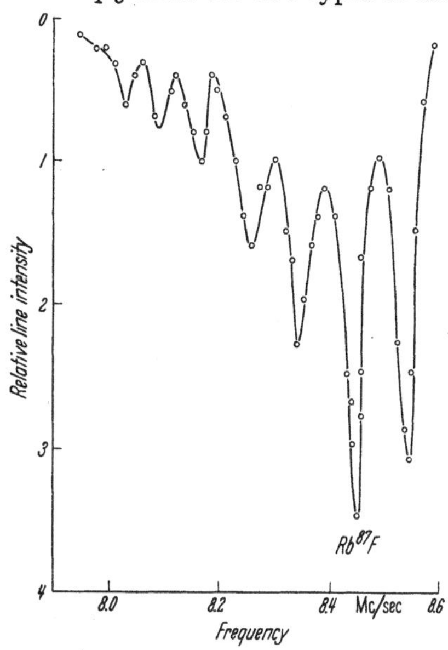

Fig. 99.

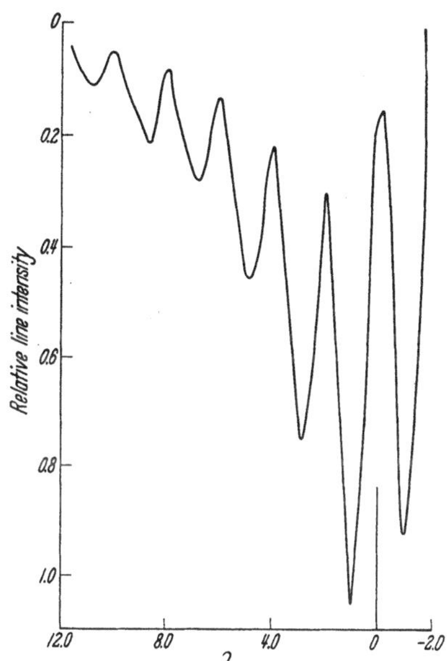

Fig. 100.

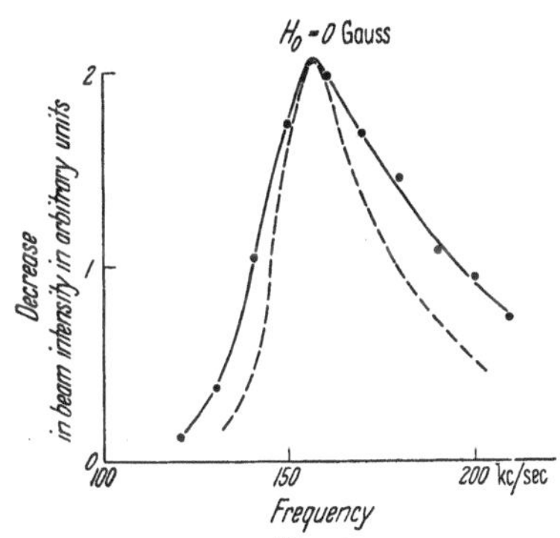

Fig. 101.

Fig. 98. Nuclear resonance spectrum of Na^{23} in NaI arising from the transitions $F=(J+\frac{3}{2}) \leftrightarrow (J+\frac{1}{2})$ and $F=(J-\frac{1}{2}) \leftrightarrow (J-\frac{3}{2})$ at zero static magnetic field.

Fig. 99. Nuclear resonance spectrum of Rb^{87} in Rb^{87}F arising from the transitions $F=(J+\frac{3}{2}) \leftrightarrow (J+\frac{1}{2})$ and $F=(J-\frac{1}{2}) \leftrightarrow (J-\frac{3}{2})$ at zero static magnetic field. The maximum change in line intensity corresponds to a 0.2% change in the total beam intensity.

Fig. 100. Theoretical line shape corresponding to the transitions observed for Fig. 99. The spectrum is computed as the superposition of transitions given by Eqs. (50.2a) and (50.2c) for the vibrational and rotational states present in the beam. The constants used are: $|e\bar{q}Q|/h = 33.96$ (2) Mc/sec; $|eq^{(v)}Q|/h = 410$ (40) kc/sec, $|eq^{(J)}Q|/h = 26$ (4) cps, as defined by Eq. (50.3). Both $q^{(v)}$ and $q^{(J)}$ have a sign opposite to that of $\bar{q}$. The quantity in parenthesis is the uncertainty in the last quoted figure. $|c|/h = 1.1$ (0.1) kc/sec. $\mathfrak{z} = h(f-f_0)/c\bar{J}$ and $f_0 = e\bar{q}Q/4h$.

Fig. 101. Nuclear resonance spectrum of Br^{79} and Br^{81} in KBr arising from the transition $F=(J+\frac{1}{2}) \leftrightarrow (J-\frac{1}{2})$ at zero static magnetic field. The solid curve is drawn through the experimental points. The dashed curve is a theoretical curve using the values $eqQ(Br^{79})/(8h) = 1670$ kc/sec; $eqQ\ (Br^{81})/(8h) = 1400$ kc/sec; $c\,(Br^{79})/h = 1.21$ kc/sec; $c\,(Br^{81})/h = 1.30$ kc/sec.

are in agreement[1]. The two transitions $F=(J+\frac{3}{2}) \leftrightarrow (J+\frac{1}{2})$ and $F=(J-\frac{1}{2}) \leftrightarrow (J-\frac{3}{2})$ should be separated in frequency by $2c\bar{J}/h$ where $\bar{J}$ is the most probable

[1] W.A. Nierenberg and N.F. Ramsey: Phys. Rev. **72**, 1075 (1947).

value of J at the oven temperature. Fig. 98 shows[1] this splitting for Na in NaI. A more complex low field spectrum observed for Rb^{87} ($I=\frac{3}{2}$) in RbF is shown in Fig. 99. The line complex is identified as arising from the transitions $F=(J+\frac{3}{2})\leftrightarrow(J+\frac{1}{2})$ and $F=(J-\frac{1}{2})\leftrightarrow(J-\frac{3}{2})$, whose frequencies vary with vibrational state because of the dependence of q on v. The theoretical line shape shown in Fig. 100 is obtained under the assumption that q depends on vibrational and rotational state:

$$q(v,J)=\bar{q}+q^{(J)}J(J+1)+q^{(v)}v \tag{50.3}$$

in which $\bar{q}$, $q^{(J)}$, and $q^{(v)}$ are constants. With the choice of constants indicated a reasonable fit to the experimental data is obtained. Analyses of this type for spectra involving Na in NaF and NaCl by different investigators have led to markedly different values for the constants[1,2]. The transition $F=(J+\frac{1}{2})\leftrightarrow(J-\frac{1}{2})$ has been observed near zero field for several alkali halide molecules[1]. The line for Br in KBr is shown in Fig. 101. The theoretical curve is obtained by using the interaction constants indicated, which agree with the known ratio $Q^{79}/Q^{81}=1.194$. In some cases—for example[1], Na in NaCl and NaBr—there is considerable discrepancy between the observed resonance frequency for the transition $F=(J+\frac{1}{2})\leftrightarrow(J-\frac{1}{2})$ and the frequency calculated from values of the nuclear-molecular interaction constants obtained from the other low field or high field lines.

II. Resolved molecular spectra by electric resonance method.

51. Stark effect. The previous discussion has dealt with the spectrum of a diatomic molecule in the absence of all external fields and in the presence of an external magnetic field only. In this section the effect of an electric field on the energy levels of a rigid rotator of moment of inertia A and electric dipole moment $\boldsymbol{\mu}$ directed along the internuclear axis[3] is considered. This characterization of the molecule is sufficient to allow the interpretation of details of the spectrum observed for a molecule in an electric field. The Schrödinger equation for this rotator in an electric field $\boldsymbol{E}$ is:

$$\left(+\frac{\hbar^2}{2A}\boldsymbol{J}^2-\boldsymbol{\mu}\cdot\boldsymbol{E}\right)\psi=W\psi. \tag{51.1}$$

Let $\boldsymbol{E}$ be in the z direction and ϑ and φ be the usual spherical polar coordinates to specify the direction of $\boldsymbol{\mu}$. Then

$$\left(+\frac{\hbar^2}{2A}\boldsymbol{J}^2-\mu E\cos\vartheta\right)\psi(\vartheta,\varphi)=W\psi(\vartheta,\varphi). \tag{51.1a}$$

This equation can be written in dimensionless form with the notations $\varepsilon=\frac{W}{(\hbar^2/2A)}$, $\lambda=\frac{\mu E}{(\hbar^2/2A)}$:

$$\left[\frac{1}{\sin\vartheta}\frac{\partial}{\partial\vartheta}\left(\sin\vartheta\frac{\partial}{\partial\vartheta}\right)+\frac{1}{\sin^2\vartheta}\frac{\partial^2}{\partial\varphi^2}\right]\psi+(\varepsilon-\lambda\cos\vartheta)\psi=0. \tag{51.1b}$$

At zero electric field the energy eigenvalues are:

$$\varepsilon^{(0)}=J(J+1),\quad\text{or}\quad W^{(0)}=J(J+1)\hbar^2/(2A). \tag{51.2}$$

The corresponding zero field eigenfunctions are:

$$\psi^{(0)}=\varphi_{J,m_J}=N_{J,m_J}P_J^{m_J}(\cos\vartheta)\,e^{+im_J\varphi} \tag{51.3}$$

[1] R.E. COTÉ and P. KUSCH: Phys. Rev. **90**, 103 (1953).
[2] H.J. ZEIGER and D.I. BOLEF: Phys. Rev. **85**, 788 (1952).
[3] P. DEBYE: Polar Molecules. New York: Dover Publishing Co. 1945.

in which $P_J^{m_J}$ is an associated Legendre function and the normalization constant N_{J,m_J} is given by:

$$N_{J,m_J} = \left[\frac{(2J+1)(J-|m_J|)!}{4\pi(J+|m_J|)!}\right]^{\frac{1}{2}}. \tag{51.3a}$$

At weak electric fields for which $\lambda \ll 1$ perturbation theory is applicable. The non-zero matrix elements of $\cos\vartheta$ in the (J, m_J) representation are[1]:

$$(J+1, m_J|\cos\vartheta|J, m_J) = \left[\frac{(J+1)^2 - m_J^2}{(2J+1)(2J+3)}\right]^{\frac{1}{2}}, \tag{51.4a}$$

$$(J-1, m_J|\cos\vartheta|J, m_J) = \left[\frac{J^2 - m_J^2}{(2J-1)(2J+1)}\right]^{\frac{1}{2}}. \tag{51.4b}$$

The lowest order contribution to the energy is an E^2 term:

$$\varepsilon^{(2)} = \frac{\lambda^2}{2}\left[\frac{J(J+1) - 3m_J^2}{J(J+1)(2J-1)(2J+3)}\right]. \tag{51.5}$$

The corresponding perturbed wavefunction is:

$$\left.\begin{aligned}\psi^{(1)} = \varphi_{J,m_J} + \lambda\Bigg[\frac{-1}{(2J+2)}\sqrt{\frac{(J+1)^2-m_J^2}{(2J+1)(2J+3)}}\,\varphi_{J+1,m_J} + \\ + \frac{1}{2J}\sqrt{\frac{J^2-m_J^2}{(2J-1)(2J+1)}}\,\varphi_{J-1,m_J}\Bigg].\end{aligned}\right\} \tag{51.6}$$

The next contribution to the energy is an E^4 term, which has been computed[2] for $J=0$, 1, and 2. The energies for these rotational states are: (For $J=0$, the energy is given to order E^6.)

$$\left.\begin{aligned}
&J=0, \quad m_J=0, \\
&\qquad \varepsilon = -\frac{1}{6}\lambda^2 + \frac{11}{1080}\lambda^4 - \frac{1}{725}\lambda^6, \\
&J=1, \quad m_J=0, \\
&\qquad \varepsilon = 2 + \frac{1}{10}\lambda^2 - \frac{73}{7000}\lambda^4, \\
&J=1, \quad m_J=\pm 1, \\
&\qquad \varepsilon = 2 - \frac{1}{20}\lambda^2 + \frac{19}{56000}\lambda^4, \\
&J=2, \quad m_J=0. \\
&\qquad \varepsilon = 6 + \frac{1}{42}\lambda^2 + \left(\sim\frac{1}{3000}\right)\lambda^4, \\
&J=2, \quad m_J=\pm 1, \\
&\qquad \varepsilon = 6 + \frac{1}{84}\lambda^2 - \frac{1}{3000}\lambda^4, \\
&J=2, \quad m_J=\pm 2, \\
&\qquad \varepsilon = 6 - \frac{1}{42}\lambda^2 + \sim\frac{1}{25000}\lambda^4.
\end{aligned}\right\} \tag{51.7}$$

---

[1] E. U. CONDON and G. H. SHORTLEY: The Theory of Atomic Spectra, Chap. III. Cambridge: Cambridge University Press 1953.

[2] F. BROUWER: Electric Polarization of a Dipole Gas. Diss. Amsterdam 1930. — H. K. HUGHES: Phys. Rev. 76, 1675 (1949).

The solution for the energy in the high field limit for which λ is large is given to order $1/\lambda^{\frac{1}{2}}$ by:

$$W = -\mu E + (2J - |m_J| + 1) \sqrt{\frac{\mu E}{\hbar^2/(4A)}} \left(\frac{\hbar^2}{2A}\right). \tag{51.8}$$

The energy eigenvalues of Eq. (51.1) are degenerate for states with $+m_J$ and $-m_J$.

Usually the electric field values used for deflection are such that the parameter λ is of the order of 1 so that neither a low nor a high field approximation is adequate, and a general solution valid for any value of E is required[1]. In the (J, m_J) representation for each value of m_J there exists a determinantal equation in ε:

$$D_0 = \begin{vmatrix} m_J(m_J+1)-\varepsilon & \lambda A(m_J, m_J) & 0 & \cdots \\ \lambda A(m_J, m_J) & (m_J+1)(m_J+2)-\varepsilon & \lambda A(m_J+1, m_J) & \cdots \\ 0 & \lambda A(m_J+1, m_J) & (m_J+2)(m_J+3)-\varepsilon & \cdots \\ 0 & 0 & \lambda A(m_J+2, m_J) & \cdots \\ \vdots & \vdots & \vdots & \vdots \end{vmatrix} = 0 \tag{51.9}$$

in which

$$A(m'_J, m_J) = -\left[\frac{(m'_J+1)^2 - u_J^2}{(2m'_J+1)(2m'_J+3)}\right]^{\frac{1}{2}}.$$

An expansion by minors, where $B(k, m_J) = [A(m_J+k, m_J)]^2$, gives:

$$D_0 = [m_J(m_J+1) - \varepsilon]\, D_1 - \lambda^2 B(0, m_J)\, D_2 = 0$$

where

$$D_1 = [(m_J+1)(m_J+2) - \varepsilon]\, D_2 - \lambda^2 B(1, m_J)\, D_3.$$

In general,

$$D_k = [(m_J+k)(m_J+k+1) - \varepsilon]\, D_{k+1} - \lambda^2 B(k, m_J)\, D_{k+2}. \tag{51.10}$$

From the expression for D_0, an equation can be obtained for ε:

$$\varepsilon = m_J(m_J+1) - \cfrac{\lambda^2 B(0, m_J)}{(m_J+1)(m_J+2) - \varepsilon - \cfrac{\lambda^2 B(1, m_J)}{D_2/D_3}}. \tag{51.11}$$

The result can be expressed in another form:

$$\varepsilon = m_J(m_J+1) - F_0 \tag{51.12}$$

where

$$F_0 = \frac{\lambda^2 B(0, m_J)}{(m_J+1)(m_J+2) - \varepsilon - F_1}$$

and in general,

$$F_k = \frac{\lambda^2 B(k, m_J)}{(m_J+k+1)(m_J+k+2) - \varepsilon - F_{k+1}}$$

where

$$B(k, m_J) = \frac{1}{4}\left[1 - \frac{4m_J^2 - 1}{4(m_J+k+1)^2 - 1}\right].$$

[1] This solution is due to W.E. LAMB jr.

The roots of Eqs. (51.11) or (51.12) correspond in the limit $\lambda \to 0$ to the energy eigenvalues $J(J+1)$ for the different J states. These formulae are useful for numerical computation of ε. Table 13 gives values of ε for the rotational states

Table 13. *Stark effect of a rigid rotator.*

Values of ε are tabulated as a function of λ.

λ	$J=0$, $m_J=0$	$J=1$, $m_J=0$	$J=1$, $m_J=1$
0.1	− 0.002	2.001	2.000
0.2	− 0.007	2.004	1.998
0.3	− 0.015	2.009	1.996
0.4	− 0.027	2.016	1.992
0.5	− 0.041	2.024	1.988
0.6	− 0.059	2.035	1.982
0.7	− 0.079	2.047	1.976
0.8	− 0.103	2.060	1.968
0.9	− 0.129	2.075	1.960
1.0	− 0.158	2.091	1.950
2.0	− 0.557	2.287	1.805
3.0	− 1.093	2.478	1.574
4.0	− 1.705	2.602	1.270
5.0	− 2.366	2.639	0.905
6.0	− 3.060	2.587	0.490
7.0	− 3.780	2.458	0.033
8.0	− 4.520	2.264	− 0.458
9.0	− 5.276	2.015	− 0.978
10.0	− 6.045	1.723	− 1.523
12.0	− 7.616	1.031	− 2.675
14.0	− 9.222	0.227	− 3.894
16.0	− 10.856	− 0.663	− 5.166
18.0	− 12.512	− 1.623	− 6.481
20.0	− 14.187	− 2.641	− 7.834

λ	$J=2$, $m_J=0$	$J=2$, $m_J=1$	$J=2$, $m_J=2$
1.0	6.024	6.012	5.976
2.0	6.098	6.043	5.905
3.0	6.228	6.084	5.789
4.0	6.413	6.123	5.629
5.0	6.645	6.149	5.428
6.0	6.909	6.154	5.189
7.0	7.182	6.131	4.915
8.0	7.444	6.076	4.609
9.0	7.681	5.989	4.273
10.0	7.879	5.869	3.912
12.0	8.141	5.535	3.117
14.0	8.221	5.088	2.243
16.0	8.138	4.543	1.300
18.0	7.922	3.914	0.300
20.0	7.597	3.212	− 0.750

λ	$J=3$, $m_J=0$	$J=3$, $m_J=1$	$J=3$, $m_J=2$	$J=3$, $m_J=3$
2	12.044	12.034	11.999	11.945
4	12.185	12.131	11.992	11.780
6	12.418	12.285	11.961	11.510
8	12.768	12.478	11.891	11.143
10	13.239	12.681	11.765	10.685
12	13.808	12.865	11.574	10.146
14	14.428	13.004	11.314	9.533
16	15.034	13.079	10.984	8.853
18	15.574	13.079	10.587	8.115
20	16.011	13.001	10.125	7.324
22	16.331	12.847	9.605	6.485
24	16.535	12.621	9.030	5.603
26	16.632	12.327	8.405	4.682
28	16.633	11.971	7.734	3.726
30	16.552	11.559	7.021	2.737
32	16.397	11.095	6.269	1.718
34	16.178	10.584	5.481	0.672
36	15.901	10.029	4.660	− 0.399
38	15.571	9.435	3.809	− 1.494
40	15.194	8.803	2.929	− 2.610

λ	$J=4$, $m_J=0$	$J=4$, $m_J=1$	$J=4$, $m_J=2$	$J=4$, $m_J=3$	$J=4$, $m_J=4$
2.0	20.026	20.022	20.010	19.991	19.964
4.0	20.111	20.097	20.040	19.962	19.855
6.0	20.237	20.199	20.088	19.911	19.676
8.0	20.424	20.350	20.147	19.832	19.428
10.0	20.671	20.547	20.213	19.721	19.114
12.0	20.983	20.779	20.276	19.574	18.738
14.0	21.366	21.037	20.328	19.386	18.302
16.0	21.827	21.325	20.360	19.155	17.811
18.0	22.366	21.616	20.365	18.880	17.269
20.0	22.975	21.901	20.335	18.560	16.678
24.0	24.309	22.400	20.155	17.787	15.364
28.0	25.595	22.736	19.800	16.846	13.895
32.0	26.635	22.869	19.269	15.751	12.290
36.0	27.349	22.790	18.572	14.517	10.568
40.0	27.749	22.513	17.723	13.158	8.742
44.0	27.883	22.057	16.737	11.688	6.824
48.0	27.796	21.441	15.625	10.119	4.825
52.0	27.524	20.686	14.403	8.459	2.753
56.0	27.094	19.804	13.081	6.719	0.615

$J=0$, 1, 2, 3, and 4 as a function of λ. Figs. 102 to 104 present plots of ε vs. λ for these rotational states. These results were used in computing the effective electric dipole moment μ_e given in Sect. 22.

52. Energy levels including hyperfine structure; selection rules. The energy associated with the vibration and rotation of the molecule [the term $F^{\mathrm{op}}_{v,J}$ in Eq. (44.1)] was discussed in Sect. 44, and the interaction energy with the external electric field in Sect. 51. The present section deals with the internal nuclear-molecular interactions. The two limiting cases in which the interaction $-\boldsymbol{\mu}\cdot\boldsymbol{E}$ with the external electric field is large compared to the internal nuclear molecular interactions (strong field case) and in which the reverse is true (weak field case) are particularly useful. Electric resonance spectra have not yet been studied in combined electric and magnetic fields, so interaction with an external magnetic field will be omitted[1].

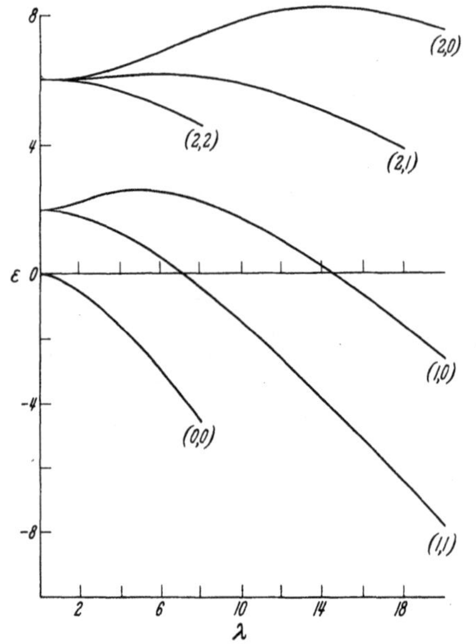

Fig. 102. Energy levels of a rigid rotator in a static electric field as computed from Eq. (51.11) for rotational states $J=0$, $J=1$, and $J=2$. $\varepsilon = W/(\hbar^2/2A)$; $\lambda = \mu E/(\hbar^2/2A)$. The numbers in parentheses designate the (J, m_J) values. The energy value is the same for states (J, m_J) and $(J, -m_J)$.

The most common case studied under weak field conditions is that in which the electric quadrupole interaction with the alkali nucleus is the dominant interaction, and the field interaction which is characterized by $\lambda\mu E$ is small compared to an internal interaction involving the halide nucleus. If nucleus 1 has an electric quadrupole moment and nucleus 2 has spin $\frac{1}{2}$, the energy levels for the relevant part of the Hamiltonian of Eq. (44.1) are given by[2]:

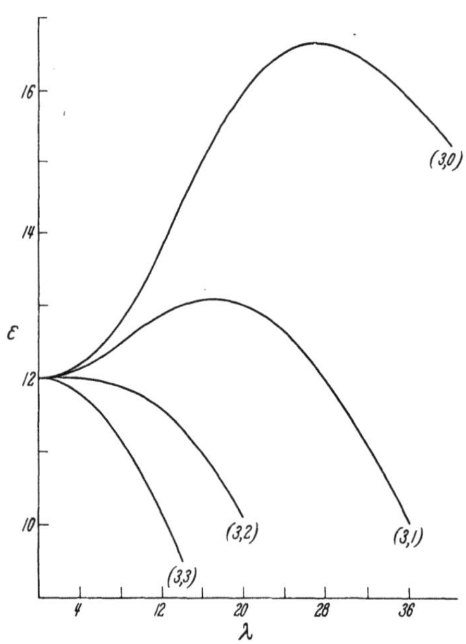

Fig. 103. Same as for Fig. 102 for rotational state $J=3$.

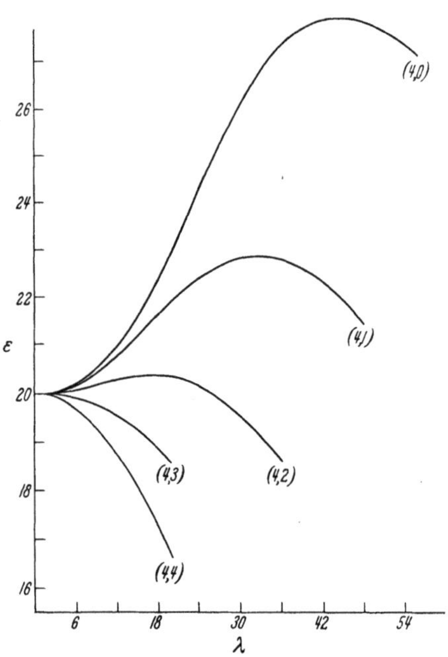

Fig. 104. Same as for Fig. 102 for rotational state $J=4$.

[1] A. M. Russell: Phys. Rev. **106**, 1100 (1957). A measurement of the rotational magnetic moments of Li^6F and Li^7F by a method in which the deflection is achieved by electric fields and the transitions occur in a magnetic field is described.

[2] V.W. Hughes and L. Grabner: Phys. Rev. **79**, 829 (1950).

$$\left.\begin{aligned}
W_{F_1, F=F_1\pm\frac{1}{2}, m} &= -\frac{e q_1 Q_1}{(2J-1)(2J+3)(2I_1)(2I_1-1)}\left\{\frac{3}{4}K_1(K_1+1) - I_1(I_1+1)J(J+1)\right\} + \\
&+ \frac{c_1 K_1}{2} \pm c_2 \frac{K_1'}{4F_1(F_1+1)}\left(F_1 + \frac{1}{2} \mp \frac{1}{2}\right) + \\
&+ \frac{g_{I_1} g_{I_2} \mu_0^2}{2R^3(2J+3)(2J-1)}\left\{\frac{3K_1 K_1'}{2F_1(F_1+1)} - \frac{J(J+1)K_1''}{F_1(F_1+1)}\right\} K - \\
&- \lambda^2 \frac{\hbar^2}{2A}\,\frac{[3(m^2+\frac{1}{4}) - F_1(F_1+1) \mp 6m^2/(2F_1+1)][3K_1'(K_1'-1) - 4F_1(F_1+1)J(J+1)]}{2J(J+1)(2J-1)(2J+3)2F_1(F_1+1)(2F_1-1)(2F_1+3)}
\end{aligned}\right\} \quad (52.1)$$

in which $\boldsymbol{F}_1 = \boldsymbol{I}_1 + \boldsymbol{J}$, $\boldsymbol{F} = \boldsymbol{F}_1 + \boldsymbol{I}_2$, m is the magnetic quantum number associated with $\boldsymbol{F}$.

$$K_1 = F_1(F_1+1) - I_1(I_1+1) - J(J+1); \quad K_1' = F_1(F_1+1) - I_1(I_1+1) + J(J+1),$$
$$K_1'' = F_1(F_1+1) - J(J+1) + I_1(I_1+1); \quad K = F(F+1) - F_1(F_1+1) - I_2(I_2+1).$$

Eq. (52.1) is valid only when:

$$e q_1 Q_1 \gg c_2 \quad \text{or} \quad \frac{g_{I_1} g_{I_2} \mu_0^2}{R^3}$$

and

$$c_2 \quad \text{or} \quad \frac{g_{I_1} g_{I_2} \mu_0^2}{R^3} \gg \lambda \mu E.$$

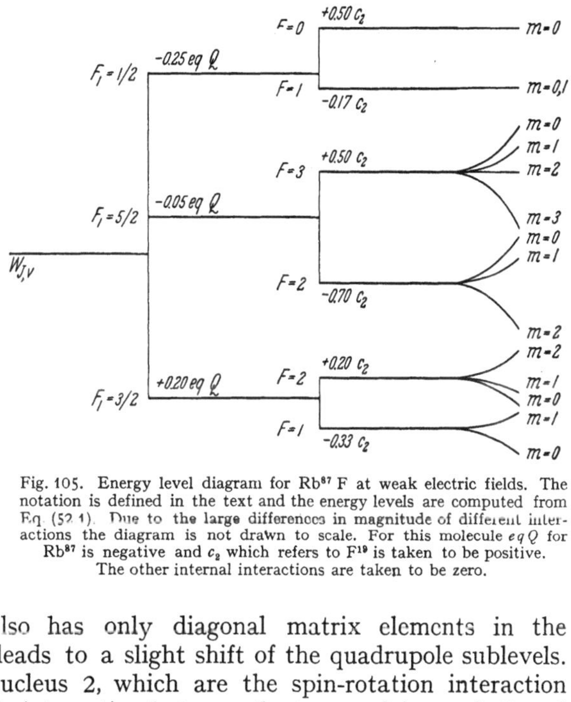

Fig. 105. Energy level diagram for $Rb^{87}F$ at weak electric fields. The notation is defined in the text and the energy levels are computed from Eq. (52.1). Due to the large differences in magnitude of different interactions the diagram is not drawn to scale. For this molecule eqQ for Rb^{87} is negative and c_2 which refers to F^{19} is taken to be positive. The other internal interactions are taken to be zero.

A level of particular v and J with energy $F_{v,J}$ [see Eq. (44.2)] is perturbed primarily by the electric quadrupole interaction with nucleus 1. This interaction couples $\boldsymbol{I}_1$ and $\boldsymbol{J}$ to form $\boldsymbol{F}_1$. Its contribution to the energy is the diagonal matrix element in an (F_1, m_{F_1}) representation, and results in the splitting of a level (v, J) into $2J+1$ (or $2I_1+1$, if $I_1 < J$) electric quadrupole sublevels. The spin-rotation interaction for nucleus 1 also has only diagonal matrix elements in the (F_1, m_{F_1}) representation and leads to a slight shift of the quadrupole sublevels. The interactions involving nucleus 2, which are the spin-rotation interaction and the magnetic dipole-dipole interaction between the two nuclei, couple $\boldsymbol{I}_2$ and $\boldsymbol{F}_1$ to form $\boldsymbol{F}$. Their contribution to the energy is computed as the diagonal matrix element in the (F_1, I_2, F, m) representation. Each F_1 level is thus split into $2I_2+1 = 2$ sublevels designated by the total angular momentum quantum number F. The contribution of the weak interaction $-\boldsymbol{\mu}\cdot\boldsymbol{E}$ with the external electric field is computed by second order perturbation theory, which involves intermediate states of different J (see Sect. 51), and splits each F level into sublevels $|m|$ having different energies. The energy level diagram for $Rb^{87}F$ shown in Fig. 105 illustrates the weak field case.

If nucleus 2 as well as nucleus 1 has an electric quadrupole moment but if $eq_2Q_2 \ll eq_1Q_1$, then the energy levels are given by Eq. (52.1) where the term in c_2 is replaced by the diagonal matrix element in the (F_1, I_2, F, m) representation and the diagonal matrix element of the electric quadrupole interaction of nucleus 2 in the (F_1, I_2, F, m) representation[1] is added. If the electric quadrupole interactions for the two nuclei are of comparable magnitude, a diagonalization of the complete secular determinant is required[2]. This solution was obtained[3] for KBr.

In the strong field case the solution can be given for a molecule in which both nuclei have electric quadrupole moments. The representation $(I_1, I_2, J, m_{I_1}, m_{I_2}, m_J)$ is used and the energy eigenvalues are given by:

$$\left.\begin{aligned} W_{m_{I_1}, m_{I_2}, m_J} &= \lambda^2 \frac{\hbar^2}{2A}\left[\frac{J(J+1)-3m_J^2}{J(J+1)(2J-1)(2J+3)}\right] - \\ &- e q_1 Q_1 \frac{[3m_J^2 - J(J+1)][3m_{I_1}^2 - I_1(I_1+1)]}{4I_1(2I_1-1)(2J-1)(2J+3)} - \\ &- e q_2 Q_2 \frac{[3m_J^2 - J(J+1)][3m_{I_2}^2 - I_2(I_2+1)]}{4I_2(2I_2-1)(2J-1)(2J+3)} + \\ &+ c_1 m_{I_1} m_J + c_2 m_{I_2} m_J + \frac{2 g_{I_1} g_{I_2} \mu_0^2}{R^3(2J-1)(2J+3)} m_{I_1} m_{I_2} [3m_J^2 - J(J+1)]. \end{aligned}\right\} \tag{52.2}$$

This equation is applicable if $|m_J| \neq 1$ or if $|m_J| = 1$ and $|m_J + m_{I_i}| = I_i$ or $I_i + 1$. If $|m_J| = 1$ and $|m_J + m_{I_i}| < I_i$, the two energy levels are obtained by the use of perturbation theory involving degenerate states[4]. In Eq. (52.2) the electric field dependent energy term is given to order E^2 [Eq. (51.5)] and contributions of the internal nuclear-molecular interactions are given by the diagonal matrix elements in the strong field representation. Fig. 106 shows the energy levels of $K^{39}F$ where, in the limiting low and high field cases, the energies can be calculated from Eqs. (52.1) and (52.2)[5].

Transitions between energy levels are induced by an oscillating electric field $\boldsymbol{E}_1 \cos\omega t$ in the C-region which contributes the time-dependent Hamiltonian term

$$V = -\mu \cdot \boldsymbol{E}_1 \cos\omega t. \tag{52.3}$$

Non-zero matrix elements of V exist only for $\Delta J = \pm 1$. Two types of transitions have been studied by the molecular beam electric resonance method: one is a transition at a microwave frequency involving a change in the rotational energy, and the other is a transition at a radiofrequency involving no change in the rotational energy. For the microwave transition the principal energy change is that associated with the molecular rotational energy; for the radiofrequency transition only changes in internal nuclear-molecular interaction energy or in the interaction energy with the external electric field occur. If $\boldsymbol{E}_1$ is parallel to the static field $\boldsymbol{E}$, $\Delta m = 0$ (σ transition); if $\boldsymbol{E}_1$ is perpendicular to $\boldsymbol{E}$, $\Delta m = \pm 1$ (π transition). In a microwave transition under weak field conditions for which the representation (I_1, I_2, J, F_1, F, m) of Eq. (52.1) is appropriate, the additional selection rules determined by the matrix elements of V are $\Delta F_1 = 0, \pm 1$ and $\Delta F = 0, \pm 1$. In a microwave transition under strong field conditions for which the representation $(I_1, I_2, J, m_{I_1}, m_{I_2}, m_J)$ of Eq. (52.2) isappropriate, the selection rules are $\Delta m_{I_1} = 0$; $\Delta m_{I_2} = 0$; $\Delta m_J = 0, \pm 1$.

---

[1] Lee, Fabricand, Carlson and Rabi: Phys. Rev. **91**, 1395 (1953).
[2] J. Bardeen and C. H. Townes: Phys. Rev. **73**, 97 (1948).
[3] Fabricand, Carlson, Lee and Rabi: Phys. Rev. **91**, 1403 (1953).
[4] V. W. Hughes and L. Grabner: Phys. Rev. **79**, 829 (1950).
[5] L. Grabner and V. W. Hughes: Phys. Rev. **79**, 819 (1950).

A radiofrequency transition occurs[1] only in the presence of a static electric field $\boldsymbol{E}$ which causes an energy eigenstate to be an admixture of different J states, as given to the first power in E by Eq. (51.6). Matrix elements of V thus occur between states having the same rotational energy and they are of the order $\lambda\,\mu\,E_1$. Under weak field conditions the selection rules are $\Delta F_1 = 0, \pm 1, \pm 2$, $\Delta F = 0, \pm 1, \pm 2, \Delta m = 0, \pm 1$; under strong field conditions $\Delta m_{I_1} = 0, \Delta m_{I_2} = 0$, and $\Delta m_J = 0, \pm 1$.

The observability of a transition requires not only that the transition be allowed by the selection rules but also that the effective electric dipole moment

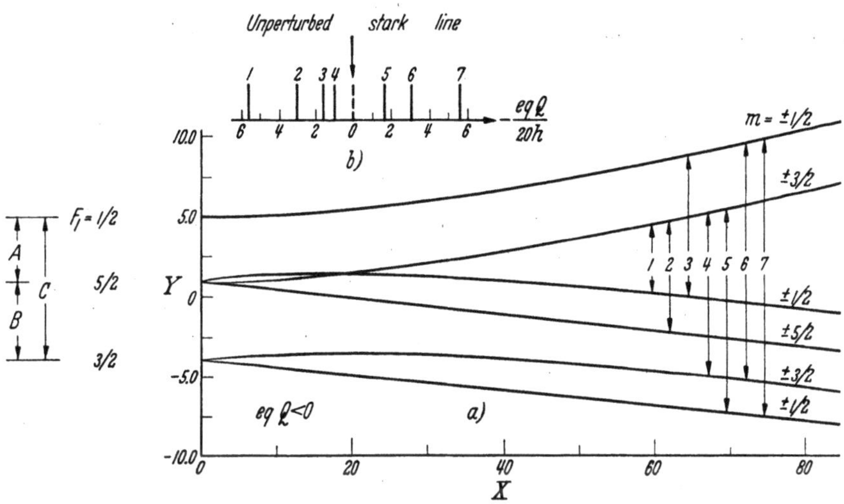

Fig. 106. Energy level diagram for the $J=1$ state of $K^{39}F$ as a function of electric field. $Y = -20\,W/(eq\,Q)$; $X = \frac{-20(\mu E)^2}{eqQ(\hbar^2/2A)}$. The only internal nuclear-molecular interaction included is the electric quadrupole interaction for K^{39} [$eq\,Q\ (K^{39}) < 0$]. The transitions observed under strong electric field conditions [see Fig. 108] are indicated by numbered arrows. A schematic diagram of the lines is shown in (b).

of the molecule change in the transition. This condition will generally be satisfied in a microwave transition for which a change in rotational state occurs. For a radiofrequency transition there is no change in rotational state and a change in effective electric dipole moment occurs only if the two states involved in the transition correspond to different values of $|m_J|$ under strong field conditions.

Absolute and even relative line intensities usually play only a minor role in the interpretation of electric resonance spectra, partly because of uncertainties about their interpretation. Several reasons for the uncertainties are: (1) The amplitude of the radiofrequency electric field is not well known throughout the C-field. (2) Non-adiabatic transitions can occur in the interfield regions and alter the distribution of the states of different space quantization. Such transitions apparently depend critically upon the field conditions in the interfield regions and have been observed to cause such gross effects as the complete disappearance of a particular line or the change from a decrease to an increase in beam intensity at the resonance frequency.

53. Molecular constants; atomic masses. From observations of the frequencies of transitions, particularly those between different rotational states, certain molecular properties of polar diatomic molecules have been determined with

[1] Multiple quantum transitions are exceptions (see Sect. 30).

high precision. To a lesser extent the frequencies of transitions within a single rotational state have also given such information. The line widths are customarily between 5 and 10 kc/sec and are determined by the transit time of the molecules through the C-region and by inhomogeneity of the electric field; the resonance frequency can be chosen to about 1 kc/sec. The molecular constants include the rotational and the rotational-vibrational constants and the electric dipole moment. Values for the ratios of the atomic masses for a number of pairs of isotopes have been derived from these molecular constants with a precision comparable to that obtained by conventional mass spectroscopy.

The simple theory of the determination of mass ratios is based on an approximate expression for the energy levels of a vibrating rotator derived from Eq. (44.4):

$$F_{v,J} = \omega_e (v + \tfrac{1}{2}) + [B_e - \alpha_e (v + \tfrac{1}{2})] J(J+1). \tag{53.1}$$

The corresponding frequency for a transition between the rotational states J and $J-1$ is:

$$\nu = (W_J - W_{J-1})/h = 2J\,[B_e - \alpha_e (v + \tfrac{1}{2})]. \tag{53.2}$$

Measurement of the frequencies of this rotational state transition for two vibrational states allows the determination of B_e and α_e. If values of B_e are obtained in this manner for two isotopic molecular species containing atomic masses m_1 and M, and m_2 and M, then since the ratio $B_{e(1)}/B_{e(2)}$ for the two isotopic molecules equals the ratio $M_{r(2)}/M_{r(1)}$ of the reduced masses, the isotope mass ratio can be determined from the equation:

$$\frac{m_1}{m_2} = \frac{(M/m_2)\,(B_{e(2)}/B_{e(1)})}{1 + M/m_2 - B_{e(2)}/B_{e(1)}}. \tag{53.3}$$

Equation (53.3) indicates that the mass ratio M/m_2 as well as $B_{e(2)}/B_{e(1)}$ is required to obtain m_1/m_2, but since $B_{e(2)}/B_{e(1)} \simeq 1$ the effect of an uncertainty in M/m_2 is small. Because terms of relative order $(B_e/\omega_e)^2$, which is about 10^{-6} for alkali halide molecules, are omitted in Eq. (53.1), the mass ratio m_1/m_2 determined from Eq. (53.3) is in error by this amount.

A more accurate treatment[1] based on Eq. (44.2) for $F_{v,J}$ provides a value of m_1/m_2 which includes the correction term of order $(B_e/\omega_e)^2$. The frequencies for the rotational state transitions $J = 0 \to J = 1$ for the lowest four vibrational states are obtained from Eq. (44.2) to be:

$$\left.\begin{aligned} \nu_0 &= 2Y_{01} + 4Y_{02} + Y_{11} + \tfrac{1}{2}Y_{21}, \\ \nu_1 &= 2Y_{01} + 4Y_{02} + 3Y_{11} + \tfrac{9}{2}Y_{21}, \\ \nu_2 &= 2Y_{01} + 4Y_{02} + 5Y_{11} + \tfrac{25}{2}Y_{21}, \\ \nu_3 &= 2Y_{01} + 4Y_{02} + 7Y_{11} + \tfrac{49}{2}Y_{21}. \end{aligned}\right\} \tag{53.4}$$

The four measured frequencies ν_0, ν_1, ν_2 and ν_3 are used to compute the three coefficients Y_{01}, Y_{11}, and Y_{21}. The smallest coefficient Y_{02} is computed from Eq. (44.3): $Y_{02} = -4B_e^3/\omega_e^2$. B_e is computed from the equation:

$$Y_{01} = B_e[1 + (B_e^2/\omega_e^2)\,\beta_{01}] \tag{53.5}$$

in which

$$\beta_{01} = Y_{10}^2\,Y_{21}/(4Y_{01}^3) + 16a_1\,Y_{20}/(3Y_{01}) - 8a_1 - 6a_1^2 + 4a_1^3 \tag{53.5a}$$

and

$$a_1 = Y_{11}\,Y_{10}/(6Y_{01}^2) - 1 \simeq -\alpha_e\omega_e/(6Y_{01}^2) - 1. \tag{53.5b}$$

[1] LEE, FABRICAND, CARLSON and RABI: Phys. Rev. **91**, 1395 (1953). — J. L. DUNHAM: Phys. Rev. **41**, 721 (1932).

The coefficients Y_{10} and Y_{20} need not be known to a high accuracy because they only appear in a small correction term, and their values are taken from results of optical spectroscopy[1]. The ratio of the reduced masses of the two isotopic molecules is obtained from:

$$\frac{Y_{01(1)}}{Y_{01(2)}} = \frac{M_{r(2)}}{M_{r(1)}}\left[1 + (\beta_{01})_{(2)}\left(\frac{B_e^2}{\omega_e^2}\right)_{(2)}\left(\frac{M_{r(2)}}{M_{r(1)}} - 1\right)\right]. \tag{53.6}$$

Finally the isotope mass ratio is given by:

$$\frac{m_1}{m_2} = \frac{(M/m_2)\,(M_{r(1)}/M_{r(2)})}{1 + (M/m_2) - (M_{r(1)}/M_{r(2)})}. \tag{53.7}$$

Table 14 presents values for isotope mass ratios obtained by the molecular beam electric resonance method with the use of Eq. (53.7). Table 15 gives the measured values of the rotational and rotational-vibrational constants Y_{lk} for alkali halide molecules.

Table 14. *Isotope mass ratios.*

Isotope ratio	Mass ratio	Reference
Cl^{35}/Cl^{37}	0.9459803(15)	53 *L1*
K^{39}/K^{41}	0.9512189(15)	53 *L1*
Br^{79}/Br^{81}	0.9753088(20)	53 *F1*
Rb^{85}/Rb^{87}	0.9770163(45)	54 *T2*

The lack of adequate knowledge about molecular wavefunctions, and hence about the contribution of the electrons to the moment of inertia of the molecule, provides at present a limit to the accuracy of the theory of mass ratio determinations which is comparable to the accuracy obtained experimentally for the transition frequencies (0.1 to 1 ppm)[2].

The electric dipole moment μ of a molecule can be determined from the observation of the frequency shift of a transition due to a known static electric field. If the electric field is small ($\lambda \ll 1$), the energy level shifts are proportional to E^2 and are given by Eq. (51.5), or by Eq. (52.1) or (52.2) if molecular hyperfine structure is considered. Measurement of the frequency shift together with a knowledge of B_e (or of A) suffices to determine μ. Transitions between different rotational states and between sublevels of a single rotational state have both been used. A typical electric field is 10 volts/cm which may result in a frequency shift of the order of 10^5 cps.; an accuracy of somewhat better than 1% is attained[3] for the value of μ.

If the value of A is not known, as is the case if a microwave transition between different rotational states has not been observed, then the measurement of the Stark shift at low fields will determine only the quantity $\mu^2 A$, which is the coefficient of E^2. A measurement at a higher field for which the E^4 term is appreciable allows the determination of $\mu^4 A^3$ as well, and hence μ and A can be calculated. Electric fields of about 300 volts/cm, which result in a frequency shift of the order of 100 Mc/sec, have been used. Eq. (51.7) or solutions as given in Table 13 are employed together with a strong field representation [see Eq. (52.2)]. Values for μ and A accurate to 5 to 10% are obtained[4]. Table 4 (Sect. 22) presents values of μ, determined for the most part by the molecular beam electric resonance method.

[1] G. HERZBERG: Spectra of Diatomic Molecules. New York: Van Nostrand 1950.

[2] GESCHWIND, GUNTHER-MOHR and TOWNES: Rev. Mod. Phys. **26**, 444 (1954). — G.C. WICK: Phys. Rev. **73**, 51 (1948). — ROSENBLUM, NETHERCOT and TOWNES: Phys. Rev. **109**, 400 (1958).

[3] LEE, FABRICAND, CARLSON and RABI: Phys. Rev. **91**, 1395 (1953).

[4] R. BRAUNSTEIN and J.W. TRISCHKA: Phys. Rev. **98**, 1092 (1955).

Table 15. *Rotational and rotational-vibrational constants of alkali halide molecules.*

The error estimates are indicated by the numbers in parentheses, which apply to the last figure given. The quantities without error estimates were obtained less directly from experimental data than were the quantities with error estimates; see original references for further details.

The values of B_e designated by stars were obtained from measurements of the Stark effect at radiofrequencies.

Molecule	Y_{01} (Mc/sec)	B_e (Mc/sec)	$-Y_{11}$ (Mc/sec)	Y_{21} (kc/sec)	$-Y_{02}$ (kc/sec)	References
Li^6F^{19}		49000 (4000) *				52 *S4*; 55 *B3*
Li^7F^{19}		44000 (4000) *				55 *B3*
Li^6Cl^{35}		16000 (6000) *				56 *M1*
Li^6Cl^{37}		16000 (6000) *				56 *M1*
Li^6Br^{81}	19161.511 (70)	19162.316 (70)	208.75	868	109	54 *H3*
Li^7Br^{79}	16650.570 (50)	16651.186 (50)	169.09 (8)	656 (40)	82	54 *H3*
Li^7Br^{81}	16650.002 (50)	16617.617 (50)	168.58 (8)	653 (40)	82	54 *H3*
Li^6I^{127}	15381.448 (80)	15381.986 (80)	152.59 (10)	610 (20)	69.0	54 *H3*
Li^7I^{127}	13286.386 (80)	13286.785 (80)	122.62 (10)	455 (50)	51.5	54 *H3*
$K^{39}F^{19}$		6060 (300) *				50 *G1*
$K^{39}Cl^{35}$	3856.370 (1)	3856.399 (7)	23.680 (1)	50 (1)		53 *L1*; 54 *H3*
$K^{39}Cl^{37}$	3746.583 (1)	3746.611 (7)	22.676 (2)	47 (1)		53 *L1*
$K^{41}Cl^{35}$	3767.394 (1)	3767.421 (7)	22.865 (1)	48 (1)		53 *L1*
$K^{39}Br^{79}$	2434.947 (1)	2434.953 (7)	12.136 (1)	23 (1)	+1	53 *F1*
$K^{39}Br^{81}$	2415.075 (1)	2415.081 (7)	11.987 (1)	22 (1)	+1	53 *F1*
$K^{39}I^{127}$	1825.006 (30)	1825.012 (30)	8.0337 (18)	12.21 (25)	1.03 (30)	54 *H3*
$K^{41}I^{127}$	1756.903 (30)					54 *H3*
$Na^{23}Cl^{35}$	6536.86 (10)	6537.07 (10)	48.28	145	8.60	54 *H3*
$Na^{23}Cl^{37}$					8.24	54 *H3*
$Na^{23}Br^{79}$	4534.51 (10)	4534.52 (10)	28.25 (10)	85 (30)	7 (3)	54 *H3*
$Na^{23}Br^{81}$	4509.34 (10)	4509.35 (10)	28.06 (10)	84	4.13	54 *H3*
$Na^{23}I^{127}$	3531.759 (35)	3531.778 (40)	19.439 (30)	46.9 (60)	2.39	54 *H3*
$Rb^{85}F^{19}$	6315.543 (9)	6315.639 (36)	45.638 (17)	94 (7)		58 *L2*
$Rb^{87}F^{19}$	6289.002 (10)	6289.098 (36)	45.350 (17)	94 (7)		58 *L2*
$Rb^{85}Cl^{35}$	2627.394 (2)	2627.414 (10)	13.601 (5)	21 (2)		54 *T2*
$Rb^{87}Cl^{35}$	2609.779 (3)		13.464 (5)	21 (2)		54 *T2*
$Rb^{85}Br^{79}$	1424.83 (2)	1424.84 (2)	5.5782 (56)	7.9 (11)	0.45 (20)	54 *H3*
$Rb^{85}Br^{81}$	1406.59 (2)		5.461 (11)			54 *H3*
$Rb^{87}Br^{79}$	1409.06 (2)		5.4744 (85)			54 *H3*
$Rb^{85}I^{127}$	984.314 (12)	984.317 (12)	3.2806 (12)	2.98 (20)	+0.234 (35)	54 *H3*
$Rb^{87}I^{127}$	970.760 (12)		3.2135 (30)			54 *H3*
$Cs^{133}F^{19}$	5527.27 (4)	5527.34 (4)	33.13 (5)	9 (12)		54 *H3*; 49 *T4*
$Cs^{133}Cl^{35}$	2161.195 (15)	2161.208 (15)	10.085 (4)	7.1 (7)		54 *H3*
$Cs^{133}Cl^{37}$	2068.761 (15)		9.46 (3)			54 *H3*
$Cs^{133}Br^{79}$	1081.34 (2)	1081.34 (2)	3.7175 (29)	3.1 (4)	+0.27 (8)	54 *H3*
$Cs^{133}Br^{81}$	1064.59 (2)		3.6313 (24)	3.1 (5)		54 *H3*
$Cs^{133}I^{127}$	708.36 (2)	708.36 (2)	2.0441 (20)	1.45 (40)	+0.152 (25)	54 *H3*
$Tl^{205}Cl^{35}$		2620.1 (25)	13 (1)			52 *C1*

54. Internal nuclear-molecular interactions. The internal nuclear-molecular interactions, which include principally electric quadrupole and spin-rotation interactions, have been measured for many alkali halide molecules by the molecular beam electric resonance method. The principal advantage of the electric resonance method as compared to the magnetic resonance method is that a single rotational state can be selected. As a result resolved spectra arising from single rotational states are obtained as compared to the unresolved magnetic resonance spectra from a superposition of many rotational states.

The internal interaction constants are usually obtained from weak field spectra, where the energy levels are given by Eq. (52.1) if one nucleus has an electric quadrupole moment and the other nucleus has spin $\frac{1}{2}$ or are calculated for other cases as discussed in Sect. 52. Occasionally strong field spectra have also been used, and the constants obtained from both weak and strong field spectra agree. Both microwave frequency transitions involving a change in rotational energy ($\Delta J = \pm 1$) and radiofrequency transitions involving no such change yield values for these interaction constants. Typically, the line widths are from 5 to 10 kc/sec, as determined by the transit time of the molecules through the C-region, and the interaction constants are determined to an accuracy of about 1 kc/sec.

An electric quadrupole interaction is usually the dominant internal nuclear-molecular interaction. The quadrupole interaction constants eqQ/h have been found to vary from 70 Mc/sec for Rb^{85} in RbF to less[1] than 10 kc/sec for Li^6 in LiF. For a given molecular species eqQ varies with vibrational state. Typically for the vibrational states studied from $v=0$ to $v=4$, eqQ varies by several percent from one vibrational state to the next; however, for KCl the unusual result is found that the Cl^{35} electric quadrupole interaction increases by as much as a factor of 3 from one vibrational state to the next higher one. No variation of eqQ with rotational state has been observed for the low rotational states of the alkali halide molecules studied. The spin-rotation interaction constants c/h have values in the range from 1 kc/sec or less up to about 35 kc/sec. Usually the nuclear magnetic dipole-dipole interaction constant, $g_{I_1} g_{I_2} \mu_0^2/(R^3 h)$, is less than 1 kc/sec, but in Li^7F it is about 13 kc/sec.

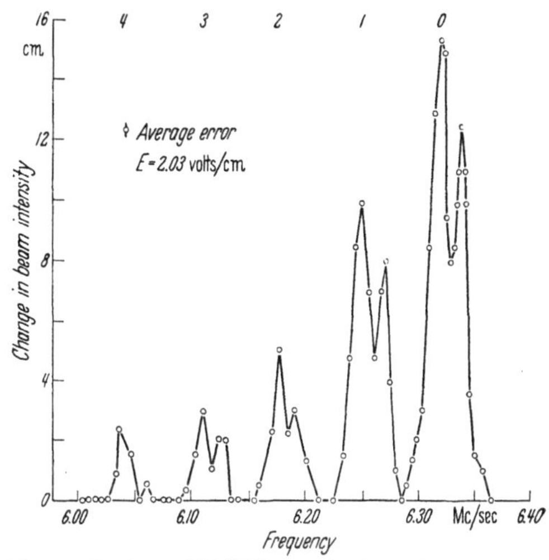

Fig. 107. Spectrum of $Rb^{85}F$ for rotational state $J = 1$ at weak electric field, showing the transition $F_1 = \frac{3}{2} \leftrightarrow \frac{7}{2}$ and the small splitting due to the spin-rotation interaction involving the F nucleus. The numbers 0 to 4 designate the vibrational states.

Typical spectra are given in Figs. 107 to 109, which show the effect of the internal nuclear-molecular interactions under weak and strong electric field conditions. Fig. 107 shows a line group from the spectrum of $Rb^{85}F$ taken at weak fields[2]. The lines arise primarily from a change in electric quadrupole interaction energy for Rb^{85} associated with the change $F_1 = \frac{3}{2} \leftrightarrow \frac{7}{2}$ $[\boldsymbol{F}_1 = \boldsymbol{I}_{Rb} + \boldsymbol{J}]$. The different lines labelled 0 to 4 come from the vibrational states $v=0$ to $v=4$, and illustrate the variation of eqQ with v. The decrease in the intensities of these lines from $v=0$ to $v=4$ is in accord with the expected population of the vibrational states in the source oven. The splitting of each of the lines is due to the spin-rotation interaction of the fluorine nucleus ($I_2 = \frac{1}{2}$). The spin-rotation interaction for Rb^{85}, the nuclear magnetic dipole-dipole interaction, and the external field interaction do not noticeably affect the line positions. Fig. 108a shows the strong field spectrum[3] of the $v=0$ vibrational state of $K^{39}F$. The lines are identified in Fig. 108b. The only internal interaction of importance is the electric quadrupole interaction with K^{39}. Fig. 109 shows[4] the spectrum of the $v=0$ vibrational state of $K^{39}Cl^{35}$ for the rotational state transition $J=0 \rightarrow J=1$.

[1] N. A. Schuster and G. E. Pake: Phys. Rev. **81**, 157 (1951). The interaction constant for Li^6F is determined from the constant for Li^7F and the ratio $|Q_6/Q_7|$.

[2] V. W. Hughes and L. Grabner: Phys. Rev. **79**, 314 (1950).

[3] L. Grabner and V. W. Hughes: Phys. Rev. **79**, 819 (1950).

[4] Lee, Fabricand, Carlson and Rabi: Phys. Rev. **91**, 1395 (1953).

The three component lines arise from the quadrupole interaction of K^{39}, and the hfs transitions are identified in the figure caption.

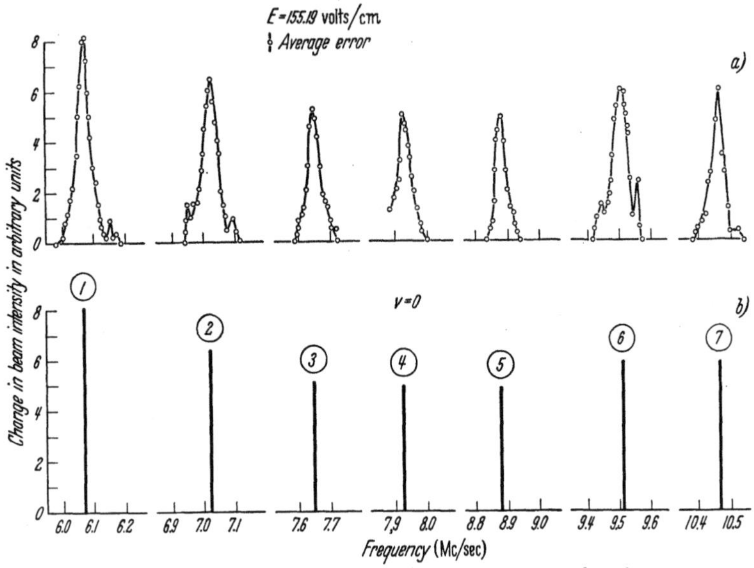

Fig. 108. Spectrum of $K^{39}F$ for rotational state $J=1$ and vibrational state $v=0$ at moderately strong electric field. The lines are identified in Fig. 106.

Because the calculation of the molecular wavefunction is a formidable task which has not yet been done for alkali halide molecules, the interaction constants eqQ cannot be used to obtain the nuclear electric quadrupole moments[1]. However, if values of eqQ are obtained for two isotopic molecular species—e.g., $Rb^{85}F$ and $Rb^{87}F$—then the ratio of the values of eqQ will give a purely nuclear constant, the ratio of Q values—e.g., $Q(Rb^{85})/Q(Rb^{87})$. This statement implies that q is the same in the two molecules. That this is true to a high accuracy despite the somewhat different vibrational properties of the molecules is indicated by the smallness of the dependence of q on vibrational state[2]. For the case of the alkali atoms nuclear quadrupole moments have been determined by atomic beam spectroscopy of optically excited states (see Table 10, Sect. 38), and hence quadrupole interaction constants for alkali nuclei in alkali halide molecules can be used to give values of q.

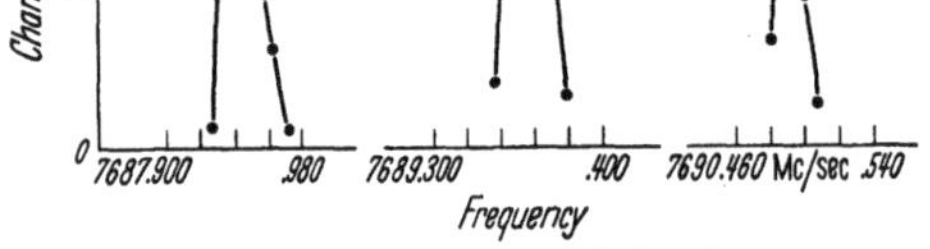

Fig. 109. Microwave spectrum of $K^{39}Cl^{35}$ for vibrational state $v=0$ at weak electric field, showing the transition $J=0\rightarrow J=1$. The F_1 values refer to the K^{39} nucleus and the $J=1$ state. The separation of the three lines is due to the K^{39} electric quadrupole interaction.

The spin-rotation interaction constants are used to determine the molecular magnetic fields H_e at the nuclei[3] and give information about the diamagnetic shielding of nuclei in molecules [24].

[1] The calculation of q in polar molecules is the subject of a series of papers by R. M. STERNHEIMER and H. M. FOLEY. See Phys. Rev. **102**, 731 (1956) and references there quoted.

[2] V. W. HUGHES and L. GRABNER: Phys. Rev. **79**, 314 (1950).

[3] J. W. TRISCHKA: Phys. Rev. **74**, 718 (1948). — R. L. WHITE: Rev. Mod. Phys. **27**, 276 1955).

Tables 16 and 17 list values for the internal nuclear-molecular interaction constants for alkali halide molecules as determined by the molecular beam electric and magnetic resonance methods. Certain derived quantities are also given.

Table 16. *Electric quadrupole interaction constants for alkali halide molecules* (eqQ/h *in* Mc/sec).

Throughout the table the numbers in parentheses give the uncertainty in the last place of the quoted value. The values of eqQ/h for the individual low vibrational states were determined by the molecular beam electric resonance method or by microwave absorption spectroscopy. For the molecular beam electric resonance experiments, eqQ/h is usually measured in the $J=1$ state and occasionally in the $J=2$ state as well; no variation of eqQ/h from $J=1$ to $J=2$ has been observed. For microwave absorption experiments higher J states are often used and again no dependence of eqQ/h on J has been observed. The values of eqQ/h for the average over vibrational and rotational states were determined by the molecular beam magnetic resonance method.

Molecule	Nucleus	Vibrational state					References
		$v=0$	$v=1$	$v=2$	$v=3$	average over vibrational and rotational states	
Li^7F^{19}	Li^7	+ 0.4120 (70)	+ 0.4120 (70)			+ 0.408	55 *B3*; 52 *L3*; 49 *K4*
Li^6Cl^{35}	Cl^{35}	− 3.071 72 (61)	− 3.4793 (17)	− 3.8730 (18)	− 4.250 (11)		56 *M1*
Li^6Cl^{37}	Cl^{37}	− 2.4199 (24)	− 2.7366 (54)				56 *M1*
Li^7Cl	Li^7					+ 0.192	52 *L3*; 49 *K4*
Li^7Br^{79}	Li^7					+ 0.184	52 *L3*; 49 *K4*
Li^7Br^{79}	Br^{79}	+ 37.20 (45)					54 *H3*
Li^7Br^{81}	Br^{81}	+ 30.71 (40)	+ 33.19 (20)				54 *H3*
Li^7I^{127}	Li^7					+ 0.172	52 *L3*; 49 *K4*
Li^7I^{127}	I^{127}	− 198.15 (30)	− 206.5 (3)	− 212.6 (15)			54 *H3*
Li^6I^{127}	I^{127}	− 199.43 (30)	− 206.8 (20)				54 *H3*
$K^{39}F^{19}$	K^{39}	− 7.938 (40)	− 7.828 (40)	− 7.744 (40)	− 7.658 (40)		50 *G1*
$K^{39}Cl^{35}$	K^{39}	− 5.656 (6)	− 5.622 (6)	− 5.571 (8)	− 5.511 (8)		53 *L1*
$K^{39}Cl^{37}$	K^{39}	− 5.660 (6)	− 5.628 (10)				53 *L1*
$K^{41}Cl^{35}$	K^{41}	− 6.899 (6)	− 6.840 (10)				53 *L1*
$K^{39}Cl^{35}$	Cl^{35}	< 0.040*	0.075 (10)*	0.237 (10)*	0.393 (10)*	−0.420	53 *L1*; 52 *L3*
$K^{39}Br^{79}$	K^{39}	− 5.003 (3)	− 4.984 (3)	− 4.915 (3)			53 *F1*
$K^{39}Br^{81}$	K^{39}	− 5.002 (3)					53 *F1*
$K^{39}Br^{79}$	Br^{79}	+ 10.244 (6)	+ 11.224 (6)	+ 12.204 (6)			53 *F1*
$K^{39}Br^{81}$	Br^{81}	+ 8.555 (6)					53 *F1*
$K^{39}I^{127}$	I^{127}	− 60 (10)					54 *H3*
$Na^{23}F^{19}$	Na^{23}					− 8.12 (2)	52 *L3*
$Na^{23}Cl$	Na^{23}					− 5.40 (2)	47 *N3*; 52 *L3*
$Na^{23}Br$	Na^{23}					− 4.68	47 *N3*; 52 *L3*
$Na^{23}Br^{79}$	Br^{79}	+ 58 (2)					54 *H3*
$Na^{23}I^{127}$	Na^{23}					− 3.88	47 *N3*; 52 *L3*
$Na^{23}I^{127}$	I^{127}	− 259.87 (60)	− 264.52 (65)	− 267.59 (40)	− 271.03 (20)		54 *H3*
$Rb^{85}F^{19}$	Rb^{85}	− 70.31 (10)	− 69.54 (10)	− 68.71 (10)	− 67.99 (10)†		50 *H2*
$Rb^{87}F^{19}$	Rb^{87}	− 34.00 (6)	− 33.63 (6)	− 33.20 (6)		−33.96 (20)	50 *H2*; 52 *B3*
$Rb^{85}Cl^{35}$	Rb^{85}	− 52.675 (5)	− 52.306 (30)	− 51.903 (40)			54 *T2*
$Rb^{85}Cl^{35}$	Cl^{35}	+ 0.774 (9)	+ 0.612 (13)	+ 0.470 (17)			54 *T2*
$Rb^{87}Cl$	Rb^{87}					25.37 (4)*	52 *B3*
$Cs^{133}F^{19}$	Cs^{133}	+ 1.240 (8)					48 *T1*
$Cs^{133}Cl^{35}$	Cs^{133}	< 4					51 *L4*
$Cs^{133}Cl^{35}$	Cl^{35}	< 3					51 *L4*
$Tl^{205}Cl^{35}$	Cl^{35}	− 15.795 (4)				−15.788 (20)	52 *C1*; 52 *Z1*
$Tl^{205}Cl^{37}$	Cl^{37}	− 12.446 (3)				−12.425 (20)	52 *C1*; 52 *Z1*

* Only the absolute values of eqQ are known for the cases with stars.

† An additional value for the $v=4$ state of $Rb^{85}F$ is eqQ (Rb^{85}) $= -67.20(10)$ Mc/sec.

Table 17. *Spin-rotation interaction constants, c/h, for alkali halide molecules in $v=0$ vibrational state by electric resonance.*

The error estimates are indicated by the numbers in parentheses, which apply to the last figure given.

Values of c/h for certain alkali halide molecules have also been obtained by the molecular beam magnetic resonance method (see Sects. 47, 49, and 50) but these values are averages over all vibrational and rotational states in the beam and in general are obtained with less accuracy than by the molecular beam electric resonance method. For a compilation of values of c/h by the magnetic resonance method see Ref. [55 *W* 2].

Molecule	Nucleus	c/h (kc/sec)	H_e (gauss per rotation quantum)	References	Molecule	Nucleus	c/h (kc/sec)	H_e (gauss per rotation quantum)	References
Li^6F^{19} *	Li^6	0 (1)		52 *S 4*;	$Rb^{87}F^{19}$ ††	F^{19}	14 (4)	3.5 (10)	50 *H 2*
				55 *K 1*	$Rb^{85}Cl^{35}$	Rb^{85}	0.3 (3)	0.7 (7)	54 *T 2*
	F^{19}	36.3 (10)	9.1 (3)			Cl^{35}	0.0 (8)		
Li^7F^{19} **	Li^7	2.2 (6)	1.3 (4)	55 *B 3*	$Cs^{133}F^{19}$	Cs^{133}	0 (1)		48 *T 1*
	F^{19}	32.9 (10)	8.2 (3)			F^{19}	16 (2)	4.0 (5)	
Li^6Cl^{35} †	Cl^{35}	2.07 (10)	5.0 (2)	56 *M 1*	$Tl^{205}Cl^{35}$	Tl^{205}	73 (2)	29.6 (8)	52 *C 1*
Li^6Cl^{37} †	Cl^{37}	1.88 (30)	5.4 (9)	56 *M 1*		Cl^{35}	1.2 (2)	2.9 (5)	
$K^{39}F^{19}$	F^{19}	0 (10)		50 *G 1*	$Tl^{205}Cl^{37}$	Tl^{205}	73 (2)	29.6 (8)	52 *C 1*
$Rb^{85}F^{19}$ ††	F^{19}	11 (3) *	2.7 (8)	50 *H 2*		Cl^{37}	1.0 (2)	2.9 (6)	

* The value of c/h for F^{19} in Li^6F^{19} in the $v=0$ state was observed in rotational states $J=1$ and $J=2$, and no difference was observed in the value of c/h in these two states within the experimental uncertainty.

** Additional values of c/h in Li^7F^{19} in the $v=1$ state are: $(c/h)_F = 32.4\,(10)$ kc/sec; $(c/h)_{Li^7} = 1.8\,(6)$ kc/sec. The values of c/h for Li^7F^{19} were obtained from observations on rotational states $J=1$ and $J=2$.

† Additional values of c/h in Li^6Cl^{35} are $(c/h)_{Cl}$: 2.22(20) kc/sec in $v=1$ state; 2.19(21) kc per sec in $v=2$ state. For Li^6Cl^{37}, $(c/h)_{Cl} = 1.60\,(34)$ kc/sec in $v=1$ state.

†† For $Rb^{85}F^{19}$ and $Rb^{87}F^{19}$ no variation of $(c/h)_F$ was observed with vibrational state for the states $v=0$, 1, 2, 3, 4. For $Rb^{85}F^{19}$ and $Rb^{87}F^{19}$ only the absolute values of $(c/h)_F$ are known; the signs are undetermined.

III. Resolved molecular spectra by magnetic resonance method.

55. Spectra of hydrogen molecules and O_2. The hydrogen molecules H_2, D_2, HD, and the oxygen molecule O_2 have been studied in low rotational and zeroth vibrational states by the molecular beam magnetic resonance method. Spectra for individual rotational states of these molecules can be obtained, because at the low source temperature of 77° (liquid nitrogen temperature) or even at 300° (room temperature) the beam is populated primarily by a few low-lying rotational states (Sect. 5). The spectra of the hydrogen molecules have been studied with particularly high resolution because the detailed theory of the internal nuclear-molecular interactions can be tested most completely for these simple molecules and because the electric quadrupole moment of the deuteron is determined from studies of these molecules.

The Hamiltonian for H_2 and D_2, which includes the internal nuclear-molecular interactions and interaction of the molecule with an external magnetic field, can be written[1]:

$$\left.\begin{aligned}\frac{\mathscr{H}}{h} = {} & \frac{5d}{(2J-1)(2J+3)}\left\{3(\boldsymbol{I}\cdot\boldsymbol{J})^2 + \frac{3}{2}(\boldsymbol{I}\cdot\boldsymbol{J}) - \boldsymbol{I}^2\boldsymbol{J}^2\right\} - c\,\boldsymbol{I}\cdot\boldsymbol{J} - \\ & - [1-\sigma_i(\boldsymbol{J})]\,a\,\boldsymbol{I}\cdot\boldsymbol{H}_0/H_0 - [1-\sigma_J(\boldsymbol{J})]\,b\,\boldsymbol{J}\cdot\boldsymbol{H}_0/H_0 - \\ & - \frac{5f}{3(2J-1)(2J+3)}\{3(\boldsymbol{J}\cdot\boldsymbol{H}_0)^2/H_0^2 - \boldsymbol{J}^2\} - g.\end{aligned}\right\} \qquad (55.1)$$

[1] N. F. RAMSEY: Phys. Rev. **85**, 60 (1952). In this article the spin-rotation interaction has heretofore been written as $+c\boldsymbol{I}\cdot\boldsymbol{J}$. The sign used in Eq. (55.1) is that used by RAMSEY who has discussed the problem of molecular hydrogen in detail [*25*].

The first term, in which I is the total nuclear spin quantum number of the molecule and J is the rotational quantum number, includes the combination of the spin-spin magnetic interaction of the two nuclei and the electric quadrupole interaction for a homonuclear molecule, and it is equivalent to the sum of terms (2), (3), and (6) in the Hamiltonian of Eq. (44.1). The constant d can be expressed as:

$$d h = \left\langle \frac{\mu_i^2}{i^2 R^3} \right\rangle \frac{I(I+1) + 4i(i+1)}{5(2I-1)(2I+3)} - \frac{eqQ}{10i(2i-1)} \left[1 - \frac{I(I+1) + 4i(i+1)}{(2I-1)(2I+3)} \right] \tag{55.2}$$

where R is the distance between the two nuclei, μ_i is the magnetic moment of one of the nuclei, i is the corresponding nuclear spin, Q is the deuteron quadrupole moment (the corresponding quantity for H_2 is zero), and q is the electric field gradient as defined for Eq. (44.1). The second term in Eq. (55.1) is the spin-rotation interaction and corresponds to terms (4) and (5) of Eq. (44.1).

The third term in Eq. (55.1) gives the interaction of the nuclear magnetic moments with the external magnetic field $\boldsymbol{H}_0$, in which a is defined by

$$a = \mu_i H_0 / (i\, h). \tag{55.3}$$

The factor $[1 - \sigma_i(\boldsymbol{J})]$ arises from the magnetic shielding of the nucleus by the molecule.

$$\sigma_i(\boldsymbol{J}) = \tfrac{2}{3}\sigma_\sigma + \tfrac{1}{3}\sigma_\pi + (\sigma_\sigma - \sigma_\pi)\left[\tfrac{2}{3}(2J-1)(2J+3)\right]\left[3(\boldsymbol{J}\cdot\boldsymbol{H}_0)^2/H_0^2 - \boldsymbol{J}^2\right] \tag{55.4}$$

in which σ_σ is the magnetic shielding constant for a magnetic field applied perpendicular to the internuclear axis and σ_π is the shielding constant for a field parallel to the internuclear axis. The fourth term in Eq. (55.1) gives the interaction of the molecular rotational magnetic moment with the external magnetic field and b is defined by

$$b = \mu_J H_0 / (J\, h) \tag{55.5}$$

where μ_J is the magnetic moment due to the rotation of the molecule in rotational state J. The corresponding terms (8), (9), and (10) of Eq. (44.1) for interaction with the external field $\boldsymbol{H}_0$ do not include the magnetic shielding effect.

The fifth and sixth terms in Eq. (55.1) give the diamagnetic interaction of the molecule with the external magnetic field. If ξ_σ is the magnetic susceptibility of the molecule for a field perpendicular to the internuclear axis and ξ_π is the susceptibility parallel to that axis,

$$f = (\xi_\sigma - \xi_\pi)\, H_0^2/(5\, h), \tag{55.6}$$

$$g = (\tfrac{1}{3}\xi_\sigma + \tfrac{1}{6}\xi_\pi)\, H_0^2/h. \tag{55.7}$$

These small diamagnetic terms were not included in Eq. (44.1).

The theory of HD has been considered[1] with a detailed Hamiltonian analogous to that of Eq. (55.1). The Hamiltonian given in Eq. (44.1) was sufficient for the interpretation of the spectra observed for H_2, D_2 and HD in earlier work.

Radiofrequency transitions with $\Delta m_I = \pm 1$, $\Delta m_J = 0$ and also with $\Delta m_I = 0$, $\Delta m_J = \pm 1$ have been observed. The original studies[2,3] were made on the $J = 0$, 1 and 2 rotational states under strong field conditions and their most important result was the discovery of the electric quadrupole moment of the deuteron through the electric quadrupole interaction in D_2 and HD. Fig. 110 shows the complete

[1] N.F. Ramsey and H.R. Lewis: Phys. Rev. **108**, 1246 (1957).
[2] Kellogg, Rabi, Ramsey and Zacharias: Phys. Rev. **56**, 728 (1939); **57**, 677 (1940).
[3] N.F. Ramsey: Phys. Rev. **58**, 226 (1940).

nuclear reorientation spectrum ($\Delta m_I = \pm 1$, $\Delta m_J = 0$) observed for D_2. The central peak arises from the $J=0$ state and the six side peaks come from the $J=1$ state. A value of Q for the deuteron was obtained after a calculation of q was made[1].

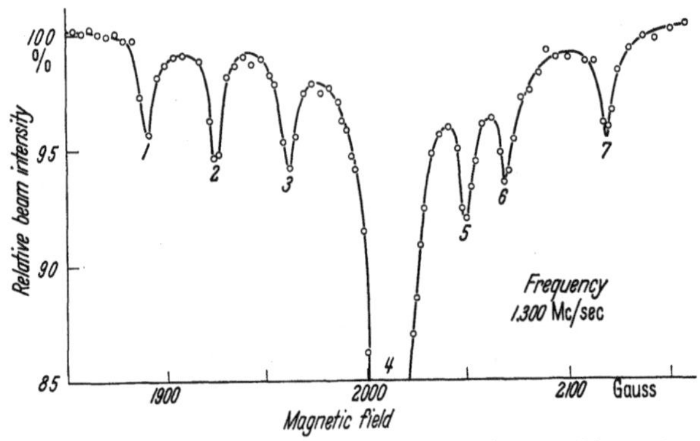

Fig. 110. Radiofrequency spectrum of D_2 for the transitions $\Delta m_I = \pm 1$, $\Delta m_J = 0$. Line 4 arises from the state $J=0$, $I=2$, and the remaining lines arise from the state $J=1$, $I=1$.

In the more recent work spectra for H_2, D_2 and HD have been[2] studied at low, intermediate, and strong fields with a greatly increased precision [25]. The transition $\Delta m_I = \pm 1$, $\Delta m_J = 0$ at strong field in the $J=0$, $I=2$ state of D_2 is shown in Fig. 53 of Sect. 29. The resonance line was obtained by the method of separated oscillating fields. It has a half-width for the central peak of about 200 cps and allows the determination of the resonance frequency to an accuracy of 10 to 20 cps.

Table 18. *Principal constants for hydrogen molecules.*

All data were taken for the $v=0$ vibrational state. For the molecule HD, c_p and c_d refer to the proton and deuteron respectively and $d_1 = \left\langle \frac{4\mu_p \mu_d}{5hR^3} \right\rangle$.

The error estimates are indicated by the numbers in parentheses, which apply to the last figure given. The results for HD are from unpublished work by N. F. RAMSEY, J. T. LATOURRETTE and others.

Quantity	H_2	D_2	HD
c	113,904 (30) cps	8,788 (25) cps	$c_p = 85{,}600$ (18) cps $c_d = 13{,}122$ (11) cps
d	57,671 (24) cps	25,237 (10) cps	$d_1 = 17{,}761$ (12) cps
$\langle r^3\rangle^{-\frac{1}{3}}$	0.74677 (10) $\times 10^{-8}$ cm		0.74604 (10) $\times 10^{-8}$ cm
eqQ/h		224,992 (100) cps	224,540 (60) cps
Q		2.738 (14) $\times 10^{-27}$ cm²	2.738 (19) $\times 10^{-27}$ cm²
σ_i	2.62 (4) $\times 10^{-5}$	263 (5) $\times 10^{-5}$	2.622 (32) $\times 10^{-5}$
$\mu_{Rot}(J=1)$	0.88291 (8) nm	0.442884 (51) nm	0.663211 (14) nm
f/H^2	$-(27.6 \pm 1.5) \times$ 10^{-6} cps-gauss^{-2}	$-(26.2 \pm 3.0) \times$ 10^{-6} cps-gauss^{-2}	$-(26.90 \pm 0.40) \times$ 10^{-6} cps-gauss^{-2}

A summary of the principal constants obtained for the hydrogen molecules is given in Table 18. In the cases where spectra have been studied under different magnetic field conditions the values of the constants obtained agree. The inaccuracy assigned to Q arises from uncertainty in the calculation of q for D_2

[1] A. NORDSIECK: Phys. Rev. **58**, 310 (1940).

[2] KOLSKY, PHIPPS, RAMSEY and SILSBEE: Phys. Rev. **87**, 395 (1952). — HARRICK, BARNES, BRAY and RAMSEY: Phys. Rev. **90**, 260 (1953). - HARRICK and RAMSEY: Phys. Rev. **88**, 228 (1952). — BARNES, BRAY and RAMSEY: Phys. Rev. **94**, 893 (1954).

and HD molecules[1]. A detailed account of the important work on the hydrogen molecules is given in Ref. [*25*].

The transitions $\Delta m_J = \pm 1$ as well as multiple quantum transitions $\Delta m_J = \pm 2, \pm 3$ in the low rotational states of O_2 have been observed[2] as resolved spectra by the molecular beam magnetic resonance method. The ratio of the g-value of the rotational magnetic moment g_K to that of the electron spin moment g_s was found to be $g_K/g_s = (6.08 \pm 0.74) \times 10^{-5}$. The ratio of the unpaired electron spin moments in O_2 to that of the free electron spin moment was measured to be $g_s(O_2)/g_s = 1 - (190 \pm 13) \times 10^{-6}$.

---

[1] G.F. Newell: Phys. Rev. **78**, 711 (1950).
[2] J.M. Hendrie and P. Kusch: Phys. Rev. **107**, 716 (1957).

Acknowledgment.

The authors gratefully acknowledge the assistance of the Air Force Office of Scientific Research, the Office of Naval Research and the U.S. Army Signal Corps, all of the U.S. Department of Defense, and of the National Science Foundation in the preparation of this article. We express our thanks to our wives, Inge Michelson Hughes and Edith McRoberts Kusch, for their forbearance during the preparation of this article and for their sympathetic understanding of the requirements of research.

General references.

[*1*] Andrew, E.R.: Nuclear Magnetic Resonance. Cambridge: Cambridge University Press 1955.
[*2*] Bethe, H.A., and E.E. Salpeter: Quantum Mechanics of One- and Two-Electron Systems. Encyclopedia of Physics. Vol. XXXV. Berlin: Springer 1957.
[3[Blatt, J.M., and V.F. Weisskopf: Theoretical Nuclear Physics. New York: John Wiley & Sons; London: Chapmann & Hall 1952.
[*4*] Condon, E.U., and G.H. Shortley: The Theory of Atomic Spectra. Cambridge: Cambridge University Press 1953.
[*5*] Debye, P.: Polar Molecules. New York: Dover Publications 1945.
[*6*] Dirac, P.A.M.: The Principles of Quantum Mechanics. Oxford: Clarendon Press 1958.
[*7*] Estermann, I.: Molecular Beam Technique. Rev. Mod. Phys. **18**, 300 (1946).
[*8*] Fraser, R.: Molecular Rays. Cambridge: Cambridge University Press 1931.
[*9*] Fraser, R.: Molecular Beams. London: Methuen & Co.; New York: John Wiley & Sons 1937.
[*10*] Gordy, W., W.V. Smith and R.F. Trambarulo: Microwave Spectroscopy. New York: John Wiley & Sons; London: Chapman & Hall 1953.
[*11*] Heitler, W.: The Quantum Theory of Radiation. Oxford: Clarendon Press 1954.
[*12*] Herzberg, G.: Spectra of Diatomic Molecules. New York: Van Nostrand 1950.
[*13*] Hund, F.: Allgemeine Quantenmechanik des Atom- und Molekelbaues. In Handbuch der Physik, Vol. XXIV/1. Berlin: Springer 1933.
[*14*] Jauch, J.M., and F. Rohrlich: The Theory of Photons and Electrons. Cambridge, Mass.: Addison-Wesley 1955.
[*15*] Jeans, J.: The Dynamical Theory of Gases. Cambridge: Cambridge University Press 1925.
[*16*] Kellogg, J.B.M., and S. Millman: Molecular Beam Magnetic Resonance Method. The Radiofrequency Spectra of Atoms and Molecules. Rev. Mod. Phys. **18**, 323 (1946).
[*17*] Kennard, E.H.: Kinetic Theory of Gases. New York-Toronto-London: McGraw-Hill 1938.
[*18*] King, J.G., and J.R. Zacharias: Some New Applications and Techniques of Molecular Beams. Advances in Electronics and Electron Physics, Vol. VIII. New York: Academic Press 1956.
[*19*] Kopfermann, H.: Kernmomente, 2. Aufl. Stuttgart 1955. — Nuclear Moments. New York: Academic Press 1958.
[*20*] Mayer, M.G., and J.H.D. Jensen: Elementary Theory of Nuclear Shell Structure. New York: John Wiley & Sons; London: Chapman & Hall 1955.
[*21*] Montgomery, C.G.: Technique of Microwave Measurements. Massachusetts Institute of Technology Radiation Laboratory Series, Vol. 11. New York-Toronto-London: McGraw-Hill 1947.

[22] MONTGOMERY, C. G., R.H. DICKE and E.M. PURCELL: Principles of Microwave Circuits. Massachusetts Institute of Technology Radiation Laboratory Series, Vol. 8. New York: McGraw-Hill 1948.
[23] PAULI, W., and A.O.G. KÄLLÉN: Two articles on the principles of quantum theory. Encyclopedia of Physics, Vol. V/1. Berlin: Springer 1958.
[24] RAMSEY, N.F.: Nuclear Moments. New York: John Wiley & Sons; London: Chapman & Hall 1953.
[25] RAMSEY, N.F.: Molecular Beams. Oxford: Clarendon Press 1956.
[26] SCHIFF, L.I.: Quantum Mechanics. New York: McGraw-Hill 1955.
[27] SMITH, K.F.: Molecular Beams. London: Methuen & Co.; New York: John Wiley & Sons 1955.
[28] TOWNES, C.H., and A.L. SCHAWLOW: Microwave Spectroscopy. New York: McGraw-Hill 1955.
[29] VAN VLECK, J.H.: The Theory of Electric and Magnetic Susceptibilities. London: Oxford University Press 1953.

References to original papers.

1900–1919

[K1] KNUDSEN, M.: Ann. Physik **28**, 999 (1909). Molecular effusion theory.
[D1] DUNOYER, L.: C. R. Acad. Sci., Paris **152**, 592 (1911). Existence of atomic beams.
[D2] DUNOYER, L.: Le Radium **8**, 142 (1911). Existence of atomic beams.
[K2] KNUDSEN, M.: Ann. Physik **48**, 1113 (1915). Molecular effusion theory.

1920

[S1] STERN, O.: Z. Physik **2**, 49 (1920). Velocity distribution, Ag.
[S2] STERN, O.: Z. Physik **3**, 417 (1920). Velocity distribution, Ag.

1921

[K1] KALLMANN, H., and F. REICHE: Z. Physik **6**, 352 (1921). Magnetic deflection of molecules.
[S1] STERN, O.: Z. Physik **7**, 249 (1921). Space quantization of magnetic moments.

1922

[E1] EINSTEIN, A., and P. EHRENFEST: Z. Physik **11**, 31 (1922). Critique Stern Gerlach experiment.
[G1] GERLACH, W., and O. STERN: Z. Physik **8**, 110 (1922). Space quantization of magnetic moments, Ag.
[G2] GERLACH, W., and O. STERN: Z. Physik **9**, 349 (1922). Space quantization of magnetic moments, Ag.
[G3] GERLACH, W., and O. STERN: Z. Physik **9**, 353 (1922). Atomic magnetic moment by deflection, Ag.

1924

[G1] GERLACH, W., and A.C. CILLIERS: Z. Physik **26**, 106 (1924). Atomic magnetic moments by deflection: Ag, Cu, Au, Tl, Pb.
[G2] GERLACH, W., and O. STERN: Ann. Physik **74**, 673 (1924). Confirmation of Debye-Sommerfeld Quantization.

1925

[G1] GERLACH, W.: Ann. Physik **76**, 163 (1925). Atomic magnetic moments by deflection: Cu, Ag, Au, Pb, Sn, Fe, Ni.

1926

[K1] KNAUER, F., and O. STERN: Z. Physik **39**, 764 (1926). (U. z. M. 2.) Molecular beam apparatus.
[K2] KNAUER, F., and O. STERN: Z. Physik **39**, 780 (1926). (U. z. M. 3.) Molecular magnetic moments by deflection ($\sim$ nm), H_2O.
[S1] STERN, O.: Z. Physik **39**, 751 (1926). (U. z. M. 1.) Review of molecular beam technique, application.
[T1] TAYLOR, J.B.: Phys. Rev. **28**, 576 (1926). Atomic magnetic moments by deflection: Na, K.

1927

[E1] ELDRIDGE, J.A.: Phys. Rev. **30**, 931 (1927). Test Maxwell velocity distribution.
[L1] LEU, A.: Z. Physik **41**, 551 (1927). (U. z. M. 4.) Atomic magnetic moments: Na, K, Tl.
[P1] PHIPPS, T.E., and J.B. TAYLOR: Phys. Rev. **29**, 309 (1927). Atomic magnetic moment by deflection, H.

[R1] RODEBUSH, W. H.: Proc. Nat. Acad. Sci. U.S.A. **13**, 50 (1927). Effect velocity distribution on deflection atoms.

[S1] STERN, O.: Z. Physik **41**, 563 (1927). (U. z. M. 5.) Intensity distribution of deflected beam.

[W1] WREDE, E.: Z. Physik **41**, 569 (1927). (U. z. M. 6) Atomic magnetic moment by deflection,

[W2] H. WREDE, E.: Z. Physik **44**, 261 (1927). (U. z. M. 7.) Molecular electric dipole moments by deflection: NaCl, KCl.

1928

[B1] BREIT, G.: Nature, Lond. **122**, 649 (1928). Relativistic theory of electron *g*-value.

[E1] ESTERMANN, I.: Z. phys. Chem. **1**, 161 (1928). Electric dipole moment of organic molecules by deflection.

[J1] JOHNSON, T. H.: Phys. Rev. **31**, 103 (1928). Pressure detector for molecular beams.

[L1] LEU, A.: Z. Physik **49**, 498 (1928). (U. z. M. 8.) Bi magnetic moment and heat of dissociation.

1929

[C1] CLARK, R. J.: Proc. Roy. Soc. Lond. **124**, 689 (1929). Electric dipole moment of As_2O_3.

[E1] ESTERMANN, J.: Z. phys. Chem. **2**, 287 (1929). Electric dipole moment of Penta-Erythrite by deflection.

[G1] GERLACH, W.: J. Phys. Radium (VI) **10**, 273 (1929). Space quantization of magnetic moments, review.

[G2] GOUDSMIT, S., and R. BACHER: Phys. Rev. **34**, 1499 (1929). Hfs theory.

[K1] KNAUER, F., and O. STERN: Z. Physik **53**, 766 (1929). (U. z. M. 10.) Molecular beam intensity measurements with Pirani gauge.

[K2] KNAUER, F., and O. STERN: Z. Physik **53**, 779 (1929). (U. z. M. 11.) Reflection and diffraction of molecular beams.

[K3] KURT, E. O., and T. E. PHIPPS: Phys. Rev. **34**, 1357 (1929). Magnetic moment of O by deflection.

[L1] LAMMERT, B.: Z. Physik **56**, 244 (1929). (U. z. M. 13.) Velocity selector; test of Maxwell velocity distribution.

[M1] MAYER, H.: Z. Physik **58**, 373 (1929). Cosine law for molecular beams.

[R1] RABI, I. I.: Nature, Lond. **123**, 163 (1929). Magnetic deflecting field.

[R2] RABI, I. I.: Z. Physik **54**, 190 (1929). (U. z. M. 12.) Magnetic deflecting field.

[T1] TAYLOR, J. B.: Z. Physik **52**, 846 (1929). (U. z. M. 9.) Magnetic moment of Li.

[T2] TAYLOR, J. B.: Z. Physik **57**, 242 (1929). (U. z. M. 14.) Surface ionization detector and beam intensity distribution.

1930

[C1] CLAUSING, P.: Ann. Physik **7**, 569 (1930). Molecular velocity distribution and cosine law of effusion.

[C2] CLAUSING, P.: Z. Physik **66**, 471 (1930). Molecular flow and beam shape.

[E1] ESTERMANN, I., and O. STERN: Z. Physik **61**, 95 (1930). (U. z. M. 15.) Diffraction of beams by crystals.

[F1] FERMI, E.: Z. Physik **60**, 320 (1930). Theory hfs.

[J1] JACKSON, L. C.: Nature, Lond. **125**, 131 (1930). Stern-Gerlach experiment on N.

[J2] JACKSON, L. C., and L. F. BROADWAY: Proc. Roy. Soc. Lond. **127**, 678 (1930). Stern-Gerlach experiment on N.

[K1] KNAUER, F., and O. STERN: Z. Physik **60**, 414 (1930). Limits of cosine law of effusion.

[Z1] ZARTMAN, I. F.: Phys. Rev. **35**, 134 (1930). Molecular velocity distribution.

1931

[B1] BREIT, G., and I. I. RABI: Phys. Rev. **38**, 2082 (1931). Breit-Rabi formula.

[E1] ELLETT, A., and R. M. ZABEL: Phys. Rev. **37**, 1112 (1931). Beam intensities.

[E2] ESTERMANN, I., R. FRISCH and O. STERN: Phys. Z. **32**, 670 (1931). Monochromatic de Broglie waves of molecular rays.

[E3] ESTERMANN, I., R. FRISCH and O. STERN: Z. Physik **73**, 348 (1931). (U. z. M. 18.) Monochromatic de Broglie waves of molecular rays.

[G1] GOUDSMIT, S.: Phys. Rev. **37**, 663 (1931). Hfs theory.

[G2] GÜTTINGER, P.: Z. Physik **73**, 169 (1931). Atoms in rotating magnetic fields.

[L1] LEWIS, L. C.: Z. Physik **69**, 786 (1931). (U. z. M. 16.) Composition of beams.

[P1] PHIPPS, T. E., and O. STERN: Z. Physik **73**, 185 (1931). (U. z. M. 17.) Non-adiabatic spin transitions.

[R1] RACAH, G.: Z. Physik **71**, 431 (1931). Hfs theory.

1932

[*B1*] BREIT, G.: Phys. Rev. **42**, 348 (1932). Theory isotope displacement in hfs.
[*C1*] CASIMIR, H.: Z. Physik **77**, 811 (1932). Theory perturbations in hfs.
[*D1*] DUNHAM, J.L.: Phys. Rev. **41**, 721 (1932). Theory vibrating rotator.
[*M1*] MAJORANA, E.: Nuovo Cim. **9**, 43 (1932). Theory non-adiabatic magnetic transitions.
[*R1*] RODEBUSH, W.H., and W.F. HENRY: Phys. Rev. **39**, 386 (1932). Beams of salts.
[*R2*] ROSENTHAL, J.E., and G. BREIT: Phys. Rev. **41**, 459 (1932). Theory isotope shift in hfs.

1933

[*B1*] BREIT, G., and L.A. WILLS: Phys. Rev. **44**, 470 (1933). Theory hfs.
[*E1*] ESTERMANN, I., and R.G.J. FRASER: J. Chem. Phys. **1**, 390 (1933). Electric deflection; electric dipole moment of HCl.
[*E2*] ESTERMANN, I., and O. STERN: Z. Physik **85**, 17 (1933). (U. z. M. 27.) Magnetic moment of proton from H_2 deflection.
[*E3*] ESTERMANN, I., and O. STERN: Z. Physik **85**, 135 (1933). (U. z. M. 28.) Intensities of molecular beams.
[*E4*] ESTERMANN, I., and O. STERN: Z. Physik **86**, 132 (1933). (U. z. M. 29.) Magnetic moment of deuteron from D_2, HD deflection.
[*E5*] ESTERMANN, I., and M. WOHLWILL: Z. phys. Chem. **20**, 195 (1933). Electric dipole moments by deflection.
[*F1*] FERMI, E., and E. SEGRÈ: Z. Physik **82**, 729 (1933). Theory hfs.
[*F2*] FRISCH, R., and E. SEGRÈ: Z. Physik **80**, 610 (1933). (U. z. M. 22.) Experiments on non-adiabatic transitions.
[*F3*] FRISCH, R.: Z. Physik **86**, 42 (1933). (U. z. M. 30.) Molecule recoil associated with optical radiation.
[*F4*] FRISCH, R., and O. STERN: Z. Physik **85**, 4 (1933). (U. z. M. 24.) Magnetic moment of proton from H_2 deflection.
[*F5*] FRISCH, R., and O. STERN: Z. Physik **84**, 430 (1933). (U. z. M. 23.) Refraction and reflection of beams from crystals.
[*F6*] FRISCH, R.: Z. Physik **84**, 443 (1933). (U. z. M. 25.) Refraction and reflection of beams from crystals.
[*G1*] GOUDSMIT, S.: Phys. Rev. **43**, 636 (1933). Theory hfs.
[*G2*] GOUDSMIT, S., and R. BACHER: Phys. Rev. **43**, 894 (1933). Theory hfs.
[*J1*] JOSEPHY, B.: Z. Physik **80**, 755 (1933). (U. z. M. 21.) Reflection of Hg atoms from crystals.
[*K1*] KNAUER, F.: Z. Physik **80**, 80 (1933). (U. z. M. 20.) Beam scattering by gases.
[*M1*] MEISSNER, W., and H. SCHEFFERS: Phys. Z. **34**, 48 (1933). Precise atomic magnetic moment by deflection: K, Li.
[*M2*] MEISSNER, W., and H. SCHEFFERS: Phys. Z. **34**, 245 (1933). Precise atomic magnetic moment by deflection: K, Li.
[*R1*] RABI, I.I., and V.W. COHEN: Phys. Rev. **43**, 582 (1933). Nuclear spin of Na^{23}.
[*S1*] SCHNURMANN, R.: Z. Physik **85**, 212 (1933). (U. z. M. 26.) O_2 magnetic moment by deflection.
[*S2*] SONKIN, S.: Phys. Rev. **43**, 788 (1933). Detection of Hg metastable atoms.
[*W1*] WICK, G.C.: Nuovo Cim. **10**, 118 (1933). Theory rotational magnetic moment of H_2.
[*W2*] WICK, G.C.: Z. Physik **85**, 25 (1933). Theory rotational magnetic moment of H_2.
[*W3*] WOHLWILL, M.: Z. Physik **80**, 67 (1933). (U. z. M. 19.) Electric dipole moment of *p*-nitranaline by deflection.

1934

[*C1*] COHEN, V.W.: Phys. Rev. **46**, 713 (1934). Nuclear spin of Cs^{133}.
[*C2*] COPLEY, M.J., and T.E. PHIPPS: Phys. Rev. **45**, 344 (1934). Surface ionization detection of K.
[*F1*] FRASER, R.G.J.: Trans. Faraday Soc. **30**, 182 (1934). Molecular beams for study of free radicals.
[*R1*] RABI, I.I., J.M.B. KELLOGG and J.R. ZACHARIAS: Phys. Rev. **46**, 157 (1934). Magnetic moment of proton by deflection.
[*R2*] RABI, I.I., J.M.B. KELLOGG and J.R. ZACHARIAS: Phys. Rev. **46**, 163 (1934). Magnetic moment of deuteron by deflection.
[*R3*] RABI, I.I., and V.W. COHEN: Phys. Rev. **46**, 707 (1934). Method spin measurement.
[*S1*] SCHEFFERS, H.: Phys. Z. **35**, 425 (1934). Electric dipole moments by deflection.
[*S2*] SCHEFFERS, H., and J. STARK: Phys. Z. **35**, 625 (1934). Electric deflection of alkali atoms.

1935

[*C1*] Casimir, H. B. G.: Physica, Haag **2**, 719 (1935). Theory hfs of Eu.
[*C2*] Copley, M. J., T. E. Phipps and J. Glasser: Rev. Sci. Instrum. **6**, 371 (1935). Ionization gauge beam detector.
[*F1*] Fox, M., and I. I. Rabi: Phys. Rev. **48**, 746 (1935). Nuclear magnetic dipole moments of Li, Na, K.
[*K1*] Kratzenstein, M.: Z. Physik **93**, 279 (1935). Angular distribution of beam intensity; "clouds".
[*M1*] Millman, S.: Phys. Rev. **47**, 739 (1935). Nuclear spins and magnetic dipole moments of K isotopes.

1936

[*C1*] Casimir, H. B. G.: Teylors' Tweede Genootschap **11**, 36 (1936). Theory hfs.
[*F1*] Fraser, R. G. J., and J. V. Hughes: J. Chem. Phys. **4**, 730 (1936). Electric dipole moments of molecules by deflection.
[*H1*] Huntoon, R. D., and A. Ellett: Phys. Rev. **49**, 381 (1936). Ionization gauge beam detector.
[*K1*] Kellogg, J. M. B., I. I. Rabi and J. R. Zacharias: Phys. Rev. **50**, 472 (1936). Magnetic dipole moment of proton and deuteron from hydrogen molecules by zero moment method.
[*M1*] Manley, J. H.: Phys. Rev. **49**, 921 (1936). K^{41} nuclear spin and magnetic dipole moment by zero moment method.
[*M2*] Millman, S., and M. Fox: Phys. Rev. **50**, 220 (1936). Rb^{85}, Rb^{87} nuclear spins and magnetic moments.
[*M3*] Motz, L., and M. Rose: Phys. Rev. **50**, 348 (1936). Theory space quantization.
[*R1*] Rabi, I. I.: Phys. Rev. **49**, 324 (1936). Theory space quantization.
[*R2*] Rodebush, W. H., L. A. Murray and M. E. Bixler: J. Chem. Phys. **4**, 372, (1936). Alkali halide electric dipole moments by deflection.
[*R3*] Rodebush, W. H.: J. Chem. Phys. **4**, 536 (1936). Alkali halide electric dipole moments by deflection.
[*S1*] Scheffers, H., and J. Stark: Phys. Z. **37**, 217 (1936). Electric deflection of H.
[*S2*] Scheffers, H.: Phys. Z. **37**, 220 (1936). Electric deflection of H.

1937

[*C1*] Cohen, V. W., and A. Ellett: Phys. Rev. **52**, 502 (1937). Velocity distribution.
[*C2*] Copley, M. J., and V. Deitz: Rev. Sci. Instrum. **8**, 314 (1937). Torsion manometer for beam force.
[*E1*] Estermann, I., O. C. Simpson and O. Stern: Phys. Rev. **52**, 535 (1937). Magnetic dipole moment of proton by deflection.
[*M1*] Manley, J. H., and S. Millman: Phys. Rev. **51**, 19 (1937). Li^6 nuclear spin, magnetic dipole moment by zero moment method.
[*M2*] Millman, S., and J. R. Zacharias: Phys. Rev. **51**, 1049 (1937). Signs of nuclear magnetic dipole moments of Li^7, Rb^{85}, Rb^{87}, Cs^{133} by zero moment method.
[*R1*] Rabi, I. I.: Phys. Rev. **51**, 652 (1937). Theory space quantization in rotating magnetic field.
[*S1*] Schwinger, J.: Phys. Rev. **51**, 648 (1937). Theory non-adiabatic transitions.
[*S2*] Stern, O.: Phys. Rev. **51**, 852 (1937). Measurement of Bohr magneton.
[*T1*] Torrey, H. C.: Phys. Rev. **51**, 501 (1937). Sign of K^{39} nuclear magnetic dipole moment.

1938

[*G1*] Gorham, J. E.: Phys. Rev. **53**, 563 (1938). Signs of Li^6, K^{41} nuclear magnetic dipole moments.
[*K1*] Kellogg, J. M. B., and N. F. Ramsey: Phys. Rev. **53**, 331 (1938). Magnetic dipole moments of Ne, A.
[*M1*] Millman, S., I. I. Rabi and J. R. Zacharias: Phys. Rev. **53**, 384 (1938). Nuclear magnetic dipole moments of In isotopes by zero moment method.
[*R1*] Rabi, I. I., J. R. Zacharias, S. Millman and P. Kusch: Phys. Rev. **53**, 318 (1938). Molecular beam magnetic resonance method.

1939

[*C1*] Caldirola, P.: Nuovo Cim. **16**, 242 (1939). Non-adiabatic processes.
[*H1*] Hamilton, D. R.: Phys. Rev. **56**, 30 (1939). In^{115} hfs and quadrupole moment by zero moment method.
[*K1*] Kellogg, J. M. B., I. I. Rabi, N. F. Ramsey and J. R. Zacharias: Phys. Rev. **56**, 728 (1939). Magnetic dipole moments of proton and deuteron by magnetic resonance on hydrogen molecules.

[K 2] KUSCH, P., S. MILLMAN and I.I. RABI: Phys. Rev. **55**, 666 (1939). Be nuclear magnetic dipole moment by magnetic resonance.

[K 3] KUSCH, P., S. MILLMAN and I.I. RABI: Phys. Rev. **55**, 1176 (1939). N^{14}, Na^{23}, K^{39}, Cs^{133} nuclear magnetic dipole moments by magnetic resonance.

[K 4] KUSCH, P., and S. MILLMAN: Phys. Rev. **56**, 527 (1939). Rb, Cl nuclear magnetic dipole moments by magnetic resonance.

[M 1] MILLMAN, S.: Phys. Rev. **55**, 628 (1939). Signs of nuclear magnetic dipole moments by magnetic resonance.

[M 2] MILLMAN, S., P. KUSCH and I.I. RABI: Phys. Rev. **56**, 165 (1939). B nuclear magnetic dipole moments by magnetic resonance.

[M 3] MILLMAN, S., and P. KUSCH: Phys. Rev. **56**, 303 (1939). Al^{27} nuclear spin and magnetic dipole moment by magnetic resonance.

[R 1] RABI, I.I., S. MILLMAN, P. KUSCH and J.R. ZACHARIAS: Phys. Rev. **55**, 526 (1939). Molecular beam magnetic resonance method for nuclear magnetic dipole moments: Li^6, Li^7, F^{19}.

[S 1] SCHEFFERS, H.: Phys. Z. **40**, 1 (1939). Electric deflection of O, H_2O.

1940

[A 1] AMDUR, I., and H. PEARLMAN: J. Chem. Phys. **8**, 7 (1940). High velocity atomic beams.

[B 1] BLOCH, F., and A. SIEGERT: Phys. Rev. **57**, 522 (1940). Theory magnetic resonance in non-rotating magnetic fields.

[K 1] KELLOGG, J.M.B., I.I. RABI, N.F. RAMSEY and J.R. ZACHARIAS: Phys. Rev. **57**, 677 (1940). Radiofrequency spectra of HD, D_2 by magnetic resonance; electric quadrupole moment of deuteron.

[K 2] KUSCH, P., S. MILLMAN and I.I. RABI: Phys. Rev. **57**, 765 (1940). Radiofrequency spectra of atoms by magnetic resonance; hfs and Zeeman effect of Li, K.

[M 1] MARGENAU, H.: Phys. Rev. **57**, 383 (1940). Relativistic theory of electron g-value in an atom.

[M 2] MILLMAN, S., and P. KUSCH: Phys. Rev. **58**, 438 (1940). Hfs of Na, Rb, Cs by magnetic resonance.

[N 1] NORDSIECK, A.: Phys. Rev. **58**, 310 (1940). Theory deuteron electric quadrupole interaction in HD, D_2.

[R 1] RAMSEY, N.F. jr.: Phys. Rev. **58**, 226 (1940). H_2, D_2, and HD rotational magnetic moments.

[R 2] RENZETTI, N.A.: Phys. Rev. **57**, 753 (1940). Ga hfs and electric quadrupole moment by zero moment method.

[S 1] SCHEFFERS, H.: Phys. Z. **41**, 89 (1940). Electric dipole moments by deflection.

[S 2] SCHEFFERS, H.: Phys. Z. **41**, 98 (1940). Electric deflection of NH_3.

[S 3] SHRADER, E.F., S. MILLMAN and P. KUSCH: Phys. Rev. **58**, 925 (1940). Nuclear moments of Cl isotopes by molecular beam magnetic resonance.

[S 4] STEVENSON, A.F.: Phys. Rev. **58**, 1061 (1940). Theory magnetic resonance transitions.

[Z 1] ZACHARIAS, J.R., and J.M.B. KELLOGG: Phys. Rev. **57**, 570 (1940). N^{15} nuclear moment from N_2 magnetic resonance.

1941

[H 1] HAMILTON, D.R.: Amer. J. Phys. **9**, 319 (1941). Molecular beam magnetic resonance method, introductory review.

[H 2] HAY, R.H.: Phys. Rev. **60**, 75 (1941). C^{13}, Ba^{135}, Ba^{137} nuclear moments by molecular beam magnetic resonance.

[L 1] LAMB jr., W.E.: Phys. Rev. **60**, 817 (1941). Theory diamagnetic shielding of nuclei in atoms.

[M 1] MILLMAN, S., and P. KUSCH: Phys. Rev. **60**, 91 (1941). Precise nuclear magnetic moments of H^1, Li^7, F^{19}, Na^{23} by molecular beam magnetic resonance.

[P 1] PHILLIPS, M.: Phys. Rev. **60**, 100 (1941). Theory atomic g_J values of alkali atoms.

[T 1] TORREY, H.C.: Phys. Rev. **59**, 293 (1941). Theory magnetic resonance transition including beam velocity distribution.

1942

[B 1] BESSEY, W.H., and O.C. SIMPSON: Chem. Revs. **30**, 239 (1942). Molecular beam experiments, review.

[C 1] CASIMIR, H.B.G., and G. KARREMAN: Physica, Haag **9**, 494 (1942). Theory magnetic octopole moment.

[D 1] DORRESTEIN, R.: Physica, Haag **9**, 433 (1942). Detection, metastable states of noble gas atoms.

[D 2] DORRESTEIN, R.: Physica, Haag **9**, 447 (1942). Detection, metastable states of noble gas atoms.

[H 1] HARDY, T.C., and S. MILLMAN: Phys. Rev. **61**, 459 (1942). Hfs and nuclear spins of In isotopes by atomic beam magnetic resonance.
[R 1] RACAH, G.: Phys. Rev. **62**, 438 (1942). Theory of complex spectra.
[Z 1] ZACHARIAS, J.R.: Phys. Rev. **61**, 270 (1942). K^{40} nuclear spin and magnetic moment by atomic beam magnetic resonance.

1943

[H 1] HEIL, H.: Z. Physik **120**, 212 (1943). Ionizer for beam detector.

1944

[C 1] COBAS, A., and W.E. LAMB jr.: Phys. Rev. **65**, 327 (1944). Theory metastable atom detection by electron ejection from a metal.

1945

[B 1] BLOCH, F., and I.I. RABI: Rev. Mod. Phys. **17**, 237 (1945). Theory atoms in variable magnetic fields.
[F 1] FELD, R.T., and W.E. LAMB jr.: Phys. Rev. **67**, 15 (1945). Theory effect nuclear electric quadrupole moment on molecular beam magnetic resonance spectra.

1946

[E 1] ESTERMANN, I.: Rev. Mod. Phys. **18**, 300 (1946). Molecular beam technique, review.
[H 1] HUGHES, H.K.: Phys. Rev. **70**, 570 (1946). Molecular beam electric resonance method.
[K 1] KELLOGG, J.B.M., and S. MILLMAN: Rev. Mod. Phys. **18**, 323 (1946). Molecular beam magnetic resonance method, review.
[N 1] NIERENBERG, W.A., N.F. RAMSEY and S.B. BRODY: Phys. Rev. **70**, 773 (1946). Nuclear electric quadrupole interactions in sodium halide molecules by magnetic resonance.

1947

[B 1] BETHE, H.A.: Phys. Rev. **72**, 339 (1947). Theory Lamb shift.
[B 2] BREIT, G.: Phys. Rev. **72**, 984 (1947). Theory anomalous magnetic moment of electron.
[B 3] BREIT, G., and R.E. MEYEROTT: Phys. Rev. **72**, 1023 (1947). Theory reduced mass correction to H hfs.
[B 4] BRODY, S.B., W.A. NIERENBERG and N.F. RAMSEY: Phys. Rev. **72**, 258 (1947). Br nuclear moments by molecular beam magnetic resonance.
[E 1] ESTERMANN, I., O.C. SIMPSON and O. STERN: Phys. Rev. **71**, 238 (1947). Atom free fall and velocity distribution.
[F 1] FELD, B.T.: Phys. Rev. **72**, 1116 (1947). Theory nuclear electric quadrupole interaction.
[F 2] FOLEY, H.M.: Phys. Rev. **71**, 747 (1947). Theory nuclear electric quadrupole interaction in homo-nuclear diatomic molecule.
[F 3] FOLEY, H.M.: Phys. Rev. **72**, 504 (1947). Theory internal nuclear-molecular interactions in diatomic molecules; pseudo-quadrupole effect.
[H 1] HUGHES, H.K.: Phys. Rev. **72**, 614 (1947). Molecular beam electric resonance method; A and μ of CsF.
[K 1] KUSCH, P., and H.M. FOLEY: Phys. Rev. **72**, 1256 (1947). Anomalous spin magnetic moment of electron from Ga g-values.
[L 1] LAMB, W.E., and R.C. RETHERFORD: Phys. Rev. **72**, 241 (1947). Fine structure of H; Lamb shift.
[M 1] MACK, J.E., and N. AUSTERN: Phys. Rev. **72**, 972 (1947). Fine structure of He^+; Lamb shift.
[N 1] NAFE, J.E., E.B. NELSON and I.I. RABI: Phys. Rev. **71**, 914 (1947). Hfs of H, D by magnetic resonance.
[N 2] NAGLE, D.E., R.S. JULIAN and J.R. ZACHARIAS: Phys. Rev. **72**, 971 (1947). Hfs of H, D by magnetic resonance.
[N 3] NIERENBERG, W.A., and N.F. RAMSEY: Phys. Rev. **72**, 1075 (1947). Na halide radiofrequency spectra by magnetic resonance.
[T 1] TOWNES, C.H.: Phys. Rev. **71**, 909 (1947). Theory nuclear electric quadrupole interaction in molecules.

1948

[B 1] BECKER, G.E., and P. KUSCH: Phys. Rev. **73**, 584 (1948). Ga hfs by magnetic resonance.
[B 2] BOHR, A.: Phys. Rev. **73**, 1109 (1948). Theory hfs anomaly in D.
[B 3] BREIT, G., and G.E. BROWN: Phys. Rev. **74**, 1278 (1948). Theory effect nuclear motion on H fine structure.

[D1] DAVIS jr., L., B.T. FELD, C.W. ZABEL and J.R. ZACHARIAS: Phys. Rev. **73**, 525 (1948). Cl hfs by magnetic resonance.
[D2] DAVIS jr., L.: Phys. Rev. **74**, 1193 (1948). Radioactive Na^{22} nuclear spin and hfs by magnetic resonance.
[F1] FANO, U.: J. Res. Nat. Bur. Stand. **40**, 215 (1948). Theory Stark effect and nuclear electric quadrupole interaction in diatomic molecules.
[F2] FOLEY, H.M., and P. KUSCH: Phys. Rev. **73**, 412 (1948). Anomalous spin magnetic moment of electron.
[K1] KUSCH, P., and H.M. FOLEY: Phys. Rev. **74**, 250 (1948). Anomalous spin magnetic moment of electron from atomic g-values.
[L1] LEW, H.: Phys. Rev. **74**, 1550 (1948). Al^{27} hfs and nuclear electric quadrupole moment by magnetic resonance.
[N1] NAFE, J.E., and E.B. NELSON: Phys. Rev. **73**, 718 (1948). H, D hfs by magnetic resonance.
[N2] NIERENBERG, W.A., I.I. RABI and M. SLOTNICK: Phys. Rev. **73**, 1430 (1948). Theory Stark effect in diatomic molecules.
[P1] PAUL, W., and G. WESSEL: Z. Physik **124**, 691 (1948). Ag beam velocity, density.
[R1] RAMSEY, N.F.: Phys. Rev. **74**, 286 (1948). Theory magnetic resonance transitions at twice Larmor frequency.
[S1] SCHWINGER, J.: Phys. Rev. **73**, 416 (1948). Theory anomalous spin magnetic moment of electron.
[T1] TRISCHKA, J.W.: Phys. Rev. **74**, 718 (1948). CsF nuclear electric quadrupole interaction by electric resonance.
[W1] WELTON, T. A.: Phys. Rev. **74**, 1157 (1948). Theory Lamb shift and electron spin magnetic moment.
[W2] WICK, G.C.: Phys. Rev. **73**, 51 (1948). Theory spin-rotation interaction in diatomic molecule.

1949

[A1] ANDERSON, H.L.: Phys. Rev. **76**, 1460 (1949). He^3 nuclear magnetic dipole moment by nuclear resonance absorption.
[B1] BITTER, F.: Phys. Rev. **76**, 150 (1949). Nuclear magnetic dipole moments of Rb isotopes by nuclear resonance absorption; hfs anomaly.
[B2] BITTER, F.: Phys. Rev. **76**, 833 (1949). Optical detection of radiofrequency resonance.
[B3] BREIT, G., G.E. BROWN and G.B. ARFKEN: Phys. Rev. **76**, 1299 (1949). Theory effect nuclear motion on H hfs.
[B4] BREIT, G., and R.E. MEYEROTT: Phys. Rev. **75**, 1447 (1949). Theory effect of nuclear motion on hfs of H.
[B5] BROSSEL, J., and A. KASTLER: C. R. Acad. Sci., Paris **229**, 1213 (1949). Magnetic resonance in excited atomic states by depolarization.
[C1] CRAWFORD, M.F., and A.L. SCHAWLOW: Phys. Rev. **76**, 1310 (1949). Theory effect of nuclear size on hfs.
[D1] DAVIS jr., L.: Phys. Rev. **76**, 435 (1949). Radioactive Cs^{137} nuclear spin and hfs by magnetic resonance.
[D2] DAVIS jr., L., B.T. FELD, C.W. ZABEL and J.R. ZACHARIAS: Phys. Rev. **76**, 1076 (1949). Cl hfs by magnetic resonance.
[D3] DAVIS jr., L., D.E. NAGLE and J.R. ZACHARIAS: Phys. Rev. **76**, 1068 (1949). Radioactive Na^{22}, K^{40}, Cs^{135} and Cs^{137} spins and hfs by magnetic resonance.
[F1] FRENCH, J.B., and V.F. WEISSKOPF: Phys. Rev. **75**, 1240 (1949). Theory Lamb shift.
[G1] GARDNER, J.H., and E.M. PURCELL: Phys. Rev. **76**, 1262 (1949). Proton magnetic dipole moment in Bohr magnetons, g_P/g_L.
[H1] HUGHES, H.K.: Phys. Rev. **76**, 1675 (1949). Theory Stark effect of polar diatomic molecules.
[K1] KNAUER, F.: Z. Physik **126**, 310 (1949). Surface ionization detection of alkali beams.
[K2] KNAUER, F.: Z. Physik **126**, 319 (1949). Velocity distribution.
[K3] KROLL, N.M., and W.E. LAMB jr.: Phys. Rev. **75**, 388 (1949). Theory Lamb shift.
[K4] KUSCH, P.: Phys. Rev. **75**, 887 (1949). Electric quadrupole moment of Li^6 by molecular beam magnetic resonance.
[K5] KUSCH, P.: Phys. Rev. **76**, 138 (1949). Sign of electric quadrupole interaction of Li^7 in Li_2 and LiBr.
[K6] KUSCH, P., and A.K. MANN: Phys. Rev. **76**, 707 (1949). Li^6, Li^7 nuclear g values by molecular beam magnetic resonance.
[K7] KUSCH, P., and H. TAUB: Phys. Rev. **75**, 1477 (1949). Alkali atoms hfs and g_J values.
[L1] LEW, H.: Phys. Rev. **76**, 1086 (1949). Al^{27} hfs and nuclear electric quadrupole moment.

[N1] NAGLE, D.E.: Phys. Rev. **76**, 847 (1949). Radioactive Cs^{135}, Cs^{137} nuclear spins and hfs by magnetic resonance.
[N2] NELSON, E.B., and J.E. NAFE: Phys. Rev. **75**, 1194 (1949). H^3 hfs by magnetic resonance.
[N3] NELSON, E.B., and J.E. NAFE: Phys. Rev. **76**, 1858 (1949). g_J in H, D ground states by magnetic resonance.
[N4] NÖLLER, H.G., G.W. OETJEN and R. JAECKEL: Z. Naturforsch. **4**a, 101 (1949). Intense beam of H_2.
[R1] RAMSEY, N.F.: Phys. Rev. **76**, 996 (1949). Theory method of separated oscillating fields.
[S1] SCHWINGER: J. Phys. Rev. **76**, 790 (1949). Theory quantum electrodynamic radiative corrections.
[T1] TAUB, H., and P. KUSCH: Phys. Rev. **75**, 1481 (1949). Proton magnetic moment by molecular beam magnetic resonance.
[T2] TOWNES, C.H., H.M. FOLEY and W. LOW: Phys. Rev. **76**, 1415 (1949). Nuclear electric quadrupole moments and shell structure.
[T3] TOWNES, C.H., and B.P. DAILEY: J. Chem. Phys. **17**, 782 (1949). Theory electric quadrupole interaction and molecular structure.
[T4] TRISCHKA, J.W.: Phys. Rev. **76**, 1365 (1949). CsF μ, A by electric resonance.
[V1] VAUTHIER, R.: C. R. Acad. Sci., Paris **228**, 1113 (1949). Electric deflection of molecules.
[W1] WEISSKOPF, V.F.: Rev. Mod. Phys. **21**, 305 (1949). Modern theory of electron, review.

1950

[B1] BERMAN, A., P. KUSCH and A.K. MANN: Phys. Rev. **77**, 140 (1950). Hfs $^2P_{\frac{1}{2}}$ state Tl^{205}, Tl^{203}.
[B2] BETHE, H.A., L.M. BROWN and J.R. STEHN: Phys. Rev. **77**, 370 (1950). Numerical evaluation of Lamb shift.
[B3] BOHR, A., and V.F. WEISSKOPF: Phys. Rev. **77**, 94 (1950). Theory hfs anomaly.
[B4] BROSSEL, J., P. SAGALYN and F. BITTER: Phys. Rev. **79**, 225 (1950), and errata **79**, 196 (1950). Optical detection of radiofrequency resonance.
[D1] DICKINSON, W.C.: Phys. Rev. **80**, 563 (1950). Theory magnetic shielding of nuclei in atoms.
[F1] FOLEY, H.M.: Phys. Rev. **80**, 288 (1950). Theory Ga, In nuclear moments from atomic magnetic resonance.
[F2] FRIEDBURG, H., and W. PAUL: Naturwiss. **37**, 20 (1950). Beam reflection by magnetic field.
[G1] GRABNER, L., and V. HUGHES: Phys. Rev. **79**, 819 (1950). $K^{39}F$ radiofrequency spectrum by electric resonance.
[H1] HIRSCHBERG jr., J.G., and J.E. MACK: Phys. Rev. **77**, 745 (1950). Lamb shift in He^+.
[H2] HUGHES, V., and L. GRABNER: Phys. Rev. **79**, 314 (1950). $Rb^{85}F$, $Rb^{87}F$ radiofrequency spectrum by electric resonance.
[H3] HUGHES, V., and L. GRABNER: Phys. Rev. **79**, 829 (1950). Theory energy levels and transitions for electric resonance with diatomic molecules.
[H4] HYLLERAAS, E., and P.R. SKAVLEM: Phys. Rev. **79**, 117 (1950). Theory magnetic shielding of nuclei in He, H_2.
[K1] KARPLUS, R., and N. KROLL: Phys. Rev. **77**, 536 (1950). Theory anomalous spin magnetic moment of electron.
[K2] KOLSKY, H.G., T.E. PHIPPS, N.F. RAMSEY and H.B. SILSBEE: Phys. Rev. **79**, 883 (1950). H_2 radiofrequency spectrum by magnetic resonance.
[K3] KOLSKY, H.G., T.E. PHIPPS, N.F. RAMSEY and H.B. SILSBEE: Phys. Rev. **80**, 483 (1950). D_2 radiofrequency spectrum by magnetic resonance.
[K4] KUHN, H., and G.W. SERIES: Proc. Roy. Soc. Lond., Ser. A **202**, 127 (1950). H_α fine structure by optical spectroscopy.
[K5] KUSCH, P.: Phys. Rev. **78**, 615 (1950). Ga nuclear magnetic moments by magnetic resonance.
[L1] LAMB, W.E., and M. SKINNER: Phys. Rev. **78**, 539 (1950). He^+ fine structure and Lamb shift.
[L2] LAMB jr., W.E., and R.C. RETHERFORD: Phys. Rev. **79**, 549 (1950). H fine structure and Lamb shift.
[L3] LOW, F.: Phys. Rev. **77**, 361 (1950). Theory hfs anomaly of deuterium.
[M1] MANN, A.K., and P. KUSCH: Phys. Rev. **77**, 427 (1950). In hfs by magnetic resonance.
[M2] MANN, A.K., and P. KUSCH: Phys. Rev. **77**, 435 (1950). Anomalous magnetic moment of electron from atomic g_J values.

[N1] NEWELL, G.F.: Phys. Rev. **78**, 711 (1950). D quadrupole moment.
[N2] NIERENBERG, W.A.: Phys. Rev. **80**, 1102 (1950). Theory effect of spin-rotation interaction on molecular spectra.
[O1] OCHS, S.A., R.A. LOGAN and P. KUSCH: Phys. Rev. **78**, 184 (1950). K hfs anomaly.
[P1] PRODELL, A.G., and P. KUSCH: Phys. Rev. **79**, 1009 (1950). H, D hfs by magnetic resonance.
[P2] PRYCE, M.H.L.: Phys. Rev. **77**, 136 (1950). Theory optical detection of radiofrequency resonance.
[R1] RAMSEY, N.F.: Phys. Rev. **77**, 567 (1950). Diamagnetic shielding of proton in H_2.
[R2] RAMSEY, N.F.: Phys. Rev. **78**, 221 (1950). Theory quadrupole moment of electron distribution in hydrogen molecules.
[R3] RAMSEY, N.F.: Phys. Rev. **78**, 695 (1950). Separated oscillating fields method.
[R4] RAMSEY, N.F.: Phys. Rev. **78**, 699 (1950). Theory magnetic shielding of nuclei in molecules.
[S1] STERNHEIMER, R.: Phys. Rev. **80**, 102 (1950). Theory electric quadrupole interaction in atoms.
[Z1] ZEIGER, H.J., D.I. BOLEF and I.I. RABI: Phys. Rev. **78**, 340 (1950). $TlCl^{35}$, $TlCl^{37}$ radiofrequency spectra by magnetic resonance.

1951

[B1] BELLAMY, E.H.: Nature, Lond. **168**, 556 (1951). Radioactive Rb^{56} nuclear spin by magnetic resonance.
[B2] BOHR, A.: Phys. Rev. **81**, 331 (1951). Theory hfs anomaly.
[B3] BRAUNSTEIN, R., and J.W. TRISCHKA: Phys. Rev. **82**, 319 (1951). Space charge detector for alkali halide molecules.
[F1] FRIEDBURG, H.: Z. Physik **130**, 493 (1951). Magnetic deflection by multipole fields.
[F2] FRIEDBURG, H., and W. PAUL: Naturwiss. **38**, 159 (1951). Magnetic deflection by multipole fields.
[G1] GRABNER, L., and V. HUGHES: Phys. Rev. **82**, 561 (1951). Two quantum transition by molecular beam electric resonance.
[G2] GUNTHER-MOHR, G. R., S. GESCHWIND and C.H. TOWNES: Phys. Rev. **81**, 289 (1951). Theory polarization of nuclei by molecular electric fields.
[G3] GUTOWSKY, H.S., and R.E. MCCLURE: Phys. Rev. **81**, 276 (1951). Measurement magnetic shielding of proton in H_2, H_2O, mineral oil.
[G4] GARDNER, J.H.: Phys. Rev. **83**, 996 (1951). Measurement of g_l/g_p.
[J1] JACCARINO, V., and J.G. KING: Phys. Rev. **83**, 471 (1951). Cl^{35}, Cl^{37} hfs by magnetic resonance.
[K1] KANTROWITZ, A., and J. GREY: Rev. Sci. Instrum. **22**, 328 (1951). Theory high intensity beam source.
[K2] KING, J.G., and V. JACCARINO: Phys. Rev. **84**, 852 (1951). Hfs Cl, $^2P_{\frac{3}{2}}$ state.
[K3] KISTIAKOWSKY, G.B., and W.P. SLICHTER: Rev. Sci. Instrum. **22**, 333 (1951). High intensity beam source, experimental.
[K4] KOENIG, S., A.G. PRODELL and P. KUSCH: Phys. Rev. **83**, 687 (1951). Anomalous magnetic moment of electron from H g-value.
[K5] KOLSKY, H.G., T.E. PHIPPS jr., N.F. RAMSEY and H.B. SILSBEE: Phys. Rev. **81**, 1061 (1951). H_2, D_2 radiofrequency spectra by magnetic resonance; quadrupole moment of D.
[K6] KORSUNSKII, M.I., and YA.M. FOGEL: J. exp. theor. Phys. USSR. **21**, 25 (1951). Magnetic deflection and focusing of beams.
[K7] KROLL, N.M., and F. POLLOCK: Phys. Rev. **84**, 594 (1951). Theory radiative corrections to H hfs.
[K8] KRUSE, U.E., and N.F. RAMSEY: J. Math. Phys. **30**, 40 (1951). Evaluation of integral useful for resonance line shape with velocity distribution.
[K9] KUSCH, P.: Physica, Haag **17**, 339 (1951). Nuclear and electron properties from atomic beam magnetic resonance experiments, review.
[L1] LAMB jr., W.E., and R.C. RETHERFORD: Phys. Rev. **81**, 222 (1951). H fine structure and Lamb shift.
[L2] LOGAN, R.A., and P. KUSCH: Phys. Rev. **81**, 280 (1951). Na^{23} magnetic moment by magnetic resonance.
[L3] LOW, F.E., and E.E. SALPETER: Phys. Rev. **83**, 478 (1951). Theory hfs of H and D.
[L4] LUCE, R.G., and J.W. TRISCHKA: Phys. Rev. **83**, 851 (1951). Molecular constants of $Cs^{133}Cl^{35}$ by electric resonance.
[P1] POHLMAN, W.B., B. BEDERSON and J.T. EISINGER: Phys. Rev. **83**, 475 (1951). K g_J value.
[R1] RAMSEY, N.F.: Physica, Haag **17**, 303 (1951). Magnetic shielding of nuclei, review.
[R2] RAMSEY, N.F.: Physica, Haag **17**, 388 (1951). H_2, D_2 radiofrequency spectra, review.

[R3] Ramsey, N.F.: Phys. Rev. **83**, 540 (1951). Theory dependence of magnetic shielding of nuclei on molecular orientation.
[R4] Ramsey, N.F., and H.B. Silsbee: Phys. Rev. **84**, 506 (1951). Theory resonance line shape from separated oscillating fields with phase shifts.
[S1] Series, G.W.: Proc. Roy. Soc. Lond., Ser. A **208**, 277 (1951). D_α fine structure by optical spectroscopy.
[S2] Smith, K.F.: Nature, Lond. **167**, 942 (1951). Radioactive Na^{24} spin by atomic beam magnetic resonance.
[S3] Sternheimer, R.: Phys. Rev. **84**, 244 (1951). Theory nuclear quadrupole interaction in atoms.

1952

[B1] Barnes, R.G., and P.J. Bray: Phys. Rev. **87**, 229 (1952). D_2 radiofrequency spectrum by magnetic resonance.
[B2] Berman, A.: Phys. Rev. **86**, 1005 (1952). Tl^{205}, Tl^{203} hfs anomalies.
[B3] Bolef, D.I., and H.J. Zeiger: Phys. Rev. **85**, 799 (1952). $Rb^{87}F$, $Rb^{87}Cl$ radiofrequency spectra by magnetic resonance.
[B4] Bray, P.J., R.G. Barnes and N.J. Harrick: Phys. Rev. **87**, 229 (1952). H, D magnetic moment ratio from magnetic resonance in hydrogen molecules.
[B5] Brossel, J., A. Kastler and J. Winter: J. Phys. Radium **13**, 668 (1952). Optical pumping.
[B6] Brossel, J., and F. Bitter: Phys. Rev. **86**, 308 (1952). Optical-radiofrequency resonances for Hg in $^3P_1$ state.
[B7] Bederson, B., and V. Jaccarino: Phys. Rev. **87**, 228 (1952). Hfs of Rb^{85} and Rb^{87}.
[C1] Carlson, R.O., C.A. Lee and B.P. Fabricand: Phys. Rev. **85**, 784 (1952). TlCl radiofrequency and microwave spectra by electric resonance.
[E1] Eisinger, J.T., B. Bederson and B.T. Feld: Phys. Rev. **86**, 73 (1952). K hfs and K^{40} magnetic moment by magnetic resonance.
[F1] Fabricand, B.P., R.O. Carlson, C.A. Lee and I.I. Rabi: Phys. Rev. **86**, 607 (1952). KCl rotational spectrum by electric resonance.
[F2] Franken, P., and S. Koenig: Phys. Rev. **88**, 199 (1952). K atomic g_J by magnetic resonance.
[F3] Frosch, R.A., and H.M. Foley: Phys. Rev. **88**, 1337 (1952). Magnetic hfs in diatomic molecules.
[H1] Harrick, N.J., and N.F. Ramsey: Phys. Rev. **88**, 228 (1952). H_2 radiofrequency spectra by magnetic resonance.
[J1] Jaccarino, V., B. Bederson and H.H. Stroke: Phys. Rev. **87**, 676 (1952). Cs^{134} nuclear spin, magnetic moment by magnetic resonance.
[K1] Karplus, R., and A. Klein: Phys. Rev. **85**, 972 (1952). Theory radiative corrections to hfs.
[K2] Karplus, R., A. Klein and J. Schwinger: Phys. Rev. **86**, 288 (1952). Theory Lamb shift.
[K3] Koenig, S.H., A.G. Prodell and P. Kusch: Phys. Rev. **88**, 191 (1952). Anomalous magnetic moment of electron from H g_J.
[K4] Kolsky, H.G., T.E. Phipps jr., N.F. Ramsey and H.B. Silsbee: Phys. Rev. **87**, 395 (1952). H_2, D_2 radiofrequency spectra by magnetic resonance.
[K5] Koster, G.F.: Phys. Rev. **86**, 148 (1952). Theory effect of configuration interaction on Ga hfs.
[K6] King, J.G., and V. Jaccarino: Phys. Rev. **87**, 228 (1952). Bromine g_J experiment.
[L1] Lamb jr., W.E.: Phys. Rev. **85**, 259 (1952). Theory H fine structure measurement.
[L2] Lamb jr., W.E., and R.C. Retherford: Phys. Rev. **86**, 1014 (1952). H fine structure and Lamb shift.
[L3] Logan, R.A., R.E. Coté and P. Kusch: Phys. Rev. **86**, 280 (1952). Sign of eqQ in diatomic molecules.
[L4] Lyons, H.: Ann. New York Acad. Sci. **55**, 831 (1952). Spectral lines for frequency standards.
[O1] Ochs, S.A., and P. Kusch: Phys. Rev. **85**, 145 (1952). Rb hfs anomaly.
[P1] Phillips, M.: Phys. Rev. **88**, 202 (1952). Theory effect of purity atomic state on g_J.
[P2] Prodell, A.G., and P. Kusch: Phys. Rev. **88**, 184 (1952). H, D hfs.
[R1] Rabi, I.I.: Phys. Rev. **87**, 379 (1952). Atomic beam resonance method for excited states.
[R2] Ramsey, N.F.: Phys. Rev. **85**, 60 (1952). Theory H_2, D_2 in magnetic fields.

[R3] RAMSEY, N.F.: Phys. Rev. **85**, 688 (1952). Proton-deuteron magnetic moment ratio from HD.
[R4] RAMSEY, N.F.: Phys. Rev. **87**, 1075 (1952). Theory vibrational and centrifugal effects in hydrogen molecules.
[R5] RAMSEY, N.F., and E.M. PURCELL: Phys. Rev. **85**, 143 (1952). Theory nuclear spin-spin interaction in molecules.
[R6] RAWSON, E.B., and R. BERINGER: Phys. Rev. **88**, 677 (1952). g_J(O) experiment.
[S1] SHERWOOD, J.E., H. LYONS, R.H. MCCRACKEN and P. KUSCH: Phys. Rev. **86**, 618 (1952). Cs^{133} hfs.
[S2] SMALLER, B., E.L. YASAITIS, E.C. AVERY and D.A. HUTCHISON: Phys. Rev. **88**, 414 (1952). Nuclear spin-spin interactions in HD.
[S3] STERNHEIMER, R.: Phys. Rev. **86**, 316 (1952). Theory effect of configuration interaction on magnetic hfs.
[S4] SWARTZ, J.C., and J.W. TRISCHKA: Phys. Rev. **88**, 1085 (1952). Li^6F^{19} radiofrequency spectrum by electric resonance.
[T1] TRISCHKA, J.W., D.T.F. MARPLE and A. WHITE: Phys. Rev. **85**, 136 (1952). Detection of halogens by electron capture on a surface.
[W1] WEINREICH, G., G. TUCKER and V.W. HUGHES: Phys. Rev. **87**, 229 (1952). Spin of Ne^{20}.
[Z1] ZEIGER, H.J., and D.I. BOLEF: Phys. Rev. **85**, 788 (1952). TlCl radiofrequency spectrum by magnetic resonance.

1953

[A1] ABRAGAM, A., and J.H. VAN VLECK: Phys. Rev. **92**, 1448 (1953). Theory g_J(O).
[B1] BARANGER, M., H.A. BETHE and R.P. FEYNMAN: Phys. Rev. **92**, 482 (1953). Relativistic theory of Lamb shift.
[B2] BELLAMY, E.H., and K.F. SMITH: Phil. Mag. **44**, 33 (1953). Radioactive Na^{24}, K^{42}, Rb^{86}, Cs^{131}, Cs^{134} spins, magnetic moments by magnetic resonance.
[B3] BROSSEL, J., B. CAGNAC and A. KASTLER: C. R. Acad. Sci., Paris **237**, 984 (1953). Magnetic resonance in optically oriented Na beam.
[B4] BRIX, P., J.T. EISINGER, H. LEW and G. WESSEL: Phys. Rev. **92**, 647 (1953). Cr g_J by magnetic resonance.
[C1] COTÉ, R.E., and P. KUSCH: Phys. Rev. **90**, 103 (1953). Low frequency resonances in diatomic molecules by magnetic resonance.
[D1] DALY jr., R.T., and J.R. ZACHARIAS: Phys. Rev. **91**, 476 (1953). Doppler effects on resonance line shape. Hfs of Rb^{87}.
[D2] DAYHOFF, E.S., S. TRIEBWASSER and W.E. LAMB jr.: Phys. Rev. **89**, 106 (1953). H fine structure and Lamb shift.
[F1] FABRICAND, B.P., R.O. CARLSON, C.A. LEE and I.I. RABI: Phys. Rev. **91**, 1403 (1953). KBr rotational spectrum by electric resonance.
[G1] GOLDBERG, J.N., J.W. TRISCHKA and P.G. BERGMANN: Phys. Rev. **89**, 278 (1953). Electric resonance transitions in tapered electric field.
[H1] HAGSTRUM, H.D.: Phys. Rev. **89**, 244 (1953). Electron ejection from metals by ions.
[H2] HARRICK, N.J., R.G. BARNES, P.J. BRAY and N.F. RAMSEY: Phys. Rev. **90**, 260 (1953). Spectra of D_2, H_2 by magnetic resonance.
[H3] HAWKINS, W.B., and R.H. DICKE: Phys. Rev. **91**, 1008 (1953). Optical polarization of Na.
[H4] HUGHES, V.W., G. TUCKER, E. RHODERICK and G. WEINREICH: Phys. Rev. **91**, 828 (1953). He g_J in $^3S_1$ state.
[H5] HUGHES, V.W., and G. WEINREICH: Phys. Rev. **91**, 196 (1953). Theory effect of nuclear structure on He^3 hfs.
[K1] KUSCH, P.: J. Chem. Phys. **21**, 1424 (1953). Dimerization in NaF.
[K2] KUSCH, P.: Phys. Rev. **92**, 268 (1953). Sign of nuclear electric quadrupole interaction of Li^6 in LiCl.
[L1] LEE, C.A., B.P. FABRICAND, R.O. CARLSON and I.I. RABI: Phys. Rev. **91**, 1395 (1953). KCl rotational spectrum by electric resonance.
[L2] LEW, H.: Phys. Rev. **91**, 619 (1953). Pr^{141} hfs by magnetic resonance.
[L3] LEW, H., and G. WESSEL: Phys. Rev. **90**, 1 (1953). Al^{27} hfs by magnetic resonance.
[L4] LUCE, R.G., and J.W. TRISCHKA: J. Chem. Phys. **21**, 105 (1953). Molecular constants of CsCl by electric resonance.
[O1] OBI, S.Y., T. ISHIDZU, H. HORIE, S. YANAGAWA, T. TANABE and M. SATO: Ann. Tokyo Astronom. Obs., Ser. II **3**, 89 (1953). Tables of Racah coefficients.
[O2] OCHS, S.A., R.E. COTÉ and P. KUSCH: J. Chem. Phys. **21**, 459 (1953). NaCl radiofrequency spectra; dimerization.
[P1] PERL, W.: Phys. Rev. **91**, 852 (1953). Relativistic theory atomic g_J values.

[P2] PERL, W., and V. W. HUGHES: Phys. Rev. **91**, 842 (1953). Relativistic theory g_J He in $^3S_1$ state.
[R1] RAMSEY, N.F.: Phys. Rev. **89**, 527 (1953). Theory pseudo-quadrupole effect in molecules.
[R2] RAMSEY, N.F.: Phys. Rev. **90**, 232 (1953). Nuclear spin interactions in molecules.
[R3] RAMSEY, N.F.: Phys. Rev. **91**, 303 (1953). Nuclear spin-spin interactions in molecules.
[S1] SALPETER, E.E.: Phys. Rev. **89**, 92 (1953). Theory H, D Lamb shift.
[S2] STERNHEIMER, R.M., and H.M. FOLEY: Phys. Rev. **92**, 1460 (1953). Theory Li_2 nuclear quadrupole interaction.
[T1] TRIEBWASSER, S., E. S. DAYHOFF and W.E. LAMB jr.: Phys. Rev. **89**, 98 (1953). H fine structure and Lamb shift.
[W1] WEINREICH, G., G.M. GROSOF and V.W. HUGHES: Phys. Rev. **91**, 195 (1953). Hfs He^3 in $^3S_1$ state.
[W2] WENESER, J., R. BERSOHN and N.M. KROLL: Phys. Rev. **91**, 1257 (1953). Theory fourth order radiative corrections to atomic energy levels.
[W3] WESSEL, G., and H. LEW: Phys. Rev. **92**, 641 (1953). Ag, Au hfs by magnetic resonance.
[W4] WESSEL, G.: Phys. Rev. **92**, 1581 (1953). B^{11} hfs by magnetic resonance.
[W5] WIMETT, T.F.: Phys. Rev. **91**, 476 (1953). Nuclear resonance absorption in hydrogen molecules.

1954

[B1] BARNES, R.G., P.J. BRAY and N.F. RAMSEY: Phys. Rev. **94**, 893 (1954). Rotational magnetic moments of H_2, D_2 by magnetic resonance.
[B2] BECKER, E.W., and K. BIER: Z. Naturforsch. **9**a, 975 (1954). Intense H_2 beam.
[B3] BENNETT, A.: Phys. Rev. **95**, 608 (1954). Beam velocity distribution.
[B4] BENNEWITZ, H.G., and W. PAUL: Z. Physik **139**, 489 (1954). Use of multiple focusing magnets for nuclear resonance.
[B5] BERINGER, R., and M.A. HEALD: Phys. Rev. **95**, 1474 (1954). Anomalous magnetic moment of electron from H g_J by microwave absorption spectroscopy.
[B6] BROSSEL, J., B. CAGNAC and A. KASTLER: J. Phys. Radium **15**, 6 (1954). Magnetic resonance in optically oriented beams.
[C1] CLENDENIN, W.W.: Phys. Rev. **94**, 1590 (1954). Relativistic theory of hfs of doublet states.
[C2] COHEN, V.W., and D.A. GILBERT: Phys. Rev. **95**, 569 (1954). Radioactive Cs^{134m} hfs, spin by magnetic resonance.
[D1] DALY jr., R.T., and J.H. HOLLOWAY: Phys. Rev. **96**, 539 (1954). Hfs of Ga in $^2P_{\frac{3}{2}}$ state; magnetic octupole moments.
[F1] FOLEY, H.M., R.M. STERNHEIMER and D. TYCKO: Phys. Rev. **93**, 734 (1954). Theory nuclear electric quadrupole interaction in polar molecules.
[G1] GOODMAN, L.S., and S. WEXLER: Phys. Rev. **95**, 570 (1954). Radioactive Cs^{134m} spin, hfs by magnetic resonance.
[G2] GORDON, J.P., H.J. ZEIGER and C.H. TOWNES: Phys. Rev. **95**, 282 (1954). Maser and NH_3 inversion spectrum.
[H1] HAWKINS, W.B.: Phys. Rev. **96**, 532 (1954). Optical alignment of Na.
[H2] HOBSON, J.P., J.C. HUBBS, W.A. NIERENBERG and H.B. SILSBEE: Phys. Rev. **96**, 1450 (1954). Radioactive Rb^{81} spin, hfs by magnetic resonance.
[H3] HONIG, A., M. MANDEL, M.L. STITCH and C.H. TOWNES: Phys. Rev. **96**, 629 (1954). Microwave spectra of alkali halides.
[H4] HEALD, M.A., and R. BERINGER: Phys. Rev. **96**, 645 (1954). g_J(N) experiment.
[J1] JACCARINO, V., J.G. KING, R.A. SATTEN and H.H. STROKE: Phys. Rev. **94**, 1798 (1954). I^{127} hfs, octopole moment.
[K1] KING, J.G., and V. JACCARINO: Phys. Rev. **94**, 1610 (1954), errata, **95**, 1706 (1954). Br hfs by magnetic resonance.
[K2] KUSCH, P.: J. Chem. Phys. **22**, 1203 (1954). $KClFeCl_2$, $KBrFeBr_2$ rotational moments by magnetic resonance.
[K3] KUSCH, P.: Phys. Rev. **93**, 1022 (1954). Multiple quantum magnetic resonance transitions.
[K4] KUSCH, P., and T.G. ECK: Phys. Rev. **94**, 1799 (1954). In^{115} hfs and octopole moment.
[K5] KAMBE, K., and J.H. VAN VLECK: Phys. Rev. **96**, 66 (1954). Theory g_J(O).
[L1] LEMONICK, A., and F.M. PIPKIN: Phys. Rev. **95**, 1356 (1954). Radioactive Ag^{111}, Cu^{64} spin, hfs.
[L2] LOUISELL, W.H., R.W. PIDD and H.R. CRANE: Phys. Rev. **94**, 7 (1954). Magnetic moment of free electron.

[R1] RABI, I. I., N. F. RAMSEY and J. SCHWINGER: Rev. Mod. Phys. **26**, 167 (1954). Theory use of rotating coordinates in magnetic resonance problems.
[S1] SAGALYN, P. L.: Phys. Rev. **94**, 885 (1954). Hfs Na^{23} in $3^2P_{\frac{3}{2}}$ state by optical radiofrequency technique.
[S2] STERNHEIMER, R. M.: Phys. Rev. **95**, 736 (1954). Theory nuclear electric quadrupole interaction in atoms.
[T1] TEUTSCH, W. B., and V. W. HUGHES: Phys. Rev. **95**, 1461 (1954). Theory hfs of He^3 in $^3S_1$ state.
[T2] TRISCHKA, J. W., and R. BRAUNSTEIN: Phys. Rev. **96**, 968 (1954). RbCl rotational spectra by electric resonance.
[T3] TATE, P. A., and M. W. P. STRANDBERG: J. Chem. Phys. **22**, 1380 (1954). Microwave spectra KCl and NaCl.
[W1] WEINREICH, G., and V. W. HUGHES: Phys. Rev. **95**, 1451 (1954). Hfs of He^3 in $^3S_1$ state.
[W2] WITTKE, J. P., and R. H. DICKE: Phys. Rev. **96**, 530 (1954). H hfs by microwave absorption spectroscopy.

1955

[A1] AULTER, S. H., and C. H. TOWNES: Phys. Rev. **100**, 703 (1955). Theory Stark effect in rapidly varying fields.
[B1] BEMSKI, G., W. A. NIERENBERG and H. B. SILSBEE: Phys. Rev. **98**, 470 (1955). Spin-rotation interactions in CsF, RbF by magnetic resonance.
[B2] BENNEWITZ, H. G., W. PAUL and C. SCHLIER: Z. Physik **141**, 6 (1955). Multipole magnetic deflecting fields.
[B3] BRAUNSTEIN, R., and J. W. TRISCHKA: Phys. Rev. **98**, 1092 (1955). Radiofrequency spectrum of LiF by electric resonance.
[B4] BROSSEL, J., J. MARGERIE and J. WINTER: C. R. Acad. Sci., Paris **241**, 556 (1955). Theory Bloch-Siegert effect in multiple quantum transitions.
[B5] BROSSEL, J., J. MARGERIE and A. KASTLER: C. R. Acad. Sci., Paris **241**, 865 (1955). Optical detection Na magnetic resonance transitions.
[D1] DEHMELT, H. G.: Phys. Rev. **99**, 527 (1955). Phosphorous g_J experiment.
[F1] FRICKE, G.: Z. Physik **141**, 166 (1955). Universal ionizer beam detector.
[F2] FRICKE, G., and H. FRIEDBURG: Z. Physik **141**, 171 (1955). Conductivity beam detector.
[G1] GILBERT, D. A., and V. W. COHEN: Phys. Rev. **97**, 243 (1955). Radioactive Cs^{134m} hfs by magnetic resonance.
[G2] GOODMAN, L. S., and S. WEXLER: Phys. Rev. **97**, 242; **99**, 192 (1955). Radioactive Cs^{134m} hfs by magnetic resonance.
[G3] GOODMAN, L. S., and S. WEXLER: Phys. Rev. **100**, 1245 (1955). Nuclear spin and magnetic moment of In^{114m}.
[G4] GOODMAN, L. S., and S. WEXLER: Phys. Rev. **100**, 1796 (1955). Nuclear spin and magnetic moment of In^{116m}.
[H1] HAWKINS, W. B.: Phys. Rev. **98**, 478 (1955). Optical orientation and alignment of Na.
[K1] KASTNER, S. O., A. M. RUSSELL and J. W. TRISCHKA: J. Chem. Phys. **23**, 1730 (1955). Radiofrequency spectrum of Li^6F by electric resonance.
[K2] KUSCH, P.: Phys. Rev. **100**, 1188 (1955). Hfs of H, D in $1^2S_{\frac{1}{2}}$ state by magnetic resonance.
[L1] LEMONICK, A., F. M. PIPKIN and D. R. HAMILTON: Rev. Sci. Instrum. **26**, 1112 (1955). Focussing atomic beam apparatus with multipole magnetic deflecting fields for radioactive atoms.
[L2] LIEB, E. H.: Phil. Mag. **46**, 311 (1955). Theory quantum electrodynamic radiative corrections to electron spin magnetic moment.
[L3] LURIO, A., and A. G. PRODELL: Phys. Rev. **99**, 613 (1955). Hfs anomaly in Ga.
[M1] MILLER, R. C., and P. KUSCH: Phys. Rev. **99**, 1314 (1955). Atomic beam velocity distributions.
[M2] MILLS, R. L., and N. M. KROLL: Phys. Rev. **98**, 1489 (1955). Theory fourth order radiative corrections to atomic energy levels.
[M3] MOELLERING, W. M., A. C. ZEMACH, A. KLEIN and F. E. LOW: Phys. Rev. **100**, 441 (1955). Theory relativistic recoil and proton structure contributions to H hfs.
[N1] NEWCOMB, W. A., and E. E. SALPETER: Phys. Rev. **97**, 1146 (1955). Relativistic mass corrections to H hfs.
[N2] NOVICK, R., E. LIPWORTH and P. F. YERGIN: Phys. Rev. **100**, 1153 (1955). He^+ fine structure.
[P1] PERL, M. L., I. I. RABI and B. SENITZKY: Phys. Rev. **97**, 838; **98**, 611 (1955). Hfs of Na^{23} in 3^2P states by optical-radiofrequency atomic beam resonance method.

[R1] RAMSEY, N.F.: Phys. Rev. **98**, 1853 (1955). Theory magnetic alignment of beam by collisions.
[R2] REYNOLDS, J.B., R.L. CHRISTENSEN, D.R. HAMILTON, A. LEMONICK, F.M. PIPKIN and H.H. STROKE: Phys. Rev. **99**, 613 (1955). Au^{198}, Au^{199} spin, hfs by magnetic resonance.
[S1] SALWEN, H.: Phys. Rev. **99**, 1274 (1955). Theory magnetic resonance transitions in rotating magnetic field.
[S2] SATO, M.: Progr. Theor. Phys. **13**, 405 (1955). Racah coefficients.
[S3] SCHWARTZ, C.: Phys. Rev. **97**, 380 (1955). Hfs theory.
[S4] SCHWARTZ, C.: Phys. Rev. **99**, 1035 (1955). Theory hfs anomaly in P states.
[S5] SESSLER, A.M., and H.M. FOLEY: Phys. Rev. **98**, 6 (1955). Theory He^{3+}, He^3 hfs including relativistic and nuclear structure effects.
[T1] TINKHAM, M., and M.W.P. STRANDBERG: Phys. Rev. **97**, 937, 951 (1955). Theory O_2 Zeeman effect.
[W1] WHITE, L.D.: Phys. Rev. **98**, 1194 (1955). Atomic g_J of H in $2^2S_{\frac{1}{2}}$ state.
[W2] WHITE, R.L.: Rev. Mod. Phys. **27**, 276 (1955). Spin-rotation interactions in $^1\Sigma$ molecules, review.
[W3] WINTER, J.: C. R. Acad. Sci., Paris **241**, 600 (1955). Theory Bloch-Siegert effect in multiple quantum transitions.
[W4] WOODGATE, G.K., and R.W. HELLWARTH: Nature, Lond. **176**, 395 (1955). Ag^{111} hfs, spin by magnetic resonance.

1956

[A1] ARAKI, G.: Phys. Rev. **101**, 1410 (1956), and errata, **103**, 1906 (1956). Theory fine structure of ^3P states of He.
[B1] BECKER, E.W., and W. HENKES: Z. Physik **146**, 320 (1956). Beam velocity distribution from Laval nozzle.
[B2] BECKER, E.W., K. BIER and W. HENKES: Z. Physik **146**, 333 (1956). Beam velocity distribution from Laval nozzle.
[B3] BROSSEL, J.: Cahiers de Phys. **65**, 59 (1956). Optical orientation methods, review.
[B4] BUCK, P., I.I. RABI and B. SENITZKY: Phys. Rev. **104**, 553 (1956). Hfs of K^{39} in 4^2P and of Cs^{133} in 6^2P states by optical-radiofrequency atomic beam magnetic resonance method.
[C1] CHRISTENSEN, R.L., D.R. HAMILTON, A. LEMONICK, F.M. PIPKIN, J.B. REYNOLDS and H.H. STROKE: Phys. Rev. **101**, 1389 (1956). Au^{198}, Au^{199} hfs, spin by magnetic resonance.
[C2] COHEN, V.W., N.R. CORNGOLD and N.F. RAMSEY: Phys. Rev. **104**, 283 (1956). Neutron magnetic moment by beam resonance method.
[C3] CHILDS, W.J., and L.S. GOODMAN: Bull. Amer. Phys. Soc. [II], **1**, 342 (1956). Spin and magnetic moment of In^{113m}.
[F1] FRANKEN, P., and S. LIEBES jr.: Phys. Rev. **104**, 1197 (1956). Proton magnetic moment in Bohr magnetons.
[G1] GOULD, G.: Phys. Rev. **101**, 1828 (1956). Hfs of Tl^{203}, Tl^{205} in metastable $6^2P_{\frac{3}{2}}$ state.
[H1] HEBERLE, J.W., H.A. REICH and P. KUSCH: Phys. Rev. **101**, 612 (1956). Hfs of H in $2^2S_{\frac{1}{2}}$ state.
[H2] HOBSON, J.P., J.C. HUBBS, W.A. NIERENBERG, H.B. SILSBEE and R.J. SUNDERLAND: Phys. Rev. **104**, 101 (1956). Radioactive Rb^{81}, Rb^{82}, Rb^{83}, Rb^{84} spins by magnetic resonance.
[H3] HUBBS, J.C., and G.M. GROSOF: Phys. Rev. **104**, 715 (1956). Ne^{21} spin by magnetic resonance.
[H4] HUBBS, J., C.W.A. NIERENBERG, H.A. SHUGART and H.B. SILSBEE: Phys. Rev. **104**, 757 (1956). Radioactive Rb^{81m} spin by magnetic resonance.
[K1] KUSCH, P.: Phys. Rev. **101**, 627 (1956). Multiple quantum transitions, and radiofrequency saturation for K^{39} magnetic resonances.
[K2] KUSCH, P.: Science, Lancaster, Pa. **123**, 207 (1956). Electron spin magnetic moment, review.
[L1] LAMB jr., W.E.: Science, Lancaster, Pa. **123**, 439 (1956). H fine structure, review.
[L2] LURIO, A., and A.G. PRODELL: Phys. Rev. **101**, 79 (1956). Hfs of Ga^{69}, Ga^{71}, Tl^{203}, Tl^{205} in $^2P_{\frac{3}{2}}$ state by magnetic resonance.
[M1] MARPLE, D.T.F., and J.W. TRISCHKA: Phys. Rev. **103**, 597 (1956). Radiofrequency spectrum of Li^6Cl by electric resonance.
[N1] NIERENBERG, W.A., H.A. SHUGART, H.B. SILSBEE and R.J. SUNDERLAND: Phys. Rev. **104**, 1380 (1956). Radioactive Cs^{127}, Cs^{129}, Cs^{130} spins by magnetic resonance.
[N2] NOVICK, R., and E. COMMINS: Phys. Rev. **103**, 1897 (1956). Hfs of singly ionized He^3 in $2^2S_{\frac{1}{2}}$ state.
[N3] NUTTER, P.B.: Phil. Mag. **1**, 587 (1956). Radioactive In^{116} hfs.

[N 4] Nierenberg, W.A., J.C. Hubbs, H.A. Shugart, H.B. Silsbee and P.O. Strom: Bull. Amer. Phys. Soc. [II], **1**, 343 (1956). Spin of Cs^{132}.

[P 1] Phillips, M.: Phys. Rev. **103**, 322 (1956). Theory Rb hfs in 5 P states.

[R 1] Reich, H.A., J.W. Heberle and P. Kusch: Phys. Rev. **104**, 1585 (1956). Hfs of D in $2^2S_{\frac{1}{2}}$ state.

[S 1] Salwen, H.: Phys. Rev. **101**, 621 (1956). Theory magnetic resonance transitions—averages over velocity distribution.

[S 2] Salwen, H.: Phys. Rev. **101**, 623 (1956). Theory K^{39} multiple quantum transitions.

[S 3] Senitzky, B., and I.I. Rabi: Phys. Rev. **103**, 315 (1956). Hfs of Rb^{85}, Rb^{87} in 5^2P states by optical-radiofrequency atomic beam magnetic resonance method.

[S 4] Shimoda, K., T.C. Wang and C.H. Townes: Phys. Rev. **102**, 1308 (1956). Maser theory.

[S 5] Sternheimer. R.M., and H.M. Foley: Phys. Rev. **102**, 731 (1956). Theory nuclear electric quadrupole interaction in polar molecules.

[S 6] Silsbee, H.B., W.A. Nierenberg, H.A. Shugart and P.O. Strom: Bull. Amer. Phys. Soc. [II], **1**, 389 (1956). Spin of Ag^{105}.

[W 1] Woodgate, G.K., and R.W. Hellwarth: Proc. Phys. Soc. Lond. A **69**, 581 (1956). Radioactive Ag^{111} hfs, spin by magnetic resonance.

[W 2] Woodgate, G.K., and R.W. Hellwarth: Proc. Phys. Soc. Lond. A **69**, 588 (1956). Theory magnetic resonance line shapes.

[W 3] Wittke, J.P., and R.H. Dicke: Phys. Rev. **103**, 620 (1956). Hfs of H by microwave absorption.

[Z 1] Zemach, A.C.: Phys. Rev. **104**, 1771 (1956). Theory effect of proton structure on H hfs.

1957

[B 1] Buck, P., and I.I. Rabi: Phys. Rev. **107**, 1291 (1957). Hfs of K^{39} in 4^2P states by optical-rf atomic beam magnetic resonance method.

[B 2] Brink, G.O., J.C. Hubbs, W.A. Nierenberg and J.L. Worcester: Phys. Rev. **107**, 189 (1957). Radioactive Tl^{197}, $Tl^{198\,m}$, Tl^{199}, Tl^{204} spins and Tl^{204} hfs by magnetic resonance.

[C 1] Christensen, R.L., H.G. Bennewitz, D.R. Hamilton, J.B. Reynolds and H.H. Stroke: Phys. Rev. **107**, 633 (1957). Nuclear spin of radioactive As^{76}.

[D 1] Drake, C., V.W. Hughes and A. Lurio: Bull. Amer. Phys. Soc. [II], **2**, 37 (1957). g_J(Ne), g_J(A) in $^3P_2$ state.

[E 1] Eck, T.G., A. Lurio and P. Kusch: Phys. Rev. **106**, 954 (1957). Hfs and hfs anomalies in $5^2P_{\frac{3}{2}}$ state of In^{115} and In^{113}.

[E 2] Eck, T.G., and P. Kusch: Phys. Rev. **106**, 958 (1957). Hfs and magnetic octupole interactions in $5^2P_{\frac{3}{2}}$ state of In^{115} and In^{113}.

[E 3] Ehrenstein, D. v., G. Fricke, H. Kopfermann and S. Penselin: Naturwiss. **44**, 255 (1957). Y^{89} hfs by magnetic resonance.

[E 4] Essen, L., and J.V.L. Parry: Phil. Trans. Roy. Soc. Lond. **250** A, 45 (1957). Cesium atomic frequency standard.

[E 5] Ewbank, W.B., L.L. Marino, H.A. Shugart and H.B. Silsbee: Bull. Amer. Phys. Soc. [II], **2**, 383 (1957). Nuclear Spins of Au^{191}, Au^{192}, and Au^{193}.

[E 6] Ewbank, W.B., W.A. Nierenberg, H.A. Shugart and H.B. Silsbee: Bull. Amer. Phys. Soc. [II], **2**, 317 (1957). Nuclear spins of radioactive Ag^{106}, Ag^{110m}.

[F 1] Friedburg, H., and H. Kuiper: Naturwiss. **44**, 487 (1957). g_J(Xe) experiment.

[G 1] Garvin, H.L., T.M. Green and E. Lipworth: Bull. Amer. Phys. Soc. [II], **2**, 344 (1957). Spin of radioactive Br^{82}.

[G 2] Garvin, H.L., T.M. Green and E. Lipworth: Bull. Amer. Phys. Soc. [II], **2**, 383 (1957). Spin of radioactive I^{124}.

[G 3] Green, T.M., H.L. Garvin, E. Lipworth and W.A. Nierenberg: Bull. Amer. Phys. Soc. [II], **2**, 383 (1957). Nuclear moments of Br^{82}.

[G 4] Goodman, L.S., and S. Wexler: Phys. Rev. **108**, 1524 (1957). Radioactive In^{114m} and In^{116m} spins and magnetic dipole moments.

[G 5] Garrison, A.K., and W. Gordy: Phys. Rev. **108**, 899 (1957). Microwave absorption spectrometer using a high-temperature molecular beam.

[G 6] Geiger, J.S., V.W. Hughes and H.E. Radford: Phys. Rev. **105**, 183 (1957). g_J(D) experiment.

[H 1] Hubbs, J.C., W.A. Nierenberg, H.A. Shugart and J.L. Worcester: Phys. Rev. **105**, 1928 (1957). Radioactive Ga^{66}, Ga^{67} spins by magnetic resonance.

[H 2] Hughes, V.W.: Phys. Rev. **105**, 170 (1957). Electron-proton charge difference by electric deflection.

[*H 3*] Hubbs, J.C., R. Marrus, W.A. Nierenberg and J.L. Worcester: Bull. Amer. Phys. Soc. [II], **2**, 316 (1957). Hfs of Pu^{239}.

[*H 4*] Hendrie, J.M., and P. Kusch: Phys. Rev. **107**, 716 (1957). Molecular O_2 Zeeman effect.

[*H 5*] Hooke, W.M., R.L. Christensen, D.R. Hamilton, J.B. Reynolds and H.H. Stroke: Bull. Amer. Phys. Soc. [II], **2**, 344 (1957). Hfs of radioactive Au^{194}.

[*H 6*] Haun jr., R.D., and J.R. Zacharias: Phys. Rev. **107**, 107 (1957). Stark effect on Cs^{133} hfs.

[*H 7*] Hubbs, J.C., W.A. Nierenberg, H.A. Shugart, H.B. Silsbee and R.J. Sunderland: Phys. Rev. **107**, 723 (1957). Radioactive Rb^{81}, Rb^{82}, Rb^{83}, and Rb^{84} hfs and magnetic moments.

[*L 1*] Lichten, W.: J. Chem. Phys. **26**, 306 (1957). Lifetime N_2 metastable states.

[*L 2*] LaTourrette, J.T., W.E. Quinn and N.F. Ramsey: Phys. Rev. **107**, 1202 (1957). Ne^{21} magnetic moment by magnetic resonance.

[*L 3*] Lipworth, E., and H.L. Garvin: Bull. Amer. Phys. Soc. [II], **2**, 316 (1957). Nuclear spins of radioactive I^{123}, I^{131}.

[*L 4*] Lipworth, E., and R. Novick: Phys. Rev. **108**, 1434 (1957). Fine structure of He^+.

[*L 5*] Lewis, H.R., A. Pery, W. Quinn and N.F. Ramsey: Phys. Rev. **107**, 446 (1957). Distortions of magnetic resonances.

[*M 1*] Marino, L.L., W.B. Ewbank, W.A. Nierenberg, H.A. Shugart and H.B. Silsbee: Bull. Amer. Phys. Soc. [II], **2**, 383 (1957). Nuclear spin of radioactive Bi^{206}.

[*M 2*] Mittleman, M.H.: Phys. Rev. **107**, 1170 (1957). Theory α^3 corrections to H hfs.

[*N 1*] Nierenberg, W.A., H.A. Shugart and H.B. Silsbee: Bull. Amer. Phys. Soc. [II], **2**, 200 (1957). Nuclear spin of radioactive Cu^{61}.

[*N 2*] Nierenberg, W.A.: Annual Rev. Nucl. Sci. **7**, 349 (1957). Review article on measurement of nuclear spins and static moments of radioactive isotopes.

[*P 1*] Prodell, A.G., and P. Kusch: Phys. Rev. **106**, 87 (1957). Hfs of H^3 in $1\,^2S_{\frac{1}{2}}$ state.

[*P 2*] Petermann, A.: Nuclear Phys. **3**, 689 (1957). Electron spin magnetic moment (α^2) theory.

[*R 1*] Russell, A.M.: Phys. Rev. **106**, 1100 (1957). Rotational magnetic moments of Li^6F, Li^7F by electric resonance.

[*R 2*] Reynolds, J.B., R.L. Christensen, W.M. Hooke, D.R. Hamilton, H.H. Stroke, W.B. Ewbank, W.A. Nierenberg, H.A. Shugart and H.B. Silsbee: Bull. Amer. Phys. Soc. [II], **2**, 317 (1957). Nuclear Spin of Au^{194}.

[*R 3*] Ramsey, N.F., and H.R. Lewis: Phys. Rev. **108**, 1246 (1957). Theory HD in magnetic fields.

[*S 1*] Schwartz, C.: Phys. Rev. **105**, 173 (1957). Hfs theory.

[*S 2*] Stroke, H.H., V. Jaccarino, D.S. Edmonds jr. and R. Weiss: Phys. Rev. **105**, 590 (1957). Cs^{133}, Cs^{134}, Cs^{135}, Cs^{137} hfs by magnetic resonance.

[*S 3*] Schlier, C.: Z. Physik **147**, 600 (1957). Radiofrequency spectrum of KF by electric resonance.

[*S 4*] Sommerfield, C.M.: Phys. Rev. **107**, 328 (1957); Ann. of Physics **5**, 26 (1958). Theory anomalous magnetic moment of electron.

[*S 5*] Smith, J.H., E.M. Purcell and N.F. Ramsey: Phys. Rev. **108**, 120 (1957). Experimental limit to neutron electric dipole moment.

[*S 6*] Silsbee, H.B., W.A. Nierenberg, H.A. Shugart and R.L. Sunderland: Bull. Amer. Phys. Soc. [II], **2**, 30 (1957). Hfs of Cs^{127} and Cs^{129}.

[*S 7*] Sternheimer, R.M.: Phys. Rev. **105**, 158 (1957). Theory effect of configuration interaction on nuclear quadrupole coupling in atoms.

[*S 8*] Schlier, C.: Fortschr. Phys. **5**, 378 (1957). Review of recent work in molecular beams.

[*T 1*] Ting, Y., and H. Lew: Phys. Rev. **105**, 581 (1957). Cu^{63}, Cu^{65} hfs by magnetic resonance.

[*T 2*] Ting, Y.: Phys. Rev. **108**, 295 (1957). La^{139} hfs by magnetic resonance.

[*W 1*] Worcester, J.L., J.C. Hubbs and W.A. Nierenberg: Bull. Amer. Phys. Soc. [II], **2**, 316 (1957). Hfs and nuclear moments of radioactive Ga^{67}.

[*W 2*] Wieder, I., and W.E. Lamb jr.: Phys. Rev. **107**, 125 (1957). Fine structure of $2\,^3P$ and $3\,^3P$ states in He.

1958

[*A 1*] Arditi, M., and T.R. Carver: Phys. Rev. **109**, 1012 (1958). Hfs of Na^{23} using optical detection.

[*C 1*] Childs, W.J., and L.S. Goodman: Bull. Amer. Phys. Soc. [II], **3**, 21 (1958). Nuclear spin of radioactive Mn^{56}.

[*D 1*] Dehmelt, H.G.: Phys. Rev. **109**, 381 (1958). g_s/g_J(Na) experiment.

[*D 2*] Drake, C., A. Lurio, V.W. Hughes and J.A. White: Bull. Amer. Phys. Soc. [II], **3**, 7 (1958). g_J (He) experiment.
[*E 1*] Ewbank, W.B., L.L. Marino, W.A. Nierenberg, H.A. Shugart and H.B. Silsbee: Bull. Amer. Phys. Soc. [II], **3**, 185 (1958). Hfs of radioactive Au^{192}.
[*E 2*] Ewbank, W.B., W.A. Nierenberg, H.A. Shugart and H.B. Silsbee: Phys. Rev. **110**, 595 (1958). Nuclear spins of radioactive Ag^{105}, Ag^{106}, and Ag^{110m}.
[*G 1*] Goodman, L.S., and W.J. Childs: Bull. Amer. Phys. Soc. [II], **3**, 21 (1958). Nuclear spin and hfs of radioactive Ga^{72}.
[*H 1*] Hubbs, J.C., and R. Marrus: Phys. Rev. **110**, 287 (1958). Nuclear spin and g_J for radioactive Np^{239}.
[*I 1*] Innes, F.R., and C.W. Ufford: Phys. Rev. **111**, 194 (1958). Theory of g_J(O).
[*L 1*] Lurio, A., C.A. Drake, V.W. Hughes and J.A. White: Bull. Amer. Phys. Soc. [II], **3**, 8 (1958). g_J (Ne) and g_J (A) experiment.
[*L 2*] Lew, H., D. Morris, F.E. Geiger jr. and J.T. Eisinger: Canad. J. Res. **36**, 171 (1958). RbF rotational spectrum by electric resonance.
[*N 1*] Novick, R., and E.D. Commins; Phys. Rev. **111**, 822 (1958). Hfs of metastable He^{3+}.
[*M 1*] Marino, L.L., G.O. Brink, W.E. Ewbank, H.A. Shugart and H.B. Silsbee: Bull. Amer. Phys. Soc. [II], **3**, 186 (1958). Nuclear spins of radioactive Tl^{200}, Tl^{201}.
[*M 2*] Marrus, R., R.G. Albridge and J.C. Hubbs: Bull. Amer. Phys. Soc. [II], **3**, 186 (1958). Nuclear spin of radioactive Np^{238}.
[*R 1*] Radford, H.E., and V.W. Hughes: Bull. Amer. Phys. Soc. [II], **3**, 8 (1958). g_J (O) experiment.
[*R 2*] Radford, H.E.: Bull. Amer. Phys. Soc. [II], **3**, 325 (1958). g_J (F) experiment.

The Vibration-rotation Energies of Molecules and their Spectra in the Infra-red[1].

By

H. H. Nielsen.

With 29 Figures.

Symbols and notation used.

X, Y, Z	space-fixed coordinate system.
X_0, Y_0, Z_0	coordinates of center of mass in space-fixed coordinate system.
X_i, Y_i, Z_i	coordinates of nuclei relative to center of mass in space-fixed coordinate system
x, y, z	body-fixed coordinate system.
ϑ, φ, ψ	Eulerian angles relating x, y, z and X, Y, Z axes.
$\alpha(\beta, \gamma)$	symbol to denote x, y, or z.
α_i', α_j'	coordinates of nuclei and electrons, respectively, relative to center of mass in body-fixed system.
α_i, α_j	$M_i^{\frac{1}{2}} \alpha_i' = \alpha_i$, $m^{\frac{1}{2}} \alpha_j' = \alpha_j$, M_i, and m being, respectively, the masses of the i-th nucleus and of an electron.
α_i^0	equilibrium value of α_i.
$I_{\alpha\alpha}$	moment of inertia about α-axis.
$I_{\alpha\beta}$	product of inertia.
$I'_{\alpha\alpha}$	effective moment of inertia about α axis. See Eq. (12.19).
$I'_{\alpha\beta}$	effective product of inertia. See Eq. (12.19).
ω_α	angular velocity about α-axis.
$Q_{s\sigma}, q_{s\sigma}$	normal coordinates. $Q_{s\sigma} = (\hbar^2/\lambda_s)^{\frac{1}{4}} q_{s\sigma}$; $\lambda_s = (2\pi c\,\omega_s)^2$.
ω_s	normal vibration frequency in cm^{-1}.
$l_{is\sigma}$	transformation coefficients relating $\delta\alpha_i$ to normal coordinates $Q_{s\sigma}$.
$p_{s\sigma}$	linear momentum conjugate to $q_{s\sigma}$.
r_t, χ_t, ϑ_t	polar coordinates used to describe degenerate vibrations. (Note difference between ϑ and ϑ_t.)
$p_{r_t}, p_{\chi_t}, p_{\vartheta_t}$	momenta conjugate to r_t, χ_t, and ϑ_t.
p_Ξ, p_H, p_Z	linear momenta conjugate to coordinates of center of mass $\Xi = \left(\sum_i M_i + Nm\right)^{\frac{1}{2}} X_0$, etc.
p_α	component of internal angular momentum of nuclei directed along α-axis.
π_α	component of angular momentum of electrons directed along α-axis.
Π_α	$\Pi_\alpha = \pi_\alpha + p_\alpha$.
s_X, s_Y, s_Z	components of spin angular momentum directed along space-fixed X, Y, Z axes.
s_x, s_y, s_z	components of spin angular momentum directed along body-fixed x, y, z axes.
$\mathfrak{S}_\alpha = \sum_j s_{j\alpha}$	component of total spin angular momentum directed along the α-axis.
M_α	$M_\alpha = \Pi_\alpha + \mathfrak{S}_\alpha$ total internal angular momentum directed along the α-axis.
P_α	component of total angular momentum directed along the α-axis.
s_{mn}	distance between two atomic nuclei m and n.
s_{mn}^0	equilibrium value of s_{mn} $(s_{mn} - s_{mn}^0 = \delta s_{mn})$.

[1] Literature references are not given in footnotes, but in a bibliography at the end of this article.

$A^{(\alpha\beta)}_{s\sigma s\sigma}$, $a^{(\alpha\beta)}_{s\sigma}$, etc., see definitions (18.24).

$\zeta^{(\alpha)}_{s\sigma s'\sigma'}$ Coriolis coupling factor. See definition (18.24).

v_s total vibration quantum number for a harmonic oscillator.

l_s quantum number of total vibrational angular momentum associated with a two- or three-fold degenerate vibration.

m_s component of vibrational angular momentum associated with a three-fold degenerate oscillator directed along an axis fixed in the molecule.

J quantum number of angular momentum of a molecule exclusive of spin.

F quantum number of angular momentum of a molecule inclusive of spin.

K quantum number associated with the component of J directed along the z-axis.

M the magnetic quantum number of rotation of a molecule.

Λ quantum number of electronic angular momentum directed along z axis of a linear molecule.

Σ quantum number of spin angular momentum directed along z axis of a molecule.

L angular momentum of the molecular framework directed along the z axis (not quantized).

g_s weight factor assuming the values 1, 2, or 3, respectively, as ω_s is one-, two-, or threefold degenerate.

$x_{ss'}$, $x_{l_s l_s}$, etc. anharmonic constants (corresponding to $\omega_e x_e$ in diatomic molecules).

D_J, D_{JK}, D_{JL}, etc. centrifugal stretching coefficients. See definitions (34.15) and (34.16).

$B^{(\alpha\alpha)}_e$ reciprocal of inertia $(h/8\pi^2 I^{(e)}_{\alpha\alpha} c)$.

$B^{(\alpha\beta)}_v$ effective reciprocal of inertia or reciprocal product of inertia.

$\lambda_{\alpha B}$ direction cosines relating body-fixed axes to space-fixed axes.

I. Introduction.

1. The spectra of substances may, in a general way, be assigned to one of three categories, namely; the continuous spectra, the bright line spectra and the band spectra. The first of these occur only in emission and are produced by bodies heated to incandescence and are incapable of resolution into lines regardless of the resolving power of the available instruments. The second type are the spectra of atoms. They may be produced as bright line emission spectra by suitable excitation of the atoms, as for example by placing them in the crater of an arc or by passing an electric discharge through their vapors. When radiation from an incandescent source is allowed to pass through an atomic vapor and is examined with a spectroscope the spectra occur in absorption as dark line spectra against a continuous bright background.

The third kind of spectra referred to above is the one which we shall be concerned with in this work. They are the spectra of molecules and derive their name from the fact that under low spectroscopic resolution they appear in the visible region, where they were first observed, as continuous bands of color. They appear with a sharp intense edge at one end and fade off in intensity toward the other end. This intense sharp edge is usually referred to as a *head* and may occur at the high frequency end of a band or at the low frequency end, depending upon circumstances. Bands are observed, in some cases, to have more than one head; in fact band spectra of polyatomic molecules frequently have many heads. Band spectra also may be observed in emission when the molecules are properly excited, or in absorption when continuous radiation is allowed to pass through a molecular vapor. The designation of such spectra as *band* spectra is, of course, anomalous since when they are examined with spectroscopic apparatus of sufficient resolving power they may be shown to consist of a large number of discrete spectral lines. Band spectra are comparable in complexity only to the most intricate of atomic line spectra.

Band spectra may be found throughout the spectral range from the very far infra-red and micro-wave region to the extreme ultra-violet region. While no clear cut distinction can be drawn between band spectra in one region and band spectra in another region they are frequently referred to as micro-wave or far infra-red band spectra (1 to 1000 cm^{-1}), near infra-red band spectra (1000 to 10000 cm^{-1}) and optical band spectra (frequencies greater than 10000 cm^{-1}).

It is observed experimentally that molecules have a spectrum in the far infra-red only when they possess a permanent electric moment. This suggests that these spectra are associated with the rotation of the molecule. This view was further advanced by BJERRUM who pointed out that the molecules in an absorption cell could be regarded as being in thermal equilibrium and that their rotational energy distribution was, therefore, given by a Maxwell-Boltzmann law. This law states, in the case of a diatomic molecule, that the number of molecules, dN, which have rotation frequencies between the values ν_r and $\nu_r + d\nu_r$ will be

$$dN(\nu_r) = (4\pi^2 I \nu_r / kT)\, N \exp(-E_r/kT)\, d\nu_r \tag{1.1}$$

where N is the total number of molecules present, I is the moment of inertia of the molecule, E_r is the rotational energy, k is the Boltzmann constant, 1.3803×10^{-16} erg deg.$^{-1}$ and T is the absolute temperature. The relation $dN(\nu_r)/d\nu_r$; has a most probable value for a certain value of ν_r. The value of ν_r, $(\nu_r)_{\max}$, for which $dN(\nu_r)/d\nu_r$ has its maximum value may be obtained by solving the relation $d\,dN(\nu_r)/d\nu_r = 0$ for ν. The frequency $(\nu_r)_{\max}$ may, moreover, be expected to be the most intense line in the spectrum since the intensity with which a frequency occurs is proportional to the number of molecules, $dN(\nu_r)/d\nu_r$, which participate in the frequency.

One may quickly arrive at the value

$$(\nu_r)_{\max} = \frac{1}{2\pi} \sqrt{\frac{kT}{I}} \tag{1.2}$$

by setting $E_r = \frac{1}{2} I \omega_r^2 = 2\pi^2 I \nu_r^2$. For a diatomic molecule $I = \mu r^2$ where μ is the reduced mass, $m_1 m_2/(m_1 + m_2)$, m_1 and m_2 being the masses, respectively, of the two nuclei and $r = r_1 + r_2$, r_1 and r_2 being, respectively, the distances of the two nuclei from the center of mass. An order of magnitude calculation is now most helpful. One may take r to be of the order of magnitude of 1 Å and the masses to be of the order of magnitude of 1.6×10^{-24} gm. At room temperature, i.e., $T \sim 300°$ K, one obtains numerically for $(\nu_r)_{\max}$ a value close to 60 cm^{-1}. The rotational frequencies of a molecule may, therefore, be expected to extend from values of a few cm^{-1} to about 500 cm^{-1} with the most intense region near 50 to 100 cm^{-1}, which is, indeed, in the far infra-red.

The near infra-red bands have been interpreted as due to the vibration of the atomic nuclei. One may be persuaded in several ways that this interpretation is a reasonable one, but the following method is probably sufficient. The forces between the two atomic nuclei in a diatomic molecule are much the same as the forces existing between the nucleus and an electron in an atom. The masses of the nuclei are, however, of the order of two thousand times greater than the masses of the electrons. When one considers that the frequencies with which a particle moves is roughly proportional to the inverse square root of the reduced masses it is indicated that the nuclear vibrations should be about one thirtieth as great as the frequencies at which line spectra of atoms are found. This may be near 2000 cm^{-1} or, indeed, in the near infra-red. The breadth of such bands was accounted for by Lord RAYLEIGH who showed that a diatomic molecule rotating with a frequency ν_r while it is vibrating should give rise to the frequencies $\nu_v \pm \nu_r$ where ν_v is the vibration frequency. Recalling the argument of BJERRUM already referred to, a most probable value of $\nu_v \pm \nu_r$ must, of course, exist. This accounts for the apparent doublet structure of the near infra-red absorption bands of diatomic molecules when they are studied under low dispersion.

The optical band spectra lie in the region of the spectrum occupied by the line spectra of atoms which are known to originate with transitions of the electrons from one electronic state to another. This suggests that optical band spectra must originate in much the same manner as line spectra, i.e., with transitions of the electrons from one electronic state in the molecule to another. It is found experimentally, moreover, that complete families of bands, known as *systems*, appear at one time under given conditions of excitation and under other conditions of excitation disappear together. This is understood to mean that the general position of such bands is determined by the electronic frequency and that the accompanying families of bands are due to a series of vibration-rotation frequencies on either side of the electronic frequency.

The first attempts at analyzing a system of bands were made by DESLANDRES who confined himself to diatomic molecules. He noted that the heads of the bands were very nearly equally spaced with respect to each other, but that the interval decreases slowly toward shorter wave lengths. He demonstrated that it was possible to arrange the band heads in *progressions*, the frequencies of these being given by the simple relation,

$$\nu = a + b v + c v^2 \tag{1.3}$$

where $b \gg c$ and where v may take positive and negative integral values.

When observed under high resolving power one observes that at one end of the band the lines may be well separated and are, in fact, diverging from each other, but as one approaches the other end the lines converge upon each other until in many instances they fold back on each other to form a virtual continuum. The frequency at which the line structure folds back on itself marks the position of the head. Again DESLANDRES observed that for diatomic molecules the positions of the lines in the spectrum can be represented by an expansion of the form

$$\nu = d + e m + f m^2 \tag{1.4}$$

where d, e and f are constants and where m is an integer. It is customary to begin numbering the lines at the band center, the higher frequency side being given by the positive values of m and the lower frequency side being determined by the negative values of m. Generally speaking e is large compared with f, in fact, in the case of the infrared bands where a head is seldom achieved, f is very small compared to e.

The formation of spectra as discrete lines was never satisfactorily coped with theoretically before the introduction of the quantum concepts. According to the quantum theory the frequency of an observed spectral line will be represented by the BOHR frequency condition

$$\nu = (E' - E'')/hc \tag{1.5}$$

where E' and E'' are respectively the *quantized* energies of the molecule in its final and initial states and where h is the universal constant of PLANCK $6{,}626 \times 10^{-27}$ erg sec and c is the velocity of light. Using the classical formulation of the quantum mechanics KRATZER calculated the energy states for a rotating-vibrating diatomic molecule with zero electronic angular momentum. The rotational energies for a diatomic molecule where the electronic angular momentum was different from zero were studied by KRAMERS and PAULI, KRATZER and KEMBLE and finally by HUND. The results of these calculations threw much light upon the structure of the band spectra of molecules, but many details were never clarified until the introduction of the new quantum mechanics of SCHRÖDINGER, HEISENBERG, BORN and DIRAC. We shall devote a large portion of this work to the calculation of the energies of molecules using this powerful tool known as the quantum mechanics.

II. The general formulation of the quantum mechanics.

2. Introduction. We shall consider the general principles of the quantum theory in this chapter, giving the development of the Schrödinger equation, a brief resumé of the symbolic algebra of states and observables and such other items as are of especial importance to us in this work. The reader will often find it useful ro refer to Chap. II of *The Theory of Atomic Spectra* by CONDON and SHORTLEY[1] for more detailed discussion of the theory.

3. The Hamiltonian in quantum mechanics. A dynamical system of particles with n degrees of freedom can be described in the classical dynamics in terms of the values of the coordinates, q_k, and the conjugate momenta, p_k, at a given time, t. To achieve this, it has frequently proved useful to express the total energy in the Hamiltonian form. The Hamiltonian is a function of the p's and the q's only and is numerically equal to the total energy for a conservative system. The usefulness of the Hamiltonian function is related to the symmetry of the Hamiltonian equations of motion:

$$\dot{q}_k = \frac{\partial H}{\partial p_k}, \qquad \dot{p}_k = -\frac{\partial H}{\partial q_k}.$$

The time variation of the *state* is thereby established.

There exists also in quantum mechanics a Hamiltonian function which formally is the same function of the p's and the q's as is the classical Hamiltonian for the corresponding dynamical system. The quantum mechanical Hamiltonian is, however, an operator since the dynamical variables are represented in the quantum mechanics by linear operators. The time dependence of a state, which also in the quantum mechanics is established through the Hamiltonian function, is determined, according to DIRAC, by the equation

$$H(q, p, t)\,\psi(q, t) = i\hbar\,\frac{\partial\psi(q, t)}{\partial t}. \tag{3.1}$$

The solution, $\psi(q_1, \dots q_n, t)$, to the Eq. (3.1) describes the state of the system and is known as a wave function. Wave functions must be single valued and continuous throughout the entire range of the variables, q_k. A wave function, ψ, has a complex conjugate which shall here be denoted by $\overline{\psi}$ and it is useful to require the integral $\int \overline{\psi}(q_1, \dots q_n, t)\,\psi(q_1, \dots q_n, t)\,d\tau$, where $d\tau$ is the element of volume in the configuration space of the q's, to be equal to unity. The integral is taken over the entire range of the variables, q_k. This stipulation may be interpreted to mean that $\overline{\psi}\psi\,d\tau$ is the probability that the system shall be in the co-

[1] E. U. CONDON and G. H. SHORTLEY: The Theory of Atomic Spectra. Cambridge: University Press 1935.

ordinate volume element $d\tau$. Since each of the variables, q_k, must have some value, the total probability must be equal to one, consistent with the requirement that

$$\int \bar{\psi}(q_1, \dots q_n, t)\, \psi(q_1, \dots q_n, t)\, d\tau = 1. \tag{3.2}$$

4. The algebra of observables. A dynamical variable is known in quantum mechanics as an *observable* since experimentally one is interested in measuring some particular quantity which will fix the value of the dynamical variable at a particular time. An observable is represented by a linear operator, say α. The only observable quantities which have a meaning are real. We may, therefore, restrict ourselves to linear operators which are Hermitian, i.e., to operators which have the property that

$$\int \bar{\psi}''(\alpha\, \psi')\, d\tau = \int \psi'(\bar{\alpha}\, \bar{\psi}'')\, d\tau. \tag{4.1}$$

An observable may be defined as a rule for acting upon a function, ψ, and converting it into another ψ-function. The result of α operating upon a ψ-function will, therefore, be another ψ-function which we shall denote by $\alpha\psi$. The complex conjugate of $\alpha\psi$ we shall denote by $\overline{\alpha\psi}$. Moreover, since $\alpha\psi$ is another ψ-function, the integral $\int \bar{\psi}'\alpha\, \psi''\, d\tau$ must be an ordinary number, where the primes and double primes are arbitrary. An obvious generalization of the interpretation that $\bar{\psi}\,\psi\, d\tau$ is the probability of finding the system in the coordinate volume element $d\tau$ will be that the integral $\int \bar{\psi}\alpha\,\psi\, d\tau$ gives the mean value of the observable α in the state characterized by ψ.

The sum of two observables we shall define in the following manner $(\alpha_1 + \alpha_2)\psi = \alpha_1\psi + \alpha_2\psi$. The product of the observables, $\alpha_1\alpha_2$, is defined as follows $(\alpha_1\alpha_2)\psi = \alpha_1(\alpha_2\psi)$. It is evident that the product $\alpha_1\alpha_2$ may, in general, be quite different from the product $\alpha_2\alpha_1$. The observables α_1 and α_2 are said to *commute* in the event that $\alpha_1\alpha_2 - \alpha_2\alpha_1 = 0$. The linearity of the operators, α, will enable us to say further that $\alpha[a'\psi' + a''\psi''] = a'\alpha\psi' + a''\alpha\psi''$ where ψ' and ψ'' are the wave functions of any two states and where a' and a'' are constants.

There are certain of the ψ-functions which, when operated upon by an observable, α, are merely multiplied by a constant. We have then that

$$\alpha\,\psi' = \alpha'\,\psi' \tag{4.2}$$

where α' is an ordinary number. We postulate that for such a state the observable, α, *has a value* α'. The various values of the ordinary numbers α' which an observable, α, may take are known as the *eigenvalues* of α. The ψ-functions which satisfy the Eq. (4.2) are known as *eigenfunctions* and the states which they denote are known as *eigenstates* of the observable, α.

It may be shown that, since the operators, α, are Hermitian the eigenvalues, α', are always real. This may be demonstrated in the following manner. When the quantities in (4.2) are replaced by their complex conjugates one obtains

$$\bar{\alpha}\,\bar{\psi}' = \bar{\alpha}'\,\bar{\psi}'. \tag{4.3}$$

Multiplying (4.2) by $\bar{\psi}'$ and (4.3) by ψ' and integrating one obtains respectively:

$$\int \bar{\psi}'(\alpha\,\psi')\, d\tau = \alpha' \int \bar{\psi}'\,\psi'\, d\tau$$

and

$$\int \psi'(\bar{\alpha}\,\bar{\psi}')\, d\tau = \bar{\alpha}' \int \psi'\,\bar{\psi}'\, d\tau$$

which according to the relation (4.1), are equal. α' is, therefore, equal to $\bar{\alpha}'$ which can be so only if α' is real.

5. The Schrödinger equation. If the function $\psi(q, t)$ in Eq. (3.1) is an eigenfunction of the Hamiltonian H, and E is its eigenvalue, then

$$H(q, p, t)\,\psi(q, t) = E\,\psi(q, t) \tag{5.1}$$

where $\psi(q, t)$ must be

$$\psi(q, t) = \psi(q) \exp(-iEt/\hbar) \tag{5.2}$$

in which $\psi(q)$ is a solution to the equation

$$H(q, p, t)\,\psi(q) = E\,\psi(q). \tag{5.3}$$

Closely associated with the Hamiltonian equations of motion in classical mechanics is the theory of the Poisson brackets. The Poisson bracket of two functions f and g may be defined as follows:

$$[f, g] = \{(\partial f/\partial q_k)(\partial g/\partial p_k) - (\partial f/\partial p_k)(\partial g/\partial q_k)\}. \tag{5.4}$$

It is evident from (5.4) that the following relationships must exist:

$$\left.\begin{aligned}
&[f, g] = -[g, f],\\
&[f + g, h] = [f, h] + [g, h],\\
&[fg, h] = f[g, h] + g[f, h],\\
&[f, [g, h]] + [g, [h, f]] + [h, [f, g]] = 0.
\end{aligned}\right\} \tag{5.5}$$

When f is a function not containing the time explicitly one may, with the aid of (5.4), derive that

$$(df/dt) = \{(\partial f/\partial q_k)(\partial H/\partial p_k) - (\partial f/\partial p_k)(\partial H/\partial q_k)\} = [f, H] \tag{5.6}$$

where H is the Hamiltonian function. It is evident that the Poisson bracket theory applied specifically to the coordinates, q_k, and the conjugate momenta, p_k, may be summarized as follows:

$$[x_k, x_l] = 0, \quad [p_k, p_l] = 0, \quad [p_k, x_l] = -\delta_{kl} \quad \text{and} \quad \dot{f} = [f, H]. \tag{5.7}$$

The laws governing quantum mechanics state, according to DIRAC, that the cartesian coordinates, $x_1, \ldots x_{3n}$ of a system of n particles and the conjugate momenta, $p_1, p_2, \ldots p_{3n}$ must satisfy the Poisson bracket relations "translated" into quantum theory. These relations are

$$\left.\begin{aligned}
&[x_k, x_l] = x_k x_l - x_l x_k = 0,\\
&[p_k, p_l] = p_k p_l - p_l p_k = 0,\\
&[x_k, p_l] = x_k p_l - p_k x_l = i\hbar\,\delta_{kl}.
\end{aligned}\right\} \tag{5.8}$$

SCHRÖDINGER's discovery was that the first two of the quantum conditions were fulfilled by adopting for the x_k a set of commuting observables and that the third condition was satisfied by setting $p_k = -i\hbar(\partial/\partial x_k)$. When p_k is replaced in (5.1) by $-i\hbar(\partial/\partial x_k)$ we obtain the Schrödinger equation

$$H(x_k, -i\hbar\,\partial/\partial x_k, t)\,\psi(x, t) = E\,\psi(x, t). \tag{5.9}$$

6. States and representations of observables. It may be demonstrated that the eigenfunctions, ψ, of an Hermitian operator, α, corresponding to different eigenvalues are orthogonal functions in the interval of the complete range of the variables. We take as two eigenfunctions of the Hermitian operator, α, the functions ψ' and ψ''. Then

$$\alpha\psi' = \alpha'\psi' \quad \text{and} \quad \alpha\psi'' = \alpha''\psi''.$$

Multiplying the first of these relations by $\overline{\psi}''$ and integrating we have

$$\int \overline{\psi}'' \alpha \psi' \, d\tau = \alpha' \int \overline{\psi}'' \psi' \, d\tau. \tag{6.1}$$

Because of the Hermitian character of α we have, according to (4.2) that

$$\int \overline{\psi}'' \alpha \psi' \, d\tau = \int \psi' \overline{\alpha} \overline{\psi}'' \, d\tau = \overline{\alpha}'' \int \psi' \overline{\psi}'' \, d\tau = \alpha'' \int \psi' \overline{\psi}'' \, d\tau. \tag{6.2}$$

since α'' is real and therefore $\overline{\alpha}'' = \alpha''$. Equating the relations (6.1) and (6.2) we have

$$(\alpha' - \alpha'') \int \overline{\psi}'' \psi' \, d\tau = 0 \tag{6.3}$$

This requires that

$$\int \overline{\psi}'' \psi' \, d\tau = 0 \tag{6.4}$$

unless $\alpha' = \alpha''$ and the functions ψ' and ψ'' are, therefore orthogonal functions. When $\alpha' = \alpha''$ a degeneracy exists. A set of ψ-functions may still be chosen in cases of this kind, however, which can be normalized and which are mutually orthogonal. We shall examine this more in detail.

When one is dealing with orthogonal functions it is possible to expand an arbitrary function in terms of these functions. Let ψ be a function which is expandible in the interval of the range of the variables in terms of a set of normalized and orthogonal functions $\psi(\alpha' r')$. Then

$$\psi = \sum_{\alpha' r'} c_{\alpha' r'} \psi(\alpha' r') \tag{6.5}$$

where the $c_{\alpha' r'}$ are constants, r' is an index running from one to $d_{\alpha'}$, $d_{\alpha'}$ being the order of degeneracy. Multiplying both sides of (6.5) by $\overline{\psi}(\alpha'' r'')$ and integrating one obtains

$$\int \overline{\psi}(\alpha'' r'') \psi \, d\tau = \sum_{\alpha' r'} \int c_{\alpha' r'} \overline{\psi}(\alpha'' r'') \psi(\alpha' r') \, d\tau. \tag{6.6}$$

All the terms on the right side of (6.6) vanish except the one where $\alpha'' = \alpha'$ and $r'' = r'$ so that

$$c_{\alpha' r'} = \int \overline{\psi}(\alpha' r') \psi \, d\tau.$$

We shall henceforth use the notation $(\alpha' r'|)$ in place of $c_{\alpha' r'}$ and write instead of (6.5)

$$\psi = \sum_{\alpha' r'} \psi(\alpha' r') (\alpha' r'|). \tag{6.7}$$

We shall follow the general practice of CONDON and SHORTLEY of reserving the right blank space to designate the ψ which is being expanded. Thus we write

$$\psi(\beta') = \sum_{\alpha' r'} \psi(\alpha' r') (\alpha' r'|\beta'). \tag{6.8}$$

Now let the observable β operate on a function $\psi(\alpha'' r'')$. The result of this operation shall be designated by $\beta\psi(\alpha'' r'')$. This may be expanded in the following manner

$$\beta \psi(\alpha'' r'') = \sum_{\alpha' r'} a_{\alpha' r'} \psi(\alpha' r'). \tag{6.9}$$

Multiplying both sides of Eq. (6.9) by $\overline{\psi}(\alpha''' r''')$ and integrating one obtains

$$\int \overline{\psi}(\alpha''' r''') \beta \psi(\alpha'' r'') \, d\tau = \int \sum_{\alpha' r'} a_{\alpha' r'} \overline{\psi}(\alpha''' r''') \psi(\alpha' r') \, d\tau. \tag{6.10}$$

All the terms on the right side of (6.10) vanish unless $\alpha''' = \alpha'$ and $r''' = r'$ so that one obtains for $a_{\alpha' r'}$

$$a_{\alpha' r'} = \int \overline{\psi}(\alpha' r') \beta \psi(\alpha'' r'') \, d\tau. \tag{6.11}$$

Here also the general notation of CONDON and SHORTLEY is adhered to and we replace $a_{\alpha' r'}$ by $(\alpha' r'|\beta|\alpha'' r'')$ so that instead of (6.9) we shall write

$$\beta\psi(\alpha'' r'') = \sum_{\alpha' r'} \psi(\alpha' r')(\alpha' r'|\beta|\alpha'' r''). \tag{6.12}$$

We may extend the above to define the product of two operators $\gamma\beta$. Using (6.12) we have

$$\gamma\beta\psi(\alpha' r') = \gamma \sum_{\alpha'' r''} \psi(\alpha'' r'')(\alpha'' r''|\beta|\alpha' r').$$

Moreover $\gamma\psi(\alpha'' r'')$ will be

$$\gamma\psi(\alpha'' r'') = \sum_{\alpha''' r'''} \psi(\alpha''' r''')(\alpha''' r'''|\gamma|\alpha'' r'')$$

so that $\gamma\beta\psi(\alpha' r')$ will be

$$\gamma\beta\psi(\alpha' r') = \sum_{\alpha'' r''} \sum_{\alpha''' r'''} \psi(\alpha''' r''')(\alpha''' r'''|\gamma|\alpha'' r'')(\alpha'' r''|\beta|\alpha' r'). \tag{6.13}$$

We shall have occasion many times in the following sections of dealing with observables which commute. Let us designate by α and β two observables which commute, i.e., $\alpha\beta = \beta\alpha$. Letting $\alpha\beta$ operate on $\psi(\alpha' r')$ we have according to relation (6.12) that

$$\begin{aligned}\alpha\beta\psi(\alpha' r') &= \alpha \sum_{\alpha'' r''} \psi(\alpha'' r'')(\alpha'' r''|\beta|\alpha' r')\\ &= \sum_{\alpha'' r''} \alpha'' \psi(\alpha'' r'')(\alpha'' r''|\beta|\alpha' r').\end{aligned}$$

Moreover

$$\beta\alpha\psi(\alpha' r') = \alpha'\beta\psi(\alpha' r') = \alpha' \sum_{\alpha'' r''} \psi(\alpha'' r'')(\alpha'' r''|\beta|\alpha' r').$$

Then since $\alpha\beta = \beta\alpha$ the coefficients of $\psi(\alpha'' r'')$ must be equal to each other and we have that $(\alpha' - \alpha'')(\alpha'' r''|\beta|\alpha' r') = 0$. Unless, therefore, $\alpha' = \alpha''$ the coefficients $(\alpha'' r''|\beta|\alpha' r')$ must vanish.

It will be important to us in many problems which we shall consider to be able to express the eigenstates of an observable, β, which commutes with another observable, α, in the representation of α. We denote the eigenfunctions of β by $\psi(\beta')$ which has the expansion (6.8). Since $\psi(\beta')$ is an eigenfunction of the operator β we have $\beta\psi(\beta') = \beta'\psi(\beta')$ of which the expansion will be

$$\sum_{\alpha' r'} \sum_{\alpha'' r''} \psi(\alpha' r')(\alpha' r'|\beta|\alpha'' r'')(\alpha'' r''|\beta') = \sum_{\alpha' r'} \beta' \psi(\alpha' r')(\alpha' r'|\beta'). \tag{6.14}$$

Equating coefficients of the functions $\psi(\alpha' r')$ and taking account of the fact that α and β are commuting observables, so that $\alpha' = \alpha''$, we obtain the homogeneous set of linear expansions for determining the coefficients $(\alpha' r'|\beta')$ in relation (6.8)

$$\sum_{r''} (\alpha' r'|\beta|\alpha' r'')(\alpha' r''|\beta') = \beta'(\alpha' r'|\beta'). \tag{6.15}$$

The determinant of the coefficients of (6.15) must vanish for these equations to coexist, except in the trivial case where all the coefficients are identically zero, i.e.,

$$\begin{vmatrix} (\alpha' 1|\beta|\alpha' 1) - \beta' & (\alpha' 1|\beta|\alpha' 2) & \cdots & (\alpha' 1|\beta|\alpha' d_{\alpha'}) \\ (\alpha' 2|\beta|\alpha' 1) & (\alpha' 2|\beta|\alpha' 2) - \beta' & \cdots & (\alpha' 2|\beta|\alpha' d_{\alpha'}) \\ \vdots & \vdots & & \vdots \\ (\alpha' d_{\alpha'}|\beta|\alpha' 1) & (\alpha' d_{\alpha'}|\beta|\alpha' 2) & \cdots & (\alpha' d_{\alpha'}|\beta|\alpha' d_{\alpha'}) - \beta' \end{vmatrix} = 0.$$

The determinant will give rise to $d_{\alpha'}$ roots, each root being an eigenvalue of the observable β identified with the α' sub-group. Each eigenstate is characterized by an eigenfunction $\psi(\beta') = \sum_{\alpha' r'} \psi(\alpha' r')(\alpha' r' | \beta')$ of which the coefficients $(\alpha' r' | \beta')$ will be the normalized cofactors of the above determinant when the roots $\beta_1, \beta_2, \ldots, \beta_{d_{\alpha'}}$ are successively introduced for β'. Since the $\psi(\beta')$ form a complete set and all the eigenvalues of β have been determined a complete set of eigenfunctions $\psi(\alpha'\beta')$ have been derived and the eigenstates they denote are eigenstates of both α and β.

7. The matrix formulation of the quantum mechanics. It may be seen from, for example, Eq. (6.13) that the observables β, γ, etc., expressed in the α representation all obey the algebra of matrices. The operators β, γ, etc., may therefore, be interpreted as matrices having the elements $(\alpha'' r'' | \beta | \alpha' r')$, $(\alpha'' r'' | \gamma | \alpha' r')$, etc., in the α representation. The observables x_k and p_k are specific examples of the above operators and as such may be represented by Hermitian matrices.

A function, f, of the observables x_k and p_k may, evidently also be represented by a Hermitian matrix which has the elements

$$(\alpha'' r'' | f | \alpha' r') = \int \overline{\psi}(\alpha'' r'') f \psi(\alpha' r') \, d\tau. \tag{7.1}$$

The Hamiltonian, $H = \sum_k \{(p_k/2m_k) + V(x_k)\}$, is a special case of such a function and according to Eqs. (5.9) and (7.1) it will be a diagonal matrix, i.e.,

$$(\alpha'' r'' | H | \alpha' r') = E(\alpha'' r'') \, \delta_{\alpha''\alpha'} \, \delta_{r''r'}.$$

Let us replace $\overline{\psi}(\alpha'' r'')$ and $\psi(\alpha' r')$ in Eq. (7.1) by their equivalents $\overline{\psi}''(x_k) \exp(iE'' t/\hbar)$ and $\psi'(x_k) \exp(-iE' t/\hbar)$. We have then that

$$\left.\begin{aligned}(\alpha'' r'' | f | \alpha' r') &= \int \overline{\psi}''(x_k) f \psi'(x_k) \exp[i(E'' - E') t/\hbar] \, d\tau \\ &= (\alpha'' r'' | f_0 | \alpha' r') \exp[i(E'' - E') t/\hbar].\end{aligned}\right\} \tag{7.2}$$

Taking the time derivative of (7.2) we have

$$i\hbar (d/dt)(\alpha'' r'' | f | \alpha' r') = (\alpha'' r'' | f | \alpha' r') E' - E''(\alpha'' r'' | f | \alpha' r')$$

which, in the matrix notation may be written

$$i\hbar \dot{f} = f H - H f. \tag{7.3}$$

Eq. (7.3) is the quantum mechanical analogue of Eq. (5.6) in the classical mechanics.

The matrix formulation of the quantum mechanics may thus be derived as a mathematical consequence of the Schrödinger theory. The matrix mechanics had, however, been developed independently by HEISENBERG who showed from correspondence principle considerations that a set of matrices satisfying the relations (5.8) and (7.3) can be derived for x_k, p_k and H. By formal analogy with the classical mechanics he was led to the general assumption that any Poisson bracket relation, $(f, g) = k$, where k is a function of p and q, must "translate" into the quantum mechanics as $fg - gf = i\hbar k$.

We refer here to the matrix form of the quantum mechanics because there are times when this formulation has advantages over the Schrödinger formulation. If, for example, it is practicable to determine uniquely a set of matrices representing x_k, p_k and H which satisfy the relations (5.8) and the relation (7.3) by an algebraic process, then all the physically important quantities have been arrived at without actually solving the Schrödinger equation.

8. The theory of perturbations. The Schrödinger Eq. (5.9) is often a differential equation which has no known mathematical solution. This is particularly true in instances, where the equation does not become separable in the coordinates. One will observe, however, in many cases that if H in (5.9) is replaced by another operator, H_0, which differs from H only by the omission of small terms, that the Schrödinger equation will assume a form which does have a well known solution, say, $\psi_0(E_0)$ and which gives rise to the eigenvalues E_0. It is possible then, in general, to take account in the energy of the small terms $H - H_0$ by the method of the perturbation theory. When H is expressed by an expansion in orders of magnitude

$$H = H_0 + \lambda H_1 + \lambda^2 H_2 + \cdots \tag{8.1}$$

the corrected energies will be given by the relations

$$E = E_0 + \lambda E_1 + \lambda^2 E_2 + \cdots \tag{8.2}$$

where

$$\lambda E_1(\Gamma) = (\Gamma | H_1 | \Gamma), \tag{8.3}$$

$$\lambda^2 E_2(\Gamma) = (\Gamma | \lambda^2 H_2 | \Gamma) + \sum_{\Gamma'} \frac{(\Gamma | \lambda H_1 | \Gamma')(\Gamma' | \lambda H_1 | \Gamma)}{E_0(\Gamma) - E_0(\Gamma')}$$

where Γ is made to represent the quantum numbers which are associated with the state $E_0(\Gamma)$. The derivations of Eq. (8.3) will not here be given since they may be found in any standard text on quantum mechanics.

The method of the perturbation theory, as it is presented here, although simple in principle becomes often times extremely laborious when many terms are involved. Another method may be found advantageous in such cases. H_0 is an operator, as we have seen, which has the eigenfunction $\psi_0(\Gamma)$ and yields the eigenvalues $E_0(\Gamma)$. Consider now the term λH_1. This portion of H may have matrix elements which are diagonal in Γ, but will certainly also have terms which lie off of the principal diagonal of the matrix. The elements non-diagonal in Γ can contribute to the energy only in the order λ^2 if the eigenvalues $E_0(\Gamma)$, $E_0(\Gamma')$, etc., are all distinct and different from each other. With this in mind one seeks to effect a contact transformation of the original Hamiltonian, H, into a new form, H', such that $\lambda H_1'$ will still give the same elements diagonal in Γ as λH_1, but will yield no elements which are non-diagonal in Γ. The terms in λH_1 which were non-diagonal in Γ will give rise to a contribution to the new $\lambda^2 H_2'$ and will have elements of the order of λ^2 which are both diagonal and non-diagonal in Γ. When one seeks a set of energy values which are valid to second order of approximation only it is necessary to take account only of the elements of $\lambda^2 H_2'$ which are diagonal in Γ.

We shall here indicate such a contact transformation by

$$H' = T H T^{-1} = H_0' + \lambda H_1' + \lambda^2 H_2' + \cdots \tag{8.4}$$

where T is an unitary transformation $T = e^{i\lambda S}$ and where $T^{-1} = e^{-i\lambda S}$ is its inverse. Expanding T in orders of magnitude one obtains

$$T = 1 + i\lambda S - \frac{\lambda^2}{2} S^2 - \frac{i\lambda^3}{6} S^2 + \cdots. \tag{8.5}$$

Carrying out the transformation indicated in (8.4) and equating terms of equal powers of λ from the two sides of the equation one obtains that

$$\left.\begin{aligned} H_0' &= H_0, \\ H_1' &= H_1 - i(H_0 S - S H_0), \\ H_2' &= H_2 + \frac{i}{2}[S(H_1 + H_1') - (H_1 + H_1') S]. \end{aligned}\right\} \tag{8.6}$$

If still higher orders of approximation to the energy are required higher orders in the expansion of H must first be obtained and then one may repeat the above process and perform a second contact transformation, this time, on H' converting it into H''. The transformation, in this latter case, must have $\lambda H_1'$ unchanged and must give rise to a new second order Hamiltonian, $\lambda^2 H_2''$, which will have the same elements diagonal in Γ as had $\lambda^2 H_2'$, but which will have no elements which are non-diagonal in Γ. The elements of $\lambda^2 H_2'$ which were non-diagonal in Γ will then manifest themselves in orders of approximation higher than the second.

The advantage of this method is that much of the work may be carried through symbolically before the actual computations and calculations are begun and far fewer terms need be evaluated in arriving at the energies to the desired approximation.

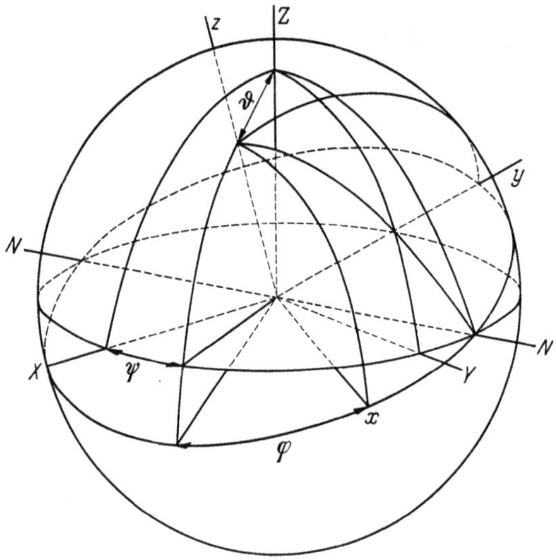

Fig. 1. Eulerian angles relating body-fixed axes, x, y, x, to space-fixed axes, X, Y, Z.

9. Angular momentum. The angular momentum of a molecule is a matter of much importance in the following sections. The angular momentum of a system of mass particles is defined, classically, by the vector equation

$$P = \sum_i m_i (\boldsymbol{r}_i \times \dot{\boldsymbol{r}}_i) \tag{9.1}$$

where $\boldsymbol{r}_i$ is the vector distance of the mass particles m_i from the origin. We have in terms of its components, that

$$P_{\mathsf{A}} = \sum_i (B_i p_{\Gamma_i} - \Gamma_i p_{B_i}) \tag{9.2}$$

where A, B, Γ take the values X, Y and Z, but where $\mathsf{A} \neq \mathsf{B} \neq \Gamma$, X, Y and Z representing a space-fixed coordinate system.

We may, with the aid of Eq. (5.4) deduce the classical Poisson brackets

$$[P_{\mathsf{A}}, P_{\mathsf{B}}] = P_{\Gamma}. \tag{9.3}$$

The operators P_{A} go over in the quantum mechanics into,

$$P_{\mathsf{A}} = -i\hbar \sum_i (\mathsf{B}_i \partial/\partial \Gamma_i - \Gamma_i \partial/\partial \mathsf{B}_i).$$

These have been developed, particularly by BREIT and by CASIMIR, for a many particle system, in terms of the Eulerian angles ϑ, φ and ψ which define the orientation of a set of coordinates x, y and z fixed in the system to the set X, Y and Z fixed in space. When the Eulerian angles are chosen as indicated in Fig. 1, the direction cosines relating these two sets of coordinates are the following

	X	Y	Z
x	$\cos\psi\cos\vartheta\cos\varphi - \sin\psi\sin\varphi$	$\sin\psi\cos\vartheta\cos\varphi + \cos\psi\sin\varphi$	$-\sin\vartheta\cos\varphi$
y	$-\cos\psi\cos\vartheta\sin\varphi - \sin\psi\cos\varphi$	$-\sin\psi\cos\vartheta\sin\varphi + \cos\psi\cos\varphi$	$\sin\vartheta\sin\varphi$
z	$\cos\psi\sin\vartheta$	$\sin\psi\sin\vartheta$	$\cos\vartheta$

(9.4)

and the expressions for P_X, P_Y and P_Z take the form

$$\left.\begin{aligned} P_X &= -i\hbar\left[\frac{\cos\psi}{\sin\vartheta}\left(\frac{\partial}{\partial\varphi} - \cos\vartheta\frac{\partial}{\partial\psi}\right) - \sin\psi\frac{\partial}{\partial\vartheta}\right], \\ P_Y &= -i\hbar\left[\frac{\sin\psi}{\sin\vartheta}\left(\frac{\partial}{\partial\varphi} - \cos\vartheta\frac{\partial}{\partial\psi}\right) + \cos\psi\frac{\partial}{\partial\vartheta}\right], \\ P_Z &= -i\hbar\frac{\partial}{\partial\psi}. \end{aligned}\right\} \tag{9.5}$$

The Poisson brackets (9.3) evidently then translate into the quantum mechanics as follows

$$[P_{\mathsf{A}}, P_{\mathsf{B}}] = (P_{\mathsf{A}} P_{\mathsf{B}} - P_{\mathsf{B}} P_{\mathsf{A}}) = i\hbar P_{\Gamma}. \tag{9.6}$$

Similarly if Q is a vector which has the components Q_{A}, as for example the direction cosines in (9.4), one may further deduce the Poisson bracket

$$[P_{\mathsf{A}}, Q_{\mathsf{B}}] = Q_{\Gamma}, \tag{9.7}$$

which in the quantum mechanics translates into

$$[P_{\mathsf{A}}, Q_{\mathsf{B}}] = (P_{\mathsf{A}} Q_{\mathsf{B}} - Q_{\mathsf{B}} P_{\mathsf{A}}) = i\hbar Q_{\Gamma}. \tag{9.8}$$

The commutation relations (9.6) and (9.8) lead to the important result that the operators P_{A} commute with the square of the total angular momentum, P^2, where, of course, $P = e_1 P_X + e_2 P_Y + e_3 P_Z$. We have seen that if P_Z and P^2 are commuting operators it is possible to find a set of functions which are, simultaneously, eigenfunctions of both of them, i.e.,

$$P^2\psi(P'^2, P_Z') = P'^2\psi(P'^2, P_Z'), \tag{9.9}$$

$$P_Z\psi(P'^2, P_Z') = P_Z'\psi(P'^2, P_Z'). \tag{9.10}$$

Similarly

$$P_Z^2\psi(P'^2, P_Z') = P_Z[P_Z'\psi(P'^2, P_Z')] = P_Z'^2\psi(P'^2, P_Z')$$

so that

$$\left.\begin{aligned} (P^2 - P_Z^2)\psi(P'^2, P_Z') &= (P_X^2 + P_Y^2)\psi(P'^2, P_Z') \\ &= (P'^2 - P_Z'^2)\psi(P'^2, P_Z'). \end{aligned}\right\} \tag{9.11}$$

All of the P_{A} are real observables so that P^2 and $P_X^2 + P_Y^2$ must necessarily be positive. This leads to the result that $|P_Z'| \leq |P'|$. The commutation relations yield further

$$\left.\begin{aligned} P_Z(P_X \pm iP_Y)\psi(P'^2, P_Z') &= (P_X \pm iP_Y)(P_Z' \pm \hbar)\psi(P'^2, P_Z') \\ &= (P_Z' \pm \hbar)(P_X \pm iP_Y)\psi(P'^2, P_Z'). \end{aligned}\right\} \tag{9.12}$$

Since P_Z' is an eigenvalue of P_Z, $P_Z' \pm \hbar$ must also be an eigenvalue of P_Z unless $(P_X \pm iP_Y)\psi(P'^2, P_Z')$ vanishes. Similarly $(P_X \pm iP_Y)\psi(P'^2, P_Z')$ must be eigenstates also of P^2 belonging to the eigenvalue P'^2.

We shall, in general, therefore, have a whole series of allowed eigenvalues

$$\cdots P_Z' - \hbar; \quad P_Z'; \quad P_Z' + \hbar \cdots \tag{9.13}$$

associated with a given value of P'^2. Since, however, $|P_Z'| \leq |P'|$ this series must have a lowest and a highest member. Indicating these by P_{Z-}' and P_{Z+}' we obtain, with the aid of (9.12), the two relations

$$\left.\begin{aligned} (P_X - iP_Y)\psi(P'^2, P_{Z-}') &= 0, \\ (P_X + iP_Y)\psi(P'^2, P_{Z+}') &= 0 \end{aligned}\right\} \tag{9.14}$$

which the two corresponding extreme states must satisfy in order that no eigenstates belonging to $P'_{Z-} - \hbar$ and $P'_{Z+} + \hbar$ shall occur. If these two equations be operated upon respectively by the operators $(P_X + i P_Y)$ and $(P_X - i P_Y)$ one obtains that

$$\left.\begin{aligned} [P'^2 - (P'_{Z-})^2 + P'_{Z-}\hbar] &= 0, \\ [P'^2 - (P'_{Z+})^2 - P'_{Z+}\hbar] &= 0 \end{aligned}\right\} \tag{9.15}$$

or combining these we have

$$(P'_{Z+} + P'_{Z-})(P'_{Z-} - P'_{Z+} - \hbar) = 0$$

which requires $P'_{Z-} = -P'_{Z+}$ since P'_{Z+} was taken greater or equal to P'_{Z-}. Eq. (9.13) states, however, that P'_{Z+} must be greater than P'_{Z-} by an integral number times $\hbar$; hence P'_{Z+} must be of the form $J\hbar$ where J is a number in one of the series 0, 1, 2, ... or $\frac{1}{2}, \frac{3}{2}, \frac{5}{2}, \ldots$. Moreover, Eqs. (9.9) and (9.10) lead to the result that

$$P'^2 = J(J+1)\hbar^2. \tag{9.16}$$

With any state belonging to P'^2 we shall then have simultaneous eigenstates belonging to the allowed values of P'_Z

$$-J\hbar, \quad (-J+1)\hbar, \ldots (J-1)\hbar, \quad J\hbar. \tag{9.17}$$

The allowed values of P'_Z shall be designated by $M\hbar$ where M takes the values $-J, -J+1, \ldots J-1, J$ or simply

$$(M|P_Z|M) = M\hbar \quad (M = -J, -J+1, \ldots J-1, J). \tag{9.18}$$

The allowed values of P_X and P_Y must, of course, be the same as those of P_Z since they enter into the commutation relations in a manner entirely similar to P_Z.

It is, further, possible by these means to derive the matrix components of P_X, P_Y and P_Z in the representation for which P^2 and P_Z are diagonal. We proved by using Eq. (9.12), in effect, that

$$(P_X \pm i P_Y)\psi(J, M) = N_\pm \psi(J, M \pm 1) \tag{9.19}$$

where P and P_Z are denoted by the quantum numbers J and M. $N_\pm$ in Eq. (9.19) may be regarded as a normalization factor for $\psi(J, M \pm 1)$, $\psi(J, M)$ being normalized in the manner indicated by Eq. (3.2). We proceed by multiplying Eq. (9.19) by its complex conjugate and integrating over configuration space

$$\left.\begin{aligned} &\int \overline{(P_X \pm i P_Y)\psi(J, M)}\,(P_X \pm i P_Y)\psi(J, M)\,d\tau \\ &\qquad = \overline{N}_\pm N_\pm \int \overline{\psi}(J, M \pm 1)\,\psi(J, M \pm 1)\,d\tau. \end{aligned}\right\} \tag{9.20}$$

Since $(P_X \pm i P_Y)$ are Hermitian, Eq. (9.20) may, using (4.1), be written

$$\left.\begin{aligned} &\overline{N}_\pm N_\pm \int \overline{\psi}(J, M \pm 1)\,\psi(J, M \pm 1)\,d\tau \\ &\qquad = \int \overline{(P_X \pm i P_Y)\psi(J, M)}\,(P_X \pm i P_Y)\psi(J, M)\,d\tau, \\ &\qquad = \int \overline{\psi}(J, M)\,(P_X \mp i P_Y)(P_X \pm i P_Y)\,\psi(J, M)\,d\tau, \\ &\qquad = \int \overline{\psi}(J, M)\,[P^2 - P_Z^2 \mp P_Z\hbar]\,\psi(J, M)\,d\tau, \\ &\qquad = [J(J+1) - M(M \pm 1)]\,\hbar^2. \end{aligned}\right\} \tag{9.21}$$

We have, therefore $\bar{N}_{\pm} N_{\pm} = N_{\pm}^2 = [J(J+1) - M(M \pm 1)]\hbar^2$ choosing the normalization factor to be real, whence

$$N_{\pm} = \sqrt{J(J+1) - M(M \pm 1)}\,\hbar = \sqrt{(J \pm M + 1)(J \mp M)}\,\hbar. \tag{9.22}$$

Therefore,

$$(P_X \pm i P_Y)\,\psi(J, M) = \sqrt{(J \pm M + 1)(J \mp M)}\,\hbar\,\psi(J, M \pm 1) \tag{9.23}$$

and from (9.23) we may at once, by integration, obtain the matrix components of P_X and P_Y. The non-vanishing matrix components will evidently be

$$\left.\begin{aligned}(J, M \pm 1\,|P_X|\,J, M) &= \pm i (J, M \pm 1\,|P_Y|\,J, M)\\ &= \frac{\hbar}{2}\sqrt{(J \mp M)(J \pm M + 1)}\,.\end{aligned}\right\} \tag{9.24}$$

It is customary in molecular problems, to refer the angular momentum operators to axes fixed in the molecule, i.e., the angular momentum is computed in reference to a system of coordinates fixed in space and then projected on axes mounted in the molecule. This is equivalent to writing

$$P_\alpha^* = \lambda_{\alpha X} P_X + \lambda_{\alpha Y} P_Y + \lambda_{\alpha Z} P_Z \tag{9.25}$$

where the $\lambda_{\alpha X}$, etc., are the direction cosines [in our case, specifically those given in (9.4)] relating the x, y, z axes to the X, Y, Z axes. The index α takes the values x, y and z. Carrying out the operation indicated in (9.25) we obtain for P_x^*, P_y^* and P_z^*, the operators

$$\left.\begin{aligned}P_x^* &= -i\hbar\left[-\frac{\cos\varphi}{\sin\vartheta}\left(\frac{\partial}{\partial\psi} - \cos\vartheta\frac{\partial}{\partial\varphi}\right) + \sin\varphi\frac{\partial}{\partial\vartheta}\right],\\ P_y^* &= -i\hbar\left[\frac{\sin\varphi}{\sin\vartheta}\left(\frac{\partial}{\partial\psi} - \cos\vartheta\frac{\partial}{\partial\varphi}\right) + \cos\varphi\frac{\partial}{\partial\vartheta}\right],\\ P_z^* &= -i\hbar\frac{\partial}{\partial\varphi}\,.\end{aligned}\right\} \tag{9.26}$$

It was discovered by KLEIN that the angular momenta, P_α^*, given in (9.26) obeyed a commutation relation similar to (9.8) except that the algebraic sign before the i was reversed, i.e.,

$$P_\alpha^* P_\beta^* - P_\beta^* P_\alpha^* = -i\hbar P_\gamma^*, \tag{9.27}$$

α, β, γ taking the values x, y, z with $\alpha \neq \beta \neq \gamma$. KLEIN justified these anomalous commutation relations by correspondence principle methods, but a method suggested more recently by VAN VLECK is, probably, more direct. VAN VLECK'S method consists in using the relations (9.25) for P_α^* and expanding the commutator $P_\alpha^* P_\beta^* - P_\beta^* P_\alpha^*$. He obtains in the particular case $P_x^* P_y^* - P_y^* P_x^*$

$$\left.\begin{aligned}P_x^* P_y^* - P_y^* P_x^* &= (\lambda_{xX} P_X + \lambda_{xY} P_y + \lambda_{xZ} P_Z)(\lambda_{yX} P_X + \lambda_{yY} P_Y + \lambda_{yZ} P_Z) -\\ &- (\lambda_{yX} P_X + \lambda_{yY} P_Y + \lambda_{yZ} P_Z)(\lambda_{xX} P_X + \lambda_{xY} P_Y + \lambda_{xZ} P_Z)\\ &= (\lambda_{xY}\lambda_{yZ} - \lambda_{xZ}\lambda_{yY})(P_Y P_Z - P_Z P_Y) - (\lambda_{xX}\lambda_{yZ} - \lambda_{xZ}\lambda_{yX})(P_Z P_X - P_X P_Z) +\\ &+ (\lambda_{xX}\lambda_{yY} - \lambda_{xY}\lambda_{yX})(P_X P_Y - P_Y P_X) - i\hbar\lambda_{xX}(-\lambda_{yZ} P_Y + \lambda_{yY} P_Z) -\\ &- i\hbar\lambda_{xY}(\lambda_{yZ} P_X - \lambda_{yX} P_Z) - i\hbar\lambda_{xZ}(-\lambda_{yY} P_X + \lambda_{yX} P_Y) +\\ &+ i\hbar\lambda_{yX}(-\lambda_{xZ} P_Y + \lambda_{xY} P_Z) + i\hbar\lambda_{yY}(\lambda_{xZ} P_X - \lambda_{xX} P_Z) +\\ &+ i\hbar\lambda_{yZ}(-\lambda_{xY} P_X + \lambda_{xX} P_Y)\,.\end{aligned}\right\} \tag{9.28}$$

The principal axes of the molecule are determined by the nuclear coordinates. The direction cosines may, therefore, apart from a normalizing factor be regarded as linear functions of the coordinates of the various nuclei. Since commutation relations are unaffected by taking linear combinations of the coordinates and the normalizing factor is an invariant which commutes with all the P_α^*'s, one may use the commutation relation (9.8) to derive another commutation relation, namely

$$\lambda_{\alpha A} P_B - P_B \lambda_{\alpha A} = i \lambda_{\alpha \Gamma}. \tag{9.29}$$

Using this relation and observing the fact that a direction cosine is equal to its cofactor in the determinant of the direction cosines, the expansion (9.28) reduces to

$$\left.\begin{aligned} P_x^* P_y^* - P_y^* P_x^* &= \lambda_{zX}(P_Y P_Z - P_Z P_X) + \lambda_{zY}(P_Z P_X - P_X P_Z) + \\ &\quad + \lambda_{zZ}(P_X P_Y - P_Y P_X) - 2i\hbar\lambda_{zX} P_X - 2i\hbar\lambda_{zY} P_Y - 2i\hbar\lambda_{zZ} P_Z \\ &= -i\hbar\lambda_{zX} P_X - i\hbar\lambda_{zY} P_Y - i\hbar\lambda_{zZ} P_Z = -i\hbar P_z^*. \end{aligned}\right\} \tag{9.30}$$

Evidently also $P_x^{*2} + P_y^{*2} + P_z^{*2} = P_X^2 + P_Y^2 + P_Z^2$ which equals P^2, the square of the total angular momentum operator, which we know has the eigenvalues $J(J+1)\hbar^2$. Moreover, the operators P_α^* and P_A all commute with each other, i.e.,

$$[P_\alpha^*, P_A] = 0. \tag{9.31}$$

It is now possible to deduce many important relations concerning the P_α^* which are analogous to those derived earlier for the P_A. Some of the most important of these follow hereafter.

It is possible, for example, to find a set of functions which are simultaneously eigenfunctions of both P^2 and P_z^* since they are commuting operators. Hence

$$\left.\begin{aligned} P^2\psi(P'^2, P_z^{*\prime}) &= P'^2\psi(P'^2, P_z^{*\prime}) = J(J+1)\hbar^2\psi(P'^2, P_z^{*\prime}), \\ P_z^*\psi(P'^2, P_z^{*\prime}) &= P_z^{*\prime}\psi(P'^2, P_z^{*\prime}) \end{aligned}\right\} \tag{9.32}$$

and

$$P_z^{*2}\psi(P'^2, P_z^{*\prime}) = P_z^*[P_z^{*\prime}(P'^2, P_z^{*\prime})] = P_z^{*\prime 2}\psi(P'^2, P_z^{*\prime}).$$

Another very important relation is one which is analogous to Eq. (9.12). This relation is

$$\left.\begin{aligned} P_z^*(P_x^* \pm iP_y^*)\psi(P'^2, P_z^{*\prime}) &= (P_x^* \pm iP_y^*)(P_z^* \mp \hbar)\psi(P'^2, P_z^{*\prime}) \\ &= (P_z^{*\prime} \mp \hbar)(P_x^* \pm iP_y^*)\psi(P'^2, P_z^{*\prime}). \end{aligned}\right\} \tag{9.33}$$

P_z' is an eigenvalue of P_z^*; $P_z^{*\prime} \mp \hbar$ must, therefore, also be eigenvalues of P_z^* unless $(P_x^* \pm iP_y^*)\psi(P'^2, P_z^{*\prime})$ vanishes. Associated with a given value of P'^2 we shall, therefore, have a whole array of permitted eigenvalues, P_z', namely;

$$\ldots\ P_z^* - \hbar;\ \ P_z^{*\prime};\ \ P_z^{*\prime} + \hbar;\ \ldots \tag{9.34}$$

where, as before, $|P_z^{*\prime}| \leq |P'|$. It is possible by arguments entirely like those applied to P_A, to show that the allowed values of P_z^* are

$$-J\hbar, (-J+1)\hbar, \ \ldots\ (J-1)\hbar, J\hbar$$

or designating $P_z^{*\prime}$ by $K\hbar$, simply

$$(K|P_z^*|K) = K\hbar \quad (K = -J, -J+1, \ \ldots\ J-1, J). \tag{9.35}$$

The relation (9.33) states, effectively, that

$$(P_x^* \pm iP_y^*)\psi(J, K) = N_\mp\psi(J, K \mp 1) \tag{9.36}$$

where we now designate P and P_z^* by the quantum numbers J and K. $N_{\mp}$, in (9.36) is a normalization factor for $\psi(J, K\mp 1)$, $\psi(J, K)$ being normalized in the usual manner. Proceeding as in (9.21) we may here write, taking into account the Hermitian character of $P_x^* \pm i P_y^*$,

$$\left.\begin{aligned}&\overline{N}_{\mp} N_{\mp} \int \overline{\psi}(J, K \mp 1)\, \psi(J, K \mp 1)\, d\tau \\ &= \int \overline{(P_x^* \pm i P_y^*)\, \psi(J, K)}\, (P_x^* \pm i P_y^*)\, \psi(J, K)\, d\tau \\ &= \int \overline{\psi}\, (J, K)\, (P_x^* \mp i P_y^*)\, (P_x^* \pm i P_y^*)\, \psi(J, K)\, d\tau \\ &= \int \overline{\psi}\, (J, K)\, [P^2 - P_z^{*\,2} \pm P_z^* \hbar]\, \psi(J, K)\, d\tau \\ &= [J(J+1) - K(K \mp 1)]\, \hbar^2.\end{aligned}\right\} \tag{9.37}$$

Therefore, if as before we choose $N_{\mp}$ to be real we have $\overline{N}_{\mp} N_{\mp} = N_{\mp}^2 = [J(J+1) - K(K \mp 1)]\, \hbar^2$. Hence

$$N_{\mp} = \sqrt{J(J+1) - K(K \mp 1)}\, \hbar = \sqrt{(J \pm K)\, (J \mp K + 1)}\, \hbar \tag{9.38}$$

so that

$$(P_x^* \pm i P_y^*)\, \psi\, (J, K) = \sqrt{(J \pm K)\, (J \mp K + 1)}\, \hbar\, \psi(J, K \mp 1). \tag{9.39}$$

We may obtain at once from (9.39) by integration the non-vanishing elements of P_x^* and P_y^*. These are

$$(K\,|\,P_x^*\,|\,K \pm 1) = \pm i (K\,|P_x^*|\,K \pm 1) = \frac{\hbar}{2} \sqrt{J \mp K)\, (J \pm K + 1)}. \tag{9.40}$$

The matrix elements for P_x^*, P_y^* and P_z^* have thus been determined.

10. The addition of angular momenta. Much of the success of the interpretation of the complex spectra of atoms must be attributed to the vector model of the atom as embraced already by the classical form of the quantum mechanics. Since we shall have occasion to experience how great is the importance of the vector model also in the discussion of molecular spectra we refer briefly to it here.

Consider, for example, a two-electron system. The classical quantum theory visualized that the two electrons were moving in independent orbits. In units of $\hbar$, the orbital angular momenta of these electrons, referred to space-fixed axes, was defined by quantum numbers, l_1 and l_2, respectively. The projection of each of these along an axis fixed in space, say the axis Z, was also quantized and was designated by quantum numbers m_1 and m_2, respectively, where m_1 and m_2 would take the $(2l_1+1)$ and $(2l_2+1)$ values $l_1, l_1-1, \cdots -l_1+1, -l_1$ and $l_2, l_2-1, \cdots -l_2+1, -l_2$, respectively. There would therefore be $(2l_1+1)\times(2l_2+1)$ states, which in the absence of external fields, would be degenerate. The theory conceived, moreover, that these two vectors, $\boldsymbol{l}_1$ and $\boldsymbol{l}_2$, might be coupled together to form a single new vector, $\boldsymbol{l}$, the absolute values of which would be $l_1+l_2, l_1+l_2-1, \ldots l_1-l_2$. The new vector $\boldsymbol{l}$ would again be space quantized, the quantum number characterizing this being designated as m, which would take the $2l+1$ values $l, l-1, \cdots -l+1, -l$. It is readily verified that also here there will be $(2l_1+1)(2l_2+1)$ states corresponding to the $(2l_1+1)(2l_2+1)$ values of m.

The vector model of the atom can be taken over and regularized by the rigorous quantum mechanics. The procedure for doing this in a system like the above of two commuting angular momentum vectors, $\boldsymbol{l}_1$ and $\boldsymbol{l}_2$, characterized by the quantum numbers γ, l_1, l_2, m_1 and m_2 and represented by the ψ function $\psi(\gamma, l_1, l_2, m_1, m_2)$ is to transform to a new representation in which $\boldsymbol{l} = \boldsymbol{l}_1 + \boldsymbol{l}_2$ and where l_Z evidently is equal to $l_{1Z} + l_{2Z}$, l_Z, l_{1Z} and l_{2Z} being respectively the projection of $\boldsymbol{l}$, $\boldsymbol{l}_1$ and $\boldsymbol{l}_2$ along the Z axis, (i.e., $m=m_1+m_2$) characterized by the quantum numbers γ, l_1, l_2, l, m and represented by a ψ function $\Phi(\gamma, l_1, l_2, l, m)$.

The problem of the vector addition in quantum mechanics has been dealt with quite generally and in much detail by CONDON and SHORTLEY. This subject lies entirely beyond the scope of this work and the reader is referred to their book, *The Theory of Atomic Spectra*, for a complete exposition of this subject. We shall here refer only to one or two of the most significant results derived by them which will be of much use to us in later sections.

Consider two angular momentum vectors $\boldsymbol{P}_1$ and $\boldsymbol{P}_2$ which commute with each other and which add up to give

$$\boldsymbol{P}_1+\boldsymbol{P}_2=\boldsymbol{P}, \tag{10.1}$$

where $\boldsymbol{P}$ is the total angular momentum. Evidently, $\boldsymbol{P}_1$ and $\boldsymbol{P}_2$ commute also with $\boldsymbol{P}$. Consider further, the interaction term, $\boldsymbol{P}_2\cdot\boldsymbol{P}$, between one of the component angular momentum vectors, $\boldsymbol{P}_2$, and the total angular momentum vector, $\boldsymbol{P}$. Rewriting Eq. (10.1) as $\boldsymbol{P}-\boldsymbol{P}_2=\boldsymbol{P}_1$, it follows that $(\boldsymbol{P}_1\cdot\boldsymbol{P}_1)=(\boldsymbol{P}\cdot\boldsymbol{P})+(\boldsymbol{P}_2\cdot\boldsymbol{P}_2)-2(\boldsymbol{P}_2\cdot\boldsymbol{P})$ or that

$$(\boldsymbol{P}_2\cdot\boldsymbol{P})=\tfrac{1}{2}\{(\boldsymbol{P}\cdot\boldsymbol{P})+(\boldsymbol{P}_2\cdot\boldsymbol{P}_2)-(\boldsymbol{P}_1\cdot\boldsymbol{P}_1)\}. \tag{10.2}$$

The eigenvalues of $(\boldsymbol{P}\cdot\boldsymbol{P})$ are known to be $J(J+1)\hbar^2$ where J, the quantum number of total angular momentum is integral. Similarly, the eigenvalues of $(\boldsymbol{P}_i\cdot\boldsymbol{P}_i)$ will be $J_i(J_i+1)\hbar^2$, the quantum numbers J_i being integral. The value of J_1, for a given value of J and J_2, however, may take only the values

$$J_1=J+J_2,\ J+J_2-1,\ \ldots\ |J-J_2|. \tag{10.3}$$

The eigenvalues of the interaction term $(\boldsymbol{P}_2\cdot\boldsymbol{P})$ are, therefore, determined.

The eigenfunction characteristic of a substate for a given value of J_2 and where J_1 takes a value, say, $J_1=J+J_2$, and where the projection of J along the space-fixed axis Z is M, will be a linear combination of the wave functions characteristic of the states where J_1 and J_2 are quantized separately with projections M_1 and M_2 respectively along the space-fixed axis Z, i.e.,

$$\psi(J_1, J_2, J, M)=\sum_{M_1}\sum_{M_2}\psi(J_1, J_2, M_1, M_2)\,(J_1, J_2, M_1, M_2|J_1, J_2, J, M), \tag{10.4}$$

retaining a notation introduced in Sect. 6. The coefficients

$$(J_1, J_2, M_1, M_2|J_1, J_2, J, M)$$

are the well known Wigner coefficients and stated explicitly by CONDON and SHORTLEY in their Tables 2^3 and 4^3.

It may happen that one of the angular momentum vectors referred to before, say $\boldsymbol{P}_2$, is the sum of two or more vectors, for example, $\boldsymbol{P}_3$, $\boldsymbol{P}_4$, etc. In such cases an extension of the foregoing is needed. Let us, however, confine ourselves to the case where

$$\boldsymbol{P}_2=\boldsymbol{P}_3+\boldsymbol{P}_4. \tag{10.5}$$

Consider, moreover, the interaction term $\zeta_3(\boldsymbol{P}_3\cdot\boldsymbol{P})+\zeta_4(\boldsymbol{P}_4\cdot\boldsymbol{P})$. This may be written, when one uses Eq. (10.1), in the form

$$\left.\begin{aligned}\zeta_3(\boldsymbol{P}_3\cdot\boldsymbol{P})+\zeta_4(\boldsymbol{P}_4\cdot\boldsymbol{P})&=\{\zeta_3(\boldsymbol{P}_3\cdot\boldsymbol{P}_1)+\zeta_4(\boldsymbol{P}_4\cdot\boldsymbol{P}_1)+\\&\quad+\zeta_3(\boldsymbol{P}_3\cdot\boldsymbol{P}_3)+\zeta_4(\boldsymbol{P}_4\cdot\boldsymbol{P}_4)+(\zeta_3+\zeta_4)(\boldsymbol{P}_3\cdot\boldsymbol{P}_4)\}.\end{aligned}\right\} \tag{10.6}$$

The term $(\boldsymbol{P}_3\cdot\boldsymbol{P}_4)$ may, however, be arrived at in a manner similar to that employed to derive (8.2), remembering that $(\boldsymbol{P}_2\cdot\boldsymbol{P}_2)=((\boldsymbol{P}_3+\boldsymbol{P}_4)\cdot(\boldsymbol{P}_3+\boldsymbol{P}_4))$. Eq. (8.5) then becomes

$$\left.\begin{aligned}\{\zeta_3(\boldsymbol{P}_3\cdot\boldsymbol{P})+\zeta_4(\boldsymbol{P}_4\cdot\boldsymbol{P})\}=\Big\{\Big(\frac{\zeta_3+\zeta_4}{2}\Big)(\boldsymbol{P}_2\cdot\boldsymbol{P}_2)+\Big(\frac{\zeta_3-\zeta_4}{2}\Big)(\boldsymbol{P}_3\cdot\boldsymbol{P}_3)-\\ -(\zeta_3-\zeta_4)(\boldsymbol{P}_4\cdot\boldsymbol{P}_4)+\zeta_3(\boldsymbol{P}_3\cdot\boldsymbol{P}_1)+\zeta_4(\boldsymbol{P}_4\cdot\boldsymbol{P}_1)\Big\}.\end{aligned}\right\}\quad(10.7)$$

The formulation by CONDON and SHORTLEY (Sect. 12[3]) quickly allows one to draw the following conclusions: (a) all nonvanishing matrix elements of the form $(\boldsymbol{P}_i\cdot\boldsymbol{P}_i)$ are diagonal in all the quantum numbers and have the values

$$J_i(J_i+1)\,\hbar^2 \qquad (J_i \text{ denoting } J,\ J_3,\ J_4 \text{ and } J_1 \text{ or } J_2); \qquad (10.8)$$

(b) all nonvanishing matrix elements of the form $(\boldsymbol{P}_3\cdot\boldsymbol{P}_1)$ or $(\boldsymbol{P}_4\cdot\boldsymbol{P}_1)$ are diagonal in the quantum numbers J, J_1, J_3 and J_4, but have non-diagonal as well as diagonal elements in the quantum number J_2; (c) that the nonvanishing matrix elements of $(\boldsymbol{P}_3\cdot\boldsymbol{P}_1)$ and $(\boldsymbol{P}_4\cdot\boldsymbol{P}_1)$ are the following ones.

$$\left.\begin{aligned}&\left.\begin{aligned}(J_2|(\boldsymbol{P}_3\cdot\boldsymbol{P}_1)|J_2)\\(J_2|(\boldsymbol{P}_4\cdot\boldsymbol{P}_1)|J_2)\end{aligned}\right\}\begin{aligned}&[\pm J_3(J_3+1)\mp J_4(J_4+1)+J_2(J_2+1)]\times\\&\times[J(J+1)-J_1(J_1+1)-J_2(J_2+1)]\times\\&\times[\hbar^2/4\,J_2(J_2+1)],\end{aligned}\\&\left.\begin{aligned}(J_2|(\boldsymbol{P}_3\cdot\boldsymbol{P}_1)|J_2-1)\\-(J_2|(\boldsymbol{P}_4\cdot\boldsymbol{P}_1)|J_2-1)\end{aligned}\right\}\begin{aligned}&[(J_2-J_3+J_4)(J_2+J_3-J_4)(J_2+J_3+J_4+1)\times\\&\times(J_3+J_4-J_2+1)(J+J_2-J_1)]^{\frac12}[(J+J_1-\\&-J_2+1)(J+J_1+J_2+1)(J_1+J_2-J)/4J_2^2\times\\&\times(4J^2-1)]^{\frac12}[\hbar^2/2].\end{aligned}\end{aligned}\right\}\quad(10.9)$$

III. The quantum mechanical Hamiltonian for a polyatomic molecule.

11. Introduction. The energies of the molecule are obtained, theoretically, by the solution of the Schrödinger equation

$$(H-E)\,\psi=0 \qquad (11.1)$$

where H is the quantum mechanical Hamiltonian operator for the vibrating-rotating molecule. We shall here derive the Hamiltonian for a vibrating and rotating polyatomic molecule where also the motions of the electrons as well as the motions of the atomic nuclei will be included. The appropriate form for the operator, H, has been derived by WILSON and HOWARD and again by DARLING and DENNISON in a slightly different form. While their form differs slightly they may be shown to be equivalent to each other. The method we shall follow here is essentially that given by WILSON and HOWARD.

12. The derivation of the classical Hamiltonian. The kinetic energy for the molecule will be

$$T=\frac{1}{2}\sum_i M_i\dot{\boldsymbol{R}}_i^2+\frac{m}{2}\sum_j\dot{\boldsymbol{R}}_j^2 \qquad (12.1)$$

where the position of an atomic nucleus or an electron measured from the origin of a right handed coordinate system X, Y, Z *fixed in space* is described by a vector $\boldsymbol{R}$. M_i in Eq. (12.1) designates the mass of the i-th atomic nucleus in the molecule and m represents the mass of an electron.

Let us denote the position of the center of mass of the molecule by the radius vector $\boldsymbol{R}_0(X_0, Y_0, Z_0)$ and the positions of the particles with respect to the center of mass by $\boldsymbol{R}'$ $(X'Y'Z')$. Evidently, if $\boldsymbol{R}_0$ is to mark the center of mass we may write that $\sum_i M_i \boldsymbol{R}'_i + m \sum_j \boldsymbol{R}'_j = 0$. Eq. (12.1) then may be written as

$$2T = \Big(\sum_i M_i + Nm\Big)\dot{\boldsymbol{R}}_0^2 + \sum_i M_i \dot{\boldsymbol{R}}_i'^2 + m\sum_j \dot{\boldsymbol{R}}_j'^2 \tag{12.2}$$

in which N is taken to denote the number of electrons associated with the molecule.

The problem which we wish to consider is much simplified if the motions of the atomic nuclei and the electrons are considered in terms of a set of *body-fixed coordinates*. We choose a set of coordinates x', y' and z' fixed in the molecule so that the origin is identified with the center of mass. In the vector notation we then have that

$$\dot{\boldsymbol{R}}' = (\boldsymbol{\omega}\times\boldsymbol{r}') + \dot{\boldsymbol{r}}' \tag{12.3}$$

where $\boldsymbol{\omega}$ denotes the angular velocity of the system of coordinates x', y', z' with respect to the coordinates X, Y, Z fixed in space and where $\boldsymbol{r}'(x', y', z')$ is a radius vector defining the position of a particle relative to the coordinate system x', y', z'. Combining Eqs. (12.2) and (12.3) and observing the fact that $\dot{\boldsymbol{r}}'(\boldsymbol{\omega}\times\boldsymbol{r}') = \boldsymbol{\omega}(\boldsymbol{r}'\times\dot{\boldsymbol{r}}')$, one obtains

$$\left.\begin{aligned} 2T = {} & \Big(\sum_i M_i + Nm\Big)\dot{\boldsymbol{R}}_0^2 + \sum_i M_i \dot{\boldsymbol{r}}_i'^2 + m\sum_j \dot{\boldsymbol{r}}_j'^2 + \\ & + \sum_i M_i(\boldsymbol{\omega}\times\boldsymbol{r}'_i)\cdot(\boldsymbol{\omega}\times\boldsymbol{r}'_i) + m\sum_j(\boldsymbol{\omega}\times\boldsymbol{r}'_j)\cdot(\boldsymbol{\omega}\times\boldsymbol{r}'_j) + \\ & + 2\boldsymbol{\omega}\sum_i M_i(\boldsymbol{r}'_i\times\dot{\boldsymbol{r}}'_i) + 2\boldsymbol{\omega} m\sum_j(\boldsymbol{r}'_j\times\dot{\boldsymbol{r}}'_j). \end{aligned}\right\} \tag{12.4}$$

The energy relations for a molecule are much simplified by identifying the coordinate system x', y', z' with the center of mass of the atomic nuclei, instead of the center of mass of the complete molecule as we have done. To introduce this simplification we must neglect certain terms in the kinetic energy which are of the order of magnitude of (m/M_i) times the ordinary rotational and vibrational energies. Evidently if this condition is to be maintained one must introduce the condition

$$\sum_i M_i\,\delta\boldsymbol{r}'_i = 0 \tag{12.5}$$

where $\delta\boldsymbol{r}'_i$ is the displacement of the i-th nucleus from its equilibrium position. The nuclei in a molecule seldom depart far from their positions of equilibrium so that $\delta\boldsymbol{r}'_i$ are, in general, small. We impose the further condition

$$\sum_i M_i(\boldsymbol{r}'_{0i}\times\dot{\boldsymbol{r}}'_i) = 0, \tag{12.6}$$

$\boldsymbol{r}_{0i}$ denoting the equilibrium position of the i-th particle relative to the moving system. Eq. (12.6) implies that the internal motions of the nuclei in the body fixed coordinate system shall generate no angular momentum of the molecule considered as a rigid framework, relative to these. Eq. (12.4) then becomes

$$\left.\begin{aligned} 2T = {} & \Big(\sum_i M_i + Nm\Big)\dot{\boldsymbol{R}}_0^2 + \sum_i M_i\,\delta\dot{\boldsymbol{r}}_i'^2 + m\sum_j \dot{\boldsymbol{r}}_j'^2 + \\ & + \sum_i M_i(\boldsymbol{\omega}\times\boldsymbol{r}'_i)\cdot(\boldsymbol{\omega}\times\boldsymbol{r}'_i) + m\sum_j(\boldsymbol{\omega}\times\boldsymbol{r}'_j)\cdot(\boldsymbol{\omega}\times\boldsymbol{r}'_j) + \\ & + 2\boldsymbol{\omega}\sum_i M_i(\delta\boldsymbol{r}'_i\times\dot{\boldsymbol{r}}'_i) + 2\boldsymbol{\omega} m\sum_j(\boldsymbol{r}'_j\times\dot{\boldsymbol{r}}'_j) \end{aligned}\right\} \tag{12.7}$$

or dropping the vector notation one has explicitly

$$\left.\begin{aligned} 2T = &\left(\sum_i M_i + Nm\right)(\dot{X}_0^2 + \dot{Y}_0^2 + \dot{Z}_0^2) + \sum_i \sum_\alpha M_i\, \delta\dot{\alpha}_i'^2 + m \sum_j \sum_\alpha \dot{\alpha}_j'^2 + \\ &+ \sum_\alpha I_{\alpha\alpha}\,\omega_\alpha^2 - \sum_{\alpha\beta} I_{\alpha\beta}\,\omega_\alpha\,\omega_\beta + 2\sum_\alpha \Omega_\alpha\,\omega_\alpha \end{aligned}\right\} \quad (12.8)$$

where

$$\left.\begin{aligned} I_{\alpha\alpha} &= \sum_i M_i\,(\beta_i'^2 + \gamma_i'^2) + m \sum_j (\beta_j'^2 + \gamma_j'^2), \\ I_{\alpha\beta} &= \sum_i M_i\, \alpha_i'\, \beta_i' + m \sum_j \alpha_j'\, \beta_j', \\ \Omega_\alpha &= -\sum_i M_i\,(\delta\dot{\beta}_i'\,\delta\gamma_i' - \delta\beta_i'\,\delta\dot{\gamma}_i') - m_j \sum_j (\dot{\beta}_j'\gamma_j' - \beta_j'\,\dot{\gamma}_j') \end{aligned}\right\} \quad (12.9)$$

in which α, β, γ are summed over x, y, z. It is important, moreover that the molecule be so oriented in the body-fixed coordinates that the axes x, y, z coincide with the principal axes of inertia when the nuclei are in their equilibrium positions. This is accomplished by requiring that

$$\sum_i M_i\, \alpha_i'^0\, \beta_i'^0 = 0. \quad (12.10)$$

It is convenient to make the following changes of variables

$$\left.\begin{aligned} &\left(\sum_i M_i + Nm\right)^{\frac{1}{2}} \dot{X}_0 = \dot{\Xi}, \quad \left(\sum_i M_i + Nm\right)^{\frac{1}{2}} \dot{Y}_0 = \dot{\mathrm{H}}, \quad \left(\sum_i M_i + Nm\right)^{\frac{1}{2}} \dot{Z}_0 = \dot{\mathrm{Z}}, \\ &\qquad m^{\frac{1}{2}}\,\dot{\alpha}_j' = \dot{\alpha}_j, \qquad M_i^{\frac{1}{2}}\,\delta\dot{\alpha}_i' = \sum_s l_{is}^{(\alpha)}\,\dot{Q}_s. \end{aligned}\right\} \quad (12.11)$$

Q_s may be identified with the normal coordinates introduced to describe the nuclear vibrations, the constants $l_{is}^{(\alpha)}$ being transformation coefficients. The manner in which the constants $l_{is}^{(\alpha)}$ are chosen in order that the coordinates Q_s are the normal coordinates will be discussed in Sect. 18. Here we shall merely anticipate that this may be done. In any event the $l_{is}^{(\alpha)}$ must be so chosen that $\sum_i \sum_\alpha (l_{is}^{(\alpha)})^2 = 1$, $\sum_i \sum_\alpha l_{is}^{(\alpha)} l_{is'}^{(\alpha)} = 0$ and $\sum_i \sum_\alpha l_{is}^{(\alpha)} l_{is}^{(\beta)} = 0$. When these changes in variables have been effected $2T$ becomes

$$2T = (\dot{\Xi}^2 + \dot{\mathrm{H}}^2 + \dot{\mathrm{Z}}^2) + \sum_s \dot{Q}_s^2 + \sum_j \sum_\alpha \dot{\alpha}_j^2 + \sum_\alpha I_{\alpha\alpha}\,\omega_\alpha^2 - \sum_{\alpha\beta} I_{\alpha\beta}\,\omega_\alpha\,\omega_\beta + 2\sum_\alpha \Omega_\alpha\,\omega_\alpha \quad (12.12)$$

where Ω_α now is

$$\Omega_\alpha = -\sum_s \mathfrak{A}_s^{(\alpha)}\,\dot{Q}_s - \sum_j (\dot{\beta}_j\gamma_j - \beta_j\dot{\gamma}_j), \quad (12.13)$$

$\mathfrak{A}_s^{(\alpha)}$ being

$$\mathfrak{A}_s^{(\alpha)} = -\sum_i \sum_{s'} (l_{is'}^{(\beta)}\, l_{is}^{(\gamma)} - l_{is}^{(\beta)}\, l_{is'}^{(\gamma)})\, Q_{s'}. \quad (12.14)$$

Eq. (12.12) must be converted to the Hamiltonian form for our purpose. This is accomplished with the aid of the relations

$$\left.\begin{aligned} p_\Xi &= \frac{\partial T}{\partial \dot{\Xi}}, \quad \text{etc.}, \quad p_{\alpha_j} = \frac{\partial T}{\partial \dot{\alpha}_j} = \dot{\alpha}_j + \gamma_j\,\omega_\beta - \beta_j\,\omega_\gamma, \\ p_s &= \frac{\partial T}{\partial \dot{Q}_s} = \dot{Q}_s - \sum_{(\alpha)} \omega_\alpha\,\mathfrak{A}_s^{(\alpha)}, \\ P_\alpha &= \frac{\partial T}{\partial \omega_\alpha} = I_{\alpha\alpha}\,\omega_\alpha - I_{\alpha\beta}\,\omega_\beta - I_{\alpha\gamma}\,\omega_\gamma - \sum_s \mathfrak{A}_s^{(\alpha)}\,\dot{Q}_s + \sum_j (\beta_j\,\dot{\gamma}_j - \dot{\beta}_j\,\gamma_j) \end{aligned}\right\} \quad (12.15)$$

which permits $2T$ to be written

$$2T = \sum_\alpha P_\alpha \omega_\alpha + \sum_s p_s \dot{Q}_s + \sum_j \sum_\alpha p_{\alpha_j} \dot{\alpha}_j + p_\Xi^2 + p_H^2 + p_Z^2. \tag{12.16}$$

Replacing $\dot{Q}_s$, $\dot{\alpha}_j$ in (12.16) by the values obtained from (12.15) Eq. (12.16) becomes

$$2T = \sum_\alpha (P_\alpha - \Pi_\alpha)\,\omega_\alpha + \sum_s p_s^2 + \sum_j \sum_\alpha p_{\alpha_j}^2 + p_\Xi^2 + p_H^2 + p_Z^2 \tag{12.17}$$

where $(P_\alpha - \Pi_\alpha) = I'_{\alpha\alpha}\,\omega_\alpha - I'_{\alpha\beta}\,\omega_\beta - I'_{\alpha\gamma}\,\omega_\gamma$ in which

$$\Pi_\alpha = -\sum_s \mathfrak{A}_s^{(\alpha)} p_s + \sum (\beta_j p_{\gamma_j} - \gamma_j p_{\beta_j}), \tag{12.18}$$

$$\left.\begin{aligned} I'_{\alpha\alpha} &= I_{\alpha\alpha} - \sum_s (\mathfrak{A}_s^{(\alpha)})^2 - \sum_j (\beta_j^2 + \gamma_j^2), \\ I'_{\alpha\beta} &= I_{\alpha\beta} + \sum_s \mathfrak{A}_s^{(\alpha)} \mathfrak{A}_s^{(\beta)} - \sum_j \alpha_j \beta_j. \end{aligned}\right\} \tag{12.19}$$

The first term in Π_α shall henceforth be denoted by p_α and the second by π_α.

An inverse transformation permits one to write

$$\omega_\alpha = \mu_{\alpha\alpha}(P_\alpha - \Pi_\alpha) + \mu_{\alpha\beta}(P_\beta - \Pi_\beta) + \mu_{\alpha\gamma}(P_\gamma - \Pi_\gamma), \tag{12.20}$$

the coefficients $\mu_{\alpha\alpha}$ and $\mu_{\alpha\beta}$ being $\mu_{\alpha\alpha} = (I'_{\beta\beta} I'_{\gamma\gamma} - I'^2_{\beta\gamma})/\Delta$ and $\mu_{\alpha\beta} = \mu_{\beta\alpha} = (I'_{\gamma\gamma} I'_{\alpha\beta} + I'_{\alpha\gamma} I'_{\gamma\beta})/\Delta$ where

$$\Delta = \begin{vmatrix} I'_{xx} & -I'_{xy} & -I'_{xz} \\ -I'_{yx} & I'_{yy} & -I'_{yz} \\ -I'_{zx} & -I'_{zy} & I'_{zz} \end{vmatrix} = \mu^{-1}. \tag{12.21}$$

This makes it possible to express the energy in the Hamiltonian form

$$2T = \sum_{\alpha\beta} \mu_{\alpha\beta}(P_\alpha - \Pi_\alpha)(P_\beta - \Pi_\beta) + \sum_s p_s^2 + \sum_j \sum_\alpha p_{\alpha_j}^2 + p_\Xi^2 + p_H^2 + p_Z^2. \tag{12.22}$$

The terms p_Ξ^2, p_H^2 and p_Z^2 in Eq. (12.22) are the squares of the momenta associated with the translational motion of the molecule as a whole. Since this type of motion holds no spectroscopic interest, they will henceforth be omitted from the discussion.

In terms of the Eulerian angles ϑ, φ and ψ already defined in (9.3) which relate the body-fixed coordinates x', y', z' to the space-fixed coordinates X', Y', Z', we obtain

$$\begin{aligned} \omega_x &= \dot{\vartheta} \sin\varphi - \dot{\psi} \sin\vartheta \cos\varphi, \\ \omega_y &= \dot{\vartheta} \cos\varphi + \dot{\psi} \sin\vartheta \sin\varphi, \\ \omega_z &= \dot{\varphi} + \dot{\psi} \cos\vartheta \end{aligned}$$

from which one obtains that

$$\left.\begin{aligned} P_x &= \sin\varphi\, p_\vartheta - \frac{\cos\varphi}{\sin\vartheta}(p_\psi - \cos\vartheta\, p_\varphi), \\ P_y &= \cos\varphi\, p_\vartheta + \frac{\sin\varphi}{\sin\vartheta}(p_\psi - \cos\vartheta\, p_\varphi), \\ P_z &= p_\varphi \end{aligned}\right\} \tag{12.23}$$

where

$$p_\vartheta = \frac{\partial T}{\partial \dot{\vartheta}} = \sum_\alpha \frac{\partial T}{\partial \omega_\alpha} \frac{\partial \omega_\alpha}{\partial \dot{\vartheta}}, \quad \text{etc.},$$

and where

$$\frac{\partial T}{\partial \omega_\alpha} = P_\alpha.$$

13. Derivation of the quantum mechanical Hamiltonian. The correct form of the quantum mechanical Hamiltonian corresponding to the classical kinetic energy expressed in terms of the momenta, p_i, conjugate to a set of generalized coordinates, q_i, has been considered by PODOLSKY. If the classical kinetic energy, expressed in terms of the generalized coordinates, q_i, is written

$$T = \tfrac{1}{2} \sum_{ij} g_{ij} \dot{q}_i \dot{q}_j$$

the classical kinetic energy written in terms of the momenta conjugate to q_i will be

$$T = \tfrac{1}{2} \sum_{ij} g^{ij} p_i p_j \tag{13.1}$$

where the g^{ij} are the elements of a matrix inverse to the matrix consisting of the elements g_{ij}. According to PODOLSKY the quantum mechanical operator, H, will be

$$H = \tfrac{1}{2} g^{\frac{1}{4}} \sum_{ij} p_i g^{ij} g^{-\frac{1}{2}} p_j g^{\frac{1}{4}} + V \tag{13.2}$$

in which $p_i = -i\hbar \partial/\partial q_i$, g is the determinant of the coefficients g^{ij} and V is the potential energy. Eq. (13.2) is subject to the requirement that its eigenfunctions normalize in the following way $\int \overline{\psi} \psi \, dq_1 \ldots dq_n = 1$.

By introducing, into Eq. (13.1), a new set of momenta, P_m, which are not conjugate to any set of coordinates, such that

$$p_i = \sum_m s_{im} P_m \tag{13.3}$$

the kinetic energy becomes transformed into

$$T = \tfrac{1}{2} \sum_{m,n} G^{m,n} P_m P_n \tag{13.4}$$

where

$$G^{m,n} = \sum_{ij} s_{im} g^{ij} s_{jn}.$$

The relation (12.22) is, indeed, of the form (13.4) and the method of WILSON and HOWARD consists in inquiring into the condition under which it is possible to express the quantum mechanical Hamiltonian in a form analogous to (13.2), i.e.,

$$H = \tfrac{1}{2} G^{\frac{1}{4}} \sum_{m,n} P_m G^{m,n} G^{-\frac{1}{2}} P_n G^{\frac{1}{4}} + V \tag{13.5}$$

in which G is the determinant of the matrix with the elements G^{mn}. To determine this condition, they substitute the transformation

$$P_m = \sum_i s^{mi} P_i, \tag{13.6}$$

which will be recognized as the inverse of the transformation (13.3), into (13.5). Observing that $G = s^2 g$, where s is the determinant of the matrix of the elements

s_{im}, then with the aid of the expression for $G^{m,n}$ one obtains

$$H = \tfrac{1}{2} g^{\frac{1}{4}} \sum_{kl} \Big[\Big(\sum_{im} s^{\frac{1}{2}} s^{mi} p_i s_{km} s^{-\frac{1}{2}}\Big) g^{kl} g^{-\frac{1}{2}} s^{-\frac{1}{2}} p_l s^{\frac{1}{2}}\Big] g^{\frac{1}{4}} + V$$

which will reduce to

$$H = \tfrac{1}{2} g^{\frac{1}{4}} \sum_{kl} \big[(s^{-\frac{1}{2}} p_k s^{\frac{1}{2}})\, g^{kl} g^{-\frac{1}{2}} (s^{-\frac{1}{2}} p_l s^{\frac{1}{2}})\big] g^{\frac{1}{4}} + V \tag{13.7}$$

only if

$$\sum_{im} s^{mi} p_i s_{km} = s^{-1} p_k s. \tag{13.8}$$

Eq. (13.7) is of the form (13.2) since the p_i in the latter may be identified with the $(s^{-\frac{1}{2}} p_k s^{\frac{1}{2}})$ occurring in (13.7).

We have already seen that the classical P_α of Eq. (12.23) may be expressed as linear combinations of the momenta p_ϑ, p_φ and p_ψ, conjugate to the coordinates ϑ, φ and ψ, and from the relation (12.18) we know that Π_α may be expressed in terms of the momenta p_s and p_{α_j}, conjugate to the coordinates Q_s and α_j. The quantities $(P_\alpha - \Pi_\alpha)$ in Eq. (12.22) may, therefore, be identified with the P_m in (13.4) and the momenta p_s, and p_{α_j} with the momenta p_i in (13.6). The quantities s^{mi} (and s_{km}) are at once available and from these one obtains that $s = (\sin\vartheta)^{-1}$. It is readily shown, moreover, that the $(P_\alpha - \Pi_\alpha)$ will conform to the condition (13.8). The quantum mechanical Hamiltonian may, therefore, be written

$$\left.\begin{aligned} H = \tfrac{1}{2} \Big[\mu^{\frac{1}{4}} \sum_{\alpha\beta} (P_\alpha - \Pi_\alpha)\, \mu_{\alpha\beta}\, \mu^{-\frac{1}{2}} (P_\beta - \Pi_\beta)\, \mu^{\frac{1}{4}}\Big] + \tfrac{1}{2} \mu^{\frac{1}{4}} \sum_s p_s \mu^{-\frac{1}{2}} p_s \mu^{\frac{1}{4}} + \\ + \tfrac{1}{2} \sum_j \sum_\alpha p_{\alpha_j}^2 + V \end{aligned}\right\} \tag{13.9}$$

in which the quantities P_α are equal to $(\sin\vartheta)^{\frac{1}{2}} P_\alpha^* (\sin\vartheta)^{-\frac{1}{2}}$, the P_α^*'s being identically the operators (9.26), and where the quantities Π_α, p_s, etc., are the following operators

$$\Pi_\alpha = p_\alpha + \pi_\alpha = \sum_i \sum_s \sum_{s'} (l_{is'}^{(\beta)} l_{is}^{(\gamma)} - l_{is}^{(\beta)} l_{is'}^{(\gamma)})\, Q_{s'} p_s + \sum_j (\beta_j p_{\gamma_j} - \gamma_j p_{\beta_j}),$$

$$p_s = -i\hbar\, \partial/\partial Q_s, \quad p_{\alpha_j} = -i\hbar\, \partial/\partial \alpha_j.$$

The function ψ which is the solution to the resulting SCHRÖDINGER equation normalizes in the following manner

$$\int \cdots \int \overline{\psi} \psi \, dV_1 \ldots dV_r = 1 \quad \text{where} \quad dV_1 \ldots dV_r = \prod_s dQ_s \prod_j \prod_\alpha d\alpha_j \, d\vartheta \, d\varphi \, d\psi.$$

14. The introduction of spin coordinates. The coordinates ϑ, φ, ψ, Q_s, α_j are positional coordinates used in the description of the atomic nuclei and the electrons in the molecule. Each electron has associated with it a spin angular momentum in addition to its orbital angular momentum. To take this into account it is necessary to introduce into Eq. (13.8) the spin coordinates, s, first used by PAULI, in addition to the positional coordinates stated above. We shall here include in V the interaction terms between the spins and the orbital motions of the electrons and it shall be regarded as legitimate to disregard the dependence of the moments of inertia on the spin orientation.

The spin angular momentum may be coupled to the angular momentum of the molecule in two essentially different manners. The limiting case in the one instance is one where a strong interaction, mainly magnetic in character, exists between the spins and the orbital angular momentum of the electrons. This limiting case corresponds to what is known as HUND'S case (a) in diatomic molecules. The limiting case in the second instance corresponds to HUND'S case (b) in diatomic molecules. Here the magnetic interaction between the spins and the orbital motion of the electrons is small compared with the interaction between the spins and the rotation of the molecule. In the former case the spins must be referred to the moving coordinate system x, y, z. Each coordinate may assume two values, A_j and B_j, and corresponding to these values the spin angular momentum for each electron will have a component along one of the rotating axes, say the z axis, which takes the values $\frac{1}{2}\hbar$ and $-\frac{1}{2}\hbar$, respectively. The spins, in the latter case, are referred to the stationary coordinate system X, Y, Z. Here also each coordinate may take two values, a_j and b_j, corresponding to which the spin angular momentum will have a component along one of the space-fixed axes, for example, the Z axis.

The solutions, ψ, to the Eq. (11.1) will, therefore, be functions of the positional coordinates $\vartheta, \varphi, \psi, Q_s, \alpha_j$, but also of the spin coordinates. The latter shall be designated by σ_j in case (a) and by s_j in case (b). There will evidently be required 2^f wave functions, for a molecule of f electrons, to describe the motion of the molecule including its spin motion.

The vector $\mathbf{s}$ associated with the spin of an electron is regarded by PAULI as an operator. The components of $\mathbf{s}$, s_X, s_Y and s_Z may, in case (b) where the spin is referred to the space-fixed axes X, Y, Z, be represented in the matrix notation by

$$s_X = \begin{vmatrix} 0 & \frac{1}{2} \\ \frac{1}{2} & 0 \end{vmatrix} \hbar, \qquad s_Y = \begin{vmatrix} 0 & -i/2 \\ i/2 & 0 \end{vmatrix} \hbar, \qquad s_Z = \begin{vmatrix} \frac{1}{2} & 0 \\ 0 & -\frac{1}{2} \end{vmatrix} \hbar. \tag{14.1}$$

The two wave functions, for a single electron, identified with the two values o the spin coordinates s shall be denoted by $\psi(a)$ and $\psi(b)$. The operators s_X, etc. operating on $\psi(a)$ and $\psi(b)$ represented in the matrix form $\begin{Vmatrix} \psi(a) \\ \psi(b) \end{Vmatrix}$, lead at once to the relations

$$\begin{aligned} s_X\,\psi(a) &= (\hbar/2)\,\psi(b), & s_X\,\psi(b) &= (\hbar/2)\,\psi(a), \\ s_Y\,\psi(a) &= (-i\,\hbar/2)\,\psi(b), & s_Y\,\psi(b) &= (i\,\hbar/2)\,\psi(a), \\ s_Z\,\psi(a) &= (\hbar/2)\,\psi(a), & s_Z\,\psi(b) &= (-\,\hbar/2)\,\psi(b). \end{aligned}$$

Similarly in a system of many electrons the spin operators have the characteristics

$$\left.\begin{aligned} s_{kX}\,\psi(a_1 \ldots a_k \ldots) &= (\hbar/2)\,\psi(a_1 \ldots b_k \ldots), \\ s_{kX}\,\psi(a_1 \ldots b_k \ldots) &= (\hbar/2)\,\psi(a_1 \ldots a_k \ldots), \\ s_{kY}\,\psi(a_1 \ldots a_k \ldots) &= (-\,i\,\hbar/2)\,\psi(a_1 \ldots b_k \ldots), \\ s_{kY}\,\psi(a_1 \ldots b_k \ldots) &= (i\,\hbar/2)\,\psi(a_1 \ldots a_k \ldots), \\ s_{kZ}\,\psi(a_1 \ldots a_k \ldots) &= (\hbar/2)\,\psi(a_1 \ldots a_k \ldots), \\ s_{kZ}\,\psi(a_1 \ldots b_k \ldots) &= (-\,\hbar/2)\,\psi(a_1 \ldots b_k \ldots). \end{aligned}\right\} \tag{14.2}$$

In case (a) where the spins are referred to the coordinates x, y, z rotating with the molecule the wave functions $\psi(\sigma_1, \ldots \sigma_k)$ are related to the wave functions $\psi(s_1 \ldots s_k)$ by the transformation $\psi(\sigma_1 \ldots \sigma_k \ldots) = S\psi(s_1 \ldots s_k \ldots)$ where S is a matrix with the elements $S(\sigma_1, \ldots \sigma_k \ldots; s_1 \ldots s_k \ldots) = S(\sigma_1; s_1) \ldots S(\sigma_k, s_k) \ldots$

in which

$$\left.\begin{aligned} S(A_k;\,a_k) &= e^{\frac{i}{2}\left(\psi+\frac{\pi}{2}\right)}\cos\frac{\vartheta}{2}\,e^{\frac{i}{2}\left(\varphi-\frac{\pi}{2}\right)},\\ S(A_k;\,b_k) &= i\,e^{-\frac{i}{2}\left(\psi+\frac{\pi}{2}\right)}\sin\frac{\vartheta}{2}\,e^{\frac{i}{2}\left(\varphi-\frac{\pi}{2}\right)},\\ S(B_k;\,a_k) &= i\,e^{\frac{i}{2}\left(\psi+\frac{\pi}{2}\right)}\sin\frac{\vartheta}{2}\,e^{-\frac{i}{2}\left(\varphi-\frac{\pi}{2}\right)},\\ S(B_k;\,b_k) &= e^{-\frac{i}{2}\left(\psi+\frac{\pi}{2}\right)}\cos\frac{\vartheta}{2}\,e^{\frac{i}{2}\left(\varphi-\frac{\pi}{2}\right)}. \end{aligned}\right\} \tag{14.3}$$

15. The Schrödinger equation for a polyatomic molecule. The wave equation for case (b) will evidently be

$$\{H(\alpha_i,\,Q_s,\,\vartheta,\,\varphi,\,\psi,\,s_{kX},\,s_{kY},\,s_{kZ}) - E\}\,\psi(\alpha_j,\,Q_s,\,\vartheta,\,\varphi,\,\psi,\,s_1,\,\ldots\,s_k\,\ldots) = 0 \tag{15.1}$$

where $H(\alpha_j,\,Q_s,\,\vartheta,\,\varphi,\,\psi,\,s_{kX},\,s_{kY},\,s_{kZ})$ is the operator (13.8). The wave equation for case (a) may be obtained in the following manner. The spin portion of the ψ function which is a solution to Eq. (15.1) in case (b) will be $\psi(s_1\,\ldots\,s_k\,\ldots)$. This will, according to the former section, be equal to $S^{-1}\psi(\sigma_1\,\ldots\,\sigma_k\,\ldots)$. When $S^{-1}\psi(\sigma_1\,\ldots\,\sigma_k\,\ldots)$ is substituted in Eq. (15.1) for $\psi(s_1\,\ldots\,s_k\,\ldots)$ and the equation is multiplied from in front by S we obtain

$$\{S\,H(\alpha_j,\,Q_s,\,\vartheta,\,\varphi,\,\psi,\,s_{kX},\,s_{kY},\,s_{kZ})\,S^{-1} - E\}\,\psi(\alpha_j,\,Q_s,\,\vartheta,\,\varphi,\,\psi,\,\sigma_k) = 0. \tag{15.2}$$

It becomes apparent that only those operators in H which depend upon the Eulerian angles will be affected by the transformation $S H S^{-1}$, i.e., nothing but the operators involving P_α and the term V will be changed. It is readily seen from the nature of the function S and the characteristics of the spin matrices (14.2) that

$$S P_\alpha S^{-1} = P_\alpha - \mathfrak{S}_\alpha \tag{15.3}$$

where $\mathfrak{S}_\alpha = \sum_j s_{j\alpha}$ in which $s_{j\alpha}$ are the same matrices in the x, y, z coordinates as are the s_{kX}, etc., in the space-fixed coordinates. The Schrödinger Eq. (15.1) becomes, after the transformation has been carried out, the following which is appropriate for case (a).

$$\left.\begin{aligned} &(H - E)\,\psi(\alpha_j,\,Q_s,\,\vartheta,\,\varphi,\,\psi,\,\sigma_1\,\ldots\,\sigma_k\,\ldots)\\ &\quad= \Big\{\tfrac{1}{2}\mu^{\frac14}\sum_{\alpha\beta}(P_\alpha-\Pi_\alpha-\mathfrak{S}_\alpha)\,\mu_{\alpha\beta}\,\mu^{-\frac12}(P_\beta-\Pi_\beta-\mathfrak{S}_\beta)\,\mu^{\frac14}+\mu^{\frac14}\sum_s p_s\mu^{-\frac12}p_s\mu^{\frac14}+\\ &\qquad+\sum_j\sum_\alpha p_{\alpha j}^2 + S V S^{-1} - E\Big\}\,\psi(\alpha_j,\,Q_s,\,\vartheta,\,\varphi,\,\psi,\,\sigma_1\,\ldots\,\sigma_k\,\ldots) = 0. \end{aligned}\right\} \tag{15.4}$$

It is, perhaps, worthwhile to make the following observation concerning the operator $(P_\alpha - \Pi_\alpha - \mathfrak{S}_\alpha)$. The Π_α and $\mathfrak{S}_\alpha$ commute in the same manner as do the operators P_A. That this should be so is natural since they are functions only of the coordinates relative to the molecular framework. They, therefore, commute with the Eulerian angles or direction cosines which relate the coordinate system x, y, z to the coordinate system X, Y, Z fixed in space. They occur in the operator $P_\alpha - \Pi_\alpha - \mathfrak{S}_\alpha$, however, with a minus sign. It is readily seen that if Π_α were replaced by $-\Pi_\alpha$ they will commute according to the rule (9.27). Evidently since P_α, Π_α and $\mathfrak{S}_\alpha$ are, respectively, described by entirely separate quantum numbers so that the order of multiplication is unimportant we have further that all components of $\boldsymbol{P}$ commute with those of $\boldsymbol{\Pi}$, i.e., $P_\alpha\,\Pi_\beta - \Pi_\beta\,P_\alpha = 0$.

The potential energy function, V, in Eq. (13.5) embraces all the terms of interaction between the electrons, between the nuclei and between the electrons and the nuclei, including in case (a) a further dependence upon the spins of the electrons. It is important to inquire in what way SVS^{-1} in Eq. (15.3) differs from V. Only two of the contributions to V depend upon the Eulerian angles. One of these arises from the magnetic fields which originate with the nuclei of the molecule as the molecule rotates. Another originates with the electrons as they are carried around with the nuclei. Evidently these two types of terms would tend to cancel each other and would become significant only for large values of the rotational quantum numbers. The function V will, therefore, be independent of the Eulerian angles if these terms are neglected and to this approximation the term SVS^{-1} will be equal to V.

16. The separation of variables. The solution to a problem in quantum mechanics is most readily achieved if a separation of variables in the Schrödinger equation can be had. Eqs. (15.1) and (15.4) are, however, not exactly separable since if functions $\psi = \Phi(\alpha_j, s_j, s_{m,n}) R(Q_s, \vartheta, \varphi, \psi)$ and $\psi = \Phi(\alpha_j, \sigma_j, s_{m,n}) \times R(Q_s, \vartheta, \varphi, \psi)$ in which $s_{m,n}$ is the distance between two atomic nuclei, are adopted as solutions to Eq. (15.1) and (15.4), the terms

$$\left.\begin{aligned} U = \frac{1}{2\Phi R} \Big\{ & \mu^{\frac14} \sum_{\alpha} (P_\alpha - p_\alpha)\, \mu_{\alpha\beta}\, \mu^{-\frac14} R\, (P_\beta - p_\beta)\, \Phi + \\ & + \mu^{\frac14} \sum_{\alpha} \left(\mu_{\alpha\beta}\, \mu^{-\frac12} (P_\beta - p_\beta)\, \mu^{\frac14} R\right) \left((P_\alpha - p_\alpha)\, \Phi\right) - \\ & - \mu^{\frac14} \sum_{\alpha\beta} M_\alpha \left[\mu_{\alpha\beta} \mu^{-\frac12} (P_\beta - p_\beta) + (P_\beta - p_\beta)\, \mu_{\alpha\beta} \mu^{-\frac12} - \mu_{\alpha\beta} \mu^{-\frac12} M_\beta\right] \mu^{\frac14} R\, \Phi + \\ & + \mu^{\frac14} \sum_{s} \left[p_s\, \mu^{-\frac14} R\, p_s\, \Phi + (\mu^{-\frac12} p_s\, \mu^{\frac14} R)\, (p_s\, \Phi)\right] \Big\} \end{aligned}\right\} \quad (16.1)$$

in which case $M_\alpha = \pi_\alpha + \mathfrak{S}_\alpha$ in case (a) and simply π_α in case (b), remain which are functions of all the coordinates.

We shall replace U by $\overline{U}$ which is the function U averaged over the electronic coordinates, i.e.,

$$\overline{U} = \sum_{\sigma_1 \ldots \sigma_j} \int \overline{\Phi}\, U\, \Phi\, \prod_j dx_j\, dy_j\, dz_j.$$

The functions, Φ, are independent of the angles ϑ, φ and ψ but do depend slightly upon the coordinates, Q_s. It is, therefore, evident that the terms in U, when averaged over the electronic coordinates, will yield two kinds of terms, one of which is a function of the Q_s only and the other which is a product of a factor, which depends only upon the Q_s, multiplying the operators, $P_\alpha - p_\alpha$. We may, therefore, write for $\overline{U}$

$$\overline{U} = \sum_{\alpha} \left[\varepsilon_\alpha(Q_s)/I^{(e)}_{\alpha\alpha}\right] (P_\alpha - p_\alpha) + \sum_{\alpha} \left[\delta_\alpha(Q_s)/I^{(e)}_{\alpha\alpha}\right]$$

where $\varepsilon_\alpha(Q_s)$ and $\delta_\alpha(Q_s)$ are functions of the Q_s only, the explicit form of which cannot be determined without a knowledge of the function, Φ. The first term in $\overline{U}$ may conveniently be combined and treated with such terms in H as $p_\alpha P_\alpha$ while the second term may best be combined with the potential energy of the vibrating molecule. Since Φ depends only slightly on the Q_s the quantity $U - \overline{U}$ will be small so that the approximate equation in which $\overline{U}$ occurs will differ from the true equation which contains U itself only by small terms.

A partial separation of the wave Eqs. (15.1) and (15.4) into the electronic coordinates and the nuclear positional coordinates are, therefore, effected and the resulting Schrödinger equations are these

$$\left\{\tfrac{1}{2}\sum_{j}\sum_{\alpha} p_{\alpha j}^2 + \left(V - E\left(s_{m,n}\right)\right)\right\} \Phi\left(\alpha_j, s_j, s_{m,n}\right) = 0 \quad \text{for case (b)}$$

$$\left\{\tfrac{1}{2}\sum_{j}\sum_{\alpha} p_{\alpha j}^2 + \left(V - E_e\left(s_{m,n}\right)\right)\right\} \Phi\left(\alpha_j, \sigma_j, s_{m,n}\right) = 0 \quad \text{for case (a)}, \tag{16.2a}$$

$$\left.\begin{aligned}\Big\{\tfrac{1}{2}\Big[\mu^{\frac{1}{4}}\sum_{\alpha\beta}\left(P_\alpha - p_\alpha\right)\mu_{\alpha\beta}\,\mu^{-\frac{1}{2}}\left(P_\beta - p_\beta\right)\mu^{\frac{1}{4}} + \mu^{\frac{1}{4}}\sum_{s} p_s\,\mu^{-\frac{1}{2}}\,p_s\,\mu^{\frac{1}{4}}\Big] + \\ + E_e\left(s_{m,n}\right) - \overline{U} - E\Big\} R\left(Q_s, \vartheta, \varphi, \psi\right) = 0.\end{aligned}\right\} \tag{16.2b}$$

$s_{m,n}$ being the distance between the m-th and the n-th nucleus.

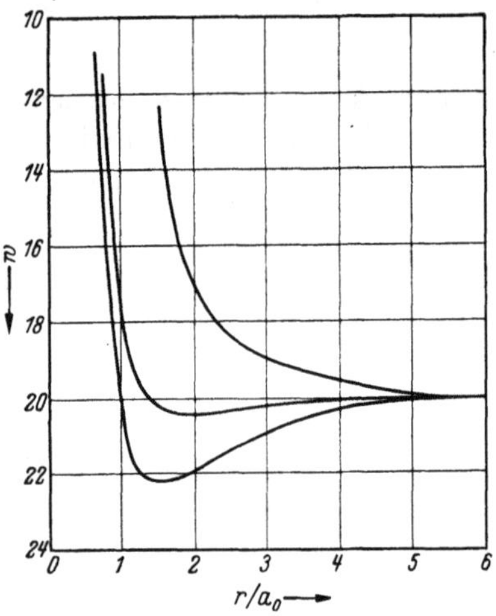

Fig. 2. Energy curves for the hydrogen molecule (in units of $e^2/2a_0$).

17. The potential energy function of a polyatomic molecule. Eq. (16.2a) is the Schrödinger equation for the motions of the electrons in the molecule. The usual manner which is followed to cope with this equation is to integrate it over the coordinates α_j, keeping the nuclei clamped, i.e., fixed with respect to each other. When this is done, the energies, E_e, will depend upon the internuclear distances, $s_{m,n}$, as parameters. The Eq. (16.2a) has been solved only in a relatively small number of instances. The case of the diatomic molecule[1] has been studied in considerable detail, particularly such molecules as the H_2^+ ion and the H_2 molecule. Here E_e is a function of only one internuclear distance, r, and it is known that for certain electronic states the curve of E_e, plotted as a function of r will have a minimum value while for other electronic states no minimum occurs in the curve of E_e for any value of r. The quantity, $E_e(r)$ may in reality be regarded as the potential energy function governing the vibration of the two atomic nuclei in the diatomic molecule with respect to each other. Those electronic states for which $E_e(r)$ has a minimum value represent the stable electronic states and the value, r_e, of r for which the curve has its minimum value represents the equilibrium value of r. The electronic states where the function $E_e(r)$ has no minimum value represent the unstable electronic states of a diatomic molecule. Fig. 2 shows a series of such curves for $E_e(r)$ in the case of the H_2 molecule.

The function $E_e(s_{mn})$ for a polyatomic molecule is, of course, much more complicated than the corresponding one for a diatomic molecule. It will, however, as before, depend upon the internuclear distances as parameters and by analogy to the diatomic case we may infer that if the electronic state is to be a stable one it is necessary that at certain values, s_{mn}^0, of the internuclear distances intermediate between $s_{mn}=0$ and $s_{mn}=\infty$ the function $E_e(s_{mn})$ must have minima. The values s_{mn}^0 at which minima in the function $E_e(s_{mn})$ occur will be the equilibrium values of the internuclear distances. Quite evidently, one may expect the equilibrium values, s_{mn}^0, of the nuclear separations in molecules to vary considerably from one electronic state to another. The function $E_e(s_{mn})$ may, as in the diatomic case, be taken to define the potential energy function

[1] Cf. the article of M. KOTANI in Part 2 of this volume.

which determines the manner in which the atomic nuclei may vibrate when they are free to move. Since it has not been practicable in general to evaluate the function $E_e(s_{mn})$ other methods must be resorted to if one is to make progress in solving the Schrödinger Eq. (16.2b) for the nuclear motions of the molecule.

It is common practice to replace the function $E_e(s_{mn})$ in Eq. (16.2b) by a function $E_e(Q_s)$ which simulates the actual function closely, particularly in the regions where, in a classical sense, the nuclei spend most of their time. The vibration states in which there will be most immediate interest are those of small energies, i.e., those states which correspond to the condition where the nuclei are oscillating about their positions of equilibrium with what in a classical sense would be regarded as simple harmonic motion. The amplitudes of oscillation will be small in such cases and it shall be considered as a legitimate approximation to replace the actual potential energy surface by a power series expansion about the equilibrium values, s^0_{mn}, of the nuclear separations. This expansion may be written

$$\left.\begin{aligned} V = V(s^0_{1,2};\, s^0_{1,3};\, \dots s^0_{mn} \dots) + \sum_{m,n} \left(\frac{\partial V}{\partial s_{mn}}\right)_0 (s_{mn} - s^0_{mn}) + \\ + \frac{1}{2} \sum_{m,n} \sum_{k,l} \left(\frac{\partial^2 V}{\partial s_{mn}\, \partial s_{kl}}\right)_0 (s_{mn} - s^0_{mn})\, (s_{kl} - s^0_{kl}) + \cdots \end{aligned}\right\} \tag{17.1}$$

where, for the purpose of studying the normal vibrations, all terms of higher order in $(s_{mn} - s^0_{mn})$ than the second may be neglected. The first term, $V(s^0_{1,2};\, s^0_{1,3};\, \dots s^0_{mn} \dots)$ is the value of V when all the s_{mn} assume their equilibrium values, s^0_{mn}. The coefficients of the second set of terms, $(\partial V/\partial s_{mn})_0$ will all vanish since at equilibrium the slopes of the potential energy curve must be zero. The coefficients, $(\partial^2 V/\partial s_{mn}\, \partial s_{kl})_0$, are then the generalized force constants which, hereafter, are denoted by K_{mnkl}. The actual function $E_e(s_{mn})$ is therefore replaced by the approximation

$$V - V(s^0_{1,2};\, s^0_{1,3};\, \dots s^0_{m\text{-}n} \dots) = \tfrac{1}{2} \sum_{m,n} \sum_{k,l} K_{mnkl}\, \delta s_{mn}\, \delta s_{kl} \tag{17.2}$$

where $\delta s_{mn} = s_{mn} - s^0_{mn}$, etc. The function (17.2) is a perfectly general function of the coordinates in which the constants are to be so determined as to give the correct positions of the fundamental vibration frequencies of the molecule.

18. The introduction of normal coordinates. The problem of obtaining the energies of the nuclear motions of a polyatomic molecule are greatly simplified by the introduction of normal coordinates. The techniques for arriving at these are sketched briefly in the following section. The kinetic energy of the atomic nuclei in the coordinates x'_i, y'_i, z'_i will be

$$T = \tfrac{1}{2} \sum_i M_i (\dot{x}'^2_i + \dot{y}'^2_i + \dot{z}'^2_i). \tag{18.1}$$

The potential energy which determines the manner in which the atomic nuclei may oscillate has already been given by Eq. (17.2). Replacing the δs_{mn}, which of course are functions of the x'_i, y'_i, z'_i, by q'_m, etc., which are linear combinations of x'_i, y'_i and z'_i Eq. (18.1) becomes

$$T = \tfrac{1}{2} \sum_{mk} a_{mk}\, \dot{q}'_m \dot{q}'_k \tag{18.2}$$

in which, for small oscillations, the a_{mk} may be regarded as constants. The values of a_{mk} are those they would assume at equilibrium, i.e., when the $q'_m = 0$.

There will be f equations of LAGRANGE for the system, one equation for each degree of freedom. These will evidently be

$$\sum_m a_{km} \ddot{q}'_m + \sum_m b_{km} q'_m = 0 \tag{18.3}$$

where b_{km} is an abbreviation for K_{klmn}. Each of the Eq. (18.3) is respectively multiplied by a constant c_k and they are then added together. The constants c_k are then to be determined so that the new equations will be of the form

$$\frac{d^2 Q}{dt^2} + \lambda Q = 0, \tag{18.4}$$

Q being a linear combination of the q'_m

$$Q = \sum_m h_m q'_m. \tag{18.5}$$

Identifying coefficients in Eq. (18.4) with equal coefficients in the above sums, it is evident that the relations which must be satisfied are these

$$\sum_k \lambda c_k a_{km} = \sum_k c_k b_{km} = \lambda h_m. \tag{18.6}$$

The relations (18.6) permit one to write

$$\sum_k (\lambda a_{km} - b_{km}) c_k = 0. \tag{18.7}$$

This set of linear equations, as is well known, can coexist, except in the trivial case of $c_k = 0$, only if the determinant

$$|\lambda a_{km} - b_{km}| = 0. \tag{18.8}$$

The determinant (18.8) will yield f roots, λ_f, which when substituted into (18.7) will give a set of equations

$$\sum_k (\lambda_f a_{km} - b_{km}) c_k = 0 \tag{18.9}$$

which are just sufficient to determine the various c_k in terms of one of these, say c_1. Another manner of achieving the same result is, of course, respectively to substitute the λ_f back into (18.8). The coefficients c_k are then the cofactors of the determinant (18.8). Once these have been determined the constants h_{mf} may be evaluated with the aid of (18.6), their absolute values being fixed by normalization. There will then be f coordinates, Q_f, each of which is a linear combination of the q'_m and each of these will satisfy the equation

$$\frac{d^2 Q_m}{dt^2} + \lambda_m Q_m = 0. \tag{18.10}$$

The Q's may then be regarded as a new set of coordinates for which (18.10) are the Lagrange equations of motion. These coordinates, Q, are known as the normal coordinates and in terms of these the kinetic and potential energies have the simple form, respectively

$$T = \tfrac{1}{2} \sum_s \dot{Q}_s^2 \quad \text{and} \quad V = \tfrac{1}{2} \sum_s \lambda_s Q_s^2. \tag{18.11}$$

The quantities, λ, must be real and positive if the equilibrium is to be stable and in this case we may set $\lambda_s = (2\pi c \omega_s)^2$ where c is the velocity of light and ω_s is the normal frequency in cm.$^{-1}$. The solution to Eq. (18.10) will be

$$Q_s = A_s \cos\left(\sqrt{\lambda_s}\, t + \alpha_s\right) = A_s \cos\left(2\pi c \omega_s t + \alpha_s\right)] \tag{18.12}$$

where A_s and α_s are arbitrary constants. The determinantal Eq. (18.8) will, as the problem has been set up, yield six roots (or five in linear cases) of λ which are equal to zero. These six roots are referred to as the *null-vibrations* and correspond to translations of the system along the three coordinate axes and to rotations of the system about the three (or two in linear cases) axes. These six roots may be eliminated if the relations (12.5) and (12.6) are introduced before the determinantal equation is set up.

Arriving at the appropriate form for the normal coordinates is in most cases a tedious and complicated task. The work is often much facilitated by introducing certain intermediate coordinates which reflect the symmetry of the model under consideration. The usefulness of such *symmetry coordinates* may be seen from the two following examples.

Consider first the planar XYZ_2 model[1] which is shown in Fig. 3. The model is so placed in the coordinate system that its plane coincides with the xz plane and the z axis lies along the axis of symmetry. There are six intermediate symmetry coordinates and these are chosen as follows

$$\left.\begin{aligned}
q_1' &= \frac{1}{\sqrt{2}}\,(\Delta d_1 + \Delta d_2)\,,\\
q_2' &= \Delta d_3\,,\\
q_3' &= \frac{1}{\sqrt{6}}\,(\Delta\alpha_1 + \Delta\alpha_2 - 2\Delta\alpha_3)\,,\\
q_4' &= \frac{1}{\sqrt{2}}\,(\Delta d_1 - \Delta d_2)\,,\\
q_5' &= \frac{1}{\sqrt{2}}\,(\Delta\alpha_1 - \Delta\alpha_2)\,,\\
q_6' &= \Delta\beta
\end{aligned}\right\} \qquad (18.13)$$

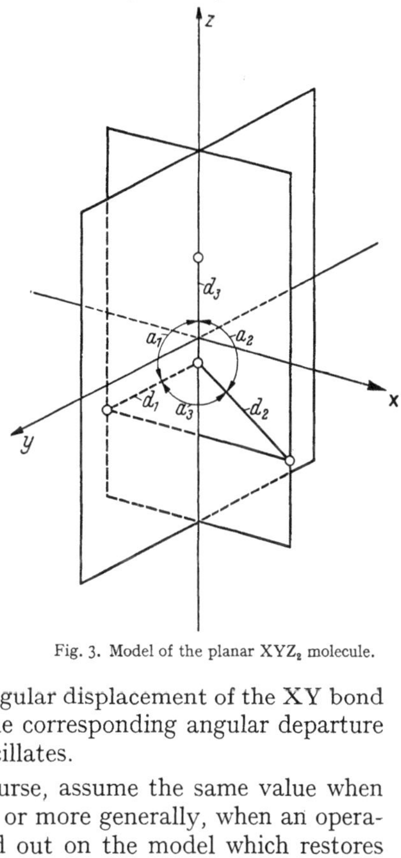

Fig. 3. Model of the planar XYZ_2 molecule.

and are illustrated in Fig. 4. q_5' measures the angular displacement of the XY bond from the z axis in the xz plane while $\Delta\beta$ is the corresponding angular departure of XY from z in the yz plane as the model oscillates.

The potential energy function must, of course, assume the same value when the two identical Z particles are interchanged, or more generally, when an operation (known as a *covering operation*) is carried out on the model which restores its original configuration. This may be accomplished here in four different ways, namely: (1), the identity operation I; (2), reflection of the model in the xz plane, $\sigma_v(x, z)$; (3), reflection of the model in the yz plane, $\sigma_v(y, z)$ and (4), rotation of the model about the z axis through an angle π, C_2. The manner in which the internal coordinates, $\Delta d_1, \Delta d_2, \Delta\alpha_1$, etc., are transformed by the above operations is shown in Table 1 where $\Delta d_1 \rightarrow \Delta d_2$, etc., may be taken to mean that d_1 goes over into d_2, etc., by the indicated operation. It is now readily verified that the three coordinates q_1', q_2' and q_3' are, through all the above operations converted, respectively, into themselves, i.e., into $+1 \cdot q_1'$, $+1 \cdot q_2'$ and $+1 \cdot q_3'$. They are, therefore, said to be symmetrical to all the above covering operations

[1] The capital letters X, Y, Z stand for chemical atom symbols here and in the following example.

and are referred to as type A_1 coordinates. Similarly one may see that the two coordinates q_4' and q_5' are converted into $+1 \cdot q_4'$ and $+1 \cdot q_5'$ respectively by the operations I and $\sigma_v(x, z)$, but are converted into $-1 \cdot q_4'$ and $-1 \cdot q_5'$ respectively by the operations C_2 and $\sigma_v(y, z)$. They are, therefore, symmetric to the two former operations, but are antisymmetric to the two latter operations and are referred to as type B_2 coordinates. Finally, q_6' is symmetric to operations I and $\sigma_v(y, z)$ and antisymmetric to C_2 and $\sigma_v(x, z)$ and is referred to as a type B_1 coordinate. The three coordinates q_1', q_2' and q_3', therefore, transform alike; similarly the coordinates q_4' and q_5' transform alike and finally the coordinate q_6' transforms by itself under the above covering operations, each of which must leave the potential energy function unchanged. Evidently, therefore, the potential energy function may have cross product terms between q_1', q_2' and q_3' and between q_4' and q_5', but no cross product terms between q_1', q_2', q_3' and q_4', q_5', q_6' or between q_4', q_5' and q_6' are permitted. Hence when V is expressed in terms of the q's it may be written

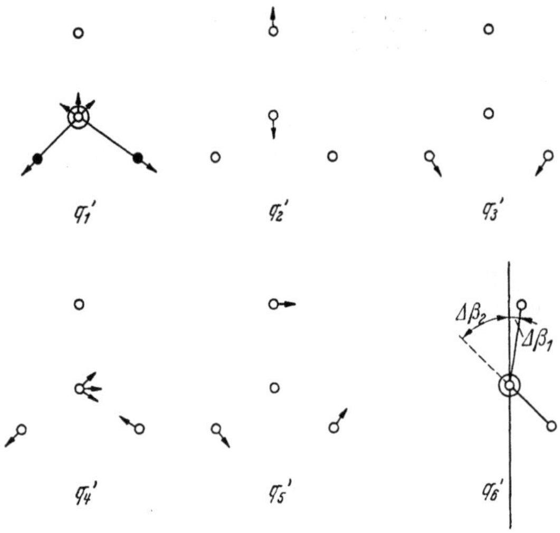

Fig. 4. Symmetry vibrational coordinates of the planar XYZ_2 molecular model.

Table 1. *Transformations of the internal coordinates by the covering operations I, $\sigma_v(x, z)$, $\sigma_v(y, z)$ and C_2.*

	I	C_2	$\sigma_v(x, z)$	$\sigma_v(y, z)$
$\Delta d_1 \to$	Δd_1	Δd_2	Δd_1	Δd_2
$\Delta d_2 \to$	Δd_2	Δd_1	Δd_2	Δd_1
$\Delta d_3 \to$	Δd_3	Δd_3	Δd_3	Δd_3
$\Delta \alpha_1 \to$	$\Delta \alpha_1$	$\Delta \alpha_2$	$\Delta \alpha_1$	$\Delta \alpha_2$
$\Delta \alpha_2 \to$	$\Delta \alpha_2$	$\Delta \alpha_1$	$\Delta \alpha_2$	$\Delta \alpha_1$
$\Delta \alpha_3 \to$	$\Delta \alpha_3$	$\Delta \alpha_3$	$\Delta \alpha_3$	$\Delta \alpha_3$
$\Delta \beta \to$	$\Delta \beta$	$-\Delta \beta$	$-\Delta \beta$	$\Delta \beta$

$$\left.\begin{aligned} 2V = {} & k_1 q_1'^2 + k_2 q_2'^2 + k_3 q_3'^2 + \\ & + 2k_{1,2} q_1' q_2' + 2k_{1,3} q_1' q_3' + \\ & + 2k_{2,3} q_2' q_3' + k_4 q'^2 + k_5 q_5'^2 + \\ & + 2k_{4,5} q_4' q_5' + k_6 q_6', \end{aligned}\right\} \tag{18.14}$$

the k's being potential energy constants. Evidently from the form of (18.13), the kinetic energy, T, will have the very simple form

$$T = \tfrac{1}{2} \sum_{s=1}^{6} a_s \dot{q}_s'^2 \tag{18.15}$$

so that the Lagrange determinant becomes simply

$$\begin{vmatrix} a_1\lambda - k_1 & -k_{1,2} & -k_{1,3} & 0 & 0 & 0 \\ -k_{1,2} & a_2\lambda - k_2 & -k_{2,3} & 0 & 0 & 0 \\ -k_{1,3} & -k_{2,3} & a_3\lambda - k_3 & 0 & 0 & 0 \\ 0 & 0 & 0 & a_4\lambda - k_4 & -k_{4,5} & 0 \\ 0 & 0 & 0 & -k_{4,5} & a_5\lambda - k_5 & 0 \\ 0 & 0 & 0 & 0 & 0 & a_6\lambda - k_6 \end{vmatrix} = 0\,. \tag{18.16}$$

The operations which may be performed on a molecule to restore its original configuration have interesting properties. Thus, for example, the identity operation followed by one of the other operations referred to above, say R, is equivalent to the operation R followed by the identity operation and is equivalent to R. It may be seen, moreover, that performing two of the above operations on the molecule, one after another, is equivalent to a third of the above operations, for example; performing first $\sigma_v(x, z)$ and then $\sigma_v(y, z)$ on the above model is equivalent to performing C_2; etc. We may note, further, that the order in which the operations are carried out is unimportant so that for example $C_2(\sigma_v(x,z)\,\sigma_v(y,z))$ is the same as $C_2\,\sigma_v(x, z)\,\sigma_v(y, z)$. Finally, for every operation R, a reciprocal R^{-1} may always be found which will give $RR^{-1}=R^{-1}R=I$, where I is the identity operator. Summarizing the above we have:

1. There exists an identity operation, I, such that $IR=RI=R$, where R is one of the permitted operations.

2. The product of two operations is equivalent to another one of the permitted operations.

3. The associative law of multiplication holds.

4. Every permitted operation has a reciprocal such that $RR^{-1}=R^{-1}R=I$.

Table 2. *Symmetry characters for the point group C_2.*

C_2	I	C_2	$\sigma_v(x, z)$	$\sigma_v(y, z)$
A_1	+1	+1	+1	+1
A_2	+1	+1	−1	−1
B_1	+1	−1	+1	−1
B_2	+1	−1	−1	+1

Quantities which have the above qualifications are known as *group elements* and the set forms what is known as a *group*[1]. The effects of the covering operations on the symmetry coordinates we have adopted for the XYZ_2 model are, in fact, described in the Table 2 which gives the *characters* for the *point group* C_2, to which this model belongs. The XYZ_2 model has no normal coordinates of the type A_2, although a rotation of the model as a whole about the z axis, R_z, would fall in this category.

It is found, in symmetric molecules, that the determinantal Eq. (18.8) will frequently give two or more roots, λ, which have the same values. This means that there are two or more normal coordinates associated with the same frequency and such vibrations are said to be two or more-fold degenerate. It is convenient, for this reason, to denote a normal coordinate by $Q_{s\sigma}$ rather than by Q_s, where s will denote the particular frequency, ω_s, to be associated with the coordinate and σ will take the values 1; 1, 2 and 1, 2, 3, depending upon whether the frequency is nondegenerate, two-, or threefold degenerate, respectively. Thus, for example, a twofold degenerate frequency of a two-dimensionally isotropic oscillator, oscillating with a frequency ω_a will have associated with it the two coordinates $Q_{a,1}$ and $Q_{a,2}$. The constants $l_{is}^{(\alpha)}$ must henceforth also be written as $l_{is\sigma}^{(\alpha)}$.

As an example of a molecule where twofold degenerate vibrations may occur, we choose the pyramidal XY_3 model which is shown in Fig. 5. The model is so placed in the coordinate system that the equilibrium values of the coordinates of the four particles have the following values: Y(1), $(-\sqrt{3}\,a_0/2,\ -a_0/2,\ -h_1)$; Y(2), $(\sqrt{3}\,a_0/2,\ -a_0/2, -h_1)$; Y(3), $(0,\ a_0,\ -h_1)$; and X(4), $(0, 0, h_4)$. The center of mass of the model coincides with the origin of the coordinate system so that $h_1=(\mu_1/3\,m_Y)\,h_0$, $h_4=(\mu_1/m_X)\,h_0$, μ_1 being $3\,m_X\,m_Y/(3\,m_Y+m_X)$ and h_0 the

[1] Cf. for mathematical details the article of G. Falk on algebra in vol. II of this Encyclopedia which contains an extensive treatment of group theory.

height of the pyramid. Evidently the z axis coincides with the axis of symmetry of the model. As before we shall introduce certain symmetry coordinates. We shall choose these to be

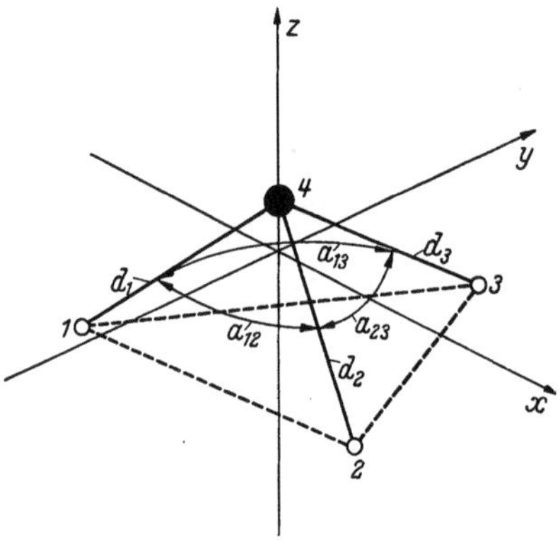

Fig. 5. The pyramidal XY_3 molecular model.

$$\left.\begin{aligned} q_1' &= (\Delta d_1 + \Delta d_2 + \Delta d_3)/\sqrt{3}\,, \\ q_2' &= (\Delta\alpha_{1,2} + \Delta\alpha_{2,3} + \Delta\alpha_{3,1})/\sqrt{3}\,, \\ q_{3,1}' &= (2\Delta d_1 - \Delta d_2 - \Delta d_3)/\sqrt{6}\,, \\ q_{3,2}' &= (\Delta d_2 - \Delta d_3)/\sqrt{2}\,, \\ q_{4,1}' &= (2\Delta_{2,3} - \Delta\alpha_{1,2} - \Delta\alpha_{3,1})/\sqrt{6}\,, \\ q_{4,2}' &= (\Delta d_{3,1} - \Delta\alpha_{1,2})/\sqrt{2}\,. \end{aligned}\right\} \tag{18.17}$$

The covering operations in this case are the following: (a), the identity operation, I; (b), rotation about the z axis through an angle $2\pi/3$, designated a $C(2\pi/3)$; (c), rotation about the z axis through an angle $4\pi/3$, designated as $C(4\pi/3)$; reflection of all particles through the plane which is perpendicular to the Y_3 plane and which contains (d), the particles 1, and 4; (e) the particles 2 and 4 and (f), the particles 3 and 4. The last three operations are respectively designated as $\sigma_v(1,4)$, $\sigma_v(2,4)$ and $\sigma_v(3,4)$. The manner in which the internal coordinates transform in the XY_3 model under these covering operations is given in Table 3. It is, at once, apparent that the two coordinates, q_1' and q_2' are symmetric to each one of the covering operations indicated in Table 3.

Table 3. *Transformations of the internal coordinates of the XY_3 model by the covering operations I, $C(2\pi/3)$, $C(4\pi/3)$, $\sigma_v(1,4)$, $\sigma_v(2,4)$ and $\sigma_v(3,4)$.*

	I	$C(2\pi/3)$	$C(4\pi/3)$	$\sigma_v(1,4)$	$\sigma_v(2,4)$	$\sigma_v(3,4)$
$\Delta d_1 \rightarrow$	Δd_1	Δd_2	Δd_3	Δd_1	Δd_3	Δd_2
$\Delta d_2 \rightarrow$	Δd_2	Δd_3	Δd_1	Δd_3	Δd_2	Δd_1
$\Delta d_3 \rightarrow$	Δd_3	Δd_1	Δd_2	Δd_2	Δd_1	Δd_3
$\Delta\alpha_{1,2} \rightarrow$	$\Delta\alpha_{1,2}$	$\Delta\alpha_{2,3}$	$\Delta\alpha_{3,1}$	$\Delta\alpha_{3,1}$	$\Delta\alpha_{2,3}$	$\Delta\alpha_{1,2}$
$\Delta\alpha_{2,3} \rightarrow$	$\Delta\alpha_{2,3}$	$\Delta\alpha_{3,1}$	$\Delta\alpha_{1,2}$	$\Delta\alpha_{2,3}$	$\Delta\alpha_{1,2}$	$\Delta\alpha_{3,1}$
$\Delta\alpha_{3,1} \rightarrow$	$\Delta\alpha_{3,1}$	$\Delta\alpha_{1,2}$	$\Delta\alpha_{2,3}$	$\Delta\alpha_{1,2}$	$\Delta\alpha_{3,1}$	$\Delta\alpha_{2,3}$

When the coordinates $q_{3,1}'$, $q_{3,2}'$, $q_{4,1}'$ and $q_{4,2}'$ are subjected to one of the above covering operations they are not converted simply into themselves or the negative of themselves, i.e., they are not simply symmetric or antisymmetric, but are converted into linear combinations of the two coordinates associated with the same frequency. Consider, for example, the operator $C(2\pi/3)$ applied to the coordinates $q_{3,1}'$ and $q_{3,2}'$. We have then that

$$(C(2\pi/3))\, q_{3,1}' = (2\Delta d_2 - \Delta d_3 - \Delta d_1)/\sqrt{6} = A q_{3,1}' + B q_{3,2}' \tag{18.18}$$

and that

$$(C(2\pi/3))\, q_{3,2}' = (\Delta d_3 - \Delta d_1)/\sqrt{2} = A' q_{3,1}' + B' q_{3,2}' \tag{18.19}$$

with A, A', B and B' constants. When one substitutes the values of $q_{3,1}'$ and $q_{3,2}'$ given in (18.17) into (18.18) and (18.19) and equates coefficients one obtains that $A = B' = -\frac{1}{2}$ and $A' = -B = -\frac{1}{2}\sqrt{3}$. We have in the matrix notation

$$(C(2\pi/3))\begin{pmatrix} q_{3,1}' \\ q_{3,2}' \end{pmatrix} = \boldsymbol{a}\begin{pmatrix} q_{3,1}' \\ q_{3,2}' \end{pmatrix}$$

where the transformation matrix is

$$\boldsymbol{a} = \begin{pmatrix} A & B \\ A' & B' \end{pmatrix} = \begin{pmatrix} -\frac{1}{2} & \frac{1}{2}\sqrt{3} \\ -\frac{1}{2}\sqrt{3} & -\frac{1}{2} \end{pmatrix}.$$

Similarly $C(4\pi/3)$ applied to $q'_{3,1}$ and $q'_{3,2}$ lead to the transformation matrix

$$\begin{pmatrix} -\frac{1}{2} & -\frac{1}{2}\sqrt{3} \\ \frac{1}{2}\sqrt{3} & -\frac{1}{2} \end{pmatrix}.$$

Moreover, applying $\sigma_v(1,4)$, $\sigma_v(2,4)$ and $\sigma_v(3,4)$ to the same coordinates we obtain, respectively, these transformation matrices

$$\begin{pmatrix} 1 & 0 \\ 0 & -1 \end{pmatrix}, \quad \begin{pmatrix} -\frac{1}{2} & -\frac{1}{2}\sqrt{3} \\ -\frac{1}{2}\sqrt{3} & +\frac{1}{2} \end{pmatrix} \quad \text{and} \quad \begin{pmatrix} -\frac{1}{2} & \frac{1}{2}\sqrt{3} \\ \frac{1}{2}\sqrt{3} & +\frac{1}{2} \end{pmatrix}.$$

So also the identity operation, I, applied to these same coordinates gives, in the matrix form

$$\begin{pmatrix} 1 & 0 \\ 0 & 1 \end{pmatrix}.$$

The character of a matrix is defined as the sum of the diagonal elements. These are evidently -1 for the transformation matrices $C(2\pi/3)$ and $C(4\pi/3)$; for σ_v, these are zero and for the identity operation the character is 2. It is readily seen, moreover, that $q'_{4,1}$ and $q'_{4,2}$ transform respectively in the same manner as $q'_{3,1}$ and $q'_{3,2}$.

The characters for the point group C_{3v} are given in Table 4. Comparison of the foregoing results with Table 4 makes it evident that the coordinates q'_1 and q'_2 transform according to the characters of type A_1 and that the coordinates $q'_{3,1}$, $q'_{3,2}$, $q'_{4,1}$ and $q'_{4,2}$ transform according to the characters of type E.

Table 4. *Characters for the point group C_{3v}.*

C_{3v}	I	$2C_3$	$3\sigma_v$
A_1	1	1	1
A_2	1	1	−1
E	2	−1	0

Since the potential energy relation must have the same value after one of the symmetry operations, it becomes evident from the foregoing that cross product terms between q'_1 and q'_2, between $q'_{3,1}$ and $q'_{4,1}$ and between $q'_{3,2}$ and $q'_{4,2}$ may be present in V, but that all others may not occur. We have then that

$$\left.\begin{aligned} 2V = k_1 q_1'^2 + k_2 q_2'^2 + 2k_{1,2} q'_1 q'_2 + k_3(q'^2_{3,1} + q'^2_{3,2}) + k_4(q^2_{4,1} + q^2_{4,2}) + \\ + 2k_{3,4}(q'_{3,1} q'_{4,1} + q'_{3,2} q'_{4,2}) \end{aligned}\right\} \tag{18.20}$$

and evidently, therefore, the Lagrange determinant (18.8) is broken up into subdeterminants, the two involving $q'_{3,1}$ and $q'_{4,1}$, and $q'_{3,2}$ and $q'_{4,2}$ being alike. The Lagrange determinant in its *irreducible* form is shown in Eq. (18.21).

$$\begin{array}{|cc|cc|cc|} \hline a_1\lambda - k_1 & -k_{1,2} & & & & \\ -k_{1,2} & a_2\lambda - k_2 & & & & \\ \hline & & a_3\lambda - k_3 & -k_{3,4} & & \\ & & -k_{3,4} & a_4\lambda - k_4 & & \\ \hline & & & & a_3\lambda - k_3 & -k_{3,4} \\ & & & & -k_{3,4} & a_4\lambda - k_4 \\ \hline \end{array} = 0. \tag{18.21}$$

The vibration of a molecule is, of course, harmonic only to zero order ot approximation. The anharmonic terms are of higher order and may be expressed in terms of the normal coordinates, $Q_{s\sigma}$, which have now been determined. We may then, to second order of approximation, write

$$E_e(Q_{s\sigma}) \approx V = V_0 + V_1 + V_2 + \cdots$$

in which

$$V_0 = \tfrac{1}{2} \sum_{s,\sigma} \lambda_s Q_{s\sigma}^2 \quad \text{where} \quad \lambda_s = (2\pi c \omega_s)^2,$$

$$V_1 = hc \sum_{s\sigma} \sum_{s'\sigma'} \sum_{s''\sigma''} k'_{ss's''} Q_{s\sigma} Q_{s'\sigma'} Q_{s''\sigma''} \tag{18.22}$$

where $s \leq s' \leq s''$ and

$$V_2 = hc \sum_{s\sigma} \sum_{s'\sigma'} \sum_{s''\sigma''} \sum_{s'''\sigma'''} k'_{ss's''s'''} Q_{s\sigma} Q_{s'\sigma'} Q_{s''\sigma''} Q_{s'''\sigma'''}$$

where $s \leq s' \leq s'' \leq s'''$. One may, as in the case of the quadratic portion of V, make use of the fact that the potential energy function must repeat itself when one of the covering operations for the model has been effected, to determine which terms in V_1 and V_2 may be present, i.e., which of the coefficients $k'_{ss's''}$ and $k'_{ss's''s'''}$ are non-vanishing.

We must, further, express the moments and the products of inertia of (12.12) in terms of the normal coordinates before we can proceed to consider the solutions to the Eq. (16.2b). It is now possible to do this since now the normal coordinates have been determined and hence the direction cosines $l_{is\sigma}^{(\alpha)}$ are also known. It may readily be verified that the quantities $I'_{\alpha\alpha}$ and $I'_{\alpha\beta}$ will be, explicitly:

$$\left.\begin{aligned} I'_{\alpha\alpha} &= I_{\alpha\alpha}^{(e)} + \sum_{s\sigma} a_{s\sigma}^{(\alpha\alpha)} Q_{s\sigma} + \sum_{s\sigma} \sum_{s''\sigma''} \Big[A_{s\sigma s''\sigma''}^{(\alpha\alpha)} - \sum_{s'\sigma'} \zeta_{s\sigma s'\sigma'}^{(\alpha)} \zeta_{s''\sigma'' s'\sigma'}^{(\alpha)}\Big] Q_{s\sigma} Q_{s''\sigma''}, \\ I'_{\alpha\beta} &= -\sum_{s\sigma} a_{s\sigma}^{(\alpha\beta)} Q_{s\sigma} - \sum_{s\sigma} \sum_{s''\sigma''} \Big[A_{s\sigma s''\sigma''}^{(\alpha\beta)} - \sum_{s'\sigma'} \zeta_{s\sigma s'\sigma'}^{(\alpha)} \zeta_{s''\sigma'' s'\sigma'}^{(\beta)}\Big] Q_{s\sigma} Q_{s''\sigma''} \end{aligned}\right\} \tag{18.23}$$

where

$$\left.\begin{aligned} a_{s\sigma}^{(\alpha\alpha)} &= 2 \sum_i M_i^{\frac{1}{2}} (\beta_i'^0 l_{is\sigma}^{(\beta)} + \gamma_i'^0 l_{is\sigma}^{(\gamma)}), \\ a_{s\sigma}^{(\alpha\beta)} &= -\sum_i M_i^{\frac{1}{2}} (\alpha_i'^0 l_{is\sigma}^{(\beta)} + \beta_i'^0 l_{is\sigma}^{(\alpha)}), \\ A_{s\sigma s''\sigma''}^{(\alpha\alpha)} &= \sum_i (l_{is\sigma}^{(\beta)} l_{is''\sigma''}^{(\beta)} + l_{is\sigma}^{(\gamma)} l_{is''\sigma''}^{(\gamma)}), \\ A_{s\sigma s''\sigma''}^{(\alpha\beta)} &= -\sum_i l_{is\sigma}^{(\alpha)} l_{is''\sigma''}^{(\beta)}, \end{aligned}\right\} \tag{18.24}$$

and where we set

$$\sum_i (l_{is\sigma}^{(\beta)} l_{is'\sigma'}^{(\gamma)} - l_{is'\sigma'}^{(\beta)} l_{is\sigma}^{(\gamma)}) = \zeta_{s\sigma s'\sigma'}^{(\alpha)}.$$

Even now the Eq. (16.2b) does not lend itself to an exact solution. One proceeds by observing that the displacement coordinates δx_i, δy_i and δz_i (and therefore also the normal coordinates $Q_{s\sigma}$) are small compared to the equilibrium values of the nuclear coordinates x_i^0, y_i^0 and z_i^0. Eq. (16.2b) may, therefore, to a good approximation be replaced by its expansion about the equilibrium values of the nuclear coordinates and an approximate solution obtained in orders of magnitude by the methods of the perturbation theory. When the quantities μ, $\mu_{\alpha\beta}$, etc., are expanded on the basis that $\delta x_i \ll x_i^0$, etc., it is possible to set down the quantum mechanical Hamiltonian for the general polyatomic molecule in orders of magnitude. One obtains the following after $Q_{s\sigma}$ is replaced by $(\hbar^2/\lambda_s)^{\frac{1}{4}} q_{s\sigma}$

$$H^{(0)} = \frac{\hbar}{2} \sum_{s\sigma} \lambda_s^{\frac{1}{2}} \left[\left(\frac{p_{s\sigma}}{\hbar} \right)^2 + q_{s\sigma}^2 \right] + \frac{1}{2} \sum_{\alpha} \frac{(P_\alpha - p_\alpha^*)^2}{I_{\alpha\alpha}^{(e)}}, \tag{18.25a}$$

$$\left. \begin{aligned} H^{(1)} = hc \sum_{s\sigma} \sum_{s'\sigma'} \sum_{s''\sigma''} k_{ss's''} q_{s\sigma} q_{s'\sigma'} q_{s''\sigma''} - {\sum_{\alpha}}^* \frac{p'_\alpha P_\alpha}{I_{\alpha\alpha}^{(e)}} - \\ - \frac{1}{2} \sum_{s\sigma} \left(\frac{\hbar^2}{\lambda_s} \right)^{\frac{1}{4}} \sum_{\alpha\beta} \frac{a_{s\sigma}^{(\alpha\beta)}}{I_{\alpha\alpha}^{(e)} I_{\beta\beta}^{(e)}} (P_\alpha - p_\alpha^*) (P_\beta - p_\beta^*) q_{s\sigma}, \end{aligned} \right\} \tag{18.25b}$$

$$\left. \begin{aligned} H^{(2)} = hc \sum_{s\sigma} \sum_{s'\sigma'} \sum_{s''\sigma''} \sum_{s'''\sigma'''} k_{ss's''s'''} q_{s\sigma} q_{s'\sigma'} q_{s''\sigma''} q_{s'''\sigma'''} - \\ - \frac{1}{2} \sum_{s\sigma} \sum_{s'\sigma'} \left(\frac{\hbar^4}{\lambda_s \lambda_{s'}} \right)^{\frac{1}{4}} \sum_{\alpha\beta} \frac{A'^{(\alpha\beta)}_{s\sigma s'\sigma'} - \sum_{\gamma} (a_{s\sigma}^{(\alpha\gamma)} a_{s'\sigma'}^{(\gamma\beta)} / I_{\gamma\gamma}^{(e)})}{I_{\alpha\alpha}^{(e)} I_{\beta\beta}^{(e)}} (P_\alpha - p_\alpha^*) (P_\beta - p_\beta^*) q_{s\sigma} q_{s'\sigma'} + \\ + {\sum_{s'\sigma'}}^* \left(\frac{\hbar^2}{\lambda_s} \right)^{\frac{1}{4}} \sum_{\alpha\beta} \frac{a_{s\sigma}^{(\alpha\beta)}}{I_{\alpha\alpha}^{(e)} I_{\beta\beta}^{(e)}} p'_\alpha P_\beta q_{s\sigma} + \frac{1}{2} {\sum_{\alpha}}^* \frac{p_\alpha^2}{I_{\alpha\alpha}^{(e)}}. \end{aligned} \right\} \tag{18.25c}$$

$A'^{(\alpha\beta)}_{s\sigma s'\sigma'}$ has been used to denote $\left(A^{(\alpha\beta)}_{s\sigma s'\sigma'} - \sum_{s''\sigma''} \zeta^{(\alpha)}_{s\sigma s''\sigma''} \zeta^{(\beta)}_{s'\sigma' s''\sigma'} \right)$ in Eq. (18.25c) and p'_α to denote $p_\alpha + \varepsilon_\alpha(q_{s\sigma})$ where the operators p_α, expressed in terms of the coordinates $q_{s\sigma}$ and the conjugate momenta, $p_{s\sigma}$, are the following

$$p_\alpha = \sum_i \sum_{s\sigma} \sum_{s'\sigma'} l_{is\sigma}^{(\beta)} l_{is'\sigma'}^{(\gamma)} \left[(\lambda_{s'}/\lambda_s)^{\frac{1}{4}} q_{s\sigma} p_{s'\sigma'} - (\lambda_s/\lambda_{s'})^{\frac{1}{4}} q_{s'\sigma'} p_{s\sigma} \right]. \tag{18.26}$$

It will be seen that the last component of Eq. (18.25b) contains a set of terms similar to the last component of Eq. (18.25c). These two sets of terms are, in fact, similar in every respect except that the p_α^* are made up of a sum of CORIOLIS operators where each operator is associated with a degenerate frequency. These terms may, therefore, contribute to the energy in second order. The significance of $\sum^*$ in Eqs. (18.25b) and (18.25c) is that the sum does not include those terms which already occur in the preceding Eqs. (18.25a) and (18.25b). The latter set of terms will contribute to second order only if resonance between certain frequencies occur. It should, moreover, be noted that the coordinates, $q_{s\sigma}$, are dimensionless. Evidently, therefore, the potential energy coefficients $k_{ss's''}$ and $k_{ss's''s'''}$ will have the dimensions cm^{-1}.

IV. The energies of a polyatomic molecule.

19. Introduction. The energies of a polyatomic molecule are obtained by solution of the Schrödinger Eq. (15.4). This equation, as we have seen, does not adapt itself to an exact solution and we shall proceed to arrive at the eigenvalues of H by successive approximations. The method of the perturbation theory, as already outlined in Chap. II, will serve our purpose best in this respect. We have replaced H by its expansion in orders of magnitude to this end in the previous section and as a first step in our work we seek a solution to the zero order Schrödinger equation $(H_0 - E_0)\,\psi_0 = 0$ where H_0 is the operator (18.25a).

The first part of (18.25a) is that portion of the zero order Hamiltonian which describes the vibrational frequencies of the molecule. We have already seen in Chap. III that the vibration frequencies may be non-degenerate, twofold degenerate and even threefold degenerate. A single coordinate, q_s, is all that is required in the first instance (i.e., σ takes only the value 1 and may therefore conveniently be omitted) to describe the frequency ω_s. We have the linear harmonic oscillator in this case and the portion of H_0 characteristic of it is, evidently

$$H_0' = (h c \omega_s/2)\,[(p_s/\hbar)^2 + q_s^2]. \tag{19.1}$$

The two-dimensionally isotropic oscillator requires two coordinates, $q_{t,1}$ and $q_{t,2}$, to describe the frequency ω_t adequately (i.e., the index σ takes the values $\sigma=1$ and $\sigma=2$). It is found advantageous in this instance to introduce cylindrically polar coordinates, i.e., to let

$$\begin{aligned}
q_{t,1} &= r_t \cos\chi_t,\\
q_{t,2} &= r_t \sin\chi_t,\\
p_{t,1} &= -i\hbar\left[\cos\chi_t \frac{\partial}{\partial r_t} - \frac{\sin\chi_t}{r_t}\frac{\partial}{\partial\chi_t}\right],\\
p_{t,2} &= -i\hbar\left[\sin\chi_t \frac{\partial}{\partial r_t} + \frac{\cos\chi_t}{r_t}\frac{\partial}{\partial\chi_t}\right].
\end{aligned}$$

Introducing these into Eq. (18.25a) we have the following as the portion of the zero order Hamiltonian characteristic of the two-dimensionally isotropic oscillator

$$H_0'' = \frac{1}{2} h c \omega_t \left[\frac{1}{r_t}\frac{p_{r_t}}{\hbar} r_t \frac{p_{r_t}}{\hbar} + \frac{1}{r_t^2}\frac{p_{\chi_t}^2}{\hbar^2} + r_t^2\right] \tag{19.2}$$

in which $p_{r_t} = -i\hbar\,\partial/\partial r_t$ and $p_{\chi_t} = -i\hbar\,\partial/\partial\chi_t$.

It is convenient, in the case of the three-dimensionally isotropic oscillator to introduce the spherical polar coordinates ϱ_u, ϑ_u and χ_u in place of the linear coordinates $q_{u,1}$, $q_{u,2}$ and $q_{u,3}$. These may be taken to be

$$\begin{aligned}
q_{u,1} &= \varrho_u \sin\vartheta_u \cos\chi_u,\\
q_{u,2} &= \varrho_u \sin\vartheta_u \sin\chi_u,\\
q_{u,3} &= \varrho_u \cos\vartheta_u.
\end{aligned}$$

The momenta, $p_{u,\sigma}$, conjugate to $q_{u\sigma}$ will then become the following operators

$$\begin{aligned}
p_{u,1} &= -i\hbar\left\{\sin\vartheta_u\cos\chi_u\frac{\partial}{\partial\varrho_u} + \frac{\cos\vartheta_u\cos\chi_u}{\varrho_u}\frac{\partial}{\partial\vartheta_u} - \frac{\sin\chi_u}{\varrho_u\sin\vartheta_u}\frac{\partial}{\partial\chi_u}\right\},\\
p_{u,2} &= -i\hbar\left\{\sin\vartheta_u\sin\chi_u\frac{\partial}{\partial\varrho_u} + \frac{\cos\vartheta_u\sin\chi_u}{\varrho_u}\frac{\partial}{\partial\vartheta_u} + \frac{\cos\chi_u}{\varrho_u\sin\vartheta_u}\frac{\partial}{\partial\chi_u}\right\},\\
p_{u,3} &= -i\hbar\left\{\cos\vartheta_u\frac{\partial}{\partial\varrho_u} - \frac{\sin\vartheta_u}{\varrho_u}\frac{\partial}{\partial\vartheta_u}\right\}.
\end{aligned}$$

Inserting these into Eq. (18.25a) the portion of H_0 characteristic of a three-dimensionally isotropic oscillator will become

$$H_0''' = \frac{1}{2} h c \omega_u \left\{\frac{1}{\varrho_u^2}\frac{p_{\varrho_u}}{\hbar}\varrho_u^2\frac{p_{\varrho_u}}{\hbar} + \frac{1}{\varrho_u^2\sin\vartheta_u}\left[\frac{p_{\vartheta_u}}{\hbar}\sin\vartheta_u\frac{p_{\vartheta_u}}{\hbar} + \frac{1}{\sin\vartheta_u}\frac{p_{\chi_u}^2}{\hbar^2}\right] + \varrho^2\right\}, \tag{19.3}$$

where $p_{\varrho_u} = -i\hbar\,\partial/\partial\varrho_u$, $p_{\vartheta_u} = -i\hbar\,\partial/\partial\vartheta_u$ and $p_{\chi_u} = -i\hbar\,\partial/\partial\chi_u$.

The second part of (18.25a) is essentially that part of the Hamiltonian which describes the rotational characteristics of the molecule. The operators, P_α, which occur have already been given in the relations (14.3) and the p_α^* are sums of the Coriolis operators, where each operator is associated with a degenerate frequency. The latter are of two types, those associated with twofold degenerate frequencies, such as commonly occur in axially symmetric molecules; and those which are associated with threefold degenerate frequencies, which occur in molecules of tetrahedral symmetry and symmetries of higher order.

Consider first the p_α^* characteristic of a twofold degenerate frequency in an axially symmetric molecule. The axis of symmetry is usually identified with the z axis so that the vibration described by r_t and χ_t will lie in the xy plane. The only p_α^* we shall need to consider will, therefore, be p_z^* which according to Eq. (18.26) will be

$$p_z^* = \sum_t \zeta_t^{(z)} [q_{t,1} p_{t,2} - q_{t,2} p_{t,1}]. \tag{19.4}$$

When the $q_{t,\sigma}$ and $p_{t,\sigma}$ are replaced by their equivalents in terms of r_t and χ_t we have for p_z^*

$$p_z^* = \sum_t \zeta_t^{(z)} p_{\chi_t}. \tag{19.5}$$

Let us consider also the p_α^* associated with a threefold degenerate oscillation in, say, a tetrahedrally symmetric molecule. We shall consider the $q_{u,\sigma}$ directed along the principal axes x, y and z. The operators p_x^*, p_y^* and p_z^* then become respectively, $\zeta_{u,2;u,3}^{(x)}(q_{u,2}p_{u,3} - q_{u,3}p_{u,2})$, $\zeta_{u,3;u,1}^{(y)}(q_{u,3}p_{u,1} - q_{u,1}p_{u,3})$ and $\zeta_{u,1;u,2}^{(z)}(q_{u,1}p_{u,2} - q_{u,2}p_{u,1})$. When the coordinates $q_{u,\sigma}$ and their conjugate momenta are replaced by their equivalents in terms of ϱ_u, ϑ_u and χ_u the operators p_α^* become

$$\left.\begin{aligned} p_x^* &= i\hbar \sum_u \zeta_u^{(x)} \left(\sin\chi_u \frac{\partial}{\partial\vartheta_u} + \cot\vartheta_u \cos\chi_u \frac{\partial}{\partial\chi_u}\right), \\ p_y^* &= -i\hbar \sum_u \zeta_u^{(y)} \left(\cos\chi_u \frac{\partial}{\partial\vartheta_u} - \cot\vartheta_u \sin\chi_u \frac{\partial}{\partial\chi_u}\right), \\ p_z^* &= -i\hbar \sum_u \zeta_u^{(z)} \frac{\partial}{\partial\chi_u}. \end{aligned}\right\} \tag{19.6}$$

The zero order Schrödinger equation $(H_0 - E_0)\psi_0 = 0$ may be solved exactly. In the case of asymmetric molecules where only non-degenerate vibrations occur so that internal angular momenta p_α^* are absent a complete separation of variables may be effected by using for ψ_0 the product $\psi_0 = \prod_s F(q_s) R(\vartheta, \varphi, \psi)$. In molecules with axial symmetry, vibrations occur which are twofold degenerate in addition to those which are non-degenerate. Here also the operator p_z^* will be present. The only term which conceivably could cause trouble is the term $p_z^* P_z$. The operators p_z^* and P_z commute with each other. They also commute with H_0'' and with P^2 and P_Z, the latter two commuting with each other. We have, therefore, from Chap. II, that the wave function ψ_0 will simultaneously be an eigenfunction of H_0'', P^2, P_Z, P_z and p_z^*. The eigenfunction will be characterized by the quantum numbers V, J, K, M and in this case also by l_t. The coordinates φ, ψ, and χ_t being ignorable, we may take for ψ_0 the product function

$$\psi_0 = \prod_s F(q_s) \prod_t e^{i l_t \chi_t} F(r_t) R(\vartheta) e^{iK\varphi} e^{iM\psi} \tag{19.7}$$

which brings about a complete separation of variables as before.

When a molecule has tetrahedral or an even higher degree of symmetry, vibrations which are threefold degnerate may also be present. Moreover, the three moments of inertia, $I_{xx}^{(e)}$, $I_{yy}^{(e)}$ and $I_{zz}^{(e)}$ are all alike. Here the operators p_x^*, p_y^* will be present as well as the operator p_z^* where it may be shown that the Coriolis coupling factor $\zeta_u^{(\alpha)}$ occurring in each of the above are all equal to each other. Here as before it will be only the terms $p_\alpha^* P_\alpha$ which can stand in the way of a complete separation of variables. p_z^* still commutes with P_z; H_0''', P^2 and P_Z also commute with each other, but neither p_z^* nor P_z commute with H_0''' any longer. A complete separation of variables is here not possible unless the terms

$p_\alpha^* P_\alpha$ are neglected. Complete separation of variables may be achieved if these are neglected by taking ψ_0 to be

$$\psi_0 = \prod_s F(q_s) \prod_u e^{i m_u \chi_u} \Theta_{l_u m_u}(\vartheta_u) F(\varrho_u) R(\vartheta) e^{i K \varphi} e^{i M \psi}. \tag{19.8}$$

It will be seen later that the terms $-p_\alpha^* P_\alpha$ actually present no real difficulty since certain linear combinations of the functions $e^{i m_u \chi_u} \Theta_{l_u m_u}(\vartheta_u) R(\vartheta) e^{i K \varphi}$ may be chosen which will simultaneously be eigenfunctions of that part of H_0''' which depends only on ϑ_u and of $\frac{1}{2}\sum_\alpha (P_\alpha - p_\alpha^*)^2/I_{\alpha\alpha}^{(e)}$. At the present time we shall, therefore, ignore the terms $-p_\alpha^* P_\alpha$ and to this extent separation of variables has been achieved.

The separation of H_0 yields a set of differential equations as follows for:

(a) the *linear* harmonic oscillator

$$\{(h c \omega_s/2) [(p_s/\hbar)^2 + q_s^2] - E_s\} F(q_s) = 0; \tag{19.9}$$

(b) the *two-dimensionally isotropic* oscillator

$$\left\{\frac{\partial^2}{\partial r_t^2} + \frac{1}{r_t}\frac{\partial}{\partial r_t} - \frac{l_t^2}{r_t^2} + [(2E/h c \omega_t) - r_t^2]\right\} F(r_t) = 0; \tag{19.10}$$

(c) the *three-dimensionally isotropic* oscillator

$$\left\{\frac{\partial^2}{\partial \varrho_u^2} + \frac{2}{\varrho_u}\frac{\partial}{\partial \varrho_u} + \left[\frac{2E}{h c \omega_u} - \varrho_u^2 - \frac{l_u(l_u+1)}{\varrho_u^2}\right]\right\} R(\varrho_u) = 0, \tag{19.11a}$$

$$\left\{\frac{1}{\sin\vartheta_u}\frac{\partial}{\partial\vartheta_u}\left(\sin\vartheta_u \frac{\partial}{\partial\vartheta_u}\right) - \frac{m_u^2}{\sin^2\vartheta_u} + k_u\right\} \Theta_{l_u m_u}(\vartheta_u) = 0 \tag{19.11b}$$

where $0 \leqq \vartheta_u \leqq \pi$ and $0 \leqq \chi_u \leqq 2\pi$ and

(d) the *rotator*

$$\left\{\tfrac{1}{2}\sum_\alpha (P_\alpha - p_\alpha^*)^2/I_{\alpha\alpha}^{(e)} - E_r\right\} R(\vartheta, \varphi, \vartheta_u) e^{i M \psi} e^{i m_u \chi_u} = 0. \tag{19.12}$$

20. The linear harmonic oscillator. The Eq. (19.9) was studied already by SCHRÖDINGER who showed that a suitable solution was

$$F(q_s) = N_{v_s} e^{-\frac{1}{2} q_s^2} H_{v_s}(q_s) \tag{20.1}$$

where H_{v_s} is the hermite polynomial

$$H_{v_s}(q_s) = (-1)^{v_s} e^{-q_s^2} \frac{d^{v_s}}{d q_s^{v_s}} e^{-q_s^2}$$

and

$$N_{v_s} = \left\{\frac{1}{\sqrt{\pi}} \frac{1}{2^{v_s} v_s!}\right\}^{\frac{1}{2}} \tag{20.2}$$

is a normalization factor determined by setting

$$\int_{-\infty}^{\infty} F_{v_s}(q_s) F_{v_s}(q_s) d q_s = 1.$$

We give here also the matrix elements of the coordinates, q_s, and their conjugate momenta p_s. These are

$$\left.\begin{aligned}(v_s | q_s | v_s - 1) &= (v_s - 1 | q_s | v_s) = (v_s | -i p_s/\hbar | v_s - 1)\\ &= (v_s - 1 | i p_s/\hbar | v_s) = [v_s/2]^{\frac{1}{2}}\end{aligned}\right\} \tag{20.3}$$

which quite evidently satisfy the commutation relation $p_s q_s - q_s p_s = -i\hbar$. The relations (20.3) will be found useful in later sections, particularly when we come to deal with the perturbation calculations, where matrix elements of certain functions, $f(q_s, p_s)$ of the coordinates, q_s, and the momenta, p_s, are required.

21. The two-dimensionally isotropic oscillator. Eq. (19.10) was first studied by DENNISON, who adopted for $F(r_t)$ a function

$$F(r_t) = e^{-\frac{1}{2}r_t^2} r_t^{l_t} Z(r_t).$$

Inserting this form for $F(r_t)$ into Eq. (19.8) it becomes transformed into

$$Z''(r_t) + \left[\frac{2l_t+1}{r_t} - 2r_t\right] Z'(r_t) + 2\left[\frac{E}{hc\omega_t} - (l_t+1)\right] Z(r_t) = 0. \tag{21.1}$$

Requirements for the single valuedness of $e^{il_t\chi_t} F(r_t)$ demands that l_t shall be an integer or zero. Eq. (21.1) is further converted into the form of the differential equation for the associated Laguerre polynomial

$$\sigma_t G''(\sigma_t) + (T_t + 1 - \sigma_t) G'(\sigma_t) + (\tau - T_t) G(\sigma_t) = 0 \tag{21.2}$$

where $\tau = \frac{1}{2}\left[(E/hc\omega_t) + l_t - 1\right] = \frac{1}{2}(v_t + l_t)$, v_t being an integer and $T_t = l_t$, if a new variable $\sigma_t = r_t^2$ is introduced. The normalized function $F(r_t)$ may, therefore, be written as

$$F(r_t) = N_{v_t, l_t} e^{-(\sigma_t/2)} \sigma_t^{\frac{l_t}{2}} L_\tau^{l_t}(\sigma_t) e^{il_t\chi_t} \tag{21.3}$$

where $L_{\tau_t}^{l_t}$ is the associated Laguerre polynomial and

$$N_{v_t, l_t} = \sqrt{2}\{[(v_t - l_t)/2]!\}^{\frac{1}{2}}/\{[(v + l_t)/2]!\}^{\frac{3}{2}}. \tag{21.4}$$

The function $e^{il_t\chi_t} F(r_t)$, in order to be a suitable solution to our problem, places the further restriction on $|l_t|$ that it may take only the integral values $v_t, v_t - 2, \ldots 1$ or 0. The energy, E_{v_t}, for the two-dimensionally isotropic oscillator is then found to be

$$E_{v_t} = (v_t + 1) hc\omega_t. \tag{21.5}$$

The matrix elements of the coordinates r_t and the momenta, p_{r_t} and p_{χ_t}, analogous to the matrix elements of q_s and p_s for the linear harmonic oscillator will be of interest and importance also here when the perturbation calculations are to be carried out. The matrix elements of $e^{\pm i\chi_t} r_t$; i.e., $q_{t,1} \pm i q_{t,2}$, have been evaluated by SHAFFER. He showed, moreover, that the matrix elements of p_{r_t} and (p_{χ_t}/r_t) are very complicated and difficult to evaluate. It is found, however, that p_{r_t} and (p_{χ_t}/r_t) occur always in the molecular problem in the following combinations, $e^{\pm i\chi_t}(p_{r_t} \pm i p_{\chi_t}/r_t)$, i.e., as $p_{t,1} \pm i p_{t,2}$. These may be thought of as corresponding to the momenta conjugate to $e^{\pm i\chi_t} r_t$ and their matrix elements may be obtained from SHAFFER'S values of the matrix components for $e^{\pm i\chi_t} r_t$ with the aid of the commutation relations

$$(p_{t,1} \pm i p_{t,2})(q_{t,1} \mp i q_{t,2}) - (q_{t,1} \mp i q_{t,2})(p_{t,1} \pm i p_{t,2}) = -2i\hbar.$$

The matrix elements of $e^{\pm i\chi_t} r_t$ and $e^{\pm i\chi_t}(p_{r_t} \pm i p_{\chi_t}/r_t)$, analogous to those of q_s and p_s given in Sect. 20 are found to be the following.

$$\left.\begin{aligned}
(v_t, l_t | e^{-i\chi_t} r_t | v_t - 1, l_t + 1) &= (v_t - 1, l_t + 1 | e^{i\chi_t} r_t | v_t, l_t) \\
&= -(v_t, l_t | i e^{-i\chi_t}(p_{r_t} - i p_{\chi_t}/r_t) | v_t - 1, l_t + 1)/\hbar \\
&= (v_t - 1, l_t + 1 | i e^{i\chi_t}(p_{r_t} + i p_{\chi_t}/r_t) | v_t, l_t)/\hbar = \sqrt{\tfrac{1}{2}(v_t - l_t)}, \\
(v_t l_t | e^{i\chi_t} r_t | v_t - 1, l_t - 1) &= (v_t - 1, l_t - 1 | e^{-i\chi_t} r_t | v_t, l_t) \\
&= -(v_t l_t | i e^{i\chi_t}(p_{r_t} + i p_{\chi_t}/r_t) | v_t - 1, l_t - 1)/\hbar \\
&= (v_t - 1, l_t - 1 | i e^{-i\chi_t}(p_{r_t} - i p_{\chi_t}/r_t) | v_t, l_t)/\hbar = -\sqrt{\tfrac{1}{2}(v_t + l_t)}.
\end{aligned}\right\} \tag{21.6}$$

22. The three-dimensionally isotropic oscillator. A suitable solution to Eq. (19.12) is, evidently, the asscciated Legendre polynomial, $P_{l_u}^{m_u}(\cos\vartheta_u)$. The normalized function, $\Theta_u(\vartheta_u)$, may, therefore, be written

$$\Theta_{l_u, m_u}(\vartheta_u) = N_{l_u, m_u} \sin^{-m_u}\vartheta_u \frac{d^{l_u - m_u} \sin^{2 l_u}\vartheta_u}{(d\cos\vartheta_u)^{l_u - m_u}} \tag{22.1}$$

where

$$N_{l_u, m_u} = (-1)^{l_u} \{\tfrac{1}{2}(2l_u + 1)(l_u + m_u)!\,(l_u - m_u)!\}^{\frac{1}{2}} (1/2^{l_u} l_u!) \tag{22.2}$$

and where, moreover, $\Theta_{l_u, -m_u}(\vartheta_u) = (-1)^{m_u}\Theta_{l_u, m_u}(\vartheta_u)$. It may be shown, further, that the constant k_u in Eq. (19.10) must be equal to $l_u(l_u+1)$ in which l_u is an integer or zero such that $l_u \geqq m_u \geqq 0$. l_u may be identified as a quantum number of angular momentum associated with the threefold degenerate oscillation of frequency, ω_u. The quantum number, m_u, may be regarded as defining the component of the angular momentum, l_u, which is directed along the body-fixed axis z.

The solution $R(\varrho_v)$ to the Eq. (19.9) may be shown to be of the form

$$R(\varrho_u) = e^{-\frac{1}{2}\varrho_u^2} \varrho_u^{l_u} Z(\varrho_u). \tag{22.3}$$

We then obtain, in place of Eq. (19.9)

$$Z''(\varrho_u) + 2\left[\frac{l_u + 1}{\varrho_u} - \varrho_u\right] Z'(\varrho_u) + \left[\frac{2E_u}{hc\omega_u} - 2l_u - 3\right] Z(\varrho_u) = 0. \tag{22.4}$$

As in Sect. 21, if we replace ϱ_u^2 by σ_u we obtain once more Eq. (21.2) where now $\tau = [E_u/2hc\omega_u) + (l_u/2) - \frac{1}{4}] = \frac{1}{2}(v_u + l_u + 1)$ and where $T_u = l_u + \frac{1}{2}$. The normalized wave function will, therefore, be

$$R_{v_u, l_u}(\sigma_u) = N_{v_u, l_u} e^{-\sigma_u/2} \sigma^{l_u/2} L_\tau^{(l_u + \frac{1}{2})}(\sigma_u) \tag{22.5}$$

where $L_\tau^{(l_u+\frac{1}{2})}(\sigma_u)$ is the associated Laguere polynomial and where the normalization factor, N_{v_u, l_u}, is the constant

$$N_{v_u, l_u} = \sqrt{2}\{[\tfrac{1}{2}(v_u - l_u)]!\}^{\frac{1}{2}} / \{[\tfrac{1}{2}(v_u + l_u + 1)]!\}^{\frac{3}{2}}. \tag{22.6}$$

The solution to the wave equation for the spatial oscillator requires that l_u may take the values $v_u, v_u - 2, \ldots 1$ or 0 only as was also the case for the two-dimensionally isotropic oscillator. It is evident from the definition of the quantity, τ_u, which occurs in the Laguere differential equation, that

$$E_{v_u} = (v_u + \tfrac{3}{2})\, hc\omega_u. \tag{22.7}$$

Matrix components of the quantities $\varrho_u \cos\vartheta_u$, $e^{\pm i\chi_u}\varrho_u \sin\vartheta_u$ which are analogous to the matrix elements of the coordinates q_s in the instance of the linear harmonic oscillator and the coordinates $e^{\pm i\chi_t} r_t$ for the two-dimensionally isotropic oscillator have been derived by SHAFFER. Also here the matrix elements of the operators $p_{\varrho_u}, p_{\vartheta_u}, p_{\chi_u}$ are complicated and difficult to derive. Fortunately these occur in the case of the molecule only as

$$-i\hbar\left(\cos\vartheta_u \frac{\partial}{\partial\varrho_u} - \frac{\sin\vartheta_u}{\varrho_u}\frac{\partial}{\partial\vartheta_u}\right)$$

and

$$-i\hbar\, e^{\pm i\chi_u}\left(\sin\vartheta_u \frac{\partial}{\partial\varrho_u} + \frac{\cos\vartheta_u}{\varrho_u}\frac{\partial}{\partial\vartheta_u} \pm \frac{i}{\varrho_u\sin\vartheta_u}\frac{\partial}{\partial\chi_u}\right),$$

and matrix elements of these may be evaluated from SHAFFER'S values for the matrix elements for $e^{\pm i\chi_u}\varrho_u \sin\vartheta_u$ and $\varrho_u\cos\vartheta_u$ if one observes that the commutation rules

$$(p_{u,1} \pm i\, p_{u,2})\,(q_{u,1} \mp i q_{u,2}) - (q_{u,1} \mp i q_{u,2})\,(p_{u,1} \pm p_{u,2}) = 2i\hbar$$

and

$$p_{u,3}\, q_{u,3} - q_{u,3}\, p_{u,3} = -i\hbar$$

must be fulfilled. These correspond again to momenta conjugate to $e^{\pm i\chi_u} \varrho_u \sin\vartheta_u$ and $\varrho_u \cos\vartheta_u$. These matrix elements are found to be

$$\left.\begin{aligned}
&(v_u, l_u, m_u \,|\, \varrho_u \cos\vartheta_u \,|\, v_u - 1, l_u - 1, m_u) = (v_u - 1, l_u - 1, m_u \,|\varrho_u \cos\vartheta_u|\, v_u, l_u, m_u)\\
&\quad = -\left(v_u, l_u, m_u \left| \cos\vartheta_u \frac{\partial}{\partial\varrho_u} - \frac{\sin\vartheta_u}{\varrho_u}\frac{\partial}{\partial\vartheta_u}\right| v_u - 1, l_u - 1, m_u\right)\\
&\quad = \left(v_u - 1, l_u - 1, m_u \left| \cos\vartheta_u \frac{\partial}{\partial\varrho_u} - \frac{\sin\vartheta_u}{\varrho_u}\frac{\partial}{\partial\vartheta_u}\right| v_u, l_u, m_u\right)\\
&\quad = -\left[\frac{v_u + l_u + 1}{2}\right]^{\frac{1}{2}} \left[\frac{(l_u - m_u)\,(l_u + m_u)}{(2\,l_u - 1)\,(2\,l_u + 1)}\right]^{\frac{1}{2}},\\
&(v_u, l_u, m_u \,|\, \varrho_u \cos\vartheta_u \,|\, v_u - 1, l_u + 1, m_u) = (v_u - 1, l_u + 1, m_u \,|\, \varrho_u \cos\vartheta_u \,|\, v_u, l_u, m_u)\\
&\quad = -\left(v_u, l_u, m_u \left| \cos\vartheta_u \frac{\partial}{\partial\varrho_u} - \frac{\sin\vartheta_u}{\varrho_u}\frac{\partial}{\partial\vartheta_u}\right| v_u - 1, l_u + 1, m_u\right)\\
&\quad = \left(v_u - 1, l_u + 1, m_u \left| \cos\vartheta_u \frac{\partial}{\partial\varrho_u} - \frac{\sin\vartheta_u}{\varrho_u}\frac{\partial}{\partial\vartheta_u}\right| v_u, l_u, m_u\right)\\
&\quad = \left[\frac{v_u - l_u}{2}\right]^{\frac{1}{2}} \left[\frac{(l_u - m_u + 1)\,(l_u + m_u + 1)}{(2\,l_u + 1)\,(2\,l_u + 3)}\right]^{\frac{1}{2}},\\
&v_u, l_u, m_u \,|\, e^{\pm i\chi_u} \varrho_u \sin\vartheta_u \,|\, v_u - 1, l_u - 1, m_u \mp 1)\\
&\quad = (v_u - 1, l_u - 1, m_u \mp 1 \,|\, e^{\mp i\chi_u} \varrho_u \sin\vartheta_u \,|\, v_u, l_u, m_u)\\
&\quad = -\left(v_u, l_u, m_u \left| e^{\pm i\chi_u}\left[\sin\vartheta_u \frac{\partial}{\partial\varrho_u} + \frac{\cos\vartheta_u}{\varrho_u}\frac{\partial}{\partial\vartheta_u} \pm \frac{i}{\varrho_u \sin\vartheta_u}\frac{\partial}{\partial\chi_u}\right]\right| v_u - 1, l_u - 1, m_u \mp 1\right)\\
&\quad = \left(v_u - 1, l_u - 1, m_u \mp 1 \left| e^{\mp i\chi_u}\left[\sin\vartheta_u \frac{\partial}{\partial\varrho_u} + \frac{\cos\vartheta_u}{\varrho_u}\frac{\partial}{\partial\vartheta_u} \mp \frac{i}{\varrho_u \sin\vartheta_u}\frac{\partial}{\partial\chi_u}\right]\right| v_u, l_u, m_u\right)\\
&\quad = \pm\left[\frac{v_u + l_u + 1}{2}\right]^{\frac{1}{2}} \left[\frac{(l_u \pm m_u)\,(l_u \pm m_u - 1)}{(2\,l_u - 1)\,(2\,l_u + 1)}\right]^{\frac{1}{2}},\\
&v_u, l_u, m_u \,|\, e^{\pm i\chi_u} \varrho_u \sin\vartheta_u \,|\, v_u - 1, l_u + 1, m_u \mp 1)\\
&\quad = (v_u - 1, l_u + 1, m_u \mp 1 \,|\, e^{\mp i\chi_u} \varrho_u \sin\vartheta_u \,|\, v_u, l_u, m_u)\\
&\quad - -\left(v_u, l_u, m_u \left| e^{\pm i\chi_u}\left[\sin\vartheta_u \frac{\partial}{\partial\varrho_u} + \frac{\cos\vartheta_u}{\varrho_u}\frac{\partial}{\partial\vartheta_u} \pm \frac{i}{\varrho_u \sin\vartheta_u}\frac{\partial}{\partial\chi_u}\right]\right| v_u - 1, l_u + 1, m_u \mp 1\right)\\
&\quad = \left(v_u - 1, l_u + 1, m_u \mp 1 \left| e^{\mp i\chi_u}\left[\sin\vartheta_u \frac{\partial}{\partial\varrho_u} + \frac{\cos\vartheta_u}{\varrho_u}\frac{\partial}{\partial\vartheta_u} \mp \frac{i}{\varrho_u \sin\vartheta_u}\frac{\partial}{\partial\chi_u}\right]\right| v_u, l_u, m_u\right)\\
&\quad = \pm\left[\frac{v_u - l_u}{2}\right]^{\frac{1}{2}} \left[\frac{(l_u \mp m_u + 1)\,(l_u \mp m_u + 2)}{(2\,l_u + 1)\,(2\,l_u + 3)}\right]^{\frac{1}{2}}.
\end{aligned}\right\} \quad (22.8)$$

23. The rotator. We shall consider first the rigid rotator, i.e., the rotator where the internal angular momenta, p^*_α, are zero. The Eq. (19.11) then reduces to

$$\left\{\tfrac{1}{2}\sum_\alpha (P^2_\alpha / I^{(e)}_{\alpha\alpha}) - E_R\right\} R(\vartheta, \varphi)\, e^{iM\psi} = 0. \quad (23.1)$$

The general case of the asymmetric rotator where $I^{(e)}_{xx} \neq I^{(e)}_{yy} \neq I^{(e)}_{zz}$ was first studied in detail by WANG who formulated the Eq. (23.1) in the following form

$$(\mathcal{H} - \varepsilon)\, R(\vartheta, \varphi)\, e^{iM\psi} = 0 \quad (23.2)$$

in which $\hbar^2\varepsilon = 2E_r/C$ and in which

$$\hbar^2 \mathcal{H} = (2H_r/C) = \{AP^2 + P_z^2\} + B(P_x^2 - P_y^2), \tag{23.3}$$

A, B and C being the following constants

$$A = (1/2C)\left[(1/I_{xx}^{(e)} + (1/I_{yy}^{(e)})\right],$$
$$B = (1/2C)\left[(1/I_{xx}^{(e)}) - (1/I_{yy}^{(e)})\right],$$
$$C = \{(1/I_{zz}^{(e)} - \tfrac{1}{2}\left[(1/I_{xx}^{(e)}) + (1/I_{yy}^{(e)})\right]\}.$$

The intermediate moment of inertia is always identified with $I_{yy}^{(e)}$ so that $-1 \leq B \leq 0$. When the two moments of inertia $I_{xx}^{(e)}$ and $I_{yy}^{(e)}$ take the same values so that B goes to zero we have the case of the symmetric rotator which has been studied in the matrix formulation of the quantum mechanics by DENNISON and by REICHE and RADEMACHER and almost simultaneously by KRONIG and RABI in the wave mechanics. The SCHRÖDINGER equation for the symmetric rotator reduces to

$$\left.\begin{aligned}\left\{\frac{1}{\sin\vartheta}\frac{\partial}{\partial\vartheta}\left(\sin\vartheta\frac{\partial}{\partial\vartheta}\right) + \frac{1}{\sin^2\vartheta}\frac{\partial^2}{\partial\psi^2} - \frac{2\cos\vartheta}{\sin^2\vartheta}\frac{\partial^2}{\partial\varphi\,\partial\psi} + \frac{1}{\sin^2\vartheta}\frac{\partial^2}{\partial\varphi^2} + \right.\\ \left. + \frac{I_{xx}^{(e)} - I_{zz}^{(e)}}{I_{zz}^{(e)}}\frac{\partial^2}{\partial\varphi^2} + 8\pi^2 I_{xx}^{(e)} E_R/h^2\right\} R = 0.\end{aligned}\right\} \tag{23.4}$$

The term in Eq. (23.4) enclosed by the first set of brackets will be recognized as P^2. Evidently $\mathcal{H}$, P^2, P_z and P_Z all commute with each other. We have from Chap. II, therefore, that the function R will simultaneously be an eigenfunction of P^2, P_z and P_Z which will be characterized by the three quantum numbers J, K and M respectively. The coordinates φ and ψ are ignorable coordinates and $R(J, K, M)$ may accordingly be taken to be

$$R(J, K, M) = Y_{J,K,M}(\vartheta)\, e^{iK\varphi} e^{iM\psi} \tag{23.5}$$

where K and M must be zero or take integral values in order to satisfy the requirement that R be single valued. The substitution of (23.5) into the Eq. (23.4) leads to the differential equation for $Y(\vartheta)$

$$Y'' + \cot\vartheta\, Y' - \left[(M - K\cos\vartheta)^2/\sin^2\vartheta\right] Y + \lambda Y = 0 \tag{23.6}$$

where $\lambda = (8\pi^2 I_{xx}^{(e)} E_R/h^2) - (I_{xx}^{(e)}/I_{zz}^{(e)})\,K^2$.

Introducing the abbreviations $s = |K+M|$, $d = |K-M|$, $t = \tfrac{1}{2}(1-\cos\vartheta)$ and letting $Y(\vartheta)$ be $Y(\vartheta) = t^{d/2}(1-t)^{s/2}F(t)$ one obtains the differential equation which F must satisfy,

$$t(1-t)F'' + [\gamma - (\alpha+\beta+1)t]F' - \alpha\beta = 0, \tag{23.7}$$

in which α, β and γ are defined in the following manner, $\gamma = 1+d$, $\alpha+\beta = 1+d+s$, $\alpha\beta = \tfrac{1}{2}(d+s)\left[\tfrac{1}{2}(d+s)+1\right] - \lambda - K^2$. This differential equation is the hypergeometric equation and F is the hypergeometric function

$$F(\alpha, \beta, \gamma; t) = 1 + \frac{\alpha\beta}{\gamma}t + \frac{\alpha(\alpha+1)\beta(\beta+1)}{2!\,\gamma(\gamma+1)}t^2 + \cdots. \tag{23.8}$$

The series (23.8) must terminate if the function R is to remain finite for all values of the coordinates. It is necessary, in order for this to be so, that α shall be a negative integer or zero, i.e., $\alpha = -p\ (p = 0, 1, \ldots)$. This enables us to determine the energy which is

$$E_R = J(J+1)(h^2/8\pi^2 I_{xx}^{(e)}) + (K^2 h^2/8\pi^2)\left[(1/I_{zz}^{(e)}) - (1/I_{xx}^{(e)})\right], \tag{23.9}$$

where J is the positive integer $J=p+\frac{1}{2}(d+s)$. The integer J is, clearly, always greater than or at least equal to $|K|$ or $|M|$ since $\frac{1}{2}(d+s)=\frac{1}{2}|K+M|+\frac{1}{2}|K-M| = |K|$ for $|K|\geq|M|$ and equal to $|M|$ for $|M|\geq|K|$. This is consistent with the classical concept that the total angular momentum (described by J) must always be greater or equal to any of its components. The normalized wave function $R_{J,K,M}(\vartheta,\varphi,\psi)$ will then be

$$R_{J,K,M}(\vartheta,\varphi,\psi)=N\,t^{d/2}(1-t)^{s/2}F(-p,1+d+s+p,1+d;t)\,e^{iK\varphi}\,e^{iM\psi} \tag{23.10}$$

or, which is the equivalent

$$R_{J,K,M}(\vartheta,\varphi,\psi)=N\,t^{-d/2}(1-t)^{-s/2}\,\frac{d!}{(d+p)!}\,\frac{d^p}{dt^p}\left(t^{d+p}(1-t)^{s+p}\right)e^{iK\varphi}\,e^{iM\psi} \tag{23.11}$$

in which N is the normalizing factor

$$N=\sqrt{\frac{(1+d+s+2p)\,(d+s+p)!\,(d+p)!}{4\pi^2 p!\,d!^2(s+p)!}}\,. \tag{23.12}$$

The terminating hypergeometric series $F(-p,1+d+s+p,1+d;t)$ is sometimes also referred to as a Jacobi polynomial of degree p, i.e.,

$$F(-p,1+d+s+p,1+d,t)=G_p(1+d+s,1+d,t).$$

24. The asymmetric rotator. The general rotator where no two of the moments of inertia are equal is more difficult to deal with. Here the operators $\mathcal{H}$ [Eq. (23.3)], P^2 and P_Z still commute with each other. Thus, if the eigenfunctions of the asymmetric rotator are designated by S, it is possible to find a set of $(2J+1)$ eigenfunctions, S_{JM}, for every set of values of the quantum numbers J and M. These were identified with the $(2J+1)$ eigenvalues of the operator P_z, $K\hbar$, in the case of the symmetric rotator. The operators, P_z and P_z^2, do not commute any longer with the Hamiltonian, $\mathcal{H}$, in the case of the asymmetric rotator where $B\neq 0$ and K is therefore no longer a suitable quantum number with which to identify the $(2J+1)$ eigenfunctions S_{JM}. The eigenfunctions S_{JM} are, indeed, much more complicated relations than the R_{JKM} of Eq. (23.11), but since P_Z still commutes with $\mathcal{H}$ (i.e., ψ remains as an ignorable coordinate) they may still be written

$$S_{JM}=F_{JM}(\vartheta,\varphi)\,e^{iM\psi}. \tag{24.1}$$

WANG's essential contribution was to show that when $I_{xx}^{(e)}\neq I_{yy}^{(e)}$ so that $B\neq 0$, $\mathcal{H}\,R_{J,K,M}^{\times}$ could be expanded into the following

$$\left.\begin{aligned}\mathcal{H}R_{J,K,M}^{\times}=B\,g(J,K+1)\,R_{J,K+2,M}^{\times}+[A\,J(J+1)+K^2]\,R_{J,K,M}^{\times}+\\+B\,g(J,K-1)\,R_{J,K-2,M}^{\times}\end{aligned}\right\} \tag{24.2}$$

where

$$g(J,K)=-\tfrac{1}{2}\left[(J-K)(J-K+1)(J+K)(J+K+1)\right]^{\frac{1}{2}}, \tag{24.3}$$

in which the $R_{J,K,M}^{\times}$ are the functions (23.10) multiplied by $(-1)^{\sigma}$ where $\sigma=K$ when $K>M$ and where $\sigma=M$ when $K<M$, i.e., $R_{J,K,M}^{\times}=R_K^{\times}=(-1)^{\sigma}R_{J,K,M}$.

The matrix elements of $\mathcal{H}$ may be calculated at once from the relation (24.2). It is evident that the matrix of these elements will be diagonal in J and M and is, in fact, independent of M. The elements of the matrix $\mathcal{H}$ are the following

$$\left.\begin{aligned}(J,K,M|\mathcal{H}|J,K,M)&=(K|\mathcal{H}|K)=A\,J(J+1)+K^2\\(J,K,M|\mathcal{H}|J,K\pm 2,M)&=(K|\mathcal{H}|K\pm 2)\\&=-\tfrac{1}{2}\left[(J\mp K)(J\mp K-1)(J\pm K+1)(J\pm K+2)\right]^{\frac{1}{2}}B.\end{aligned}\right\} \tag{24.4}$$

The matrix $\mathcal{H}$ will, therefore, split up into submatrices, one set for each value of J. Since the energy does not depend upon M, each of the above submatrices is repeated $(2J+1)$ times, once for each value of M. Moreover since K takes all integral values from $-J$ to $+J$ each submatrix will have $(2J+1)$ rows and columns. The energies are then obtained by diagonalizing the matrix with the elements (24.4). This is most readily achieved by setting the secular determinant of the matrix $|(K|\mathcal{H}|K') - \varepsilon\,\delta_{KK'}|$ equal to zero and solve for the roots. The secular determinant given by WANG is

$$\begin{vmatrix} J^2-W & 0 & (J|\mathcal{H}|J-2) & & & & & & \\ 0 & (J-1)^2-W & 0 & & & & & & \\ (J-2|\mathcal{H}|J) & 0 & (J-2)^2-W & & & & & & \\ & & & \ddots & & & & & \\ & & & 1-W & 0 & (1|\mathcal{H}|-1) & & & \\ & & & 0 & -W & 0 & & & \\ & & & (-1|\mathcal{H}|1) & 0 & 1-W & & & \\ & & & & & & \ddots & & \\ & & & & & & (J-2)^2-W & 0 & (-J+2|\mathcal{H}|-J) \\ & & & & & & 0 & (J-1)^2-W & 0 \\ & & & & & & (-J|\mathcal{H}|-J+2) & 0 & J^2-W \end{vmatrix} = 0 \quad (24.5)$$

where $W=\varepsilon - AJ(J+1)$ and $\varepsilon = 8\pi^2 E_r/h^2 C$. The energy of the rotator will, therefore, be

$$E_r = (h^2 C/8\pi^2)\,[A\,J(J+1) + W]. \qquad (24.6)$$

It has been pointed out by MULLIKEN that it is advantageous to use symmetrized wave functions, $R^s_{K,\gamma}$, so defined that

$$\left.\begin{aligned} R^s_{K,\gamma} &= \frac{1}{\sqrt{2}}\,[R^\times_K + (-1)^\gamma R^\times_{-K}] \qquad (K = 1, 2, \ldots, J)\\ R^s_{0,0} &= R^\times_0 \qquad\qquad\qquad\qquad\quad (\gamma = 0, 1)\end{aligned}\right\} \qquad (24.7)$$

rather than the $R^\times_K$ in that by so doing the energy matrix will split up into submatrices, each submatrix being associated with functions $R^s_{K,\gamma}$ belonging only to one symmetry species.

We shall examine the symmetry properties of the eigenfunctions $R^s_{K,\gamma}$. It may be seen that there are three covering operations which in addition to the identity operation, I, will restore the rotational ellipsoid to a position where it can not be distinguished from its original position. These are the rotations through angles π of the ellipsoid about any one of its principal axes of inertia, designated respectively by $C_2^{(x)}$, $C_2^{(y)}$ and $C_2^{(z)}$. These operations cause the Eulerian angles to transform in the following manner.

$$\begin{array}{lll} C_2^{(x)} & C_2^{(y)} & C_2^{(z)} \\ \vartheta \to \pi - \vartheta & \vartheta \to \pi - \vartheta & \vartheta \to \vartheta \\ \varphi \to 2\pi - \varphi & \varphi \to \pi - \varphi & \varphi \to \pi + \varphi \\ \psi \to \pi + \psi & \psi \to \pi + \psi & \psi \to \psi. \end{array}$$

Since the product of any two of the $C_2^{(\alpha)}$ is equal to another $C_2^{(\beta)}$, the $C_2^{(\alpha)}$ and the identity operator, evidently, form a group. A covering operation can, of course, have no effect upon $\mathcal{H}$ since it must leave the energy unchanged. Applying a covering operation to the Schrödinger equation $C_2^{(\alpha)}(\mathcal{H}-E_K)\psi_K=0$ will, therefore, be tantamount to applying the operation to the wave function directly, i.e., $C_2^{(\alpha)}(\mathcal{H}-E)R_{K,\gamma}^s=(\mathcal{H}-E_K)C_2^{(\alpha)}R_{K,\gamma}^s$ where now $C_2^{(\alpha)}R_{K,\gamma}^s$ also may be regarded as a solution to the wave equation with the eigenvalue E_K.

Examination of the functions (23.11) will show that the covering operations $C_2^{(\alpha)}$ cause the $R_{K,\gamma}^s$ to transform in the following manner

$$\left.\begin{aligned} C_2^{(x)}R_{K,\gamma}^s &= (-1)^{J+\gamma}R_{K,\gamma}^s \\ C_2^{(y)}R_{K,\gamma}^s &= (-1)^{J+K+\gamma}R_{K,\gamma}^s \\ C_2^{(z)}R_{K,\gamma}^s &= (-1)^{K}R_{K,\gamma}^s. \end{aligned}\right\} \tag{24.8}$$

The relation of the symmetry species of the $R_{K,\gamma}^s$ to the parity of $(J+\gamma)$ and K is probably most readily shown in a tabular form as has been done by VAN WINTER. In Table 5 these parity relations are shown and the symmetry species A, B_x, B_y and B_z of the symmetrized wave functions are defined in Table 6 in terms of the parities of the rotation operators.

Table 5.

Parity of K	Parity of $J+\gamma$	Symmetry species of $R_{K,\gamma}^s$
e	e	A
e	o	B_z
o	e	B_x
o	o	B_y

Table 6. *Symmetry species of symmetrized wave functions defined in terms of the parities for the identity and rotation operators.*

Symmetry species	Operators I	$C_2^{(x)}$	$C_2^{(y)}$	$C_2^{(z)}$
A	+1	+1	+1	+1
B_x	+1	+1	−1	−1
B_y	+1	−1	+1	−1
B_z	+1	−1	−1	+1

One may now derive relations corresponding to those derived by WANG, here using the symmetrized wave functions $R_{K,\gamma}^s$. It is readily verified that these will become

$$\left.\begin{aligned} \mathcal{H}R_{K,\gamma}^s = Bg(J,K+1)R_{K+2,\gamma}^s &+ [AJ(J+1)+K^2]R_{K,\gamma}^s \\ &+ Bg(J,K-1)R_{K-2,\gamma}^s. \end{aligned}\right\} \tag{24.9}$$

The relations (24.9) are valid in general. For values of $K<3$, examination will show, however, that they may be expressed in a somewhat more useful form, namely;

$$\left.\begin{aligned} \mathcal{H}R_{1,\gamma}^s &= \{(-1)^\gamma Bg(J,0)+[AJ(J+1)+1^2]\}R_{1,\gamma}^s+Bg(J,2)R_{3,\gamma}^s \\ \mathcal{H}R_{2,0}^s &= \sqrt{2}Bg(J,1)R_{00}^s+[AJ(J+1)+2^2]R_{2,0}^s+Bg(J,3)R_{4,0}^s \\ \mathcal{H}R_{0,0}^s &= [AJ(J+1)+0^2]R_{0,0}^s+\sqrt{2}Bg(J,1)R_{2,0}^s \\ \mathcal{H}R_{2,1}^s &= [AJ(J+1)+2^2]R_{2,1}^s-Bg(J,3)R_{4,1}^s. \end{aligned}\right\} \tag{24.10}$$

It is clear then that the symmetrized matrix, $\mathcal{H}^s$, which consists of the elements, $(K|\mathcal{H}^s|K')$, will split up into the submatrices $\mathcal{H}_0^s$ and H_1^s to be identified respectively with $\gamma=0$ and $\gamma=1$. Using the notation introduced in

(24.4) we have

$$\left.\begin{aligned}
\mathcal{H}_0^s &= \left|\begin{matrix}
(0|\mathcal{H}|0) & 0 & \sqrt{2}(0|\mathcal{H}|2) & 0 & 0 & \cdots \\
0 & (1|\mathcal{H}|1)+(-1|\mathcal{H}|1) & 0 & (1|\mathcal{H}|3) & 0 & \cdots \\
\sqrt{2}(2|\mathcal{H}|0) & 0 & (2|\mathcal{H}|2) & 0 & (2|\mathcal{H}|4) & \cdots \\
0 & (3|\mathcal{H}|1) & 0 & (3|\mathcal{H}|3) & 0 & \cdots \\
0 & 0 & (4|\mathcal{H}|2) & 0 & (4|\mathcal{H}|4) & \cdots \\
\cdots & \cdots & \cdots & \cdots & \cdots & \cdots
\end{matrix}\right| \\
\mathcal{H}_1^s &= \left|\begin{matrix}
(1|\mathcal{H}|1)-(-1|\mathcal{H}|1) & 0 & (1|\mathcal{H}|3) & 0 & 0 & \cdots \\
0 & (2|\mathcal{H}|2) & 0 & (2|\mathcal{H}|4) & 0 & \cdots \\
(3|\mathcal{H}|1) & 0 & (3|\mathcal{H}|3) & 0 & (3|\mathcal{H}|5) & \cdots \\
0 & (4|\mathcal{H}|2) & 0 & (4|\mathcal{H}|4) & 0 & \cdots \\
0 & 0 & (5|\mathcal{H}|3) & 0 & (5|\mathcal{H}|5) & \cdots \\
\cdots & \cdots & \cdots & \cdots & \cdots & \cdots
\end{matrix}\right|
\end{aligned}\right\} \quad (24.11)$$

Arranging the rows and columns of $\mathcal{H}_0^s$ and $\mathcal{H}_1^s$ so that the elements of even values of K stand together and the elements of odd values of K stand together the submatrix $\mathcal{H}_0^s$ is further factorized into the submatrix E^+ consisting of even rows and columns, and into the submatrix O^+ consisting of odd rows and columns. Similarly, $\mathcal{H}_1^s$ may be factored into E^- and O^-.

Each of the submatrices may now be diagonalized independently to obtain the energies in the same manner as the matrix (24.5) was diagonalized, i.e., by setting the secular determinant $|(K|P^s|K')_\gamma - E\,\delta_{KK'}| = 0$ and solving for the roots, where $(K|P^s|K')_\gamma$ represents the elements of any of the submatrices E^+, E^-, O^+ and O^-. The eigenfunctions, (ψ_{JM}^s), for a given state in the rotator, will be a linear combination of the symmetrized wave functions $R_{K\gamma}^s$ belonging to one symmetry species only and will itself, therefore, belong to this symmetry species. Hence we may write

$$(\psi_{JM}^s) = \sum_K a_{K\gamma}^s R_{K\gamma}^s \tag{24.12}$$

where the $a_{K\gamma}^s$ are the normalized cofactors of the determinants of the submatrices E^+, E^-, O^+ and O^-. We shall summarize in Table 7 the essential information we have derived concerning the rotator, i.e., the rank of the submatrices, $\mathcal{H}_\gamma^s$, and the symmetry species to which the eigenfunctions $(\psi_{JM}^s)_\gamma$ belong.

Table 7.

Submatrix	Parity of K	γ	Rank		Symmetry species of ψ^s	
			J even	J odd	J even	J odd
E^+	e	0	$\frac{1}{2}J+1$	$\frac{1}{2}(J+1)$	A	B_z
E^-	e	1	$\frac{1}{2}J$	$\frac{1}{2}(J-1)$	B_z	A
O^+	o	0	$\frac{1}{2}J$	$\frac{1}{2}(J+1)$	B_x	B_y
O^-	o	1	$\frac{1}{2}J$	$\frac{1}{2}(J+1)$	B_y	B_x

A discussion of the asymmetric rotator is incomplete without a reference to the method of KLEIN which undoubtedly represents the most elegant of all the approaches to the solution of the rotator problem. We have already indicated in Chap. II that this method was one of translating the classical rotator problem into quantum mechanics. The essential elements of his analysis have already been discussed and need not be repeated here. Using as the matrix elements of P_x, P_y and P_z those stated in (9.40) he derives the following elements for P_x^2

and P_y^2

$$\left.\begin{aligned}(K|P_x^2|K) &= (K|P_y^2|K) = (\hbar^2/2)\,[J(J+1)-K^2],\\ (K|P_x^2|K\pm 2) &= -(K|P_y^2|K\pm 2)\\ &= (\hbar/2)^2\,[(J\mp K)(J\mp K+1)(J\pm K+1)(J\pm K+2)]^{\frac{1}{2}}\end{aligned}\right\}\qquad(24.13)$$

from which he deduces that $(K|P^2|K')=J(J+1)\,\hbar^2\,\delta_{KK'}$, $\delta_{KK'}$ being the Kronecker symbol. These permit writing, at once, the matrix elements for $\mathcal{H}$. These are

$$\left.\begin{aligned}(K|\mathcal{H}|K) &= (\hbar^2/2)\,\{J(J+1)\,\tfrac{1}{2}[(1/I_{xx}^{(e)})+(1/I_{yy}^{(e)})]+K^2[(1/I_{zz}^{(e)})-\\ &\quad -\tfrac{1}{2}\left((1/I_{xx}^{(e)})+(1/I_{yy}^{(e)})\right)]\},\\ (K|\mathcal{H}|K\pm 2) &= (\hbar/2)^2\,\{\tfrac{1}{2}\,[(1/I_{xx}^{(e)})-(1/I_{yy}^{(e)})]\}\times\\ &\quad\times\{(J\mp K)(J\mp K-1)(J\pm K+1)(J\pm K+2)\}^{\frac{1}{2}}.\end{aligned}\right\}\qquad(24.14)$$

The matrix elements (24.14) here given may be regarded as entirely equivalent to those stated in (24.4) if the $g(J, K)$ notation introduced earlier is retained.

Methods for evaluating the energies of the rotator have been studied in much detail by KING, HAINER and CROSS. They follow the method of RAY and write for the energy of the rotator

$$E_\tau^J(a,b,c) = \frac{a+c}{2}\,J(J+1)+\frac{a-c}{2}\,E_\tau^J(\varkappa)\qquad(24.15)$$

where a, b and c are, respectively, $\hbar^2/2I_a$, $\hbar^2/2I_b$ and $\hbar^2/2I_c$ and where the moments of inertia are so chosen that $I_a \leq I_b \leq I_c$. $E_\tau^J(\varkappa)$ in Eq. (24.15) is one of the $(2J+1)$ roots of the secular determinant which is similar to that of WANG given in (24.5). The quantity $\varkappa$ is a parameter of asymmetry, $\varkappa=(2b-a-c)/(a-c)$, which varies from $\varkappa=1$ for the oblate symmetric rotator ($b=a$) to $\varkappa=-1$ for the prolate symmetric rotator ($b=c$). The index τ assumes the $(2J+1)$ integral values $-J\leq\tau\leq J$ such that $\tau=J$ identifies the level of highest energy for a given value of J while $\tau=-J$ identifies the level of lowest energy for a given value of J. To evaluate the eigenvalues, $E_\tau^J(\varkappa)$, of the rotator, KING, HAINER and CROSS exploit the fact, observed earlier, that the energy matrix may be reduced to the four submatrices E^+, E^-, O^+ and O^-. The submatrices E^+, E^-, O^+ and O^- which are used by KING, HAINER and CROSS are similar to those given in (24.4) although the constants are not the same. Actually the four submatrices used by them are the following[1]

$$\left.\begin{aligned}E^\pm\colon\ E &= \begin{Vmatrix} FJ(J+1) & \sqrt{2}H\sqrt{f(J,1)} & 0 & \cdots\\ \cdots & \cdots & \cdots & \cdots\\ \sqrt{2}H\sqrt{f(J,1)} & 4(G-F)+FJ(J+1) & H\sqrt{f(J,3)} & \cdots\\ 0 & H\sqrt{f(J,3)} & 16(G-F)+FJ(J+1) & \cdots\\ \cdots & \cdots & \cdots & \cdots\end{Vmatrix},\\ O^\pm\colon\ E &= \begin{Vmatrix} (G-F)+FJ(J+1)\pm H\sqrt{f(J,0)} & H\sqrt{f(J,2)} & 0 & \cdots\\ H\sqrt{f(J,2)} & 9(G-F)+FJ(J+1) & H\sqrt{f(J,4)} & \cdots\\ 0 & H\sqrt{f(J,4)} & 25(G-F)+FJ(J+1) & \cdots\\ \cdots & \cdots & \cdots & \cdots\end{Vmatrix},\end{aligned}\right\}\qquad(24.16)$$

[1] E^- and E^+ have the same elements. In the case of E^- the first row and first column are omitted as indicated by the dotted line.

where $2\sqrt{f(J, K)}$ is the same quantity, $g(J, K)$, used by WANG, given earlier in (24.3) and where F, G and H are constants. The values of the constants F, G and H depend upon the manner in which the a, b and c axes are identified with the axes x, y and z. Six such identifications are possible, but if we confine ourselves to the two cases referred to earlier, i.e., letting the axis b be always identified with the y axis and then (1) letting the axis a be identified with x and the axis c with z and (2) letting the axis a be identified with the axis z and the axis c with x, we have respectively that $F=\frac{1}{2}(\varkappa+1)$, $G=-1$, $H=\frac{1}{2}(\varkappa-1)$ and $F=\frac{1}{2}(\varkappa-1)$, $G=1$, $H=\frac{1}{2}(\varkappa+1)$.

KING, HAINER and CROSS have found that in solving for the roots of (24.16) which are of the form

$$E=\left\|\begin{matrix} k_0 & b_1^{\frac{1}{2}} & 0 & \cdots \\ b_1^{\frac{1}{2}} & k_1 & b_2^{\frac{1}{2}} & \cdots \\ 0 & b_2^{\frac{1}{2}} & k_3 & \cdots \\ \cdots & \cdots & \cdots & \cdots \end{matrix}\right\| \tag{24.17}$$

it was advantageous to transform them to the form

$$D=L^{-1}EL=\left\|\begin{matrix} k_0 & 1 & 0 & \cdots & \cdots \\ b_1 & k_1 & 1 & \cdots & \cdots \\ 0 & b_2 & k_2 & 1 & \cdots \\ \cdots & \cdots & \cdots & \cdots & \cdots \end{matrix}\right\| \tag{24.18}$$

where the transformation function L is the following

$$L=\left\|\begin{matrix} 1 & 0 & 0 & \cdots & \cdots \\ 0 & b_1^{-\frac{1}{2}} & 0 & \cdots & \cdots \\ 0 & 0 & b_1^{-\frac{1}{2}}\, b_2^{-\frac{1}{2}} & \cdots \end{matrix}\right\|. \tag{24.19}$$

The four secular determinants $|D-\lambda I|=0$ of the four submatrices are then expanded into terminating continued fractions written in the form

$$\left.\begin{aligned} \lambda_m = k_m - b_m/\{(k_{m-1}-\lambda_m) - [b_{m-1}/(k_{m-2}-\lambda_m) - \cdots]\} - \\ - b_{m+1}/\{(k_{m+1}-\lambda_m) - [b_{m+2}/(k_{m+2}-\lambda_m) - \cdots]\}. \end{aligned}\right\} \tag{24.20}$$

Numerical values for $E_\tau^J(\varkappa)$ have thus been calculated throughout the range of asymmetry $\varkappa=-1$ to $\varkappa=0$, $\varkappa$ varying by steps of 0.1 and tables of these have been prepared. With the observation of RAY that $E_\tau^J(\varkappa)=-E_\tau^J(-\varkappa)$ the tables are applicable to a rotator of any degree of asymmetry, i.e., from $\varkappa=1$ to $\varkappa=-1$. The tables of KING, HAINER and CROSS, therefore, represent a very convenient means of approximating to the energies of any rigid asymmetric rotator. The values of $E_\tau^J(\varkappa)$ have been recomputed on the EDVAC electronic computer by TURNER, HICKS and REITWEISNER at the Ballistic Research Laboratories, Aberdeen Proving Grounds, Md., with $\varkappa$ varying from $\varkappa=-1$ to $\varkappa=0$ in steps of 0.01. Recently tables of the same type have been computed for values of $J=12$ to $J=40$ on the BESK electronic computer by ERLANDSSON, at Royal Institute of Technology, Stockholm.

MULLIKEN has proposed an alternate method of identifying an energy level for an asymmetric rotator to that employing the index τ. This method consists in attaching as indices to every value of $E^J_\tau(\varkappa)$ the absolute values of the quantum numbers K by which the level would be characterized in the limiting symmetric rotator instances of $\varkappa=-1$ and $\varkappa=+1$. These indices are denoted respectively, by K_{-1} and K_1. It is quickly verified that the index τ will be $\tau=K_{-1}-K_1$. The method of identifying a level by the indices K_{-1} and K_1 has the advantage that it can be made to indicate the symmetry species of the level in question. Thus, if K_1 is even, the level must according to Table 5, where now $B_z=B_c$ and $B_x=B_a$, have A or B_c symmetry. Table 5 may become applicable to $\varkappa=-1$ simply by setting $B_z=B_a$ and $B_x=B_c$. If K_{-1} is also even the level must now according to Table 5 have either A or B_a symmetry. The symmetry common to K_1 even and K_{-1} even is A and the level characterized by K_1 and K_{-1} both even must, therefore, have A symmetry. We may arrive at corresponding results for K_1 even, K_{-1} odd; etc. The results are tabulated in Table 7.

Table 7. *Symmetry classification of energy levels for an asymmetric rotator.*

Symmetry species of ψ^s	A	B_a	B_b	B_c
Parities of $K_{-1}K_1$	ee	eo	oo	oe

The information contained in Table 7 together with the fact that $\tau=K_{-1}-K_1$ enables one further to produce the Table 8 which gives the symmetry classification of the levels in sequential order of the eigenvalues of $E^J_\tau(\varkappa)$ for a given

Table 8. *Symmetry classification of asymmetric rotator levels in sequential order of the eigenvalues $E^J_\tau(\varkappa)$.*

		J	$J-1$	$J-2$	$J-3$	$J-4$	...
Symmetry species of ψ^s	J even	A	B_a	B_b	B_c	A	...
	J odd	B_c	B_b	B_a	A	B_c	...

J value. The information in Table 8 together with that given in Table 5 make it clear that the eigenvalues $E^J_J(\varkappa)$, $E^J_{J-1}(\varkappa)$, $E^J_{J-2}(\varkappa)$, ... for a given J value are obtained successively from E^+, O^-, O^+, E^-, E^+, etc.

Recently GOLDEN has demonstrated that for large values of J the energy matrix for the asymmetric rotator assumes a form which asymptotically becomes very similar to that obtained from the characteristic value problem of the Mathieu differential equation. The characteristic values of the Mathieu equation may, therefore, serve as a basis for a good approximation to the energies of the asymmetric rotator, particularly those levels which for a given value of J correspond to small values of K in the limiting symmetric case.

25. The selection rules for the symmetric rotator. The selection rules for the rotator may be arrived at from a study of the non-vanishing components of the electric moment; the actual transition probabilities being given by the squares of the modulus of the matrix elements of the electric moment. We shall denote by μ the electric moment of the molecule with the components μ_α directed along the body fixed system of coordinates. These components may consist of two parts, the permanent electric moment, $\mu_\alpha^{(0)}$, and a varying part induced by molecular vibration, $\sum_{s\sigma}(\partial\mu_\alpha/\partial q_{s\sigma})_0\, q_{s\sigma}+\cdots$. We may, accordingly, write for μ_α

$$\mu_\alpha=\mu_\alpha^{(0)}+\sum_{s\sigma}\left(\frac{\partial\mu_\alpha}{\partial q_{s\sigma}}\right)_0 q_{s\sigma}+\cdots.$$

The quantities μ_α are, evidently, functions only of the vibrational coordinates and their matrix elements will be diagonal in all the rotational quantum numbers, i.e., $(v_{s''}, J, K|\mu_\alpha| v_{s'}, J, K) = (v_{s''}|\mu_\alpha| v_{s'})$. The components of μ along the space-fixed axes may be written

$$\mu_A = \lambda_{Ax}(v_{s''}|\mu_x| v_{s'}) + \lambda_{Ay}(v_{s''}|\mu_y| v_{s'}) + \lambda_{Az}(v_{s''}|\mu_z| v_{s'}) \tag{25.1}$$

where the $\lambda_{A\alpha}$ are the direction cosines relating the x, y, z axes to the X, Y, Z axes and may be taken to be those given in Eq. (9.4). It becomes necessary then to inquire into what are the matrix elements of the $\lambda_{A\alpha}$. A convenient method for accomplishing this has been given by CASIMIR. This method is an extension of a method by BORN and JORDAN which depends upon computations involving the commutation relations (9.6) and (9.7). The relation (9.7) may, for our purpose, better be stated as $[P_{\mathsf{A}}, \lambda_{\mathsf{B}\alpha}] = i\hbar\lambda_{\Gamma\alpha}$. Replacing $\lambda_{X\alpha} \pm i\lambda_{Y\alpha}$ by $\lambda_{\mathrm{I}\alpha}$ and $\lambda_{\mathrm{II}\alpha}$ respectively the matrix elements for $\lambda_{\mathrm{I}\alpha}$ and $\lambda_{\mathrm{II}\alpha}$, given by BORN and JORDAN, are the following

$$\left.\begin{aligned}
(J, M|\lambda_{\mathrm{I}\alpha}| J, M-1) &= \sqrt{(J+M)(J-M+1)}\,(J, K|a| J, K'),\\
(J, M|\lambda_{\mathrm{II}\alpha}| J, M+1) &= \sqrt{(J+M+1)(J-M)}\,(J, K|a| J, K'),\\
(J, M|\lambda_{Z\alpha}| J, M) &= M(J, K|a| J, K'),\\
(J, M|\lambda_{\mathrm{I}\alpha}| J-1, M-1) &= \sqrt{(J+M)(J+M-1)}\,(J, K|a| J-1, K'),\\
(J, M|\lambda_{\mathrm{II}\alpha}| J-1, M+1) &= -\sqrt{(J-M-1)(J-M)}\,(J, K|a| J-1, K'),\\
(J, M|\lambda_{Z\alpha}| J-1, M) &= -\sqrt{(J+M)(J-M)}\,(J, K|a| J-1, K'),\\
(J, M|\lambda_{\mathrm{I}\alpha}| J+1, M-1) &= \sqrt{(J-M+1)(J-M+2)}\,(J, K|a| J+1, K'),\\
(J, M|\lambda_{\mathrm{II}\alpha}| J+1, M+1) &= -\sqrt{(J+M+1)(J+M+2)}\,(J, K|a| J+1, K'),\\
(J, M|\lambda_{Z\alpha}| J+1, M) &= \sqrt{(J+M+1)(J-M+1)}\,(J, K|a| J+1, K'),
\end{aligned}\right\} \tag{25.2}$$

or in general,

$$(J, K, M|\lambda_{A\alpha}| J' K' M') = (J, K|a| J' K')\,(J, M|F_A| J' M') \tag{25.3}$$

where the $(J, K|a| J'K')$ are arbitrary when $J - J' = \pm 1$ or 0, but are otherwise zero.

If one confines oneself to the symmetric rotator one may similarly make use of the commutation relations (9.27) and (9.6), the latter of which is here best stated $[P_\alpha, \lambda_{\beta A}] = -i\hbar\,\lambda_{\gamma A}$, to derive analagous set of matrix elements for $\lambda_{\mathrm{I}A}$ and $\lambda_{\mathrm{II}A}$ where these are respectively $\lambda_{xA} \pm i\lambda_{yA}$. Interchanging rows and columns (i.e., here α and A), we may then write

$$(J, K, M|\lambda_{A\alpha}| J', K', M') = (J, M|b| J' M')\,\overline{(J, K|F_\alpha| J' K')} \tag{25.4}$$

where, as before, the coefficients $(J, M|b| J'M')$ are arbitrary when $J - J' = \pm 1$ or 0, but are otherwise zero. The absolute values of the $\overline{(J, K|F_\alpha| J'K')}$ are the same as those stated in (25.2) and (25.3) where the quantum number M is everywhere replaced by K. The combination of the relations (25.3) and (25.4) give

$$(J, K, M|\lambda_{A\alpha}| J', K', M') = (J|C|J)\,(J, M|F_A| J', M')\,\overline{(J, K|F_\alpha| J', K')}, \tag{25.5}$$

where $(J|C|J')$ is zero except when $J - J' = \pm 1$ or 0.

The coefficient $(J|C|J)$ is evaluated by first writing

$$P_A = \tfrac{1}{2}\lambda_{A\mathrm{I}}(P_x - iP_y) + \tfrac{1}{2}\lambda_{A\mathrm{II}}(P_x + iP_y) + \lambda_{Az} P_z \tag{25.6}$$

where $\lambda_{A\,\mathrm{I}}$ and $\lambda_{A\,\mathrm{II}}$ are respectively $\lambda_{Ax} \pm i\lambda_{Ay}$ and then multiplying the matrices together and adding. One obtains in this way that

$$(J\,|\,C\,|\,J) = [J(J+1)]^{-1}. \tag{25.7}$$

Exploiting next the fact that

$$(\lambda_{\mathrm{II}}\,\lambda_{\mathrm{II\,I}} - \lambda_{\mathrm{II\,I}}\,\lambda_{\mathrm{II}}) = 0 \quad \text{and} \quad (\lambda_{\mathrm{II}}\,\lambda_{\mathrm{II\,II}} - \lambda_{\mathrm{II\,II}}\,\lambda_{\mathrm{II}}) = 0 \tag{25.8}$$

where $\lambda_{\mathrm{II}} = \lambda_{A\,\mathrm{I}} + i\lambda_{B\,\mathrm{I}}$, $\lambda_{\mathrm{II\,I}} = \lambda_{A\,\mathrm{I}} - i\lambda_{B\,\mathrm{I}}$, etc., one further obtains through matrix multiplication and addition that

$$(J\,|\,C\,|\,J)^2 - (2J-1)\,|(J\,|\,C\,|\,J-1)|^2 + (2J+3)\,|(J\,|\,C\,|\,J+1)|^2 = 0$$

and

$$\left.\begin{aligned} &J(J+1)\,(J\,|\,C\,|\,J)^2 + J(J-1)\,(2J-1)\,|(J\,|\,C\,|\,J-1)|^2 - \\ &\qquad - (J+1)\,(J+2)\,(2J+3)\,|(J\,|\,C\,|\,J+1)|^2 = 0. \end{aligned}\right\} \tag{25.9}$$

We get from these with the aid of (25.7) that

$$(J\,|\,C\,|\,J-1) = (J-1\,|\,C\,|\,J) = J^{-1}[(2J-1)\,(2J+1)]^{-\frac{1}{2}}. \tag{25.10}$$

The general quantum mechanical selection rule $\Delta J = \pm 1$ or 0 is reflected here in the fact that the coefficients $(J\,|\,C\,|\,J')$ vanish unless $J - J' = \pm 1$ or 0. A study of the quantities $(J, M\,|F_A|\,J', M')$ and $(J, K\,|F_\alpha|\,J', K')$ which occur in the Eq. (25.5) show that these vanish in all cases except where $\Delta M = \pm 1$ or 0 and $\Delta K = \pm 1$ or 0, respectively.

The energies of the rotator do not depend upon M in the absence of external electric or magnetic fields, i.e., each rotation level, $E_{J,K}$, is $(2J+1)$-fold degenerate in the quantum number M. It is convenient, for this reason, in making intensity calculations to have available the squares of the moduli of the total electric moment produced when the molecule makes a transition from the $(2J+1)$ degenerate states, $E_{J,K}$, to the $(2J'+1)$ degenerate states, $E_{J'K'}$. Such a quantity, $(v_{s''} J, K\,|\mu^2|\,v_{s'} J', K')$, may be obtained by summing the $\mu_X^2 + \mu_Y^2 + \mu_Z^2$ over M in the following manner

$$\left.\begin{aligned} (v_{s''}, J, K\,|\mu^2|\,v_{s'}, J', K') = \sum_\alpha \{&(J, K, M\,|\,\lambda_{z\alpha}|\,J'\,K'\,M)^2 \\ &+ \tfrac{1}{2}(J, K, M\,|\,\lambda_{x\alpha} + i\,\lambda_{y\alpha}|\,J', K', M-1)^2 \\ &+ \tfrac{1}{2}(J, K, M\,|\,\lambda_{x\alpha} - i\,\lambda_{y\alpha}|\,J', K', M+1)^2\}\,(v_{s''}|\,\mu_\alpha|\,v_{s'})^2. \end{aligned}\right\} \tag{25.11}$$

When the electric moment lies in the xy plane of the molecule only μ_x and μ_y are different from zero. When this is so we have the following quantum mechanical amplitudes squared of the electric moment

$$\left.\begin{aligned} (J, K\,|\mu^2|\,J, K\mp 1) &= \frac{(J\pm K)\,(J\mp K+1)}{4J(J+1)}\,[(v_{s''}|\,\mu_x^2|\,v_{s'}) + (v_{s''}|\,\mu_y^2|\,v_{s'})], \\ (J, K\,|\mu^2|\,J-1, K\mp 1) &= \frac{(J\pm K)\,(J\pm K-1)}{4J(2J+1)}\,[(v_{s''}|\,\mu_x^2|\,v_{s'}) + (v_{s''}|\,\mu_y^2|\,v_{s'})], \\ (J, K\,|\mu^2|\,J+1, K\mp 1) &= \frac{(J\mp K+1)\,(J\mp K+2)}{4(J+1)\,(2J+1)}\,[(v_{s''}|\,\mu_x^2|\,v_{s'}) + (v_{s''}|\,\mu_y^2|\,v_{s'})]. \end{aligned}\right\} \tag{25.12}$$

When, on the other hand, the electric moment lies along the axis of symmetry, z, of the molecule only μ_z will be different from zero. We have, in this case, the following relations for the quantum mechanical amplitudes squared of the electric moment

$$\left.\begin{aligned}(J,K|\mu^2|J,K) &= \frac{K^2}{J(J+1)}(v_{s''}|\mu_z^2|v_{s'}),\\ (J,K|\mu^2|J-1,K) &= \frac{(J+K)(J-K)}{J(2J+1)}(v_{s''}|\mu_z^2|v_{s'}),\\ (J-1,K|\mu^2|J,K) &= \frac{(J+K)(J-K)}{J(2J-1)}(v_{s''}|\mu_z^2|v_{s'}).\end{aligned}\right\} \tag{25.13}$$

Whenever a symmetric molecule has a permanent electric moment it will always lie along the z axis.

26. Selection rules for the asymmetric rotator. The energy levels of the asymmetric rotator are characterized, no longer by K, but by an index τ. It is evident, therefore, that the relations (25.12) and (25.13) are no longer applicable. The modification which must be made is to evaluate the matrix elements $(J,\tau|F_\alpha|J',\tau')$ rather than the elements $(J,K|F_\alpha|J',K')$ used in Sect 25. These may be shown to be

$$(J,\tau|F_\alpha|J',\tau') = \sum_{K',K''}(\tau|S|K')(J,K'|F_\alpha|J',K'')(K''|S|\tau') \tag{26.1}$$

where the $(\tau|S|K')$ are the elements of the WANG transformation matrix, i.e., the matrix whose elements are the normalized cofactors of the secular determinant (24.5) when W has been replaced by the specific values of W, W_τ. Evidently,

$$\sum_{K'}(\tau|S|K')(K'|S|\tau') = \delta_{\tau\tau'} \quad \text{and} \quad (K|S|\tau) = \overline{(\tau|S|K)}.$$

A complete discussion of the evaluation of the line strenghts for the asymmetric rotator is beyond the scope of this article. It is, nevertheless, worthwhile to inquire into the selection rules for the asymmetric rotator. The elements of S are evidently diagonal in the quantum numbers J and M. The selection rules for J and M must, for this reason, be the same for the asymmetric rotator as for the symmetric rotator, namely; $\Delta J = \pm 1, 0$ and $\Delta M = \pm 1, 0$. The quantum number K is no longer a good quantum number and it is meaningless to speak of ΔK. Consider, however, the case where the electric moment lies along the body fixed axis x so that λ_{Ax} only is different from zero. The relations (25.1) reduce now to $\mu_A = \lambda_{Ax}\mu_x$. The quantum mechanical amplitudes may then be written

$$\mu_A = \int \overline{\psi}_{JM\tau}\lambda_{Ax}\mu_x\psi_{J'M'\tau'}\,dv \tag{26.2}$$

where A takes the values X, Y and Z.

At least one of the integrals (26.2) must be different from zero if a transition is to occur between the two energy states E_τ and $E_{\tau'}$. This is tantamount to saying that under the covering operations $C_2^{(\alpha)}$ at least one of the integrands of (26.2) must not change sign. The covering operation $C_2^{(x)}$ applied to the direction cosines λ_{Ax} causes them to go over into themselves. Since the μ_{Ax} remain unchanged by these operations $\psi_{J,M,\tau}$ and $\psi_{J'M'\tau'}$ must have the same symmetry. On the other hand the covering operations, $C_2^{(y)}$ and $C_2^{(z)}$ applied to the direction cosines λ_{Ax} cause them to reverse sign. Evidently then for the same reason as before the functions $\psi_{J,M,\tau}$ and $\psi_{J',M',\tau'}$ must here belong to the opposite symmetries. We have from Table 6 (p. 219), in this case that the allowed transitions are the following $A \leftrightarrow B_x$ and $B_y \leftrightarrow B_z$.

Similarly when $\mu_y \neq 0$ with $\mu_x = \mu_y = 0$. The covering operation $C_2^{(y)}$ leaves the direction cosines unchanged in this case while the operations $C_2^{(x)}$ and $C_2^{(z)}$

cause them to reverse sign. The functions $\psi_{J,M,\tau}$ and $\psi_{J'M'\tau'}$ must, in these instances, be respectively of the same symmetry and of different symmetries. This establishes the allowed transitions to be $A \leftrightarrow B_y$ and $B_x \leftrightarrow B_z$. When $\mu_z \neq 0$ with $\mu_x = \mu_y = 0$ the allowed transitions will be found to be $A \leftrightarrow B_z$ and $B_x \leftrightarrow B_y$.

It is, of course, immaterial whether one identifies the axes x, y, z in the order a, b, c or in the order c, b, a. In the first case μ_x becomes the component of the electric moment, μ_a, directed along the axis of the least moment of inertia. The transition rule $A \leftrightarrow B_x$ and $B_y \leftrightarrow B_z$, stated above, becomes $A \leftrightarrow B_a$ and $B_b \leftrightarrow B_c$. μ_z, accordingly, becomes the component of the electric moment, μ_c, directed along the axis of the largest moment of inertia and the rule for permitted transitions $A \leftrightarrow B_z$ and $B_x \leftrightarrow B_y$ becomes $A \leftrightarrow B_c$ and $B_a \leftrightarrow B_b$. In the second choice μ_x becomes the component of the electric moment, μ_c, directed along the axis of the largest moment of inertia and the selection rules $A \leftrightarrow B_x$ and $B_y \leftrightarrow B_z$ become $A \leftrightarrow B_c$ and $B_b \leftrightarrow B_a$. μ_z, then, becomes μ_a, the component of the electric moment directed along the axis of the least moment of inertia and the selection rules $A \leftrightarrow B_z$ and $B_x \leftrightarrow B_y$ become $A \leftrightarrow B_a$ and $B_c \leftrightarrow B_b$. μ_y would, in either choice, be identified with μ_b and the transition rules $A \leftrightarrow B_y$ and $B_x \leftrightarrow B_z$ would become $A \leftrightarrow B_b$ and $B_a \leftrightarrow B_c$. The two choices evidently give identical results. The selection rules for the asymmetric rotator as derived in the preceding paragraphs are summarized in Table 9.

Table 9.

Dipole Moment	μ_a	μ_b	μ_c
Allowed transitions	$A \leftrightarrow B_a$ $B_b \leftrightarrow B_c$	$A \leftrightarrow B_b$ $B_a \leftrightarrow B_c$	$A \leftrightarrow B_c$ $B_a \leftrightarrow B_b$

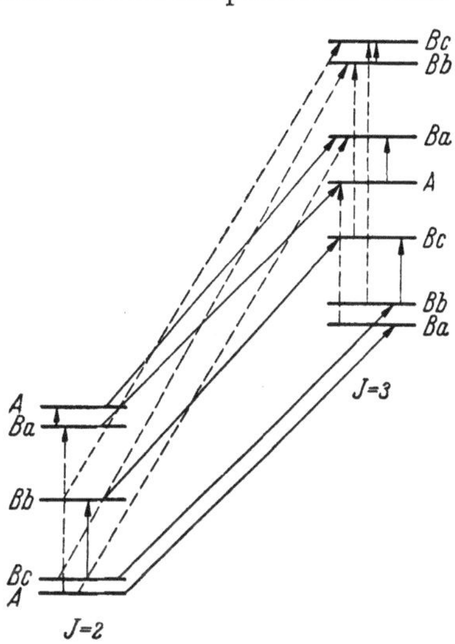

Fig. 6. Permitted transitions for an asymmetric rotator involving the levels with $J = 2$ and $J = 3$.

The transitions permitted for an asymmetric rotator between the levels characterized by $J = 2$ and $J = 3$ are shown in Fig. 6 where the electric moment lies along the axis of a.

27. The axially symmetric rotator with internal angular momentum. The case of the axially symmetric rotator with internal angular momentum originating from the twofold degenerate oscillations is almost a trivial one. The twofold degenerate oscillations must, because of symmetry, lie in the xy plane of the rotator. This determines that the vector p_z^*, which is equal to $-i\hbar \sum_t \zeta_t^{(z)} \frac{\partial}{\partial \chi_t}$, must be parallel or antiparallel to the axis of symmetry, z, of the molecule. The Schrödinger equation for this rotator will evidently be

$$\tfrac{1}{2}\{[(P_x^2 + P_y^2)/I_{xx}^{(e)}] + [(P_z - p_z^*)^2/I_{zz}^{(e)}] - E_{JKl_t}\}\psi_0 = 0 \tag{27.1}$$

where we have seen in Sect. 19 that ψ_0 may be taken to be

$$\psi_0 = \prod_t R(\vartheta)\, e^{iM\psi}\, e^{iK\varphi}\, e^{il_t\chi_t} \tag{27.2}$$

where M, K and l_t all are integers. This gives

$$\tfrac{1}{2}\left\{[(P^2 - P_z^2)/I_{xx}^{(e)}] + \left[\left(P_z - \sum_t \zeta_t^{(z)} l_t\right)^2 \Big/ I_{zz}^{(e)}\right] - E_{J,K,l_t}\right\}\psi_0 = 0. \qquad (27.3)$$

The eigenvalues of (27.3) will clearly be the following

$$E_{J,K,l_t} = [J(J+1) - K^2]\,(h^2/8\pi^2 I_{xx}^{(e)}) + \left(K - \sum \zeta_t^{(z)} l_t\right)^2 (h^2/8\pi^2 I_{zz}^{(e)}). \qquad (27.4)$$

Since K and l_t may both assume positive and negative values a degeneracy is clearly removed, manifesting itself in the term, $-2K\sum_t \zeta_t^{(z)} l_t (h^2/8\pi^2 I_{zz}^{(e)})$, which is a part of the second term in (27.4). The selection rules as stated in Sect. 25 are evidently incomplete since they take no account of the quantum numbers l_t. This selection rule is quickly determined if we confine ourselves to harmonic oscillations. When, for example, a symmetrical molecule makes a transition from the normal vibration state where $v_t = l_t = 0$ to the first excited state of a *perpendicular* vibration with $v_t = l_t = 1$ the components of the electric moment induced are μ_x and μ_y while the component of the electric moment μ_z remains zero. One then replaces μ_x and μ_y by $\sum_t e\,r_t \cos\chi_t$ and $\sum_t e\,r_t \sin\chi_t$ respectively. Eq. (25.1) for the space fixed components of the classical electric moment, μ_A, then becomes

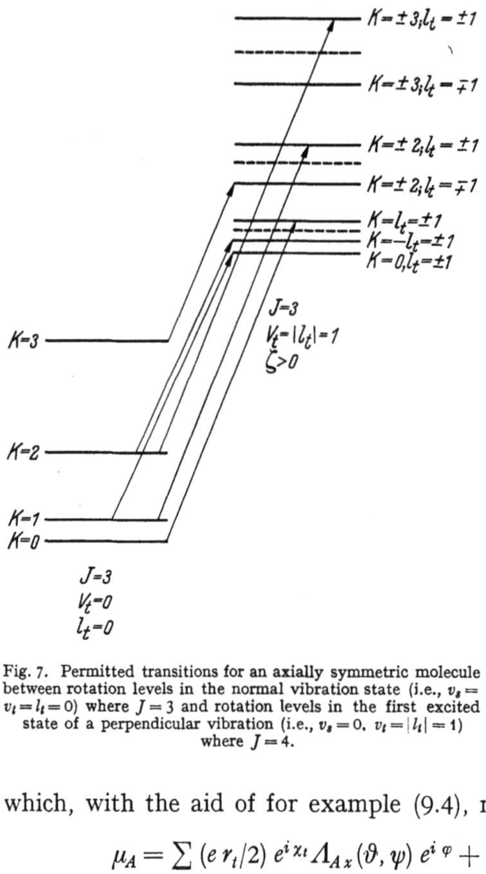

Fig. 7. Permitted transitions for an axially symmetric molecule between rotation levels in the normal vibration state (i.e., $v_s = v_t = l_t = 0$) where $J = 3$ and rotation levels in the first excited state of a perpendicular vibration (i.e., $v_s = 0$, $v_t = |l_t| = 1$) where $J = 4$.

$$\left.\begin{aligned} \mu_A = {} & \sum_t (e\,r_t/2)\,e^{i\chi_t}(\lambda_{Ax} - i\lambda_{Ay}) + \\ & + \sum_t (e\,r_t/2)\,e^{-i\chi_t}(\lambda_{Ax} + i\lambda_{Ay}) \end{aligned}\right\} \qquad (27.5)$$

which, with the aid of for example (9.4), may further be written to be

$$\mu_A = \sum_t (e\,r_t/2)\,e^{i\chi_t}\Lambda_{Ax}(\vartheta,\psi)\,e^{i\varphi} + \sum_t (e\,r_t/2)\,e^{-i\chi_t}\bar{\Lambda}_{Ax}(\vartheta,\psi)\,e^{-i\varphi}. \qquad (27.6)$$

$\Lambda_{Ax}(\vartheta,\psi)$ and $\bar{\Lambda}_{Ax}(\vartheta,\psi)$ being functions only of ϑ and ψ. The matrix elements for μ_A corresponding to (26.2) may be obtained using the wave functions stated in (27.2). This leads at once to the result that μ_A is zero except when $\Delta l_t = \Delta K = \pm 1$. The selection rule for J and M is evidently unchanged. The above selection rule is due to TELLER and TISZA.

Spectral frequencies are sometimes observed in the far infrared or in the microwave region which arise from transitions between the rotation levels of a molecule when it is in an excited perpendicular vibration state where $v_t = |l_t| = 1$. It is the permanent electric moment, in such cases, which is important and as we have already noted it is the component μ_z which then is different from zero while $\mu_x = \mu_y = 0$. The direction cosines are in such cases independent of both χ_t and φ so that $\Delta l_t = \Delta K = 0$.

Fig. 7 shows two sets of energy levels for $J=3$; one set where all the v's are zero except v_t and l_t, which are taken to be one, and a second set where all the v's are zero. The transitions permitted, by the selection rules given above, to take place between these levels are also indicated in the figure.

28. The spherical rotator with internal angular momentum. A molecule must have a high degree of spatial symmetry (i.e., tetrahedral or higher symmetries) in order that three dimensionally isotropic oscillations may occur. Such molecules will have the three moments of inertia equal and the Coriolis coupling factors, $\zeta_u^{(\alpha)}$, will all be alike. The Hamiltonian for such a rotator may, therefore, be written

$$H = \tfrac{1}{2}\{(P_x - p_x^*)^2 + (P_y - p_y^*)^2 + (P_z - p_z^*)^2\}/I^{(e)} \tag{28.1}$$

which evidently will be

$$\left.\begin{aligned} H = \tfrac{1}{2}\{(P^2 + p^{*2}) - (p_x^* + i p_y^*)(P_x - i P_y) - \\ - (p_x^* - i p_y^*)(P_x + i P_y) - 2 p_z^* P_z\}/I^{(e)} \end{aligned}\right\} \tag{28.2}$$

where $p^{*2} = p_x^{*2} + p_y^{*2} + p_z^{*2}$ and which for a given frequency ω_u is equal to

$$\{(1/\sin\vartheta_u)(\partial/\partial\vartheta_u)(\sin\vartheta_u\,\partial/\partial\vartheta_u) - (m_u^2/\sin^2\vartheta_u) + l_u(l_u+1)\}\,\hbar^2.$$

The operator, H, is independent of the variable ϱ_u and must, therefore, be diagonal in the total vibration quantum number v_t. The operators P^2 and p^{*2} commute with each other and commute also with the components of the total and the internal angular momenta P_α and p_α^*. The P_α commute among themselves according to the relation (9.3) while the p_α^* commute among themselves according to the relation (9.6). Evidently therefore H is diagonal in the quantum numbers J, M and l_u. It is equally evident that H will not be diagonal in K and m_u, but choosing a representation in which P_z and p_z^* are diagonal may indeed have matrix elements of both the type $(K, m_u|H|K, m_u)$ and the type $(K, m_u|H|K\pm 1, m_u\pm 1)$. It may, in fact, be shown that if $\psi^0_{l_u m_u, J, K, M} = \psi^0_{m_u, K} = R_{J,K,M}(\vartheta)\, e^{iM\psi}\, e^{iK\varphi}\, \Theta_{l_u, m_u}(\vartheta_u)\, e^{i m \chi_u}$ be taken as the basic wave function, that

$$\left.\begin{aligned} H\psi^0_{m_u,K} = \{(l_u - m_u + 1)(J - K + 1)\,(l_u + m_u)(J + K)\}^{\frac{1}{2}}(2B\zeta hc)\,\psi^0_{m_u-1,K-1} + \\ + \{J(J+1) + l_u(l_u+1)\zeta^2 - 2m_u K\zeta\}\,B\psi_{m_u,K} + \\ + \{(l_u - m_u)(l_u + m_u + 1)(J-K)(J+K+1)\}^{\frac{1}{2}}(2B\zeta hc)\,\psi^0_{m_u+1,K+1} \end{aligned}\right\} \tag{28.3}$$

where $B = h/8\pi^2 I_e c$. The matrix elements of H may be calculated directly from the relation (28.3) and they are found to be

$$\left.\begin{aligned} (K, m_u|H|K, m_u) &= [J(J+1) + l_u(l_u+1)\zeta^2 - 2m_u K\zeta]\,B_e hc, \\ (K, m_u|H|K+1, m_u+1) &= (K+1, m_u+1|H|K, m_u) \\ &= \{(l_u - m_u)(l_u + m_u + 1)(J-K)(J+K+1)\}^{\frac{1}{2}}\zeta B_e hc. \end{aligned}\right\} \tag{28.4}$$

The eigenvalues are then obtained, as in similar cases, by diagonalizing the matrix with the elements (28.4). One achieves this by setting the determinant

$$|(K, m_u|H|K', m_u') - E\,\delta_{KK'}\,\delta_{m_u m_u'}|$$

equal to zero and solving for the roots. The secular determinant will be one of $(2l_u+1)(2J+1)$ rows and columns since there are $2l_u+1$ values of m_u and $2J+1$ values of K. The determinant will split up, however, into subdeterminants and if one divides all the elements (28.4) by $2B_e\zeta hc$ the general form of such a subdeterminant will be

$$\begin{array}{r|cccc|}
 & \Theta(l,l)\,R(K+l) & \Theta(l,l-1)\,R(K+l-1)\ldots & \Theta(l,-l+1)\,R(K-l+1) & \Theta(l,-l)\,R(K-l) \\
\hline
\Theta(l,l)\,R(K+l) & -l(K+l)-\varepsilon & (l/2)^{\frac{1}{2}}[(J+K+l)(J-K-l+1)]^{\frac{1}{2}}\ldots & 0 & 0 \\
\Theta(l,l-1)\,R(K+l-1) & (l/2)^{\frac{1}{2}}[(J+K+l)(J-K-l+1)]^{\frac{1}{2}} & -(l-1)(K+l-1)-\varepsilon\ldots & 0 & 0 \\
\vdots & \vdots & \vdots & \vdots & \vdots \\
\Theta(l,-l+1)\,R(K-l+1) & 0 & 0\ldots & (l-1)(K-l+1)-\varepsilon & (l/2)^{\frac{1}{2}}[(J-K+l)(J+K-l+1)]^{\frac{1}{2}} \\
\Theta(l,-l)\,R(K-l) & 0 & 0 & (l/2)^{\frac{1}{2}}[(J-K+l)(J+K-l+1)]^{\frac{1}{2}} & l(K-l)-\varepsilon \\
\hline
\end{array} = 0 \quad (28.5)$$

where $E_{J,K,l_u,m_u} = [J(J+1) + l_u(l_u+1)\,\zeta^2]\; B_e hc + 2B_e\zeta\, hc\,\varepsilon$. The eigenfunctions, ψ, for a given substate, $E_{J,l_u,\gamma}$ will be a linear combination of the wave functions $\Theta(l_u, m_u)\,R(K)$, the coefficients multiplying these being, in each case, the cofactors of the determinant (28.5) where ε has successively been replaced by the various roots ε_γ.

As an example of the foregoing, consider the excited vibration state where $v_u = l_u = 1$. The quantum number, m_u, may here take the values $+1, 0, -1$. The secular determinant (28.5) then becomes merely

$$\begin{vmatrix} -K-1-\varepsilon & [\frac{1}{2}(J-K)(J+K+1)]^{\frac{1}{2}} & 0 \\ [\frac{1}{2}(J-K)(J+K+1)]^{\frac{1}{2}} & -\varepsilon & [\frac{1}{2}(J+K)(J-K+1)]^{\frac{1}{2}} \\ 0 & [\frac{1}{2}(J+K)(J-K+1)]^{\frac{1}{2}} & K-1-\varepsilon \end{vmatrix} = 0 \qquad (28.6)$$

which gives the three values for ε, $\varepsilon_- = J$, $\varepsilon_0 = -1$ and $\varepsilon_+ = -(J+1)$. There will be three eigenfunctions ψ_-, ψ_0 and ψ_+ and these will be found to be

$$\left.\begin{aligned}
\psi_+ &= \left\{\frac{(J-K)(J-K+1)}{2(J+1)(2J+1)}\right\}^{\frac{1}{2}} \psi(1,1)\,R(K+1) + \left\{\frac{(J-K+1)(J+K+1)}{(J+1)(2J+1)}\right\}^{\frac{1}{2}} \times \\
&\quad \times \psi(1,0)\,R(K) + \left\{\frac{(J+K)(J+K+1)}{2(J+1)(2J+1)}\right\}^{\frac{1}{2}} \psi(1,-1)\,R(K-1), \\
\psi_0 &= -\left\{\frac{(J-K)(J+K+1)}{2J(J+1)}\right\}^{\frac{1}{2}} \psi(1,1)\,R(K+1) - \left\{\frac{K^2}{J(J+1)}\right\}^{\frac{1}{2}} \times \\
&\quad \times \psi(1,0)\,R(K) + \left\{\frac{(J+K)(J-K+1)}{2J(J+1)}\right\}^{\frac{1}{2}} \psi(1,-1)\,R(K-1), \\
\psi_- &= \left\{\frac{(J+K)(J+K+1)}{2J(2J+1)}\right\}^{\frac{1}{2}} \psi(1,1)\,R(K+1) - \left\{\frac{(J+K)(J-K)}{J(2J+1)}\right\}^{\frac{1}{2}} \times \\
&\quad \times \psi(1,0)(RK) + \left\{\frac{(J-K)(J-K+1)}{2J(2J+1)}\right\}^{\frac{1}{2}} \psi(1,-1)\,R(K-1).
\end{aligned}\right\} \qquad (28.7)$$

The wave functions (28.7) permit the determination of the selection rules for transitions from the normal vibration state to the first excited vibration state. One finds them to be $\Delta J = +1$, 0 or -1 as the transition takes place to the component states ε_+, ε_0 and ε_- respectively. The transitions permitted as the molecule makes a vibrational transition from the normal vibration state to the first excited state of a three dimensionally isotropic vibration is illustrated in Fig. 8 with $J = 1$, 2 and 3 in the normal state and in the excited state.

The results which were just derived may also be obtained by the method of vector addition in quantum mechanics referred to earlier in Sect. 10. We indicated there that this consisted in transforming from a system of representation in which two angular momentum vectors, P_1 and P_2 are independently space quantized to a system where $P_1 + P_2 = P$ is space quantized and where P^2 is diagonal. The Hamiltonian, in our case, is of the form $\sum_\alpha (P_\alpha - p_\alpha^*)^2/I_{\alpha\alpha}^{(e)}$ rather than of the form $\sum_\alpha (P_\alpha + p_\alpha^*)^2/I_{\alpha\alpha}^{(e)}$ as would seem consistent with the formulation of Sect. 10. A word of clarification is, therefore, indicated as to why the method of Sect. 10 is directly applicable.

The operators P_α commute according to the commutation relation (9.30) with the anomalous algebraic sign before i. The operators p_α^*, however, commute according to the commutation relation (9.6) with the usual sign before i. The fact that p_α^* commute according to relation (9.6) rather than (9.30) is related to the fact that they are computed in a system of coordinates fixed in the molecular framework rather than in space. The quantities p_α^* may be regarded as measured relative to the axes mounted in the molecule and are not, as are the P_α, simply

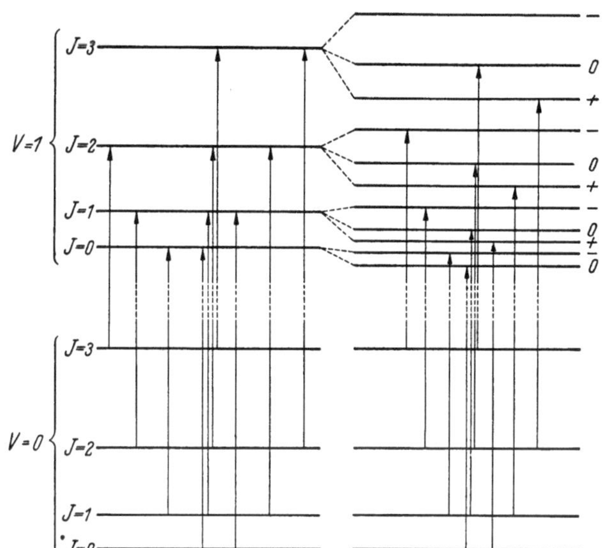

Fig. 8. a) Permitted transitions for a tetrahedrally symmetric XY_4 molecule between rotation levels in the norma vibration state (i.e., $v_s = v_u = l = m_u = 0$) and rotation levels in the first excited state of a three-fold degenerate vibration (i.e., $v_s = 0$, $v_u = l = 1$, $m_u = 1, 0, -1$) where no Coriolis interaction is present. b) Permitted transitions for a tetrahedrally symmetric XY_4 molecule between rotation levels in the normal vibration state and rotation levels in the first excited state of a three-fold degenerate vibration where CORIOLIS interaction is present.

referred to these axes. It is, however, true that the operators $-p_\alpha^*$ (and indeed $-\Pi_\alpha$) conform to the same commutation relation (9.30) as do the P_α. Hence, as we have noted already in Sect. 21, the entire operators $(P_\alpha - p_\alpha^*)$ [and indeed the operators $(P_\alpha - \Pi_\alpha - \mathfrak{S}_\alpha)$] also commute according to relation (9.30). The significant fact is that P_α and $-p_\alpha^*$ commute in the same manner. Since evidently P_α and $-p_\alpha^*$ also commute the conditions of Sect. 10 are met since the sign assigned to i is unimportant so long as consistency in the use of the sign is maintained. We may, therefore, write instead of (28.1),

$$\left.\begin{aligned} H &= \tfrac{1}{2}\sum_\alpha [P_\alpha + (-p_\alpha^*)]^2/I^{(e)} = \tfrac{1}{2}[P + (-p^*)]^2/I^{(e)} \\ \text{or}\qquad H &= \tfrac{1}{2}[P^2 + p^{*2} + 2(-p^*)\cdot P]/I^{(e)}. \end{aligned}\right\} \tag{28.8}$$

We then proceed to identify P with the total angular momentum operator P of Sect. 10 and p^* with the operator ζP_2. The angular momentum operator, P_1, which will be $P - P_2$, may be identified with the angular momentum of the molecular framework. The eigenvalues of P^2 and p^{*2} are, as we know, $J(J+1)\hbar^2$ and $l_u(l_u+1)\zeta^2\hbar^2$ respectively, where l_u, according to Sect. 22 may take the values $v_u, v_u - 2, \ldots 1$ or 0. We obtain as the eigenvalue of $(-p^*)\cdot P/I^{(e)}$ the

relation
$$\varepsilon = (-p^*)\cdot P/\hbar^2\zeta = -\tfrac{1}{2}\{J(J+1) + l_u(l_u+1) - L(L+1)\} \tag{28.9}$$
where L takes the values $J+l_u, J+l_u-1, \ldots, J-l_u$. Applying this to the case where $v_u=1$ and where accordingly $l_u=1$ only we have for the values of ε, remembering, of course, that L may take the values $J+1$, J and $J-1$, $\varepsilon_-=J$, $\varepsilon_0=-1$ and $\varepsilon_+=-(J+1)$. The result is the same as that stated before.

The eigenfunctions for the three component states are given by Eq. (10.4) when the proper notation is introduced. The eigenfunctions may be written
$$\begin{Vmatrix}\psi_+\\ \psi_0\\ \psi_-\end{Vmatrix} = \begin{Vmatrix}a_+ & b_+ & c_+\\ a_0 & b_0 & c_0\\ a_- & b_- & c_-\end{Vmatrix}\cdot\begin{Vmatrix}\psi(1,1)\,R(K+1)\\ \psi(1,0)\,R(K)\\ \psi(1,-1)\,R(K-1)\end{Vmatrix} \tag{28.10}$$
where the transformation matrix above is that given by CONDON and SHORTLEY in their Table 2[3]. Since the role of J and J_1 has been interchanged here with respect to their notation, J_1 in their case must be replaced by J. Moreover, because the operators P and $-p^*$ may be regarded as referred to the axes x, y and z rather than X, Y and Z and particularly because of the negative sign before p^*, their quantum number m_2 must here be replaced by $-K$. The eigenfunctions (28.10) will be seen to be identical to (28.6).

The results for the second excited state of ν_3, i.e., $2\nu_3$, are also quickly arrived at by this method. Here $v_u=2$ and l_u may, according to Sect. 22 take the two values $l_u=2$ and $l_u=0$. The quantum number L will then take the five values $J+2$, $J+1$, J, $J-1$, $J-2$ where $l_u=2$ and the value J only where $l_u=0$. Eq. (28.9) gives at once the five values for ε; $\varepsilon_1=2J$, $\varepsilon_2=(J-2)$, $\varepsilon_3=-3$, $\varepsilon_4=-(J+3)$, $\varepsilon_5=-2(J+1)$ in the instance where $l_u=2$ and the single value $\varepsilon_0=0$ when $l_u=0$.

The general form for the basic wave functions corresponding to those for the singly excited state ν_3 will here be
$$\left.\begin{aligned}\chi_0 &= \Theta(0,0)\,R(K),\\ \chi_1 &= \Theta(2,2)\,R(K+2),\\ \chi_2 &= \Theta(2,1)\,R(K+1),\\ \chi_3 &= \Theta(2,0)\,R(K),\\ \chi_4 &= \Theta(2,-1)\,R(K-1),\\ \chi_5 &= \Theta(2,-2)\,R(K-2)\end{aligned}\right\} \tag{28.11}$$
where $R(K)$ is a normalized wave function of the symmetric rotator. One may quite readily show that the eigenfunctions which go with the values of ε stated above are these
$$\left.\begin{aligned}l_u &= 0 \quad \psi_0(\varepsilon_0) = \chi_0\\ l_u &= 2 \quad \psi_i(\varepsilon_i) = \|A\|\cdot\|\chi_i\|\end{aligned}\right\} \tag{28.12}$$
where $\|A\|$ is the transformation matrix given by CONDON and SHORTLEY in their Table 4[3], J and J_1 being interchanged as before and their m_2 being here replaced by $-K$.

The selection rules have also been determined in this case and are as follows: (*a*) Transitions from the normal state to the state ε_0 may not take place; (b) $\Delta J=0$, ± 1 are permitted from the normal state to the state ε_1, ε_2, ε_3, ε_4, and ε_5.

When the internal angular momentum stems from the molecules being in a state where two or more threefold degenerate oscillations are excited the problem becomes, at once, much more complicated. One may, here also, make use of the method of vector addition as delineated in Sect. 10, specifically the relation (10.7), to arrive at the energies although no such simple relation as (28.9) may be written down. We shall here confine ourselves to the example of a combination state where two three dimensionally isotropic oscillations, say ν_3 and ν_4, are excited.

Referring again to Eq. (28.8) we still have that the eigenvalue of P^2 is $J(J+1)\hbar^2$. The eigenvalues of p^{*2} are arrived at in the following manner. One identifies p^* with $\zeta_3 P_3 + \zeta_4 P_4$ in this case so that $p^{*2} = [\zeta_3^2 (P_3 \cdot P_3) + \zeta_4^2 (P_4 \cdot P_4) + 2\zeta_3\zeta_4 (P_3 \cdot P_4)]$. We have, from Sect. 10 however, that $P_3 \cdot P_4 = \frac{1}{2}[(P_2 \cdot P_2) - (P_3 \cdot P_3) - (P_4 \cdot P_4)]$ whence $p^{*2} = \zeta_3\zeta_4 (P_2 \cdot P_2) + \zeta_3(\zeta_3 - \zeta_4)(P_3 \cdot P_3) + \zeta_4(\zeta_3 - \zeta_4)(P_4 \cdot P_4)$. The eigenvalues of p^{*2} will, therefore, be

$$\{\zeta_3\zeta_4 J_2(J_2+1) + (\zeta_3 - \zeta_4)[\zeta_3 l_3(l_3+1) - \zeta_4 l_4(l_4+1)]\}\hbar^2. \tag{28.13}$$

We identify the term $-p^* P$ with $-\zeta_3(P_3 \cdot P) - \zeta_4(P_4 \cdot P)$, the matrix components of which have been given in (10.7) and, formally, the problem is solved.

Specifically we shall consider the case of the state $\nu_3 + \nu_4$ where ν_3 and ν_4 are both once excited, i.e., where $v_3 = l_3 = 1$ and $v_4 = l_4 = 1$ at the same time. As in (28.13) we identify l_3 with J_3 (the quantum number associated with P_3) and l_4 with J_4 in (10.7). The last two terms in (10.7) evidently cancel each other out in our example so as to give

$$-\{\zeta_3(P_3 \cdot P) + \zeta_4(P_4 \cdot P)\} = -\left\{\frac{(\zeta_3+\zeta_4)}{2}(P_2 \cdot P_2) + \zeta_3(P_3 \cdot P_1) + \zeta_4(P_4 \cdot P_1)\right\}. \tag{28.14}$$

Eq. (28.14) reduces to

$$-\{\zeta_3(P_3 \cdot P) + \zeta_4(P_4 \cdot P)\} = -\left\{\frac{(\zeta_3+\zeta_4)}{2} J_2(J_2+1)\hbar^2 + \zeta_3(P_3 \cdot P_1) + \zeta_4(P_4 \cdot P_1)\right\} \tag{28.15}$$

since the eigenvalue of $(P_2 \cdot P_2) = J_2(J_2+1)\hbar^2$.

We must, in order to obtain the eigenvalues of the rotator, determine the values now of the matrix components $(P_3 \cdot P_1)$ and $(P_4 \cdot P_1)$ of Sect. 10 for our example where $l_3 = l_4 = 1$. We have for our specific case where J_2 may take the values $J_2 = 0$, 1 and 2, the following:

$$\left.\begin{aligned} &J_1 = J+2 \qquad (J_2 = 2)\\ &(2|(P_3 \cdot P_1)|2) = (2|(P_4 \cdot P_1)|2) = -(J+3)\hbar^2 \end{aligned}\right\} \tag{28.16a}$$

$$\left.\begin{aligned} &J_1 = J+1 \qquad (J_2 = 2, 1)\\ &(2|(P_3 \cdot P_1)|2) = (2|(P_4 \cdot P_1)|2) = -(J+4)\hbar^2\\ &(1|(P_3 \cdot P_1)|1) = (1|(P_4 \cdot P_1)|1) = -(J+2)\hbar^2\\ &(2|(P_3 \cdot P_1)|1) = -(2|(P_4 \cdot P_1)|1) = -\sqrt{J(J+2)}\,\frac{\hbar^2}{2} \end{aligned}\right\} \tag{28.16b}$$

$$\left.\begin{aligned} &J_1 = J \qquad (J_2 = 2, 1, 0)\\ &(2|(P_3 \cdot P_1)|2) = (2|(P_4 \cdot P_1)|2) = -\frac{3}{2}\hbar^2\\ &(1|(P_3 \cdot P_1)|1) = (1|(P_4 \cdot P_1)|1) = -\frac{1}{2}\hbar^2\\ &(0|(P_3 \cdot P_1)|0) = (0|(P_4 \cdot P_1)|0) = 0\\ &(2|(P_3 \cdot P_1)|1) = -(2|(P_4 \cdot P_1)|1) = \sqrt{\frac{1}{3}}\sqrt{(2J-1)(2J+3)}\,\frac{\hbar^2}{2}\\ &(1|(P_3 \cdot P_1)|0) = -(1|(P_4 \cdot P_1)|0) = \sqrt{\frac{2}{3}}\sqrt{J(J+1)}\,\hbar^2 \end{aligned}\right\} \tag{28.16c}$$

$$\left.\begin{aligned} &J_1 = J-1 \qquad (J_2 = 2, 1)\\ &(2|(P_3\cdot P_1)|2) = (2|(P_4\cdot P_1)|2) = (J-3)\frac{\hbar^2}{2}\\ &(1|(P_3\cdot P_1)|1) = (1|(P_4\cdot P_1)|1) = (J-1)\frac{\hbar^2}{2}\\ &(2|(P_3\cdot P_1)|1) = -(2|(P_4\cdot P_1)|1) = \sqrt{(J-1)(J+1)}\frac{\hbar^2}{2} \end{aligned}\right\} \quad (28.16\text{d})$$

$$\left.\begin{aligned} &J_1 = J-2 \qquad (J_2 = 2)\\ &(2|(P_3\cdot P_1)|2) = (2|(P_4\cdot P_1)|2) = (J-2)\hbar^2 \end{aligned}\right\} \quad (28.16\text{e})$$

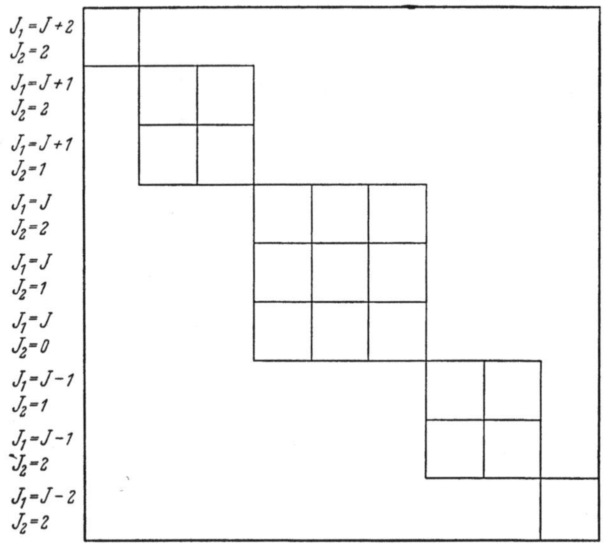

Fig. 9. Form of the sub-matrix for the vibration-rotation energies of a tetrahedrally symmetric molecule for $\nu_3+\nu_4$.

The elements (28.16) permit setting up the matrix of H. The elements may be so arranged as indicated in Fig. 9. We are aware that the elements (28.10) are diagonal in the quantum number J_1 so that the matrix will factor into steps, a separate step for each value of J_1. The roots of the secular determinant of H in which each element has been divided by $2B_e hc$ are quickly found to be

$$\left.\begin{aligned} &J_1 = J+2\\ &\varepsilon_1 = J(\zeta_3+\zeta_4) \end{aligned}\right\} \quad (28.17\text{a})$$

$$\left.\begin{aligned} &J_1 = J+1\\ &\varepsilon_{3,4} = \tfrac{1}{2}\{(J-1)(\zeta_3+\zeta_4) \pm [(J+1)^2(\zeta_3-\zeta_4)^2 + 4\zeta_3\zeta_4]^{\frac{1}{2}}\} \end{aligned}\right\} \quad (28.17\text{b})$$

$$\left.\begin{aligned} &J_1 = J\\ &\varepsilon^3 + 2(\zeta_3+\zeta_4)\,\varepsilon^2 + [\tfrac{3}{4}(\zeta_3+\zeta_4)^2 - (J^2+J-\tfrac{1}{4})(\zeta_3-\zeta_4)^2]\,\varepsilon -\\ &\qquad\qquad - J(J+1)(\zeta_3+\zeta_4)(\zeta_3-\zeta_4)^2 = 0 \end{aligned}\right\} \quad (28.17\text{c})$$

$$\left.\begin{aligned} &J_1 = J-1\\ &\varepsilon_{5,6} = -\tfrac{1}{2}\{(J+2)(\zeta_3+\zeta_4) \mp [J^2(\zeta_3-\zeta_4)^2 + 4\zeta_3\zeta_4]^{\frac{1}{2}}\} \end{aligned}\right\} \quad (28.17\text{d})$$

$$\left.\begin{aligned} &J_1 = J-2\\ &\varepsilon_2 = -(J+1)(\zeta_3+\zeta_4) \end{aligned}\right\} \quad (28.17\text{e})$$

where

$$E_{J,K,l_2,l_4,J_2} = \{J(J+1)+\zeta_3\zeta_4 J_2(J_2+1)+(\zeta_3-\zeta_4)[\zeta_3 l_3(l_3+1)-\zeta_4 l_4(l_4+1)]+2\varepsilon\} B_e\, hc.$$

In a manner analogous to the state where $v_u=2$, $l_u=2$ and $l_u=0$ we adopt here as preliminary wave functions, the following

$$\left.\begin{aligned}
&\left.\begin{matrix} J_2= & 0 \\ M_2= & 0\end{matrix}\right\} \chi_0 = \frac{1}{\sqrt{6\pi}}\{2\Theta_3(1,1)\,\Theta_4(1,1)\cos(\chi_3-\chi_4)-\Theta_3(1,0)\,\Theta_4(1,0)\}\,R(K)\\
&\left.\begin{matrix} J_2= & 1 \\ M_2= & 1\end{matrix}\right\} \chi_{1,1} = \frac{1}{\sqrt{4\pi}}\{\Theta_3(1,1)\,\Theta_4(1,0)\,e^{i\chi_3}-\Theta_3(1,0)\,\Theta_4(1,1)\,e^{i\chi_4}\}\,R(K+1)\\
&\left.\begin{matrix} J_2= & 1 \\ M_2= & 0\end{matrix}\right\} \chi_{1,0} = \frac{-i}{\sqrt{\pi}}\Theta_3(1,1)\,\Theta_4(1,1)\sin(\chi_3-\chi_4)\,R(K)\\
&\left.\begin{matrix} J_2= & 1 \\ M_2= & -1\end{matrix}\right\} \chi_{1,-1} = \frac{-1}{\sqrt{4\pi}}\{\Theta_3(1,1)\,\Theta_4(1,0)\,e^{-i\chi_3}-\Theta_3(1,0)\,\Theta_4(1,1)\,e^{-i\chi_4}\}\,R(K-1)\\
&\left.\begin{matrix} J_2= & 2 \\ M_2= & 2\end{matrix}\right\} \chi_{2,2} = \frac{1}{\sqrt{2\pi}}\Theta_3(1,1)\,\Theta_4(1,1)\,e^{i(\chi_3+\chi_4)}\,R(K+2)\\
&\left.\begin{matrix} J_2= & 2 \\ M_2= & 1\end{matrix}\right\} \chi_{2,1} = \frac{1}{\sqrt{4\pi}}\{\Theta_3(1,1)\,\Theta_4(1,0)\,e^{i\chi_3}+\Theta_3(1,0)\,\Theta_4(1,1)\,e^{i\chi_4}\}\,R(K+1)\\
&\left.\begin{matrix} J_2= & 2 \\ M_2= & 0\end{matrix}\right\} \chi_{2,0} = \frac{1}{\sqrt{3\pi}}\{\Theta_3(1,1)\,\Theta_4(1,1)\cos(\chi_3-\chi_4)+\Theta_3(1,0)\,\Theta_4(1,0)\}\,R(K)\\
&\left.\begin{matrix} J_2= & 2 \\ M_2= & -1\end{matrix}\right\} \chi_{2,-1} = \frac{1}{\sqrt{4\pi}}\Theta_3(1,1)\,\Theta_4(1,0)\,e^{-i\chi_3}+\Theta_3(1,0)\,\Theta_4(1,1)\,e^{-i\chi_4}\}\,R(K-1)\\
&\left.\begin{matrix} J_2= & 2 \\ M_2= & -2\end{matrix}\right\} \chi_{2,-2} = \frac{1}{\sqrt{2\pi}}\Theta_3(1,1)\,\Theta_4(1,1)\,e^{-i(\chi_3+\chi_4)}\,R(K-2)
\end{aligned}\right\} \qquad (28.18)$$

The basic wave functions used in setting up the matrix in Fig. 8 will be linear combinations of the preliminary wave functions (28.18). It may be shown that these are

$$\|\psi\|\,(J_2=2)=\|A\|\cdot\|\chi\|\,(J_2=2) \qquad (28.19)$$

where $\|\psi\|\,(J_2=2)=\|\psi_{\mathrm{I}},\psi_{\mathrm{II}},\psi_{\mathrm{III}},\psi_{\mathrm{IV}},\psi_{\mathrm{V}}\|$, $\|\chi\|\,(J_2=2)=\|\chi_{2,2},\chi_{2,1}\,\chi_{2,0},\chi_{2,-1}\,\chi_{2,-2}\|$ and the transformation matrix is identically that given by CONDON and SHORTLEY in their table 4[3], their J_1 and m_2 being replaced in our case respectively by J and $-K$;

$$\|\psi\|\,(J_2=1)=\|B\|\cdot\|\chi\|\,(J_2=1) \qquad (28.20)$$

where $\|\psi\|\,(J_2=1)=\|\psi_{\mathrm{VI}},\psi_{\mathrm{VII}},\psi_{\mathrm{VIII}}$, $\|\chi\|\,(J_2=1)=\|\chi_{1,1},\chi_{1,0},\chi_{1,-1}\|$ and the transformation matrix $\|B\|$ is that given by CONDON and SHORTLEY in their Table 2[3], J_1 and m_2 being once more replaced in our case by J and $-K$; and

$$\psi_{\mathrm{IX}}=\chi_0. \qquad (28.21)$$

The actual eigenfunctions associated with the values of ε stated earlier will be further linear combinations of the functions $\psi(J_2=2)$, $\psi(J_2=1)$ and $\psi(J_2=0)$, i.e.,

$$\psi(\varepsilon_i) = A_i \psi(J_2 = 2) + B_i \psi(J_2 = 1) + C_i \psi(J_2 = 0) \tag{28.22}$$

where A_i, B_i and C_i are the normalized cofactors of the determinant of the matrix in Fig. 8.

The wave functions (28.22) will permit the determination of the allowed transitions from the normal vibration state to the state $\nu_3 + \nu_4$. They have been found to be simply the following: $\Delta J = 0, \pm 1$ from the normal state to all the component states.

29. Summary of Chap. IV. We may summarize the results of Chap. IV briefly in the following manner. The vibration-rotation energies of a polyatomic molecule, *to the extent that the oscillations are harmonic and the rotator is semirigid,* may be stated as the sum of two terms, one of which is strictly the harmonic energy of vibration and the other which may be regarded as the energy of rotation of the molecule, i.e.,

$$E(v, r) = E(v_s) + E(r). \tag{29.1}$$

The vibrational energy, to this approximation, will be a sum of terms, each term giving the harmonic energy of vibration of the molecule in one of its normal modes. It may be written

$$E(v_s) = h c \sum_s \omega_s (v_s + \tfrac{1}{2} g_s) \tag{29.2}$$

where ω_s is the frequency of oscillation expressed in cm.$^{-1}$, v_s is the vibration quantum number and g_s is a weight factor, numerically equal to the number of dimensions of isotropy of the oscillator.

The form of the rotational energy will vary from molecule to molecule. Thus for an asymmetric molecule where there will be no internal angular momentum of vibration, $E(r)$ may, to this approximation, be written

$$E(r) = \{J(J+1)\,[\tfrac{1}{2}(B_e^{(xx)} + B_e^{(yy)})] + W\,[B_e^{(zz)} - \tfrac{1}{2}(B_e^{(xx)} + B_e^{(yy)})]\}\, h c \tag{29.3}$$

in which W is one of the $(2J+1)$ roots of the Wang secular determinant (24.5). The energy may most conveniently be calculated after the method of HAINER, KING and CROSS who write the energy as

$$(E(r) = J(J+1)\frac{(a+c)}{2} + \frac{(a-c)}{2} E(\varkappa) \tag{29.4}$$

where $a = h^2/8\pi^2 I_a$ and $c = h^2/8\pi^2 I_c$ and where $a > c$. $E(\varkappa)$ is one of the roots of the secular determinant appropriate for their problem and corresponds to the quantity W for the Wang description. The quantities $E(\varkappa)$ may be obtained from tables prepared by them for various degrees of asymmetry, the degree of asymmetry being defined by a parameter equal to $\frac{2b-a-c}{a-c}$ with $b = h^2/8\pi^2 I_b$, I_b being the intermediate moment of inertia. The root $E(\varkappa)$ may be related to W in the following manner

$$E(\varkappa) = J(J+1)\frac{b-c}{a-c} + W\frac{2c-a-b}{a-c}.$$

An axially symmetric molecule has two equal moments of inertia, i.e., $B_e^{(xx)} = B_e^{(yy)}$. It may, moreover, have two dimensionally isotropic oscillations, ω_t, which give rise to angular momenta of magnitude $l_t \zeta_t \hbar$, ζ_t being the modulus of the angular momentum vector, directed along the axis of symmetry. The quantum number l_t takes the values v_t, $v_t-2, \dots 1$ or 0 where, of course, v_t is the total vibration quantum number. The rotational energy of such a molecule is

$$E(r) = \{[J(J+1) - K^2]\, B_e^{(xx)} + L^2 B_e^{(zz)}\}\, h\, c \tag{29.5}$$

in which J and K are respectively the quantum number of total angular momentum and of the total angular momentum of the rotator along the axis of symmetry and where $L = K - \sum_t l_t \zeta_t$.

Molecules of higher symmetry have also three dimensionally isotropic oscillations. The rotational energy is here more complicated, but for a molecule in which there are only two three dimensionally isotropic oscillations it may be written

$$\left.\begin{aligned} E(r) = \{&J(J+1) + \zeta_3\zeta_4 J_2(J_2+1) + \\ &+ (\zeta_3 - \zeta_4)\,[\zeta_3 l_3(l_3+1) - \zeta_4 l_4(l_4+1)] + 2\varepsilon\}\, B_e\, h\, c, \end{aligned}\right\} \tag{29.6}$$

J being as before the quantum number of total angular momentum of the molecule. l_3 and l_4 are the quantum numbers of angular momentum associated with the two oscillations ω_3 and ω_4 and the ζ's are again the moduli of the angular momentum vectors. The quantum number J_2 takes the values l_3+l_4, l_3+l_4-1, $\dots l_3-l_4$. The ε's are the roots of the interaction term $-p^* P$ of which special cases only have been discussed.

30. Intensities in infrared bands. The intensity with which a band occurs in the infrared spectrum of a molecule depends on several factors, but is particularly related to the strength of the electric moment μ which has the components

$$\mu_\alpha = \mu_\alpha^{(0)} + \sum_{s\sigma} \left(\frac{\partial \mu_\alpha}{\partial q_{s\sigma}}\right)_0 q_{s\sigma} + \cdots \quad (\alpha = x, y, z) \tag{30.1}$$

directed along the body fixed axes x, y and z. The first term in μ_α, we recall from Sect. 25, is the permanent electric moment while the remaining terms are those induced by molecular vibration, the $q_{s\sigma}$ being the normal coordinates.

The theoretical intensity of absorption corresponding to the molecule executing a transition from a state τ'' to a state τ' has been studied by TOLMAN who has derived the following relation

$$(\tau''\,|I|\,\tau') = (8\pi^3 \nu\, g_{\tau'} N_{\tau''}/3\, h\, c\, g_{\tau''})\,(\tau'|\mu|\,\tau'')^2\,(1 - e^{-h\nu/kT}) \tag{30.2}$$

where $g_{\tau''}$ and $g_{\tau'}$ are, respectively, the statistical weights of the states τ'' and τ', $N_{\tau'}$ is the number of molecules per cm.^{3} in the state τ' and $(\tau'|\mu|\tau'')$ are the $(\tau'|\tau'')$ matrix elements of the electric moment determined quantum mechanically. When the squares of the matrix elements of μ are expressed in terms of the squares of the matrix elements of its components, then $(\tau'|\mu|\tau'')^2 =$

$$(\tau'|\mu_x|\tau'')^2 + (\tau'|\mu_y|\tau'')^2 + (\tau'|\mu_z|\tau'')^2.$$

The first term in (30.1), when different from zero, is diagonal in the quantum numbers v_s and gives rise to the rotation spectrum of the molecule. The second and higher order terms of (30.1) give rise to the near infrared bands in the spectrum of the molecule. Regarding the oscillational motion of the molecule in its several modes as harmonic it is readily shown that the non-vanishing matrix elements of the second terms in (30.1) are these

$$\left.\begin{aligned} (\tau'|\mu_\alpha|\,\tau'') &= (v_1, v_2, \dots v_s, \dots v_n\,|\mu_\alpha|\, v_1, v_2 \dots v_s - 1, \dots v_n) \\ &= \sqrt{\frac{v_s \hbar}{4\pi\omega_s}} \left(\frac{\partial \mu_\alpha}{\partial q_{s\sigma}}\right)_0. \end{aligned}\right\} \tag{30.3}$$

Eq. (30.3) is equivalent to stating that if the oscillational motion is harmonic $\Delta v_s = 1$ and only the fundamental absorption will occur in the spectrum. Actually, of course, the force fields are such that the oscillational motion is not harmonic. The relation (30.1) for μ will have terms of higher order than the first in $q_{s\sigma}$ and the wave functions will not be those of an harmonic oscillator so that one may also expect overtones and combination bands to occur.

When the initial state of the molecule is the normal vibration state (i.e., $v_s - 1 = 0$) the relation (30.2) applied to fundamental bands will become

$$(0, 0, \ldots, 1, \ldots 0 \,|I|\, 0, 0, \ldots 0, \ldots 0) = \sum_\alpha \left(\frac{\partial \mu_\alpha}{\partial q_{s\sigma}}\right)_0^2 \frac{N_0 \pi g_1}{3 c g_0} \left(1 - e^{-h c \omega_s / k T}\right). \quad (30.4)$$

Measurements on the absorption spectra of molecules are usually made at room temperatures. At these temperatures nearly all the molecules are in the normal state so that N_0 may be replaced by N the total number of molecules per cubic centimeter. Moreover, most of the vibration frequencies, ω_s, are greater than 500 cm^{-1} in which case the quantity $e^{-h c \omega_s / k T}$ is so small that it may be neglected. Eq. (30.4) then becomes simply

$$(0, 0, \ldots, 1, \ldots 0 \,|I|\, 0, 0, \ldots 0, \ldots 0) = \sum_\alpha \left(\frac{\partial \mu_\alpha}{\partial q_{s\sigma}}\right)_0^2 \frac{N \pi g_1}{3 c g_0}. \quad (30.5)$$

It is necessary to know the values of $(\partial \mu_\alpha / \partial q_{s\sigma})_0$ if one is to compare the actual intensities of absorption bands and while various assumptions have been made concerning these factors no really adequate method is known for calculating them theoretically.

Experimentally the intensity of an absorption band is arrived at by evaluating the integral

$$I = \int k(\nu)\, d\nu \quad (30.6)$$

where $k(\nu)$ is the absorption coefficient that occurs in LAMBERT'S law for the percent transmission

$$T = (I_\nu / I_{0\nu}) = e^{-k(\nu) l} \quad (30.7)$$

in which l is the length of the absorbing column. The integral (30.7) may be equated to the relation (30.5) and the coefficient $(\partial \mu_\alpha / \partial q_{s\sigma})_0$ evaluated if it is feasible to determine reliable values of $k(\nu)$ experimentally. The constant $(\partial \mu_\alpha / \partial q_{s\sigma})_0$ is the slope of the curve of the dipole moment along the coordinate $q_{s\sigma}$ when the molecule is in equilibrium. It is therefore, possible in principle to calculate the absolute change of dipole moment associated with a vibration since the amplitude of the vibration may readily be determined.

The transmission of radiation through a gaseous medium does not, unfortunately, often obey LAMBERT'S law. The most frequent failure of LAMBERT'S law stems from the difficulty of making reliable measurements on the absorption coefficient, $k(\nu)$. The difficulty which arises is the following. An absorption band is composed of a group of rotation lines which have a natural breadth of from 0.1 to 0.25 cm^{-1}. It would be necessary to explore the transmission across an absorption line with a spectrometer of slit widths which subtend frequency intervals in the spectrum which are small when compared to the line width if reliable values of the absorption coefficient, $k(\nu)$ are to be obtained. Such slit widths would have to be equivalent to something of the order of 0.01 cm^{-1} and until the present time it has been impracticable to operate with slit widths of this size. The difficulty encountered here may be circumvented by making the measurements on lines broadened by placing the gas under pressure or by using small amounts of gas so that $k(\nu) l$ is small. In the latter case the absorption becomes proportional to the path length and LAMBERT'S law may be said to be obeyed. Relatively few band spectra have been measured under either of these conditions and the matter of intensities of bands may be said to be in a rather unsatisfactory state.

It is of more immediate interest to consider the relative intensity distribution of rotation lines within a band in the spectrum of a molecule where it is possible to resolve the vibration bands into rotational structure. Here one must take account of the fact that the molecules are distributed among the rotational levels of a vibration state according to the law $N_{J''\tau''} = (N_0/Z)\, g_{J''\tau''} e^{-E_{J''\tau''}/kT}$ where $Z = \sum_{J''} g_{J''\tau''} e^{-E_{J''\tau''}/kT}$. Moreover, the electric moment must be expressed in terms of their projections along the space fixed axes, i.e., the matrix elements must be those of μ_A already discussed in Sect. 25 and 26. Making these observations one obtains from (30.2) that

$$\left.\begin{aligned}(v'', J'', \tau'' \,|\, I \,|\, v', J', \tau') = (8\pi^3 \omega_s N_0/3hcZ)(v' |\mu_\alpha| v'')^2 (1 - e^{-hc\omega_0/kT}) \times \\ \times (J' \tau' \,|\, A \,|\, J'' \tau'')^2 g_{J'\tau'} e^{-E_{J''\tau''}/kT}\end{aligned}\right\} \quad (30.8)$$

where $(J'\tau' | A | J''\tau'')$ represents the coefficient of μ_α in (25.12), (25.13) and (26.1) which depends entirely on the rotational quantum numbers. The frequency ω_s is in general much larger than the rotational frequency so that ω does not vary much percentage wise from one extremity of the bands to the other. It is therefore a good approximation to replace ν in (30.2), as we have done, simply by the vibration frequency ω_s.

We shall apply Eq. (30.8) to derive explicitly the intensity relations for the symmetric rotator. The statistical weight here is g_{JK} and may be shown to be equal to $(2J+1)$ where $K=0$ and $2(2J+1)$ where $K \neq 0$.

Consider first the instance where the electric moment lies along the axis of symmetry, i.e., where $\mu_z \neq 0$ but $\mu_x = \mu_y = 0$. The resulting band is usually referred to as a *parallel* band and the selection rules according to Sect. 25 are $\Delta J = \pm 1, 0$; $\Delta K = 0$. The squares of the quantum mechanical amplitudes are given in (25.13) for this case. Inserting these and the values of g_{JK} into (30.8) we have the three intensity relations

$$\left.\begin{aligned}(v_s', J, K \,|\, I \,|\, v_s'', J, K) &= \text{const} \cdot [(2J+1) K^2/J(J+1)]\, e^{-E(J,K)/kT}, \\ (v_s', J, K \,|\, I \,|\, v_s'', J-1, K) &= \text{const} \cdot [\gamma (J^2 - K^2)/2J]\, e^{-E(J,K)/kT}, \\ (v_s', J-1, K \,|\, I \,|\, v_s'', J, K) &= \text{const} \cdot [\gamma (J^2 - K^2)/2J]\, e^{-E(J-1,K)/kT}\end{aligned}\right\} \quad (30.9)$$

where $E(J, K) = [J(J+1) B_e^{(xx)} + K^2 (B_e^{(zz)} - B_e^{(xx)})]\, hc$ and where the constant $\gamma = 1$ when $K = 0$, but $\gamma = 2$ when $K \neq 0$.

We shall now turn to the case where $\mu_z = 0$, but where μ_x and μ_y are different from zero. The resulting band is here usually referred to as a *perpendicular* band and according to Sect. 25 the selection rules are $\Delta J = \pm 1, 0$; $\Delta K = \Delta l = \pm 1$. The squares of the quantum mechanical amplitudes are stated in (25.12). As before three intensity relations result when these and the g_{JK} are inserted into (30.8). These are the following

$$\left.\begin{aligned}&(v_s', J, K, l \,|\, I \,|\, v_s'', J-1, K-1, l-1) \\ &\quad = \text{const} \cdot [(J+K-1)(J+K)/4J]\, e^{-E(J,K)/kT}, \\ &(v_s', J, K, l \,|\, I \,|\, v_s'', J, K-1, l-1) \\ &\quad = \text{const} \cdot [(2J+1)(J-K+1)(J+K)/4J(J+1)]\, e^{-E(J,K)/kT}, \\ &(v_s', J-1, K, l \,|\, I \,|\, v_s'', J, K-1, l-1) \\ &\quad = \text{const} \cdot [(J-K+1)(J-K)/4J]\, e^{-E(J-1,K)/kT}.\end{aligned}\right\} \quad (30.10)$$

The general structure of parallel type bands and perpendicular bands will be discussed in some detail in Chap. VI which deals with the interpretation of infrared spectra of polyatomic molecules.

No simple relation like those stated in (30.9) and (30.10) can be derived for the intensities of lines in the bands of asymmetrical molecules. Tables of the squares of the quantum mechanical amplitudes for molecules of varying degrees of asymmetry, i.e., for values of varying from $\varkappa=0$ to $\varkappa=1$, have been prepared by HAINER, CROSS and KING and may conveniently be used together with (30.8).

V. The improved energies of a polyatomic molecule.

31. Introduction. We shall investigate the corrections to the energy of a polyatomic molecule as stated in Chap. IV, in this section, arising from terms in the Hamiltonian arising from interaction between vibration and rotation and from the anharmonic terms in the potential energy functions, i.e., from $H^{(1)}$ and $H^{(2)}$. We have indicated already in Chap. II that we shall use the techniques of the perturbation theory to approximate to the energy to second order. We shall, moreover, use the method of the contact transformation discussed in that chapter. We find this method advantageous here because it is evident that none of the terms in $H^{(1)}$ can contribute to the energy in first order, but will contribute to the energy only in second and higher orders. This means we may transform H in such a manner that the transformed first order Hamiltonian, $H^{(1)\prime}$, of Chap. II will reduce to zero. The second order transformed Hamiltonian, $H^{(2)\prime}$, will therefore become $H^{(2)\prime}=H^{(2)}+\frac{i}{2}[S H^{(1)}-H^{(1)} S]$. The second order energies will, therefore, be merely the diagonal elements of $H^{(2)\prime}$ and the problem is, formally, reduced to a first order calculation. The method of the contact transformation is convenient also because it permits much of the calculation to be carried out as symbolic matrix algebra, thereby facilitating the final addition of the contributions due to $H^{(2)\prime}$.

32. The contact transformation function. The transformation function $T(\lambda)$ was taken, in Chap. II, to be $T(\lambda)=e^{i\lambda S}$ (with its inverse equal to $T^{-1}(\lambda)=e^{-i\lambda S}$), λ being a parameter of smallness. Expanded in orders of magnitude, this will be

$$T=1+i\lambda S-\frac{\lambda^2}{2}S^2-\frac{i\lambda^3}{6}S^3+\cdots. \tag{32.1}$$

The function T, i.e., the operators S, are so determined that $H^{(1)\prime}=0$ which requires

$$H^{(1)}=i\,(H^{(0)}S-S\,H^{(0)}). \tag{32.2}$$

$H^{(1)}$ consists of a sum of terms each of which is a function of $q_{s\sigma}$ (or the conjugate momenta, $p_{s\sigma}$, or both) multiplied by a coefficient. These coefficients are usually constants, but in some terms they are functions of the operators, P_α. The S function will accordingly consist of a sum of terms, $S=\sum_\varrho S_\varrho$, each S_ϱ so chosen that it will remove a single term in $H^{(1)}$.

When the coefficient multiplying a term in $H^{(1)}$ is a constant an S_ϱ can be found which will exactly remove that term from $H^{(1)}$. Consider, for example, a term $a q_{s\sigma}^3$. It is quickly seen that choosing $S_s=-2(a/hc\omega_s)[(p_{s\sigma}^3/3\hbar^3)+q_{s\sigma}(p_{s\sigma}/\hbar)\,q_{s\sigma}/2]$ will give a result $i\,(H^{(0)}S_s-S_s H^{(0)})=a q_{s\sigma}^3$ since S_s commutes with the P_α's. On the other hand if the coefficient of the term in $H^{(1)}$ is a function of P_α the term can, generally, only be removed from $H^{(1)}$ to higher orders of approximation than the first. Consider, for example, a term $b q_{s\sigma} P_x^2$ where b

is of the order of magnitude $(B^{(xx)}/\omega_s (I_e^{(xx)})^2)^{\frac{1}{2}}$. Choosing for S_s the operator $-(b/hc\omega_s)P_x^2(p_{s\sigma}/\hbar)$ we obtain for $i(H^{(0)}S_s - S_s H^{(0)})$ the following, with the aid of the commutation relation (9.27),

$$\left.\begin{aligned} i(H^{(0)}S_s - S_s H^{(0)}) &= b\,q_{s\sigma}P_x^2 - \\ &- \frac{1}{\sqrt{2}}\left\{\left(\frac{2B^{\overline{(xx)}}}{\omega_s}\right)^{\frac{3}{2}}\left(\frac{2B^{(yy)}}{\omega_s}\right) hc\omega_s[P_y(P_xP_z + P_zP_x) + (P_xP_z + P_zP_x)P_y]/\hbar^3 - \right. \\ &\left. - \left(\frac{2B^{(xx)}}{\omega_s}\right)^{\frac{1}{2}}\left(\frac{2B^{(zz)}}{\omega_s}\right) hc\omega_s[P_x(P_zP_y + P_yP) + (P_zP_y + P_yP_z)P_x]/\hbar^3\right\}(p_{s\sigma}/\hbar). \end{aligned}\right\} \quad (32.3)$$

The second term in (32.3) may safely be neglected, however, if we confinc ourselves to corrections of second order since $(B^{(\alpha\alpha)}/\omega_s)$ is of the order of λ. Should one wish to extend calculations of this kind to higher orders of approximation, say the fourth, a refinement of S_s would have to be made which would remove this second term in (32.3). We shall, however, confine ourselves to corrections of second order magnitude and for our purpose the coefficients involving P_α may be treated as constants.

The transformation functions required for our purpose have been studied by SHAFFER, NIELSEN and THOMAS and again by HERMAN and SHAFFER. The functions required to remove the terms from $H^{(1)}$ are given in Table 10 together with the corresponding terms which they remove, i.e., $H^{(1)} = i(H^{(0)}S - SH^{(0)})$.

Table 10. *Transformation functions S required to remove types of terms, $H_\varrho^{(1)}$, occurring in $H^{(1)}$, together with the corresponding values of $i(H^{(0)}S_\varrho - S_\varrho H^{(0)}) = H^{(1)}$.*

ϱ	S_ϱ	$H_\varrho^{(1)} = i(H^{(0)}S_\varrho - S_\varrho H^{(0)})$
1	$-(a_1/hc\omega_s)(p_{s\sigma}/\hbar)$	$a_1 q_{s\sigma}$
2	$(a_2/2\pi c\omega_s)q_{s\sigma}$	$a_2 p_{s\sigma}$
3	$-(2a_3/hc\omega_s)[\frac{1}{3}(p_{s\sigma}/\hbar)^3 + \frac{1}{2}q_{s\sigma}(p_{s\sigma}/\hbar)q_{s\sigma}]$	$a_3 q_{s\sigma}^3$
4	$-[(a_4/2\pi c)/(\omega_s^2 - \omega_{s'}^2)][\omega_{s'}q_{s\sigma}q_{s'\sigma'} + \omega_s(p_{s'\sigma'}/\hbar)(p_{s\sigma}/\hbar)]$	$a_4 q_{s\sigma}p_{s'\sigma'}$
5	$-[(a_5/hc\omega_{s'})/(4\omega_s^2 - \omega_{s'}^2)][(2\omega_s^2 - \omega_{s'}^2)q_{s\sigma}^2(p_{s'\sigma'}/\hbar) +$ $+\omega_s\omega_{s'}(p_{s\sigma}q_{s\sigma} + q_{s\sigma}p_{s\sigma})q_{s'\sigma'}/\hbar + 2\omega_s^2 p_{s\sigma}^2 p_{s'\sigma'}/\hbar^3]$	$a_5 q_{s\sigma}^2 q_{s'\sigma'}$
6	$-[(a_6/2\pi c)[(\omega_s\omega_{s'})^{\frac{1}{2}}(\omega_s^2 - \omega_{s'}^2)]^{-1}[(\omega_s^2 + \omega_{s'}^2)q_{s\sigma}q_{s'\sigma'} +$ $+ 2\omega_s\omega_{s'}p_{s\sigma}p_{s'\sigma'}/\hbar^2]$	$a_6[(\omega_{s'}/\omega_s)^{\frac{1}{2}}q_{s\sigma}p_{s'\sigma'} -$ $-(\omega_s/\omega_{s'})^{\frac{1}{2}}q_{s'\sigma'}p_{s\sigma}]$
7	$-(a_7D_{ss's''}/hc)[\omega_s(\omega_s^2 - \omega_{s'}^2 - \omega_{s''}^2)(p_{s\sigma}/\hbar)q_{s'\sigma'}q_{s''\sigma''} +$ $+\omega_{s'}(\omega_{s'}^2 - \omega_s^2 - \omega_{s''}^2)q_{s\sigma}(p_{s'\sigma'}/\hbar)q_{s''\sigma''} + \omega_{s''}(\omega_{s''}^2 - \omega_s^2 - \omega_{s'}^2)\times$ $\times q_{s\sigma}q_{s'\sigma'}(p_{s''\sigma''}/\hbar) - 2\omega_s\omega_{s'}\omega_{s''}(p_{s\sigma}p_{s'\sigma'}p_{s''\sigma''}/\hbar^3)]$ $D_{ss's''} = [(\omega_s + \omega_{s'} + \omega_{s''})(\omega_s - \omega_{s'} - \omega_{s''})\times$ $\times(\omega_s - \omega_{s'} + \omega_{s''})(\omega_s + \omega_{s'} - \omega_{s''})]^{-1}$	$a_7 q_{s\sigma}q_{s'o'}q_{s''\sigma'}$

It will be seen from Table 10 that certain of the basic S functions contain denominators of the kind $(\omega_s - \omega_{s'})$ and $(4\omega_s - \omega_{s'}^2)$. The first kind occur in the removal from $H^{(1)}$ of CORIOLIS interaction terms of the kind $p_x P_\alpha$. It may happen that two such frequencies ω_s and $\omega_{s'}$ or $2\omega_s$ and $\omega_{s'}$ are accidentally degenerate (i.e., $\omega_s \approx \omega_{s'}$ or $2\omega_s \approx \omega_{s'}$). In such instances $\omega_s - \omega_{s'}$ and $2\omega_s - \omega_{s'}$ approach zero respectively and resonance between the frequencies is said to occur. When resonance between two such frequencies exists their contributions to the energy become of first order significance and the methods of perturbation theory as here employed fail. The transformation method may still be employed, but other transformation functions, S_ϱ, than those given in Table 10 must be used in which the resonance denominators are absent. $H^{(1)}$, in addition to contributing

Tabelle 11. *Tabulation of the individual terms $H^{(2)\prime}_{\varrho\tau}$ transformed into $H^{(2)\prime}$ together with their matrix elements diagonal in the vibration quantum numbers.*

ϱ	τ	$H^{(1)}_\varrho$	$H^{(2)\prime}_{\varrho,\tau} = (i/2)\,[S_\varrho H^{(1)}_\tau - H^{(1)}_\tau S_\varrho]$	$(v\,\vert H^{(2)}_{\varrho,\tau}\vert\, v)$
1	1	$a_1 q_s$	$-(a_1^2/2hc\omega_s)$	$-(a_1^2/2hc\omega_s)$
1	1	$a_1' q_s$	$-(a_1 a_1' + a_1' a_1)/4hc\omega_s$	$-(a_1 a_1' + a_1' a_1)/4hc\omega_s$
1	3	$a_3 q_s^3$	$-(3a_1 a_3 q_s^2/2hc\omega_s)$	$-(3a_1 a_3/2hc\omega_s)(v_s+\tfrac{1}{2})$
1	5	$a_5' q_s q_{s'}^2$	$-(a_1 a_5' q_{s'}^2/2hc\omega_s)$	$-(a_1 a_5'/2hc\omega_s)(v_{s'}+\tfrac{1}{2})$
2	2	$a_2(p_s/\hbar)$	$-(a_2^2/4\pi c\omega_s)$	$-(a_2^2/4\pi c\omega_s)$
3	1	$a_1 q_s$	$-(3a_1 a_3 q_s^2/2hc\omega_s)$	$-(3a_1 a_3/2hc\omega_s)(v_s+\tfrac{1}{2})$
3	3	$a_3 q_s^3$	$-(a_3^2/hc\omega_s)[\tfrac{3}{2}(p_s^2 q_s^2 + q_s^2 p_s^2)/\hbar^2 +$ $+ 1 + \tfrac{3}{2} q_s^4]$	$-(a_3^2/hc\omega_s)[(\tfrac{15}{4})(v_s+\tfrac{1}{2})^2 +$ $+ \tfrac{7}{16}]$
3	5	$a_5' q_s q_{s'}^2$	$-(3a_3 a_5' q_s^2 q_{s'}^2/2hc\omega_s)$	$-(3a_3 a_5'/2hc\omega_s)(v_s+\tfrac{1}{2})\times$ $\times(v_{s'}+\tfrac{1}{2})$
4	4	$a_4 q_s p_{s'}/\hbar$	$(a_4^2/4\pi c)(\omega_s^2-\omega_{s'}^2)^{-1}[\omega_{s'} q_s^2 -$ $-\omega_s p_{s'}^2/\hbar^2)]$	$(a_4^2/4\pi c)(\omega_s^2-\omega_{s'}^2)^{-1}\times$ $\times[\omega_{s'}(v_s+\tfrac{1}{2})-\omega_s(v_{s'}+\tfrac{1}{2})]$
5	1	$a_1' q_{s'}$	$-(a_1' a_5 q_s^2/2hc\omega_{s'})$	$-(a_1' a_5/2hc\omega_{s'})(v_s+\tfrac{1}{2})$
5	3	$a_3' q_{s'}^3$	$-[(a_3' a_5/2hc\omega_{s'})(4\omega_{s'}^2-\omega_{s'}^2)^{-1}]\times$ $\times[(2\omega_s^2-\omega_{s'}^2)q_s^2+2\omega_s^2(p_s/\hbar)^2]q_{s'}^2$	$-(3a_3' a_5/2hc\omega_{s'})(v_s+\tfrac{1}{2})\times$ $\times(v_{s'}+\tfrac{1}{2})$
5	5	$a_5 q_s^2 q_{s'}$	$-(a_5^2/2hc\omega_{s'})(4\omega_s^2-\omega_{s'}^2)^{-1}\times$ $\times[(2\omega_s^2-\omega_{s'}^2)q_s^4 + 4\omega_s\omega_{s'}\times$ $\times q_s^2 q_{s'}^2 + (\omega_s^2/\hbar^2)(p_s^2 q_s^2 + q_s^2 p_s^2)]$	$-(a_5^2/4hc\omega_{s'})E_{ss'}(v_s+\tfrac{1}{2})^2+$ $+(3a_5^2/16hc)H_{ss'}-(a_5^2/hc)\times$ $\times F_{ss'}(v_s+\tfrac{1}{2})(v_{s'}+\tfrac{1}{2})$
5	5	$a_5'' q_{s''}^2 q_{s'}$	$-(a_5 a_5'' q_{s''}^2 q_s^2/2hc\omega_{s'})$	$-(a_5 a_5''/2hc\omega_{s'})(v_{s''}+\tfrac{1}{2})\times$ $\times(v_s+\tfrac{1}{2})$
6	4	$a_4 q_s p_{s'}/\hbar$	$(a_4 a_6\hbar/2hc)(\omega_s-\omega_{s'})^{-1}[q_s^2 -$ $-p_{s'}^2/\hbar^2]$	$(a_4 a_6\hbar/2hc)(\omega_s-\omega_{s'})^{-1}\times$ $\times(v_s-v_{s'})$
6	6	$a_6(q_s p_{s'} - q_{s'} p_s)/\hbar$	$(a_6^2\hbar/2hc)(\omega_s-\omega_{s'})^{-1}[(q_s^2 +$ $+ p_s^2/\hbar^2) - (q_{s'}^2 + p_{s'}^2/\hbar^2)]$	$(a_6^2\hbar/hc)(\omega_s-\omega_{s'})^{-1}(v_s-v_{s'})$
7	7	$a_7[(\omega_{s'}/\omega_s)^{\frac{1}{2}} q_s p_{s'} -$ $-(\omega_s/\omega_{s'})^{\frac{1}{2}} q_{s'} p_s]/\hbar$	$a_7^2\hbar[2hc(\omega_s^2-\omega_{s'}^2)]^{-1}[(3\omega_s^2 +$ $+\omega_{s'}^2)(q_s^2/\omega_s) - (3\omega_{s'}^2+\omega_s^2)\times$ $\times(q_{s'}^2/\omega_{s'}^2)]$	$(a_7^2\hbar/2hc)[\omega_s^{-1} J_{ss'}(v_s+\tfrac{1}{2})+$ $+\omega_{s'}^{-1} J_{s's}(v_{s'}+\tfrac{1}{2})]$
8	8	$a_8 q_s q_{s'} q_{s''}$	$-(a_8^2 D_{ss's''}/2hc)[\omega_s(\omega_s^2-\omega_{s'}^2 -$ $-\omega_{s''}^2)q_{s'}^2 q_{s''}^2 + \omega_{s'}(\omega_{s'}^2-\omega_{s''}^2 -$ $-\omega_s^2)q_s^2 q_{s''}^2 + \omega_{s''}(\omega_{s''}^2-\omega_s^2 -$ $-\omega_{s'}^2)q_s^2 q_{s'}^2 + \omega_s\omega_{s'}\omega_{s''}/2]$	$-(a_8^2/hc)[C_{ss's''}(v_{s'}+\tfrac{1}{2})\times$ $\times(v_{s''}+\tfrac{1}{2}) + C_{s's''s}\times$ $\times(v_s+\tfrac{1}{2})(v_{s''}+\tfrac{1}{2}) +$ $+C_{s''ss'}(v_s+\tfrac{1}{2})(v_{s'}+\tfrac{1}{2})+$ $+(\omega_s\omega_{s'}\omega_{s''}D_{ss's''}/4)]$
			$C_{ss's''} = \omega_s(\omega_s^2-\omega_{s'}^2-\omega_{s''}^2)\times$ $\times D_{ss's''}/2$	$F_{ss'} = 2\omega_s(4\omega_s^2-\omega_{s'}^2)^{-1}$
			$D_{ss's''} = [(\omega_s+\omega_{s'}+\omega_{s''})(\omega_s -$ $-\omega_{s'}-\omega_{s''})(\omega_s-\omega_{s'} +$ $+\omega_{s''})(\omega_s+\omega_{s'}-\omega_{s''})]^{-1}$	$H_{ss'} = \omega_{s'}(4\omega_s^2-\omega_{s'}^2)^{-1}$
			$E_{ss'} = (8\omega_s^2-3\omega_{s'}^2)(4\omega_s^2-\omega_{s'}^2)^{-1}$	$J_{ss'} = (3\omega_s^2+\omega_{s'}^2)(\omega_s^2-\omega_{s'}^2)^{-1}$

to $H^{(2)\prime}$ by the transformation will also yield a first order transformed Hamiltonian, $H^{(1)\prime}$, which is not zero. Cases of this kind will be treated in Chap. VII.

33. The transformed Hamiltonian. The terms in $H^{(1)}$ which by T are transformed to become a part of $H^{(2)\prime}$ are, according to Chap. II, equal to $H^{(2)}_{\varrho,\tau} = (i/2)\,[S_\varrho H^{(1)}_\tau - H^{(1)}_\tau S_\varrho]$ where τ is an index indicating a type of term in $H^{(1)}$ which the function S_ϱ is operating upon. It should be noted that the S_ϱ required to remove a term $H^{(1)}_\varrho$ from $H^{(1)}$ will operate, in general, on all other terms in $H^{(1)}$ and transform them into new terms which are a part of $H^{(2)\prime}$. Table 11 gives the effect of the transformation S_ϱ operating on components of $H^{(1)}$, $H^{(1)}_\tau$, the resulting terms being identified as $H^{(2)\prime}$. The last column of Table 11 gives the diagonal matrix elements of the terms $H^{(2)}$†. Only those terms which are transformed from $H^{(1)}$ into $H^{(2)}$ are given which contribute to second order of approximation.

When the S function has been applied as indicated earlier to transform the Hamiltonian for the molecule one obtains the following expression for H' which is set down in orders of magnitude

$$H^{(0)\prime} = H^{(0)}, \tag{33.1}$$

$$H^{(1)\prime} = 0, \tag{33.2}$$

$$\begin{aligned}
H^{(2)\prime} = {} & \frac{1}{2}\sum_{\alpha}\frac{p_\alpha^2}{I^{(e)}_{\alpha\alpha}} - \sum_{s\sigma}\frac{\lambda_s}{8}\left\{\sum_{\alpha\beta}\frac{a^{(\alpha\beta)}_{s\sigma}}{I^{(e)}_{\alpha\alpha} I^{(e)}_{\beta\beta}}\,(P_\alpha - p^*_\alpha)\,(P_\beta - p^*_\beta)\right\}^2 - \\
& - \sum_{\alpha\beta}\frac{\hbar^{\frac{1}{2}}}{2 I^{(e)}_{\alpha\alpha}}\sum_{s\sigma}\left\{\sum_{s'\sigma'}\frac{\hbar^{\frac{1}{2}}}{\lambda_s^{\frac{1}{4}}\lambda_{s'}^{\frac{1}{4}} I^{(e)}_{\beta\beta}}\left[A'^{(\alpha\beta)}_{s\sigma s'\sigma'} - \sum_{\gamma}\frac{a^{(\alpha\gamma)}_{s\sigma}\, a^{(\gamma\beta)}_{s'\sigma'}}{I^{(e)}_{\gamma\gamma}} - \sum_{s''\sigma''}\frac{\zeta^{(\alpha)}_{s\sigma s''\sigma''}\zeta^{(\beta)}_{s'\sigma' s''\sigma''}(\lambda_s + \lambda_{s''})}{\lambda_s - \lambda_{s''}}\right] q_{s'\sigma'} - \right. \\
& \left. - 2\pi c\left[2 k_{sss}\frac{a^{(\alpha\beta)}_{s\sigma}}{I^{(e)}_{\beta\beta}\lambda_s^{\frac{3}{4}}} + \sum_{s'\sigma'}{}' k_{sss'}\frac{a^{(\alpha\beta)}_{s'\sigma'}\,(3\lambda_s - \lambda_{s'})}{I^{(e)}_{\beta\beta}\lambda_{s'}^{\frac{3}{4}}\,(4\lambda_s - \lambda_{s'})}\right] q_{s\sigma}\right\} q_{s\sigma}(P_\alpha - p^*_\alpha)(P_\beta - p^*_\beta) + \\
& + \sum_{\alpha\beta}\frac{\hbar^{\frac{1}{2}}}{2 I^{(e)}_{\alpha\alpha}}\sum_{s\sigma}\left\{\sum_{s'\sigma'}\sum_{s''\sigma''}\frac{2\hbar^{\frac{1}{2}}}{\lambda_s^{\frac{1}{4}}\lambda_{s'}^{\frac{1}{4}} I^{(e)}_{\beta\beta}}\,\frac{\zeta^{(\alpha)}_{s\sigma s''\sigma''}\zeta^{(\beta)}_{s'\sigma' s''\sigma''}\lambda_s^{\frac{1}{2}}\lambda_{s'}^{\frac{1}{2}}}{\lambda_s - \lambda_{s''}}\,\frac{p_{s'\sigma'}}{\hbar} + \right. \\
& \left. + 2\pi c\left[k_{sss}\frac{a^{(\alpha\beta)}_{s\sigma}}{I^{(e)}_{\beta\beta}\lambda_s^{\frac{3}{4}}} + \sum_{s'\sigma'}{}' k_{sss'}\frac{a^{(\alpha\beta)}_{s'\sigma'}\lambda_s}{I^{(e)}_{\beta\beta}\lambda_{s'}^{\frac{3}{4}}\,(4\lambda_s - \lambda_{s'})}\right]\frac{p_{s\sigma}}{\hbar}\right\}\frac{p_{s\sigma}}{\hbar}\,(P_\alpha - p^*_\alpha)\,(P_\beta - p^*_\beta) + \\
& + hc\sum_{s\sigma}\sum_{s'\sigma'}\sum_{s''\sigma''}\sum_{s'''\sigma'''} k_{ss's''s'''}\, q_{s\sigma} q_{s'\sigma'} q_{s''\sigma''} q_{s'''\sigma'''} - h^2 c^2\left\{\sum_{s\sigma}\frac{1}{\hbar\lambda_s^{\frac{1}{2}}} k^2_{sss}\times\right. \\
& \times\left[\frac{3}{2\hbar^2}\,(p^2_{s\sigma} q^2_{s\sigma} + q^2_{s\sigma} p^2_{s\sigma}) + 1 + \frac{3}{2}\, q^4_{s\sigma}\right] + \frac{1}{2}\sum_{s\sigma}\sum_{s'\sigma'}{}'\frac{1}{\hbar\lambda_{s'}^{\frac{1}{2}}\,(4\lambda_{s'} - \lambda_{s'})} k^2_{sss'}\times \\
& \times\left[(2\lambda_s - \lambda_{s'})\, q^4_{s\sigma} + 4\lambda_s^{\frac{1}{2}}\lambda_{s'}^{\frac{1}{2}} q^2_{s\sigma} q^2_{s'\sigma'} + 4\frac{\lambda_s}{\hbar^2}\,(p_{s\sigma} q_{s\sigma} + q_{s\sigma} p_{s\sigma})\, p_{s'\sigma'} q_{s'\sigma'} + 2\frac{\lambda_s}{\hbar^2} q^2_{s\sigma} p^2_{s\sigma}\right] + \\
& + \sum_{s\sigma}\sum_{s'\sigma'}{}'\frac{1}{\hbar\lambda_s^{\frac{1}{2}}\,(4\lambda_s - \lambda_{s'})} k_{sss} k_{ss's'}\left[(5\lambda_{s'} - 2\lambda_s)\, q^2_{s\sigma} q^2_{s'\sigma'} + (4\lambda_{s'} - \lambda_s)\,\frac{p^2_{s\sigma}}{\hbar^2}\times\right. \\
& \left.\times q^2_{s\sigma} + 3\lambda_{s'} q^2_{s\sigma}\frac{p^2_{s'\sigma'}}{\hbar^2}\right] + \sum_{s\sigma}\sum_{s'\sigma'}{}'\sum_{s''\sigma''}{}''\frac{1}{2\hbar\lambda_{s'}^{\frac{1}{2}}\,(4\lambda_s - \lambda_{s'})} k_{sss'} k_{s's''s''}\times \\
& \times\left[(2\lambda_s - \lambda_{s'})\, q^2_{s\sigma} + \frac{2\lambda_s}{\hbar^2} p^2_{s\sigma}\right] q^2_{s''\sigma''} + \sum_{s\sigma}\sum_{s'\sigma'}{}'\sum_{s''\sigma''}{}''\frac{1}{2\hbar N} k^2_{ss's''}\left[2\frac{\lambda_s^{\frac{1}{2}}\lambda_{s'}^{\frac{1}{2}}\lambda_{s''}^{\frac{1}{2}}}{\hbar^2}\times\right. \\
& \times(p_{s\sigma} q_{s\sigma} p_{s'\sigma'} q_{s'\sigma'} + q_{s'\sigma'} p_{s'\sigma'} q_{s''\sigma''} p_{s''\sigma''} + q_{s''\sigma''} p_{s''\sigma''} p_{s\sigma} q_{s\sigma}) - \lambda_s^{\frac{1}{2}}(\lambda_s - \lambda_{s'} - \lambda_{s''})\times \\
& \left.\times q^2_{s'\sigma'} q^2_{s''\sigma''} - \lambda_{s'}^{\frac{1}{2}}(\lambda_{s'} - \lambda_{s''} - \lambda_s)\, q^2_{s''\sigma''} q^2_{s\sigma} - \lambda_{s''}^{\frac{1}{2}}(\lambda_{s''} - \lambda_s - \lambda_{s'})\, q^2_{s\sigma} q^2_{s'\sigma'}\right] + H^*_V + H^*_R.
\end{aligned} \tag{33.3}$$

† The values given are for a non-degenerate vibration or for a single component of a degenerate vibration. σ has, therefore, been omitted consistently in Table 11.

The denominator N in Eq. (33.3) is the following quantity

$$(\lambda_s^{\frac{1}{2}} + \lambda_{s'}^{\frac{1}{2}} + \lambda_{s''}^{\frac{1}{2}})\,(\lambda_s^{\frac{1}{2}} + \lambda_{s'}^{\frac{1}{2}} - \lambda_{s''}^{\frac{1}{2}})\,(\lambda_s^{\frac{1}{2}} - \lambda_{s'}^{\frac{1}{2}} + \lambda_{s''}^{\frac{1}{2}})\,(\lambda_s^{\frac{1}{2}} - \lambda_{s'}^{\frac{1}{2}} - \lambda_{s''}^{\frac{1}{2}}).$$

The term H_V^* contains potential energy members of the variety $q_{s\sigma}^3 q_{s'\sigma'}$ and $q_{s\sigma}\, q_{s'\sigma'}\, q_{s''\sigma''}^2$ multiplied by constant coefficients of the order of magnitude, $(h c k_{s s' s''}^2/\omega_s)$. These terms will have no elements diagonal in all the vibrational quantum numbers and will, therefore, not contribute to the energy in second order of approximation unless certain frequencies accidentally become degenerate. Similarly the quantity, H_R^*, which is a correction to the rotational energy, cannot contribute to the energy in second order unless certain frequencies accidentally become degenerate. The two quantities H_V^* and H_R^* are given by the following equations.

$$\left.\begin{aligned}
H_V^* = -(\pi h c^2) \sum_{s\sigma} \sum_{s'\sigma'} \Bigg\{ & \frac{2}{\lambda_s^{\frac{1}{2}}} k_{sss} k_{sss'} \left[\left(\frac{p_{s\sigma}^2}{\hbar^2} q_{s\sigma} + q_{s\sigma} \frac{p_{s\sigma}^2}{\hbar^2}\right) + q_{s\sigma}^3\right] q_{s'\sigma'} + \\
& + \frac{6}{\lambda_{s'}^{\frac{1}{2}} (4\lambda_s - \lambda_{s'})} k_{sss} k_{sss'} \left[(\lambda_s \lambda_{s'})^{\frac{1}{2}} q_{s\sigma}^3 q_{s'\sigma'} + \lambda_s \left(\frac{p_{s\sigma}}{\hbar} q_{s\sigma}^2 + q_{s\sigma}^2 \frac{p_{s\sigma}}{\hbar}\right)\right] \frac{p_{s'\sigma'}}{\hbar} + \\
& + \frac{4}{\lambda_{s'}^{\frac{1}{2}} (4\lambda_s - \lambda_{s'})} k_{sss'}^2 \lambda_s \frac{p_{s'\sigma'}}{\hbar} q_{s'\sigma'} \left(\frac{p_{s\sigma}}{\hbar} q_{s\sigma} + q_{s\sigma} \frac{p_{s\sigma}}{\hbar}\right) + \\
& + \frac{2}{\lambda_{s'}^{\frac{1}{2}} (4\lambda_s - \lambda_{s'})} k_{sss'} k_{s's's} \Big[(2\lambda_s - \lambda_{s'})\, q_{s\sigma}^3 q_{s'\sigma'} + (\lambda_s \lambda_{s'})^{\frac{1}{2}} q_{s'\sigma'}^3 q_{s\sigma} + \\
& \qquad + 2\lambda_s \left(q_{s'\sigma'} \frac{p_{s\sigma}^2}{\hbar^2} q_{s\sigma} + q_{s'\sigma'}^2 \frac{p_{s'\sigma'}}{\hbar} \frac{p_{s\sigma}}{\hbar}\right)\Big] + \\
& + \sum_{s''\sigma''}{}' \frac{1}{\lambda_{s''}^{\frac{1}{2}} (4\lambda_s - \lambda_{s''})} k_{sss''} k_{ss's''} \Big[(2\lambda_s - \lambda_{s''})\, q_{s\sigma}^3 q_{s'\sigma'} + \\
& \qquad + \lambda_s \left(\frac{p_{s\sigma}^2}{\hbar^2} q_{s\sigma} + q_{s\sigma} \frac{p_{s\sigma}^2}{\hbar^2}\right) q_{s'\sigma'}\Big] + \\
& + \sum_{s''\sigma''}{}' \frac{\lambda_s}{\lambda_{s''}^{\frac{1}{2}} N} k_{sss''} k_{ss's''} \Big[(\lambda_{s''} - \lambda_s - \lambda_{s'})\, q_{s\sigma}^3 q_{s'\sigma'} + \\
& \qquad + (\lambda_s \lambda_{s''})^{\frac{1}{2}} \left(\frac{p_{s\sigma}}{\hbar} q_{s\sigma}^2 + q_{s\sigma}^2 \frac{p_{s\sigma}}{\hbar}\right) \frac{p_{s'\sigma'}}{\hbar}\Big]\Bigg\},
\end{aligned}\right\} \quad (33.4\text{a})$$

where the prime attached to $\sum\limits_{s''\sigma''}{}'$ indicates that $s''\sigma'' \neq s, \sigma$ and s', σ'. Moreover certain terms in H_V^* which involve $p_{s\sigma}$ and/or $q_{s\sigma}$ to the fourth power have been omitted because they would contribute only to the energies of levels involving three or four vibration frequencies.

$$\left.\begin{aligned}
H_R^* = \sum_{\alpha\beta} \frac{\hbar^{\frac{1}{2}}}{2 I_{\alpha\alpha}^{(e)}} \sum_{s\sigma} \sum_{s'\sigma'} 2\pi c \Bigg\{ & k_{sss'} \frac{a_{\sigma s}^{(\alpha\beta)}}{I_{\beta\beta}^{(e)} \lambda_s^{\frac{3}{4}} (4\lambda_s - \lambda_{s'})} \Big[(5\lambda_s - \lambda_{s'})\, q_{s\sigma} q_{s'\sigma'} + \\
& + \frac{2\lambda_s^{\frac{3}{2}}}{\lambda_{s'}^{\frac{1}{2}}} \frac{p_{s\sigma}}{\hbar} \frac{p_{s'\sigma'}}{\hbar}\Big] + \sum_{s''\sigma''} k_{ss's''} \frac{a_{s\sigma}^{(\alpha\beta)}}{2 I_{\beta\beta}^{(e)} \lambda_s^{\frac{3}{4}}} \Big[\left(1 + \frac{\lambda_s(\lambda_s - \lambda_{s'} - \lambda_{s''})}{N}\right) q_{s'\sigma'} q_{s''\sigma''} - \\
& - \frac{2\lambda_s^{\frac{1}{2}} (\lambda_s \lambda_{s'} \lambda_{s''})^{\frac{1}{2}}}{N} \frac{p_{s'\sigma'}}{\hbar} \frac{p_{s''\sigma''}}{\hbar}\Big]\Bigg\} (P_\alpha - p_\alpha^*)(P_\beta - p_\beta^*).
\end{aligned}\right\} \quad (33.4\text{b})$$

H_R^* contains also certain terms which have as coefficients P_α and $P_\alpha P_\beta P_\gamma$. These are regarded as unimportant and are omitted.

The Hamiltonian for the vibration-rotation energies of a polyatomic molecule as stated in relations (33.1), (33.2) and (33.3) is very convenient for successive approximations to the eigenvalues.

34. Second order corrections to the energy. The development of the Hamiltonian has been entirely general and it is, therefore possible, in principle, to evaluate the higher order contributions to the energy for any molecular model. These higher order corrections to the energies of the molecule which originate with the transformed second order Hamiltonian are obtained by carrying out a first order perturbation calculation using $H^{(2)'}$ as the perturbing potential. The wave functions (19.7) and (28.7) to (28.12) which are the eigenfunctions of $H^{(0)}$ are used in making these calculations.

The eigenfunctions (28.7) to (28.12) which pertain to molecules in which occur threefold degenerate oscillations are, as we have seen, in Chap. III, very special linear combinations of the functions (19.8) where the coefficients depend in a complicated manner upon J and K. The actual evaluation of the corrections to the energies due to $H^{(2)'}$ becomes, in such cases, a most laborious task. The matrix of $H^{(2)'}$, in such cases, could also be set up in terms of the basic wave functions (19.8) rather than in terms of the eigenfunctions (28.7) to (28.12) in which case elements non-diagonal in both m_u and K would have to be evaluated because of the degeneracy in m_u and K. The actual diagonalization of the resulting matrix would be tantamount to the calculation using the functions (28.7), etc. and would again be a very tedious and laborious task. We shall not discuss such molecules further here because they do not lend themselves to a convenient formulation of their energies, but shall rather regard them as molecules which, together with certain other molecules such as those in which inversion, internal rotation or other anomalies occur are best treated individually. These might best be treated as special cases.

The terms in the transformed second order Hamiltonian are essentially of two kinds, one kind contributing only to the vibration energy and the other kind contributing essentially what are refinements to the rotation energy. The first set of terms originate with the anharmonic portion of the potential energy and are quantities in $p_{s\sigma}$ and $q_{s\sigma}$ only. They will contribute, to this approximation, only to the elements of H which are diagonal in the vibrational and the rotational quantum numbers. The second kind consists of terms which are quartic in the operators P_α, which represent the centrifugal distortion of the molecule, and of terms quadratic in P_α, multiplied by coefficients depending quadratically on $p_{s\sigma}$ and $q_{s\sigma}$. The coefficients of this latter set of terms may be regarded as refinements to the moments of inertia.

The wave functions $R^\times(J, K, M)$ defined in (23.11) are, we know, not eigenfunctions of the asymmetric rotator. $H^{(2)'}$ will, therefore, contribute also to the elements of H which lie off of the principal diagonal. The elements will, however, be diagonal in all quantum numbers except the quantum number K. The non-vanishing elements of H are the following $(K|K)$, $(K|K\pm 1)$, $(K|K\pm 2)$, $(K|K\pm 3)$ and $(K|K\pm 4)$. The elements $(K|K\pm 1)$ and $(K|K\pm 3)$ arise only in molecules where complete absence of symmetry exists. The coefficients are always small, being in both instances of the order of the centricugal distortion, i.e., of the magnitude $B^{(e)}(B^{(e)}/\omega_s)$ where $B^{(e)}$ equals $h/8\pi^2 I^{(e)} c$ and ω_s is a vibration frequency. They are almost certainly of small importance in nearly all cases and can usually be neglected. They will, therefore, not be given here.

The actual evaluation of the matrix components of $H^{(2)'}$ is tedious and hinges upon deducing the matrix elements of a number of functions of $q_{s\sigma}$ and $p_{s\sigma}$. This may be accomplished by methods suggested in Sect. 20 and 21 or by reference, in many cases, to Table 11. It is important, where one is dealing with two dimensionally isotropic oscillations, to replace the two coordinates, $q_{t,1}$ and $q_{t,2}$, and their conjugate momenta, $p_{t,1}$ and $p_{t,2}$, by their equivalents in terms of r_t and χ_t as indicated in Sect. 20.

Before collating the results stated in Sect. 29 with the corrections to the energy arising out of $H^{(2)\prime}$ it is important also to investigate the nature of the contribution from $\overline{U}$ with which it was necessary to replace U in Chap. III in order to effect a separation of variables in the Schrödinger equation. We saw that this term consisted of two types of terms, namely; $\varepsilon_\alpha(q_{s\sigma})(P_\alpha - p_\alpha)/I_{\alpha\alpha}^{(e)}$ and $\delta_\alpha(q_{s\sigma})/I_{\alpha\alpha}^{(e)}$ in which ε_α and δ_α are coefficients which in HUND's case (b) may be regarded as functions of the positional coordinates only, but in case (a) would contain also a component due to the spins of the electrons. The quantity $\overline{U}$ is, in general, small because of the fact that Φ is so nearly independent of the $q_{s\sigma}$, and with the exceptions of the third and fourth terms we shall neglect it. The third term in $\overline{U}$ may, however, give a contribution which is significant when the electrons are in states where their angular momenta are different from zero. The fourth term may also contribute significantly to the energies in the instance of linear molecules and may thereby significantly influence the positions of the vibrational states. This term will be considered separately at a later point.

We may write, for our present purpose, the term $\overline{U}$ simply as

$$\overline{U} = + \frac{1}{2\Phi R} \sum_\alpha \mu_{\alpha\alpha} \left\{\overline{M_\alpha^2} - 2\,\overline{M_\alpha (P_\alpha - p_\alpha^*)}\right\} \Phi R. \tag{34.1}$$

The quantity $\overline{M_\alpha^2}$ may be written $\overline{M_\alpha^2} = \overline{\pi}_\alpha^2 + 2\overline{\pi}_\alpha \overline{\mathfrak{S}}_\alpha + \overline{\mathfrak{S}}_\alpha^2$. The cross terms, $2\overline{\pi}_\alpha \overline{\mathfrak{S}}_\alpha$ will vanish when α is equal to x or y, giving the result $\overline{\pi}_\alpha^2 + \overline{\mathfrak{S}}_\alpha^2$ in these instances. When $\alpha = z$, however, the cross product $2\overline{\pi}_z \overline{\mathfrak{S}}_z$ may not necessarily vanish since $\overline{\pi}_z$, while not quantized, will not necessarily be equal to zero and $\mathfrak{S}_z$ will also not, in general, be non-vanishing.

It may be established with the aid of the spin matrices given in Chap. III that $\mathfrak{S}_z$ has only the diagonal elements equal to $\hbar\Sigma$ where Σ takes the values $S, S-1, \cdots -S+1, -S$; S being the quantum number of total spin angular momentum. We write for these

$$(S, \Sigma | \mathfrak{S}_z | S, \Sigma) = \hbar \Sigma. \tag{34.2}$$

The operators $\mathfrak{S}_x$ and $\mathfrak{S}_y$, on the other hand, while still diagonal in S, are non-diagonal in Σ. One may quickly verify that the non-vanishing matrix elements are these

$$(S, \Sigma | \mathfrak{S}_x | S, \Sigma \pm 1) = \mp i (S, \Sigma | \mathfrak{S}_y | S, \Sigma \pm 1) = (\hbar/2)\,[S(S+1) - \Sigma(\Sigma \pm 1)]^{\frac{1}{2}}. \tag{34.3}$$

$S^2 = \sum_\alpha \mathfrak{S}_\alpha^2$ will evidently be a diagnonal matrix. $\mathfrak{S}_z$ and, therefore $\mathfrak{S}_z^2$, are also diagonal matrices. Hence $\mathfrak{S}_x^2 + \mathfrak{S}_y^2 = S^2 - \mathfrak{S}_z^2$ will also be diagonal. We have, therefore, that

$$\left.\begin{aligned} (S, \Sigma | \mathfrak{S}_z^2 | S, \Sigma) &= \hbar^2 \Sigma^2, \\ (S, \Sigma | \mathfrak{S}^2 | S, \Sigma) &= \hbar^2 S(S+1), \\ (S, \Sigma | \mathfrak{S}_x^2 + \mathfrak{S}_y^2 | S, \Sigma) &= \hbar^2 [S(S+1) - \Sigma^2]. \end{aligned}\right\} \tag{34.4}$$

Using Eq. (34.4) we may write as the contribution due to the first term in (34.1)

$$\left.\begin{aligned}\frac{1}{2\Phi R}\sum_{\alpha}\mu_{\alpha\alpha}\overline{M}^2\,\Phi R &= (\bar{\pi}_x^2 B_e^{(xx)} + \bar{\pi}_y^2 B_e^{(yy)}) + \\ &+ [S(S+1) - \Sigma^2]\,\frac{1}{2}\,(B_e^{(xx)} + B_e^{(yy)}) + (\bar{\pi}^2 + 2\bar{\pi}_z\Sigma + \Sigma^2)\,B_e^{(zz)}.\end{aligned}\right\}\qquad(34.5)$$

The second terms in (34.1) will vanish on the average except in the case where $\alpha = z$. When $\alpha = z$ we have that $\overline{M}_z = \bar{\pi}_z + \overline{\mathfrak{S}}_z$ which need not be zero since $\bar{\pi}_z$, while not quantized, is not necessarily zero and, of course, $\mathfrak{S}_z$ is not, in general, equal to zero. Moreover $(\overline{P_z - p_z^*})$ is not zero, but has the value $\left(K - \sum_t l_t \zeta_t^{(z)}\right)\hbar$. Hence the second term in (34.1) becomes

$$-\frac{1}{\Phi R}\,\mu_{zz}\,M_z(P_z - p_z^*)\,\Phi R = -2(\bar{\pi}_z + \Sigma)\left(K - \sum_t l_t \zeta_t^{(z)}\right) B_{zz}^{(e)}. \qquad (34.6)$$

We have, therefore, for the contribution to the energy due to $\overline{U}$, the following

$$\left.\begin{aligned}(v_s, \Sigma, K\,|\,\overline{U}\,|\,v_s, \Sigma, K) &= (\bar{\pi}_x^2 B_e^{(xx)} + \bar{\pi}_y^2 B_e^{(yy)}) + [S(S+1) - \Sigma^2]\times \\ &\times \tfrac{1}{2}\,(B_e^{(xx)} + B_e^{(yy)}) + (\bar{\pi}_z^2 + 2\bar{\pi}_z\Sigma + \Sigma^2)\,B_e^{(zz)} - \\ &- 2(\bar{\pi}_z + \Sigma)\left(K - \sum_t l_t \zeta_t^{(z)}\right) B_e^{(zz)}.\end{aligned}\right\}\qquad(34.7)$$

Combining the results stated in Sect. 29 with the corrections to the energy arising out of $H^{(2)\prime}$ and the results stated in Eq. (34.7) one obtains for the elements of the entire energy matrix, corrected to second order of approximation, the following:

$$\left.\begin{aligned}(K|H|K) = hc\{(E_v/hc) &+ \tfrac{1}{2}(B_v^{(xx)} + B_v^{(yy)})\,[J(J+1) + S(S+1) - K^2 - \Sigma^2] + \\ &+ B_v^{(zz)} L^2 - D_J\,[J(J+1) + S(S+1) - K^2 - \Sigma^2]^2 - \\ &- D_{JL}\,[J(J+1) + S(S+1) - K^2 - \Sigma^2]\,L^2 - D_L L^4\},\end{aligned}\right\}\qquad(34.8)$$

$$\left.\begin{aligned}(K|H|K\pm 2) = hc\{&[J(J+1) - K(K\pm 1)]\,[J(J+1) - (K\pm 1)(K\pm 2)]^{\frac{1}{2}}\}\times \\ &\times \{\tfrac{1}{4}(B_v^{(xx)} - B_v^{(yy)} \pm 2i\,B_v^{(xy)}) - (D_1 \pm i D_1')\,J(J+1) + \\ &+ (D_2 \pm i D_2')\,[K^2 + (K\pm 2)^2]\},\end{aligned}\right\}\qquad(34.9)$$

$$\left.\begin{aligned}(K|H|K\pm 4) = hc\{&[J(J+1) - K(K\pm 1)]\,[J(J+1) - \\ &- (K\pm 1)(K\pm 2)]\,[J(J+1) - (K\pm 2)(K\pm 3)]\times \\ &\times [J(J+1) - (K\pm 3)(K\pm 4)]\}^{\frac{1}{2}}\,(D_3 \pm i D_3').\end{aligned}\right\}\qquad(34.10)$$

$$\left.\begin{aligned}(E_v/hc) = (E_0/hc) + \sum_s \omega_s (v_s + g_s/2) &+ \sum_s\sum_{s'} x_{ss'}(v_s + g_s/2)\times \\ &\times (v_{s'} + g_{s'}/2) + \sum_t\sum_{t'} x_{l_t l_{t'}}\, l_t\, l_{t'}\end{aligned}\right\}\qquad(34.11)$$

in which g_s is the weight factor introduced in Sect. 29 and is equal to 1 or 2 depending upon whether ω_s is nondegnerate or doubly degenerate and in which E_0/hc is a constant term depending in a complicated manner on the potential energy constants and embracing also the first term of Eq. (34.5). The coefficients

$x_{ss'}$ which occur in (34.9) are the following

$$\left.\begin{aligned}
x_{ss} &= (1/4)\Big\{6k_{ssss} - 15\,(k_{sss}^2/\omega_s) - \sum_{s'}(k_{sss'}^2/\omega_{s'})\,(8\omega_s^2 - 3\omega_{s'}^2)/(4\omega_s^2 - \omega_{s'}^2)\Big\},\\
x_{ss'} &= (1/2)\Big\{k_{sss's'} - 6\,(k_{sss}\,k_{ss's'}/\omega_s) - 4k_{sss'}^2\,[\omega_s/(4\omega_s^2 - \omega_{s'}^2)] - \sum_{s''}(k_{sss''}\,k_{s''s's'}/\omega_{s''}) -\\
&\quad - \sum_{s''}\frac{k_{ss's''}^2}{2\,(g_s + g_{s'} - 2)!}\,\omega_{s''}\,(\omega_{s''}^2 - \omega_s^2 - \omega_{s'}^2)/[(\omega_s + \omega_{s'} + \omega_{s''})\,(\omega_s + \omega_{s'} - \omega_{s''})\,\times\\
&\quad \times (\omega_s - \omega_{s'} + \omega_{s''})(\omega_s - \omega_{s'} - \omega_{s''})] + (2/g_s\,g_{s'})\sum_{\alpha}\sum_{\sigma\sigma'}(\zeta_{s\sigma s'\sigma'}^{(\alpha)})^2\,B_e^{(\alpha\alpha)}\,(\omega_{s'}/\omega_s)\Big\},\\
x_{l_t l_t} &= -(1/4)\Big\{2k_{tttt} + \sum_{s'} k_{tts'}^2\,[\omega_{s'}/(4\omega_t^2 - \omega_{s'}^2)]\Big\},\\
x_{l_t l_{t'}} &= (k_{tt's''}^2/2)\,[\omega_t\omega_{t'}\omega_{s''}/(\omega_t + \omega_{t'} + \omega_{s''})\,(\omega_t + \omega_{t'} - \omega_{s''})\,(\omega_t - \omega_{s'} + \omega_{s''})\,\times\\
&\quad \times (\omega_t - \omega_{t'} - \omega_{s''})] + \sum (B_e^{(\alpha\alpha)}/2)\,(\zeta_{t1t'1}^{(\alpha)}\,\zeta_{t2t'2}^{(\alpha)} - \zeta_{t1t'2}^{(\alpha)}\,\zeta_{t2t'1}^{(\alpha)})\,^*.
\end{aligned}\right\}\quad (34.12)$$

t being those values of s for which the vibrations are two-fold degenerate. The $B_v^{(\alpha\beta)}$ which occur in Eq. (34.6) and (34.7) are the following

$$\left.\begin{aligned}
B_v^{(\alpha\alpha)} &= B_e^{(\alpha\alpha)} - b_0^{(\alpha\alpha)} - \sum_{s\sigma} b_{s\sigma}^{(\alpha\alpha)}\,(v_s + g_s/2), \quad (\alpha = x, y, z)\\
B_v^{(xy)} &= -\sum_{s\sigma} b_{s\sigma}^{(xy)}\,(v_s + g_s/2)
\end{aligned}\right\}\quad (34.13)$$

in which

$$\left.\begin{aligned}
b_0^{(xx)} &= b_0 - 2\gamma_0 - 10\beta_0,\\
b_0^{(yy)} &= b_0 - 2\gamma_0 + 10\beta_0,\\
b_0^{(zz)} &= \frac{-3}{2}\,b_0 - 8\,\gamma_0,\\
b_0 &= -\frac{h^3}{256\pi^6c^3}\sum_{s\sigma}\frac{1}{8\omega_s^2}\,[(a_{s\sigma}^{(xx)}/(I_{xx}^{(e)})^2 - a_{s\sigma}^{(yy)}/(I_{yy}^{(e)})^2)^2 + 4\,(a_{s\sigma}^{(xy)}/I_{xx}^{(e)}\,I_{yy}^{(e)})^2],\\
\beta_0 &= -\frac{h^3}{256\pi^6c^3}\sum_{s\sigma}\frac{1}{8\omega_s^2}\,[(a_{s\sigma}^{(xz)}/I_{xx}^{(e)}\,I_{zz}^{(e)})^2 - (a_{s\sigma}^{(yz)}/I_{yy}^{(e)}\,I_{zz}^{(e)})^2],\\
\gamma_0 &= -\frac{h^3}{256\pi^6c^3}\sum_{s\sigma}\frac{1}{8\omega_s^2}\,[(a_{s\sigma}^{(xz)}/I_{xx}^{(e)}\,I_{zz}^{(e)})^2 + (a_{s\sigma}^{(yz)}/I_{yy}^{(e)}\,I_{zz}^{(e)})^2],\\
b_{s\sigma}^{(\alpha\beta)} &= B_e^{(\alpha\alpha)}\,(\hbar/I_{\beta\beta}^{(e)}\,g_s^2\,\lambda_s)^{\frac12}\Big\{(\hbar/I_{\beta\beta}^{(e)})^{\frac12}\Big[A_{s\sigma s\sigma}^{(\alpha\beta)} - \sum_{\gamma} a_{s\sigma}^{(\alpha\gamma)}\,a_{s\sigma}^{(\gamma\beta)}/I_{\gamma\gamma}^{(e)} -\\
&\quad - 4\sum_{s'\sigma'}\zeta_{s\sigma s'\sigma'}^{(\alpha)}\,\zeta_{s\sigma s'\sigma'}^{(\beta)}\,\lambda_s/(\lambda_s - \lambda_{s'}) - \zeta_{ss}^{(\alpha)}\,\zeta_{ss}^{(\beta)}\Big] -\\
&\quad - 2\pi c\Big[3\,k_{sss}\,((a_{s\sigma}^{(\alpha\beta)})^2/I_{\beta\beta}^{(e)})^{\frac12}/\lambda_s^{\frac14} + \sum_{s'\sigma'} k_{sss'}\,((a_{s'\sigma'}^{(\alpha\beta)})^2/I_{\beta\beta}^{(e)})^{\frac12}(\lambda_s^2/\lambda_{s'}^3)^{\frac14}\Big]\Big\}.
\end{aligned}\right\}\quad (34.14)$$

---

* It was found by HANSEN and DENNISON that a term $hck_{4,8,12}\,q_4(q_{8,1}\,q_{12,2} - q_{8,2}\,q_{12,1})$ was needed in the potential energy function of the C_2H_6 molecule in order to explain certain perturbations in the rotational structure of the perpendicular band, ν_8, in the spectrum of that molecule. When the $q_{s,\sigma}$ are replaced by their equivalents in polar coordinates, this takes the form $hck_{4,8,12}\,q_4\,r_8,\,r_{12}\sin(\chi_8 - \chi_{12})$. More recently KELLER has suggested that much better and consistent agreement among the anharmonic potential energy constants can be achieved in the case of the C_2H_2 molecule if one includes, in addition to the usual ones, a similar term coupling ν_3, ν_4 and ν_5, i.e., a term $hck_{345}\,q_3(q_{4,1}\,q_{5,2} - q_{4,2}\,q_{5,1})$ which is equivalent to $hck_{345}\,q_3\,r_4\,r_5\sin(\chi_4 - \chi_5)$. The above would imply that when one perpendicular vibration, ν_s, is excited a second perpendicular vibration, $\nu_{s'}$ cannot be expected to occur in a completely isotopic force field since there now is a dependence upon the angle $\chi_s - \chi_{s'}$.

The above terms are of the form $hck_{ss's''}\,q_{s\sigma}\,q_{s'\sigma'}\,q_{s''\sigma''}$ and contribute to the constants $x_{ss'}$, as indicated in the relations (34.12). It introduces, further, a term

$$x_{l_t l_{t'}} = (k_{tt's''}^2/2)\,[\omega_t\omega_{t'}\omega_{s''}/(\omega_t + \omega_{t'} + \omega_{s''})\,(\omega_t + \omega_{t'} - \omega_{s''})\,(\omega_t - \omega_{t'} + \omega_{s''})\,(\omega_t - \omega_{t'} - \omega_{s''})].$$

KELLER has computed the constant $x_{l_4 l_5}$ in his doctoral dissertation and he has informed us that it agrees with the value given above for $x_{l_t l_{t'}}$.

Other constants which are present in Eqs. (34.6) to (34.8) are these

$$
\left.\begin{aligned}
D_J &= \frac{h^3}{512\pi^6 c^3}\sum_{s\sigma}\frac{1}{8\omega_s^2}\left[3(a_{s\sigma}^{(xx)})^2/(I_{xx}^{(e)})^4 + 3(a_{s\sigma}^{(yy)})^2/I_{yy}^{(e)})^4 + \right.\\
&\quad \left. + 2\left(a_{s\sigma}^{(xx)}\,a_{s\sigma}^{(yy)} + 2(a_{s\sigma}^{(xy)})^2\right)/(I_{xx}^{(e)})^2\,(I_{yy}^{(e)})^2\right],\\
D_L &= \frac{h^3}{64\pi^6 c^3}\sum_{s\sigma}\frac{1}{8\,\omega_s^2}\left[(a_{s\sigma}^{(zz)})^2/(I_{zz}^{(e)})^4\right],\\
D_{JL} &= \frac{h^3}{64\pi^6 c^3}\sum_{s\sigma}\frac{1}{8\,\omega_s^2}\left[\left(a_{s\sigma}^{(zz)}\,a_{s\sigma}^{(xx)} + 2(a_{s\sigma}^{(zx)})^2\right)/(I_{xx}^{(e)})^2\,(I_{zz}^{(e)})^2 + \right.\\
&\quad \left. + \left(a_{s\sigma}^{(yy)}\,a_{s\sigma}^{(zz)} + 2(a_{s\sigma}^{(yz)})^2\right)/(I_{yy}^{(e)})^2\,(I_{zz}^{(e)})^2\right],\\
D_1 &= \frac{h^3}{256\pi^6 c^3}\sum_{s\sigma}\frac{1}{8\omega_s^2}\left[\left((a_{s\sigma}^{(xx)})^2/(I_{xx}^{(e)})^4\right) - \left((a_{s\sigma}^{(yy)})^2/(I_{yy}^{(e)})^4\right)\right],\\
D_1' &= \frac{-h^3}{128\pi^6 c^3}\sum_{s\sigma}\frac{1}{8\omega_s^2}\left\{\left[\left(a_{s\sigma}^{(xx)}/(I_{xx}^{(e)})^2\right) - \left(a_{s\sigma}^{(yy)}/(I_{yy}^{(e)})^2\right)\right] a_{s\sigma}^{(xy)}/I_{xx}^{(e)}\,I_{yy}^{(e)}\right\},\\
D_2 &= \frac{h^3}{512\pi^6 c^3}\sum_{s\sigma}\frac{1}{8\omega_s^2}\left\{\left(a_{s\sigma}^{(xx)}/(I_{xx}^{(e)})^2\right)^2 - \left(a_{s\sigma}^{(yy)}/(I_{yy}^{(e)})^2\right)^2 - 2\left(a_{s\sigma}^{(xx)}\,a_{s\sigma}^{(zz)} + \right.\right.\\
&\quad \left.\left. + 2(a_{s\sigma}^{(xz)})^2\right)/(I_{xx}^{(e)})^2\,(I_{zz}^{(e)})^2 + 2\left(a_{s\sigma}^{(yy)}\,a_{s\sigma}^{(zz)} + 2(a_{s\sigma}^{(yz)})^2\right)/(I_{yy}^{(e)})^2\,(I_{zz}^{(e)})^2\right\},\\
D_2' &= \frac{-h^3}{256\pi^6 c^3}\sum_{s\sigma}\frac{1}{8\omega_s^2}\left\{\left(a_{s\sigma}^{(xx)}\,a_{s\sigma}^{(xy)}/(I_{xx}^{(e)})^3\,(I_{yy}^{(e)})\right) + \left(a_{s\sigma}^{(yy)}\,a_{s\sigma}^{(xy)}/(I_{xx}^{(e)})\,(I_{yy}^{(e)})^3\right) - \right.\\
&\quad \left. - 2\left(a_{s\sigma}^{(zz)}\,a_{s\sigma}^{(xy)} + 2a_{s\sigma}^{(yz)}\,a_{s\sigma}^{(xz)}\right) I_{xx}^{(e)}\,I_{yy}^{(e)}\,(I_{zz}^{(e)})^2\right\},\\
D_3 &= -\frac{h^3}{1024\pi^6 c^3}\sum_{s\sigma}\frac{1}{8\omega_s^2}\left\{\left[\left(a_{s\sigma}^{(xx)}/(I_{xx}^{(e)})^2\right) - \left(a_{s\sigma}^{(yy)}/(I_{yy}^{(e)})^2\right)\right]^2 - \right.\\
&\quad \left. - 4(a_{s\sigma}^{(xy)})^2/(I_{xx}^{(e)})^2\,(I_{yy}^{(e)})^2\right\},\\
D_3' &= \frac{h^3}{256\pi^6 c^3}\sum_{s\sigma}\frac{1}{8\,\omega_s^2}\left\{\left(a_{s\sigma}^{(xx)}\,a_{s\sigma}^{(xy)}/(I_{xx}^{(e)})^3\,I_{yy}^{(e)}\right) - \left(a_{s\sigma}^{(yx)}\,a_{s\sigma}^{(yx)}/I_{xx}^{(e)}\,(I_{yy}^{(e)})^3\right)\right\}.
\end{aligned}\right\} \tag{34.15}
$$

It has been customary in earlier works to use as centrifugal distortion constants, D_K and D_{JK} instead of D_L and D_{JL} as we are doing. We believe the use of D_L and D_{JL} to be more consistent with the actual physical picture, but in order to be able to tie in with earlier works the relationships between D_L, D_{JL} and D_K and D_{JK} are given here

$$D_{JK} = D_{JL} - 2D_J; \qquad D_K = D_L - D_{JL} + D_J. \tag{34.16}$$

The number L takes on a slightly altered connotation here from that assigned to it in Sect. 4. Here it becomes

$$
\left.\begin{aligned}
L &= \left|K - \sum_t l_t\,\zeta_t^{(z)} - \bar{\pi}_z - \Sigma\right|\\
&\text{and}\\
L &= \left|K - \sum_t l_t\,\zeta_t^{(z)} - \bar{\pi}_z\right|.
\end{aligned}\right\} \tag{34.17}
$$

respectively for cases (a) and (b).

It will be noted that the $(K|K\pm 2)$ and $(K|K\pm 4)$ matrix elements have complex coefficients. Complex matrix components were first encountered in connection with the bent XYZ molecular model. It was found that these arose out of the lack of symmetry of the model. The imaginary parts are never greater than of the order of $b_{s\sigma}^{(\alpha)}$ and are, therefore, small. It is readily demonstrated that because of the Hermitian character of the energy matrix, the imaginary parts can probably not contribute to the energy in second order of approximation.

It is interesting to note that although $I_{xx}^{(e)} + I_{yy}^{(e)} = I_{zz}^{(e)}$ in planar molecules, where $I_{xx}^{(e)} < I_{yy}^{(e)} < I_{zz}^{(e)}$, this is not the case for the effective moments of inertia, but there does, indeed, exist a relation $\Delta = I_v^{(xx)} + I_v^{(yy)} - I_v^{(zz)}$ which is different from zero. The quantity is, to this approximation, independent of the anharmonic constants for planar molecules. That this is so, is readily verified simply by setting all the $z_i^0 = 0$ in Eq. (18.24). The coefficients of the anharmonic constants in $I_v^{(xx)} + I_v^{(yy)} - I_v^{(zz)}$, $a_{s\sigma}^{(xx)} + a_{s\sigma}^{(yy)} - a_{s\sigma}^{(zz)}$, then become zero, i.e., $a_{s\sigma}^{(xx)} + a_{s\sigma}^{(yy)} - a_{s\sigma}^{(zz)} = 0$. Such a relation for Δ does not exist, in general, for non-planar molecules.

35. Contribution to the energy from $U - \overline{U}$. When the energy difference between two electronic states corresponding to two orientations of Σ are sufficiently large one may neglect the difference between U and $\overline{U}$. This is not always the case and when two such states are close together it may be necessary to correct for this difference. The terms which are of interest are the terms in U which may be written

$$2\mu_{xx}\mathfrak{S}_x(P_x - p_x^*) + 2\mu_{yy}\mathfrak{S}_y(P_y - p_y^*) \tag{35.1}$$

which we have seen will vanish on the average. The terms $2\mu_{xx}\mathfrak{S}_x(P_x - p_x^*) + 2\mu_{yy}\mathfrak{S}_y(P_y - p_y^*)$ have the non-vanishing matrix components $(\Sigma, K | \Sigma \pm 1, K \pm 1)$. The values of these are readily obtained by use of Eq. (34.3) and (24.13) and they are found to be

$$\left.\begin{aligned} &(v_s, \Sigma, K | U - \overline{U} | v_s, \Sigma \pm 1, K \pm 1) \\ &\quad = \tfrac{1}{2}(B_e^{(xx)} + B_e^{(yy)}) \sqrt{[S(S+1) - \Sigma(\Sigma \pm 1)][J(J+1) - K(K \pm 1)]}. \end{aligned}\right\} \tag{35.2}$$

When the electronic term values corresponding to the two orientations of Σ are rather widely separated (i.e., the deviation from case (a) is not great), the contribution of these terms to the energy may be calculated by the ordinary perturbation theory to give

$$(\Delta E/hc) = [(B_e^{(xx)} + B_e^{(yy)})/\Delta\nu]\{[J(J+1) - K^2]A + KB\} \tag{35.3}$$

where A and B are constants. The effect of the perturbation under these conditions is effectively to cause the $B_v^{(\alpha\alpha)}$ values to depend upon the multiplet level. When the frequency intervals between the components of the spin multiplet levels are small, on the other hand, the correction to the energy, $(\Delta E/hc)$, given above is no longer valid and one must resort to the use of the degenerate perturbation theory in such cases. A special case which may be solved exactly, is that of the doublet state where $S = \frac{1}{2}$. The secular determinant to be solved for a symmetric molecule where $B_e^{(xx)} = B_e^{(yy)}$ is readily found from Eq. (34.8) and (35.2) to be

	$K = K,\ \Sigma = \frac{1}{2}$	$K = K - 1,\ \Sigma = -\frac{1}{2}$	
$K = K$, $\Sigma = \frac{1}{2}$	$E_{e,v,\frac{1}{2}} + [(J+\frac{1}{2})^2 - k(k+1)] \times B_v^{(xx)} + L^2 B_v^{(zz)} - \varepsilon$	$[(J+\frac{1}{2})^2 - k^2]\, B_v^{(xx)}$	$= 0$ (35.4)
$K = K - 1$, $\Sigma = -\frac{1}{2}$	$[(J+\frac{1}{2})^2 - k^2]\, B_v^{(xx)}$	$E_{e,v,-\frac{1}{2}} + [(J+\frac{1}{2})^2 - k(k-1)] \times B_v^{(xx)} + L^2 B_v^{(zz)} - \varepsilon$	

where $K = k + \Sigma$. The roots are similar to those obtained for a diatomic molecule and are

$$\left.\begin{aligned}\varepsilon = \tfrac{1}{2}(E_{e,v,\frac{1}{2}} + E_{e,v,-\frac{1}{2}}) + \{[(J+\tfrac{1}{2})^2 - k^2]\, B_v^{(xx)} + L^2 B_v^{(zz)}\} \pm \\ \pm \tfrac{1}{2}\{\Delta(\Delta - 4k B_e^{(xx)}) + 4(J+\tfrac{1}{2})^2 (B_e^{(xx)})^2\}^{\frac{1}{2}}\end{aligned}\right\} \tag{35.5}$$

where $\Delta = (E_{e,v,\frac{1}{2}} - E_{e,v,-\frac{1}{2}})$. This leads to the result (34.8) when S is set equal to $\frac{1}{2}$ in the limiting instance of case (a) coupling where $\Delta \gg B_e^{(xx)}$. The equation for the energies in the other limiting case where $\Delta \ll B_e^{(xx)}$ gives $E = E_{e,v} + \{[F(F+1) - k^2]\, B_v^{(xx)} + L^2 B_v^{(zz)}\}$ which may be regarded as the equivalent of Eq. (34.8) for case (b) coupling where $S = \frac{1}{2}$. F is equal to $J + \frac{1}{2}$ and in both limiting cases the centrifugal distortion has been neglected.

The terms $2\mu_{xx}\,\pi_x\, P_x + 2\mu_{yy}\,\pi_y\, P_y$, which also vanish on the average, may usually be neglected. It has, never the less, been observed in certain microwave spectra that their contribution in second order of approximation is not entirely insignificant. To estimate the magnitude of this contribution, one might proceed much as was done to determine the magnitude of the contribution due to the term $2\mu_{xx} p_x P_x + 2\mu_{yy} p_y P_y$ considered earlier, i.e., to remove these terms from the first order Hamiltonian by a contact transformation whereby they reappear, transformed, as a part of the second order Hamiltonian. As in the latter case the term $2\mu_{xx}\pi_{xx} P_x + 2\mu_{yy}\pi_y P_y$ will produce a term in $H^{(2)}$ proportional to $P_x^2 + P_y^2$. The contribution to the energy will be proportional to $hc \sum_\alpha \sum_j Be^{(\alpha\alpha)} \cdot (Be^{(\alpha\alpha)}/\Delta\nu_j) \times [J(J+1) - K^2]$ where $\Delta\nu_j$ is the frequency between the electronic state in question and the other interacting electronic states. This term may then be added to corresponding terms in (34.8). The effect of the above contribution is to alter slightly the effective $Be^{(xx)}$ and $Be^{(yy)}$ values for a given electronic state and can be computed accurately only with a knowledge of the electronic wave function, Φ. An estimate of the magnitude of this correction in the normal electronic state has been given by TOWNES. It has, moreover, been computed by ESPE on the basis that the electronic motion may be represented by an irrotational flow and more recently by KURTZ who used as wave functions the atomic orbitals given by WANG and by KEMBLE and ZENER. The effect is extremely small and can, in a great many instances, be neglected.

VI. The interpretation of the infrared spectra of molecules.

36. Introduction. We shall attempt, in this section, to interpret the general aspects of the infrared spectra of molecules in the light of what has gone before, i.e., in terms of the energy relations of the previous section, the selection rules and the intensity relations (30.9) and (30.10). The position of each line in the spectrum is determined by the Bohr frequency relation

$$\nu = [E(v', J', \tau') - E(v'', J'', \tau'')]/hc \tag{36.1}$$

where the $E(v'', J'', \tau'')$ and $E(v', J', \tau')$ are the energies, stated earlier, for the molecules in their normal and excited states respectively. Our discussion shall be divided into two parts essentially—one part dealing with the spectra of axially symmetric molecules, of which the linear and the tetrahedrally symmetric molecules shall be regarded as special cases. The discussion concerning the spectra of tetrahedrally symmetric molecules will be confined to a few general remarks. The second part of our discussion will concern itself with the spectra of asymmetric molecules. We shall, in all cases throughout this section, confine ourselves to molecules which are in states where the electronic and the spin angular momenta are equal to zero.

37. The spectra of axially symmetric molecules. The rather simple arrangements which the atoms assume in an axially symmetric molecule introduce certain simplifications in the energies. First of all, the two moments of inertia, $I_{xx}^{(e)}$ and $I_{yy}^{(e)}$ are alike although $I_{zz}^{(e)}$ remains, in general, different from these. In the second place the symmetry which is inherent of such a molecule brings about that all the matrix elements of the energy which are non-diagonal in K must vanish. The energy of the axially symmetric molecules becomes, therefore, only

$$\left.\begin{aligned}(E/hc) = (E_v/hc) + B_v\,[J(J+1) - K^2] + C_v L^2 - D_J\,[J(J+1) - K^2]^2 - \\ - D_{JL}\,[J(J+1) - K^2]\,L^2 - D_L L^4\end{aligned}\right\} \quad (37.1)$$

in which (E_v/hc) is the vibrational term value and where

$$\left.\begin{aligned}B_v^{(xx)} &= B_v^{(yy)} = B_v = B_e - \sum_s (v_s + g_s/2)\,\alpha_s,\\ B_e &= (h/8\pi^2 I_{xx}^{(e)} c),\\ B_v^{(zz)} &= C_v = C_e - \sum_s (v_s + g_s/2)\,\gamma_s,\\ C_e &= (h/8\pi^2 I_{zz}^{(e)} c),\\ \alpha_s &= \tfrac{1}{2}\sum_\sigma (b_{s\sigma}^{(xx)} + b_{s\sigma}^{(yy)}), \qquad \gamma_s = \sum_\sigma b_{s\sigma}^{(zz)}.\end{aligned}\right\} \quad (37.2)$$

Since the electrons are regarded to be in states of zero orbital and spin angular momentum, L, is merely $L = \left|K - \sum_t l_t \zeta_t\right|$.

The inherent symmetry of an axially symmetric molecule is such that the electric moments induced by vibration lie always either parallel to the axis of symmetry or in a plane perpendicular to it. The former are often referred to as *parallel* vibrations and according to Chap. IV the selection rules are $\Delta J = \pm 1, 0$; $\Delta K = 0$. The latter are usually known as *perpendicular* vibrations and the selection rules in this case were found to be $\Delta J = \pm 1, 0$; $\Delta K = \pm 1$.

Consider first the parallel type band. Vibrations of this type are non-degenerate and the number L becomes simply the quantum number K, i.e., $L = K$. When the selection rules $\Delta J = \pm 1, 0$; $\Delta K = 0$ are used in the Bohr frequency relations (36.1) using as the $E(v, J, K)$ the energy relations (37.1), one obtains the following two expressions for the positions of lines in the spectrum:

$$\left.\begin{aligned}&\text{for } \Delta J = 0, \quad \Delta K = 0,\\ &\nu = [\nu_v + (B_v' - B_v'')] + (B_v' - B_v'')\,J(J+1) + [(C_v' - C_v'') - (B_v' - B_v'')]\,K^2\\ &\text{and for } \Delta J = \pm 1, \quad \Delta K = 0\\ &\nu = [\nu_v + (B_v' - B_v'')] \pm J(B_v'' + B_v') + J^2(B_v' - B_v'') + K^2[(C_v' - C_v'') - \\ &\qquad - (B_v' - B_v'')] \mp 4D_J J(J^2 - K^2) \mp 2D_{JL} J K^2.\end{aligned}\right\} \quad (37.3)$$

The numbers J and K may also be regarded as the ordinal numbers of the lines in the branches of the band, J beginning with $J = K + 1$ and K beginning with $K = 0$. The number J in (37.3) may also be identified as the J value in the lower state (i.e., J'') for lines in the P branch and as the J value in the upper state (i.e., J') for lines in the R branch. J' and J'' are, of course, alike in the Q branch.

Such a parallel band may be regarded as a composite of many sub-bands, a sub-band for each value of K. This is illustrated in Fig. 10 where each sub-band has been drawn separately and the resultant effect collated below. The figure has been drawn for a molecule where B_v'' and B_v' and C_v'' and C_v', respectively, are slightly different, but where the D values have been neglected. If one makes the approximation that $B_v'' = B_v'$ and $C_v'' = C_v'$ the lines in the several sub-bands coalesce so that the composite lines in the P and R branches are separated from each other by equal frequency intervals of $2B$. It will be seen, moreover, that the Q branch in the sub-band $K=0$ is absent. This is required by the intensity relation (30.9) which states that the intensity of the Q branch is proportional to K^2. The components of Q branches of other sub-bands fall off rapidly, their intensities being inversely proportional to $J(J+1)$.

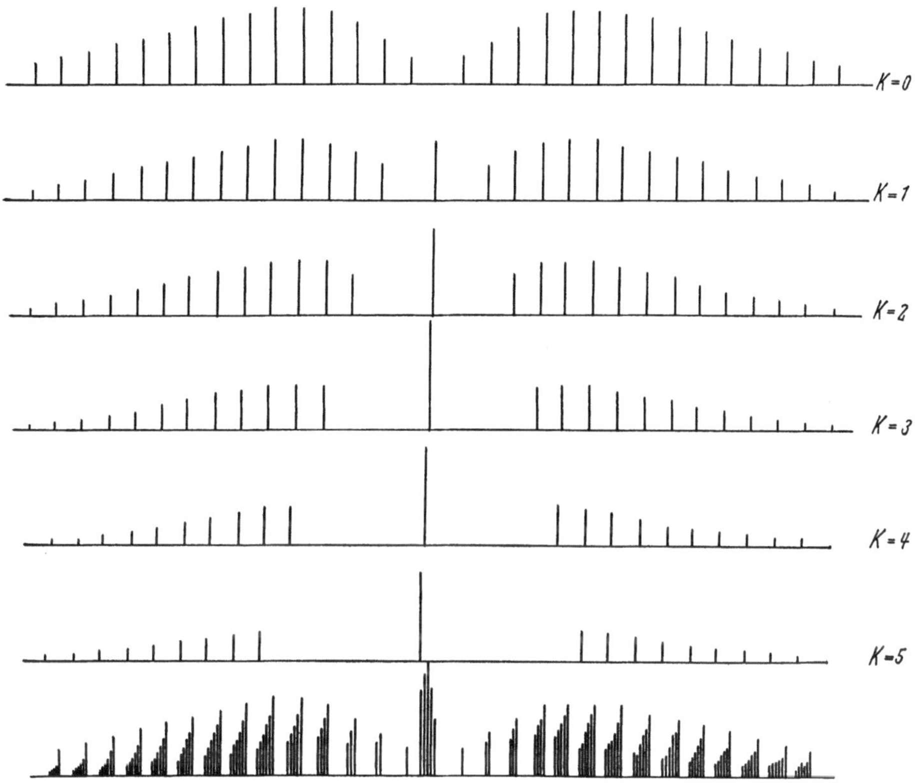

Fig. 10. The sub-band structure of a "parallel" band in the spectrum of an axially symmetric molecule.

The resolution of a parallel band into its "K" structure has rarely been achieved experimentally. An example where such resolution has been affected is the band ν_2 in the spectrum of PH_3 where strong interaction between ν_2 and ν_4 causes the B's and the C's to change by rather large amounts (see Chap. VII). Fig. 26 shows these bands in the spectrum of PH_3 and the K structure can be seen resolved in the P branch which is not overlapped by other bands.

The pure rotation spectrum of an axially symmetric molecule may be regarded as the R branch of a parallel band since in all symmetric molecules the permanent electric moment, when different from zero, lies along the axis of symmetry.

A perpendicular absorption band arises when an axially symmetric molecule makes transitions from a vibration state where the vibration angular momentum quantum numbers add up to zero, i.e., $\sum_t l_t = 0$, to an excited state where $\sum_t l_t = 1$. The selection rules we have seen are $\Delta J = \pm 1, 0$ and $\Delta K = \Delta l = \pm 1$.

This type of band, like the parallel type band, may be regarded as a composite of a series of sub-bands, a separate sub-band on each side of the center of the band for each value of K. The positions of the Q branches of these sub-bands are determined by the selection rule $\Delta J=0$. When these selection rules are used with the energies (37.1) in the frequency condition (36.1) the frequency positions of these are given by

$$\left.\begin{aligned}\nu = \nu_0 - B'_v + \zeta_t^{(z)}(\zeta_t^{(z)} - 2)\, C'_v \pm 2K\,[(1-\zeta_t^{(z)})\, C'_v - B'_v] + \\ + J(J+1)\,(B'_v - B''_v) + K^2[(C'_v - C''_v) - (B'_v - B''_v)]\end{aligned}\right\} \tag{37.4}$$

when one neglects centrifugal stretching terms. K may again be regarded as the ordinal number of the line in the band or, on the high frequency side of the band K takes the values of the quantum number K in the normal state (i.e., $K=0, 1, 2, \ldots$) while on the low frequency side it takes the values of the quantum numbers K in the upper vibration state (i.e., $K=1, 2, 3, \ldots$).

The R branches of the high frequency sub-bands and the P branches of the low frequency sub-bands arise when $\Delta J=\Delta K=+1$ and $\Delta J=\Delta K=-1$ respectively. With these selection rules and the energies (37.1) the Bohr frequency condition leads to the following frequency positions for these lines if centrifugal distortion terms are neglected

$$\left.\begin{aligned}\nu = \nu_0 - B'_v + \zeta_t^{(z)}(\zeta_t^{(z)} - 2)\, C'_v \pm 2K\,[(1-\zeta_t^{(z)})\, C'_v - B'_v] \pm J(B'_v + B''_v) + \\ + J^2(B'_v - B''_v) + K^2\,[(C'_v - C''_v) - (B'_v - B''_v)].\end{aligned}\right\} \tag{37.5}$$

The values which must be assigned to K are the same as for the relation (37.4) and J may take the values $K+1, K+2, \ldots$ on the high frequency side of the band and the values $K, K+1, \ldots$ on the low frequency side.

Similarly the R branches of the low frequency sub-bands and the P branches of the high frequency sub-bands arise when $\Delta J=-\Delta K=+1$ and $\Delta J=-\Delta K=-1$, respectively. With the aid of (37.1) and (36.1) and these selection rules the following relation for the frequency positions of the lines is obtained, neglecting centrifugal stretching terms as earlier,

$$\left.\begin{aligned}\nu = \nu_0 - B'_v + \zeta_t^{(z)}(\zeta_t^{(z)} - 2)\, C'_v \pm 2K\,[(1-\zeta_t^{(z)})\, C'_v - B'_v] \mp J(B'_v + B''_v) + \\ + J^2(B'_v - B''_v) + K^2[(C'_v - C''_v) - (B'_v - B''_v)].\end{aligned}\right\} \tag{37.6}$$

As in earlier instances K takes the values $K=0, 1, 2, \ldots$ on the high frequency side and the values $K=1, 2, 3, \ldots$ on the low frequency side of the band. The values which J is permitted to assume will be $J=K+2, K+3, \ldots$ on the high frequency side and $J=K+1, K+2, \ldots$ on the low frequency side.

The structure of a perpendicular band is shown in Fig. 11 where a series of such sub-bands has been drawn for a molecule where $C'_v \gg B'_v$ and where B'_v and B''_v and C'_v and C''_v are respectively only slightly different. Each sub-band is drawn separately and the integrated effect is shown below.

Should one make the approximation that $C'_v=C''_v$ and $B'_v=B''_v$ one would find a series of sub-bands of which the Q branches are equally spaced with respect to each other, the separation being

$$\Delta\nu = 2\,[(1-\zeta_t^{(z)})\, C_v - B_v]. \tag{37.7}$$

The frequency separation between lines in the P and R branches accompanying a given sub-band will be the same as for any parallel band, namely $\Delta\nu'=2B$. The complete structure of a perpendicular band is seldom completely resolved or even sufficiently well resolved to make a thorough analysis. What usually is seen is a series of Q branches protruding above a background of partially

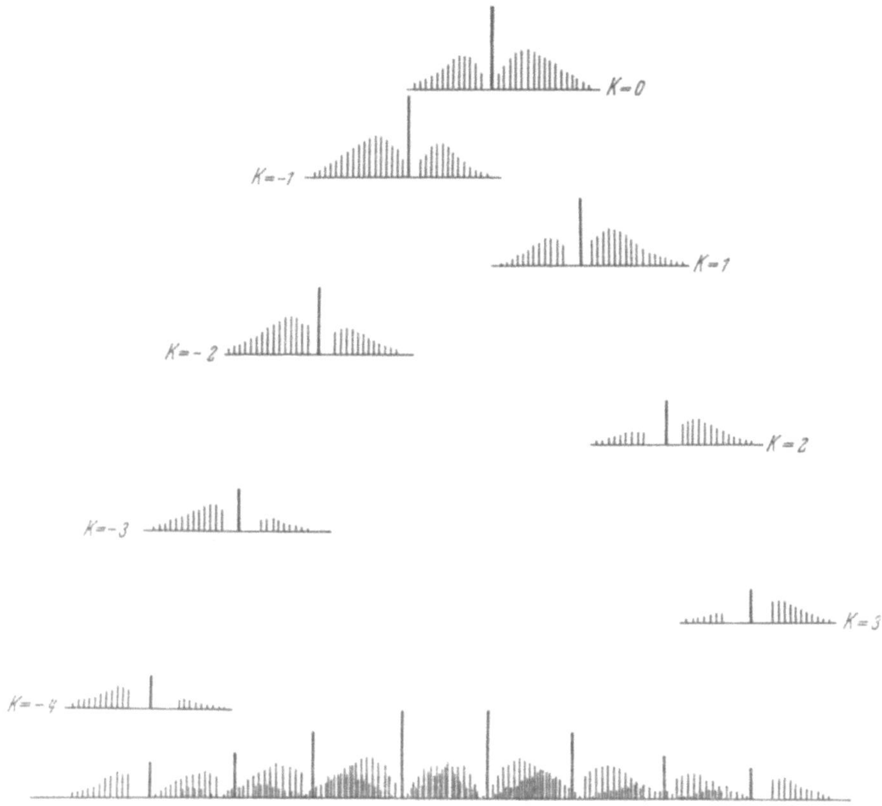

Fig. 11. The sub-band structure of a "perpendicular" band in the spectrum of an axially symmetric molecule.

resolved P and R lines. Fig. 12 is an illustration of such a band from the spectrum of CH_3F. It is, of course, apparent from the frequency relation (37.4) that a

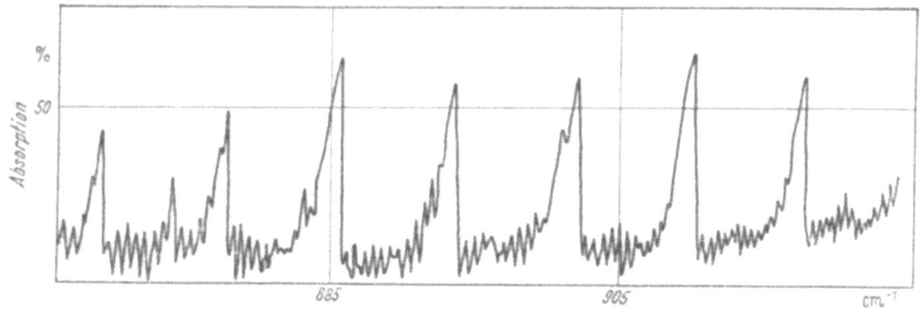

Fig. 12. Perpendicular absorption band from the spectrum of CH_3F (courtesy of G. B. B. M. SUTHERLAND).

perpendicular band may present a wide variety of appearances because the frequency separation between the Q branches depends upon the difference between C_v and B_v and upon the constant $\zeta_t^{(z)}$ which we recall from Chap. III is related to the nature of the normal vibration. While it is common that the frequency separation (37.7) is a relatively large positive quantity, examples are known where $\Delta\nu$ is very nearly zero and in some cases (viz., the pyramidal molecules NH_3, PH_3, AsH_3, PbH_3) it is actually less than zero.

Moreover since the factor, $\zeta_t^{(z)}$, depends upon the nature of the particular normal coordinates, $q_{t,1}$ and $q_{t,2}$, associated with the vibration it may be expected to vary from perpendicular band to perpendicular band and accordingly also the respective frequency intervals $\Delta\nu$ between the Q branches. This fact is indeed consistent with observation.

No direct measurement can be made on the factor $\zeta_t^{(z)}$ and unless the normal frequencies of the molecule are very accurately known (which is seldom so) it can not be computed with any degree of precision. Very little useful information can therefore be derived from a single perpendicular band in the spectrum of an axially symmetric molecule. It may be shown, however, that the sums of the $\zeta_t^{(z)}$ over the perpendicular vibrations of a molecule belonging to a given symmetry class are independent of the force fields. When they are evaluated they lead to simple relations between C_e and B_e. Thus, if it is practicable to observe all the fundamental perpendicular bands and determine all the frequency intervals between the Q branches so that the sum $\sum_t \Delta\nu_t$ can be obtained a simple relation between C'_v and B'_v may be arrived at.

The values of the $\zeta^{(z)}$-sums for symmetric molecules of virtually all symmetry species have been studied by LORD and MERRIFIELD. Their argument is based on the establishment of two theorems, namely:

I. A set of equivalent nuclei lying on the symmetry axis will contribute one unit to the $\zeta^{(z)}$ sum of a given degenerate symmetry species for each vibration the set contributes to that species.

II. A set of equivalent nuclei not lying on the figure axis will contribute a net of zero to the $\zeta^{(z)}$ sum for the vibrations of every symmetry species.

By a set of equivalent nuclei is meant sets of identical nuclei that can be transformed into one another by the symmetry operations permitted by the molecule. Thus, for the pyramidal YZ_3 molecule the 3Z atoms constitute such a set and the Y atom forms another set. Theorems I and II, which are here presented without proof, permit general expressions for the values of the sums of $\zeta^{(z)}$ to be formulated for the various degenerate species of all point groups of axial symmetry, including, however, the contributions from translation and rotation as well as those due to pure vibration. One may readily convince oneself that the doubly-degenerate translation along x and y contribute one unit (i.e., $+1$) to the value of the $\zeta^{(z)}$ sum. It may, moreover, be shown that rotations about the x and y axes contribute an amount $1 - I_{zz}^{(e)}/2I_{xx}$ to the ζ-sum where $I_{xx}^{(e)}$ and $I_{yy}^{(e)}$ are the two principal moments of inertia, $I_{zz}^{(e)}$ being that about the symmetry axis.

Let us apply the above argument to the pyramidal YZ_3 molecule. Such a molecule has only planes of symmetry passing through a threefold axis and accordingly may be shown to have only one degenerate species of vibration, namely the E species. There are two vibrations belonging to this species, but only one of these may be said to be contributed by the Y atom. According to theorem I, then the Y atom contributes an amount $+1$ to the $\zeta^{(z)}$ sum. The 3Z atoms, which lie off of the symmetry axis, contribute a net amount of zero to the $\zeta^{(z)}$ sum according to theorem II. The total $\zeta^{(z)}$ sum according to theorems I and II will, therefore, be $\Sigma\zeta^{(z)} = +1$. The E species embraces a translation as well as a rotation, the sum of which will be $2 - I_{zz}^{(e)}/2I_{xx}^{(e)}$. The part of the $\zeta^{(z)}$ sum which can be identified with the two degenerate E vibration frequencies must, therefore, be $+1 - 2 + I_{zz}^{(e)}/2I_{xx}^{(e)}$ or $I_{zz}^{(e)}/2I_{xx}^{(e)} - 1$. This argument may be extended, at once, to other molecules of the C_{3v} variety. Thus $\Sigma \zeta_{\text{vib}}^{(z)} = I_{zz}^{(e)}/2I_{xx}^{(e)}$ for the molecular model XYZ_3, $\Sigma \zeta_{\text{vib}}^{(z)} = (I_{zz}^{(e)}/2I_{xx}^{(e)}) + 1$ for the molecular model $WXYZ_3$, etc. Quite in general, the value of the $\zeta^{(z)}$-sum for C_{3v} molecules may be written

$$\Sigma \zeta_{\text{vib}}^{(z)} = m_0 - 2 + I_{zz}^{(e)}/2I_{xx}^{(e)} \tag{37.8}$$

where m_0 indicates the number of sets of equivalent nuclei lying on all symmetry elements. Table 12 reproduces the results of LORD and MERRIFIELD and states the $\zeta^{(z)}$-sums for fundamental vibrations of axially symmetric molecules belonging virtually to all point groups. m_p, in this table, is the number of sets of equivalent nuclei lying on a p-fold axis of symmetry but not on all the elements.

Table 12. *The values of the $\zeta^{(z)}$-sums for fundamental vibrations of axially symmetric molecules.*

Point groups	Species	$\Sigma \zeta_{\text{vib}}^{(z)}$	Point groups	Species	$\Sigma \zeta_{\text{vib}}^{(z)}$
C_3, C_4, C_{3v}, C_{4v}	E	$m_0 - 2 + I_{zz}^{(e)}/2I_{xx}^{(e)}$	C_{3h}, D_{3h}	E'	$m_p + m_0 - 1$
C_5, C_6, C_{5v}, C_{6v}	E_1	$m_0 - 2 + I_{zz}^{(e)}/2I_{xx}^{(e)}$		E''	$m_p - 1 + I_{zz}^{(e)}/2I_{xx}^{(e)}$
	E_2	0	C_{4h}, D_{4h}	E_g	$m_p - 1 + I_{zz}^{(e)}/2I_{xx}^{(e)}$
D_3, D_{2d}, S_4, D_4	E	$2m_p + m_0 - 2 + I_{zz}^{(e)}/2I_{xx}^{(e)}$		E_u	$m_p + m_0 - 1$
D_5, D_6	E_1	$2m_p + m_0 - 2 + I_{zz}^{(e)}/2I_{xx}^{(s)}$	C_{5h}, D_{5h}	E_1'	$m_p + m_0 - 1$
	E_2	0		E_1''	$m_p - 1 + I_{zz}^{(e)}/2I_{xx}^{(e)}$
				E_2'	0
D_{3d}, S_6	E_g	$m_p - 1 + I_{zz}^{(e)}/2I_{xx}^{(e)}$		E_2''	0
	E_u	$m_p + m_0 - 1$	C_{6h}, D_{6h}	E_{1g}	$m_p - 1 + I_{zz}^{(e)}/2I_{xx}^{(e)}$
D_{4d}	E_1	$m_p + m_0 - 1$		E_{1u}	$m_p + m_0 - 1$
	E_2	0		E_{2g}	0
	E_3	$m_p - 1 + I_{zz}^{(e)}/2I_{xx}^{(e)}$		E_{2u}	0

Let us apply the value of the $\zeta^{(z)}$-sum for the YZ_3 molecular model to equation (37.7). As we have seen, this model has two degenerate vibrations. Identifying these as ω_3 and ω_4, we have approximately

$$\left.\begin{aligned} \sum_t \Delta\nu_t &= 2\{[(1-\zeta_3^{(z)})\,C_e - B_e] + [(1-\zeta_4^{(z)})\,C_e - B_e]\} \\ &= 4\{(C_e - B_e) - \tfrac{1}{2}(\zeta_3^{(z)} + \zeta_4^{(z)})\,C_e\}. \end{aligned}\right\} \tag{37.9}$$

We have just seen, however, that $\zeta_3^{(z)} + \zeta_4^{(z)} = I_{zz}^{(e)}/2I_{xx}^{(e)} - 1$ or $(B_e/2C_e - 1)$. Inserting this value into Eq. (37.9) we obtain

$$\sum_t \Delta\nu_t = (6C_e - 5B_e). \tag{37.10}$$

Hence if all the fundamental perpendicular bands in the spectrum of an axially symmetric molecule can be observed so that the average spacings between rotation lines can be determined a good estimate of the B and C values for the molecules can be obtained.

38. The spectra of linear molecules. The linear molecule may be regarded as a special case of the axially symmetric molecule where the angular momentum of the molecular framework about the internuclear axis, z, is zero, i.e., $(P_z - \Pi_z) = 0$. The eigenvalue of $P_z - \Pi_z$, L, must accordingly also be equal to zero, i.e., $L = 0$. The condition that $P_z - \Pi_z$ must be equal to zero has the effect upon the Hamiltonian and in turn upon the energies that all terms containing $I_{zz}^{(e)}$, $a_{s\sigma}^{(xz)}$ and $a_{s\sigma}^{(yz)}$

are absent. Setting $L=0$ is tantamount to stipulating that the quantum number $K=\sum_t l_t \zeta_t^{(z)}$†. The complete cylindrical symmetry of the linear molecule about its internuclear axis requires, however, that the direction cosines $l_{it1}^{(x)}$ must be equal to the direction cosines $l_{it2}^{(y)}$. It is readily seen, from the definition of the $\zeta_t^{(z)}$ and the normalization requirements imposed upon the $l_{it\sigma}$, that the $\zeta_t^{(z)}$ are all equal to unity. Hence $K=\sum_t l_t$.

We shall now consider the spectra of linear molecules. A parallel band arises here, as in the general symmetric molecule, when the molecule makes transitions between two vibration levels of the same values of $\sum_t l_t$. The selection rules are $\Delta J=\pm 1, 0$; $\Delta K=\Delta\sum_t l_t=0$ where K has the one value $\sum_t l_t$. We shall expect to observe, therefore, only one of the sub-bands which normally go into making up a parallel band of a symmetric molecule. Conversely one may state that a parallel band in the spectrum of a linear molecule consists of a single sub-band. When the initial state of the molecule is the ground state a parallel vibration is from a state where $\sum l_t=0$ to another state where $\sum_t l_t=0$. We should expect, accordingly, to observe the $K=0$-th sub-band of Fig. 10 as, indeed, we do. Such a band from the spectrum of CO_2 is shown in Fig. 13 which shows the characteristic absence of a Q branch. Low lying perpendicular vibration states where $\sum_t l_t=1$ are, in some cases, sufficiently populated at normal temperatures so that bands originating with molecules in such states are observed. When a parallel transition results from a state where the molecule is in such a state to another state where $\sum_t l_t=1$ we should expect to observe the $K=1$-st sub-band of Fig. 10. Such a band, observed in the spectrum of C_2H_2 is shown in Fig. 14 and it exhibits the characteristic weak Q branch and the absence of the first line in the P and R branches. Transitions from vibration states of $\sum_t l_t>1$ have not been observed, but will undoubtedly be seen in future investigations.

A perpendicular band arises here, as in the general axially symmetric molecule, when the molecule makes transitions between two vibration states with values

† The electrons in a linear molecule move in cylindrically symmetric force fields when the molecule is in vibration states where $\sum_t l_t=0$. π_z is in this instance quantized as may be seen in the following manner. The coordinates x_i und y_i are replaced by cylindrical coordinates $\varrho_i\cos\alpha_i$ and $\varrho_i\sin\alpha_i$ with z_i as the axis and the angles α_i measured from the xz plane. It is advantageous, further, to introduce the relative coordinates $\beta_2=\alpha_2-\alpha_1$, $\beta_3=\alpha_3-\alpha_1$, etc., instead of the angles α_2, α_3, etc. By so doing $\overline{U}$ will not contain α_1 explicitly, but only derivatives with respect to α_1. That this must be so is seen in that the force field is isotropic and the energy must be independent of a rotation of the molecule about z. While the potential energy will depend upon the β_i it will be independent of α_1. Since α_1 is an ignorable coordinate the wave function Φ will have the form $\Phi=e^{i\Lambda\alpha_1}\Phi'(q_{s\sigma},\beta_2,\beta_3\ldots\beta_j,\varrho_j)$ where the single valuedness requirement demands that Λ must be an integer or zero. The component of electron orbital angular momentum along the z axis will then be $\pi_z=i\hbar\,\partial/\partial\alpha_1$ from which one obtains $\pi_z\Phi=\Lambda\hbar\Phi$. We have further for M_z, the total angular momentum directed along z, including spins $M_z=i\hbar(\partial/\partial\alpha_1)+\mathfrak{S}_z$. This gives $M_z\Phi=(\Lambda+\Sigma)\,\Phi$ where Σ takes the values $S, S-1, \ldots, -S$. So long as $\sum_t l_t=0$, $\pi_z=M_z$ which for linear molecules must equal P_z. We have therefore that $K=(\Lambda+\Sigma)$ for case (a) and $K=\Lambda$ for case (b) for linear molecules in a Σ vibration state. The corresponding rotation energies are $(E_{\text{rot}}/hc)=[J(J+1)+S(S+1)-2K\Sigma]\,B_v-[J(J+1)+S(S+1)-2K\Sigma]^2 D$ for case (a) and just $(E_{\text{rot}}/hc)=[J(J+1)-\Lambda^2]\,B_v-[J(J+1)-\Lambda^2]^2 D$.

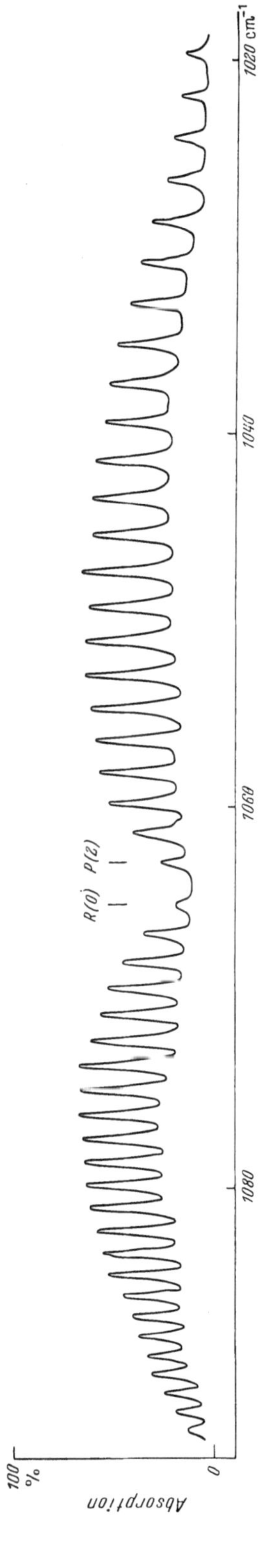

Fig. 13. Parallel band ($K = 0$-th sub-band) from the infrared spectrum of CO_2.

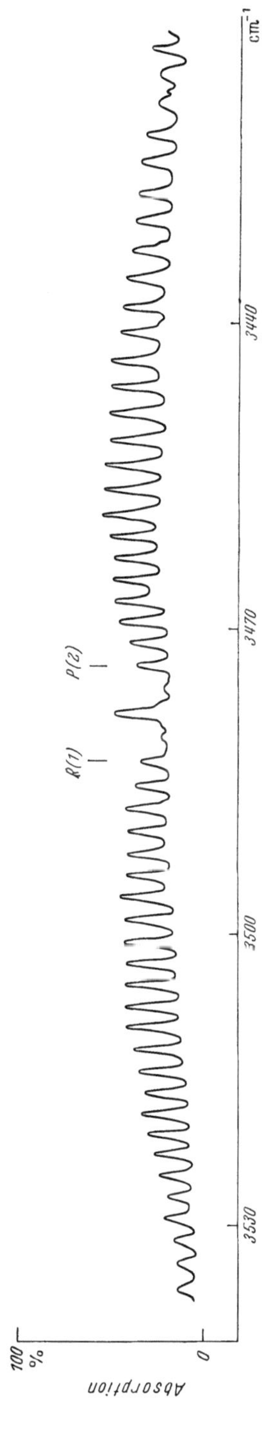

Fig. 14. Parallel band (K = first sub-band; $\Pi - \Pi$ transition) from the infrared spectrum of C_2H_2 (courtesy of A. H. Nielsen).

of $\sum_t l_t$ differing by one unit. The selection rules are $\Delta J = \pm 1, 0$; $K = \sum_t l_t \rightarrow K \pm 1 = \left(\sum_t l_t\right) \pm 1$. As in the parallel band we shall here expect to observe, for a given vibration transition, only one sub-band of the kind which normally constitutes a perpendicular band in the spectrum of a symmetric molecule, or stated conversely, a perpendicular band in the spectrum of a linear molecule consists of only one sub-band. When the molecule is in the normal state initially where all the

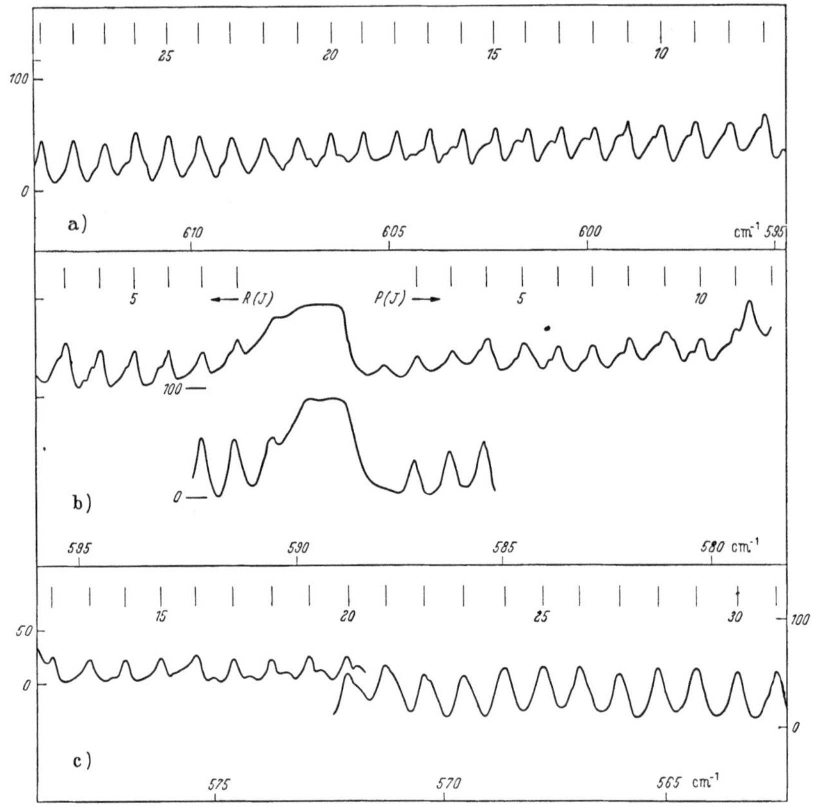

Fig. 15. Perpendicular band ($K =$ first negative sub-band; $\Pi - \Sigma$ transition) from the infrared spectrum of N_2O.

$v_s = 0$, and accordingly $\sum_t l_t = 0$, a perpendicular band will consist of transitions to a final state where $\sum_t l_t = 1$. We should accordingly expect to observe the $K = 0$-th sub-band of Fig. 11. Transitions from low lying states where $\sum_t l_t = 1$ to higher vibration states where $\sum_t l_t = 0$ have been observed in several instances. The sub-band which one would observe in such cases would be the $K =$ first negative sub-band. Such a band, observed in the spectrum of N_2O is shown in Fig. 15. The illustration shows the characteristic absence of the first line in the P branch. Perpendicular transitions of the kind $\sum_t l_t = 1 \rightarrow \sum_t l_t = 2$ have also been observed in the spectrum of CO_2, but in this case the characteristic structure is partially obscured by overlapping with other bands. Thus the simplicity of the bands in the spectrum of linear molecules provide a verification of the predicted sub-band structure of the more complicated infrared bands of axially symmetric molecules.

39. The spectra of spherical molecules. The spherical molecule is the special instance of the symmetric molecule where all three moments of inertia are equal to each other and, as we have seen in Chap. IV, the rotational energy of such molecules becomes independent of the quantum number K for this reason. The only variety of spherical molecules which have been studied extensively are the tetrahedrally symmetric (spherically symmetric only in zero order of approximation) YZ_4 type of which CH_4, CD_4, SiH_4 and GeH_4 are examples. We shall confine ourselves to this type of molecule and restrict the discussion of the structure of the infrared bands in the spectra of such molecules to the fundamental bands. We shall, moreover, interpret the structure of these bands in the light of the energy relations stated for such molecules in Eq. (29.6).

The tetrahedrally symmetric YZ_4 type of molecule has four fundamental modes of vibration. One of these, ω_1, is non-degenerate and may be described as a symmetric expansion and contraction of the tetrahedron where all the Z nuclei move along the YZ bonds in the same phase. This frequency is frequently referred to as the breathing frequency. The molecule retains its tetrahedral symmetry in this mode and remains, therefore, a spherical rotator. A second fundamental vibration, ω_2, is two dimensionally isotropic and may be visualized as a motion where the four Z nuclei move in ellipses on the surface of a sphere, the radius of which is equal to the equilibrium YZ distance. While this frequency is two dimensionally isotropic, the internal angular momentum $\zeta_2 \hbar$ associated with it will be zero. This is readily seen as follows. As a Z particle rolls on the sphere about its equilibrium position it gives rise to an angular momentum which we may represent by a vector directed along the YZ bond when the molecule is in equilibrium. Since all four Z particles are participating in the motion, four such angular momentum vectors are produced all directed away from (or toward) the center of the tetrahedron along a YZ bond. Their sum evidently will be zero. Both the frequencies ω_1 and ω_2 induce zero electric moment and may be expected to be optically inactive.

The remaining two frequencies, ω_3 and ω_4, are alike in character and in both instances the Y nucleus will move in a three dimensionally isotropic force field. These fundamental vibrations will, therefore, be threefold degenerate and may be expected to produce bands in the infrared spectrum of the molecule. Moreover, the internal angular momentum of vibration will be different from zero and the Coriolis coupling factors, ζ_3 and ζ_4, will obey the following ζ-sum, $\zeta_3+\zeta_4=\frac{1}{2}$. We have seen in Chap. IV that each rotation level in the first excited state of ω_3 and ω_4 is split into three components E_+, E_0 and E_- and that the selection rules permit the following transitions to be made to these respectively, $J\to J+1$, $J\to J$ and $J\to J-1$. These selection rules give rise to the following relations for the positions of lines in the P and R branches and for the Q branch, assuming the B_v value in the upper and lower states to be the same:

$$\nu = \nu_0 \pm 2J(1-\zeta_t)\, B_v \qquad \text{(see Fig. 8, p. 231)},$$

$$\nu = \nu_0. \tag{39.1}$$

The spherical symmetry is retained to first order of approximation, but in higher orders the distortion of the molecule due to these vibrations and interaction terms between vibration and rotation gives rise to a term in the energy proportional to K^2. The spherical molecule in such an excited state behaves like an axially symmetric molecule where one of the moments of inertia now is different from

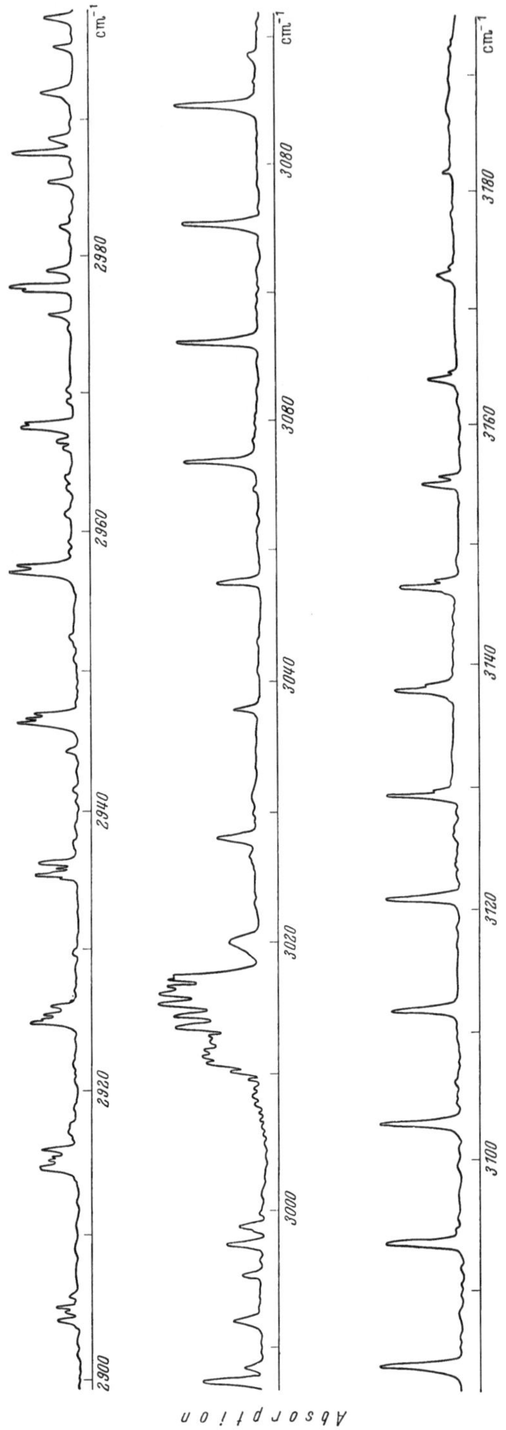

Fig. 16. The infrared band, ν_3, in the spectrum of CH_4.

the other two by a very small amount. Fig. 16 is a record of the band, ω_3, in the infrared spectrum of CH_4. Essentially it shows a series of lines nearly equally spaced with respect to each other, on either side of the central Q line. On closer inspection it can be seen that each line is partially resolved into K-structure.

40. The spectra of asymmetric molecules. The lack of symmetry of asymmetric molecules precludes the presence of degenerate vibration frequencies. This introduces the simplification that there are no p_α^* operators in the Hamiltonian and therefore the number L becomes equal to K. This simplification is more than offset, however, by the fact that the $(K|K\pm 2)$ and $(K|K\pm 4)$ elements are no longer equal to zero. The $(K|K\pm 2)$ elements in the non-rigid rotator contain not only the contributions from the asymmetry but now also contributions arising from centrifugal stretching. The $(K|K\pm 4)$ elements are unique for the non-rigid rotator and are due only to centrifugal stretching. The secular determinant of the elements given in Eq. (24.11) will factor into four sub-determinants, in spite of the $(K|K\pm 4)$ elements just as in the rigid rotator case if one neglects the imaginary part of the off-diagonal matrix elements. The roots of the secular determinants will, therefore, be obtained in much the same manner as in the case of the rigid rotator but will now be influenced by the presence of the centrifugal stretching terms in the Hamiltonian. The secular determinants for the values of $J=0$ to $J=3$ have been expanded into algebraic equations to illustrate the foregoing. They are the following

$$\left.\begin{aligned}
&J=0 \quad W=0,\\
&J=1 \quad W=0,\\
&\qquad\; W=(1+\varrho_3)\pm(\varrho_2+2\varrho_4)f,\\
&J=2 \quad W=4+16\varrho_3-\varrho_5 f(f-2),\\
&\qquad\; W=(1+\varrho_3)\pm(\varrho_2+2\varrho_4)f,\\
&\qquad\; W=\tfrac{1}{2}\left[4+16\varrho_3+\varrho_5 f(f-2)\right]\pm\\
&\qquad\qquad \pm\tfrac{1}{2}\{[4+4\varrho_3+\varrho_5 f(f-2)]^2+8f(f-2)(\varrho_2+4\varrho_4)^2\}^{\frac{1}{2}},\\
&J=3 \quad W=4+16\varrho_3-\varrho_5 f(f-2),\\
&\qquad\; W=(1+\varrho_3)\pm(\varrho_2+2\varrho_4)f,\\
&\qquad\; W=\tfrac{1}{2}\left[4+16\varrho_3+\varrho_5 f(f-2)\right]\pm\\
&\qquad\qquad \pm\tfrac{1}{2}\{[4+4\varrho_3+\varrho_5 f(f-2)]^2+8f(f-2)(\varrho_2+4\varrho_4)^2\}^{\frac{1}{2}},\\
&\qquad\; W=\tfrac{1}{2}\left[10+82\varrho_3\mp(\varrho_2+2\varrho_4)f\right]\pm\tfrac{1}{2}\{[8+80\varrho_3\pm\\
&\qquad\qquad \pm(\varrho_2\pm2\varrho_4)f^2]+4(f-2)(f-6)(\varrho_2+10\varrho_4+f\varrho_5)^2\}^{\frac{1}{2}}
\end{aligned}\right\}\quad(40.1)$$

in which $f=J(J+1)$. A notation $\varrho_i=R_i/R_1$, $(i\neq1)$ has been introduced in (40.1) where R_1 is used to denote $\{B_v^{(zz)}-\frac{1}{2}(B_v^{(xx)}+B_v^{(yy)})-(D_{JL}-2D_J)\,J(J+1)\}$, R_2 is used in place of $\{\frac{1}{4}(B_v^{(xx)}-B_v^{(yy)})-D_1\,J(J+1)\}$, R_3 to denote $-(D_L-D_{JL}+D_J)$, R_4 to signify D_2 and where R_5 is used to designate D_3. The relationship between the roots of the secular equation, W, and the energies E are given by the equation

$$W=(2E/\hbar^2-R_0)/R_1, \tag{40.2}$$

where R_0 is used to designate $(E_v/hc)+\frac{1}{2}(B_v^{(xx)}+B_v^{(yy)})\,J(J+1)-D_J\,J^2(J+1)^2$. It may be verified, that if the centrifugal distortion terms are neglected, the Eqs. (40.1) reduce to the corresponding relations for the rigid rotator.

While this is satisfactory, in principle, it has the limitation that the energy of the non-rigid asymmetric rotator can not, like the energy of the non-rigid symmetric rotator, be expressed as the sum of the rigid rotator terms plus the centrifugal stretching terms. This is particularly disturbing because tables such as those produced by King, Hainer and Cross are not applicable then and the preparation of new tables to cover all cases of centrifugal distortion, is of course not practicable.

The problem of expressing the energy of the non-rigid rotator as the sum of the energy of a rigid asymmetric rotator where the tables of King, Hainer and Cross are applicable plus energy terms which are entirely due to centrifugal stretching has been dealt with by Kivelson and Wilson. The derivation of their results, which is an approximation method, is entirely beyond the scope of this section, but we shall state their conclusions here. They obtain for the energy of the non-rigid rotator

$$\left.\begin{aligned}
E_{J,\tau}&=E_{J,\tau}^{(0)}+A_1(E_{J,\tau}^{(0)})^2+A_2(E_{J,\tau}^{(0)})\,J(J+1)+A_3\,J^2(J+1)^2+\\
&\quad+A_4\,J(J+1)\,\langle P_z\rangle^2+A_5\langle P_z^4\rangle+A_6(E_{J,\tau}^{(0)})\,\langle P_z^2\rangle
\end{aligned}\right\}\quad(40.3)$$

where $E^{(0)}_{J,\tau}$ is the energy of the rigid asymmetric rotator and where the constants A_i are defined in the following manner in terms of the constants stated in Eq. (34.15)

$$\left.\begin{aligned}
A_1 &= 16D_3/(\beta-\gamma)^2,\\
A_2 &= -[16D_3(\beta+\gamma)/(\beta-\gamma)^2+4D_1/(\beta-\gamma)],\\
A_3 &= -D_J+2D_3+16D_3\beta\gamma/(\beta-\gamma)^2+2D_1(\beta+\gamma)/(\beta-\gamma),\\
A_4 &= -[D_{JK}-2D_1\sigma-16D_3(\alpha^2-\beta\gamma)/(\beta-\gamma)^2+4D_3\sigma^2+\\
&\qquad\qquad\qquad\qquad +4D_2(\gamma+\beta)/(\beta-\gamma)],\\
A_5 &= -[D_K+4D_2\sigma+2D_3-4D_3\sigma^2],\\
A_6 &= [8D_2-16D_3\sigma]/(\beta-\gamma)
\end{aligned}\right\}\tag{40.4}$$

in which $\alpha=(\hbar^2/2I^{(e)}_{zz})$, $\beta=(\hbar^2/2I^{(e)}_{xx})$, $\gamma=(\hbar^2/2I^{(e)}_{yy})$ and $\sigma=[2\alpha-\beta-\gamma]/(\beta-\gamma)$. The symbols $\langle P_z^2\rangle$ and $\langle P_z^4\rangle$ which occur in (40.3) are used to indicate the diagonal elements of P_z^2 and P_z^4 respectively for the asymmetric rotator, i.e., the values of P_z^2 and P_z^4 averaged over the wave functions for the asymmetric rotator. They give the formula for these

$$\left.\begin{aligned}
\langle\Pi\rangle &= \{1+R_{m-1}[1+R_{m-2}(1+\cdots)]+R'_{m+1}[1+R'_{m+2}(1+\cdots)]\}^{-1}\times\\
&\quad\times\{\Pi_m+R_{m-1}[\Pi_{m-1}+R_{m-2}(\ldots)]+R'_{m+1}[\Pi_{m+1}+R'_{m+2}(\ldots)]\}
\end{aligned}\right\}\tag{40.5}$$

where the quantities R and R' are defined as follows

$$\left.\begin{aligned}
R_n &= b_{n+1}/[k_n-\lambda-b_n/(k_{n-1}-\lambda-\cdots)]^2,\\
R'_n &= b_n/[k_n-\lambda-b_{n+1}/(k_{n+1}-\lambda-\cdots)]^2.
\end{aligned}\right\}\tag{40.6}$$

The quantities, λ, in the R_n and R'_n are the roots of the KING, HAINER and CROSS secular determinant referred to in Chap. IV and the k_m and b_m may be obtained by reference to the secular determinant (24.17).

The procedure then is to evaluate the rotational constants, I_{xx}, I_{yy} and I_{zz} which, occur in the rigid rotator from energy levels of low quantum numbers J and τ where the centrifugal stretching can be neglected. Using these constants so determined, the other constants occuring in (40.3) are evaluated from energy levels of successively higher and higher J and τ. If a trend in the centrifugal distortion constants develops it may be necessary to return and adjust the rigid rotator constants and by an iterative process arrive at a set of consistent results.

We have seen in Chap. IV that there are three different sets of selection rules for the asymmetric rotator depending upon whether the electric moment lies along the axis of the smallest, intermediate or largest moment of inertia. We have, in fact, seen that when the electric moment lies along the axis of the smallest moment of inertia (i.e., $\mu_a\neq 0$, $\mu_b=\mu_c=0$) that the allowed transitions are between levels of the symmetries $A\leftrightarrow B_a$, $B_b\leftrightarrow B_c$; when the electric moment lies along the axis of the intermediate moment of inertia (i.e., $\mu_b\neq 0$, $\mu_a=\mu_c=0$) the allowed transitions are between levels of the symmetries $A\leftrightarrow B_b$, $B_a\leftrightarrow B_c$ and when the electric moment lies along the axis of the largest moment of inertia (i.e., $\mu_c\neq 0$, $\mu_a=\mu_b=0$) the allowed transitions are between levels of symmetry $A\leftrightarrow B_c$, $B_a\leftrightarrow B_b$.

Corresponding to these three types of transitions three characteristic types of bands are obtained which are commonly referred to as types A, B and C, respectively. The sub-band structure which was so helpful in describing the appearance of bands in the spectra of axially symmetric molecules may still be regarded as present in the bands of the spectra of asymmetric molecules, but because of the complexity of the energy level pattern and the far less rigid selection rules many more sub-bands spring up at irregular intervals so that much

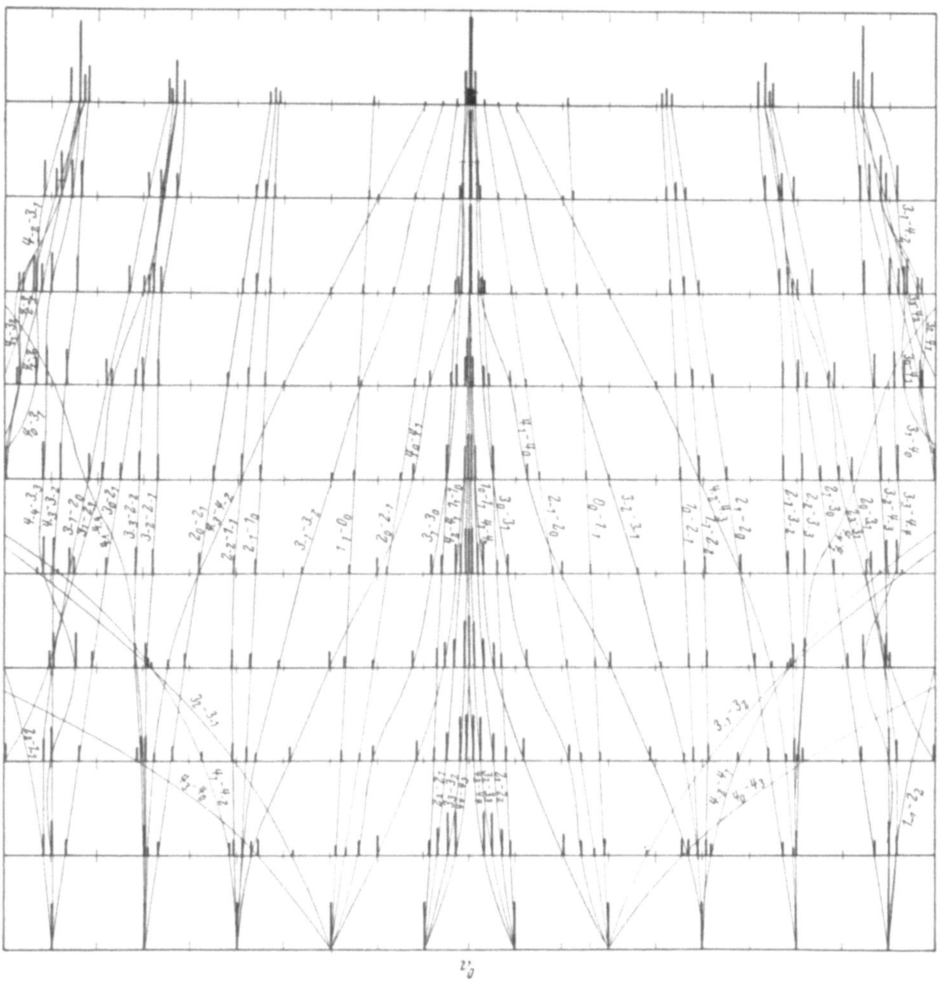

Fig. 17. The structure of an infrared absorption band for an asymmetric molecule where $\mu_a \neq 0$, $\mu_b = \mu_c = 0$.

of the effectiveness of using these to describe the bands of an asymmetric molecule is lost. No simple method of representing the three types of bands is practicable, in reality, because the band structure will depend upon the values of the three moments of inertia taken as parameters.

A glimpse of the condition which prevails has nevertheless been given by DENNISON who considered a set of perfectly hypothetical models of the planar variety where $I_a^{(e)} + I_b^{(e)} = I_c^{(e)}$. He has given two figures to illustrate the type A and the type B bands, each figure consisting of ten sets of lines, the heights of which indicate their intensities, placed one above the other corresponding to an asymmetry parameter, $\varrho = (I_a^{(e)}/I_b^{(e)})$, adopted by him ranging from $\varrho = 1.0$ to $\varrho = 0.01$. (The relationship between ϱ and $\varkappa$ may be verified to be $\varkappa = [(2\varrho - 1) \times a - c]/(a - c)$. He assumed the moments of inertia to have the same values in the upper and the lower vibration states and no account was taken of any symmetry properties of the molecules which might exist in computing the intensities. The lines are, therefore, arranged in a completely symmetrical pattern about

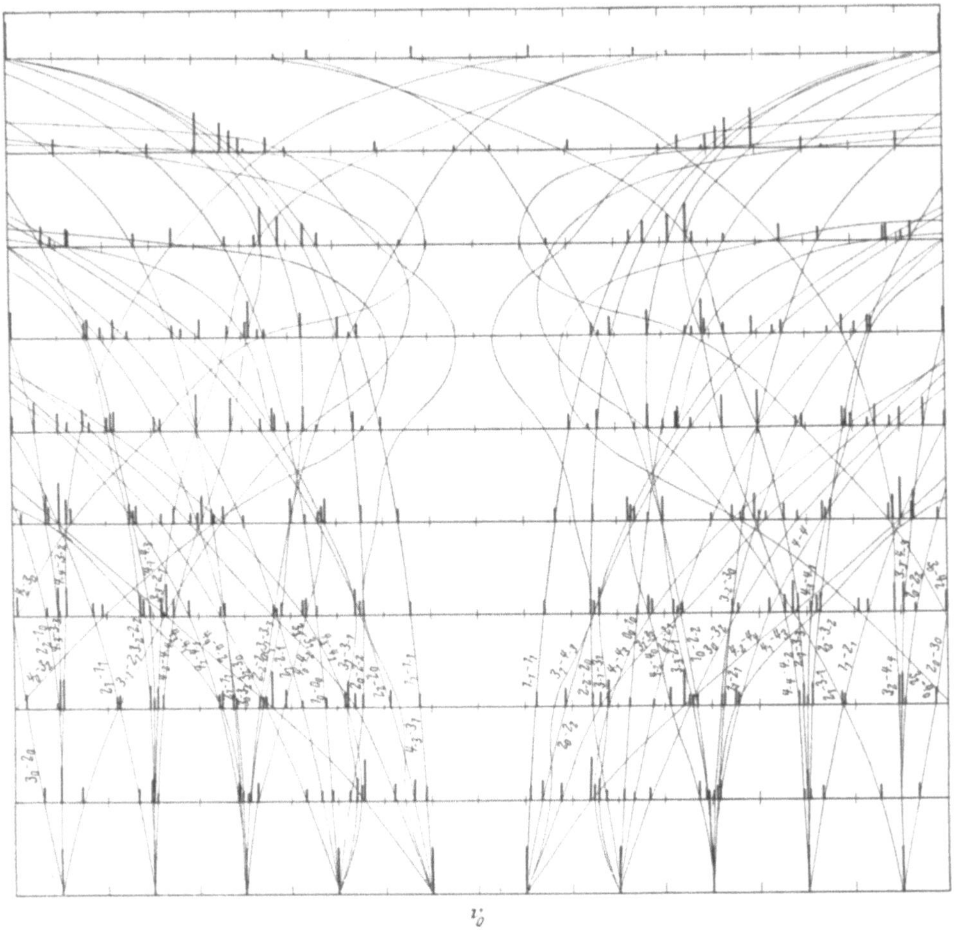

Fig. 18. The structure of an infrared absorption band for an asymmetric molecule where $\mu_a = \mu_c = 0$, $\mu_b \neq 0$.

the band center, ν_0; i.e., wherever a line occurs on the high frequency side of ν_0 corresponding to a transition $(J''_{\tau''} \to J'_{\tau'})$, a similar line of equal intensity occurs on the low frequency side of ν_0 corresponding to $(J'_{\tau'} \to J''_{\tau''})$. The lines for $\varrho = 1.0$ are plotted in both figures with an equal arbitrary height of five units. DENNISON confined himself to transitions between levels of J not greater than $J = 4$ and in all cases the Boltzmann factor was taken to be equal to unity. Many of the most significant lines in the case of small molecules, for example, hydrides where the moments of inertia are determined principally by the hydrogen atoms, particularly near the band center, will be accounted for by these transitions. Actually, of course, the Boltzmann factor would decrease rapidly in such cases and for this reason the intensities might be altered considerably from those shown in the diagrams. The lines for a given transition are connected throughout the sets by fine lines so as to allow tracing the evolution in band structure as ϱ varies from the value $\varrho=1.0$ to the value $\varrho=0.10$. DENNISON's figures describing the structures of type A and type B bands are shown respectively in Figs. 17 and 18.

Following the above scheme adopted by DENNISON, the three types A, B and C were further studied by NIELSEN in the region $\varrho=0.2$ to $\varrho=0.02$, who produced diagrams depicting their transformation throughout this interval of

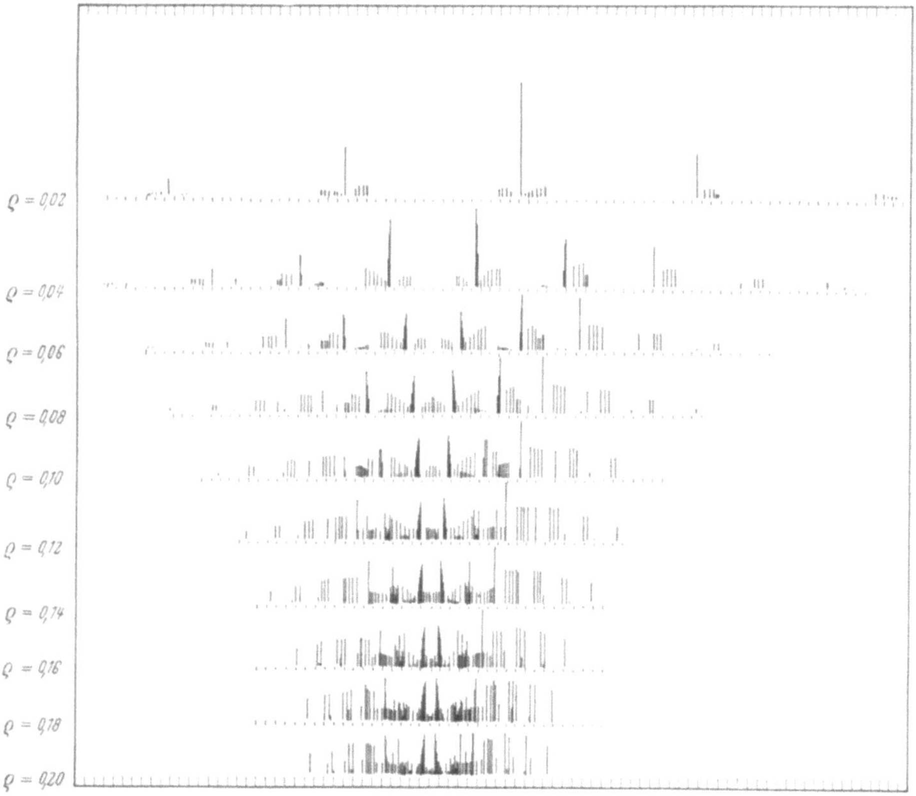

Fig. 19. The structure of an infrared absorption band for an asymmetric molecule where $\mu_a=\mu_b=0$, $\mu_c\neq 0$.

ϱ including lines arising from transitions up to $J=6$ and using a Boltzmann factor corresponding to $I_c^{(e)}=2.0\times10^{-39}$ g cm². Fig. 19 reproduces NIELSEN's diagram for a type C band in the interval $\varrho=0.2$ to $\varrho=0.02$. One may infer from it that as ϱ increases to unity the central region will build up to produce a "gathered" Q branch, the residual lines producing P and R branches respectively on opposite sides of the Q branch as is consistent with the selection rules of the oblate symmetric rotator.

It should be emphasized that no claim for completeness is made for this representation of the three types of bands for an asymmetric rotator since the molecule adopted is a purely hypothetical one and, moreover, the appearance of the band in an actual case might be strongly influenced by the variations in the values of the moments of inertia between the upper and the lower vibration states.

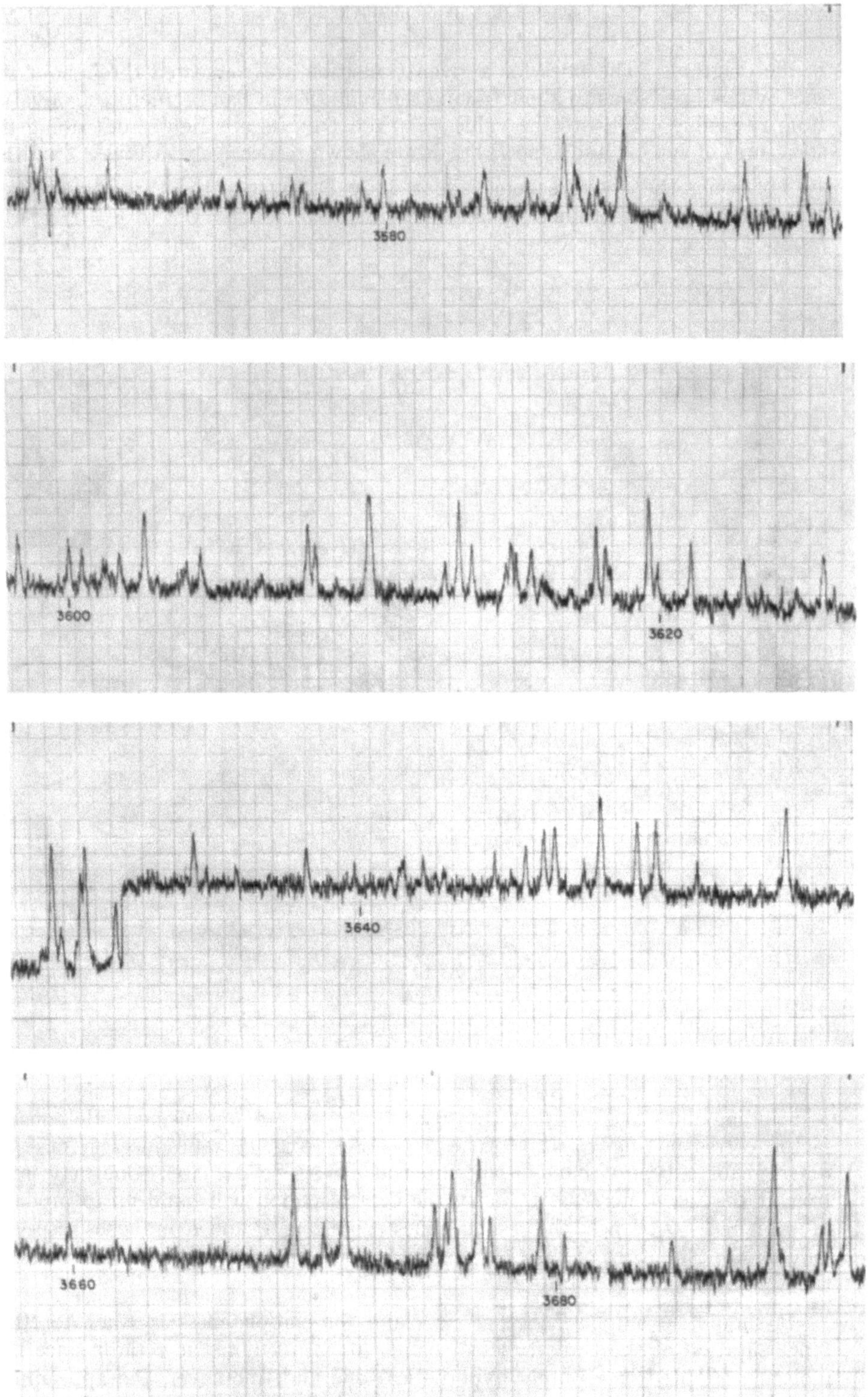

a b c d

Fig. 20 a—d.

Fig. 20 a—l. The 2.7 μ absorption region in the spectrum of H_2O vapor. This region embraces the two bands $\nu_1\,(\mu_b \neq 0)$ and $\nu_3\,(\mu_a \neq 0)$ where ν_3 is much the more intense.

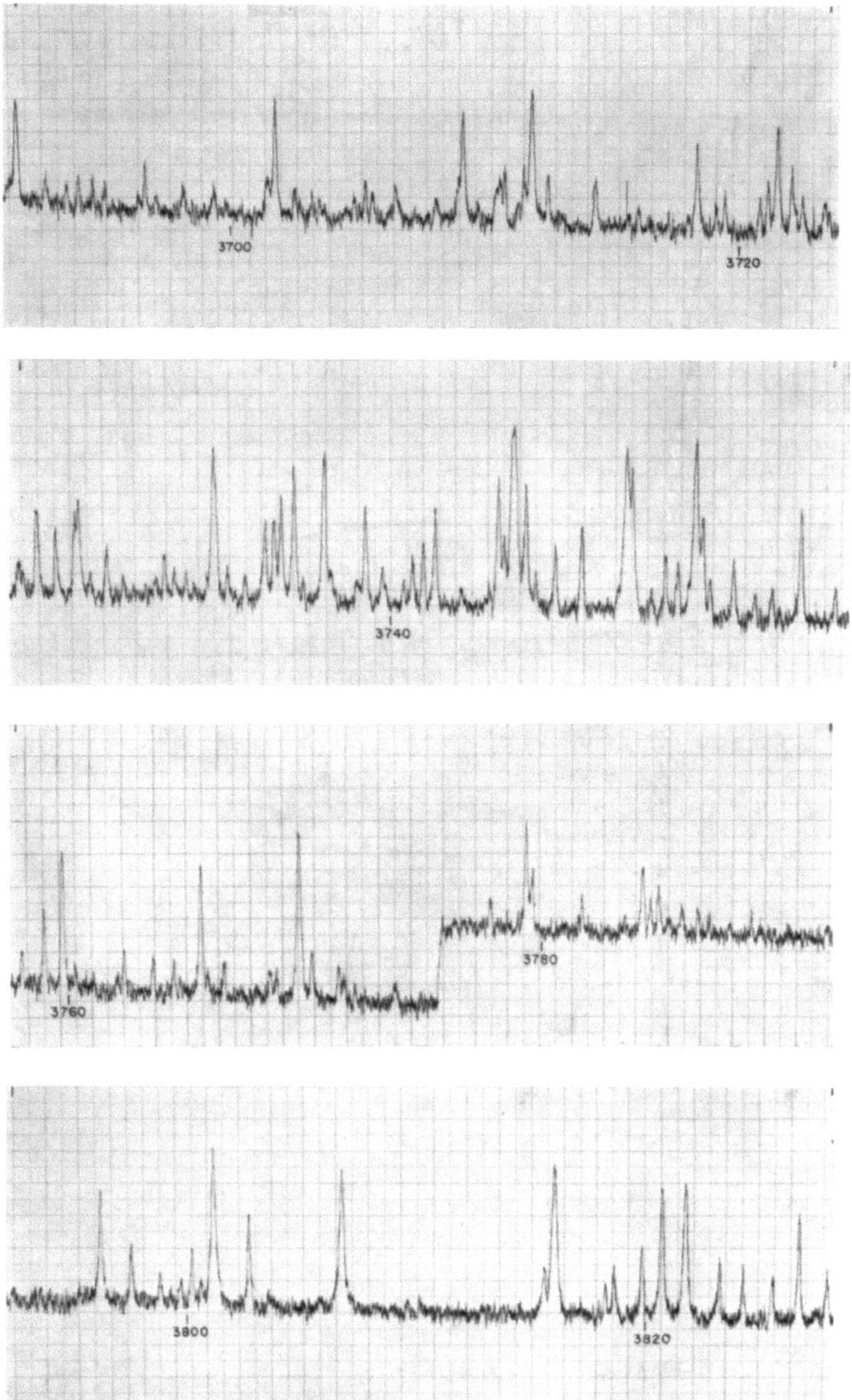

e f g h

Fig. 20 e—h.

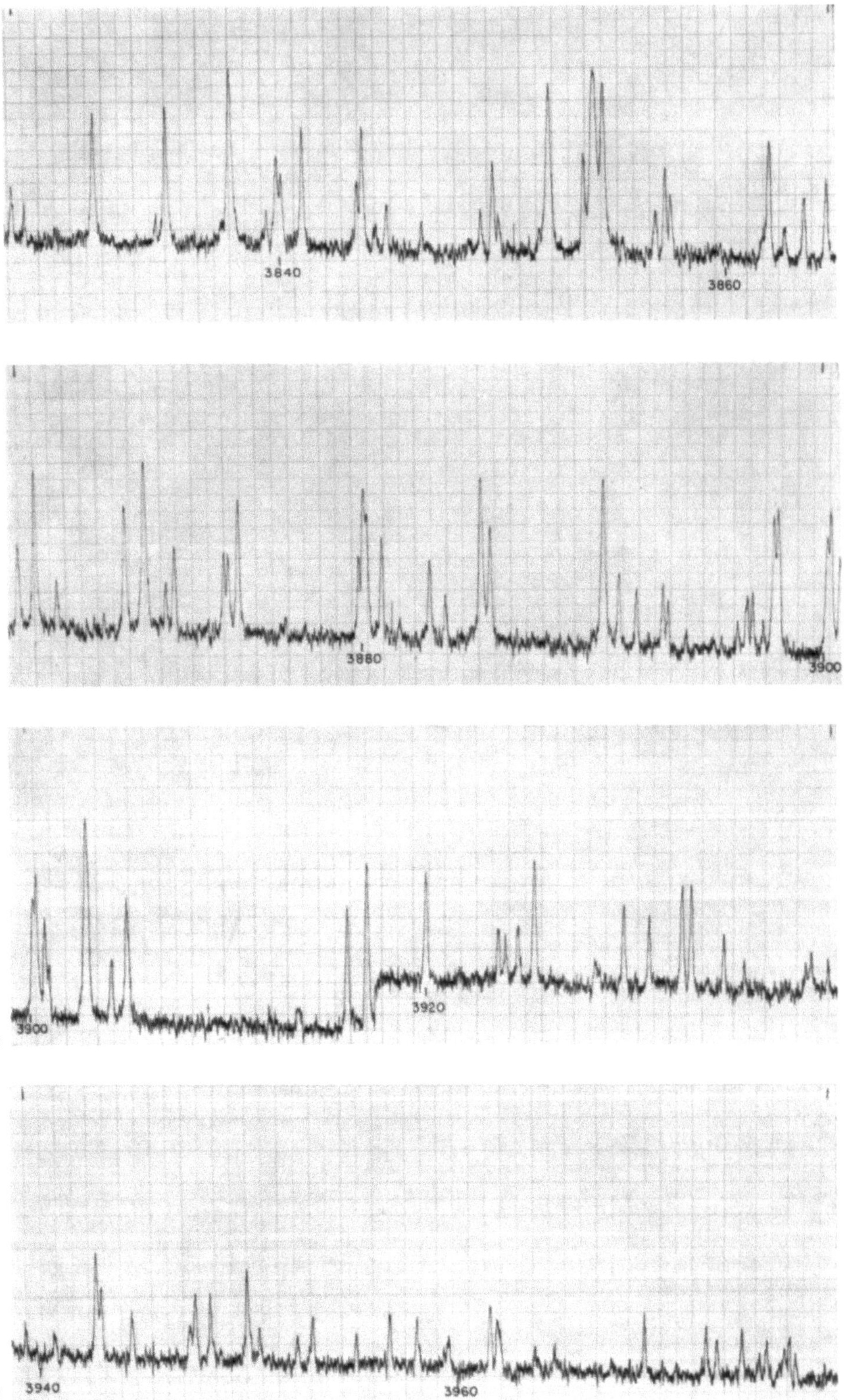

i
j
k
l

Fig. 20 i—l.

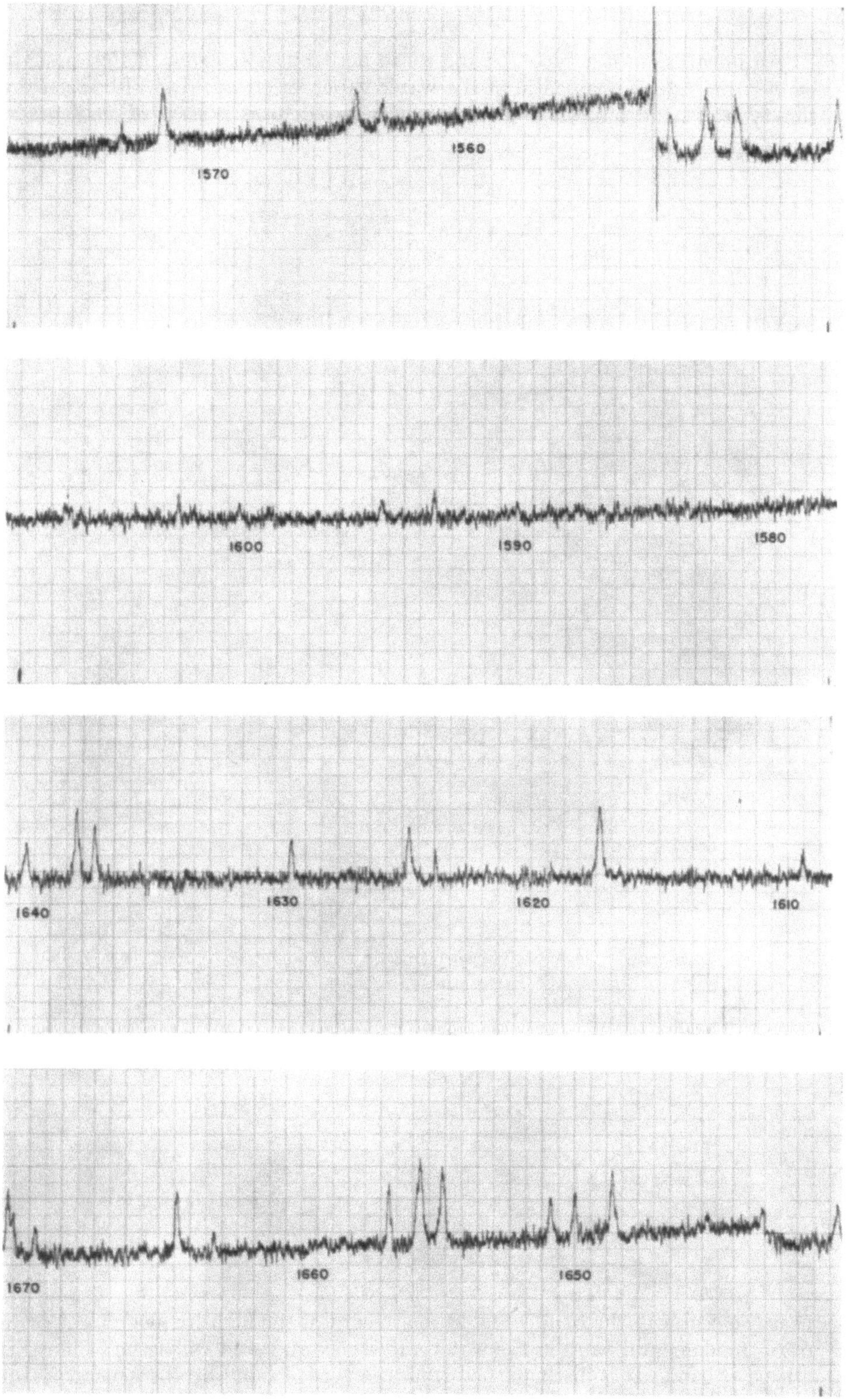

Fig. 21 a—d.

Fig. 21 a—g. The 6.26 μ absorption region in the spectrum of H_2O vapor. This region is the band $\nu_2 (\mu_b \neq 0)$.

An illustration of a type A band in an actual case is given in Fig. 20. This band is the 2.7μ fundamental of the water vapor spectrum. It will be seen that this type of band has what commonly is referred to as a *gathered* "Q" branch

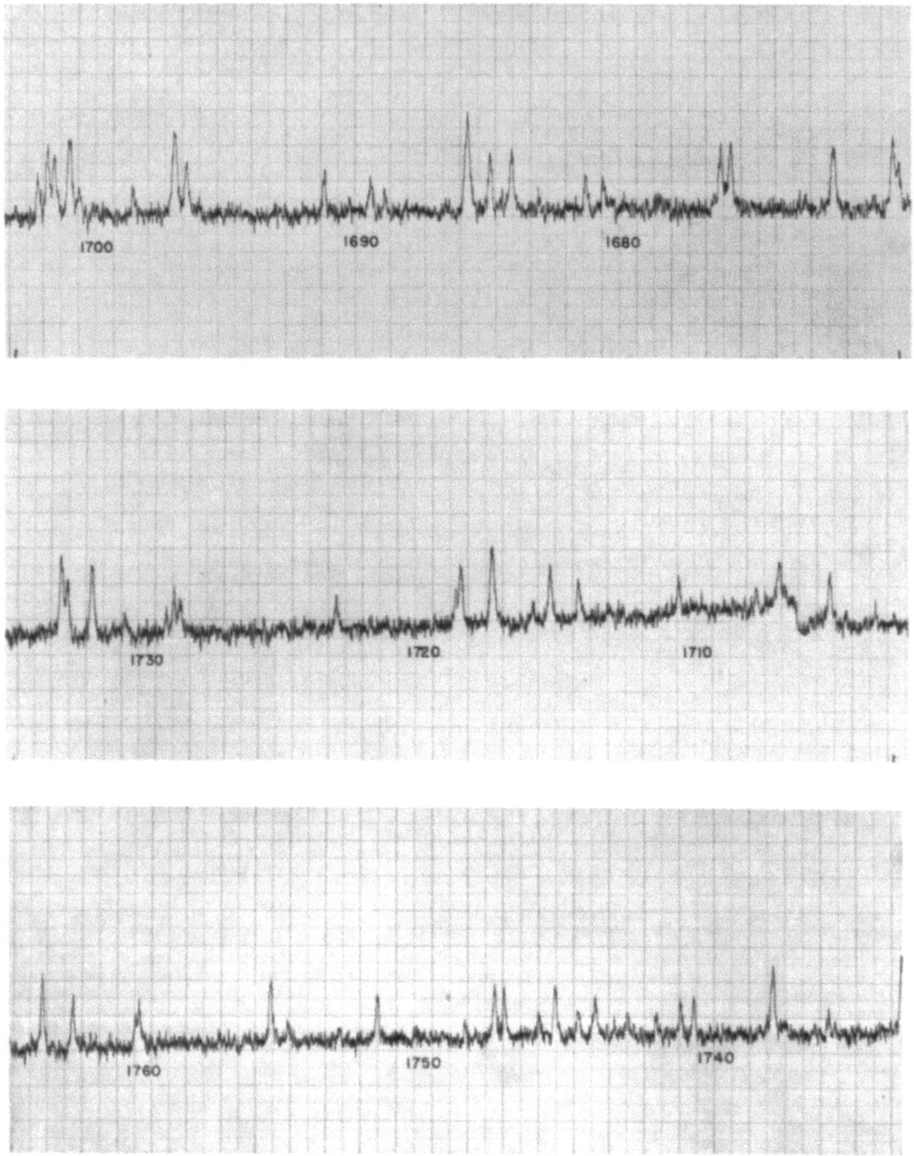

e f g

Fig. 21 e—g.

near the center of the band. Fig. 21 is a record of the high frequency side of the 6.26μ fundamental band in the water vapor spectrum and will serve to illustrate a type B band. It is distinguished from the type A band by the general feature that no *gathered* Q branch is found at the center of the band. The Q structure spreads, in fact, away from the middle of the band leaving a line free gap there.

Water vapor has evidently no type C bands. The three moments of inertia for water vapor are roughly the following $I_a = 1.05 \times 10^{-40}$ g cm^2, $I_b = 1.96 \times 10^{-40}$ g cm^2 and $I_c = 3.06 \times 10^{-40}$ g cm^2.

VII. Anomalies in the infrared spectra of molecules.

41. Introduction. The energy relations given in Chap. V are sufficient to account for many of the details in the infrared spectra of polyatomic molecules. It is an interesting fact, nevertheless, that in virtually every case that has been examined certain details are found to exist which are not accounted for by what has gone before. The anomalies arise from the fact that the assumptions which we have made concerning the Hamiltonian, namely that it may be expanded about the equilibrium values of the nuclear positional coordinates in orders of magnitude and that, to zero order of approximation, complete separation of variables in the wave equation may be effected, are idealized and in some instances are not valid. This fact is reflected in the transformation function T of Chap. V where it may be seen from Table 10 that certain of the basic S functions contain denominators of the type $(\lambda_s - \lambda_{s'})$ and $(\lambda_s + \lambda_{s'} - \lambda_{s''})$. Resonance is said to occur between the frequencies when, for example, either $\lambda_s \approx \lambda_{s'}$ or $\lambda_s + \lambda_{s'} \approx \lambda_{s''}$ so that the denominators approach zero and the usual methods of the perturbation theory fail and the problem must be treated by the degenerate perturbation methods. The resonances referred to above are first order effects, but second and even third order resonances are known to exist. Other effects which make themselves evident are those of l-type doubling and Λ-type doubling and certain higher order contributions from $U - \overline{U}$. We shall consider these effects in Chap. VII and the anomalies they give rise to in the spectra of molecules.

42. l-type doubling in axially symmetric molecules. It was seen already in Chap. IV that the degeneracy in the quantum numbers l and K was partially removed by the degenerate Coriolis operators $p_\alpha^* P_\alpha / I_e^{(\alpha\alpha)}$. Each Coriolis component, however, remained twofold degenerate. There are terms in the Hamiltonian, however, which will further remove these degeneracies. Such a doubling occurs in diatomic molecules when the electrons are in electronic states such that their components of angular momentum along the internuclear axis is not zero (i.e., $\Lambda \neq 0$). The doubling originates with a Coriolis interaction between the rotational motion of the nuclei and the motion of the electrons. This doubling is known as *Λ-type* doubling. An entirely analagous type of doubling occurs in linear and axially symmetric polyatomic molecules when the electronic motions are neglected, but where one or more of the perpendicular two-dimensionally isotropic oscillations are excited. This type of doubling originates with an interaction between the vibrational motion of the nuclei and the rotational motion of the molecule and is known as l-type doubling. Since the linear molecule may be regarded as a special case of the axially symmetric molecule and since the l-type doubling is entirely similar in linear and axially symmetric molecules we shall consider them as two examples of the same phenomenon.

We have noted in Chap. V that the terms of $H^{(2)'}$ which have P_x^2, P_y^2 and $i(P_x P_y + P_y P_x)$ as coefficients can, in general, contribute to the energy in second order only when $s = s'$ and $\sigma = \sigma'$. An exception to this occurs in the case of twofold degenerate oscillations where two coordinates, $q_{t,1}$ and $q_{t,2}$, are required to describe the frequency ω_t. Here $s = s' = t$, but σ may take the two values $\sigma = 1$ and $\sigma = 2$. When $q_{t,1}$ and $q_{t,2}$ and their conjugate momenta are replaced by their equivalents in r_t and χ_t as stated earlier and use is made of the fact that

$$A^{(xx)}_{s\sigma s\sigma} - \left((a^{(xx)}_{s\sigma})^2/I^{(e)}_{xx}\right) - \left((a^{(xy)}_{s\sigma})^2/I^{(e)}_{yy}\right) - \left((a^{(xz)}_{s\sigma})^2/I^{(e)}_{zz}\right)$$
$$= A^{(yy)}_{s\sigma' s\sigma'} - \left((a^{(xy)}_{s\sigma'})^2/I^{(e)}_{xx}\right) - \left((a^{(yy)}_{s\sigma'})^2/I^{(e)}_{yy}\right) - \left((a^{(yz)}_{s\sigma'})^2/I^{(e)}_{zz}\right)$$

where σ equals 1 and 2 with σ' taking the values 2 and 1 respectively, the portion of $H^{(2)\prime}$ referred to which is not diagonal in the quantum numbers l and K and therefore not already accounted for in the energy may be written:

$$\left.\begin{aligned}
(H^{(2)\prime}/hc) = \cdots - \sum_s \sum_{-}^{+} \frac{B_e^2}{4\omega_s} \Bigg\{\Bigg[(A^{(xx)}_{s1s1} - A^{(xx)}_{s2s2}) - \\
- \frac{(a^{(xx)}_{s1})^2 + (a^{(xy)}_{s1})^2 - (a^{(xx)}_{s2})^2 - (a^{(xy)}_{s2})^2}{I^{(e)}_{xx}} - \frac{(a^{(xz)}_{s1})^2 - (a^{(xz)}_{s2})^2}{I^{(e)}_{zz}} + (A^{(xy)}_{s1s2} + A^{(xy)}_{s2s1}) - \\
- \frac{a^{(xx)}_{s1} a^{(xy)}_{s2} + a^{(xx)}_{s2} a^{(xy)}_{s1} + a^{(yy)}_{s1} a^{(yx)}_{s2} + a^{(yy)}_{s2} a^{(yx)}_{s1}}{I^{(e)}_{xx}} - \\
- \frac{a^{(xz)}_{s1} a^{(zy)}_{s2} + a^{(xz)}_{s2} a^{(zy)}_{s1}}{I^{(e)}_{zz}} \Bigg] (e^{\pm i\chi_s} r_s)^2 + 4 \sum_{s'} (\zeta^{(x)}_{ss'})^2 \times \frac{\lambda_s}{\lambda_s - \lambda_{s'}} \times \\
\times (e^{\pm i\chi_s})^2 \left[r_s^2 + \left(\frac{p_{r_s}}{\hbar} \pm \frac{i}{r_s} \frac{p_{r_s}}{\hbar} \right)^2 \right] \Bigg\} \left(\frac{P_x}{\hbar} \mp i \frac{P_y}{\hbar} \right)^2 \Big/ I^{(e)}_{xx} - \sum_s \sum_{-}^{+} \left(\frac{B_e^2}{4\omega_s} \right) \times \\
\times \Bigg\{\Bigg[(A^{(xx)}_{s1s1} - A^{(xx)}_{s2s2}) - \frac{(a^{(xx)}_{s1})^2 + (a^{(xy)}_{s1})^2 - (a^{(xx)}_{s2})^2 - (a^{(xy)}_{s2})^2}{I^{(e)}_{xx}} - \\
- \frac{(a^{(xz)}_{s1})^2 - (a^{(xz)}_{s2})^2}{I^{(e)}_{zz}} - (A^{(xy)}_{s1s2} + A^{(xy)}_{s2s1}) + \\
+ \frac{a^{(xx)}_{s1} a^{(xy)}_{s2} + a^{(xx)}_{s2} a^{(xy)}_{s1} + a^{(yy)}_{s1} a^{(yx)}_{s2} + a^{(yy)}_{s2} a^{(yx)}_{s1}}{I^{(e)}_{xx}} + \\
+ \frac{a^{(xz)}_{s1} a^{(zy)}_{s2} + a^{(xz)}_{s2} a^{(zy)}_{s1}}{I^{(e)}_{zz}} \Bigg] (e^{\pm i\chi_s} r_s)^2 \left(\frac{P_x}{\hbar} \pm i \frac{P_y}{\hbar} \right)^2 \Big/ I^{(e)}_{xx}.
\end{aligned}\right\} \quad (42.1)$$

The non-vanishing matrix elements of Eq. (42.1) are readily seen to be these only $(v_s, l_s, K | v_s, l_s \pm 2, K \pm 2)$, $(v_s, l_s, K | v_s, l_s \pm 2, K \mp 2)$, $(v_s, l_s, K | v_s \pm 2, l_s \pm 2, K \pm 2)$ and $(v_s, l_s, K | v_s \pm 2, l_s \pm 2, K \mp 2)$. The latter two of these sets of elements which are non-diagonal in v_s can not contribute to the energy in this approximation and are therefore of no interest to us here. The former two sets may be seen to take the following values

$$\left.\begin{aligned}
&(v_s, l_s, K | H^{(2)}/hc | v_s, l_s \pm 2, K \pm 2) \\
&= \left(\frac{B_e^2}{4\omega_s} \right) \Bigg\{ (A^{(xx)}_{s1s1} - A^{(xx)}_{s2s2}) - \frac{(a^{(xx)}_{s1})^2 + (a^{(xy)}_{s1})^2 - (a^{(xx)}_{s2})^2 - (a^{(xy)}_{s2})^2}{I^{(e)}_{xx}} - \\
&\quad - \frac{(a^{(xz)}_{s1})^2 - (a^{(xz)}_{s2})^2}{I^{(e)}_{zz}} + (A^{(xy)}_{s1s2} + A^{(xy)}_{s2s1}) - \\
&\quad - \frac{a^{(xx)}_{s1} a^{(xy)}_{s2} + a^{(xx)}_{s2} a^{(xy)}_{s1} + a^{(yy)}_{s1} a^{(yx)}_{s2} + a^{(yy)}_{s2} a^{(yx)}_{s1}}{I^{(e)}_{xx}} - \\
&\quad - \frac{a^{(xz)}_{s1} a^{(zy)}_{s2} + a^{(xz)}_{s2} a^{(zy)}_{s1}}{I^{(e)}_{zz}} + 8 \sum_{s'} (\zeta^{(x)}_{ss'})^2 \frac{\lambda_s}{\lambda_s - \lambda_{s'}} \Bigg\} \{[J(J+1) - K(K \pm 1)] \times \\
&\quad \times [J(J+1) - (K \pm 1)(K \pm 2)] [(v_s \mp l_s)(v_s \pm l_s + 2)]\}^{\frac{1}{2}} \\
&= q_a^0 \{[J(J+1) - K(K \pm 1)] [J(J+1) - (K \pm 1)(K \pm 2)] \times \\
&\quad \times [(v_s \mp l_s)(v_s \pm l_s + 2]\}^{\frac{1}{2}}.
\end{aligned}\right\} \quad (42.2)$$

$$\left.\begin{aligned}
&(v_s, l_s, K|H^{(2)}/hc|v_s, l_s \pm 2, K \mp 2)\\
&= \left(\frac{B_e^2}{4\omega_s}\right)\Big\{(A_{s1s1}^{(xx)} - A_{s2s2}^{(xx)}) - \frac{(a_{s1}^{(xx)})^2 + (a_{s1}^{(xy)})^2 - (a_{s2}^{(xx)})^2 - (a_{s2}^{(xy)})^2}{I_{xx}^{(e)}} - \frac{(a_{s1}^{(xz)})^2 - (a_{s2}^{(xz)})^2}{I_{zz}^{(e)}} -\\
&\quad - \big[(A_{s1s2}^{(xy)} + A_{s2s1}^{(xy)}) + \frac{a_{s1}^{(xx)}a_{s2}^{(xy)} + a_{s2}^{(xx)}a_{s1}^{(xy)} + a_{s1}^{(yy)}a_{s2}^{(yx)} + a_{s2}^{(yy)}a_{s1}^{(yx)}}{I_{xx}^{(e)}} +\\
&\quad + \frac{a_{s1}^{(xz)}a_{s2}^{(zy)} + a_{s2}^{(xz)}a_{s1}^{(zy)}}{I_{zz}^{(e)}}\Big\} \times \{[J(J+1) - K(K \mp 1)]\,[J(J+1) - (K\mp 1)(K\mp 2)] \times\\
&\quad \times [(v_s \mp l_s)(v_s \pm l_s + 2)]\}^{\frac{1}{2}} = q_b^0\{[J(J+1) - K(K\mp 1)] \times\\
&\quad \times [J(J+1) - (K\mp 1)(K \mp 2)]\,[(v_s \mp l_s)(v_s \pm l_s + 2)]\}^{\frac{1}{2}}.
\end{aligned}\right\}\text{(42}$$

Also these would ordinarily not be taken account of in this approximation, but when $|l_s| = 1$ they will link the component states where $l_s = +1$ and $l_s = -1$ which are degenerate in l_s and K. They are, therefore, significant and will remove the further degeneracy in l and K giving rise to what is known as l-type doubling.

The quantity $\overline{U}$ contains certain terms of the form $\sum_{\alpha\beta} g_\alpha(q_{s\sigma}) P_\beta$ (α and β taking the values x and y) where the function $g_\alpha(q_{\sigma s})$ is essentially $\sum_\alpha (p_\alpha \Phi)$ and is, therefore, a function of $q_{s\sigma}$ alone. Such terms might, conceivably, also contribute to the l-type doubling. The influence of these terms on the energy cannot reliably be estimated without a knowledge of the dependence of the electronic wave function, Φ, upon the coordinates, $q_{s\sigma}$. It is assumed here that their effect is negligible compared to the term stated in (42.1) because of the insensitiveness of Φ to $q_{s\sigma}$.

We shall examine how the matrix elements (42.2) and (42.3) affect the energy levels of a symmetric molecule in a Π vibration state (i.e., where $|l_s| = 1$). We shall confine ourselves to the levels where $J = 2$ and where K takes the values 2, 1, 0, -1, -2 while $l_s = \pm 1$. Designating the two values of the relation (37.1) corresponding to the two values of L, $L = K + \zeta_s$ and $L = K - \zeta_s$, for K different from zero by E_{+K} and E_{-K} respectively, the secular determinant for the matrix of the perturbed energies factors into the following four sub-determinants where α, β, γ and δ are the matrix elements $(l_s = \pm 1, K = \pm 2 | l_s = \mp 1, K = 0)$, $(l_s = \pm 1, K = \mp 2 | l_s = \mp 1, K = 0)$, $(l_s = \pm 1, K = \pm 1 | l_s = \mp 1, K = \mp 1)$ and $(l_s = \pm 1, K = \mp 1 | l_s = \mp 1, K = \pm 1)$ of $H^{(2)}/hc$.

	$K = \pm 2$, $l_s = \pm 1$	$K = 0$, $l_s = \mp 1$	$K = \mp 2$, $l_s = \pm 1$
$K = \pm 2$, $l_s = \pm l$	$(E_{+2}/hc) - \varepsilon$	α	0
$K = 0$, $l_s = \mp 1$	α	$(E_0/hc) - \varepsilon$	β
$K = \mp 2$, $l_s = \pm 1$	0	β	$(E_{-2}/hc) - \varepsilon$

$= 0$, (42.4)

	$K = +1$, $l_s = +1$	$K = -1$, $l_s = -1$
$K = +1$, $l_s = +1$	$(E_{+1}/hc) - \varepsilon$	γ
$K = -1$, $l_s = -1$	γ	$(E_{+1}/hc) - \varepsilon$

$= 0$,

	$K = +1$, $l_s = -1$	$K = -1$, $l_s = +1$
$K = +1$, $l_s = -1$	$(E_{-1}/hc) - \varepsilon$	δ
$K = -1$, $l_s = +1$	δ	$(E_{-1}/hc) - \varepsilon$

$= 0$. (42.5)

One sees from (42.4), which occurs twice, that while the levels E_{+2}, E_0 and E_{-2} are, indeed, altered by the perturbation the degeneracies of E_{+2} and E_{-2} are not removed. The levels $K=l_s=\pm 1$ and $K=-l_s=\pm 1$, on the other hand, are both split into two components, their centers of gravity remaining at the positions of the unperturbed levels. The correction will be proportional to $(v_s+1)\,J(J+1)$ where $|K|=1$ and $|l_s|=1$ so that the splitting $\Delta\nu$, between the component levels will be $\Delta\nu=2q_a^0(v_s+1)\;J(J+1)$ and $\Delta\nu'=2q_b^0(v_s+1)\;J(J+1)$ respectively. The terms $q_a^0(v_s+1)$ and $q_b^0(v_s+1)$ may, therefore, be combined with the coefficient of $J(J+1)$ in the unperturbed energy. This is equivalent to regarding the two components of the l-doublets as having slightly different B values.

It may be shown that for virtually all axially symmetric molecules except those which belong to point groups C_{4v} and V_d the coefficient $\big[(A_{s1s1}^{(xx)}-A_{s2s2}^{(xx)})-$

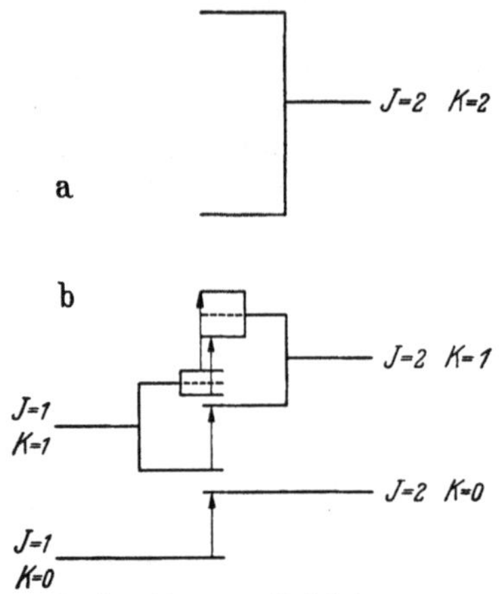

Fig. 22 a and b. Transitions permitted between components of l-type splitting in an axially symmetric molecule.

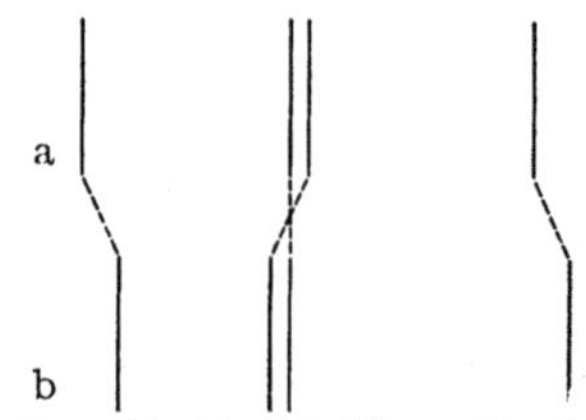

Fig. 23 a and b. l-type doubling pattern for energy levels in an axially symmetric molecule when $J=2$.

$((a_{s1}^{(xx)})^2-(a_{s2}^{(xx)})^2+(a_{s1}^{(xy)})^2-(a_{s2}^{(xy)})^2)/I_{xx}^{(e)}-((a_{s1}^{(xz)})^2-(a_{s2}^{(xz)})^2)/I_{zz}^{(e)}\big]$ is equal to the coefficient $\big[(A_{s1s2}^{(xy)}+A_{s2s2}^{(xy)})-((a_{s1}^{(xx)}a_{s2}^{(xy)}+a_{s2}^{(xx)}a_{s1}^{(xy)}+a_{s1}^{(yy)}a_{s2}^{(yx)}+a_{s2}^{(yy)}a_{s1}^{(yx)})/I_{xx}^{(e)})$ $-((a_{s1}^{(xz)}a_{s2}^{(zy)}+a_{s2}^{(xz)}a_{s1}^{(zy)})/I_{zz}^{(e)})\big]$, i.e., the coefficients of $(e^{\pm i\chi_s})^2\,(P_x\pm iP_y)^2$ vanish. This is tantamount to stating that the matrix elements $(v_s, l_s, K\,|\,v_s, l_s\pm 2, K\mp 2)$ will vanish for such molecules so that only the one Cortolis component, namely the one where $l_s=K$, i.e., the component where the framework of the rotator is at rest, will be split.

Molecules which have low lying perpendicular vibration levels may have an appreciable population in such states. Such molecules will have spectra in the microwave region associated with transitions between rotation levels within such a state. These spectra depend upon the permanent electric moment which is directed along the axis of symmetry and they have been used to study l-type doubling in axially symmetric molecules. The selection rules in such cases will, evidently, be $\Delta J=+1$, $\Delta K=0$. It is readily shown, moreover, that a selection rule exists which restricts the transitions to be from a given l-doubling component of $|l_s|=1$, $|K|=1$ for a given J value to the corresponding l-doubling component of $|l_s|=1$, $|K|=1$ for the value of J equal to $J+1$. This is illustrated in Fig. 22 in a case where the transitions are between levels where $J=1$ and $J=2$ and for a molecule of the C_{3v} symmetry species where only the one Coriolis component is split by l-type doubling. Three lines would be observed if centrifugal stretching is neglected, the central one being double and spaced exactly midway between the other two lines. This is shown in Fig. 23a. The central two lines are due to transitions between the levels where $K=0$ and the degenerate Coriolis components where $K=1$. The two other lines at equal distances on either side of the central two are due to transitions from the two doubling components for $|l_s|=$

$|K|=1$ where $J=1$ to the two l-doubling components for $|l_s|=|K|=1$ where $J=2$. The centrifugal distortion term $J(J+1)L^2 D_{JL}$ in the energy has the effect of shifting the positions of the two outer components by an amount $8D_{JL}\zeta_s^{(z)}$ toward higher or lower frequencies, depending upon the algebraic sign of D_{JL} and ζ, and the one of the two central lines by an equal amount in the opposite sense. The second central line will also be shifted by centrifugal stretching, but by an amount different from the foregoing since it involves a different value of L. The actual pattern as it is observed in the spectrum of CH_3CN is shown in Fig. 23b. Vibration-rotation bands in the near infrared showing anomalies due to l-type doubling have been observed only in the case of linear molecules and such bands will be discussed in the next paragraphs.

43. l-type doubling in linear molecules. The l-type doubling relations are much simplified in the case of linear molecules. The normalization requirements placed on the direction cosines, $l_{is\sigma}^{(\alpha)}$ and the definitions of $A_{s\sigma s\sigma}^{(\alpha\beta)}$ and $a_{s\sigma}^{(\alpha\beta)}$ as stated in Eq. (18.24) lead to the result that $(A_{s1s1}^{(xx)}-A_{s2s2}^{(xx)})=-1$ and that the $a_{s\sigma}^{(\alpha\beta)}$, $(\alpha\neq\beta)$, are equal to zero. Moreover, for linear molecules the $\zeta_s^{(z)}$ are all unity so that according to the relation (34.16) $K=\sum_s l_s^{(z)}$. The relation (42.2) therefore reduces to

$$\left.\begin{aligned}(v_s, l_s = \pm 1, \Lambda = 0, K = \pm 1\,|\,H^{(2)}/hc\,|\,v_s, l_s = \mp 1, \Lambda = 0, K = \mp 1)\\ = -B_e(B_e/2\omega_s)\left\{1 + 4\sum_{s'}(\zeta_{ss'}^{(x)})^2\,\omega_s^2/(\omega_{s'}^2-\omega_s^2)\right\}(v_s+1)\,J(J+1)\\ = q_s^0(v_s+1)\,J(J+1)\end{aligned}\right\}\qquad(43.1)$$

for linear molecules in a Π vibration state. It is of interest to note also that for linear molecules the relation $\sum_s(\zeta_{ss'}^{(x)})^2=1$ may be shown to exist.

We shall compute the magnitude of the l-type doubling for a linear molecule in a Π vibration state v_s, $l_s=1$ and where $\Lambda=0$. The secular determinant for the matrix of the energies will be

$$\begin{array}{c|c|c|}
 & l_s=1,\Lambda=0 & l_s=-1,\ \Lambda=0 \\
\hline
l_s=1,\ \Lambda=0 & (E_0/hc)-\varepsilon & q_s^0(v_s+1)\,J(J+1) \\
\hline
l_s=-1,\ \Lambda=0 & q_s^0(v_s+1)\,J(J+1) & (E_0/hc)-\varepsilon \\
\hline
\end{array} = 0 \qquad (43.2)$$

where the terms E_0/hc are the term values for the state in question and may be obtained from Eq. (37.1) by setting $L=0$. The roots of (43.2) are $(E_0/hc)\pm q_s^0(v_s+1)J(J+1)$. Hence the actual splitting of the levels will be $\Delta\nu=2q_s^0(v_s+1)\,J(J+1)$. As in axially symmetric molecules we may regard the quantity $q_s^0(v_s+1)\,J(J+1)$ as a correction to the B value of the level so that effectively the two l-doubling components have slightly different B values.

The selection rules for l-type doubling in linear molecules may be summarized in the following fashion. Suppose the initial and final states for a vibrational transition are a Σ^+ and a Π vibration state. The rotation levels in a Σ^+ state may be classified as $+$ for even values of J and $-$ for odd values of J. There are two components in a Π state; namely, Π^+ and Π^-. The first value of J in a Π state is $J=1$ since J must always be greater than or equal to $|l_s|$ and for the Π^+ component the rotation levels may be classified as $+$ and $-$ for even and odd values of J, respectively. Similarly for the Π^- component, odd and even values of J

may be classified as + and − respectively. The general transition rule in quantum mechanics which states that + rotation levels can combine only with − rotation levels and *vice versa* requires that for $\Delta J = 0$, the transitions must be from Σ^+ to Π^- while for $\Delta J = \pm 1$ the transitions take place between Σ^+ and Π^+. When a $\Pi - \Pi$ transition takes place one will have for $\Delta J = 0$ that transitions must be between Π^- and Π^+; but for $\Delta J = \pm 1$ the selection rule states that Π^+ combine with Π^+ and Π^- with Π^-.

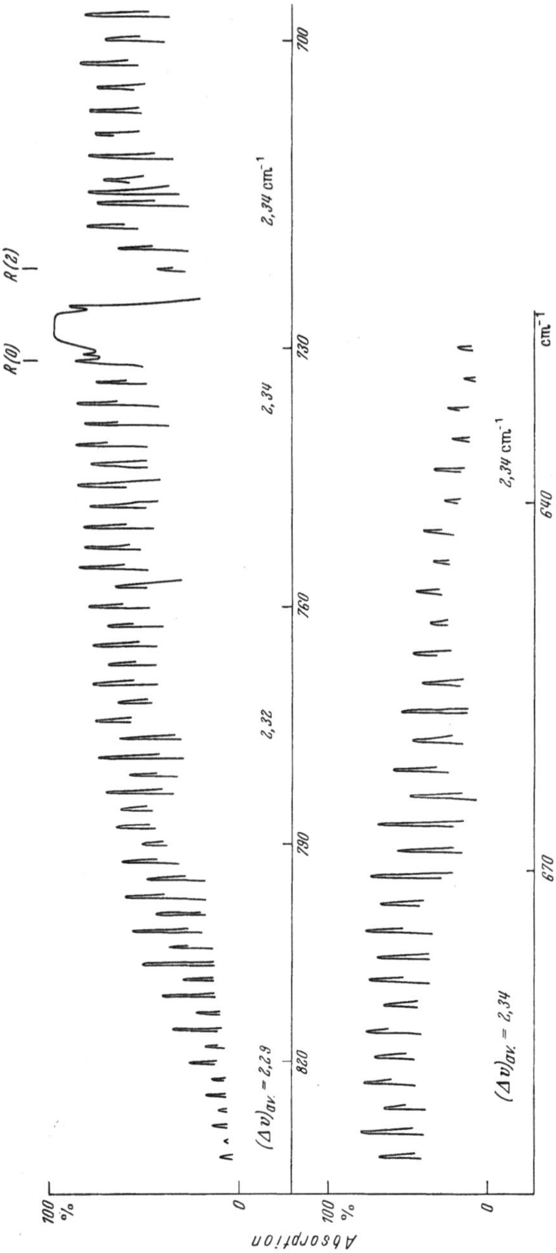

Fig. 24. The band ν_5 in the spectrum of C_2H_2 showing the effects of *l*-type doubling. Note the convergence in the *P* and *R* branches is toward higher frequencies while the *Q* branch converges toward lower frequencies.

It is evident from the foregoing that the effective values of $(B' - B'')$ will not be the same for the Q branch as for the P and R branches since the effective values of B' will be different for the $\Delta J = 0$ transitions from the $\Delta J = \pm 1$ transitions. Cases are, in fact, known where the Q branch converges in one sense while the P and R branches converge in the opposite sense. This is illustrated in Fig. 24 which is a record of ν_5, a perpendicular band in the spectrum of C_2H_2.

Measurements in the microwave region of the spectrum on the *l*-type doubling in certain molecules, particularly the HCN molecule, have shown that the quantity q_s^0 is not quite constant, but has a slight variation proportional to $J(J+1)$. This effect has also been described as due to a centrifugal distortion of the molecule. The effect can be justified theoretically, but involves a complicated fourth order perturbation calculation which can not here be discussed. An improved value of q_s^0 which is probably sufficiently good in practically all cases is the following one

$$q_s^0 = -B_e(B_e/2\omega_s)\Big[1 + 4\sum_{s'}(\zeta_{ss'}^{(x)})^2\,\omega_s^2/(\omega_{s'}^2 - \omega_s^2)\Big]\big[1 - 4(B_e/2\omega_s)^2 J(J+1)\big]. \quad (43.3)$$

An interesting example of the influence of the matrix elements (42.2) occurs in linear molecules when $v_t = 2$ and where accordingly $l_t = \pm 2$ and $l_t = 0$, all other $l_{t'}$ being equal to zero. We have in this instance a Δ-vibration state and a Σ vibration state. The non-vanishing off-diagonal matrix elements, which connect these states, are the following

$$(v_t = 2,\, l_t = \pm 2 \,|\, H^{(2)}/hc \,|\, v_t = 2,\, l_t = 0) = (v_t = 2,\, l_t = 0 \,|\, H^{(2)}/hc \,|\, v_t = 2,\, l_t = \pm 2)$$
$$= -2\sqrt{2}\, B_e (B_e/\omega_t) \Big\{1 + 4 \sum_s (\zeta_{ts}^{(x)})^2 \omega_t^2/(\omega_s^2 - \omega_t^2)\Big\} [f(f-2)]^{\frac{1}{2}} = A$$

where $f = J(J+1)$. Denoting by $\overline{T}_2$ and $\overline{T}_0$, respectively, the term values of the states $v_t = 2$, $l_t = \pm 2$ and $l_t = 0$ for a linear molecule, the secular determinant for the *l*-type splitting will be for this case

	$v_t = l_t = 2$	$v_t = 2,\, l_t = 0$	$v_t = -l_t = 2$	
$v_t = l_t = 2$	$\overline{T}_2 - \varepsilon$	A	0	
$v_t = 2$, $l_t = 0$	A	$\overline{T}_0 - \varepsilon$	A	$= 0.$
$v_t = -l_t = 2$	0	A	$\overline{T}_2 - \varepsilon$	

The roots of this will be

$$\varepsilon = \overline{T}_2$$

and

$$\varepsilon = \tfrac{1}{2}\{(\overline{T}_2 + \overline{T}_0) \pm [(\overline{T}_2 - \overline{T}_0)^2 + 8A^2]^{\frac{1}{2}}\}.$$

It is readily shown that $(\overline{T}_2 - \overline{T}_0) = 4(x_{l_t l_t} - B_e)$. Moreover, the magnitude of $(\overline{T}_2 - \overline{T}_0)$ is, in general, large compared to $B_e(B_e/\omega_t)\Big\{1 + 4\sum_s (\zeta_{st}^{(x)})^2 \omega_t^2/(\omega_s^2 - \omega_t^2)\Big\}$ so that the expressions for the latter two roots stated above may be expanded to give approximately

$$\varepsilon_+ = \overline{T}_2 + \Big\{4(B_e^2/\omega_t)^2 \Big[1 + 4\sum_s (\zeta_{st}^{(x)})^2 \omega_t^2/(\omega_s^2 - \omega_t^2)\Big]^2 \Big/ (x_{l_t l_t} - B_e)\Big\} J^2(J+1)^2$$

and

$$\varepsilon_- = \overline{T}_0 - \Big\{4(B_e^2/\omega_t)^2 \Big[1 + 4\sum_s (\zeta_{st}^{(x)})^2 \omega_t^2/(\omega_s^2 - \omega_t^2)\Big]^2 \Big/ (x_{l_t l_t} - B_e)\Big\} J^2(J+1)^2.$$

The result is that in the two states ε_+ and ε_- the centrifugal stretching constants are effectively altered, their values being

$$(D_{\text{eff}})_\mp = D \pm 4(B_e^2/\omega_t)^2 \Big[1 + 4\sum_s (\zeta_{st}^{(x)})^2 \omega_t^2/(\omega_s^2 - \omega_t^2)\Big]^2 \Big/ (x_{l_t l_t} - B_e).$$

This effect might better be referred to as *l*-type *resonance* than *l*-type doubling.

Λ-type doubling in linear polyatomic molecules.

When a linear molecule is in a Σ vibration state, i.e., $\sum_s l_s = 0$, the degeneracy in the quantum number Λ, if it is different from zero, may be removed by certain terms in the quantity $U - \overline{U}$ of the form $\sum_{\alpha\beta} \Pi_\alpha P_\beta$ (α and β taking the values x and y) when the spins are neglected, which have hitherto not been considered. These terms have only the matrix components $(\Lambda|\Lambda\pm 1)$ linking two different electronic states. Since $\sum_{\alpha\beta}\Pi_\alpha P_\beta$ have no diagonal matrix elements they cannot contribute to the energy in first order. These terms can, as was done in Chap. V, be transformed by a contact transformation to become a part of the transformed second order HAMILTONian, $H^{(2)\prime}$, where they will produce terms having $B_e(B_e/\nu_e)\,(P_\alpha/\hbar)^2$ and $i\,B_e(B_e/\nu_e)\,(P_\alpha P_\beta+P_\beta P_\alpha)/\hbar^2$ as coefficients, ν_e being the electronic frequency. The first of these may have matrix elements which are diagonal in Λ. These will be proportional to $[J(J+1)-\Lambda^2]\hbar^2$ and may be incorporated with the equilibrium value of the moment of inertia for this particular electronic state. All of these have elements of the variety $(\Lambda|\Lambda\pm 2)$ and these give rise to Λ-type doubling. We shall here refer only to a singlet Π state where the doubling is proportional to $J(J+1)\,\hbar^2$. We obtain the following elements of the energy matrix, non-diagonal in the quantum number Λ for a singlet Π electronic state with $\sum_s l_s=0$.

$$\left.\begin{aligned}(v_s,\ l_s=0,\ \Lambda=\pm 1\,|\,H^{(2)\prime}/hc\,|\,v_s,\ l_s=0,\ \Lambda=\mp 1,\ K=\mp 1)\\ =\text{const}\ B_e(B_e/\nu_e)\,J(J+1).\end{aligned}\right\}\tag{43.4}$$

l-type doubling may, of course, also occur in states other than vibration states where $l_s>1$. Such splittings have never been observed and it is possible to show for singlet states that the l-type splitting will be of the order of magnitude $B_e(B_e/\omega_s)^{2l_s-1}$. Correspondingly Λ-type doubling will be of the order of magnitude $B_e(B_e/\nu_e)^{2\Lambda-1}$. The splitting may, therefore, be expected to diminish rapidly with l_s or Λ.

The energy levels of a linear molecule, where the electrons are in a state of non-zero angular momentum (i.e., $\Lambda\neq 0$) and where at the same time the vibration state is one where $\sum_s l_s\neq 0$, will also suffer a splitting which in a general way corresponds to simultaneous l and Λ type doubling. The splitting is, however, considerably more complicated as we shall see. The simple theory of l-type doubling and Λ-type doubling does not apply when both Λ and $\sum_s l_s$ are different from zero.

We have noted in a footnote in Sect. 38 that because of the complete cylindrical symmetry of a linear molecule in a Σ vibration state the angle α_1 does not occur in the Hamiltonian except as a derivative and is, therefore, an ignorable coordinate. The function, Φ, is, as a consequence, expressible as

$$\Phi=e^{i\Lambda\alpha_1}\,\Phi'(q_{s\sigma},\beta_2,\beta_3,\cdots\beta_j,\varrho_j)\tag{43.5}$$

where Λ is an integer and $\beta_j=\alpha_j-\alpha_1$. When a perpendicular oscillation occurs the nuclei are displaced at right angles to the internuclear axis and the displacements are best described in terms of a radial coordinate, r, which makes an angle, χ, with the xz plane. The force field in which an electron finds itself will no longer be cylindrically symmetric, but will now depend upon the angle $\alpha_1-\chi$. A perturbing term, $\sigma(r)=(er/\varrho^2)\cos(\alpha_1-\chi)$, where e may be thought of as effective nuclear charge, r, the displacement and ϱ, the distance of the first electron to the dipole. It is apparent how similar the problems of HANSEN and DENNISON and of KELLER referred to in Chap. V in which one excited perpendicular vibration destroys the complete isotropy of the force field for a second perpendicular vibration are to this problem.

The angle, α_1, is no longer an ignorable coordinate and the function Φ transforms now as $e^{i\Lambda(\alpha_1-\chi)}$ rather than as $e^{i\Lambda\alpha_1}$. The perturbing term, $\sigma(r)$, has no diagonal elements in Λ and will therefore yield no first order contribution to the energy. $\sigma(r)$ does, however, have $(\Lambda|\Lambda \pm 1)$ matrix elements which would connect such states as Σ and Π-states, Π- and Δ-states, etc., so that a second order effect may result. The second order contribution to the energy matrix will, as is well known, be proportional to $(\Lambda|\sigma(r)|\Lambda \pm 1) \cdot (\Lambda \pm 1|\sigma(r)|\Lambda)$ and $(\Lambda|\sigma(r)|\Lambda \pm 1) \times (\Lambda \pm 1|\sigma(r)|\Lambda \pm 2)$. The first of these terms may produce a displacement of the energy levels, E_e, but can not produce a splitting. The second term gives no contribution, to this approximation, except when $\Lambda = \pm 1$, i.e., for Π electronic states, in which case elements exist in the energy matrix connecting $\Lambda = +1$ and $\Lambda = -1$. Thus the degeneracy in Λ is removed and the electronic energies may be written as $E_e^{\pm} = E_e \pm \varepsilon$, where ε is, according to the foregoing, proportional to r^2. This function is further augmented by the quadrupole associated with the nuclear displacement.

The degeneracy in Λ, for a Δ-electronic state where Φ transforms as $e^{\pm 2i(\alpha_1 - \chi)}$ is not removed before in fourth order of approximation, since the off-diagonal elements will have to be of the variety $(2|\sigma(r)|1)(1|\sigma(r)|0)(0|\sigma(r)|-1)(-1|\sigma(r)|-2)$. Here the splitting will be proportional to r^4 as will be the quadrupole moment associated with r. One may demonstrate by this method of reasoning that the degeneracy in E_e is removed in an order 2Λ. The magnitude of the perturbation in a Π-electronic state is, therefore, of the order of magnitude of the vibrational energy, but diminishes rapidly for electronic states of higher values of Λ. Corresponding to the two energies, E_+^e and E_e^-, there will now be two electronic wave functions $\Phi^+ = \cos(\alpha_1 - \chi)\Phi'$ and $\Phi^- = \sin(\alpha_1 - \chi)\Phi'$.

It was assumed in Chap. III that the dependence of Φ on the nuclear displacement coordinates was small so that separation of variables could be achieved by adopting a simple product ΦR for ψ. Moreover, the last term of the quantity U becomes important, as we shall see, under the conditions we have just described. The last term in U may be taken to be

$$U_1 = -(1/2\Phi R) \sum_t \{R p_t^2 \Phi + 2(p_t \Phi)(p_t R)\} \tag{43.6}$$

where t are the values of s for which the vibration is normal to the internuclear axis. When the operators, p_t, are replaced by their equivalents in r_t and χ_t and account is taken of the fact that Φ is still insensitive to r (i.e., $\partial\Phi/\partial r_t \approx 0$), the term, U_1, becomes

$$U_1 = \cdots \sum_t (\hbar^2/2\Phi R)\{(R/r_t^2)(\partial^2\Phi/\partial\chi_t^2) + (2/r_t^2)(\partial\Phi/\partial\chi_t)(\partial R/\partial\chi_t)\}. \tag{43.7}$$

It is evident that U_1 may give rise to a contribution to the energy of the same order of magnitude as the term $(1/r_t^2)(\partial^2 R/\partial\chi_t^2)$ in Eq. (19.2).

This problem has been studied by RENNER who has shown that the simple product relation must now be replaced by a function

$$\Phi^+ R^+ + \Phi^- R^- \tag{43.8}$$

where R^+ and R^- are such functions of r and χ that (43.8) is a suitable solution to the Eq. (19.2). A separation of variables, using the function (43.8) as a solution to Eq. (19.2) may still be carried out much as before, but now instead of Eq. (19.2) there will be two equations, namely:

$$\left\{\frac{\partial^2}{\partial r^2} + \frac{1}{r}\frac{\partial}{\partial r} + \frac{1}{r^2}\frac{\partial^2}{\partial\chi^2} - \frac{\Lambda^2}{r^2} + [2(E - E_e \mp \varepsilon(r))/\hbar\lambda^{\frac{1}{2}}] - r^2\right\} R^{\pm} \pm \frac{2}{r^2}\Lambda\frac{\partial R^{\mp}}{\partial\chi} = 0, \tag{43.9}$$

the equations associated with the upper and lower algebraic signs being identified with the electronic states E_e^- and E_e^+ respectively. Evidently the quantities E_e

and ε in Eq. (43.9) may be written $E_e = \frac{1}{2}(E_e^+ + E_e^-)$ and $\varepsilon = \frac{1}{2}(E_e^+ - E_e^-)$ respectively. The effect of the perturbation is, therefore, to give rise to two electronic states with two somewhat different potential energy functions. The two equations indicated in (43.9) are coupled to each other through the terms $\pm(2/r^2)\Lambda(\partial R^{\mp}/\partial\chi)$ and to arrive at a solution, the two equations must be solved simultaneously.

When the function ε is made to go to zero, the two Eq. (43.9) must become the same and degenerate into that of the two dimensional oscillator. They may, in fact, be reduced to the form of Eq. (19.2) when ε is set equal to zero by introducing for R^+ and R^- in Eq. (43.9) the functions $R^+ = e^{\pm i\Lambda\chi} F(r, \chi)$ and $R^- = \mp i e^{\pm i\Lambda\chi} F(r, \chi)$, respectively where $F(r, \chi)$ is the function (21.3) which has already been determined to be the solution to the two dimensionally isotropic oscillator.

Renner studied the Π electronic states where $\Lambda = \pm 1$ so that $\varepsilon = a r^2$ and in particular the states where $(l_s + \Lambda) = 0$ in which case the coupling term $(2/r^2)\Lambda(\partial R^{\mp}/\partial\chi)$ drops out. An exact solution may be obtained in this case and one obtains for $E_s^{\pm}$ the following

$$E_s^{\pm} = h c \omega_s (1 \pm a)^{\frac{1}{2}} (v_s + 1). \tag{43.10}$$

The energies for other values of $l_s + \Lambda$ may be arrived at by perturbation methods, but this lies beyond the scope of this section. It is sufficient to say that the splitting may be such that the energy level pattern for perpendicular vibrations will not necessarily any longer form a simple arrangement in linear molecules where $\Lambda \neq 0$[1].

44. First order anharmonic resonance interactions. Two vibration frequencies, say $\omega_s + \omega_{s'}$ and $\omega_{s''}$, will be coupled through a term in the cubic portion of the potential energy $h c k_{s s' s''} q_{s\sigma} q_{s'\sigma} q_{s''\sigma''}$. Such a term may normally be removed from the first order Hamiltonian by means of a contact transformation which, however, introduces a second order term with the denominator $\omega_s + \omega_{s'} - \omega_{s''}$. When the frequency $\omega_s + \omega_{s'}$ is nearly equal to $\omega_{s''}$ resonance sets in and the term $h c k_{s s' s''} q_{s\sigma} q_{s'\sigma'} q_{s''\sigma''}$ may produce a first order contribution to the energy of the molecule. It is then no longer a legitimate procedure to use the transformation function, S, given in Chap. V to remove this term from $H^{(1)}$. It is further evident that the constants x_{ss}, $x_{ss'}$, $x_{l_s l_s}$, $B_v^{(\alpha\alpha)}$, etc., must be modified when resonance occurs since they are a product of the second order perturbation calculation outlined in Chap. V and in the form they occur in that section they all contain contributions containing resonance denominators such as $\omega_s + \omega_{s'} - \omega_{s''}$. We shall consider here the modifications that resonance introduces into the vibration-rotation energies of a polyatomic molecule, specifically the effect upon the quantities x_{ss}, $x_{ss'}$, $x_{l_s l_s}$, $B_v^{(\alpha\alpha)}$, etc.

An examination of the problem will show that in a conventional second order perturbation calculation it would here be only those contributions involving the matrix elements $(v_{s\sigma}, v_{s'\sigma'}, v_{s''\sigma''} | v_{s\sigma} \pm 1, v_{s'\sigma'} \pm 1, v_{s''\sigma''} \mp 1)$ that would offer

[1] It has been noted by KELLER that much better and more consistent agreement among the anharmonic potential energy constants is achieved in the C_2H_2 molecule if one assumes that, in addition to the terms permitted by the $D_{\infty}h$ symmetry of the model, the term $k_{345} q_3 (q_{4,1} q_{5,1} + q_{4,2} q_{5,2})$ is also included. This term becomes $k_{345} q_3 r_4 r_5 \cos(\chi_4 - \chi_5)$ when polar coordinates are introduced instead of $q_{4,\sigma}$ and $q_{5,\sigma}$. A similar term of this variety occurs in the quartic portion of the potential energy function. Such terms would contribute to the constants $x_{3,4}$, $x_{3,5}$, $x_{4,5}$ and would introduce a new term $x_{l_4 l_5} l_4 l_5$ into the energy expressions. This would imply that when one perpendicular vibration is excited, a second perpendicular vibration can not be expected to vibrate in a completely isotropic force field, but will depend upon the angle $\chi_4 - \chi_5$. The work of KELLER would indicate that a situation entirely analogous to that described by RENNER occurs when two or more perpendicular bands are simultaneously excited.

difficulties since these terms would be the only ones containing the resonance denominators $\omega_s + \omega_{s'} - \omega_{s''}$. This suggests that a transformation function, S^*, might be adopted which will transform $H^{(1)}$ into $H^{(1)\prime}$ where $H^{(1)\prime}$ will have the same $(v_{s\sigma}, v_{s'\sigma'}, v_{s''\sigma''} | v_{s\sigma} \pm 1, v_{s'\sigma'} \pm 1, v_{s''\sigma''} \pm 1)$ elements as the term $hc k_{ss's''} q_{s\sigma} q_{s'\sigma'} q_{s''\sigma''}$, but will have all the other matrix components equal to zero, instead of the function, S, given in Table 10. Evidently such a function, S^*, would have no resonance denominator $\omega_s + \omega_{s'} - \omega_{s''}$. The following function, S^*,

$$\left.\begin{aligned} S^* = {} & \frac{2\pi c\, k_{ss's''}}{4(\lambda_s^{\frac{1}{2}} + \lambda_{s'}^{\frac{1}{2}} + \lambda_{s''}^{\frac{1}{2}})(\lambda_s^{\frac{1}{2}} - \lambda_{s'}^{\frac{1}{2}} + \lambda_{s''}^{\frac{1}{2}})(\lambda_s^{\frac{1}{2}} - \lambda_{s'}^{\frac{1}{2}} - \lambda_{s''}^{\frac{1}{2}})} \times \\ & \times \Big\{ (\lambda_s + \lambda_{s'} + \lambda_{s''} - 2\lambda_s^{\frac{1}{2}}\lambda_{s'}^{\frac{1}{2}} + 2\lambda_s^{\frac{1}{2}}\lambda_{s''}^{\frac{1}{2}} + 2\lambda_{s'}^{\frac{1}{2}}\lambda_{s''}^{\frac{1}{2}}) \left(\frac{p_{s\sigma}}{\hbar}\,\frac{p_{s'\sigma'}}{\hbar}\,\frac{p_{s''\sigma''}}{\hbar}\right) - \\ & - (\lambda_s + \lambda_{s'} - 3\lambda_{s''} - 2\lambda_s^{\frac{1}{2}}\lambda_{s'}^{\frac{1}{2}} - 2\lambda_s^{\frac{1}{2}}\lambda_{s''}^{\frac{1}{2}} - 2\lambda_{s'}^{\frac{1}{2}}\lambda_{s''}^{\frac{1}{2}}) \left(q_{s\sigma}\, q_{s'\sigma'}\, \frac{p_{s''\sigma''}}{\hbar}\right) + \\ & + (\lambda_s - 3\lambda_{s'} + \lambda_{s''} + 2\lambda_s^{\frac{1}{2}}\lambda_{s'}^{\frac{1}{2}} + 2\lambda_s^{\frac{1}{2}}\lambda_{s''}^{\frac{1}{2}} - 2\lambda_{s'}^{\frac{1}{2}}\lambda_{s''}^{\frac{1}{2}}) \left(q_{s\sigma}\, \frac{p_{s'\sigma'}}{\hbar}\, q_{s''\sigma''}\right) + \\ & + (-3\lambda_s + \lambda_{s'} + \lambda_{s''} + 2\lambda_s^{\frac{1}{2}}\lambda_{s'}^{\frac{1}{2}} - 2\lambda_s^{\frac{1}{2}}\lambda_{s''}^{\frac{1}{2}} + 2\lambda_{s'}^{\frac{1}{2}}\lambda_{s''}^{\frac{1}{2}}) \left(\frac{p_{s\sigma}}{\hbar}\, q_{s'\sigma'}\, q_{s''\sigma''}\right) \Big\}. \end{aligned}\right\} \quad (44.1)$$

will, indeed, accomplish this and the first order Hamiltonian will be transformed into

$$H^{(1)} = (hc k_{ss's''}/4)\,\{q_{s\sigma} p_{s'\sigma'} + p_{s\sigma} q_{s'\sigma'})\,(p_{s''\sigma''}/\hbar^2) - [(p_{s\sigma} p_{s'\sigma'}/\hbar^2) - q_{s\sigma} q_{s'\sigma'}]\, q_{s''\sigma''}\}. \quad (44.2)$$

Since the frequencies ω_s, $\omega_{s'}$ and $\omega_{s''}$ are non-degenerate frequencies in nearly all instances where a term $hc k_{ss's''} q_{s\sigma} q_{s'\sigma'} q_{s''\sigma''}$ may occur no ambiguity will arise by omitting the σ's entirely.

One proceeds by considering $\lambda_{s''}^{\frac{1}{2}} \approx \lambda_s^{\frac{1}{2}} + \lambda_{s'}^{\frac{1}{2}}$ (i.e., $\omega_{s''} \approx \omega_s + \omega_{s'}$) so that $\lambda_{s''}^{\frac{1}{2}}$ may be written $\lambda_{s''}^{\frac{1}{2}} = \lambda_s^{\frac{1}{2}} + \lambda_{s'}^{\frac{1}{2}} + \delta$ where δ is small. We write that portion of $H^{(0)}$ and $H^{(1)}$ which concerns the frequencies $\lambda_s^{\frac{1}{2}}$, $\lambda_{s'}^{\frac{1}{2}}$ and $\lambda_{s''}^{\frac{1}{2}}$ as

$$H^{(0)} = (\hbar/2)\,\{\lambda_s^{\frac{1}{2}}[(p_s/\hbar)^2 + q_s^2] + \lambda_{s'}^{\frac{1}{2}}[(p_{s'}/\hbar)^2 + q_{s'}^2] + (\lambda_s^{\frac{1}{2}} + \lambda_{s'}^{\frac{1}{2}})\,[(p_{s''}/\hbar)^2 + q_{s''}^2]$$

and

$$\begin{aligned} H^{(1)\prime\prime} = {} & (\delta\hbar/2)\,[(p_{s''}/\hbar)^2 + q_{s''}^2] + \\ & + (hc k_{ss's''}/4)\,\{(q_s p_{s'} + p_s q_{s'})\,(p_{s''}/\hbar^2) - [(p_s p_{s'}/\hbar^2) - q_s q_{s'}]\, q_{s''}\}. \end{aligned}$$

The zero order term value, of E_0/h_c, will evidently be

$$(E^0/hc) = \omega_s(v_s + v_{s''} + 1) + \omega_{s'}(v_{s'} + v_{s''} + 1). \quad (44.5)$$

The first order Hamiltonian has the following non-vanishing matrix elements

$$\left.\begin{aligned} & (v_s, v_{s'}, v_{s''} | H^{(1)\prime\prime}/hc | v_s, v_{s'}, v_{s''}) = (v_{s''} + \tfrac{1}{2})\,(\delta/2\pi c) \\ & (v_s, v_{s'}, v_{s''} | H^{(1)\prime\prime}/hc | v_s + 1, v_{s'} + 1, v_{s''} - 1) \\ & \quad = (v_s + 1, v_{s'} + 1, v_{s''} - 1 \,|\, H^{(1)\prime\prime}/hc | v_s, v_{s'}, v_{s''}) \\ & \quad = k_{ss's''}\,\{[(v_s + 1)/2]\,[(v_{s'} + 1)/2]\,[v_{s''}/2]\}^{\frac{1}{2}}. \end{aligned}\right\} \quad (44.6)$$

The transformed second order Hamiltonian, $H^{(2)\prime}$, will remain unaltered except in so far as the terms are concerned which have $k_{ss's''}$ as coefficients. Only the diagonal elements of $H^{(2)\prime}$ are of interest and it is readily verified that these are the same as before except in so far as the constant $x_{ss'}$ is concerned. The term in $x_{ss'}$ which has $k_{ss's''}^2$ as a coefficient will now be

$$(-k_{ss's''}^2/8)\,\{1/(\omega_s + \omega_{s'} + \omega_{s''}) + 1/(\omega_s - \omega_{s'} - \omega_{s''}) + 1/(\omega_s - \omega_{s'} - \omega_{s''})\}. \quad (44.7)$$

When the frequency $\lambda_s^{\frac{1}{2}}$ is equal to $\lambda_{s'}^{\frac{1}{2}}$ the case arises which is generally referred to as Fermi-Dennison resonance. When $\lambda_s^{\frac{1}{2}}$ and $\lambda_{s''}^{\frac{1}{2}}$ are both non-degenerate, the term in the first order Hamiltonian which permits the resonance to occur will be $hck_{sss'}q_s^2 q_{s''}$. We may proceed exactly as before, the appropriate transformation being achieved by using the function (44.1) where $s=s'$ (σ being omitted) and where the order of the operators is preserved. This time one sets $\lambda_{s''}^{\frac{1}{2}} = 2\lambda_s^{\frac{1}{2}} + \delta$, where δ is a small frequency interval. Omitting the zero order Hamiltonian here and writing only the first order Hamiltonian, we have

$$H^{(1)\prime\prime} = (\delta\hbar/2)[(p_{s''}/\hbar)^2 + q_{s''}^2] + (hck_{sss'}/4)\{(p_s q_s + q_s p_s)(p_{s''}/\hbar^2) - [(p_s/\hbar)^2 - q_s^2]\,q_{s''}\}. \tag{44.8}$$

The zero order energy will, evidently be

$$(E^0/hc) = \omega_s(v_s + 2v_{s''} + 3/2) \tag{44.9}$$

and the non-vanishing matrix component of the first order Hamiltonian (44.8) may be verified to be

$$\left.\begin{aligned}(v_s, v_{s''}|H^{(1)\prime\prime}/hc|v_s, v_{s''}) &= (v_{s''} + \tfrac{1}{2})(\delta/2\pi c),\\ (v_s, v_{s''}|H^{(1)\prime\prime}/hc|v_s+2, v_{s''}-1) &= (v_s+2, v_{s''}-1|H^{(1)\prime\prime}/hc|v_s, v_{s''})\\ &= (k_{sss''}/2)\,[(v_s+1)(v_s+2)(v_{s''}/2)].\end{aligned}\right\} \tag{44.10}$$

The frequency $\lambda_s^{\frac{1}{2}}$ is twofold degenerate in many cases, particularly in linear and axially symmetric molecules. When this is the situation the index, σ, takes the two values, $\sigma=1$ and $\sigma=2$ and resonance between $2\lambda_s^{\frac{1}{2}}$ and $\lambda_{s''}^{\frac{1}{2}}$ (i.e., $2\omega_s$ and $\omega_{s''}$) may then be induced when they are coupled through a potential energy term $hck_{sss''}(q_{s1}^2 + q_{s2})q_{s''}$. The terms above are transformed as before by use of two transformation functions of the variety S^*, Eq. (44.1), one for each value of σ, where s has been set equal to s' and where, as in the foregoing example, the order of the operators must be preserved. Setting $\lambda_{s''}^{\frac{1}{2}} = 2\lambda_s^{\frac{1}{2}} + \delta$, δ being a small frequency interval, and replacing $q_{s,1}\,q_{s,2}$ and the conjugate momenta, $p_{s,1}$ and $p_{s,2}$ by their equivalents in r_s and χ_s, one obtains for $E^{(0)}/hc$, $(E^{(0)}/hc) = (v_s + 2v_{s''} + 2)\,\omega_s$ and for the non-vanishing matrix elements of $(H^{(1)\prime\prime}/hc)$

$$\left.\begin{aligned}(v_s, l_{s'}, v_{s''}|H^{(1)\prime\prime}/hc|v_s, l_s, v_{s''}) &= (v_{s''} + \tfrac{1}{2})(\delta/2\pi c),\\ (v_s, l_s, v_{s''}|H^{(1)\prime\prime}/hc|v_s+2, l_s, v_{s''}-1) &= (v_s+2, l_s, v_{s''}-1|H^{(1)\prime\prime}/hc|v_s, l_{s'}, v_{s''})\\ &= (-k_{sss''}/2)\,[(v_s+2)^2 - l_s^2]^{\frac{1}{2}}(v_{s''}/2)^{\frac{1}{2}}\end{aligned}\right\} \tag{44.11}$$

where l_s takes the values $v_s, v_s-2, \ldots 1$ or 0.

The only matrix elements of $H^{(2)\prime}$ which are taken into account are those diagonal in the v_s. These remain the same as in Sect. V except in so far as the terms are concerned which have $k_{sss''}^2$ as coefficients. These contribute only to x_{ss}, $x_{ss''}$ and $x_{l_s l_s}$ in which the contributions containing $k_{sss''}^2$ as coefficients are changed from $(-k_{sss''}^2/4\omega_s)(8\omega_s^2 - 3\omega_{s''}^2)/(4\omega_s^2 - \omega_{s''}^2)$ to $(-k_{sss''}^2)\,[(1/2\omega_{s''}) + 1/8(2\omega_s + \omega_{s''})]$ in x_{ss}, from $-2k_{sss''}^2\,\omega_s/(4\omega_s^2 - \omega_{s''}^2)$ to $(-k_{sss''}^2/2)(2\omega_s + \omega_{s''})$ in $x_{ss''}$ and from $(-k_{sss''}^2/4)\,\omega_{s''}/(4\omega_s^2 - \omega_{s''}^2)$ to $(k_{sss''}^2/8)/(2\omega_s + \omega_{s''})$ in $x_{l_s l_s}$.

The term values themselves are obtained by diagonalizing the matrix (H/hc) which has the elements (44.6), (44.10) or (44.11). This is accomplished by setting the secular determinant of the matrix (H/hc) equal to zero and solving for the roots. We shall consider two examples. The first example will be the evaluation of the vibration-rotation energies of a linear molecule where we shall assume that Fermi resonance exists between the first overtone, $2\omega_s$, of a non-degenerate vibration and the fundamental frequency, $\omega_{s''}$, of another non-degenerate vibration. The second example will be the determination of the vibration energies of a linear molecule where Fermi resonance exists between the first overtone, $2\omega_s$, of a two-fold degenerate vibration and the fundamental frequency, $\omega_{s''}$, of a non-degenerate vibration. Both examples are similar and may, to a point, be considered together.

The elements of (H/hc) are, in both instances, diagonal in all the quantum numbers, rotational and vibrational, except v_s and $v_{s''}$. The zero order energy does not any longer depend upon v_s and $v_{s''}$ independently, but upon $v_s + 2v_{s''}$. There is, therefore, a degeneracy, the degree of which is determined by the number of ways $v_s + 2v_{s''}$ can be made to add up to a given value. The states we have chosen to study require that $v_s + 2v_{s''}$ shall add up to two. This can happen in our first example only if $v_s = 2$ with $v_{s''} = 0$ or if $v_s = 0$ with $v_{s''} = 1$ and in our second example only if $v_s = 2$, $l_s = 0$ with $v_{s''} = 0$ or if $v_s = 0$, $l_s = 0$ with $v_{s''} = 1$. The resonating levels form, therefore, a diad in both instances. The elements of the matrix (H/hc) will all lie along the principal diagonal except those associated with these particular levels. These will, respectively, form a step matrix of two rows and two columns grouped symmetrically about the principal diagonal. The diagonalization may, therefore, be made complete simply by diagonalizing this step matrix by itself.

Consider now the first example in detail. The quantity (E_0/hc) for each of the resonating components will be $7\omega_s/2$. The secular determinant of the step matrix for the levels referred to in that example is shown in (44.12) where $F(v_s, v_{s''})$ is the rotational term value

$$F(v_s, v_{s''}) = J(J+1)\, B(v_s, v_{s''}) - J^2(J+1)^2 D$$

$$\begin{array}{c} \\ \\ v_s = 2 \\ v_{s''} = 0 \\ \\ v_s = 0 \\ v_{s''} = 1 \end{array}
\begin{array}{cc}
\begin{array}{c} v_s = 2 \\ v_{s''} = 0 \end{array} & \begin{array}{c} v_s = 0 \\ v_{s''} = 1 \end{array} \\
\left|\begin{array}{c} (7\omega_s/2) + (\delta/4\pi c) + \\ + F(v_s = 2, v_{s''} = 0) - \varepsilon \end{array}\right. & \left.\begin{array}{c} (k_{sss''}/2) \end{array}\right| \\
\left|\begin{array}{c} (k_{sss''}/2) \end{array}\right. & \left.\begin{array}{c} (7\omega_s/2) + (3\delta/4\pi c) + \\ + F(v_s = 0, v_{s''} = 1) - \varepsilon \end{array}\right|
\end{array} = 0. \qquad (44.12)$$

Adopting a notation $\Delta^2 = [k_{sss''}^2 + \Delta_0^2]$ and $\Delta_0 = (\delta/2\pi c)$, the secular determinant (44.12) factors into the roots

$$\left.\begin{aligned} \varepsilon = {} & [(3\omega_s/2) + \omega_{s''} \pm (\Delta/2)] + F(v_s = 0, v_{s''} = 0) + (\alpha/2\Delta)(\Delta \mp \Delta_0) + \\ & + (\beta/2\Delta)(\Delta \pm \Delta_0) \pm (\alpha - \beta)^2 (\Delta^2 - \Delta_0^2)/4\Delta^3 \end{aligned}\right\} \quad (44.13)$$

where $\alpha = F(v_s = 2,\ v_{s''} = 0) - F(v_s = 0,\ v_{s''} = 0)$ and $\beta = F(v_s = 0,\ v_{s''} = 1) - F(v_s = 0,\ v_{s''} = 0)$ if it is assumed that Δ is large compared with α and β.

Eq. (44.13) shows that while the primary effect of the resonance is, indeed, to make the vibration levels take on anomalous values, the corresponding rotational constants are also altered so as to take on hybrid values of the rotational constants in their unperturbed states. The third and fourth terms together with $F(v_s = 0, v_{s''} = 0)$ in (44.13) give rise to the hybrid B values. Recently experimental techniques have been improved to the extent that accurate determinations of the centrifugal distortion constants, D, can be carried out. It has been found experimentally by COURTOY and HERZBERG that the D values in a polyad of states perturbed by Fermi resonance are different from the D value in the normal state. The term $F(v_s=0, v_{s''}=0)$ and the last term in (44.13) confirms theoretically this experimental fact, the effective D values in the two states considered being $D \pm [B(2,0) - B(0,1)]^2 [(\Delta^2 - \Delta_0^2)/4\Delta^3]$.

We turn now to our second example. Here the situation, though similar, is somewhat more complicated. The reason for this complication is that the state $v_s=2$, $l_s=0$ which is coupled to the state $v_{s''}=0$ by Fermi resonance is further coupled to the sub-states $v_s=2$, $l_s=\pm 2$ through the l-type doubling operator (42.1). This is tantamount to stating that in such cases the FERMI resonance and the l-type doubling resonance can not be treated independently.

The quantity (E_0/hc) will, according to the foregoing, be $4\omega_s$ for each of the levels involved. The diagonal elements of the matrix would, therefore, be $4\omega_s + (\delta/4\pi c)$ where $v_s=2$, $l_s=0$, $v_{s''}=0$ and $4\omega_s + (3\delta/4\pi c)$ where $v_s=0$, $l_s=0$, $v_{s''}=1$. The levels $v_s=2$, $l_s \pm 2$, $v_{s''}=0$ and $v_s=2$, $l_s=0$, $v_{s''}=0$ which are coupled by the l-type doubling operator do not, however, have entirely the same values as may be seen from the relation (34.11). This degeneracy is removed in second order of approximation and although the difference between the levels is small it is large as compared to the l-type doubling parameter, q. It is, therefore, necessary to replace the above diagonal values respectively, by $E_0(v_s=0, l_s=0, v_{s''}=1)/hc$, $E_0(v_s=2, l_s=+2, v_{s''}=0)/hc$, $E_0(v_s=2, l_s=0, v_{s''}=0)/hc$ and $E_0(v_s=2, l_s=-2, v_{s''}=0)/hc$ in the step matrix for these levels. The secular determinant of the step matrix appropriate for this example is shown in (44.14) where $F(v_s, l_s, v_{s''})$ is the rotational term value

$$F(v_s, l_s, v_{s''}) = [J(J+1) - l_s^2]\, B(v_s, l_s, v_{s''}) - [J(J+1) - l_s^2]^2 D$$

and where here $a = -k_{sss''}/\sqrt{2}$ and $b = 2\sqrt{2}\, q^0 \{J(J+1)[J(J+1)-2]\}^{\frac{1}{2}}$.

	$v_s=0,\ l_s=0$ $v_{s''}=1$	$v_s=2,\ l_s=2$ $v_{s''}=0$	$v_s=2,\ l_s=0$ $v_{s''}=0$	$v_s=2,\ l_s=-2$ $v_{s''}=0$
$v_s=0$ $l_s=0$ $v_{s''}=1$	$E(v_s=0, l_s=0, v_{s''}=1)/hc +$ $+ F(v_s=0, l_s=0, v_{s''}=1)$ $-\varepsilon$	0	a	0
$v_s=2$ $l_s=2$ $v_{s''}=0$	0	$E(v_s=2, l_s=2, v_{s''}=0)/hc +$ $+ F(v_s=2, l_s=2, v_{s''}=0)$ $-\varepsilon$	b	0
$v_s=2$ $l_s=0$ $v_{s''}=0$	a	b	$E(v_s=2, l_s=0, v_{s''}=0)/hc +$ $+ F(v_s=2, l_s=0, v''=0)$ $-\varepsilon$	b
$v_s=2$ $l_s=-2$ $v_{s''}=0$	0	0	b	$E(v_s=2, l_s=2, v_{s''}=0)/hc +$ $+ F(v_s=2, l_s=2, v_{s''}=0)$ $-\varepsilon$

$= 0.$ (44.14)

The determinant (44.14) has four roots, one of which is $\varepsilon = E(v_s=2, l_s=2, v_{s''}=0)/hc$ and is of no further interest here. Only two of the remaining three roots shall concern us in this discussion, namely those two roots which correspond to the two roots which would have been obtained if only the Fermi resonance had been taken into account and the l-type resonance had been neglected (i.e., when $b=0$). The two roots we wish to obtain we shall approximate to by means of perturbation theory methods since the l-type resonance is small compared to the Fermi resonance. To this end we shall solve the problem of Fermi resonance for these levels first. The secular determinant for the matrix of the elements

(44.11) only will be the following

	$v_s=2,\ l_s=0$ $v_{s''}=0$	$v_s=0,\ l_s=0$ $v_{s''}=1$	
$v_s=2,\ l_s=0$ $v_{s''}=0$	$4\omega_s+(\delta/4\pi c)+$ $+F(v_s=2, l_s=0, v_{s''}=0)$ $-\varepsilon$	$-k_{sss''}/\sqrt{2}$	$=0$
$v_s=0,\ l_s=0$ $v_{s''}=1$	$-k_{sss''}/\sqrt{2}$	$4\omega_s+(3\,\delta/4\pi c)+$ $+F(v_s=0, l_s=0, v_{s''}=1)$ $-\varepsilon$	

where $F(v_{s'}, v_{s''})$ is the rotational term value

$$F(v_{s'}, v_{s''})=[J(J+1)-l_s^2]\,B(v_{s'}, v_{s''})-[J(J+1)-l_s^2]^2 D.$$

Adopting a notation similar to that in the former case where $\Delta^2=2k_{sss''}^2+\Delta_0^2$ and $\Delta_0=(\delta/2\pi c)$ the secular determinant factors into the roots

$$\left.\begin{aligned}\varepsilon\pm &= (2\omega_s+\omega_{s''}\pm\Delta/2)+F(v_s=0, l_s=0, v_{s''}=0)+\\ &+(\alpha/2\Delta)(\Delta\mp\Delta_0)+(\beta/2\Delta)(\Delta\pm\Delta_0)\pm(\alpha-\beta)^2(\Delta^2-\Delta_0^2)/4\Delta^3.\end{aligned}\right\}\quad(44.15)$$

α and β have the same significance here as in the earlier example.

The stabilized wave functions for the states ε_+ and ε_- may be obtained in the usual manner and they will be found to be

$$\Psi_{\mathrm{I}}=[(\Delta+\Delta_0)/2\Delta]^{\frac12}\,\psi(1, 0_0)+[(\Delta-\Delta_0)/2\Delta]^{\frac12}\,\psi(0, 2_0)$$

and

$$\Psi_{\mathrm{II}}=[(\Delta-\Delta_0)/2\Delta]^{\frac12}\,\psi(1, 0_0)-[(\Delta+\Delta_0)/2\Delta]^{\frac12}\,\psi(0, 2_0).$$

We shall take as the wave functions for the two perturbing states ($v_s=2, l_s=\pm 2, v_{s''}=0$) the following $\psi_{\mathrm{III}}=\frac{1}{\sqrt{2}}[\Psi(0, 2_2)-\Psi(0, 2_{-2})]$ and $\psi_{\mathrm{IV}}=\frac{1}{\sqrt{2}}[\psi(0, 2_2)+\psi(0, 2_{-2})]$. With these four wave functions the perturbed values of ε_+ and ε_- are found to be

$$\varepsilon'_+=\varepsilon_++[(\mathrm{I}\,|\,\mathrm{III})^2/(\varepsilon_+-E_{2_2})]+[(\mathrm{I}\,|\,\mathrm{IV})^2/(\varepsilon_+-E_{2_2})]$$

and

$$\varepsilon'_-=\varepsilon_-+[(\mathrm{II}\,|\,\mathrm{III})^2/(\varepsilon_--E_{2_2})]+[(\mathrm{II}\,|\,\mathrm{IV})^2/(\varepsilon_--E_{2_2})]$$

where (I|III), (I|IV), etc., are the matrix elements of the l-type doubling operator (42.1) where the corresponding functions Ψ_{I}, Ψ_{II}, etc. have been used. (It is quickly verified that (I|III) = (II|III) = 0 and that (I|IV) $=[(\Delta-\Delta_0)/\Delta]^{\frac12}\times$ $0, 2_0|0, 2_2)=\sqrt{2}[(\Delta-\Delta_0)/\Delta]^{\frac12}q^0 J(J+1)$ and (II|IV) $=[(\Delta+\Delta_0)/\Delta]^{\frac12}(0, 2_0|0, 2_2)=\sqrt{2}[(\Delta+\Delta_0)/\Delta]^{\frac12}q^0 J(J+1)$. Moreover, it is easy to show that $\varepsilon_+-E_{2_2}=\frac12(\Delta+\Delta_0)-4x_{ll}$ and that $\varepsilon_--E_2=-\frac12(\Delta-\Delta_0)-4x_{ll}$. One obtains, therefore, for ε'_+ and ε'_- the following relations

$$\left.\begin{aligned}\varepsilon'_+&=\varepsilon_++\{16[(\Delta-\Delta_0)/(\Delta_0+\Delta-8x_{ll})]\,(q^{0\,2}/\Delta)\}\,J^2(J+1)^2,\\ \varepsilon'_-&=\varepsilon_-+\{16[(\Delta+\Delta_0)/(\Delta_0-\Delta-8x_{ll})]\,(q^{0\,2}/\Delta)\}\,J^2(J+1)^2.\end{aligned}\right\}\quad(44.16)$$

The effect of the l-type resonance has been not to contribute to the hybridization of the B values [i.e., to the terms in the energy proportional to $J(J+1)$] but to add a term to the effective centrifugal distortion coefficient [i.e., the term proportional to $J^2(J+1)^2$]. The effective D value, D_{eff}, will be seen, from (44.15) and (44.16), to be

$$\left.\begin{aligned} D_{+\text{eff}} &= D - (\alpha-\beta)^2\left[(\Delta^2-\Delta_0^2)/4\Delta^3)\right] - 16\left[(\Delta-\Delta_0)/(\Delta_0+\Delta-8x_{ll})\right](q^{02}/\Delta) \\ &\text{and} \\ D_{-\text{eff}} &= D + (\alpha-\beta)^2\left[(\Delta^2-\Delta_0^2)/4\Delta^3\right] + 16\left[(\Delta+\Delta_0)/(\Delta_0-\Delta-8x_{ll})\right](q^{02}/\Delta). \end{aligned}\right\} \quad (44.17)$$

When the roots of the secular determinant of the matrix of the energies, in a particular instance of Fermi-Dennison resonance, have been obtained, the stabilized wave functions for the component states may readily be obtained, since these are linear combinations of the wave functions of the unperturbed states, the coefficients multiplying each term being the normalized cofactors of the secular determinant.

45. Simultaneous Fermi resonance and l-type doubling. Levels in which l-type doubling occurs may in certain instances be coupled by Fermi resonance as well. A particular case of interest is the triad

$$\left\{\begin{matrix} 0, & 5_1, & 1 \\ 1, & 3_1, & 1 \\ 2, & 1_1, & 1 \end{matrix}\right\}$$

studied experimentally in the spectrum of CO_2 by GOLDBERG, MOHLER and PIERCE. We shall, in such cases, have elements lying off of the principal diagonal of the varieties of both (42.3) and (44.11). The secular determinant for the energies in the case of the above triad is readily shown to be

	$v_1=0,\ v_2=5,\ l_2=1$	$v_1=1,\ v_2=3,\ l_2=1$	$v_1=2,\ v_2=1,\ l_2=1$	$v_1=2,\ v_2=1,\ l_2=-1$	$v_1=1,\ v_2=3,\ l_2=-1$	$v_1=0,\ v_2=5,\ l_2=-1$
$v_1=0,\ v_2=5,\ l_2=1$	$T(0,5_1,1)+F(0_1,5_1,1)-\varepsilon$	$-k_{122}\sqrt{3}$	0	0	0	$6q^0J(J+1)$
$v_1=1,\ v_2=3,\ l_2=1$	$-k_{122}\sqrt{3}$	$T(1,3_1,1)+F(1,3_1,1)-\varepsilon$	$-k_{122}\sqrt{2}$	0	$4q^0J(J+1)$	0
$v_1=2,\ v_2=1,\ l_2=1$	0	$-k_{122}\sqrt{2}$	$T(2,1_1,1)+F(2,1_1,1)-\varepsilon$	$2q^0J(J+1)$	0	0
$v_1=2,\ v_2=1,\ l_2=-1$	0	0	$2q^0J(J+1)$	$T(2,1_{-1},1)+F(2,1_{-1},1)-\varepsilon$	$-k_{122}\sqrt{2}$	0
$v_1=1,\ v_2=3,\ l_2=-1$	0	$4q^0J(J+1)$	0	$-k_{122}\sqrt{2}$	$T(1,3_{-1},1)+F(1,3_{-1},1)-\varepsilon$	$-k_{122}\sqrt{3}$
$v_1=0,\ v_2=5,\ l_2=-1$	$6q^0J(J+1)$	0	0	0	$-k_{122}\sqrt{3}$	$T(0,5_{-1},1)+F(0,5_{-1},1)-\varepsilon$

$= 0$ (45.1)

where $T(v_1, v_2, v_3)$ are the vibration term values and where $F(v_1, v_2, v_3)$ are the rotational term values (44.12) and where l-type resonance has been neglected.

Eq. (45.1) factors into the following two determinants of three rows and columns

$$0=\begin{vmatrix} T(0,5_1,1)+F(0,5_1,1)\pm \\ \pm 6q^0 J(J+1)-\varepsilon_\pm & -k_{122}\sqrt{3} & 0 \\ -k_{122}\sqrt{3} & T(1,3_1,1)+F(1,3_1,1)\pm \\ & \pm 4q^0 J(J+1)-\varepsilon_\pm & -k_{122}\sqrt{2} \\ 0 & -k_{122}\sqrt{2} & T(2,1_1,1)+F(2,1_1,1)\pm \\ & & \pm 2q^0 J(J+1)-\varepsilon_\pm \end{vmatrix}. \tag{45.2}$$

Denoting the roots of the secular determinant (45.2) by $\varepsilon_\pm^0$, the l-type doubling of a given level will be given by the difference $\varepsilon_+^0, -\varepsilon_-^0$, where the E^0's are obtained in the conventional manner.

An alternate manner of computing the l-type doubling in such cases is to observe that the l-splitting is, in general, small compared to the splitting due to Fermi resonance, i.e., that the l-type doubling may be regarded as perturbation on the levels already affected by Fermi resonance. The l-doubling is, then, totally ignored at first and a secular determinant involving the Fermi resonance only is solved. The stabilized wave functions for a given state may be denoted by $R_\pm$ which are

$$R_\pm(\tau)=[C_1 R(2,1_1,1)+C_2 R(1,3_1,1)+C_3 R(0,5_1,1)]\,e^{\pm i\chi} \tag{45.3}$$

where the $R(v_1, v_2, v_3)$ are the wave functions (19.7) and the C's are the normalized cofactors of the secular determinant. Using the wave functions (45.3) to compute the l-type doubling, from the first order Hamiltonian (42.1) we have

$$\left.\begin{aligned}\big(R_\pm(\tau)\,|H^{(1)\prime}/hc|\,R_\mp(\tau)\big) &= C_1^2\big(R(2,1_1,1)\,e^{i\chi}\,|H^{(1)\prime}/hc|\,R(2,1_1,1)\,e^{-i\chi}\big)+\\ &+2C_2^2\big(R(1,3_1,1)\,e^{i\chi}\,|H^{(1)\prime}/hc|\,R(1,3_1,1)\,e^{-i\chi}\big)+\\ &+C_3^2\big(R(0,5_1,1)\,e^{i\chi}\,|H^{(1)\prime}/hc|\,R(0,5_1,1)\,e^{-i\chi}\big)\end{aligned}\right\} \tag{45.4}$$

bearing in mind, once more, that terms not diagonal in $R(v_1, v_2, v_3)$ are of no interest here. The result is that the l-type doubling coefficient in (43.1), $q^0(v_s+1)$, will be replaced, in this case, by

$$q=2q^0(C_1^2+2C_3^2+3C_3^2). \tag{45.5}$$

This alternate method of solution is somewhat simpler than the foregoing and is quite satisfactory so long as the l-type doubling is small compared to the Fermi resonance splitting and this is probably always the case for the values of J which are of interest.

It has been found by TAYLOR, BENEDICT and STRONG that the above approximations to the Fermi-Dennison resonance splitting are not entirely satisfactory in all instances, for example to describe certain energy levels in the CO_2 molecule. They have proposed a semi-empirical modification which replaces $(k_{122}/\sqrt{2})$ by

$$W=(k_{122}/\sqrt{2})\,\{1-\beta_2[(v_s+2)^2-l_2^2]^{\frac{1}{2}}\,v_1^{\frac{1}{2}}-2\beta_2-\beta_3 v_3\}^{\frac{1}{2}}$$

in the relation (44.11) where β_2 and β_3 are constants. This question has been investigated by AMAT and GOLDSMITH who have shown that the correct form for this effective value of $(k_{122}/\sqrt{2})$ is not the one proposed by TAYLOR, BENEDICT and STRONG, but rather the following

$$W=(k_{122}/\sqrt{2})\,\{1-\beta_0'-\beta_1' v_1-\beta_2' v_2-\beta_{l_2}' l_2-\beta_3' v_3\} \tag{45.6}$$

where β_0', β_1', etc. are constants depending in a complicated manner upon the coefficients which occur in the first, second and third order Hamiltonian of a molecule. MAES has recently shown theoretically that β_{l_2}' in Eq. (45.6) is equal to zero, a fact which is borne out by the experiments of COURTOY on the spectrum of CO_2. MAES has, however, shown that $\beta_0' = 2\beta_2' + \frac{1}{2}\beta_3'$.

46. Second order anharmonic resonance interactions. The Hamiltonian for the molecule has been transformed by a contact transformation in Sect. V in such a manner that the transformed first order Hamiltonian, $H^{(1)\prime}$, is zero while the second order Hamiltonian, $H^{(2)}$, becomes transformed into $H^{(2)\prime}$ where the latter includes now quadratic contributions stemming from the original first order Hamiltonian, $H^{(1)}$. The energies to second order of approximation, involve only the diagonal elements of $H^{(2)\prime}$ except in such cases where resonance occurs. Such resonances are of second order and they shall be referred to as second order resonances. Further examination of $H^{(2)\prime}$, including H_v^*, reveals that the following elements which will be off of the principal diagonal exist;

$$(v_s, v_{s'}, v_{s''} | v_s \pm 1, v_{s'} \pm 1, v_{s''}), \qquad (v_s, v_{s'}, v_{s''} | v_s \pm 1, v_{s'} \pm 1, v_{s''} \pm 2),$$
$$(v_s, v_{s'}, v_{s''} | v_s \pm 2, v_{s'} \pm 2, v_{s''}), \qquad (v_s, v_{s'}, v_{s''} | v_s \pm 3, v_{s'} \pm 1, v_{s''}),$$

etc. Normally these elements would enter into the energy quadratically in a higher order and would be multiplied by a coefficient containing the denominator $E(v_s, v_{s'}, v_{s''}) - E(v_s', v_{s'}', v_{s''}')$ if the energies were to be estimated to approximations beyond the second order. Certain of the above matrix elements would introduce terms with resonance denominators and these might prove troublesome. Examples of these would be

$$(v_s, v_{s'}, v_{s''} | v_s \pm 1, v_{s'} \mp 1, v_{s''}), \qquad (v_s, v_{s'}, v_{s''} | v_s \pm 2, v_{s'} \mp 2, v_{s''})$$

and

$$(v_s, v_{s'}, v_{s''} | v_s \pm 3, v_{s'} \mp 1, v_{s''})$$

which would, respectively, introduce the denominators $(\omega_s - \omega_{s'})$, $(2\omega_s - 2\omega_{s'})$ and $(3\omega_s - \omega_{s'})$. As in Sect. 44 these denominators will approach zero when $\omega_s \approx \omega_{s'}$ or $3\omega_s \approx \omega_{s'}$, i.e., when resonance between these frequencies set in, and the corresponding term in the energy might become indefinitely large. It is necessary then, as in the case of the first order anharmonic resonance, to regard such frequencies as degenerate and to use the degenerate form of the perturbation theory so as to include in the sub-matrix of the energies of these states also the elements which are non-diagonal in v_s and $v_{s'}$. The analogy here to that of first order resonance is complete in that the matrix will be diagonal in all the vibration quantum numbers except v_s and $v_{s'}$ to this approximation, the new elements being grouped about the principal diagonal in a little box to form a sub-matrix. These sub-matrices may then be diagonalized individually to make the entire matrix diagonal.

Most important of the second order anharmonic resonance is the one usually referred to as the Darling-Dennison type of resonance which has been observed experimentally in the water vapor spectrum. Here it is the elements $(v_1, v_2, v_3 | v_1 \pm 2, v_2, v_3 \mp 2)$ which are significant and these are important only when $2\omega_3 \approx 2\omega_1$ and have no effect on the fundamentals themselves. The Darling-Dennison resonance is the only second order anharmonic resonance which has definitely been established and we shall here state explicitly the values of the matrix elements which are involved in this case. Other such resonances are almost certain to exist and the appropriate matrix elements may rather readily be evaluated in such special instances as they become important by referring to $H^{(2)\prime}$ in Chap. V including, however, also H_v^*. The matrix elements of importance to calculate the Darling-Dennison resonance are readily found to be these

$$\left.\begin{aligned}&(v_s, v_{s'}, v_{s''}|H^{(2)\prime}/hc|\, v_s-2, v_{s'}+2, v_{s''})\\&\quad=\big[(k_{s s s' s'}/4)-(\pi c k_{s s s'}^2/\lambda_s^{\frac{1}{2}})-(\pi c k_{s s' s'}^2/\lambda_{s'}^{\frac{1}{2}})+(\pi c k_{s s s}\, k_{s s' s'}/2\lambda_s^{\frac{1}{2}})+\\&\quad+(\pi c k_{s s s'}\, k_{s' s' s'}/2\lambda_{s'}^{\frac{1}{2}})\big]+(\pi c k_{s s s''}\, k_{s' s' s''}/2\lambda_{s''}^{\frac{1}{2}})\,(\lambda_{s''}/(4\lambda_s-\lambda_{s''}))+\\&\quad+(\pi c k_{s s s''}\, k_{s' s' s''}/2\lambda_{s''}^{\frac{1}{2}})\,(\lambda_{s''}/(4\lambda_s-\lambda_{s''}))-\\&\quad-(\pi c k_{s' s s''}^2/4\lambda_{s''}^{\frac{1}{2}})\,(\lambda_{s''}/(4\lambda_s-\lambda_{s''}))\big]\,[v_s(v_s-1)\,(v_{s'}+1)\,(v_{s'}+2)]^{\frac{1}{2}}.\end{aligned}\right\}\quad(46.1)$$

Darling and Dennison's work referred specifically to the water vapor molecule where the two vibration frequencies ω_1 and ω_3 are nearly alike so that resonance between $2\omega_1$ and $2\omega_3$ may occur. When cognizance is taken of the fact that the non-vanishing anharmonic constants in the potential energy are these: $k_{1133}, k_{133}, k_{111}, k_{233}$, the results of Eq. (46.1) reduce to

$$(v_1, v_2, v_3|H^{(2)\prime}/hc|\, v_1-2, v_2, v_3+2)=(\gamma/2)\,[v_1(v_1-1)\,(v_3+1)(v_3+2)]^{\frac{1}{2}}\quad(46.2)$$

where

$$\gamma=(k_{1133}/2)-(2\pi c k_{331}^2/\lambda_3^{\frac{1}{2}})+(\pi c k_{111}\, k_{133}/\lambda_1^{\frac{1}{2}})+(\pi c k_{322}\, k_{111}/\lambda_2^{\frac{1}{2}})\,(\lambda_2/(4\lambda_3-\lambda_2)).$$

Setting $\lambda_1^{\frac{1}{2}}=\lambda_3^{\frac{1}{2}}$ it is evident that (46.2) reduces to the value derived by Darling and Dennison. One sees also, by inspection of $H^{(2)\prime}$, that in the case of water vapor the elements $(v_1, v_2, v_3|\, v_1\pm 1, v_2, v_3\mp 1)$ which would influence the positions of the fundamental bands ω_1 and ω_3 will be equal to zero.

It may be verified by reference to $H^{(2)\prime}$ including H^* that terms exist in it which may permit resonance to occur, in much the same manner as the case we have just considered, between ω_s and $\omega_{s'}$; $3\omega_s$ and $\omega_{s'}$; $\omega_s+\omega_{s'}$ and $2\omega_{s''}$; ω_s and $\omega_{s'}+2\omega_{s''}$; ω_s and $\omega_{s'}-2\omega_{s''}$; and between ω_s and various combinations of $\omega_{s'}$, $\omega_{s''}$ and $\omega_{s'''}$. Experimental evidence on such types of resonance have not been presented.

47. Anharmonic resonance interactions of higher order than the second. A few isolated instances of anharmonic resonance of order higher than the second are on record. The case reported by Douglas and Sharma dealing with two pairs of levels in the HCN molecule is of particular interest. It is reported in this work that resonance exists between the levels $(4, 0_0, 2)$ and $(1, 0_0, 4)$ and again between the levels $(0, 0_0, 4)$ and $(3, 0_0, 2)$. It is readily shown that these levels do, indeed, possess suitable symmetry properties for an anharmonic resonance, to occur and one may quickly establish that the resonance is a third order effect in this case. The matrix elements which apply here have been studied by Hanson and Nielsen and are found, by a third order perturbation calculation, to be

$$(v_1, v_2, v_3|H^{(3)\prime}/hc|\, v_1-3, v_2, v_3+2)=\gamma\left[\frac{v_1(v_1-1)\,(v_1-2)\,(v_3+1)\,(v_3+2)}{2}\right]^{\frac{1}{2}}\quad(47.1)$$

where γ is a complicated relation between the coefficients of the cubic, the quartic and the quintic terms in the potential energy function.

An interesting feature of this type of resonance is that while it is observed in the above mentioned instances it appears not to occur between levels in other cases which bear the same symmetry relations to each other. This may be understood in the following manner. Consider the example of the levels $(3, 0_0, 2)$

and $(0, 0_0, 2)$. The secular determinant of the matrix for these levels is shown in (47.2)

$$\begin{array}{c|c|c|} & (3, 0_0, 0) & (0, 0_0, 2) \\ (3, 0_0, 0) & T(3, 0_0, 0) - \varepsilon & \sqrt{6}\gamma \\ (0, 0_0, 2) & \sqrt{6}\gamma & T(0, 0_0, 2) - \varepsilon \end{array} = 0 \tag{47.2}$$

which when expanded has the roots

$$= \frac{1}{2}[T(3, 0_0, 0) + T(0, 0_0, 2)] \pm \left\{\left[\frac{T(3, 0_0, 0) - T(0, 0_0, 2)}{2}\right]^2 + 6\gamma^2\right\}^{\frac{1}{2}}. \tag{47.3}$$

The term values $T(3, 0_0, 0)$ and $T(0, 0_0, 2)$ differ, in general, by amounts o. the order of magnitude of the quantities $x_{ss'}$, i.e., by terms of second order im portance. The off diagonal elements, however, are only of third order significance and therefore the term within the radical in (47.3) in γ^2 will, in general, be so small in comparison to $\left|\frac{T(3, 0_0, 0) - T(0, 0_0, 2)}{2}\right|^2$ that it can be neglected. No resonance will occur in such cases. The term in γ^2 within the radical becomes the dominant term, however, if $T(3, 0_0, 0)$ and $T(0, 0_0, 2)$ are accidentally almost exactly equal to each other. The roots, then, are

$$\varepsilon_{\pm} = \left[\frac{T(3, 0_0, 0) + T(0, 0_0, 2)}{2}\right] \pm \sqrt{6}\gamma. \tag{47.4}$$

The levels $(4, 0_0, 2)$ and $(1, 0_0, 4)$ and the levels $(0, 0_0, 4)$ and $(3, 0_0, 2)$ studied by DOUGLAS and SHARMA are two components of two triad levels, the third components being $(7, 0_0, 0)$ and $(6, 0_0, 0)$ respectively. If the levels $(4, 0_0, 2)$ and $(1, 0_0, 4)$ and the levels $(0, 0_0, 4)$ and $(3, 0_0, 2)$ of the two triads were respectively exactly degenerate, but significantly different from the respective third components of the diads the theory would predict that the resonance would produce a splitting in the former pair twice as large as the splitting in the latter pair. This is very much borne out by experiment. The third components, which have not been observed, would not be effected significantly by the perturbation.

What has been said about the third order resonance interactions may be said also for resonance interactions of still higher order. Such resonances have thus far not been observed with great certainty and it is evident from the foregoing discussion that only when the unperturbed levels are completely coincident would there be a chance to observe them.

48. First order rotational resonance interactions. We studied the resonances produced by coupling of vibration frequencies through terms in the potential energy in the former sections of this chapter. We shall here consider resonance between vibration frequencies facilitated by the presence of certain terms in the kinetic energy, in particular the Coriolis operator

$$-\sum_{\sigma}\sum_{\sigma'}[(\lambda_{s'}/\lambda_s)^{\frac{1}{4}} q_{s\sigma} p_{s'\sigma'} - (\lambda_s/\lambda_{s'})^{\frac{1}{4}} q_{s'\sigma'} p_{s\sigma}] \sum_{\alpha}(\zeta_{ss'}^{(\alpha)} P_\alpha / I_{\alpha\alpha}^{(e)}) \tag{48.1}$$

in which $\zeta_{ss'}^{(\alpha)}$ are the Coriolis coupling factors which depend in a complex manner upon the nature of the normal coordinates associated with the frequencies ω_s and $\omega_{s'}$. It is not every pair of vibrations, ω_s and $\omega_{s'}$, which may contribute to the Coriolis operator because of symmetry reasons. The general rule for determining when two vibrations may interact through Coriolis forces has been given by JAHN. The argument is somewhat as follows. The operators, p_α, must be of the species of a rotation about one of the axes x, y or z. Moreover, the species of the operators, $p_{s,\sigma}$, in (48.1) are the same as the species of $q_{s,\sigma}$. Therefore, two vibrations may perturb each other through Coriolis interaction only if the product of the species contains the species of a rotation. As an example to illustrate this we shall refer to the planar XYZ_2 molecule which belongs to the C_{2v} point group and concerning which we studied the normal modes of vibration in Sect. 18. The rotation R_α, about an axis, α, will transform as the quantity $\beta\dot{\gamma}-\gamma\dot{\beta}$. It is readily verified with the aid of information given in Sect. 18 that R_x, R_y and R_z transform, respectively as

$$R_x \sim B_2, \quad R_y \sim B_1, \quad R_z \sim A_2.$$

We need, therefore, investigate only the product representation of the normal coordinates. We have seen in Chap. III that the normal coordinates, q_1, q_2, q_3 transform as A_1; q_4, q_5 as B_2 and q_6 as B_1. We have as the products, therefore

$$\begin{aligned} A_1\times A_1 &= A_1, & A_1\times B_1 &= B_1, & &(R_y)\\ B_1\times B_1 &= A_1, & A_1\times B_2 &= B_2, & &(R_x)\\ B_2\times B_2 &= A_1, & B_1\times B_2 &= A_2, & &(R_z). \end{aligned}$$

Since the Coriolis operator is of the form $-\sum_\alpha p_\alpha P_\alpha/I^{(e)}_{\alpha\alpha}$ in this approximation the Coriolis operator reduces to

$$\begin{aligned} -\Big\{&\sum_{s=1}^{3}\sum_{s'=4}^{5}\big[(\lambda_{s'}/\lambda_s)^{\frac14} q_s p_{s'} - (\lambda_s/\lambda_{s'})^{\frac14} q_{s'} p_s\big]\,\zeta^{(x)}_{s,s'}\,P_x/I^{(e)}_{xx}\\ &+\sum_{s=1}^{3}\big[(\lambda_6/\lambda_s)^{\frac14} q_s p_6 - (\lambda_s/\lambda_6)^{\frac14} q_6 p_s\big]\,\zeta^{(y)}_{s,6}\,P_y/I^{(e)}_{yy}\\ &+\sum_{s=4}^{5}\big[(\lambda_6/\lambda_s)^{\frac14} q_s p_6 - (\lambda_s/\lambda_6)^{\frac14} q_6 p_s\big]\,\zeta^{(z)}_{s,6}\,P_z/I^{(e)}_{zz}\Big\}. \end{aligned}$$

When a pair of frequencies ω_s and $\omega_{s'}$ which are allowed to perturb each other through Coriolis interaction and have quite different numerical values the term (48.1) may be removed from the first order Hamiltonian by a contact transformation described in Sect. V. The transformation introduces a term in the transformed second order Hamiltonian which depends upon p_α in a quadratic manner and which contains the denominator $(\lambda_s^{\frac12}-\lambda_{s'}^{\frac12})$ (i.e., $\omega_s-\omega_{s'}$). When $\omega_s\approx\omega_{s'}$ the same difficulty encountered with anharmonic resonance arises here also, it being this time the elements $(v_{s\sigma}, v_{s'\sigma'}|\,v_{s\sigma}\pm 1, v_{s'\sigma'}\mp 1)$ that present difficulty, and the contact transformation method described earlier is not at once valid. We follow the lead of the preceding paragraphs and use instead of the usual transformation function, S, a new function S_R^* which is the following

$$S_R^* = -\frac{1}{2}\,\frac{\sqrt{\lambda_{s'}}-\sqrt{\lambda_s}}{\sqrt{\lambda_{s'}}+\sqrt{\lambda_s}}\,\lambda_s^{-\frac14}\lambda_{s'}^{-\frac14}\big[q_{s\sigma}q_{s'\sigma'} - p_{s\sigma}p_{s'\sigma'}/\hbar^2\big]\sum_\alpha\big(\zeta^{(\alpha)}_{ss'}P_\alpha/I^{(e)}_{\alpha\alpha}\big). \qquad (48.2)$$

The function S_R^* transforms $H^{(1)}$ into

$$H^{(1)\prime} = -\frac{1}{2}\sum_\sigma\sum_{\sigma'}\frac{\sqrt{\lambda_{s'}}+\sqrt{\lambda_s}}{\lambda_{s'}^{\frac14}\lambda_s^{\frac14}}\big[q_{s\sigma}p_{s'\sigma'} - q_{s'\sigma'}p_{s\sigma}\big]\sum_\alpha\big(\zeta^{(\alpha)}_{ss'}P_\alpha/I^{(e)}_{\alpha\alpha}\big) \qquad (48.3)$$

which has the same $(v_{s\sigma}v_{s'\sigma'}|\,v_{s\sigma}\pm 1,\ v_{s'\sigma'}\mp 1)$ elements as the Coriolis operator (48.1), but has all others equal to zero. By analogy to the function S^* of Sect. 42. The function S_R^* here introduced does not evidently, contain the denominator $\omega_s-\omega_{s'}$.

One proceeds now precisely as before. We shall consider first the instance where ω_s and $\omega_{s'}$ are non-degenerate frequencies. We set $\lambda_{s'}^{\frac{1}{2}}$ equal to $\lambda_s^{\frac{1}{2}}+\delta$, where δ is small so that for the zero order term value involving these frequencies one obtains $(E_0/hc)=(v_s+v_{s'}+1)\omega_s$. The first order Hamiltonian is now

$$H^{(1)''}=(\delta\hbar/2)\,[(p_{s'}/\hbar)^2+q_{s'}^2]+H^{(1)'} \tag{48.4}$$

which has the following non-vanishing matrix components

$$\left.\begin{aligned}
&(\ldots v_s, v_{s'},\ldots K\,|\,H^{(1)''}/hc\,|\ldots v_s, v_{s'},\ldots K)=(v_{s'}+\tfrac{1}{2})\,(\delta/2\pi c),\\
&(\ldots v_s, v_{s'},\ldots K\,|\,H^{(1)''}/hc\,|\ldots v_s+1, v_{s'}-1,\ldots K)\\
&\qquad=-(\ldots v_s+1, v_{s'}-1\ldots K\,|\,H^{(1)''}/hc\,|\ldots v_s, v_{s'},\ldots K)\\
&\qquad=-i\zeta_{ss'}^{(z)}B_e^{(zz)}K\,[(\lambda_s^{\frac{1}{2}}+\lambda_{s'}^{\frac{1}{2}})/(\lambda_s^{\frac{1}{4}}\lambda_{s'}^{\frac{1}{4}})]\,[(v_s+1)\,v_{s'}]^{\frac{1}{2}},\\
&(\ldots v_s, v_{s'},\ldots K\,|\,H^{(1)''}/hc\,|\ldots v_s+1, v_s-1,\ldots K\pm 1)\\
&\qquad=(\ldots v_s+1, v_{s'}-1\ldots K\pm 1\,|\,H^{(1)''}/hc\,|\ldots v_s, v_{s'},\ldots K)\\
&\qquad=\mp\tfrac{1}{2}\,[\zeta_{ss'}^{(y)}B_e^{(yy)}\pm i\zeta_{ss'}^{(x)}B_e^{(xx)}]\,[J(J+1)-K(K\pm 1)]^{\frac{1}{2}}\times\\
&\qquad\quad\times[(\lambda_s^{\frac{1}{2}}+\lambda_{s'}^{\frac{1}{2}})/(\lambda_s^{\frac{1}{4}}\lambda_{s'}^{\frac{1}{4}})]\,[(v_s+1)\,v_{s'}]^{\frac{1}{2}}.
\end{aligned}\right\} \tag{48.5}$$

The transformed second order Hamiltonian will, of course, also be altered since we have used the transformation function, S_R^*, instead of the transformation function given in Chap. V. It will be altered, however, only in so far as the coefficients of P_α^2 containing $(\zeta_{ss'}^{(\alpha)})^2$ are concerned. These terms contribute to the reciprocals of inertia, $B_v^{(\alpha\alpha)}$. It is only the diagonal elements of this part of $H^{(2)'}$ which are of interest to this approximation, and these will still yield exactly the same contribution to the reciprocals of inertia as before except that the quantities in $\lambda_s/(\lambda_s-\lambda_{s'})$ which are found in the constants, $b_{s\sigma}^{(\alpha\alpha)}$, occurring in the $B_v^{(\alpha\alpha)}$ will be absent and in their place will be found $(\zeta_{ss'}^{(\alpha)})^2 B_e^{(\alpha\alpha)}(B_e^{(\alpha\alpha)}/\omega_s^{\frac{1}{2}}\omega_{s'}^{\frac{1}{2}})\,[(\lambda_s^{\frac{1}{2}}-\lambda_{s'}^{\frac{1}{2}})^2/(\lambda_s^{\frac{1}{2}}+\lambda_{s'}^{\frac{1}{2}})]$.

As in the case of anharmonic resonance, the energy values themselves must be obtained by solving for the roots of the secular determinant of the matrix which contains the elements (48.5). These are diagonal in all of the vibrational quantum numbers except v_s and $v_{s'}$ to second order of approximation. The zero order energies are degenerate in ω_s and $\omega_{s'}$ since they do not depend upon v_s and $v_{s'}$ independently, but rather upon $v_s+v_{s'}$. The degree of the degeneracy is determined by the number of ways in which $v_s+v_{s'}$ can be made to add up to a given value. The elements associated with the terms coupling the frequencies will be grouped about the principal diagonal to form a sub-matrix of H, the sub-matrix containing as many rows and columns as the degree of the degeneracy. The matrix may be completely diagonalized by diagonalizing these sub matrices individually. When diagonalization is complete the stabilized wave functions of the component levels, the coefficients being the normalized cofactors of the secular sub-determinant, may be set up.

Coriolis resonance was first identified between the two fundamental frequencies ω_5 and ω_6 in the spectrum of formaldehyde vapor. One of these frequencies, ω_5, is a deformation frequency in the plane of the molecule; the other, ω_6, is a deformation frequency normal to the plane of the molecule. The example of the planar XYZ_2 molecule was studied theoretically by SILVER who chose the axis of symmetry to be the z axis. The Coriolis coupling factors $\zeta_{ss'}^{(x)}$ and $\zeta_{ss'}^{(y)}$ will both be zero here so that the only non-vanishing of the matrix elements given in (48.5) will be the $(v_5, v_6, K\,|\,v_5\pm 1, v_6\mp 1, K)$ elements and the $(v_5, v_6, K\,|\,v_5, v_6, K)$ elements. Using the wave functions, $R_{K,\gamma}^s$, and $R_{0,0}^s$, defined in (24.7) we shall, as an example, set up the secular determinant of the sub-matrix for the first excited states ω_5 and ω_6, (i.e., where $v_5=1$, $v_6=0$ and where $v_5=0$, $v_6=1$) for the value of $J=1$. It is readily verified from earlier sections that the determinant will factor into the two subdeterminants

$$\left.\begin{array}{l}
\begin{array}{c|c|c|}
 & v_5=1,\ v_6=0,\ R^s_{1,0} & v_5=0,\ v_6=1,\ R^s_{1,0} \\ \hline
\begin{array}{c} v_5=1,\ v_6=0, \\ R^s_{1,0} \end{array} & \begin{array}{l} hc\,[R_0+R_1+R_3+ \\ \quad +2(R_2+2R_4)+ \\ \quad +\delta/4\pi c]-\varepsilon \end{array} & i\Delta_1/2 \\ \hline
\begin{array}{c} v_5=0,\ v_6=1, \\ R^s_{1,0} \end{array} & -i\Delta_1/2 & \begin{array}{l} hc\,[R'_0+R'_1+R'_3+ \\ \quad +2(R'_2+2R'_4)+ \\ \quad +3\delta/4\pi c]-\varepsilon \end{array} \\ \hline
\end{array} = 0, \\[2ex]
\begin{array}{c|c|c|}
 & v_5=1,\ v_6=0,\ R^s_{1,1} & v_5=0,\ v_6=1,\ R^s_{1,1} \\ \hline
\begin{array}{c} v_5=1,\ v_6=0, \\ R^s_{1,1} \end{array} & \begin{array}{l} hc\,[R_0+R_1+R_3- \\ \quad -2(R_2+2R_4)+ \\ \quad -\delta/4\pi c]-\varepsilon \end{array} & i\Delta_1/2 \\ \hline
\begin{array}{c} v_5=1,\ v_6=0, \\ R^s_{1,1} \end{array} & -i\Delta_1/2 & \begin{array}{l} hc\,[R'_0+R'_1+R'_3- \\ \quad -2(R'_2+2R'_4)+ \\ \quad +3\delta/4\pi c]-\varepsilon \end{array} \\ \hline
\end{array} = 0
\end{array}\right\} \qquad (48.6)$$

and the roots $\varepsilon=hcR_0$ and $\varepsilon=hcR'_0$ where the R_i are respectively the values of the R_i defined in Sect. VI where $v_5=1$, $v_6=0$ and where $v_5=0$, $v_6=1$ and where Δ_1 is used to denote the quantity $2B_e^{(zz)}\,\zeta_{ss'}^{(zz)}\,(\lambda_s^{\frac{1}{2}}+\lambda_{s'}^{\frac{1}{2}})/(\lambda_s^{\frac{1}{4}}\lambda_{s'}^{\frac{1}{4}})$.

Let us consider the effect of the above Coriolis resonance on the appearance of the interacting bands in the spectrum. The formaldehyde molecule is only slightly asymmetric and we shall for simplicity neglect the asymmetry and set $I_{xx}^{(e)}=I_{yy}^{(e)}$ and set the $(K|K\pm 2)$ and $(K|K\pm 4)$ elements equal to zero. We shall further assume, for simplicity, that the rotational constants in the initial and final states have the same values. The solution of the secular determinant of the matrix elements of the energies yield the following energy values for the two states perturbed by Coriolis interaction

$$\left.\begin{aligned} E(+)/hc = {} & (\omega_5+\omega_6)+\{(\Delta_0/2)^2+K^2(\Delta_1/2)^2\}^{\frac{1}{2}}+\sum_{s''}{}'(\omega_{s''}/2)+ \\ & +J(J+1)\,B_e^{(xx)}+K^2(B_e^{(zz)}-B_e^{(xx)}) \end{aligned}\right\} \qquad (48.7)$$

where Δ_0 is used to designate $\omega_5-\omega_6$. The bands ω_5 and ω_6 in the CH_2O spectrum, which are non-degenerate perpendicular bands, arise when transitions take place from the normal vibration state to one of the two levels (48.7), observing at the same time the selection rules $\Delta K=\pm 1$, $\Delta J=0$ and $\Delta K=\pm 1$, $\Delta J=\pm 1$. Only the first set are impotant to us since the other set gives rise to lines which are not readily resolved spectroscopically. Subtracting the energies of the molecule when it is in its normal vibration state, where all the v's are zero, from the energies given in (48.7) and observing the selection rules the following relation for the frequency positions of the Q branches in the two bands is obtained.

$$\left.\begin{aligned} \omega(\pm) = {} & [(\omega_5+\omega_6)/2]-(B_e^{(zz)}-B_e^{xx)})\pm \\ & \pm\{(\Delta_0/2)^2+K^2(\Delta_1/2)^2\}^{\frac{1}{2}}\pm 2K(B_e^{(zz)}-B_e^{(xx)}) \end{aligned}\right\} \qquad (48.8)$$

where K takes the values 0, 1, 2, ...

The relation (48.8) shows that when Δ_0 is large compared to $K\Delta_1$ so that the latter may be neglected the frequency interval between two Q branches approach the normal value of $2(B_e^{(zz)} - B_e^{(xx)})$. As K increases, however, the spacings between Q branches change, approaching asymptotically the values $2(B_e^{(zz)} - B_e^{(xx)}) \pm (\Delta_1/2) = 2\{(1 \pm \zeta_{ss'}^{(zz)}) B_e^{(zz)} - B_e^{(xx)}\}$ the upper sign to be taken with the lines in the high frequency side of the lower frequency band and *vice versa*. The wave functions $\psi(\pm)(J, K, M)$ associated with the states $E(\pm)/hc$ are readily determined to be these linear combinations of the unperturbed wave functions

$$\left.\begin{aligned}\psi(+) &= [(\Delta - \Delta_0)^{\frac{1}{2}}/(2\Delta)^{\frac{1}{2}}]\,\psi(v_5 = 1, v_6 = 0) + \\ &\quad + i\,[(\Delta + \Delta_0)^{\frac{1}{2}}/(2\Delta)^{\frac{1}{2}}]\,\psi(v_5 = 0, v_6 = 1)\\ \psi(-) &= i\,[(\Delta + \Delta_0)^{\frac{1}{2}}/(2\Delta)^{\frac{1}{2}}]\,\psi(v_5 = 1, v_6 = 0) - \\ &\quad - [(\Delta - \Delta_0)^{\frac{1}{2}}/(2\Delta)^{\frac{1}{2}}]\,\psi(v_5 = 0, v_6 = 1)\end{aligned}\right\} \tag{48.9}$$

and

where

$$\Delta^2 = (\Delta_0^2 + K^2\Delta_1^2).$$

These degenerate to the wave functions of two independent linear harmonic oscillators when Δ_0 is large enough so that $K\Delta_1$ may be neglected in comparison with it. When Δ_0 is small compared with $K\Delta_1$, on the other hand, the above functions become the wave functions of a two dimensionally isotropic oscillator. The intensities, in the first instance, are therefore those for an ordinary perpendicular band. The intensities in the second limiting case are once more those for a perpendicular band, but this time a two fold degenerate perpendicular band where the TELLER-TISZA selection rule applies which states that transitions from $K = K' - 1$ in the normal state to $K = K'$ in the upper component state $E(-)/hc$ are forbidden. The result for the intermediate case may be summarized somewhat in the following manner. As the molecule makes transitions from the normal vibration state to the state $E(+)/hc$, the probability of the rotational transition $K = K' + 1$ in the lower state to $K = K'$ in the upper state increases at the expense of the transition $K = K' - 1$ to $K = K'$ as K increases. Similarly, when the molecule makes a transition from the ground state to the component state $E(-)/hc$, the transition $K = K' - 1$ in the initial state to $K = K'$ in the final state will increase in intensity with K at the expense of the transition $K = K' + 1$ to $K = K'$.

We shall consider also the example when Coriolis resonance occurs between two adjacent perpendicular bande ω_s and $\omega_{s'}$ in an axially symmetric molecule. As usual the z axis is identified with the axis of symmetry and we shall assume that the Coriolis coupling factors $\zeta_{ss'}^{(x)}$ and $\zeta_{ss'}^{(y)}$ may once more be set equal to zero. The perpendicular frequencies ω_s and $\omega_{s'}$ are two fold degenerate in the axially symmetric molecule, but the resonance may, nevertheless, be treated in the same manner we have just described, it being the x component of ω_s which interacts with the y component of $\omega_{s'}$ and *vice versa*. It is more convenient, however, to obtain the matrix components in terms of v_s, l_s and $v_{s'}$, $l_{s'}$, i.e., by treating the problem in polar coordinates, replacing $q_{s\sigma}$ and $q_{s'\sigma'}$ and their conjugate momenta by their equivalents in r_s, χ_s and $r_{s'}$, $\chi_{s'}$. The work is straight forward and the matrix components may readily be obtained with the aid of the relations (21.6), taking into account the fact that $\zeta_{s,1,s',2}^{(z)} = -\zeta_{s,2,s',1}^{(z)}$ [see for example Eq. (18.24)]. The matrix elements are then the following where as before we set $\lambda_{s'}^{\frac{1}{2}} = \lambda_s^{\frac{1}{2}} + \delta$:

$$\left.\begin{aligned}&(v_s, l_s, v_{s'}, l_{s'}, K|\, H^{(1)''}/hc\,|\, v_s, l_s, v_{s'}, l_{s'}, K) = (v_{s'} + 1)\,\delta/2\pi c\,,\\ &(v_s - 1, l_s \pm 1, v_{s'}, l_{s'}, K\,|\, H^{(1)''}/hc\,|\, v_s, l_s, v_{s'} - 1, l_{s'} \mp 1, K)\\ &\quad = (v_s, l_s, v_{s'} - 1, l_{s'} \mp 1, K\,|\, H^{(1)''}/hc\,|\, v_s - 1, l_s \pm 1, v_{s'}, l_{s'}, K)\\ &\quad = \pm\,[(\lambda_s^{\frac{1}{2}} + \lambda_{s'}^{\frac{1}{2}})/(\lambda_s^{\frac{1}{4}}\,\lambda_{s'}^{\frac{1}{4}})]\,[\zeta_{ss'}^{(z)}\,K B_e^{(zz)}/2]\,[(v_s \mp l_s)\,(v_{s'} \mp l_{s'})]^{\frac{1}{2}}.\end{aligned}\right\} \tag{48.10}$$

When the secular determinant of the matrix which embraces the elements (48.10) is solved for the energies of the perturbed states where

$$v_s = |l_s| = 1, \quad v_{s'} = |l_{s'}| = 0 \quad \text{and} \quad v_s = |l_s| = 0, \quad v_{s'} = |l_{s'}| = 1$$

the following relations are obtained

$$\left.\begin{aligned} E(\pm)/hc = 3(\omega_s + \omega_{s'})/2 + \sum_{s''} \omega_{s''}/2 + J(J+1) B_e^{(xx)} + K^2 (B_e^{(zz)} - B_e^{(xx)}) - \\ - K B_e^{(zz)} (\zeta_{ss}^{(z)} + \zeta_{s's'}^{(z)}) \pm \sqrt{(\varDelta_0/2)^2 + (K\varDelta_1/2)^2} \end{aligned}\right\} \quad (48.11)$$

where $\varDelta_0$ is equal to $\delta/2\pi c + 2K(\zeta_{ss}^{(z)} - \zeta_{s's'}^{(z)}) B_e^{(zz)}$, J and K are the rotational quantum numbers and as before $\varDelta_1 = 2\zeta_{ss'}^{(z)} B_e^{(zz)} [(\lambda_s^{\frac{1}{2}} + \lambda_{s'}^{\frac{1}{2}})/\lambda_s^{\frac{1}{4}} \lambda_{s'}^{\frac{1}{4}}]$. Moreover, we still retain the notation $\varDelta^2 = (\varDelta_0^2 + K^2 \varDelta_1^2)$. Subtracting the energies of the molecule in its normal state from the relations (48.11), observing the appropriate selection rules, including the TELLER-TISZA selection rule for each band, the following relations for the frequency positions for the Q lines in the two bands are obtained

$$\left.\begin{aligned} \omega(\pm) = [(\omega_s + \omega_{s'})/2] - (B_e^{(zz)} - B_e^{(xx)}) \pm \{(\varDelta_0/2)^2 + (K\varDelta_1/2)^2\}^{\frac{1}{2}} \pm \\ \pm 2K \{[1 - (\zeta_{ss}^{(z)} + \zeta_{s's'}^{(z)})/2] B_e^{(zz)} - B_e^{(xx)}\}. \end{aligned}\right\} \quad (48.12)$$

As in the earlier example, when $\varDelta_0 \gg K\varDelta_1$ the relations (48.12) for the line positions go over into those for two independent two fold degenerate perpendicular bands. As the quantity $K\varDelta_1$ increases with respect to $\varDelta_0$ the spacings between Q branches approach $2\{[1 - (\zeta_{ss}^{(z)} + \zeta_{s's'}^{(z)})/2] B_e^{(zz)} - B_e^{(xx)}\} \pm (\varDelta_1/2)$ or $2\{[1 - (\zeta_{ss}^{(z)} + \zeta_{s's'}^{(z)})/2 \pm \zeta_{ss'}^{(z)}] \times B_e^{(zz)} - B_e^{(xx)}\}$, the upper sign to be taken with the line on the high frequency side of the higher frequency band and with the lines on the low frequency side of the lower frequency band and *vice versa*.

When the secular determinant of the matrix whose elements are (48.10) has been solved the stabilized wave functions, $\psi(\pm)(J, K, M)$ associated with the states $E(\pm)/hc$ may be found. They are the following linear combinations of the unperturbed wave functions

$$\begin{aligned} \psi(\pm) = [(\varDelta \mp \varDelta_0)^{\frac{1}{2}}/(2\varDelta)^{\frac{1}{2}}]\, \psi(v_s = |l_s| = +1, v_{s'} = |l_{s'}| = 0) + \\ \pm [(\varDelta \pm \varDelta_0)^{\frac{1}{2}}/(2\varDelta)^{\frac{1}{2}}]\, \psi(v_s = |l_s| = 0, v_{s'} = |l_{s'}| = +1). \end{aligned}$$

It becomes evident from these that while for each perpendicular vibration individually the TELLER-TISZA selection rule still holds, there will be a modified intensity distribution throughout the two bands.

The general effect of the Coriolis resonance in the two cases we have considered is to produce a convergence of the Q branches, the lines spreading apart on the high and low frequency sides, respectively, of the higher and lower frequency bands and the lines crowding together on the high and low frequency sides, respectively, of the lower and higher frequency bands. The convergence is of a special nature, however, in that the spacings between lines approach limiting values and heads are not reached. There is in both examples we have considered, further an anomalous intensity distribution of the spectral lines established. Fig. 25 shows the two perpendicular bands, ν_9 and ν_{10} in the spectrum of allene. These are undoubtedly the best known examples of the type of resonance we have just discussed.

A more complicated form of Coriolis interaction between vibration frequencies may arise in axially symmetric molecules when a parallel vibration, ω_s, resonates with a two-fold degenerate (isotropic) perpendicular vibration, ω_t. Here the coupling between frequencies is introduced through the operators $\frac{1}{2}(p_x \pm i p_y)(P_x \mp i P_y)$ where p_x and p_y are the x and y components of the incipient

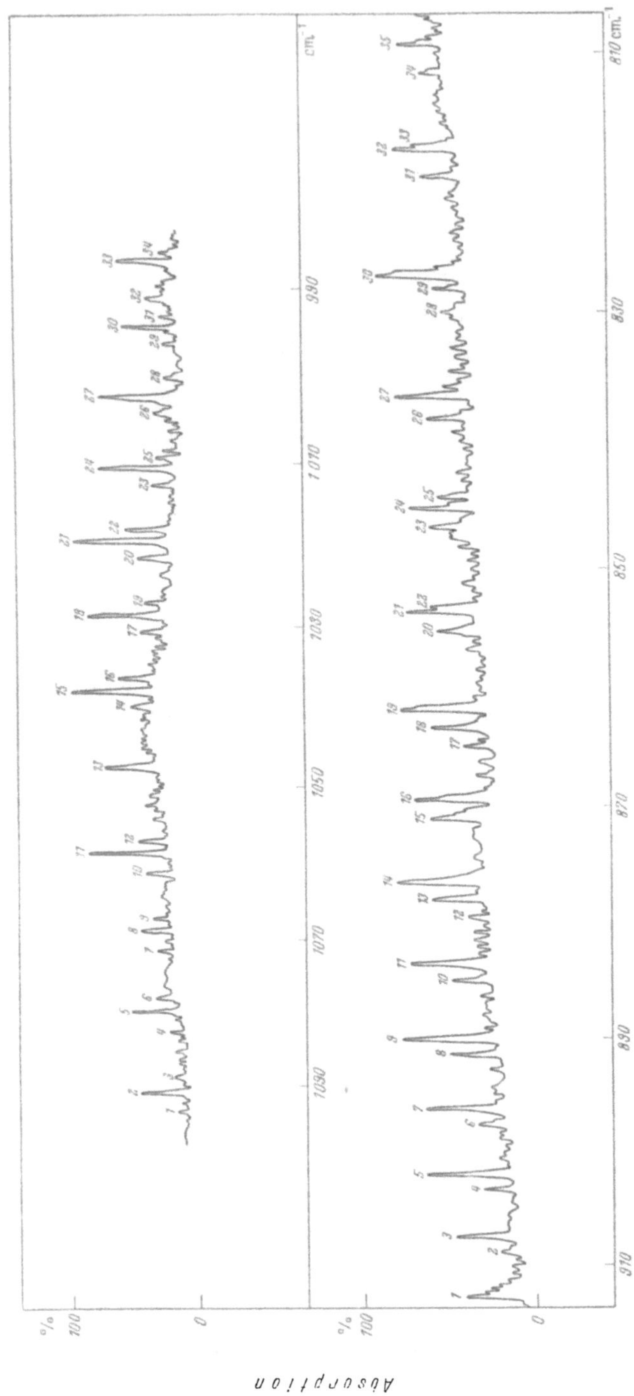

Fig. 25a. The perpendicular bands ν_9 and ν_{10} in the infrared spectrum of allene (high dispersion). Note the convergence is in the opposite sense in the two bands due to Coriolis resonance interaction.

internal angular momentum resulting from oscillation and P_x and P_y are the x and y components of the angular momentum of the molecule[1]. The Coriolis

[1] The problem which here is being considered bears a marked similarity in principle to that of the coupling of the electron spins to the rotation of the molecule through the oper-

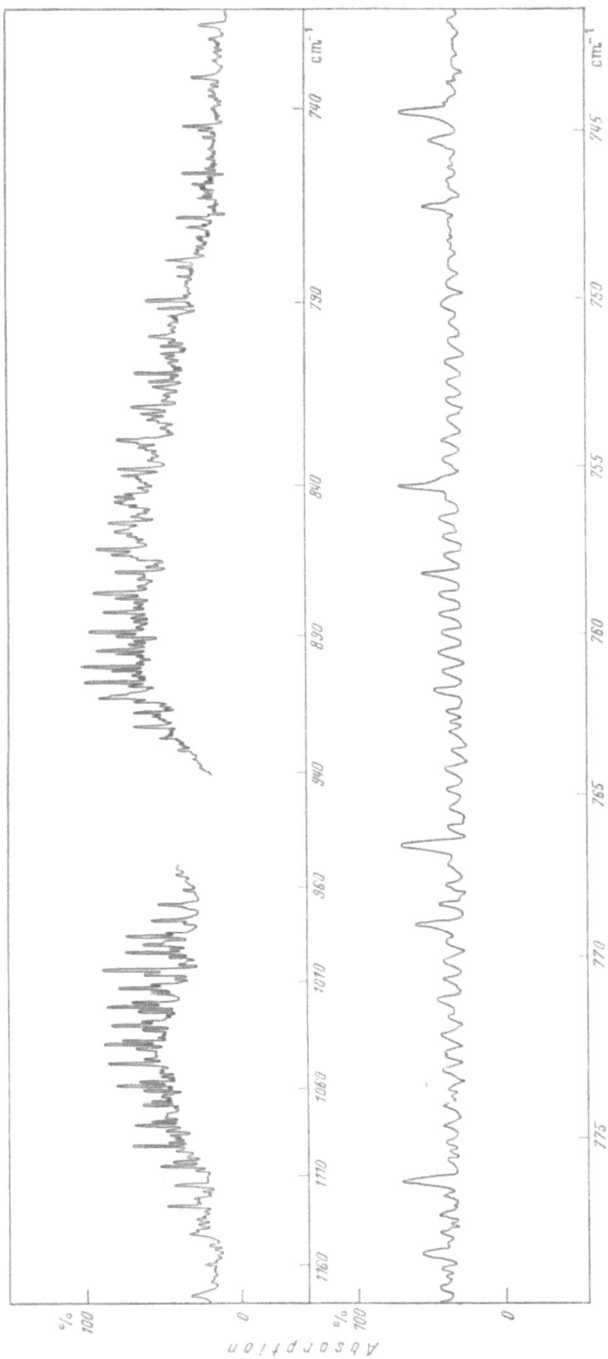

Fig. 25b. Upper part: The perpendicular bands ν_9 and ν_{10} in the infrared spectrum of allene (medium dispersion). Note the convergence is in the opposite sense in the two bands due to Coriolis resonance interaction. Lower part: A portion of lower frequency band examined under high resolution.

coupling factor, $\zeta^{(z)}_{s,t}$ vanishes under these circumstances and the factors $\zeta^{(x)}_{s,t}$ and $\zeta^{(y)}_{s,t}$ are this time different from zero. The resonance may, again, be treated by

---

ator $2\mu_{xx}\mathfrak{S}_x(P_x - p_x^*) + 2\mu_{yy}\mathfrak{S}_y(P_y - p_y^*)$, discussed in Sect. 35, where the frequency intervals between the components of the spin multiplet levels are small.

using the matrix elements (48.5) since ω_s will interact with each of the components of the frequency, ω_t, separately. It is more convenient, however, again to introduce the polar coordinates, r_t and χ_t, instead of the coordinates $q_{t,1}$ and $q_{t,2}$. When one takes into account that only two of the Coriolis coupling factors, namely; $\zeta_{s;t,2}^{(x)}$ and $\zeta_{s;t,1}^{(y)}$, will be different from zero and that the relation $\zeta_{s;t,2}^{(x)} = -\zeta_{s;t,1}^{(y)} = \zeta_{s,t}^{(x)}$ exists between them the operators $(p_x \pm i p_y)\,(P_x \mp i P_y)$ may be shown to take the form

$$\left.\begin{aligned}(p_x \pm i p_y)\,(P_x \mp i P_y) = \sum_s \sum_t \zeta_{s,t}^{(x)}\, e^{\pm i\chi_t}\{(\lambda_s/\lambda_t)^{\frac14}\, q_s\,[p_{r_t} \pm (i/r_t)\, p_{\chi_t}] - {}\\ - (\lambda_t/\lambda_s)^{\frac14}\, r_t\, p_s\}\,\{P_x \mp i P_y\}.\end{aligned}\right\}\qquad(48.14)$$

One may proceed much as before by setting $\lambda_t^{\frac12} = \lambda_s^{\frac12} + \delta$, to determine the significant elements to the energy matrix arising from this type of interaction. They are found to be

$$\left.\begin{aligned}&(v_s, v_t, l_t, K\,|\,H^{(1)\prime\prime}/hc\,|\,v_s, v_t, l_t, K) = (v_t + 1)\,\delta/2\pi c;\\ &(v_s - 1, v_t, l_t, K\,|\,H^{(1)\prime\prime}/hc\,|\,v_s, v_t - 1, l_t \pm 1, K \pm 1)\\ &\quad = (v_s, v_t - 1, l_t \pm 1, K \pm 1\,|\,H^{(1)\prime\prime}/hc\,|\,v_s - 1, v_t, l_t, K)\\ &\quad = [(\lambda_s^{\frac12} + \lambda_{s'}^{\frac12})/\lambda_s^{\frac14}\,\lambda_{s'}^{\frac14}]\,[\zeta_{s,t}^{(x)}\,B_e^{(xx)}/2]\,[v_s\,(v_t \mp l_t)]^{\frac12}\,[(J \mp K)\,(J \pm K + 1)]^{\frac12}.\end{aligned}\right\}\qquad(48.15)$$

The secular determinant of the energies of the two perturbing fundamental states $v_s = 1$, $v_t = l_t = 0$ and $v_s = 0$, $v_t = 1$, $l_t = \pm 1$ will have $3\,(2J + 1)$ rows and columns, but will break up into sub determinants. The general form of these subdeterminants is shown in (48.16)

$$\begin{array}{c|c|c|c|}
 & \begin{matrix}v_t=1, l_t=1,\\ v_s=0, K=K+1\end{matrix} & \begin{matrix}v_t=l_t=0,\\ v_s=1, K=K\end{matrix} & \begin{matrix}v_t=1, l_t=-1,\\ v_s=0, K=K-1\end{matrix}\\ \hline
\begin{matrix}v_t=1, l_t=1\\ v_s=0\; K=K+1\end{matrix} & -K-1-\varepsilon & -[\frac12(J-K)(J+K+1)]^{\frac12}\alpha & 0\\ \hline
\begin{matrix}v_t=l_t=0\\ v_s=1, K=K\end{matrix} & -[\frac12(J-K)(J+K+1)]^{\frac12}\alpha & \Delta-\varepsilon & -[\frac12(J+K)(J-K+1)]^{\frac12}\alpha\\ \hline
\begin{matrix}v_t=1, l_t=-1\\ v_s=0, K=K-1\end{matrix} & 0 & -[\frac12(J+K)(J-K+1)]^{\frac12}\alpha & K-1-\varepsilon\\ \hline
\end{array} = 0 \qquad (48.1$$

where

$$\Delta = [(\delta/2\pi c) - (B_e^{(zz)} - B_e^{(xx)})]/2[(1 - \zeta_{tt}^{(z)})\,B_e^{(zz)} - B_e^{(xx)}],$$

$$\alpha = \{[(\lambda_s^{\frac12} + \lambda_{s'}^{\frac12})/\lambda_s^{\frac14}\,\lambda_{s'}^{\frac14}]\,\zeta_{s,t}^{(x)}\,(B_e^{(xx)}/2)\}/[(1 - \zeta_{tt}^{(z)})\,B_e^{(zz)} - B_e^{(xx)}]$$

and

$$\varepsilon = \{-\varepsilon' + T(v) + J(J+1)\,B_e^{(xx)} + (K^2 - 1)\,(B_e^{(zz)} - B_e^{(xx)})\}/2\,[(1 - \zeta_{tt}^{(z)})\,B_e^{(zz)} - B_e^{(xx)}],$$

$T(v)$ being the vibration term value where $v_t = l_t = 1$ and where ε' is the perturbed energy. The value of $K - 1$ may not exceed J; neither may $K + 1$ be less than $-J$. Evidently, then, there will be two identical single roots obtained by setting $K + 1 = -J$ and $K - 1 = J$ in (48.16). There will be two other identical sub-determinants of two rows and columns each obtained by setting $K + 1 = -J + 1$ and $K - 1 = J - 1$ in (48.16). Finally there will be $(2J - 1)$ other sub-determinants of three rows and columns each. $J - 1$ of the latter are repeated twice and one (i.e., where $K + 1 = 1$) is unique. This last instance of the above sub-determinant is of particular interest because it removes the degeneracy of the levels $K=l_t=1$ and $K=l_t=-1$. The splitting is entirely similar to the type of splitting discussed earlier and referred to as l-type doubling. The essential difference is that here the perturbing frequencies are close enough together so that resonance exists. The doubling is, for this reason, a first order effect. Such a large l-type doubling has recently been observed by GARING, RAO and NIELSEN in the perpendicular band, ν_3, in the spectrum of NH_3.

Diagonalization of these sub matrices is simple in principle, but depends upon the solution, in general, of a cubic equation which is awkward in practice. One may observe, however, that if K is set equal to $J-1$ (i.e., $K+1=J$) in (48.16) two of the elements off of the principal will reach their minimum value $\sqrt{J}$. This corresponds to the classical case where the molecule is rotating entirely about the axis of symmetry. Classically then $P_z=P$ and $P_x=P_y=0$. Correspondingly, $p_x P_x + p_y P_y$ should be zero so that the perturbation should vanish. Quantum mechanically this condition is not quite achieved, but the uncoupling is so complete that the determinant of the submatrix may be expanded by some suitable method, for example, by means of continued fractions. Expanding to the first term only one obtains for the energy term value of the state where $K+1=J$

$$\left.\begin{aligned} E/hc \approx T(v) + J(J+1)\, B_e^{(xx)} + J^2 (B_e^{(zz)} - B_e^{(xx)}) - \\ - 2J\zeta_{tt}^{(z)} B_e^{(zz)} \left[1 - 2(\zeta_{s,t}^{(x)})^2 (B_e^{(xx)})^2/\zeta_{tt}^{(z)}\, \delta\, B_e^{(zz)}\right]. \end{aligned}\right\} \tag{48.17a}$$

Expanding (48.16) in the same manner for $K+1=J-1$ one obtains the somewhat less exact formula for E/hc:

$$\left.\begin{aligned} E/hc \approx T(v) + J(J+1)\, B_e^{(xx)} + (J-1)^2 (B_e^{(zz)} - B_e^{(xx)}) - \\ - 2(J-1)\,\zeta_{tt}^{(z)} B_e^{(zz)} \left[1 - 4(\zeta_{s,t}^{(x)})^2 (B_e^{(xx)})^2/\zeta_{tt}^{(z)}\, \delta\, B_e^{(zz)}\right]. \end{aligned}\right\} \tag{48.17b}$$

The term values (48.17a) and (48.17b) are evidently term values belonging to a perpendicular vibration state, ω_t. The conclusion to be drawn from these relations is that the usual degenerate Coriolis coupling factor, $\zeta_{tt}^{(z)}$, must be replaced by an effective ζ which depends upon the value K.

The lines in a perpendicular band on which the perturbation will have the most striking effect will depend upon the structure of the molecule. It has been the ${}^R R$ lines which have been of most interest in the case of pyramidal molecules like PH_3, AsH_3 and SbH_3 where the moments of inertia are such that the rotator approaches a spherical top. The ${}^P P$ lines might have been of equal interest except for the fact that they are badly overlapped by the perturbing band. The ${}^R R$ frequencies where the above term values are involved are readily shown to be

$$\nu = \nu_0 + 2JB_e^{(xx)} + 2J\left[(1-\zeta_{tt}^{(z)})\, B_e^{(zz)} - B_e^{(xx)}\right] + 4J(\zeta_{s,t}^{(x)})^2 (B_e^{(xx)})^2/\delta \tag{48.18a}$$

and

$$\left.\begin{aligned} \nu = \nu_0 + 2JB_0^{(xx)} + 2(J-1)\left[(1-\zeta_{tt}^{(z)})\, B_e^{(zz)} - B_e^{(xx)}\right] + \\ + 4(2J-1)\,(\zeta_{s,t}^{(x)})^2 (B_e^{(xx)})^2/\delta. \end{aligned}\right\} \tag{48.19}$$

The interval between two lines $\nu(J+1, K+1=J+1)$ and $\nu(J, K=J)$ will then be equal to $2[(1-\zeta_{tt}^{(z)})\, B_e^{(zz)} + 2(\zeta_{s,t}^{(x)})^2 (B_e^{(xx)})^2/\delta]$ and between two lines $\nu(J+1, K+1=J)$ and $\nu(J, K=J-1)$ the spacing will be $2[(1-\zeta_{tt}^{(z)})\, B_e^{(zz)} + 4(\zeta_{s,t}^{(x)})^2 (B_e^{(xx)})^2/\delta]$. The constants $\zeta_{tt}^{(z)} B_e^{(zz)}$ and $(\zeta_{s,t}^{(x)})^2$ can then be determined if it is possible to identify the lines indicated above.

The perturbation produces anomalies also in the structure of the parallel band, ν_s. The effect of the perturbation as it manifests itself in the rotation levels of the first excited state, ω_s, is best studied where $K=0$. These states correspond classically to the situation where the molecule does not spin at all

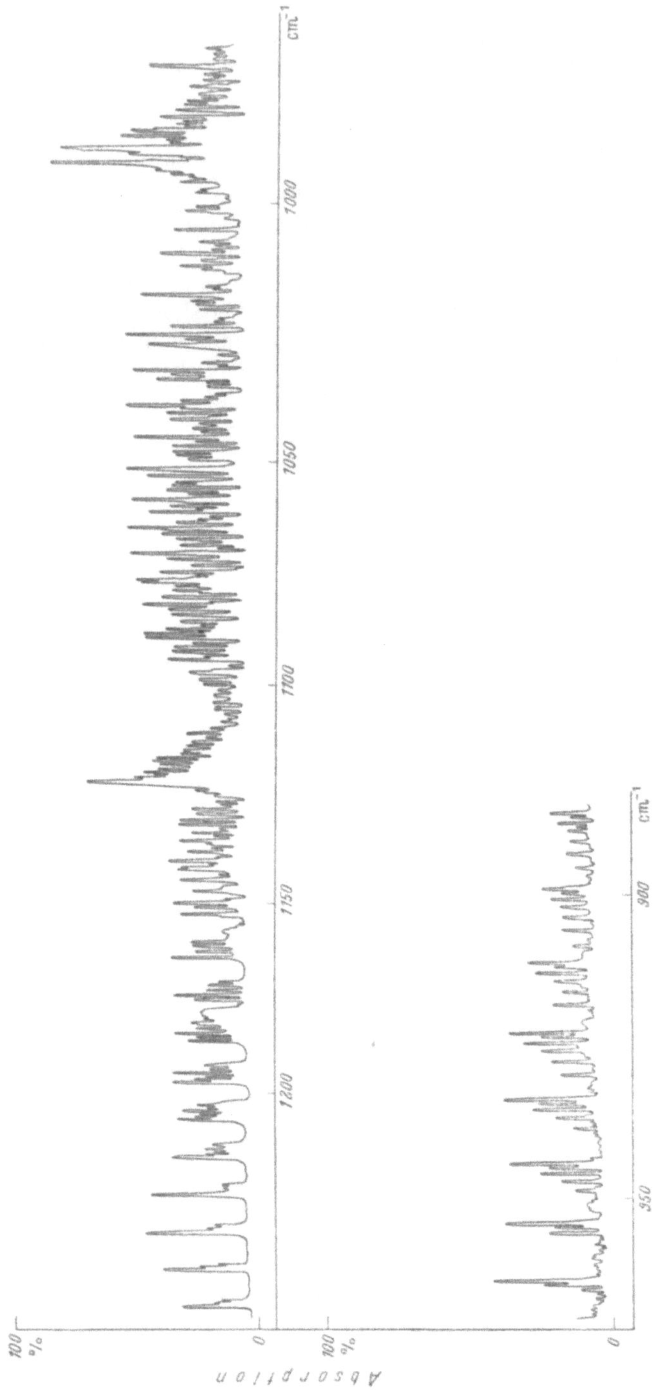

Fig. 26. The parallel and perpendicular bands, ν_2 and ν_4, in the infrared spectrum of PH_3 showing the effects of Coriolis resonance interaction.

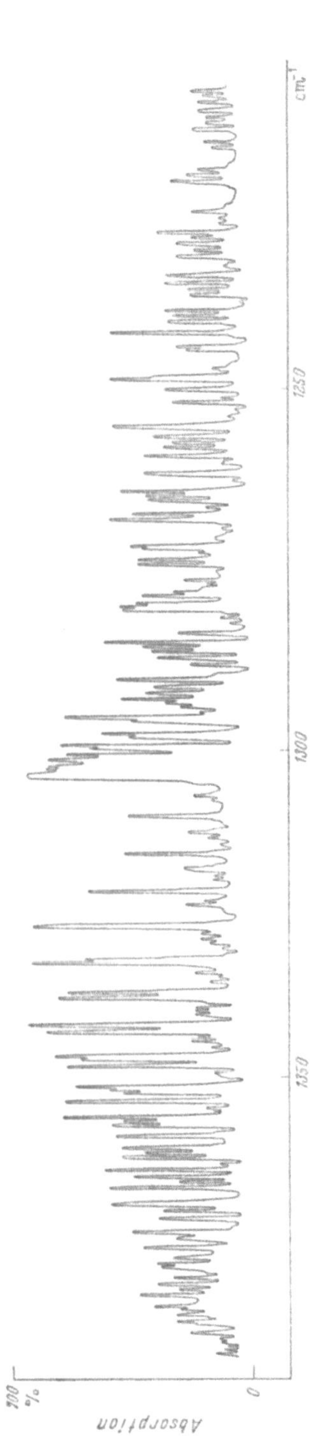

Fig. 27. The band ν_4 in the infrared spectrum of CH_4. The complex structure in the P and R branches is a splitting into K structure due to Coriolis resonance interaction of ν_4 with ν_2.

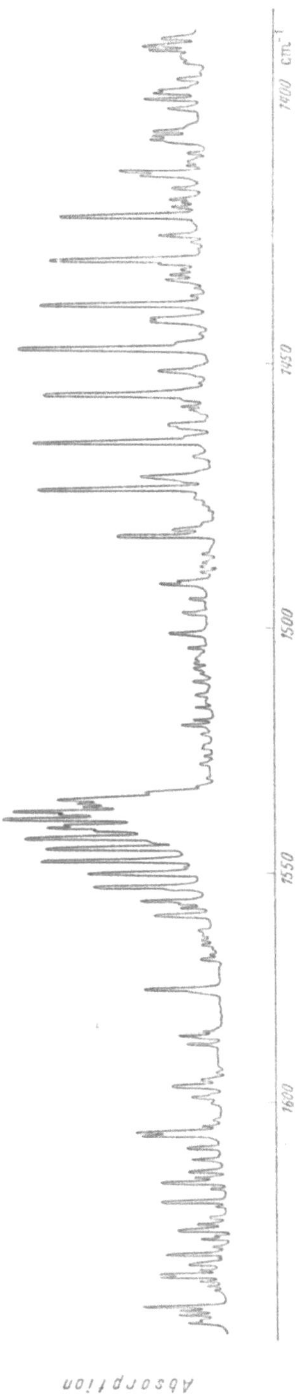

Fig. 28. The band ν_2 in the infrared spectrum of CH_4. This band becomes optically active due to Coriolis resonance interaction with the frequency ν_4.

about its axis of symmetry and is *tumbling* only. When $K=0$ the submatrix (48.16) factors into a single root and a double root. One of the double roots applies to the state ω_s and it leads to the term value.

$$\left.\begin{aligned} E/hc \approx T'(v) + (3\omega_t + \omega_s)/2 + J(J+1)\, B_e^{(xx)} + \\ + [(\delta/2)^2 + 4(\zeta_{t,s}^{(x)})^2 (B_e^{(xx)})^2 J(J+1)]^{\frac{1}{2}}. \end{aligned}\right\} \tag{48.19}$$

It has been the P branch of such bands in the case of pyramidal molecules, like PH_3, AsH_3 and SbH_3, which have been of interest, the R branch being overlapped by the perturbing band. The positions of the lines in the P branch of the sub-bands where $K=0$ are given by the following expression

$$\nu \approx (\omega_t + \omega_s/2) - 2J B_e^{(xx)} - [(\delta/2)^2 + 4J(J-1)(\zeta_{s,t}^{(x)})^2 (B_e^{(xx)})^2]^{\frac{1}{2}}. \tag{48.20}$$

As one might expect the relation (48.19) resembles in many respects the relation (48.8) except that $J(J-1)$ replaces K^2.

The parallel band, ν_2, and the perpendicular band, ν_4, in the spectrum of PH_3 which perturb each other through a Coriolis resonance of the kind just referred to are shown in Fig. 26.

The most famous instance of Coriolis resonance is undoubtedly that between the three-fold degenerate vibration frequency, ω_4, and the two-fold degenerate frequency, ω_2, in the CH_4 molecule. This problem was studied by CHILDS and JAHN who gave a theoretical interpretation to the band, ν_4, studied experimentally by NIELSEN and NIELSEN. A more complete discussion of this example is beyond the scope of this section and the reader is referred to the original papers dealing with this problem. An interesting corallary to this example is the fact that the frequency, ω_2, which in the unperturbed case is optically inactive, becomes spectroscopically observable due to the resonance with the frequency, ω_4. The two bands, ν_2 and ν_4 from the infrared spectrum of CH_4 are shown in Figs. 27 and 28.

49. Second order rotational resonance interactions. The interactions discussed in Sect. 48 are the only rotational resonance interactions on record although other types of higher order are, indeed, possible. The transformed second order Hamiltonian, $H^{(2)\prime\prime}$, contains terms in $P_\alpha P_\beta$ which have as coefficients quantities involving $A_{s,\sigma,s'\sigma'}^{(\alpha\beta)}$, $a_{s\sigma}^{(\alpha\beta)}$, etc., multiplied by $q_{s\sigma} q_{s'\sigma'}$ and $p_{s\sigma} p_{s'\sigma'}/\hbar^2$. The contributions to the energy of such terms have already been taken account of when $s=s'$ and $\sigma=\sigma'$ and these are already included in the energy expressions given in Chap. V. When $s=s'$, but $\sigma\neq\sigma'$ these terms contribute further to what has been referred to as l-type doubling and these corrections to the energy have been evaluated in Chap. VI. When $s\neq s'$ and $\sigma=\sigma'$ these terms do not contribute to the energy in second approximation except in such cases as where $\omega_s\approx\omega_{s'}$. The same is essentially the situation concerning the terms stated in Chap. V and designated as H_R^*. The first term in H_R^* might at first glance be thought to contribute to l-type doubling when $s=s'$ as well as the terms in $H^{(2)\prime}$ referred to earlier. One is quickly convinced that this is not so since coefficients containing k_{sss} as a factor can not occur in terms contributing to l-type doubling for symmetry reasons. Quite in general H_R^* will not contribute to the energy in second order of approximation except when $\omega_s\approx\omega_{s'}$ or when $\omega_s\approx\omega_{s''}$.

When resonance sets in between two frequencies ω_s and $\omega_{s'}$ one would proceed as earlier and set $\lambda_{s'}^{\frac{1}{2}}=\lambda_s^{\frac{1}{2}}+\delta$ and develop the result by means of the degenerate perturbation theory. The matrix elements which would be important would be the following, $(v_s, v_{s'}, K|v_s\pm1, v_{s'}\mp1, K')$ where K' might, in general, take the values $K, K \pm 1$ or $K \pm 2$. When the contribution for $H^{(2)\prime}$ and H_R^* referred to above are gathered together one may write as their matrix components

$$\left.\begin{aligned}
&(v_s, v_{s'}, K \,|\, H^{(2)}/hc \,|\, v_s + 1,\, v_{s'} - 1,\, K') = (v_s + 1,\, v_{s'} - 1,\, K' \,|\, H^{(2)}/hc \,|\, v_s,\, v_{s'} K) \\
&\quad = -\frac{1}{2} \sum_{\alpha\beta} B_e^{(\alpha\alpha)} \left(\frac{\hbar}{I_{\beta\beta}^{(e)} \lambda_s}\right)^{\frac{1}{2}} \left\{ \left(\frac{\hbar}{I_{\beta\beta}^{(e)}}\right)^{\frac{1}{2}} \left[A_{s s'}^{(\alpha\beta)} + A_{s' s}^{(\alpha\beta)} - \sum_{\gamma} \frac{a_s^{(\alpha\gamma)} a_{s'}^{(\gamma\beta)} + a_{s'}^{(\alpha\gamma)} a_s^{(\gamma\beta)}}{I_{\gamma\gamma}^{(e)}} - \right.\right. \\
&\quad\quad \left. - 2 \sum_{s''\sigma''}{}' \left(\zeta_{s s''}^{(\alpha)} \zeta_{s' s''}^{(\beta)} + \zeta_{s' s''}^{(\alpha)} \zeta_{s s''}^{(\beta)}\right) \right] - \\
&\quad\quad - 2\pi c \left[2 k_{s s s'} \frac{a_s^{(\alpha\beta)}}{(I_{\beta\beta}^{(e)} \lambda_s^{\frac{1}{2}})^{\frac{1}{2}}} + 2 k_{s s' s'} \frac{a_{s'}^{(\alpha\beta)}}{(I_{\beta\beta}^{(e)} \lambda_s^{\frac{1}{2}})^{\frac{1}{2}}} + \right. \\
&\quad\quad \left.\left. + \sum_{\substack{s \\ s \neq s' \neq s''}} k_{s s' s''} \frac{a_{s''}^{(\alpha\beta)}}{(I_{\beta\beta}^{(e)} \lambda_{s''}^{\frac{1}{2}})^{\frac{1}{2}}} \left(\frac{\lambda_s}{\lambda_{s''}}\right)^{\frac{1}{2}} \right] \right\} [(v_s + 1)\, v_{s'}]^{\frac{1}{2}} \,(K \,|\, P_\alpha P_\beta/\hbar^2 \,|\, K')
\end{aligned}\right\} \tag{49.1}$$

where the prime attached to the $\sum\limits_{s''\sigma''}$ indicates that $s''\sigma'' \neq s, \sigma$ and $s'\sigma'$, so long as ω_s and $\omega_{s'}$ are non-degenerate frequencies. The above elements in the energy matrix associated with v_s and $v_{s'}$ will then as before form a submatrix which may be diagonalized by itself. No resonances have been observed of this kind. Their magnitudes should be of the size of $b_{s\sigma}^{(\alpha\beta)}$ and they might in some instances give rise to anomalous values of $B_v^{(\alpha\alpha)}$.

50. Anomalous intensity distribution in the infrared bands of molecules. Experiments have shown that the intensity distribution of rotation lines in a band in the spectrum of a molecule is not exactly that predicted by the Eqs. (30.9) and (30.10). This may be understood if one takes the interaction between rotation and vibration of the molecule into account and recognizes the fact that the zero order wave functions are not sufficient for the computation of the matrix elements of the electric moment $(\tau|\mu|\tau'')$ in the relation (30.2).

Three kinds of intensity anomalies are known. The first of these was known already in the days of the classical quantum theory. The effect has been studied in the rigorous quantum mechanics, first by OPPENHEIMER and more recently by HERMAN and WALLIS and again by HERMAN and RUBIN who have confined themselves to diatomic molecules. The anomaly originates, in our theory, with the first order term in the Hamiltonian,

$$-\tfrac{1}{2} \sum_{s\sigma} (\hbar^2/\lambda_s)^{\frac{1}{4}} \left\{ \sum_{\alpha} \sum_{\beta} a_{s\sigma}^{(\alpha\beta)} (P_\alpha - p_\alpha^*)(P_\beta - p_\beta^*)/I_{\alpha\alpha}^{(0)} I_{\beta\beta}^{(0)} \right\} q_{s\sigma}.$$

A second type of intensity anomaly may arise in the case of asymmetric molecules. This effect may be understood in the following manner. The rotational part of the wave function for an asymmetric molecule, $(\psi_{J,M}^s)_\gamma$ is expandible in terms of the wave functions of the symmetric rotator, $R_{K,\gamma}^s$, as

$$(\psi_{J,M}^s)_\gamma = \sum_K a_{K,\gamma}^s R_{K,\gamma}^s$$

where the coefficients $a_{K,\gamma}^s$ are the normalized cofactors of the determinants of the submatrices E^+, E^-, O^+ and O^- and will, therefore, depend upon the asymmetry of the molecule (i.e., upon the value of $\varkappa$). It is a well known fact, however, that the effective moments of intertia are not the same in the initial and the final states and that, therefore, the effective asymmetry must be different as well. This carries with it the fact that the coefficients, $a_{K,\gamma}^s$, will be different in the initial and final states. The line strengths computed and tabulated by

HAINER, CROSS and KING on the basis that $\varkappa$ does not change will, therefore, not strictly speaking be applicable to an actual spectrum. This effect has been studied by ALLEN and while noticeable variations in line strengths arise from this effect it is probably small compared to the former effect since it originates in the second order Hamiltonian with the terms

$$-\tfrac{1}{2}\sum_{s\sigma}\sum_{s'\sigma'}(\hbar^4/\lambda_s\,\lambda_{s'})^{\frac{1}{4}}\Big\{\sum_{\alpha,\beta}\Big[\big(A^{(\alpha\beta)}_{s\sigma s'\sigma'}-\sum_{s''\sigma''}\zeta^{(\alpha)}_{s\sigma s''\sigma''}\,\zeta^{(\beta)}_{s'\sigma' s''\sigma''}\big)-$$
$$-\sum_{\gamma}\big(a^{(\alpha\gamma)}_{s\sigma}\,a^{(\gamma\beta)}_{s'\sigma'}/I^{(e)}_{\gamma\gamma}\big)\Big]\,(P_\alpha-p^*_\alpha)\,(P_\beta-p^*_\beta)/I^{(e)}_{\alpha\alpha}\,I^{(e)}_{\beta\beta}\Big\}\;q_{s\sigma}\,q_{s'\sigma'}$$

and from cross products between the terms in the first order Hamiltonian This effect we shall not discuss further.

A third type of intensity anomaly has been discussed by the author and it originates also with the first order Hamiltonian. The terms of importance are the Coriolis operators $-\sum_{\alpha} p_\alpha P_\alpha/I^{(e)}_{\alpha\alpha}$.

It is beyond the scope of this section to give a complete treatment of the intensity anomalies for a polyatomic molecule. We shall, however consider the first type, mentioned above, for a diatomic molecule and, as well, a naive example of the third type of intensity anomaly referred to. We shall proceed to do this, using here also the contact transformation method since it is more readily adaptible to future work with polyatomic molecules than the ordinary perturbation methods. We present the following argument to do this. The Schrödinger equation for a molecule will be

$$(H-E)\,\psi=0\,.$$

This equation may be written

$$T(H-E)\,T^{-1}\,T\psi=0 \tag{50.1}$$

where T is the contact transformation discussed in Chap. II and applied to the molecular problem in Chap. V. Eq. (50.1) is evidently equivalent to

$$(H'-E)\,\psi'=0$$

where ψ' is $T\psi$ which is the eigenfunction of the transformed Hamiltonian, H'. We have also that

$$\int\overline{\psi}_\tau\,\psi_{\tau''}\,dv=\int\overline{\psi}_\tau\,T^{-1}\,T\,\psi_{\tau''}\,dv=\delta_{\tau\tau''}.$$

Now the matrix elements of the electric moment, $(\tau|\mu|\tau'')$, are determined from the integrals

$$(\tau|\mu|\tau'')=\int\overline{\psi}_\tau\,E_A\,\psi_{\tau''}\,dv=\int\overline{\psi}_\tau\,T^{-1}(TE_A\,T^{-1})\,T\,\psi_{\tau''}\,dv \tag{50.2}$$

where E_A is the component of the classical electric moment directed along the X, Y and Z axes. Eq. (50.2) may be interpreted as

$$(\tau|\mu|\tau'')=\int\overline{\psi}'_\tau\,E'_A\,\psi'_{\tau''}\,dv\,,$$

E'_A being the transformed expression for the classical electric moment.

The advantage to this method comes from the fact that the first order Hamiltonian, H_1, vanishes on the average and may, therefore, be removed by a contact transformation, i.e., be made equal to zero. The wave function, corrected to first order, is known from perturbation theory, to be

$$\psi'_\tau=\psi'^{\,0}_\tau+\sum_{\tau''}\frac{(\tau|H'_1|\tau'')}{E^0_\tau-E^0_{\tau''}}\,\psi'^{\,0}_{\tau''}. \tag{50.3}$$

Since the transformed first order Hamiltonian, H_1, is zero, the correction to the wave function is zero and the wave function ψ'_τ remains the same to this approximation as ψ^0_τ. The transformation function, T, has already been discussed in detail and is well known. It may, for our purpose, be expanded into $T = (1 + i\lambda S \ldots)$ [and T^{-1} into $(1 - i\lambda S \ldots)$] and the transformed expression for the electric moment, E'_A, expanded, to the desired approximation, as

$$E'_A = E_A + i\lambda (S E_A - E_A S) + \cdots. \tag{50.4}$$

We shall proceed now to consider the first type of intensity anomaly discussed earlier for the fundamental band in the spectrum of a diatomic molecule. The zero and first order Hamiltonians for a diatomic molecule may, by reference to Chap. III, be shown to be

$$H_0 = (hc\omega/2)\,[(p/\hbar)^2 + q^2] + B_e hc (P/\hbar)^2,$$
$$H_1 = -hc(2B_e)(2B_e/\omega)^{\frac{1}{2}}(P/\hbar)^2 q + hckq^3$$

and dropping terms beyond the linear term the electric moment will evidently be

$$E_A = \lambda_{\alpha A}(a_0 + a_1 q + \cdots) \tag{50.5}$$

where $\lambda_{\alpha A}$ is as usual the direction cosine giving the projection of E along the X, Y and Z axes. We may determine from Chap. V that the function λS occurring in T will be the following:

$$S = (2B_e/\omega)^{\frac{3}{2}}(P/\hbar)^2(p/\hbar) \tag{50.6}$$

if we leave out of consideration the second term in H_1. Expanding the transformed electric moment as in (50.4) we obtain

$$E'_A = \lambda_{\alpha A} a_0 + \lambda_{\alpha A} a_1 q + i(2B/\omega)^{\frac{3}{2}} a_0 (p/\hbar)\,[P^2\lambda_{\alpha A} - \lambda_{\alpha A} P^2]/\hbar^2 + \cdots. \tag{50.7}$$

Computing now the matrix element of E'_A for a fundamental transition, we have

$$(v, J\,|E'_A|\,v+1, J') = a_1\left(\frac{v+1}{2}\right)^{\frac{1}{2}}(J\,|\lambda_{\alpha A}|\,J') +$$
$$+ \sum_{J''}(2B/\omega)^{\frac{3}{2}} a_0\left(\frac{v+1}{2}\right)^{\frac{1}{2}}[(J\,|P^2|\,J'')(J''\,|\lambda_{\alpha A}|\,J') - (J\,|\lambda_{\alpha A}|\,J'')(J''\,|P^2|\,J'')]/\hbar^2.$$

Recalling that P^2 is diagonal in J and that $(J|\lambda_{\alpha A}|J') \propto \delta_{J,J\pm 1}$ for a diatomic molecule we derive

$$(v_1 J\,|E'_A|\,v+1, J\pm 1) = a_1\left(\frac{v+1}{2}\right)^{\frac{1}{2}}(J\,|\lambda_{\alpha A}|\,J\pm 1) +$$
$$+ (2B/\omega)^{\frac{3}{2}} a_0\left(\frac{v+1}{2}\right)^{\frac{1}{2}}[J(J+1)(J\,|\lambda_{\alpha A}|\,J\pm 1) - (J|\lambda_{\alpha A}|\,J\pm 1)(J\pm 1)(J+1\pm 1)],$$

which may be written

$$a_1\left(\frac{v+1}{2}\right)^{\frac{1}{2}}[1 + 2m(2B_e/\omega)^{\frac{3}{2}}(a_0/a_1)]\,(J\,|\lambda_{\alpha A}|\,J\pm 1), \tag{50.8}$$

m being $-(J+1)$ and J for the upper and lower signs in Eq. (50.8) respectively. This relation is, to the approximation we have used, equivalent to that derived by HERMAN and WALLIS. It states that the matrix element of the electric moment is equal to the unperturbed value, $(J|\lambda_{\alpha A}|J\pm 1)$, multiplied by a factor which increases on one side of the band center and which decreases on the other side. It is interesting to note that the correction factor contains the ratio of the two constants a_0 and a_1. Similar considerations applied to overtone bands would introduce relationships between a_2, a_3, etc., and a_0 so that in principle one has a means at one's command for actually determining the value of the electric moment, E.

Consider now the third type of intensity anomaly, referred to earlier, in the case of fundamental bands of a triatomic molecule of the variety XY_2 where it is assumed that the molecule is accidentally symmetrical so that the two moments of inertia, $I_{xx}^{(e)}$ and $I_{yy}^{(e)}$, are equal (This is very nearly so in the H_2Se and H_2Te molecules). We shall orient the molecule in the x, y and z coordinate system so that the unique axis which is normal to the plane of the molecule is identified with the z axis and so that the normal vibrations, q_1 and q_2, lie along the y axis while the normal vibration, q_3, lies along the x axis. The portion of the first order Hamiltonian which is of interest to us if we confine ourselves to the Coriolis type of intensity anomaly is the following

$$H_1 = \cdots - 2hc\, B_e^{(zz)} \sum_{s=1}^{2} \zeta_{s,3}^{(z)} [(\omega_s/\omega_3)^{\frac{1}{2}} q_3 (p_s/\hbar) - (\omega_3/\omega_s)^{\frac{1}{2}} q_s (p_3/\hbar)]\, (P_z/\hbar) + \cdots . \qquad (50.9)$$

The transformation function which is required to remove (50.9) from the first order Hamiltonian may be obtained from reference to Chap. V and is the following

$$S = 2 B_e^{(zz)} \sum_{s=1}^{2} \{\zeta_{s,3}^{(z)} [(\omega_s\, \omega_3)^{\frac{1}{2}} (\omega_3^2 - \omega_s^2)]^{-1} [(\omega_3^2 + \omega_s^2)\, q_3 q_s + 2\omega_3 \omega_s p_s\, p_3/\hbar^2]\} (P_z/\hbar) . \qquad (50.10)$$

The electric moment induced by vibration may be written

$$\left.\begin{aligned} E_x &= a_3 q_3 + \cdots, \\ E_y &= a_0 + a_1 q_1 + a_2 q_2 + \cdots . \end{aligned}\right\} \qquad (50.11)$$

The components along the X, Y and Z axes will be given by

$$E_A = \lambda_{Ax} E_x + \lambda_{Ay} E_y$$

or more conveniently

$$\left.\begin{aligned} E_X \pm i E_Y &= (\lambda_{Xx} \pm i \lambda_{Yx})\, E_x + (\lambda_{Xy} \pm i \lambda_{Yy})\, E_y , \\ E_Z &= \lambda_{Zx} E_x + \lambda_{Zy} E_y . \end{aligned}\right\} \qquad (50.12)$$

The transformed form of the electric moment, $T(E_X \pm i E_Y)\, T^{-1}$, will, to the approximation desired here, be

$$\left.\begin{aligned} (E'_X \pm i E'_Y) \approx{} & (E_X \pm i E_Y) - 4 a_3 \sum_{s=1}^{2} B_e^{(zz)} \zeta_{s,3}^{(z)} [(\omega_3 \omega_s)^{\frac{1}{2}}/(\omega_3^2 - \omega_s^2)] (p_s/\hbar) \times \\ & \times (P_z/\hbar)\, (\lambda_{Xx} \pm i \lambda_{Yx}) - 4 \sum_{s=1}^{2} a_s B_e^{(zz)} \zeta_{s,3}^{(z)} [(\omega_3 \omega_s)^{\frac{1}{2}}/(\omega_3^2 - \omega_s^2)] \times \\ & \times (p_3/\hbar)\, (P_z/\hbar)\, (\lambda_{Xy} \pm i \lambda_{Yy}) , \\ E'_Z \approx{} & E_Z - 4 a_3 \sum_{s=1}^{2} B_e^{(zz)} \zeta_{s,3}^{(z)} [(\omega_3 \omega_s)^{\frac{1}{2}}/(\omega_3^2 - \omega_s^2)]\, (p_s/\hbar)\, (P_z/\hbar)\, \lambda_{Zx} - \\ & - 4 \sum_{s=1}^{2} a_s B_e^{(zz)} \zeta_{s,3}^{(z)} [(\omega_3 \omega_s)^{\frac{1}{2}}/(\omega_3^2 - \omega_s^2)]\, (p_3/\hbar)\, (P_z/\hbar)\, \lambda_{Zy} . \end{aligned}\right\} \qquad (50.13)$$

Confining ourselves to the fundamental bands we find for the matrix elements of the electric moment

$$\left.\begin{aligned}&(v_s, v_3, J, K\,|v_s+1, v_3, J', K') = \left(\frac{v_s+1}{2}\right)^{\frac{1}{2}} \{a_s\,(J, K\,|\,(\lambda_{Xy} \pm i\,\lambda_{Yy})|\,J' K') - \\ &\quad -4a_3\, B_e^{(zz)}[(\omega_3\omega_s)^{\frac{1}{2}}/(\omega_3^2-\omega_s^2)]\,(J, K\,|P_z/\hbar|\,J, K)\,(J, K\,|\,(\lambda_{Xx}\pm i\lambda_{Yx})|\,J' K')\} \\ &(s=1 \text{ and } 2) \text{ and} \\ &(v_s, v_3, J, K\,|\,v_s, v_3+1, J', K') = \left(\frac{v_3+1}{2}\right)^{\frac{1}{2}} \Big\{a_3\,(J, K\,|\,(\lambda_{Xx}\pm i\,\lambda_{Yx})|\,J', K') - \\ &\quad -\sum_{s=1}^{2} 4a_s B_e^{(zz)}[(\omega_3\omega_s)^{\frac{1}{2}}/(\omega_3^2-\omega_s^2)]\,(J, K\,|P_z/\hbar|\,J, K)\,(J, K\,|(\lambda_{Xy}\pm i\lambda_{Yy})|\,J', K')\Big\}.\end{aligned}\right\}\quad(50.14)$$

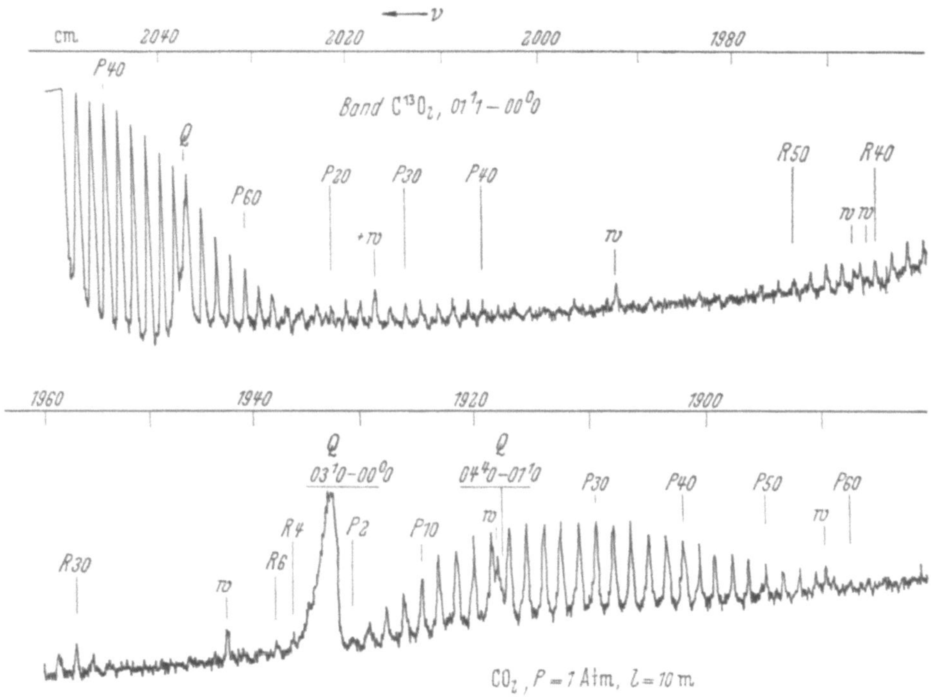

Fig. 29. The band, $\nu_s + \nu_3$, in the spectrum of CO_2 showing the intensity anomaly due to Coriolis interaction (courtesy of W. Benedict).

The above matrix components may more conveniently be expressed

$$\left.\begin{aligned}&(v_s, v_3, J, K\,|\,v_s+1, v_3, J', K\pm 1) = \left(\frac{v_s+1}{2}\right)^{\frac{1}{2}} \times \\ &\qquad \times \{a_s \mp 4a_3 B_e^{(zz)}[(\omega_3\omega_s)^{\frac{1}{2}}/(\omega_3^2-\omega_s^2)]\,K\}\,|(J, K\,|(\lambda_{Xy}\pm i\lambda_{Yy})|\,J', K\pm 1) \\ &\text{and} \\ &(v_s, v_3, J, K\,|\,v_s, v_3+1, J', K\pm 1) = \left(\frac{v_3+1}{2}\right)^{\frac{1}{2}} \times \\ &\times \Big\{a_3 \mp 4\sum_{s=1}^{2} a_s B_e^{(zz)}[(\omega_3\omega_s)^{\frac{1}{2}}/(\omega_3^2-\omega_s^2)]\,K\Big\}\,|(J, K\,|\,(\lambda_{Xx}\pm i\lambda_{Yx})|\,J', K\pm 1)|.\end{aligned}\right\}\quad(50.15)$$

The result, qualitatively, is much the same as in the preceding example, namely, that the matrix element of the electric moment in the perturbed case is that of the unperturbed case multiplied by a factor. This factor will increase on one side of the band center and decrease on the other thereby increasing the relative intensity of the lines on one side at the expense of the lines on the other side. The extent to which this takes place depends strongly upon the relative values of the constants a_1, a_2 and a_3. This is illustrated in Fig. 29.

51. Conclusion. The preceding chapters have considered the status of the theory of the vibration rotation energies of polyatomic molecules. The energies have been derived by methods of successive approximation since the Schrödinger equation, $(H-E)\psi=0$, is not one which lends itself to exact solution. The method rests upon the assumption that the atomic nuclei never depart far from their positions of equilibrium in the molecule and that, therefore, the Hamiltonian may be replaced by its expansion in terms of the coordinates about their equilibrium values. The energies are then evaluated by solving the expanded Hamiltonian to higher and higher approximations, using a form of the perturbation theory which is highly suited to the molecular problem. The relations given here have been evaluated to second order of approximation only, but are adequate to describe most of the observed phenomena so long as the quantum numbers v_s, J and K (or τ) do not become too large. The vibration energy is, for example, adequately expressed by Eq. (34.11), in general, so long as the quantities $x_{ss'}(v_s+g_s/2)(v_{s'}+g_{s'}/2)$ are small compared to $\omega_s(v_s+g_s/2)$. Similarly the rotational energy is satisfactorily described by Eqs. (34.8), (34.9) and (34.10) so long as the centrifugal distortion terms remain small compared to the energy of the rigid rotator. When this is no longer true, deviations between the predictions from the relations stated in Chap. V and the experimentally observed details may begin to become apparent so that approximations to higher orders than the second may be necessary. The inherent weakness of the method we have used is, obviously, that it rests upon the assumption that the Hamiltonian may be expanded in orders of magnitude in terms of the coordinates about their equilibrium values. The relative magnitudes of terms in the various orders of the Hamiltonian is not always obvious, particularly for large values of v_s, J and K and in such circumstances where resonance sets in.

As the techniques in molecular spectroscopy improve more and more, deviations from the spectra predicted from the foregoing discussion are being observed. Examples of such deviations are the following: the influence of centrifugal stretching proportional to the 6th and 8th power of J and K on the water vapor molecule term values as observed by BENEDICT; the effect of centrifugal distortion on l-type doubling; the variation of D values with anharmonic resonance; the splitting of the $K=3$ levels in the normal state of NH_3; resonances of higher order than the second; and the TAYLOR, BENEDICT and STRONG evidence for a refinement in the Fermi resonance terms.

These effects have been dealt with only rather superficially in the literature and can be treated more accurately only by extending the reasoning here presented to still higher approximations as is currently being done by AMAT and GOLDSMITH, and by AMAT, GOLDSMITH and NIELSEN.

It would appear on the surface that an extension of the foregoing beyond the fourth order of magnitude would be rather unrewarding. Thus far, none of the effects which have been observed to deviate from the foregoing results, even by microwave techniques, have been beyond the fourth order of magnitude. Even in the case where resonance sets in this would seem to be true. The argument for this has been given in Sect. 34, but we shall briefly review it. Consider a perturbing term, $\lambda^n H^{(n)}$ of order of magnitude n which is capable of producing resonance. When the frequencies are far separated the contribution to the energy may be calculated by perturbation theory and it will contain a resonance denominator. When, on the other hand, the frequency difference becomes small and resonance sets in one must use the degenerate form of the perturbation theory and the undiagonal elements of the perturbing term, $(v_s, v_{s'}, v_{s''} \ldots | \lambda^n H^{(n)} | v'_s, v'_{s'}, v'_{s''} \ldots)$ may now be of n-th order importance. These elements may be grouped about the principal diagonal in a little block forming a sub-matrix which may be diagonalized by itself. Consider a case where the block has only two rows and columns. We shall then have

$$\begin{array}{c|c|c|} & v_s, v_{s'}, v_{s''} \ldots & v'_s, v'_{s'}, v'_{s''} \ldots \\ \hline v_s, v_{s'}, v_{s''} \ldots & E_0 - \varepsilon & \alpha \\ \hline v'_s, v'_{s'}, v'_{s''} \ldots & \alpha & E_0 + \Delta - \varepsilon \\ \hline \end{array} = 0 \qquad (51.1)$$

where Δ is taken to represent the difference in energy between the resonating states and where $\alpha = (v_s, v_{s'}, v_{s''} \ldots | \lambda^n H^{(n)} | v'_s, v'_{s'}, v'_{s''} \ldots)$. The roots of the determinant (51.1) will be seen to be

$$= (E_0 + \Delta/2) \pm \tfrac{1}{2}\sqrt{\Delta^2 + 4\alpha^2}. \qquad (51.2)$$

The argument now is that Δ is of the order of $x_{ss'}$ and will therefore be of second order while α will be of the n-th order of approximation and will, in general, therefore be negligible compared to Δ. Only in cases where Δ is accidentally so small that it is no larger than α will a significant contribution be given to the energy by the resonance. We may, therefore, conclude that the second order of approximation to the vibration-rotation energies is one which is sufficiently good in many instances to describe the observed details and is certainly a convenient point to pause and determine what measures should be taken to arrive at results more consistent with experiment.

Acknowledgments.

The author wishes to acknowledge his indebtedness to his colleagues and students who have been helpful in making this contribution possible. He is especially grateful to Dr. K. Narahari Rao, Professor Alvin H. Nielsen, Professor G. B. B. M. Sutherland, Dr. John Shaw, Dr. William Benedict, Mr. Kurt Rossmann, Dr. William Dalby and Mrs. K. Lakshmi who have contributed illustrations of the band spectra discussed. Discussions with Dr. Gilbert Amat, Dr. Mark Goldsmith, Dr. Robert Herman, Mr Stewart Kurtz and Dr. Fred Keller have also been most helpful and are acknowledged with thanks.

Bibliography.

Chapter I.

Bjerrum, Niels: Nernst-Festschrift, Halle 1912.
Deslandres, H.: C. R. Acad. Sci. Paris **103**, 375 (1886).
Drude, P. K. L.: Ann. Physik **4**, 677 (1904).
Heurlinger, T.: Ark. Mat. Astron. Fysik **12**, 1 (1917).
— Phys. Z. **20**, 188 (1919).
— Z. Physik **1**, 82 (1920).
Hund, F.: Z. Physik **36**, 657 (1926).
Kemble, E. C.: J. Amer. Opt. Soc. **12**, 1 (1926).
Kramers, H. A., and W. Pauli: Z. Physik **13**, 351 (1923).
Kratzer, A.: Z. Physik **3**, 289 (1920).
Lord Rayleigh: Phil. Mag. **34**, 410 (1892).

Chapter II.

Casimir, H. B. G.: Rotation of a Rigid Body in Quantum Mechanics. Groningen, Den Haag, Batavia: J. B. Walters Uitgevers-Maatschappij 1931.
Condon, E. U., and G. H. Shortley: The Theory of Atomic Spectra, Cambridge: University Press 1935.
Kemble, E. C.: Fundamental Principles of Quantum Mechanics. New York, N.Y.: McGraw-Hill Book Co. Inc. 1937.
Vleck, J. H. Van: Rev. Mod. Phys. **23**, 213 (1951).

Chapter III.

BORN, M., and R. OPPENHEIMER: Z. Physik **84**, 457 (1927).
—, and P. JORDAN: Elementare Quantenmechanik. Berlin: Springer 1930.
BURRAU, O.: Kgl. danske Vid. Selsk. Mat.-fys. Medd. **7**, 14 (1927).
DARLING, B. T., and D. M. DENNISON: Phys. Rev. **57**, 128 (1940).
EYRING, H., J. WALTER and G. E. KIMBALL: Quantum Chemistry. New York: John Wiley Sons, Inc. 1944.
HEITLER, W., and F. LONDON: Z. Physik **44**, 455 (1927).
HUND, F.: Z. Physik **36**, 657 (1926); **42**, 93 (1927).
KRONIG, R. DE L.: Band Spectra and Molecular Structure. London: Cambridge University Press 1930.
MEISTER, A. G., and F. F. CLEVELAND: Amer. J. Phys. **14**, 13 (1946).
NIELSEN, H. H.: Rev. Mod. Phys. **23**, 90 (1951).
PAULI jr., W.: Z. Physik **43**, 601 (1927).
PODOLSKY, B.: Phys. Rev. **32**, 812 (1928).
VLECK, J. H. VAN: Phys. Rev. **33**, 467 (1929).
WANG, S.: Phys. Rev. **31**, 579 (1928).
WHITTAKER, E. T.: A Treatise on the Analytical Dynamics of Particles and Rigid Bodies, 3rd edition, p. 77. London: Cambridge University Press 1927.
WILSON jr., E. B., and J. B. HOWARD: J. Chem. Phys. **4**, 262 (1936).

Chapter IV.

BORN, M., and P. JORDAN: Elementare Quantenmechanik. Berlin: Springer 1930.
BOURGIN, D. G.: Phys. Rev. **29**, 794 (1927).
CASIMIR, H. B. G.: Rotation of a Rigid Body in Quantum Mechanics. Groningen, Den Haag, Batavia: J. B. Walters Uitgevers-Maatschappij 1931.
CHILDS, W. H. J., and H. A. JAHN: Proc. Roy. Soc. Lond., Ser. A **169**, 451 (1939).
CONDON, E. U., and G. H. SHORTLEY: The Theory of Atomic Spectra. London: Cambridge University Press 1935.
CROSS, P. C., R. M. HAINER and G. W. KING: J. Chem. Phys. **12**, 210 (1944).
DENNISON, D. M.: Phys. Rev. **28**, 318 (1926).
— Phys. Rev. **31**, 503 (1928).
— Rev. Mod. Phys. **3**, 280 (1931).
— Phys. Rev. **44**, 304 (1932).
DUNHAM, J. L.: Phys. Rev. **34**, 438 (1929).
ERLANDSON, G.: Ark. Fysik **10**, 65 (1955).
GOLDEN jr., S.: J. Chem. Phys. 16, 78 (1948).
HAINER, R. M., P. C. CROSS and G. W. KING: J. Chem. Phys. **17**, 826 (1949).
JOHNSTON, M., and D. M. DENNISON: Phys. Rev. **48**, 868 (1935).
KEMBLE, E. C., and D. G. BOURGIN: Nature, Lond. **117**, 789 (1926).
KLEIN, O.: Z. Physik **58**, 730 (1929).
KING, G. W., R. M. HAINER and P. C. CROSS: J. Chem. Phys. **11**, 27 (1943).
KRAMERS, H. A., and G. P. ITTMANN: Z. Physik **53**, 553 (1929); **58**, 217 (1929); **60**, 663 (1930).
KRONIG, R. DE L., and J. J. RABI: Phys. Rev. **29**, 262 (1927).
RAY, B. S.: Z. Physik **78**, 74 (1932).
REICHE, F., and H. RADEMACHER: Z. Physik **39**, 444 (1926).
SCHRÖDINGER, E.: Ann. Physik **79**, 361 (1926).
SHAFFER, W. H., H. H. NIELSEN and L. H. THOMAS: Phys. Rev. **56**, 895 (1939).
— — — Phys. Rev. **56**, 1051 (1939).
— Rev. Mod. Phys. **16**, 245 (1944).
TELLER, E.: Hand- und Jahrbuch der Chemischen Physik, vol. 9, p. 125. 1934.
TOLMAN, R. C.: Statistical Mechanics with Applications to Physics and Chemistry. New York 1927.
VLECK, J. H. VAN: Phys. Rev. **33**, 467 (1929).
— Rev. Mod. Phys. **23**, 213 (1951).
WANG, S. C.: Phys. Rev. **34**, 243 (1929).
WINTER, C. VAN: Physica, Haag **20**, 274 (1954).
WITMER, E. E.: Proc. Nat. Acad. Sci. U.S.A. **13**, 60 (1927).

Chapter V.

HANSEN, G. E., and D. M. DENNISON: J. Chem. Phys. **20**, 313 (1952).
HERMAN, R. C., and W. H. SHAFFER: J. Chem. Phys. **16**, 453 (1948).
KELLER, FRED: Dissertation for PH. D. degree, University of Tennessee, Knoxville, Tenn., 1955.

KEMBLE, E. C.: Fundamental Principles of Quantum Mechanics. New York, N. Y.: McGraw-Hill Book Co., Inc. 1937.
KRONIG, R. DE L.: Band Spectra and Molecular Structure. London: Cambridge University Press 1930.
NIELSEN, H. H.: Phys. Rev. **60**, 794 (1941).
— Rev. Mod. Phys. **23**, 90 (1951).
SHAFFER, W. H., H. H. NIELSEN and L. H. THOMAS: Phys. Rev. **56**, 895 (1939).
VLECK, J. H. VAN: Phys. Rev. **33**, 467 (1929).

Chapter VI.

DENNISON, D. M.: Astrophys. J. **62**, 84 (1925).
— Rev. Mod. Phys. **3**, 280 (1931).
KAYLOR, H. M. and A. H. NIELSEN J. Chem. Phy. **23**, 2139 (1955).
KIVELSON, D., and E. B. WILSON jr.: J. Chem. Phys. **20**, 1575 (1952).
LORD, R. C., and R. E. MERRIFIELD: J. Chem. Phys. **20**, 1348 (1952).
NIELSEN, A. H., and H. H. NIELSEN: Phys. Rev. **48**, 864 (1935).
—, H. H.: Phys. Rev. **59**, 565 (1941).
SAYVETZ, A.: J. Chem. Phys. **7**, 282 (1939).
STRALEY, J. W., C. H. TINDAL and H. H. NIELSEN: Phys. Rev. **62**, 161 (1942).
TELLER, E.: Hand- und Jahrbuch der Chemischen Physik, vol. 9, p. 125. 1934.
TINDAL, C. H., J. W. STRALEY and H. H. NIELSEN: Phys. Rev. **62**, 151 (1942).
WINTER, C. VAN: Physica, Haag **20**, 274 (1954).

Chapter VII.

AMAT, G., and M. GOLDSMITH: J. Chem. Phys. **23**, 6, 1171—1172 (1955).
— — and H. H. NIELSEN: J. Chem. Phys. **27**, 838 (1957).
—, and H. H. NIELSEN: J. Chem. Phys. **27**, 845 (1957); **29**, 665 (1958).
COLE, GOOD and HUGHES: Phys. Rev. **79**, 224A (1950).
COURTOY, C. P.: Canad. J. Phys. **35**, 608 (1957).
DARLING, B. T., and D. M. DENNISON: Phys. Rev. **57**, 128 (1950),
DENNISON, D. M.: Phys. Rev. **41**, 304 (1932).
EBERS, E. S., and H. H. NIELSEN: J. Chem. Phys. **5**, 822 (1937).
FERMI, E.: Z. Physik **71**, 251 (1931).
FUNG, L. W., and E. F. BARKER: Phys. Rev. **45**, 238 (1934).
GARING, J., K. NARAHARI RAO and H. H. NIELSEN: J. Mol. Spect. **1958** (in Press).
GOLDSMITH, M., G. AMAT and H. H. NIELSEN: J. Chem. Phys. **24**, 1178 (1956).
HANSEN, G. E., and D. M. DENNISON: J. Chem. Phys. **20**, 313 (1952).
HANSON, H., and H. H. NIELSEN: J. Chem. Phys. **25**, 591 (1956).
HEER, J. DE, and H. H. NIELSEN: J. Chem. Phys. **20**, 101 (1952).
HERMAN, R. C., and R. J. RUBIN: Astrophys. J. **121**, 533 (1955).
HERMAN, R. C., and R. F. WALLIS: J. Chem. Phys. **23**, 631 (1955).
HERZBERG, G.: Rev. Mod. Phys. **14**, 219 (1942).
JAHN, H.: Phys. Rev. **56**, 680 (1939).
KELLER, FRED: Dissertation for Ph. D. degree, University of Tennessee, Knoxville, Tenn. 1955.
KEMBLE, J. C.: Phys. Rev. **25**, 1 (1925).
KESSLER, RING, TRAMBARULO and GORDY: Phys. Rev. **79**, 54 (1950).
KRONIG, R. DE L.: Z. Physik **50**, 347 (1928).
MAES, S.: Thèse de doctorat (to be published).
McCONAGHIE, V. M., and H. H. NIELSEN: J. Chem. Phys. **21**, 1836 (1953).
MILLER, C. H., and H. W. THOMPSON: Proc. Roy. Soc. Lond., Ser. A **200**, 1 (1949).
NIELSEN, H. H.: J. Chem. Phys. **5**, 818 (1937).
— Phys. Rev. **68**, 181 (1945).
— Phys. Rev. **75**, 1961 (1949).
— Phys. Rev. **77**, 130 (1950).
— J. Chem. Phys. **21**, 142 (1953).
— J. Chem. Phys. **22**, 1383 (1954).
—, and W. H. SHAFFER: J. Chem. Phys. **11**, 140 (1943).
OPPENHEIMER, J. R.: Proc. Cambridge Phil. Soc. **23**, 327 (1926).
RAO, K. NARAHARI, and A. H. NIELSEN: In press.
RENNER, R.: Z. Physik **95**, 172 (1934).
SILVER, S.: J. Chem. Phys. **9**, 565 (1941).

The Collisions of Electrons with Molecules.

By

J. D. Craggs and H. S. W. Massey.

With 76 Figures..

Introduction.

There are a number of new features introduced when the collisions of electrons with molecules instead of with single atoms are considered. Thus, as far as elastic collisions are concerned, additional diffraction effects are introduced due to the interference of the waves scattered by the individual atoms. This has been made the basis of a powerful method for determining the geometrical configurations of polyatomic molecules.

Inelastic collisions now include the possibility of molecular dissociation as well as excitation of molecular vibration and rotation. The possibilities arising even for inelastic collisions with diatomic molecules are very numerous. Thus we may have, for collisions with a molecule AB, the following possibilities

$$
\begin{aligned}
\mathrm{AB} + e \rightarrow & \ \mathrm{AB}' + e, \\
\rightarrow & \ \mathrm{AB}^* + e, \\
\rightarrow & \ \mathrm{A}' + \mathrm{B}'' + e, \\
\rightarrow & \ \mathrm{AB}^+ + 2e, \\
\rightarrow & \ \mathrm{A} + \mathrm{B}^+ + 2e, \\
\rightarrow & \ \mathrm{A} + \mathrm{B}^-, \\
\rightarrow & \ \mathrm{A}^+ + \mathrm{B}^- + e,
\end{aligned}
$$

in which the presence of electronic excitation is denoted by one or two primes (′ or ″) and of nuclear vibration and/or vibration by an asterisk (*). The list is by no means exhaustive as it does not include double ionization processes. It is obvious that for polyatomic molecules many more reactions occur.

Electronic excitation of states which are not metastable may be studied in the same way as for atoms (see Vol. XXXVI) by observing the light emitted due to electron impact. Dissociation into normal atoms is more difficult to study because of the relative insensitivity of methods for detecting the atomic products. Ionization processes may be studied as for atoms (see Vol. XXXVI) but much more emphasis must be placed on mass spectrograph analysis of the products. Many molecules have been studied in this way, and quick methods of analysis of hydrocarbon mixtures have been devised which depend on knowledge of the relative probability of producing different ionic fragments by impact with electrons of chosen energy.

In this article we shall describe results obtained in the study of electron collisions with molecules and, whenever possible, their theoretical interpretation. Many of the experimental techniques employed have already been described in earlier articles (see Vols. XXXVI and XXI) but an account will be given of any which have either not been described elsewhere in the Encyclopedia or have only been described cursorily.

We shall deal first with elastic collisions and next with collisions which lead to excitation, electronic, vibrational or rotational, but not ionization. The final sections deal with the rather extensive data available on the ionization of diatomic and of polyatomic molecules.

A. General theoretical considerations.

As discussed earlier in this volume the wave function of a molecular state can be written to a high degree of accuracy in the form

$$\psi(\boldsymbol{r}, \boldsymbol{R})\, \varphi(\boldsymbol{R}, \Theta)$$

where $\boldsymbol{R}$ denotes the co-ordinates defining the configuration of the nuclei, Θ those defining their orientation as a whole and $\boldsymbol{r}$ those of the electrons relative to the nuclei. The electronic wave function $\psi(\boldsymbol{r}, \boldsymbol{R})$ can be thought of as calculated by solving the Schrödinger equation for the motion of the electrons under the influence of their mutual repulsion and of the field of the nuclei frozen in the configuration defined by $\boldsymbol{R}$. For each such configuration there will be, for a given electronic state, a proper energy $E(\boldsymbol{R})$. The nuclei may now be allowed to move under the influence of this energy, together with that of their mutual repulsion, so that, for each electronic state, there is a corresponding effective potential energy $V(\boldsymbol{R})$ for the nuclear motion. The functions φ are then obtained by solving the Schrödinger equation for motion of the nuclei under the influence of this potential energy.

In diatomic molecules there is only one nuclear configuration co-ordinate, the nuclear separation R, and the variation of $V(R)$ with R for a given electronic state can be fully represented by a plane curve which is usually referred to as the potential energy curve for the given electronic state. For polyatomic molecules three or more configurational co-ordinates are required and the variation of the effective potential energy with configuration can only be expressed geometrically in terms of a hypersurface. Because of this, much more can be done in interpreting reactions involving diatomic molecules and most of the theoretical discussion will be concerned with such cases.

The potential energy curves for a diatomic molecule AB tend at large nuclear separations to a limit which is the sum of the energies of two states of the respective atoms A and B or of the ions A^+, B^- or A^-, B^+. A general classification of the curves into two main types may be based in the first instance on the relation of the energy at intermediate separations to that at infinite separation. In one type the curve approaches the limit at large R from below and possesses a minimum at some separation R_0 (see Fig. 1 a), whereas in the other the curve approaches the limit at large R from above and always lies above this limit (see Fig. 1 b). At small R both types exhibit the same behaviour, the curves rising rapidly as R decreases because of the dominating effect of the nuclear repulsion.

With the first type of curve there is an equilibrium position at $R = R_0$ corresponding to a stable molecular state so that we shall refer to such curves as *attractive*. No equilibrium position exists at all in the other type which therefore corresponds to an unstable molecular state. These curves will be referred to as *repulsive*.

The broad distinction we have made between the two main types is not universally applicable. In a number of cases the curves may exhibit a deep minimum, below the separated atom limit, and yet approach the limit from above due to the presence of a subsidiary maximum (see Fig. 1 c). From our point of view, we still regard this type of curve as attractive although at large separations the force between the interacting atoms in the particular state concerned is repulsive. Equally, a predominantly repulsive-type curve may nevertheless approach the limit from below due to the presence of a subsidiary minimum at large distances (see Fig. 1d). In particular, any curve which represents a state dissociating into ions must be attractive at large distances[1].

For a state with an attractive potential energy curve the nuclear motion consists of vibration about the equilibrium position and rotation about an axis through the centre of mass perpendicular to the nuclear axis. The total energy of the molecule relative to that at infinite separation (the sum of the energies of the dissociation products) can be written

$$E = -D + E_n$$

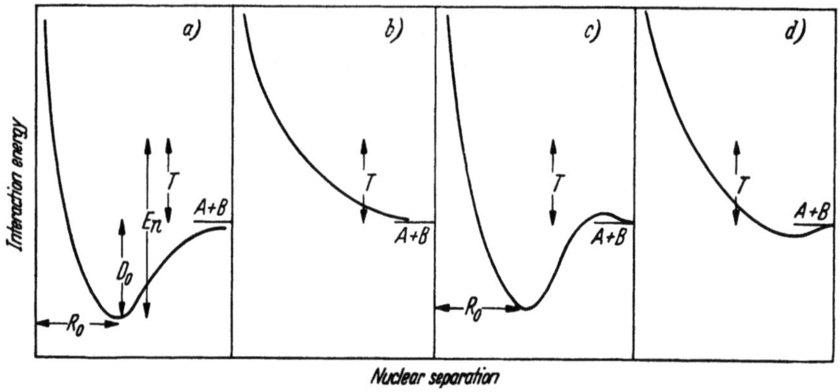

Fig. 1 a—d. Typical shapes of potential energy curves for states of diatomic molecules.

where D is the depth of the minimum and E_n is the energy of the nuclear motion. If $E_n < D$, the vibration-rotation levels form a discrete set representing bound molecular states; for $E_n > D$, the relative motion of the nuclei is unbounded, and the nuclear levels form a continuum. Under these conditions the molecule is dissociated into atoms moving relative to each other with kinetic energy $T = E_n - D$. If M_1, M_2 are the masses of the atoms, their respective kinetic energies will be

$$\frac{M_2}{M_1 + M_2} T, \quad \frac{M_1}{M_1 + M_2} T.$$

With a repulsive interaction the nuclear levels always form a continuum as all states are dissociated states of the molecule.

For polyatomic molecules it is not possible to think in such simple geometrical terms. Some progress may be made, however, by considering the variation of effective potential energy in particularly significant plane sections of the potential energy hypersurface. A less drastic procedure is to consider the energy variation in restricted configurations specified by two co-ordinates only, so that the energy surface is reduced to a two-dimensional surface. Thus for linear configurations of a molecule ABC the configurational co-ordinates may be chosen as the AB and BC distances. Very little has yet been attempted on these lines for the interpretation of the collision reactions of electrons with polyatomic molecules.

---

[1] Strictly speaking the curves will always be attractive at very large distances due to dynamical polarisation (van der Waals) forces but for most purposes with which we are concerned in this review these forces may be ignored.

1. Nuclear level changes associated with molecular electronic transitions. The Franck-Condon principle. We now seek an answer to the following question. Initially a molecule is in a given state of electronic and nuclear motion. A transition takes place to a new electronic state. What will be the probability of finding the nuclear motion in any one of the allowed states corresponding to the new electronic state?

This question is answered, at least partially, by the Franck-Condon principle, which states that, in an electronic transition, the probability is small that any change of nuclear configuration should take place. Qualitatively this is based again on the relatively great inertia of the nuclei compared with the electrons. In more precise terms it may be expressed as follows.

The probability amplitude associated with a given electronic transition can, as far as averaging over the electronic co-ordinates is concerned, be represented as a slowly varying function $f(\boldsymbol{R})$ of the nuclear configurational co-ordinates (see $\boldsymbol{R}$). The probability amplitude for a change from a nuclear state i, say, associated with the initial electronic state, to one, j, associated with the final state will then be proportional to

$$\int f(\boldsymbol{R})\, \varphi_i^* \varphi_j\, d^3\boldsymbol{R} \quad (1.1)$$

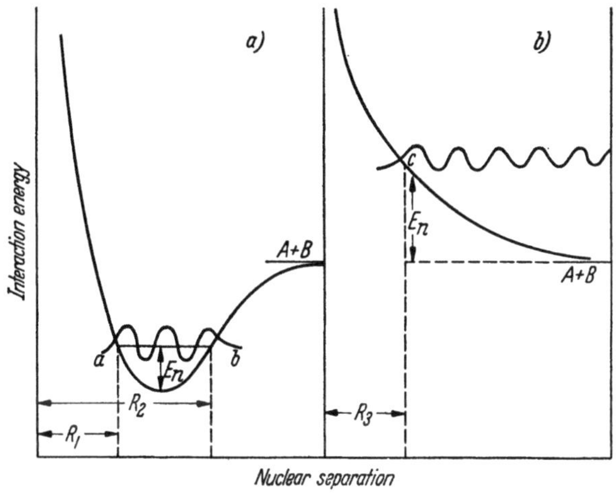

Fig. 2a and b. Illustrating classical turning points for nuclear motion (a) for an attractive electronic state, (b) for a repulsive electronic state. The general shapes of the corresponding nuclear wave functions are also indicated.

where φ_i, φ_j are the initial and final wave functions for the nuclear motion. Because of the large masses of the nuclei, their relative motion is very close to classical, so that the functions φ_i, φ_j are only large for those configurations which are probable in the classical motion. Thus, returning to the diatomic case in which the nuclear motion concerned is purely one of vibration about an equilibrium position, the most probable separations are those which correspond to the extreme ranges of the classical motion, i.e. to those points for which

$$V(R) = E_n \quad (1.2)$$

where $V(R)$ is the potential energy measured from the minimum as zero and E_n is the energy of the particular vibration concerned. This is illustrated in Fig. 2a where the points a, b represent the most probable nuclear separations R_1 and R_2 for the vibrational level of energy above the minimum. In a similar way, for unbound nuclear motion the most probable separation is the classical turning point (i.e. c in Fig. 2b). Furthermore, in both cases, the probability of finding a separation which is classically inaccessible (i.e. $<R_1$ or $>R_2$ in Fig. 2a or $<R_3$ in Fig. 2b) is very small. Reference to the general forms of the nuclear wave functions sketched in Figs. 2a and b will make these results clearer. It follows that for a transition from a bound state of a diatomic molecule, the function φ_i will have maxima near the classical turning points R_{i1} and R_{i2} say. For separations which are classically inaccessible (i.e. for $R<R_{i1}$ or $>R_{i2}$)

φ_i will be very small. Hence for a large overlap with φ_f the final state must have a classical turning point close to R_{i1} or R_{i2} and in any case not at a point $< R_{i1}$ or $> R_{i2}$. Since $f(R)$ is slowly varying a large overlap between φ_i and φ_f ensures a large probability amplitude.

The nearly classical behaviour of the nuclear motion thus provides the justification for the Franck-Condon principle in the diatomic case. Similar arguments apply to polyatomic molecules.

The application of the Franck-Condon principle to actual cases presents no difficulty in principle as far as diatomic molecules are concerned. It is only necessary to draw the potential energy curves for the two electronic states as in Figs. 3a—c. For a given initial vibrational energy the classical turning

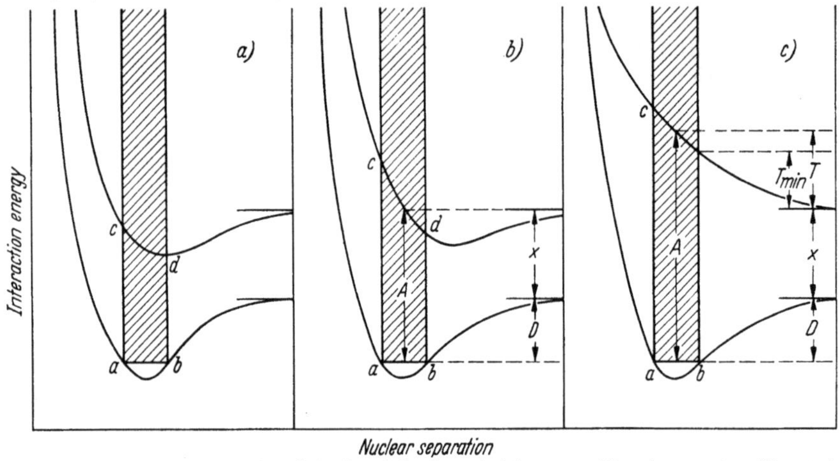

Fig. 3 a—c. Illustrating the application of the Franck-Condon principle to transitions between two different electronic states of a diatomic molecule.

points a, b are marked on the curve for the initial state. Vertical lines drawn from a and b to intersect the curve for the final electronic state at c and d give the turning points for the most probable final states of relative nuclear motion. The probability of a final nuclear energy denoted by a point between c and d will be quite appreciable, but that for an energy denoted by a point outside cd will usually be very small. Such transitions could only arise from initial separations which were classically inaccessible.

The consequences of the transition depend on the energies of the final molecular states relative to the energy at infinite separation in the final electronic level. These different cases are illustrated in Figs. 3a—c respectively. In case (a) the final states will be below the dissociation limit, so that the process is one of excitation of a stable excited electronic state, the threshold energy being given by the length bd. On the other hand, in case (c) all the final states lie above the dissociation limit. The process then leads always to dissociation. It can be seen from the Fig. 3c that the energy which must be communicated to the molecule to produce atoms moving with relative kinetic energy T is given by

$$A = D + X + T, \tag{1.3}$$

where D is the dissociation energy of the ground state of the molecule, X the excitation energy of the dissociated atoms. Thus if T and A can be measured and X determined in some way, D may be obtained. Alternatively from a knowledge of T, D and A we may obtain X. The minimum value $T_{\min}$ of the kinetic energy is indicated in Fig. 3c.

Case (b) is an intermediate one in which the end product may either be a stable excited molecule or two dissociated atoms. It may be distinguished from case (c) in that the minimum relative kinetic energy of these atoms is not finite but zero, so that the threshold energy for dissociation is

$$A = D + X. \tag{1.4}$$

We are concerned here specifically with cases in which the energy is transferred to the molecule from the kinetic energy of a colliding electron. The above considerations may be applied not only when the final molecular state is neutral, but also when it is one of a positive or negative molecular ion. For a process such as

$$\mathrm{AB} + e \rightarrow \mathrm{AB}^+ + 2e \tag{1.5}$$

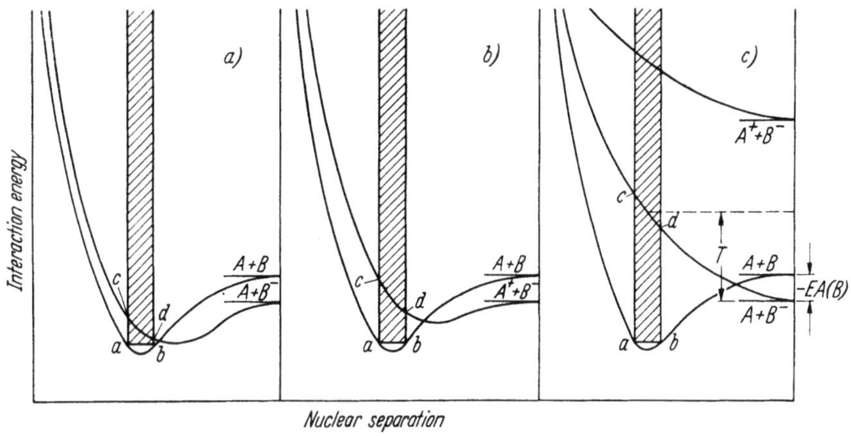

Fig. 4 a—c. Illustrating processes of negative ion formation.

the potential energy curves are as in Fig. 3a, the upper curve being that of a stable state of the ion AB^+. The threshold energy in this case is known as the vertical ionization energy of the molecule.

In a dissociative ionization process such as

$$\mathrm{AB} + e \rightarrow \mathrm{A} + \mathrm{B}^+ + 2e \tag{1.6}$$

the situation corresponds either to that of Figs. 3b or c. In case (c) if M_A, M_B are the masses of atoms A and B respectively, the relation (1.3) may be written in the form

$$A = D + X + I(\mathrm{B}) + \frac{M_A + M_B}{M_A}\, T(\mathrm{B}^+) \tag{1.7}$$

where $T(\mathrm{B}^+)$ is the kinetic energy of the ion B^+, which may be directly measured. X is now the sum of the excitation energies, if any, of A and of B^+, and $I(\mathrm{B})$ the ionization energy of B.

For processes which lead to formation of negative ions, the only important difference is that the limit of dissociation of a state of AB^- into A and B^- lies below that for the normal state of the molecule into A and B by an amount $EA(\mathrm{B})$, the electron affinity of B. Fig. 4 illustrates the possibilities corresponding to Figs. 4a—c for transitions to states of AB^-. In case (a) capture of an electron with energy between ac and bd leads to a molecular ion AB^-, but if left alone, this will reject the electron by a process of Auger type. The molecule AB^- can only be stabilised against this if it can give up enough energy, either by radiation or in collision with other molecules, before the Auger process occurs. In Fig. 4a the energy given up must be sufficient to bring the energy of AB^- below that indicated by *d*.

In case (c) the process which results is one of dissociative attachment

$$AB + e \rightarrow A + B^- \tag{1.8}$$

and will only occur with appreciable probability for electrons with energies between *ac* and *bd* in Fig. 4c. When the incident electron energy is A the kinetic energy $T(B^-)$ of the ion B^- will be given by

$$A = D - EA(B) + X + \frac{M_A + M_B}{M_A} T(B^-) \tag{1.9}$$

where X is the excitation energy of A. In most instances no allowance need be made for excitation energy of B^- as few stable excited states of negative ions are known.

As for positive ions, case (b) is distinguished by the possibility of forming negative ions with zero kinetic energy.

Finally, a repulsive state of the neutral molecule may be excited which dissociates not to neutral atoms but to ions A^+, B^- (see upper curve Fig. 4c), so that we have

$$AB + e \rightarrow A^+ + B^- + e, \tag{1.10}$$

and the kinetic energy T of relative motion of the resulting ions is given in terms of the incident electron energy by

$$A = D + I(A) - EA(B) + X + T. \tag{1.11}$$

Examples of these various processes and of the information about dissociation energies and electron affinities which have been obtained from their experimental study will be discussed in Sects. 17 to 23.

Although the same general considerations apply to transitions in polyatomic molecules so many possibilities arise that analysis is difficult and usually ambiguous. Thus in the dissociative ionization of a molecule ABC to yield C^+ ions, it is only possible to measure the kinetic energy of these ions. It is not possible to deduce from this the total kinetic energy of all fragments, for this depends on whether the process is of the form

$$ABC + e \rightarrow AB + C^+ + e, \tag{1.12}$$

or

$$ABC + e \rightarrow A + B + C^+ + e. \tag{1.13}$$

If the molecule is quite complex, such as SF_6, the excited ion SF_6^- formed by electron capture may have a lifetime, before Auger ejection of the electron, very much longer than for a diatomic ion AB^-. This is because the excess energy after capture is soon distributed over a great number of internal degrees of freedom and must reconcentrate in that mode which leads to the Auger effect before the surplus electron can be ejected. This means that stable, undissociated molecular negative ions are much more likely to result when complex molecules are bombarded by slow electrons (see Sect. 35α).

B. The elastic scattering of electrons by molecules.

Owing to the fact that the molecular force field is not spherically symmetrical, the calculation of cross sections even for elastic scattering of electrons by molecules is often much more difficult than for atoms. Under certain conditions, however, it is possible to treat the individual atoms in the molecule as independent scattering centres so that the amplitude for scattering by the molecule can be obtained by adding the amplitudes for scattering by the individual atom with proper allowance for phase differences. The resulting cross sections must then be averaged over all molecular orientations to give observable data.

This approximation will only be valid if multiple scattering of an electron within the molecule is negligible and if the distortion of the atomic fields by the valence forces is also unimportant. Both these conditions are likely to be well satisfied for impacts with fast electrons for which BORN's first approximation gives an adequate representation of the atomic scattering (Vol. XXXVI, p. 238). It is not so obvious that, when BORN's first approximation is inadequate, the assumption of independent scattering will ever be satisfactory, but there is evidence that it does have a range of validity extending down to scattering of electrons with velocities much too slow for BORN's first approximation to be applicable. On the other hand, for very slow electrons, with wavelengths comparable with the atomic separations in the molecule, there is no doubt that the independent scattering approximation is no longer valid and the calculation of the scattering presents much greater difficulties. We shall therefore discuss first the scattering of fast electrons.

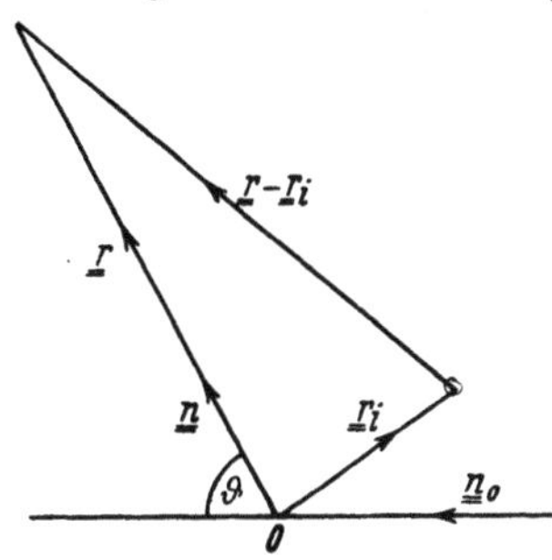

Fig. 5.

2. The scattering of fast electrons. We consider the scattering of electrons by a molecule containing n atoms which act as independent scattering centres. Let O be some point of reference (Fig. 5) within the molecule with respect to which the position vector of the i-th atom is $\boldsymbol{r}_i$. The incident electron beam is represented by the plane wave e^{ikz} where z is measured from O in the direction of incidence and k is the electron wave number. With respect to O the spherical wave scattered from the i-th atom can be written at large distances in the form

$$f_i(\vartheta)\,\frac{e^{ik|\boldsymbol{r}-\boldsymbol{r}_i|}}{|\boldsymbol{r}-\boldsymbol{r}_i|}\,e^{ikz_i} \tag{2.1}$$

where $f_i(\vartheta)$ is the scattered amplitude in the direction between ϑ and $\vartheta+d\vartheta$ relative to Oz and the factor e^{ikz_i} allows for the phase difference due to the eccentric scattering centre. Since $r \gg r_i$

$$|\boldsymbol{r}-\boldsymbol{r}_i| \approx r - \boldsymbol{n}\cdot\boldsymbol{r}_i, \tag{2.2}$$

where $\boldsymbol{n}$ is a unit vector in the direction of $\boldsymbol{r}$. The contribution to the scattered amplitude due to the i-th atom may therefore be written

$$e^{ik(\boldsymbol{n}_0-\boldsymbol{n})\cdot\boldsymbol{r}_i} f_i(\vartheta), \tag{2.3}$$

where $\boldsymbol{n}_0$ is a unit vector in the direction of incidence. According to the independent scattering approximation the differential cross section for scattering by the molecule in a fixed orientation will now be given by

$$I(\vartheta)\,d\omega = \left|\sum_{i=1}^{n} e^{ik(\boldsymbol{n}_0-\boldsymbol{n})\cdot\boldsymbol{r}_i} f_i(\vartheta)\right|^2 d\omega, \tag{2.4}$$

$$= \sum_i \sum_j f_i\, f_j^*\, e^{ik(\boldsymbol{n}_0-\boldsymbol{n})\cdot\boldsymbol{r}_{ij}}, \tag{2.5}$$

where $\boldsymbol{r}_{ij}=\boldsymbol{r}_i-\boldsymbol{r}_j$.

It remains to average over all molecular orientations. This is conveniently done by choosing $\boldsymbol{n}_0 - \boldsymbol{n}$ as polar axis and averaging over all orientations of $\boldsymbol{r}_{ij}$ relative to this axis. We obtain then for the averaged differential cross section

$$\overline{I(\vartheta)}\, d\omega = \sum_i \sum_j f_i f_j^* \frac{\sin s r_{ij}}{s r_{ij}} \tag{2.6}$$

where

$$s = k\,|\boldsymbol{n}_0 - \boldsymbol{n}| = 2k \sin \tfrac{1}{2}\vartheta. \tag{2.7}$$

If BORN's approximation is valid for the calculation of the atomic amplitudes we have[1]

$$f_i(\vartheta) = \frac{2m\varepsilon^2}{\hbar^2} \frac{Z_i - F_i(\vartheta)}{s^2} \tag{2.8}$$

where $Z_i \varepsilon$ is the nuclear charge and $F_i(\vartheta)$ the atom form factor for the i-th atom.

The formulae (2.6) and (2.8) have been made the basis of an important method of determining the atomic separations and arrangements in molecules. This consists in observing the angular distribution of electrons with several tens of kev energy scattered singly by the appropriate vapour. Given $F_i(\vartheta)$ for each of the atoms concerned, it is possible to work back from the observed data, partly by trial and error, to obtain the r_{ij}. A summary of results obtained in this way up till 1949 has been given by ALLEN and SUTTON[2].

3. Independent scattering model for medium energy electrons. The first calculations carried out using the formula (2.6), but without assuming that the scattering amplitudes $f_i(\vartheta)$ are given by BORN's approximation, were those of BULLARD and MASSEY[3]. They considered the elastic scattering of electrons of medium energy (30 to 780 ev) by N_2. For this case (2.6) takes the form

$$\bar{I}(\vartheta)\, d\omega = 2 I_a(\vartheta) \left(1 + \frac{\sin sR}{sR}\right) d\omega \tag{3.1}$$

where R is the nuclear separation. $I_a(\vartheta)\, d\omega$ is the differential cross section for scattering by a single nitrogen atom which was calculated, not by BORN's approximation, but by solution of the Schrödinger equation for the motion of the electrons in the self-consistent field of the atom (see Sect. 4 and also Vol. XXXVI, p. 268) and use of the Faxén-Holtsmark formula (4.7). Even at 780 ev $I_a(\vartheta)$ obtained in this way is appreciably different from that derived from (2.8) and at the lowest energy considered the differences were very marked. Although the application of (3.1) under these circumstances is not obvious, reasonably good agreement with the observed angular distributions was obtained.

A more remarkable application was made by HILL and WOODCOCK[4] for scattering of electrons with energies between 18 and 42 ev by the carbon tetrahalides. For all but CF_4 the scattering from the carbon atom may be neglected and

$$\bar{I}(\vartheta)\, d\omega = I_h(\vartheta) \left(1 + \frac{3 \sin sR}{sR}\right) d\omega \tag{3.2}$$

where R is the distance between the halogen atoms and $I_h(\vartheta)\, d\omega$ the atomic halogen cross section. I_h was calculated accurately for scattering by the self-consistent fields of the appropriate atoms. Good agreement was obtained with the observed

---

[1] N. F. MOTT and H. S. W. MASSEY: The Theory of Atomic Collisions, 2nd ed. Oxford: The Clarendon Press 1949.

[2] P. W. ALLEN and L. E. SUTTON: Acta crystallogr. **3**, 46 (1950).

[3] E. C. BULLARD and H. S. W. MASSEY: Proc. Cambridge Phil. Soc. **29**, 511 (1933).

[4] S. HILL and A. H. WOODCOCK: Proc. Roy. Soc. Lond., Ser. A **55**, 231 (1936).

angular distributions for scattering by CCl_4 and CF_4 but was quite poor for CBr_4. This failure is not surprising as for such large atoms as bromine there must be considerable distortion of the atomic fields in the molecule and in addition, owing to the high scattering power, multiple scattering within the molecule is likely to be important.

A detailed calculation on much the same lines was carried out by HOERNI and IBERS[1] for the scattering of electrons with 11 and 40 kev energy by UF_6, the only difference being the use of the Fermi-Thomas statistical field for the uranium atom, no self-consistent field being available. The agreement with observation was again good although the scattering by the uranium atom is very far from that given by BORN's first approximation.

Similar calculations have been carried out and applied to observed data with moderate success for CH_4, C_2H_4, C_2H_2 and Br_2.

A related phenomenon is the fine structure which appears in the X-ray absorption edges in molecular gases. This is ascribed to diffraction of the ejected photo-electron by the surrounding atoms. The theory has been worked out by KRONIG[2] on exactly similar lines to the above treatment of scattering and makes the same assumptions. The most detailed calculations have been carried out for $GeCl_4$[3] and the agreement with observation, though not perfect, is surprisingly good in view of the uncertain foundation of the assumptions. Some application has also been made to Br_2[4] and $AsCl_3$[5].

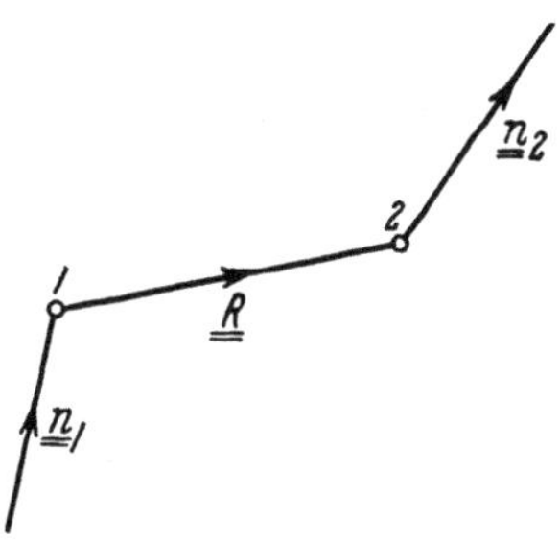

Fig. 6.

The only attempt which has been made to provide theoretical justification is that of HOERNI[6] who estimated the importance of the double scattering correction for the 40 kev electrons scattered by two uranium atoms 2 Å apart. Referring to Fig. 6, 1 and 2 represent the centres of the scattering atoms and $\boldsymbol{R}$ is the vector joining 1 and 2. We consider a scattering process in which the wave is incident in the direction of the unit vector $\boldsymbol{n}_1$ and scattered in the direction of the unit vector $\boldsymbol{n}_2$, the wave number being k throughout. The amplitude of the wave scattered from a single atom can, at a large distance from the atom, be written in the form

$$f(\boldsymbol{n}_1, \boldsymbol{n}_2) \frac{e^{ikr}}{r}. \tag{3.3}$$

According to the simple independent scattering model the scattered amplitude at a large distance from the molecule can, as we have seen, be written in the form

$$f(\boldsymbol{n}_1, \boldsymbol{n}_2) \left[1 + e^{-ik(\boldsymbol{n}_1 - \boldsymbol{n}_2)\cdot \boldsymbol{R}}\right] \frac{e^{ikr}}{r}. \tag{3.4}$$

We now consider the contribution which arises from a scattering process in which the electron is first scattered by one atom along the vector $\boldsymbol{R}$ in such a sense as to collide with the other atom which scatters it into the final direction $\boldsymbol{n}_2$. If we assume that the distance R is large compared with the electron wavelength

[1] J. A. HOERNI and J. A. IBERS: Phys. Rev. **91**, 1182 (1953).
[2] R. DE L. KRONIG: Z. Physik **75**, 468 (1932).
[3] D. R. HARTREE, R. DE L. KRONIG and H. PETERSEN: Physica, Haag **1**, 895 (1934).
[4] T. SNYDER and C. SHAW: Phys. Rev. **57**, 881 (1940).
[5] D. COSTER and G. KLAMER: Physica, Haag **1**, 899 (1934).
[6] J. A. HOERNI: Phys. Rev. **102**, 1530 (1956).

we may calculate the amplitude associated with this double scattering by taking the first scatterer as the source of a plane wave in the direction $\boldsymbol{R}$ of amplitude $f\left(\boldsymbol{n}_1, \pm\frac{\boldsymbol{R}}{R}\right) R^{-1} e^{ikR}$ the alternate signs arising from the fact that the first scattering may occur at either atom. This gives for the contribution from double scattering

$$\left[f\left(\boldsymbol{n}_1, \frac{\boldsymbol{R}}{R}\right) f\left(\frac{\boldsymbol{R}}{R}, \boldsymbol{n}_2\right) e^{-ik\boldsymbol{n}_2\cdot\boldsymbol{R}} + f\left(\boldsymbol{n}_1, -\frac{\boldsymbol{R}}{R}\right) f\left(-\frac{\boldsymbol{R}}{R}, \boldsymbol{n}_2\right) e^{ik\boldsymbol{n}_1\cdot\boldsymbol{R}}\right] \frac{e^{ikr}}{r}. \quad (3.5)$$

To complete the calculation (3.5) must be added to (3.4), the square of the modulus of the sum taken and then averaged over all orientations of the molecule. The resulting scattered intensity can be written in the form

$$\bar{I}(\vartheta) = 2|f|^2\left(1 + \frac{\sin sR}{sR}\right) + \delta I \quad (3.6)$$

where δI is the contribution from double scattering. For numerical estimation of the importance of δI HOERNI assumed that it would be sufficient to substitute in (3.5) the amplitude functions f given by BORN's first approximation, for scattering by a potential $-Z\varepsilon^2 r^{-1} e^{-r/a}$ with $Z = 92$ and $a = 0.468 Z^{-\frac{1}{3}}$ Å, the effective range given by the Fermi-Thomas statistical model. Exact atomic scattering amplitudes were still retained in (3.4). Terms involving R^{-2} were neglected.

Table 1. *Illustrating the effect of double scattering in the collisions of* 40 kev *electrons with a representative molecule consisting of two uranium atoms* 2 Å *apart.*

Angle of scattering (degrees)	Direct atomic scattering $2\|f\|^2$	Molecule interference scattering/sin sR $2\|f\|^2/sR$	Correction due to double scattering δI
0	565		0.227
2	59.2	8.17	0.178
4	9.25	0.638	0.0953
6	2.79	0.128	0.0471
8	1.13	0.0390	0.0205
10	0.566	0.0157	0.00927
12	0.325	0.00749	0.00498
16	0.128	0.00222	0.00192
20	0.064	0.00089	0.00087

Table 1 gives the results obtained. δI is seen to be very small at all angles compared with the direct atomic scattering $2|f|^2$. On the other hand, it is by no means negligible compared with the molecular interference term. However, it varies smoothly with angle and so will not have much effect in modifying the location of the maxima and minima of the diffraction pattern due to the molecular structure. On the whole, therefore, HOERNI's calculation provides some theoretical evidence that the independent scattering model may be used for electrons of intermediate energy even though BORN's first approximation is quite inadequate for the calculation of the atomic scattering amplitudes.

EKSTEIN[1] has discussed formally the procedure for calculating multiple scattering effects by an iterative procedure but no further applications to molecular problems have yet been made.

4. Slow electrons. Introductory remarks. A great deal of experimental work has been done on the scattering of slow electrons by molecular gases and vapours. The experimental technique used in these measurements has been described in Vol. XXXIV of this Encyclopedia, so we are mainly concerned here with the theoretical interpretation. It is important to realise, however, that the earlier work has been supplemented by data obtained by microwave methods which yield values of the elastic cross sections for electrons with thermal energies. Some further mention of this technique will be made in Sect. 14 in which its application to the study of energy losses of very slow electrons in molecular collisions will be described.

[1] H. EKSTEIN: Phys. Rev. **89**, 490 (1953).

The problem of developing a theory of the scattering of slow electrons by molecules is naturally much more difficult than for atoms and so far has been limited almost completely to scattering by homonuclear diatomic molecules.

It has been shown in the article on *The Theory of Atomic Collisions* in Vol. XXXVI of this Encyclopedia that the calculation of the elastic scattering of slow electrons by atoms can be reduced, to a good approximation, to the problem of solving integro-differential equations of the form

$$\nabla^2 F + [k^2 - U(r)]\, F = \int K(\boldsymbol{r}, \boldsymbol{r}')\, F(\boldsymbol{r}')\, d^3\boldsymbol{r}'. \tag{4.1}$$

This is a Schrödinger equation for the motion of an electron of mass m and energy $\frac{k^2\hbar^2}{2m}$ under the influence of an interaction which consists partly of a potential interaction $\frac{\hbar^2}{2m} U(r)$ and partly of a velocity dependent interaction expressed in terms of the interaction kernel $K(\boldsymbol{r}, \boldsymbol{r}')$. $\frac{\hbar^2}{2m} U(\boldsymbol{r})$ is the mean interaction of the electron with the atom, considered unperturbed in its ground state. $K(\boldsymbol{r}, \boldsymbol{r}')$ arises from the possibility of electron exchange between the atom and incident electron. The nucleus of the atom is taken as origin of the polar coordinates.

A solution of (4.1) is required which is well behaved throughout space and has the asymptotic form

$$F \sim e^{ikr\cos\vartheta} + f_0(\vartheta)\, \frac{e^{ikr}}{r}. \tag{4.2}$$

The differential cross section for elastic collisions can then be expressed in terms of $|f_0(\vartheta)|^2$.

Since $U(r)$ is a function of r only, and $K(\boldsymbol{r}, \boldsymbol{r}')$ depends only on r, r' and the angle between $\boldsymbol{r}$ and $\boldsymbol{r}'$, it is possible to proceed by expanding F in the form

$$F = \frac{1}{r} \sum F_l(r)\, P_l(\cos\vartheta) \tag{4.3}$$

so as to obtain uncoupled integro-differential equations in one variable for the F_l viz:

$$\frac{d^2 F_l}{dr^2} + \left[k^2 - U(r) - \frac{l(l+1)}{r^2}\right] F_l = \int K_l(r, r')\, F_l(r')\, dr', \tag{4.4}$$

where

$$K(\boldsymbol{r}, \boldsymbol{r}') = \frac{1}{4\pi r r'} \sum_{l=0}^{\infty} (2l+1)\, P_l\!\left(\frac{\boldsymbol{r}\cdot\boldsymbol{r}'}{r r'}\right) K_l(r, r'). \tag{4.5}$$

If a solution of (4.4) is obtained which vanishes at $r=0$, is well-behaved everywhere and has the asymptotic form

$$F_l \sim \sin\left(kr - \tfrac{1}{2} l\pi + \eta_l\right) \tag{4.6}$$

it may be shown that

$$f(\vartheta) = \frac{1}{2ik} \sum (2l+1)\left(e^{2i\eta_l} - 1\right) P_l(\cos\vartheta). \tag{4.7}$$

Provided $\lim_{r\to\infty} r^2 U(r) = 0$ the series converges for all ϑ. For a given k, in general η_l decreases as l exceeds a value l_0 which becomes smaller as k decreases. When k is sufficiently small only η_0 is important.

The total elastic cross section Q_0 is given in terms of the phase shifts by

$$Q_0 = \frac{4\pi}{k^2} \sum_l (2l+1)\sin^2\eta_l. \tag{4.8}$$

5. Scattering by homonuclear diatomic molecules—phenomenological analysis. Turning now to scattering by diatomic molecules we see that in the equations corresponding to (4.4) U will depend in general on ϑ and φ as well as r or at best, if the co-ordinates axis is taken along the molecular axis, on r and ϑ. The exchange kernel K will also be more complicated. It follows that, if the function F were expanded in spherical harmonics

$$F = \frac{1}{r} \sum_l \sum_m F_l^m(r)\, P_l^m(\cos\vartheta)\, e^{\pm i m \varphi} \tag{5.1}$$

the integro-differential equations for the F_l^m would not be uncoupled as the Eq. (4.4) would no longer be separable in polar co-ordinates.

For any actual molecule no choice of co-ordinate system would make (4.4) separable. The following procedure has therefore been adopted for the analysis of observed data. Instead of working from calculated molecular fields an attempt is made to obtain an effective axially symmetrical scattering field which gives theoretical cross sections in agreement with observation, a procedure very similar to that adopted by ALLIS and MORSE[1] for the analysis of data on low energy electron scattering by atoms. For this purpose it is not necessary to include the interaction kernel $K(\boldsymbol{r}, \boldsymbol{r}')$ in (4.4) as it is assumed that the effective interaction determined empirically will allow for this as well as for the true direct interaction $U(\boldsymbol{r})$. The most natural co-ordinate system to choose is a prolate spheroidal one in which the two nuclei are taken as foci. We then have as co-ordinates ξ, η, φ where

$$\xi = \cosh\mu\,, \qquad \eta = \cos\lambda\,,$$

$$x = R\sinh\mu\sin\lambda\cos\varphi\,, \qquad y = R\sinh\mu\sin\lambda\sin\varphi\,, \qquad z = R\cosh\mu\cos\lambda\,, \tag{5.2}$$

R being the nuclear separation and the molecular axis being taken as the z-axis.

The Eq. (4.4) with $K(\boldsymbol{r}, \boldsymbol{r}') = 0$, is separable if, in terms of these co-ordinates, U is chosen to have the form

$$U = \frac{g(\xi)}{\xi^2 - \eta^2} \tag{5.3}$$

where $\lim_{\xi\to\infty} \xi^2 g(\xi) = 0$. With suitable choice of $g(\xi)$ it is possible to obtain a quite adequate representation of the effective molecular scattering field.

With this form for U the required well-behaved solution of (4.1) has the form

$$F = \sum_{ml} \sum a_{ml}\, S_{ml}(R, \eta)\, T_{ml}(R, \xi)\cos\{m(\varphi - \alpha_{ml})\} \tag{5.4}$$

where S_{ml}, T_{ml} satisfy the equations

$$\frac{d}{d\xi}\left\{(\xi^2 - 1)\frac{d\,T_{ml}}{d\xi}\right\} + \left[A_{ml} + k^2R^2\xi^2 - \frac{m^2}{\xi^2 - 1} + \frac{1}{4}k^2R^2 g(\xi)\right] T_{ml} = 0\,, \tag{5.5}$$

$$\frac{d}{d\eta}\left\{(\eta^2 - 1)\frac{d\,S_{ml}}{d\eta}\right\} + \left[A_{ml} + k^2R^2\eta^2 - \frac{m^2}{\eta^2 - 1}\right] S_{ml} = 0\,. \tag{5.6}$$

The parameters A_{ml} are such that the equation for S_{ml} has well-behaved solutions. m and l are either zero or positive integers.

---

[1] W. ALLIS and P. M. MORSE: Z. Physik **70**, 567 (1931).

The functions S_{ml} are spheroidal[1] harmonics which may be expressed in the form

$$S_{ml} = \sum_n{}' d_n^l P_{m+n}^m(\cos\lambda) \tag{5.7}$$

where the prime indicates that the summation is over even or odd n according as l is even or odd respectively. As $k \to 0$, $d_n^l \to \delta_{nl}$.

When $g(\xi) = 0$, $T_{ml} \to R_{ml}$ which may be expanded in the form

$$R_{ml}(kR, \xi) = \frac{(\xi^2-1)^{\frac{1}{2}m}}{\xi^m \sum\limits_{n=0}^{\infty}{}' d_n^l \dfrac{(n+2m)!}{n!}} \sum_n{}' i^{l-n} d_n^l \frac{(n+2m)!}{n!} j_{m+n}(R\xi) \tag{5.8}$$

where j_{m+n} is the spherical Bessel function,

$$j_{m+n}(z) = \left(\frac{\pi}{2z}\right)^{\frac{1}{2}} J_{m+n+\frac{1}{2}}(z). \tag{5.9}$$

R_{ml} has the asymptotic form

$$R_{ml} \sim \frac{1}{kR\xi} \sin\left\{R\xi - \frac{1}{2}(l+m)\pi\right\}. \tag{5.10}$$

A plane wave of wave number k propagating in the direction (χ, ψ) can be expanded in terms of the functions R_{ml}, S_{ml} in the form

$$\left.\begin{aligned} e^{ikZ} = 2\sum \cos\{m(\varphi-\psi)\}\, i^{m+l} \frac{2-\delta_{0,m}}{N_{ml}} \left[\sum_{s=0}^{\infty}{}' d_s^l \frac{(s+2m)!}{s!}\right] \times \\ \times\, S_{ml}(kR, \cos\chi) \qquad S_{ml}(kR, \cos\lambda) \qquad R_{ml}(kd, \xi). \end{aligned}\right\} \tag{5.11}$$

N_{ml} is given by

$$N_{ml} = 2 \sum_{n=0,1}^{\infty}{}' \left\{\frac{d_n^l(n+2m)!}{n!}\right\}^2 \frac{1}{2n+2m+1}. \tag{5.12}$$

The asymptotic form for large ξ of the well-behaved solution of (5.5) can be written

$$T_{ml}(kR, \xi) \sim \frac{1}{kR\xi} \sin\left\{kR\xi - \frac{1}{2}(l+m)\pi + \zeta_{m,l}\right\}. \tag{5.13}$$

It follows by arguments similar to those used in dealing with scattering by a spherically symmetrical scattering potential that, if we take

$$a_{ml} = \frac{2i^{m+l}(2-\delta_{0,m})}{N_{ml}} \left[\sum_{s=0}^{\infty}{}' d_s^b \frac{(s+2m)!}{s!}\right] e^{2i\zeta_{m,l}}, \tag{5.14}$$

F has the asymptotic form, for large ξ,

$$F \sim e^{ikz} + \frac{e^{ik\xi}}{R\xi} f(\lambda, \varphi; \chi, \psi), \tag{5.15}$$

where

$$\left.\begin{aligned} f(\lambda, \varphi; \chi, \psi) = \frac{1}{ik} \sum \cos m(\varphi-\psi) \frac{2-\delta_{0,m}}{N_{m,l}} \left[\sum_{s=0,1}^{\infty}{}' d_s^l \frac{(s+2m)!}{s!}\right] \times \\ \times\, S_{ml}(kR, \cos\chi)\, S_{ml}(kR, \cos\lambda)\, (e^{2i\zeta_{m,l}} - 1). \end{aligned}\right\} \tag{5.16}$$

[1] Spheroidal Wave Functions by J. A. STRATTON, P. M. MORSE, L. J. CHU, J. D. C. LITTLE and F. J. CORBATÓ, p. 39. New York: John Wiley & Sons and Technology Press, M.I.T. 1956.

Since, at large distances from the scatterer $\cos\lambda \to \cos\vartheta$, where ϑ is the polar angle in the spherical polar co-ordinate system with axis that of the molecule, the differential cross section for an elastic collision in which the electron is incident on the molecule in the direction (χ, ψ) and is scattered in to the direction (ϑ, φ) into the solid angle element $d\omega$, is

$$|f(\vartheta, \varphi; \chi, \psi)|^2\, d\omega. \tag{5.17}$$

The observed differential cross section for scattering into the solid angle $d\Omega$ about the direction (Θ, Φ) relative to the incident direction as axis will be given by

$$\frac{1}{4\pi}\int_0^\pi\int_0^{2\pi} |f(\vartheta, \varphi; \chi, \psi)|^2 \sin\chi\, d\chi\, d\psi\, \frac{d\omega}{d\Omega}\, d\Omega \tag{5.18}$$

in which (ϑ, φ) must be expressed in terms of (Θ, Φ) and (χ, ψ). This integral cannot be evaluated in closed terms but may be evaluated numerically, using the expansions (5.7) for the $S_{m,l}$.

The total elastic cross section may be expressed quite simply in terms of the $\zeta_{m,l}$. For this may be written

$$Q_0 = \frac{1}{4\pi}\int_0^\pi\int_0^{2\pi} |f(\vartheta, \varphi; \chi, \psi)|^2\, d\omega \tag{5.19}$$

so that use of the orthonormal properties of the S_{ml} gives

$$Q_0 = \sum_{l,m} q_{m,l} \tag{5.20}$$

where

$$q_{ml} = \frac{4\pi}{k^2}(2 - \delta_{0,m}) \sin^2\zeta_{m,l}. \tag{5.21}$$

This may be compared with the corresponding formula (4.8) for the spherically symmetrical case. The similarity is clear, the essential difference being the dependence of the phase shift on m. If m were restricted for a given l so that $m \leq l$ and $\zeta_{m,l}$ became independent of m, (5.21) would agree with (4.8).

Just as for the spherically symmetrical case the series (5.20) converges rapidly when kd is small and in the limit $kd \to 0$, only q_{00} remains finite. This implies further that the observed angular distributions as given by (5.18) will also be isotropic in this limit.

6. Application to analysis of observed cross sections. The first application of the foregoing theory was made by STIER[1] for the analysis of the scattering of electrons by nitrogen molecules. He took in (5.3)

$$\left.\begin{aligned} g(\xi) &= -\frac{2Z\varepsilon^2\xi}{R}\left(\frac{\xi_0-\xi}{\xi_0-1}\right)^2, &\quad \xi &< \xi_0,\\ &= 0, & \xi &> \xi_0 \end{aligned}\right\} \tag{6.1}$$

where $R = 1.1$ Å for N_2 and ξ_0 and Z are adjustable parameters.

A comprehensive analysis of all the data on homonuclear diatomic molecules was carried out later by FISK[2] who took

$$\left.\begin{aligned} g(\xi) &= -\frac{2Z\varepsilon^2}{R}\left[1-\left(\frac{\xi_0}{\xi_0-1}\right)^2\left(\frac{\xi-1}{\xi}\right)^2\right], &\quad \xi &< \xi_0,\\ &= 0, & \xi &> \xi_0, \end{aligned}\right\} \tag{6.2}$$

[1] H. C. STIER: Z. Physik 76, 439 (1932).
[2] J. B. FISK: Phys. Rev. 49, 167 (1936).

Z and ξ_0 being again adjustable parameters. Calculations were carried out giving the partial cross sections as functions of $x = \xi_0 k R$ for different values of $\beta = \frac{1}{2}\sqrt{ZR\xi_0}$. It was found that the results were insensitive to the value taken for ξ_0 provided it was between 1.75 and 3.0, so that the detailed calculations were confined to the cases in which $\xi_0 = 2$. Just as for the corresponding calculations of ALLIS and MORSE[1] for scattering by atoms, the partial cross sections, for a fixed x, vary quasi-periodically with β, the parameter defining the strength of the molecular field. Results obtained are illustrated in Figs. 7, 9, 10 and 11.

For hydrogen (Fig. 11) q_{00} is alone important at electron energies below about 12 ev. On the whole, the choice of $\beta = 0.6$ gives perhaps the best average agreement with the experimental data up to energies of 25 ev, but this agreement is

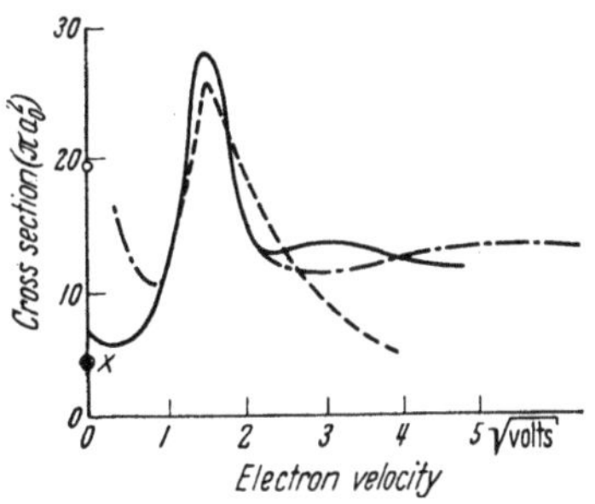

Fig. 7.

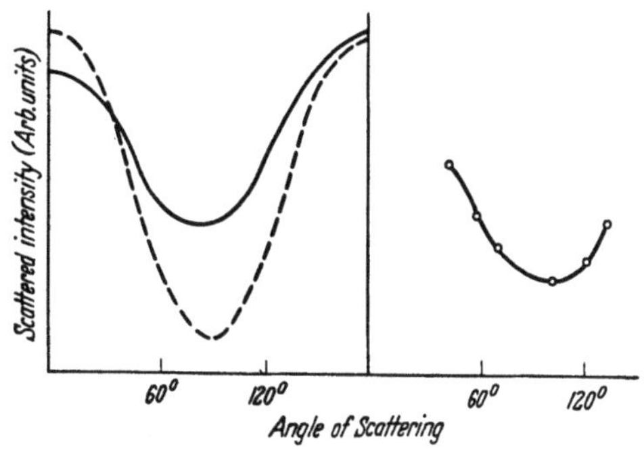

Fig. 8.

Fig. 7. Calculated and observed cross sections for the elastic scattering of electrons in N_2. ——— calculated by FISK; – – – calculated by STIER; —·—·— observed by NORMAND; ● observed by GOULD and BROWN (microwave method); o observed by ANDERSON and GOLDSTEIN (microwave method); x observed by CROMPTON and SUTTON (diffusion method).

Fig. 8. Observed and calculated angular distributions of 2.3 volt electrons scattered by N_2. ——— calculated from FISK's potential; – – – calculated from STIER's potential; o observed by BULLARD and MASSEY.

not by any means close in detail. It is somewhat surprising that the observed cross section is so large at energies greater than 7 ev. There would seem to be little prospect of modifying the field sufficiently to increase the contributions of higher order partial cross sections and even though above 9 ev or so there will be an appreciable contribution to the observed cross section from inelastic processes, it is not at all likely that they would be large enough to account for the discrepancy (see Sect. 15α). It may be that the interaction (6.2), being cut off for $\xi > \xi_0$, is of too short a range and that important contributions come from the molecular quadrupole moment at greater distances, but this is a matter for further investigation.

At low electron energies FISK's cross section is somewhat too large in the light of present data. It is of course possible that if this data had been available at the time a somewhat better choice of the parameter β could have been made as far the low energy side is concerned. Further remarks about the theory of the scattering by H_2 are given in Sect. 7 below.

FISK obtained quite a good fit to the observed cross section for nitrogen over the energy range from 1 to 25 ev by taking $\beta = 1.3$. This is illustrated in Fig. 7. According to this interpretation the sharp maximum for 2.2 ev electrons is due to the behavior of q_{10}. On the other hand, STIER, using the interaction (6.1) ascribed it to q_{01}. His values for the total cross section are also

[1] W. ALLIS and P.M. MORSE: Z. Physik **70**, 567 (1931).

given in Fig. 7. It would seem possible to discriminate between these interpretations by comparison of observed and calculated angular distributions of the scattered electrons, using (5.7) and (5.18). This comparison is illustrated in Fig. 8 and can only be regarded as indecisive. It is also worth noting that, if in Fisk's analysis β were reduced to about 1.0, a similar conclusion to that of Stier would have been obtained.

The behaviour of the N_2 cross section for very low electron energies is still uncertain. Recent measurements for electrons with thermal energies (about 0.04 ev) using microwave methods for measurement of collision frequency in a nitrogen afterglow in which atomic nitrogen is effectively absent, have not given consistent results. Phelps, Fundingsland and Brown[1], by measuring the microwave conductivity in the afterglow obtain a cross section which is rising with electron energy and is of magnitude $4.87\pi a_0^2$, whereas Anderson and

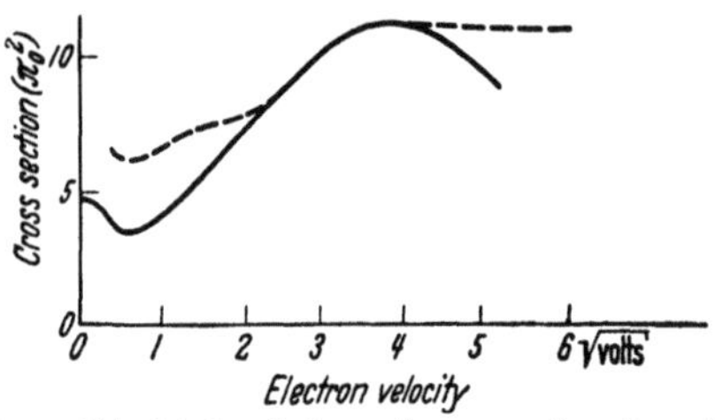

Fig. 9. Calculated and observed cross sections from the elastic scattering of electrons in O_2. ——— calculated by Fisk; – – – observed by Ramsauer and Kollath.

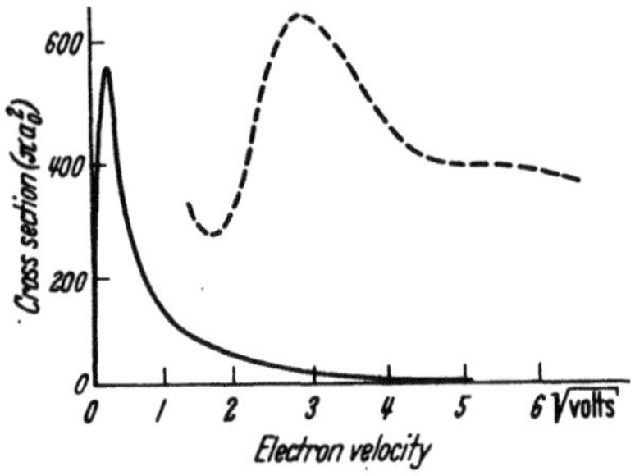

Fig. 10. Calculated and observed cross sections for the elastic scattering of electrons in Cl_2. ——— calculated by Fisk; – – – observed by Fisk.

Goldstein[2], using microwave interaction methods (see Sect. 14) obtain values decreasing with electron energy from about $19.5\,\pi a_0^2$ for thermal electrons.

The observations of Crompton and Sutton[3], using the Townsend-Bailey[4] electron drift method modified to take advantage of recent technical improvements, favour the former results, but as the lowest mean electron energy reached in the drift experiments is about 0.2 ev, this is not decisive. No decision between these conflicting results can be based on Fisk's calculations as the behaviour at very low energies is likely to depend very much on the detailed nature of the interaction.

Results for O_2, illustrated in Fig. 9, show that Fisk's interaction gives quite good results if $\beta = 1.35$. For this case there are no angular distribution data available. Once again the behaviour of Fisk's theoretical cross section in the low energy limit cannot be relied upon to give any indication of the true cross section for thermal electrons.

It is not possible to fit the observed cross sections[5] for Cl_2 in terms of the empirical interactions (Fig. 10), but this may be due to experimental rather than to theoretical difficulties.

7. Calculations for hydrogen based on the actual molecular structure. It is only for molecular hydrogen that any calculations of the scattering of slow electrons have been carried out which actually proceed from the calculated electron structure of the molecule.

---

[1] A. V. Phelps, A. Fundingsland and S. C. Brown: Phys. Rev. **84**, 559 (1951).
[2] J. M. Anderson and L. Goldstein: Phys. Rev. **102**, 388 (1956).
[3] R. W. Crompton and D. J. Sutton: Proc. Roy. Soc. Lond., Ser. A **215**, 467 (1952).
[4] J. S. Townsend and V. A. Bailey: Phil. Mag. **46**, 657 (1923).
[5] J. B. Fisk: Phys. Rev. **51**, 25 (1937).

NAGAHARA[1] calculated the mean potential interaction $V(\xi, \eta, \varphi)$ of an electron with the molecule by taking for the molecular electron wave function[2]

$$\psi(\boldsymbol{r}_1, \boldsymbol{r}_2) = N\, \mathrm{e}^{-\alpha(\xi_1+\xi_2)} \cosh\beta(\eta_1 - \eta_2) \tag{7.1}$$

with $\alpha = 1.75$, $\beta = 1.375$, $R = 1.42$ Å, so that

$$V(\xi, \eta, \varphi) = -\frac{2e^2}{R}\frac{\xi}{\xi^2-\eta^2} + e^2 \iint \left(\frac{1}{|\boldsymbol{r}-\boldsymbol{r}_1|} + \frac{1}{|\boldsymbol{r}-\boldsymbol{r}_2|}\right) |\psi(\boldsymbol{r}_1, \boldsymbol{r}_2)|^2 d^3\boldsymbol{r}_1, d^3\boldsymbol{r}_2. \tag{7.2}$$

Exchange effects are neglected, i.e. $K(\boldsymbol{r}, \boldsymbol{r}')$ in (4.4) is taken as zero. Even with this approximation the equation which results when (7.2) is substituted for V is not separable in spheroidal co-ordinates. NAGAHARA nevertheless calculates the scattering to be expected by expanding the solution of (4.1) in spheroidal harmonics in the form (5.4) which leads to an infinite set of coupled ordinary differntial equations for the T_{ml}. These may be solved approximately when the electrons are slow. The results obtained in this way give as good an agreement with the observed data as the results obtained with FISK's empirical interaction.

The only calculations which have been carried out so far which include the effects of exchange between the incident and molecular electrons are those of MASSEY and RIDLEY[3] using the HULTHÉN and KOHN variational procedure (see Vol. XXXVI, p. 258).

If $\psi(\boldsymbol{r}_1, \boldsymbol{r}_2)$ is the wave function for the ground state of the molecule, the direct interaction U which appears in (4.4) is given by (7.2) and the kernel K is given by

$$\left.\begin{aligned} K(\boldsymbol{r}, \boldsymbol{r}') = \int \Big[[\psi(\boldsymbol{r}_1, \boldsymbol{r})\, \nabla'^2 \psi(\boldsymbol{r}', \boldsymbol{r}_1) + \psi(\boldsymbol{r}', \boldsymbol{r}_1)\, \nabla^2 \psi(\boldsymbol{r}', \boldsymbol{r}_1) + \\ + \psi(\boldsymbol{r}', \boldsymbol{r}_1)\, \nabla^2 \psi(\boldsymbol{r}_1, \boldsymbol{r}) + \frac{2me^2}{\hbar^2}\left(\frac{2}{r} + \frac{2}{r'} + \frac{2}{r_1} - \frac{1}{|\boldsymbol{r}-\boldsymbol{r}'|} - \right. \\ \left. - \frac{1}{|\boldsymbol{r}-\boldsymbol{r}_1|} - \frac{1}{|\boldsymbol{r}_1-\boldsymbol{r}'|} + E\right) \psi(\boldsymbol{r}', \boldsymbol{r}_1)\, \psi(\boldsymbol{r}_1, \boldsymbol{r}) \Big]\, d^3\boldsymbol{r}_1. \end{aligned}\right\} \tag{7.3}$$

This may be obtained by exactly similar methods as those used for the corresponding three-electron scattering problem (scattering by helium, see Vol. XXXVI, p. 376). MASSEY and RIDLEY obtained an approximate solution of the resulting integro-differential equation by choosing as the trial form for F,

$$\left.\begin{aligned} F(\boldsymbol{r}) = (1 + a^2)^{-\frac{1}{2}} \{c(\xi - 1)\}^{-1} [\sin c(\xi - 1) + \{a + \\ + b\, \mathrm{e}^{-\gamma(\xi-1)} (1 - \mathrm{e}^{-\gamma(\xi-1)}\} \cos c(\xi - 1)], \quad c = \tfrac{1}{2} kR. \end{aligned}\right\} \tag{7.4}$$

The parameters a and b were determined from the usual variational conditions (see Vol. XXXVI, p. 258) and the phase ζ_{00} is then given approximately by

$$\zeta_{00} = (\arctan a) - c. \tag{7.5}$$

The form assumed for the molecular wave function in this work was the self-consistent form given by COULSON[4]. The calculations are only likely to give good results for electrons with energies less than 10 ev. The results obtained are illustrated in Fig. 11 and in view of the fact that no adjustable parameters are involved the agreement with observation is not unsatisfactory. It will be seen also that, if exchange is neglected in these calculations, the cross section obtained is

[1] S. NAGAHARA: J. Phys. Soc. Japan 9, 52 (1954).
[2] T. INUI: Proc. Phys.-Math. Soc. Japan 20, 790 (1938).
[3] H. S. W. MASSEY and R. O. RIDLEY: Proc. Phys. Soc. Lond. A 69, 659 (1956).
[4] C. A. COULSON: Proc. Cambridge Phil. Soc. 34, 204 (1938).

much too large, a result similar to that found in the theory of the scattering of slow electrons by helium atoms.

There are many further possibilities for theoretical study of the scattering by simple molecules, including particularly H_2. As the appropriate Eq. (4.1) is no longer separable in spheroidal co-ordinates when actual molecular fields are considered, there would seem to be little advantage in working with these co-ordinates. It may then be preferable to work in terms of the more familiar and more flexible spherical polar co-ordinates.

8. Scattering by highly symmetrical molecules. The critical field of a molecule with tetrahedral symmetry, such as methane, is not far from being spherically symmetrical. BUCKINGHAM, MASSEY and TIBBS[1] have calculated self-consistent wave functions for the methane molecule by averaging the proton distribution over all orientations. Using these wave functions a mean potential interaction of an electron with a methane molecule may be derived. If exchange effects are neglected, the cross sections for elastic scattering may be derived in the form (4.8) by substituting the mean interaction in (4.4) and following the procedure outlined for determination of the phase shifts η_l. Calculations on these lines were carried out by BUCKINGHAM, MASSEY and TIBBS[1] and they obtained the interesting result that, for electrons with energies below about 10 ev, the cross sections should be very similar to those for argon, calculated in the same way, exchange being neglected. This is in agreement with observation, both as regards the similarity of angular distribution and of total cross section as a function of electron velocity. However, when exchange is neglected, neither of the calculated results agree closely with the corresponding observed ones. It is not unreasonable to suppose though, that if the exchange were properly allowed for in both cases, the similarity would persist.

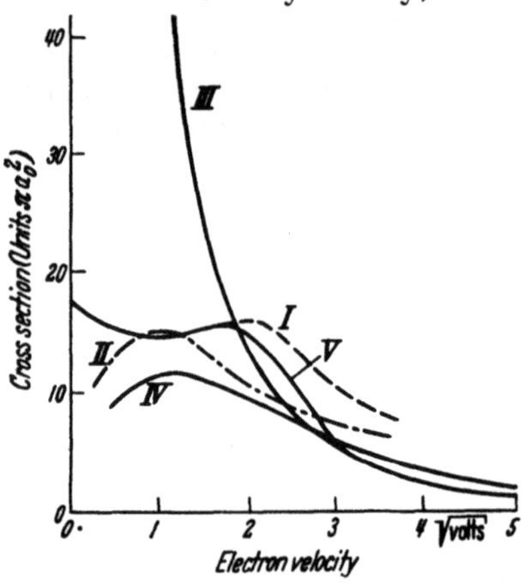

Fig. 11. Calculated and observed cross sections for the elastic scattering of electrons in H_2. *I* observed by Ramsauer method. *II* observed by swarm and diffusion methods (ordinate is the mean cross section and the abscissa is the mean electron velocity in $\sqrt{\text{volts}}$). *III* calculated by the variation method, ignoring exchange. *IV* calculated by the variation method, including exchange. *V* calculated by FISK using the semi empirical molecular scattering potential.

Comparison of observed and calculated electron scattering cross sections for highly symmetrical molecules should provide a useful check on further calculations of the structure of molecules which possess tetrahedral or octahedral symmetry.

C. Inelastic collisions of electrons with molecules.

9. Further theoretical considerations. In an elastic collision of an electron with a molecule there may be a transfer of translational energy of the electron not only to internal energy of the molecular electrons but to nuclear vibration and rotation.

Consider, for simplicity, a collision with a diatomic molecule. The initial state of the molecule may be characterised by a set of quantum numbers n specifying the state of electronic motion relative to the nuclei, the vibrational

[1] R. A. BUCKINGHAM, H. S. W. MASSEY and S. TIBBS: Proc. Roy. Soc. Lond., Ser. A **178**, 119 (1941).

quantum number v and the rotational quantum numbers J, M. The wave function for this state may be written to a good approximation in the form

$$\Psi_{n v J M} = \psi(n \mid \boldsymbol{r}_n, R)\, \varphi(v \mid R)\, \chi(J, M \mid \Theta, \Phi) \tag{9.1}$$

where $\boldsymbol{r}_m$ denotes the co ordinates of the molecular electrons relative to the nuclei, (Θ, Φ) are the polar co-ordinates of the molecular axis with respect to an axis fixed in space and R is the nuclear separation.

According to BORN's first approximation (Vol. XXXVI, p. 280) the differential cross section for exciting the molecule to the state characterised by the quantum numbers n', v', J', M' in which the momentum of the incident electron is changed from $\boldsymbol{k}\hbar$ to $\boldsymbol{k}'\hbar$ is given by

$$\left.\begin{aligned} I_{n v J M}^{n' v' J' M'}\, d\omega &= \frac{4\pi^2 m^2 e^4}{h^4} \frac{k'}{k} \times \\ &\times \left| \iiint \left(\sum \frac{1}{|\boldsymbol{r}_n - \boldsymbol{r}|} \right) e^{i(\boldsymbol{k}-\boldsymbol{k}')\cdot \boldsymbol{r}}\, \Psi_{n v J M}\, \Psi^*_{n' v' J' M'}\, d^3\boldsymbol{R}\, d^3\boldsymbol{r}\, \Pi_m\, d^3\boldsymbol{r}_m \right|^2 d\omega, \end{aligned}\right\} \tag{9.2}$$

e and m being respectively the charge and mass of the electron.

By use of the formula

$$\int \frac{e^{i(\boldsymbol{k}-\boldsymbol{k}')\cdot \boldsymbol{r}}}{|\boldsymbol{r} - \boldsymbol{r}_m|}\, d^3\boldsymbol{r} = \frac{4\pi}{|\boldsymbol{k} - \boldsymbol{k}'|^2}\, e^{i(\boldsymbol{k}-\boldsymbol{k}')\cdot \boldsymbol{r}_m} \tag{9.3}$$

and the form (9.1) for the molecular wave function, we may write

$$\left.\begin{aligned} I_{n v J M}^{n' v' J' M'} &= \left| \int \varphi(v \mid R)\, \varphi^*(v' \mid R)\, \chi(J, M \mid \Theta, \Phi) \times \right. \\ &\times \chi^*(J', M' \mid \Theta, \Phi)\, M(R, \Theta, \Phi)\, d^3\boldsymbol{R} \big|^2 \end{aligned}\right\} \tag{9.4}$$

where

$$\left.\begin{aligned} M(R, \Theta, \Phi) &= \frac{8\pi^2 m e^2}{h^2} \left(\frac{k'}{k}\right)^{\frac{1}{2}} \frac{1}{|\boldsymbol{k} - \boldsymbol{k}'|^2} \int \psi(n \mid \boldsymbol{r}_m, R) \times \\ &\times \left(\sum_m e^{i(\boldsymbol{k}-\boldsymbol{k}')\cdot \boldsymbol{r}_m} \right) \psi^*(n' \mid \boldsymbol{r}_m, R)\, d^3\boldsymbol{r}_m. \end{aligned}\right\} \tag{9.5}$$

For most purposes we need not consider separately the cross sections for excitation of the separate rotational states, but merely that summed over all final rotational states. Since

$$\left.\begin{aligned} &\sum_{J'M'} \left| \int \chi(J, M \mid \Theta, \Phi)\, G(\Theta, \Phi)\, \chi^*(J', M' \mid \Theta, \Phi) \sin\Theta\, d\Theta\, d\Phi \right|^2 \\ &\quad = \int |\chi(J, M \mid \Theta, \Phi)\, G(\Theta, \Phi)|^2 \sin\Theta\, d\Theta\, d\Phi \end{aligned}\right\} \tag{9.6}$$

and the energy differences between the rotational states are all small compared with $\frac{\hbar^2}{2m}(k^2 - k'^2)$, we have

$$\left.\begin{aligned} \sum_{J'M'} I_{n v J M}^{n' v' J' M'} &= \int \left| \int \varphi(v \mid R)\, \varphi(v' \mid R)\, M(R, \Theta, \Phi)\, R^2\, dR \right|^2 \times \\ &\times |\chi(J, M \mid \Theta, \Phi)|^2 \sin\Theta\, d\Theta\, d\Phi. \end{aligned}\right\} \tag{9.7}$$

Whereas the electronic wave functions ψ are slowly varying functions of the nuclear separation, the product $\varphi(v \mid R)\, \varphi(v' \mid R)$ will have a strong maximum for $R \sim R_0$, the initial equilibrium nuclear separation. This gives

$$\left.\begin{aligned} \sum_{J'M'} I_{n v J M}^{n' v' J' M'} &= \int |M(R_0, \Theta, \Phi)|^2\, |\chi(J, M \mid \Theta, \Phi)|^2 \times \\ &\times \sin\Theta\, d\Theta\, d\Phi \cdot \left| \int \varphi(v \mid R)\, \varphi(v' \mid R)\, R^2\, dR \right|^2 \end{aligned}\right\} \tag{9.8}$$

If the energy differences of the vibrational states are small compared with the final energy of the colliding electron so that k' is effectively independent of v' we have the following important result: for a given electronic transition the relative probability of any vibrational transition is approximately independent of the energy of the exciting electron and is proportional to the square of the overlap integral

$$|\int \varphi(v|R)\, \varphi(v'|R)\, R^2\, dR|^2. \tag{A}$$

Furthermore, since

$$\sum_{v'} \left|\int \varphi(v|R)\, \varphi(v'|R)\, R^2\, dR\right|^2 = 1,$$

we have

$$\sum_{v'J'M'} I_{n v J M}^{n' v' J' M'} = \int |M(R_0, \Theta, \Phi)|^2\, |\chi(J M|\Theta, \Phi)|^2 \sin\Theta\, d\Theta\, d\Phi. \tag{9.9}$$

As $|\chi(J M|\Theta\Phi)|^2$ will in general be independent of Θ, Φ this means that the sum of the differential cross sections over all final states of nuclear motion is simply the average, over all orientations of the nuclear axis, of that calculated for the electronic excitation with the nuclei fixed at this equilibrium separation.

These results have been obtained on the assumption of the validity of BORN'S first approximation. Provided k' is still effectively independent of v' they would still be expected to hold if the interaction between the incident and molecular electrons can no longer be treated as small. The only difference is that $M(R, \Theta, \Phi)$ can no longer be represented by the expression (9.5) but by some other slowly varying function of R.

We shall consider the experimental and theoretical work which has been carried out on inelastic collisions under three headings—electronic excitation, excitation of nuclear motion without electronic excitation and, finally, ionizing collisions. This division is appropriate because of the different experimental techniques employed in each case.

D. Electronic excitation.

The experimental methods which have been used to study electron excitation are essentially the same as those employed for atomic collisions. Thus the excitation of states from which radiation in the visible, the near ultraviolet and infra-red is emitted in optically allowed transitions may be studied by the optical methods described in Vol. XXXVI, pp. 320—329, of this Encyclopedia. Measurements of this kind have been carried out for excitation of certain levels of H_2, N_2 and CO. No attempts have been made up to the time of writing to study cross sections for excitation of metastable states, with the exception of the work which has been carried out concerning the dissociation of H_2 into two neutral atoms by electron impact—a process which proceeds largely by excitation of the lowest triplet state of the molecule from the ground singlet state.

10. Excitation of nitrogen. It will be convenient to begin by discussing the studies of the excitation of N_2. The most prominent bands in the nitrogen spectrum excited in a discharge are the First and Second Positive and First Negative Bands. The first two arise from transitions between states of the neutral molecule, the last from transitions between states of N_2^+. Fig. 12 illustrates the relative location of the energy minima of the electronic states concerned with these transitions as well as the ground state. It will be seen that the First and Second Positive Bands arise from transitions within the triplet system of levels. The lowest level for the First Positive Bands is the lowest triplet level usually referred to as $A^3\Sigma_u^+$. The upper level for these bands, which is also the lower level for the Second Positive Bands, is $B^3\Pi_g$. Finally, the upper level for the Second Positive Bands is $C^3\Pi_u$. Since the ground state of the molecule is a singlet state ($X\,^1\Sigma_g^+$), optical excitation of the Positive Bands by electron impact involves a change of multiplicity and hence takes place mainly by electron exchange. The optical excitation function for both band systems is therefore likely to be of the same general form as for the triplet lines for helium (see Vol. XXXVI, p. 327) i.e. a sharp maximum would be expected for electron energies near the threshold followed by a rapid fall off in intensity as the energy increases further. On the other hand, rather weak Lyman-Birge-Hopfield Bands arise from a transition between the first excited ($a\,^1\Pi_g$) state and the ground $^1\Sigma_g^+$ state. Although the transiton violates the $g \to u$ selection rule for an optically allowed transition, it does not involve a change of multiplicity. The excitation function for these bands would therefore be expected to resemble that for the corresponding atomic case (say the excitation of a ^1D level of helium from the ground ^1S level). Owing, however, to the weakness of the bands, experimental observation of the optical excitation function would be very difficult.

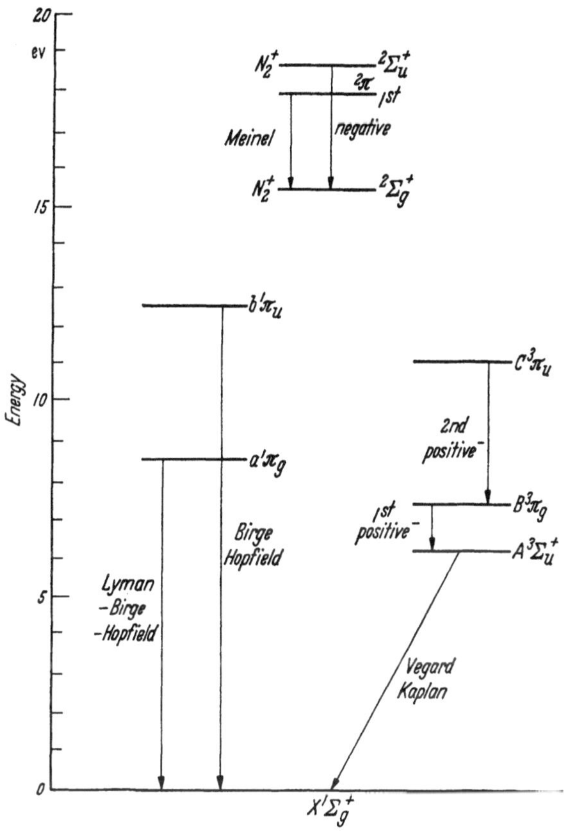

Fig. 12. Some electronic energy levels of N_2 and N_2^+ with origins of band systems indicated.

Bands, known as the Vegard-Kaplan system, arising from the intercombination $A\,^3\Sigma_u^+ - X\,^1\Sigma_g^+$ transition, have been observed, particularly in the polar aurora. The transition probability for these bands is so very low, that it is out of the question to investigate the optical excitation function by optical emission methods. It should be possible, however, to measure the variation with electron energy of the cross section for excitation of $A\,^3\Sigma_u^+$ using an optical absorption method similar to that used by WOUDENBERG and MILATZ[1] (Vol. XXXVI p. 335) for studying the impact excitation of the metastable states of helium and of neon.

The First Negative Bands arise from the transition between the $^2\Sigma_u^+$ and the $^2\Sigma_g^+$ ground state of N_2^+, so that the optical excitation function is determined by the cross section for production of N_2^+ ions in the ground state by single electron impact from ground state N_2. This band system is a prominent feature of the

[1] J. P. WOUDENBERG and J. M. MILATZ: Physica, Haag 8, 871 (1941).

auroral spectrum. A second band system of N_2^+ was discovered in this spectrum by MEINEL[1] and is usually refered to as the Meinel system. It arises from a transition between a $^2\Pi_u$ state and the ground $^2\Sigma_g^+$ state. Excitation of the upper levels of both the First Negative and Meinel Bands represents an optically allowed transition from the ground state of N_2 and the corresponding excitation function should therefore resemble those for similar transitions in atoms (as for example the $3\,^1P$ level of helium, Vol. XXXVI, p. 327).

Experimental studies of the optical excitation functions for N_2 have been carried out by LANGSTROTH[2], THIEME[3], THOMSON and WILLIAMS[4], BERNARD[5] and STEWART[6], the techniques being essentially the same as those already described in Vol. XXXVI, p. 320. Essentially an electron beam of nearly homogeneous energy is fired through the gas and the intensity of the particular band emission from a selected region of the beam measured as a function of electron energy. To ensure that the excitation is a result of single electron impact, the intensity should be proportional to the beam intensity and to the gas pressure.

LANGSTROTH studied the excitation functions of the First Negative and Second Positive Bands with the special aim of verifying (A) of Sect. 9, p. 334. He found that the intensity ratio of the $0\to 1$ and $1\to 2$ vibration bands in the First Negative System was indeed independent of electron energy over the range investigated (21 to 50 ev) and equal to 7.2. According to the discussion of Sect. 9 the intensity of the $v'\to v''$ band due to impact excitation from a state in which the vibrational quantum number is v is proportional to

$$p_{v'v''}\left|\int \varphi(v\,|\,R)\,\varphi^*(v'\,|\,R)\,R^2\,dR\right|^2 \tag{10.1}$$

where $p_{v'v''}$ is the optical transition probability from v' to v''. This in turn is proportional to

$$\nu_{v'v''}^4 \int\left|\varphi(v'\,|\,R)\,\varphi^*(v''\,|\,R)\,R^2\,dR\right|^2 \tag{10.2}$$

where $\nu_{v'v''}$ is the frequency of the transition concerned. Over an electronic band system the variation of ν with v' and v'' is small so that the intensity is practically proportional to the product of the squares of the $v v'$ and $v' v''$ overlap integrals. LANGSTROTH calculated these integrals approximately by using a method due to HUTCHINSON[7] and found for the $0\to 2/1\to 2$ intensity ratio a value in very close agreement with that observed. The overlap integrals were recalculated many years later by BATES[8] using more accurate vibrational wave functions. These were determined by representing the potential energy curves for the $X\,^1\Sigma_g^+$, $^2\Sigma_u^+$ and $^2\Sigma_g^+$ states analytically by MORSE's function[9] and using the best available data on the equilibrium nuclear separations and vibration frequencies. He obtained the intensity distribution among the vibrational transitions given in Table 2. It will be seen that the $0\to 2/1\to 2$ ratio according to Bates should be $16/22 = 7.3$ which agrees very well both with LANGSTROTH's observations and calculations.

---

[1] A. B. MEINEL: C. R. Acad. Sci. Paris **231**, 1049 (1950).
[2] G. O. LANGSTROTH: Proc. Roy. Soc. Lond., Ser. A **146**, 166 (1934).
[3] O. THIEME: Z. Physik **78**, 412 (1932).
[4] R. THOMPSON and S. E. WILLIAMS: Proc. Roy. Soc. Lond., Ser. A **147**, 583 (1934).
[5] R. BERNARD: C. R. Acad. Sci. Paris **205**, 793 (1937).
[6] D. T. STEWART: Proc. Phys. Soc. Lond. A **68**, 404 (1955).
[7] E. HUTCHINSON: Phys. Rev. **36**, 410 (1930).
[8] D. R. BATES: Proc. Roy. Soc. Lond., Ser. A **196**, 217 (1949).
[9] P. M. MORSE: Phys. Rev. **34**, 57 (1929).

LANGSTROTH[1] only measured the excitation fonction up to electron energies of 50 ev, but the work of THIEME[2], BUNDY[3] and BERNARD[4] has extended the observations up to 500 ev. Fig. 13 illustrates the results obtained in these investigations, showing that the excitation function is of the expected form.

The situation is not so satisfactory for the Second Positive Bands. LANGSTROTH found that the intensity ratios of different vibrational transitions were only independent of electron energy when this exceeded about 30 ev. The observed

Table 2. *Calculated intensity distribution among vibrational transitions of the Negative Bands arising from electron impact excitation of N_2 molecules in their ground vibrational states.*

v'	v''			
	0	1	2	3
0	59	16	2.5	0.1
1	3.3	2.2	2.2	0.5
2	0.0	0.1	0.0	0.0

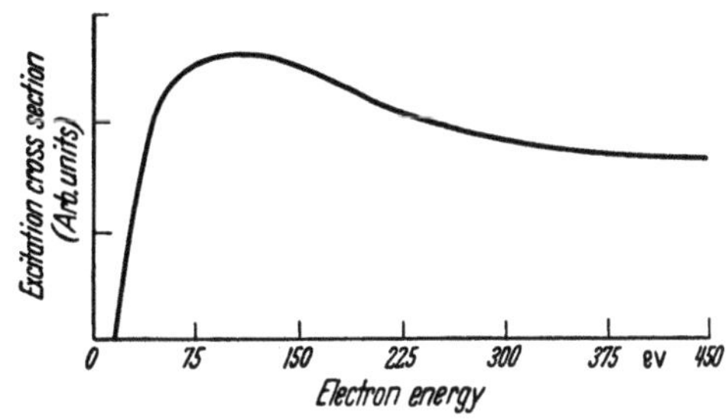

Fig. 13. Observed excitation function for the N_2^+ Negative Bands.

relative intensities for such electron energies are given in Table 3 and compared with the theoretical predictions obtained by BATES using the same method as for the Negative Bands. The observed results are normalized so that the total intensities of all the $0 \to v''$ bands observed agree with that calculated.

Table 3. *Observed and calculated intensity distribution among vibrational transitions of the second positive bands arising from electron impact excitation of $X\,^1\Sigma_g^+$ N_2 molecules in their ground vibrational states. Observed values are indicated by brackets.*

v'	v''							
	0	1	2	3	4	5	6	7
0	23 (19)	14 (16)	5.1 (6.2)	1.3 (1.9)	0.2	0.0		
1	17	1.0 (0.4$_8$)	4.6 (5.6)	3.7 (5.2)	1.4 (2.7)	0.3	0.0	
2	3.3	4.7	0.2 (0.3$_6$)	0.7 (0.7)	0.9 (2.7)	0.5	0.1	0.0
3	0.9	2.1	0.9	0.2	0.3	0.1	0.1	0.0

It will be seen that, while the agreement is fair, there are considerable discrepancies. Taken together with the observed variation of the intensity ratio with electron energys below 30 ev, these suggest that some secondary process was contributing to the observed emission. No evidence in support of this is yet forthcoming.

There is evidence that secondary processes are difficult to avoid when studying the impact excitation of the First Positive Bands. THOMPSON and WILLIAMS[5] and, much later, STEWART[6] found that much of the emission in these bands comes from outside the beam. Similar effects in the excitation of certain helium lines due to transitions from ^3D levels were observed by SKINNER and LEES[7] and

[1] G.O. LANGSTROTH: Proc. Roy. Soc. Lond., Ser. A **146**, 166 (1934).
[2] O. THIEME: Z. Physik **78**, 412 (1932).
[3] F.P. BUNDY: Phys. Rev. **52**, 698 (1937).
[4] R. BERNARD: C. R. Acad. Sci. Paris **205**, 193 (1951).
[5] W. THOMPSON and S.E. WILLIAMS: Proc. Roy. Soc. Lond., Ser. A **147**, 583 (1934)
[6] D.T. STEWART: Proc. Phys. Soc. Lond. A **68**, 404 (1935).
[7] H. W. B. SKINNER and J.H. LEES: Proc. Roy. Soc. Lond., Ser. A **37**, 186 (1932).

ascribed to collisions of the kind

$$\mathrm{He}(n^1P) + \mathrm{He}(1\,^1S) \rightarrow \mathrm{He}(1\,^1S) + \mathrm{He}(n\,^3D) \tag{10.1}$$

in which the excited He$(n\,^1P)$ atoms are formed outside the beam by resonance absorption of $n\,^1P - 1\,^1S$ radiation excited in the beam. THOMPSON and WILLIAMS suggest that the process responsible for the spreading of the light in the First Positive Bands is the analogous one

$$\mathrm{N}_2(a\,^1\Pi_g) + \mathrm{N}_2(X\,^1\Sigma_g^+) \rightarrow \mathrm{N}_2(B\,^3\Pi_g) + \mathrm{N}_2(X\,^1\Sigma_g^+). \tag{10.2}$$

The molecules in the $a\,^1\Pi_g$ state would be produced by absorption of Lyman-Birge-Hopfield band radiation. These molecules would not be converted into molecules in the upper $C\,^3\Pi_g$ state of the Second Positive Bands by a process

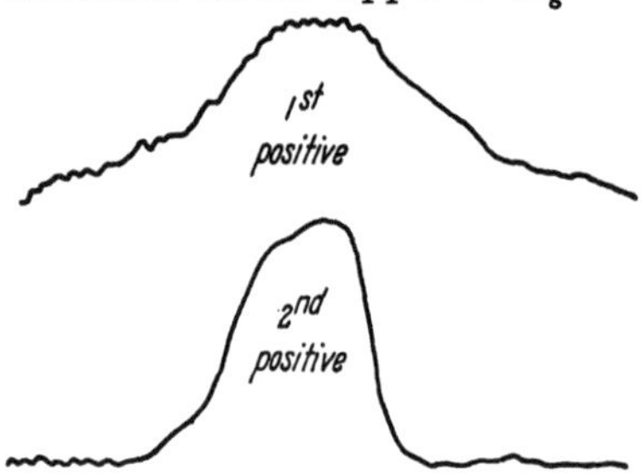

Fig. 14.

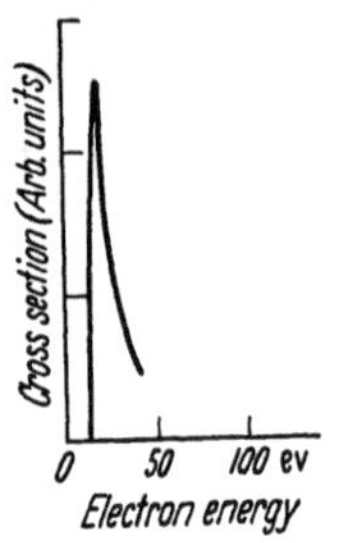

Fig. 15.

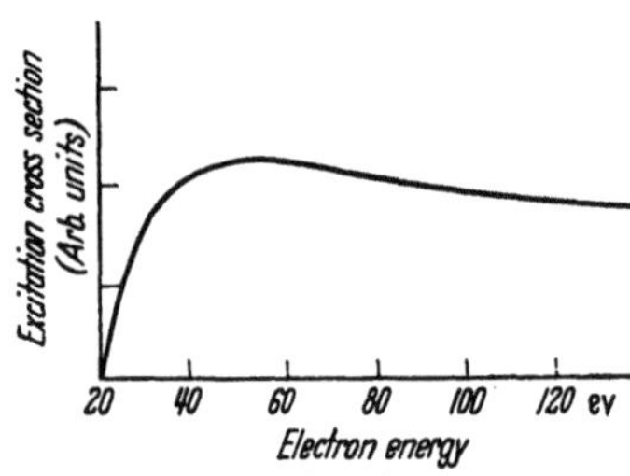

Fig. 16.

Fig. 14. Photometer traces for the First and Second Positive Bands of nitrogen taken perpendicular to the dispersion. The spreading of the First Positive trace is noticeable.

Fig. 15. Observed excitation function for the Second Positive Bands of nitrogen.

Fig. 16. Observed excitation function for the Meinel Bands of N_2^+.

similar to (10.2) unless the initial excited molecules were in a higher singlet state, such as the $b\,^1\Pi$ upper state of the Birge-Hopfield bands (see Fig. 12). As these bands are much weaker than the Lyman-Birge-Hopfield system, it would then be natural to expect that the spreading effect would be much less for the Second than for the First Positive Bands. This is confirmed from the observations of THOMPSON and WILLIAMS. Fig. 14 reproduces two typical photometric records taken across the beam for the two band systems under similar conditions of electron current and gas pressures. However, the whole mechanism of spreading cannot yet be regarded as properly understood and it is not out of the question that some secondary process was contributing to the intensities observed by LANGSTROTH. There is room for further experiment here.

The fact that the excitation function for the Second Positive Bands is indeed of the form expected for a singlet-triplet excitation will be apparent by reference to Fig. 15 in which THIEME's observed function is illustrated.

STEWART's experiments were directed in the first instance to a search for a laboratory method for exciting the Meinel Band system which had previously been observed only in auroral spectra. Fig. 16 illustrates the excitation function obtained. As expected, this is of the form typical of optically allowed transitions and is very similar to that for the First Negative Bands. No spreading of the light emitted from these bands was observed and the observed intensities were proportional to both beam current and gas pressure.

STEWART[1] also observed the excitation function for the NI lines resulting from the $3s\,^4P - 3p\,^4P^0$ transition. The intensity of these lines were found to

[1] D. T. STEWART: Proc. Phys. Soc. Lond. A **68**, 404 (1935).

be proportional to beam current and gas pressure (below 10^{-2} mm Hg) showing that the N atoms in the $3p\ ^4P^0$ state are produced in single impact of electrons with N_2 molecules, in their ground states. The appearance potential for the process was found to be below 24 volts.

11. Excitation of hydrogen. The approximate potential energy curves for some of the lowest electronic states of molecular hydrogen are illustrated in Fig. 17.

As the molecule is a two-electron system the levels fall into a singlet and triplet set. The ground level is a $(1s\sigma)^2\ ^1\Sigma_g^+$ state which is attractive with a dissociation energy, into two normal hydrogen atoms, of 4.44 ev. No other singlet state dissociates in the limit of large nuclear separations into two normal atoms. The three singlet states arising from one normal H atom and one excited to a 2-quantum level are, in ascending order of energy at the equilibrium separation of the ground state, $^1\Sigma_u^+$, $^1\Pi_u$ and $^1\Sigma_g^+$ states, usually denoted as the B, C and E states. The designation of these states by quantum numbers referring to the united and separated atom limits to which they tend offers no difficulty as far as the C and E states are concerned. The correlations are:

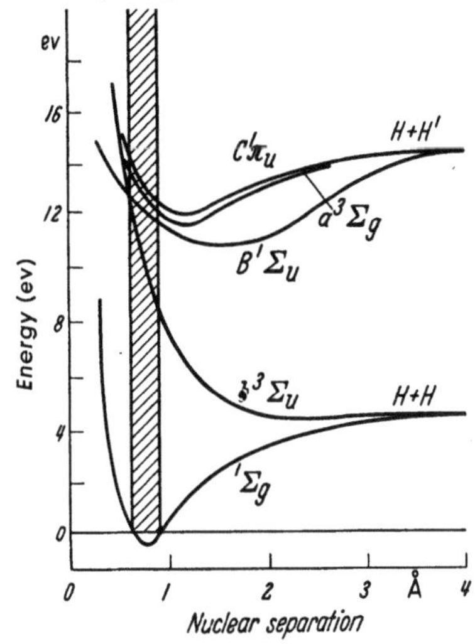

Fig. 17. Potential curves for some of the lower electronic states of H_2.

United atom limit						Separated atom limit	
$He(2^1P)$	$1s\sigma$	$2p\pi$	$\rightarrow$	C	$\rightarrow$	$\sigma_g 1s$	$\pi_u 2p$
$He(2^1S)$	$1s\sigma$	$2s\sigma$	$\rightarrow$	E	$\rightarrow$	$\sigma 1s$	$\sigma 2s$.

The order of the energy levels for these states is the same as for the $2p\pi$ and $2s\sigma$ states of H_2^+. On the same correlation scheme for the B state we would have

$$He(3^1P)\quad 1s\sigma\quad 3p\sigma \rightarrow B \rightarrow \sigma_g 1s\quad \sigma_u 2s,$$

but in this case the order of the level is not the same as for the $3p\sigma$ state of H_2^+, which, at the equilibrium distances concerned, lies above $2s\sigma$ and $2p\pi$. The complication in the B state may be ascribed to its possessing a large amount of polar character so that the energy near the equilibrium separation is much less than would be expected without allowance for this. Thus, although a $1s\sigma$, $2p\sigma$ united atom configuration dissociating into two normal H atoms can only give a $^1\Sigma_g$ or $^3\Sigma_u$ state, the same configuration can give rise to a $^1\Sigma_u$ state dissociating into normal H^- and a proton. We may therefore regard the B state as arising from $He(2^1P)\ 1s\sigma\ 2p\sigma$ which tends to dissociate into $H^-(^1S)$ and a proton. Owing to interaction between potential energy curves at very large separations, it probably dissociates into $\sigma_g 1s\ \sigma_u 2s$.

The lowest triplet state $b\ ^3\Sigma_u^+$ is repulsive in character, but the next in order of energy, $a\ ^3\Sigma_g^+$, is attractive. The correlations for these two states are as follows.

$$He(2^3P)\quad 1s\sigma\quad 2p\sigma \rightarrow b\ ^3\Sigma_u^+ \rightarrow \sigma_g 1s\quad \sigma_u 1s,$$
$$He(2^3S)\quad 1s\sigma\quad 2s\sigma \rightarrow a\ ^3\Sigma_g^+ \rightarrow \sigma_g 1s\quad \sigma_g 2s.$$

According to the Franck-Condon principle the final molecular states which will arise with appreciable probability from excitation of normal molecules in the lowest vibrational state will lie within the shaded area in Fig. 17. Thus the excitation potentials for the different states are approximately as in Table 4.

We may now discuss briefly the effects which are likely to be associated with impact excitation of the hydrogen molecule. Electrons with energy of the order of 100 ev or more can only excite with appreciable probability directly, but slow electrons also excite through exchange with the molecular electrons. We would expect from the correlations with the united atom limit that the excitation functions for the B and C states should be very similar to that of He (2^1P), while that for the E state should resemble that of He (2^1S). The B and C states should therefore be excited with comparable probability by medium and high energy electrons, the E state with much smaller probability. ROSCOE[1] has applied BORN's approximation to calculate the differential cross sections for excitation of these three states. He used molecular wave functions first suggested and used by MACDONALD[2] which are consistent in form with the above correlations. From his published data it is possible to estimate the total excitation cross sections for the three states for a few electron energies. The values obtained are given in Table 5. Included among these are also some values for the excitation of the so-called D state, which is the $3p\pi\,{}^1\Pi_u$ state with an excitation function very similar in shape to that for the B and C states.

Table 4. *Approximate excitation potentials of states of molecular hydrogen.*

State	Excitation potential (volts)
$b\,{}^3\Sigma_u$	8.8
$B\,{}^1\Sigma_u$	11.4
$a\,{}^3\Sigma_g$	11.8
$C\,{}^1\Pi_u$	12.4
$E\,{}^1\Sigma_g$	12.4

Table 5. *Cross sections for excitation of B, C, D and E states of molecular hydrogen, calculated by Born's first approximation.*

Electron energy (ev)	Cross section in units πa_0^2			
	B	C	D	E
100	0.27	0.21	0.03_5	0.08
200	0.16	0.13_5	0.02_5	0.04
300	0.14	0.10_5	0.02	0.03

Electrons with energies of 100 ev or more will be very ineffective in exciting the triplet system, but this will no longer apply for much slower electrons. Particular interest attaches to the excitation of the $b\,{}^3\Sigma_u$ and $a\,{}^3\Sigma_g$ states as in each case dissociation of the molecule into normal atoms will result. This will follow immediately on excitation of the $b\,{}^3\Sigma_u$ state as it is repulsive in character. Thus referring to Fig. 17 it will be seen that, if the Franck-Condon principle is closely satisfied, excitation of the b state will yield two normal atoms moving with relative kinetic energies, at infinite separation, between 2.2 and 4.5 ev.

Although excitation of the $a\,{}^3\Sigma_g$ state will not lead directly to dissociation, this will occur via allowed radiation transitions from this state to the unstable $b\,{}^3\Sigma_u$ state. The radiation emitted in these transitions is the source of the well-known continuous spectrum of molecular hydrogen.

The variation with electron energy of the cross section for dissociation of the molecule by either of the two excitation processes will be characteristic of that for a singlet-triplet excitation, having a sharp maximum close to the threshold and then falling off quite rapidly at higher energies. No reliable theoretical estimates of these cross sections are at present available. For the same reasons as those discussed in Vol. XXXVI, p. 369 of this Encyclopedia the Born-Oppenheimer approximation is not valid for low electron energies. It has been calculated

[1] R. ROSCOE: Phil. Mag. **31**, 349 (1941).

[2] J. K. L. MACDONALD: Proc. Roy. Soc. Lond., Ser. A **136**, 528 (1932).

for the excitation of both $a\,{}^3\Sigma_g$ and $b\,{}^3\Sigma_u$ states and gives, near the threshold, results greater than the maxima permitted by the conservation of particles. The distorted wave approximation, allowing fully for exchange distortion, might give results of adequate accuracy, but is very difficult to work out because the distortion is due to a two-centres field.

Experimental investigation of the excitation and dissociation of molecular hydrogen by electron impact is still fragmentary. The first evidence of the existence of a process of impact dissociation into neutral atoms came from experiments in which the atomic hydrogen, produced by firing an electron beam of definite energy through hydrogen gas at low pressure, was detected by chemical means. Thus GLOCKLER, BAXTER and DALTON[1] oxidized the inner walls of a copper tube containing the hydrogen so that any atomic hydrogen reduced the oxide and the resulting water vapour was frozen out on a liquid air trap. The occurrence of dissociation was then manifest by a fall in pressure in the experimental chamber. HUGHES and SKELLET[2] froze the atomic hydrogen out directly on liquid-air cooled tungsten film and again observed the pressure fall. They studied the variation of atom yield with gas pressure and obtained evidence that the dissociation process was a primary one. Both sets of investigators observed dissociation by electrons with energies as low as 11.5 ev, showing that the process did not involve ionization for which the threshold is 15.5 ev (see Sect. 21). DORSCH and KALLMANN[3], by using a more sensitive method of detection were able to observe dissociation by electrons with energies as low as 8.0 ev, which is somewhat smaller even than the threshold given by the Franck-Condon principle. They detected the formation of atomic hydrogen by the blackening of a deposit of PbCl due to the reaction

$$\mathrm{PbCl} + \mathrm{H} \rightarrow \mathrm{Pb} + \mathrm{HCl}. \tag{11.1}$$

It was also observed that the dissociation cross section rose rapidly as the electron energy increased from 10 to 15 ev.

KRUITHOF and ORNSTEIN[4] obtained information about the variation of the dissociation cross section with electron energy by studying the variation with pressure of the intensities of certain lines of molecular hydrogen excited by an electron beam of controlled energy. The technique employed was essentially similar to that used in the study of the optical excitation functions of the lines of various atoms (see Vol. XXXVI, p. 323 of this Encyclopedia).

Interpretation of the data makes it necessary to consider processes of molecular dissociation and of atom recombination. If the dissociation occurs in single collisions of molecules with electrons the rate at which atoms are produced can be written, at pressure p, in the form

$$A\,p\,\frac{1-m}{1+m} \tag{11.2}$$

where m is the degree of dissociation and A is proportional to the rate of impact dissociation per molecule. In equilibrium this would be equal to the rate of recombination. If this occurs by impact of two atoms with the wall, its rate will be proportional to

$$p^2\,\frac{4m^2}{(1+m)^2} \tag{11.3}$$

---

[1] G. GLOCKLER, W. BAXTER and R. DALTON: J. Amer. Chem. Soc. **49**, 58 (1927).
[2] A.L. HUGHES and A.M. SKELLET: Phys. Rev. **30**, 11 (1927).
[3] K. DORSCH and H. KALLMANN: Z. Physik **34**, 389 (1925).
[4] A.A. KRUITHOF and L.S. ORNSTEIN: Physica, Haag **2**, 611 (1935).

so that

$$A p \frac{1-m}{1+m} = p^2 \frac{4m^2}{(1+m)^2}. \tag{11.4}$$

Hence

$$\frac{1}{m^2} = 1 + \frac{4p}{A}. \tag{11.5}$$

The intensity of a molecular line at a pressure p will be proportional to

$$p \frac{1-m}{1+m} \tag{11.6}$$

so that from the observed variation with p, m may be determined for each p.

Fig. 18 shows how well the observed pressure variation of the intensities of two molecular lines follows that given by (11.5) and (11.6) with a suitably chosen value of A.

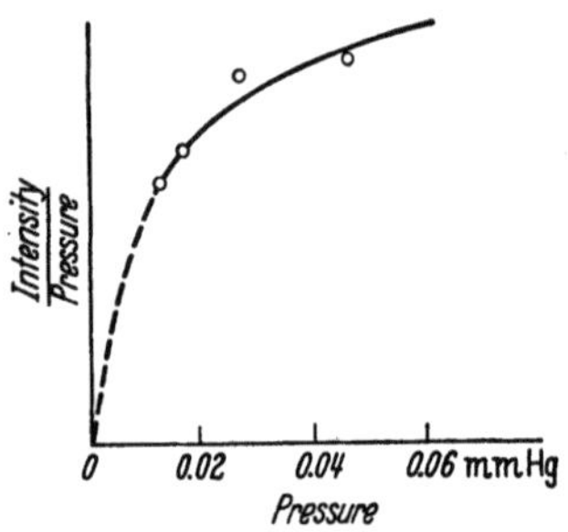

Fig. 18. Observed variation with pressure of the intensities of the lines of H_2 excited by electron impact. ○ exptl. points; — calculated from the relation (11.5) with a suitable value for A.

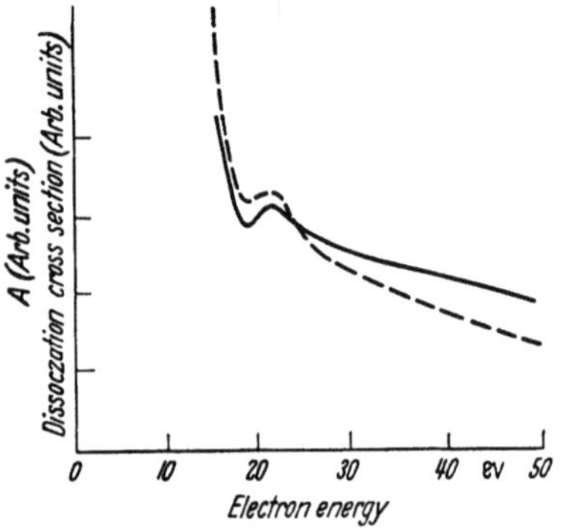

Fig. 19. Variation with electron energy of the probability of dissociating a hydrogen molecule by electron impact. ——— variation of the constant A of (11.5); - - - variation of dissociation cross section.

In this way the variation of A with electron energy may be studied, the observed results being illustrated in Fig. 19. A is proportional to the rate of dissociation per molecule and hence to the product of dissociation cross section and electron velocity. Allowance for the latter gives the energy variation of the dissociation cross section, also illustrated in Fig. 19. It will be seen that at 15 ev the cross section is still rising rapidly as the energy falls. Taken in conjunction with the observations of DORSCH and KALLMANN[1] this indicates a sharp maximum quite close to the threshold, characteristic of an intercombination excitation. The subsidiary maximum which appears near 20 ev may possibly be due to the excitation of the $a^3\Sigma_g$ and higher triplet states.

Some evidence concerning the magnitude of the cross section for dissociation by impact of slow electrons comes from the experiments of RAMIEN[2] who applied the Hertz diffusion method (Vol. XXXVI, p. 332 of this Encyclopedia) to study inelastic collisions in molecular hydrogen. He found the energy loss corresponding to excitation of the $b^3\Sigma_u$ state for electrons with initial energy slightly above 9 ev and it was observed to increase in probability as the initial energy increased to 11.7 ev, the highest studied. At this energy the excitation cross section came out to be about $\frac{1}{40}$ of the total cross section. Since the latter is known to be $12\pi a_0^2$ (see Fig. 11) the former must be about $0.3\pi a_0^2$.

[1] K. DORSCH and H. KALLMANN: Z. Physik **34**, 389 (1925).
[2] H. RAMIEN: Z. Physik **70**, 353 (1931).

JONES and WHIDDINGTON[1] also observed the energy loss due to excitation in an electron beam, initially homogeneous, after passage through hydrogen. The relative probability of this energy loss was found to reach a maximum at about 16 ev incident energy. It was then about $\frac{1}{4}$ as probable as the 12.6 loss due to excitation of the $C\,^1\Pi_u$ state. The total cross section for the latter process at 16 ev energy is probably much less than $0.3\,\pi a_0^2$, but this does not necessarily mean that there is any inconsistency with RAMIEN'S observations. As the excitation of the C state is an optically allowed transition, the angular distribution of the electrons which excite it is strongly concentrated in the forward direction, even at energies quite close to the threshold. On the other hand, for an excitation which can only take place through electron exchange the angular distribution, at energies near the threshold, is close to isotropic. As JONES and WHIDDINGTON were only observing electrons which had suffered very small deviations, their experiments indicate that the *total* cross section for excitation of the $b\,^3\Sigma_u$ state by 16 ev electrons is actually much bigger than that for excitation of the C state.

The only other experiments which provide evidence on the excitation of the $b\,^3\Sigma_u$ and $a\,^3\Sigma_g$ states involve in some way the study of discharges in hydrogen.

POOLE[2] studied the energy efficiency of atom production in the positive column of a striated glow discharge as a function of the ratio F/p of the field strength F in the column to the gas pressure p. If $f(E)\,dE$ is the number of electrons in the column with energies between E and $E+dE$, the number of molecules dissociated per second is given in terms of the dissociation cross section Q_d by

$$N(2/m)^{\frac{1}{2}} \int\limits_0^\infty Q_d E^{\frac{1}{2}} f(E)\, dE \tag{11.7}$$

where N is the number of molecules per cm³. The energy communicated to the electrons by the field per second is

$$e F u \tag{11.8}$$

where u, the drift velocity, is the mean distance traversed by an electron, of charge e, in the direction of the field, per second. The energy efficiency η, which may be defined by the number of dissociations produced per unit energy input to the electrons, is now given by

$$\eta = \frac{N}{e F u} \left(\frac{2}{m}\right)^{\frac{1}{2}} \int\limits_0^\infty Q_d E^{\frac{1}{2}} f(E)\, dE. \tag{11.9}$$

This formula may be applied in practice provided the gas is predominantly molecular. The power input may be obtained from the discharge current and potential fall down the column and the rate of dissociation by allowing the hydrogen issuing from the discharge to recombine in a continuous flow calorimeter. The ratio μ of these two quantities is not equal in general to η, because the degree of dissociation will usually be high. To allow for this in his experiments POOLE measured μ at a fixed pressure p, for a number of values of the hydrogen flow rate U and power input W. He then extrapolated to find the limit for vanishing W and $1/U$ and this was taken as η.

As F and N may be measured directly and u obtained from previous experiments (see Sect. 13) it is possible to derive a value for the integral in (11.7) for different mean energies of the electrons. If, further, the form of $f(E)$ were known it would be possible to obtain information about Q_d. Unfortunately, in a striated

[1] H. JONES and R. WHIDDINGTON: Phil. Mag. **6**, 889 (1928).

[2] H. POOLE: Proc. Roy. Soc. Lond., Ser. A **163**, 404, 415, 424 (1937).

hydrogen discharge $f(E)$ may be quite complicated and vary with position along the column. Not too much significance should therefore be attached to the fact that, if it is assumed that Q_d is the sum of the cross sections Q_u, Q_g for excitation of the $b\,^3\Sigma_u^+$ and $a\,^3\Sigma_g$ states respectively and that

$$\begin{aligned} Q_u &= 0.45\,\pi a_0^2, & 9.5\ \text{ev} < E < 19\ \text{ev},\\ &= 0, & E > 19\ \text{ev},\\ Q_g &= 0.15\,\pi a_0^2, & 12.0\ \text{ev} > E > 24\ \text{ev},\\ &= 0, & E > 24\ \text{ev} \end{aligned}$$

quite good agreement is obtained with POOLE's observed variation of η with F/p if $f(E)$ is taken to be Maxwellian about the mean energy for the appropriate value of F/p. It does, however, suggest that the estimated values for the cross sections are not too far wrong.

The excitation function for any wavelength band of the continuous spectrum of hydrogen will, apart from cascade effects which lead to population of the $a\,^3\Sigma_g^+$ state from above, be of the same form as that for excitation of that state. No measurements of this kind have been made, though FINKELNBURG[1] did find that the intensity varied with the exciting electron energy in the predicted way. LUNT, MEEK and SMITH[2] investigated the efficiency of excitation of the continuous spectrum as a function of F/p by a steady discharge in streaming hydrogen. This bears the same relation to Q_g as η in (11.9) does to Q_{diss} and the same difficulty arises in analyzing the data to obtain information about the relevant cross section.

COOLIDGE, JAMES and PRESENT[3] worked out the intensity-distribution in the continuous spectrum due to excitation by electrons with energy very close to the threshold energy. By choosing a mean electron energy very close to the threshold one vibrational level is excited with much greater probability than any other. The probability of other levels being excited could then be estimated with sufficient accuracy in terms of the square of the overlap integral for approximate initial and final vibrational wave functions. The intensity distribution is then determined largely by the overlap of the wave function for the predominantly excited $^3\Sigma_g^+$ vibrational level with wave functions for different energies of relative motion of the two atoms in the repulsive $^3\Sigma_u$ state. These were calculated numerically from a $^3\Sigma_u$ potential curve determined by an accurate variational method. The final results agreed very well with observations made by COOLIDGE[4].

Very few measurements of optical excitation functions for other electronic levels have been made. KRUITHOF and ORNSTEIN[5] measured the excitation functions of the lines arising from the $3d\sigma\,^1\Sigma_g^+ \rightarrow B$ and $4p\pi\,^3\Pi_u \rightarrow a\,^3\Sigma_g^+$ transitions. As expected they had a similar form to those for transitions from ^1D and ^3P upper states, respectively, in helium (see Vol. XXXVI, p. 327 of this Encyclopedia).

E. Excitation of molecular vibration and rotation.

12. Methods of observation. The energy losses suffered by electrons which excite nuclear motions only are so small that special techniques must usually be employed to study the cross sections for inelastic collisions of this kind.

---

[1] H. FINKELNBURG: Z. Physik **62**, 624 (1930).

[2] R. W. LUNT, C. A. MEEK and E. C. SMITH: Proc. Roy. Soc. Lond., Ser. A **158**, 729 (1937).

[3] A. S. COOLIDGE, H. JAMES and R. PRESENT: J. Chem. Phys. **4**, 187 (1936).

[4] A. S. COOLIDGE: Phys. Rev. **65**, 236 (1944).

[5] A. A. KRUITHOF and L. S. ORNSTEIN: Physica, Haag **2**, 611 (1935).

It is true that RAMIEN[1] and HARRIES[2] have applied the diffusion method of HERTZ to determine probabilities for excitation of the first vibrational levels of hydrogen and nitrogen, but all other experimental information comes either from the study of the motion of a swarm of electrons in the gas under the action of a uniform electric field at different gas pressures or, quite recently, from microwave interaction effects in a plasma formed in the gas. We shall begin by describing the principles on which these methods are based.

13. Diffusion of an electron swarm in a gas under a constant electric field. We consider a swarm of electrons diffusing in a gas at pressure p under the influence of a uniform electric field F in the x-direction. When a steady state has been reached so that the energy gained by the swarm from the field in diffusing a distance δx in the direction of the field is equal to that lost by collision with the gas molecules, the mean kinetic energy ε and velocity u of drift in the x-direction will be constants independent of time and position. It is easy to see in general terms how these two quantities may be related to the mean free path l of the electrons in the gas and to the mean fractional energy loss λ suffered by an electron on collision with a gas molecule.

We first introduce the root mean square velocity c of agitation of the electrons, defined by

$$\tfrac{1}{2} m c^2 = \varepsilon \tag{13.1}$$

where m is the electron mass. The mean time between collisions of one electron with a gas molecule will be given, apart from a factor of order unity, by l/c so that, it all directions of motion after collision are equally probable, the distance traversed in the x-direction during this interval will be $u l/c$. Since the acceleration of the electron in this direction during the interval will be $F e/m$, it follows that

$$\frac{1}{2} \frac{F e}{m} \frac{l^2}{c^2} = \frac{u l}{c}$$

so that

$$l = \frac{2 u c m}{F e}. \tag{13.2}$$

To allow for factors of order unity introduced by the averaging processes, we may write this

$$l = f_1 \frac{u c m}{F e} \tag{13.3}$$

where f_1 is of order unity.

A second equation follows from energy balance considerations. Again ignoring factors of order unity, we have that an electron in diffusing a distance x in the direction of the field will actually have moved, on the average, a distance of order $x c/u$, during which it will have made $c x/u l$ collisions. The energy lost due to these collisions will be $\lambda \varepsilon c x/u l$ so that in the steady state

$$\frac{\lambda \varepsilon c x}{u l} = F e x \tag{13.4}$$

or, since $\varepsilon = \frac{1}{2} m c^2$,

$$\lambda = \frac{2 F e l u}{m c^3}. \tag{13.5}$$

Allowing as before for averaging by introducing a factor f_2 of order unity

$$\lambda = \frac{2 F e l u}{m c^3} f_2. \tag{13.6}$$

[1] H. RAMIEN: Z. Physik **70**, 353 (1931).
[2] W. HARRIES: Z. Physik **42**, 26 (1927).

It follows that, if the factors f_1 and f_2 can be calculated, both l and λ may be determined from measurements of ε and u. f_1 and f_2 depend on the form of the velocity distribution of the electrons. If the mean energy ε is not much greater than that of the gas molecules, this distribution will be nearly Maxwellian and then

$$f_1 = 1.08_5, \quad f_2 = 1.21. \tag{13.7}$$

A second case, which probably represents a good approximation when λ is not too great, prevails at higher mean energies when the electrons only suffer elastic collisions and the elastic cross section is independent of electron velocity. This is the distribution obtained by DRUYVESTYN and it gives

$$f_1 = 1.18, \quad f_2 = 1.05. \tag{13.8}$$

In any case of interest in the present context, neither of these distributions are strictly applicable, but the true distributions will not be very different and the difference between the two sets of values (13.7) and (13.8) shows that the results obtained for l and λ are not at all sensitive to the distribution.

The drift velocity u may be measured directly by shutter methods in which case the interpretation of the observations is quite independent of the electron velocity distribution. An alternative method first introduced by TOWNSEND[1] and used recently in a modified form by HUXLEY and ZAAZOU[2] depends on the Hall effect produced when a transverse magnetic field is applied to the diffusing swarm. The extraction of the drift velocity from the observations made in their experiments depends on the assumed velocity-distribution, but only to the limited extent involved in the formulae (13.7) and (13.8). Similar considerations apply also to the determination of the mean energy which depends on the measurement of the lateral diffusion of the swarm under appropriate geometrical conditions. The most accurate measurements have been those carried out by HUXLEY and his collaborators employing modern vacuum technique.

14. Wave interaction studies. The principle of these experiments was suggested from the radio-interaction phenomenon known as the Luxemburg effect[3], observed first in 1933 in Holland and in England. Signals received from many medium wave transmitters in Western Europe were found to be modulated by the powerful long-wave transmitter (252 kc) at Luxemburg. The interference producing the modulation takes place in the ionosphere—a result established by several investigations.

The theory of the effect was first given by BAILEY and MARTYN[4]. According to this the interference takes place through a change in collision frequency of the electrons in the ionosphere due in turn to an increase in their temperature by absorption of energy from the Luxemburg radio transmissions. The change in collision frequency changes the absorption of the medium waves and it is through this that the modulation is effected. It is possible to simulate the effect in the laboratory using a discharge plasma in place of the ionosphere and two microwave signals in place respectively of the "wanted" medium wave and "disturbing" long wave radio signals. From an analysis of the observations made in such experiments on the lines of BAILEY and MARTYN's theory, it is possible to obtain the mean fractional energy loss λ per collision of the plasma electrons with the ambient gas molecules.

[1] J. S. TOWNSEND: Electricity in Gases, p. 100. Oxford 1914.
[2] L. G. HUXLEY and A. A. ZAAZON: Proc. Roy. Soc. Lond., Ser. A **196**, 402 (1949).
[3] B. D. TELLEGEN: Nature, Lond. **131**, 840 (1933).
[4] V. A. BAILEY and D. F. MARTYN: Phil. Mag. **18**, 369 (1934).

Let ε and ε_0 be the mean energies of the plasma electrons and gas molecules respectively.

The mean energy lost per second by an electron due to collisions with gas molecules will be $\lambda\nu\varepsilon$ where ν is the collision frequency. Hence if P_e is the mean energy supplied to an electron due to a disturbing microwave field applied to the plasma

$$\frac{d\varepsilon}{dt} + \lambda\nu\varepsilon = P_e. \tag{14.1}$$

We now consider the particular conditions of the microwave experiments by supposing that, in the absence of the disturbing field, the plasma electrons are in thermal equilibrium with the gas molecules so that $\varepsilon = \varepsilon_0$. Since λ vanishes in this limit it is convenient to write

$$\lambda = G\left(1 - \frac{\varepsilon_0}{\varepsilon}\right) \tag{14.2}$$

where G is a quantity which may depend on ε, but is such that $\lambda \to 0$ as $\varepsilon \to \varepsilon_0$.

If the disturbing electric field is given by $E_0 e^{i\omega t}$ the total rate at which power is dissipated in the plasma is given by

$$P_e = \tfrac{1}{2}\int_V |E_0|^2 \sigma_r \, dV \tag{14.3}$$

where V is the volume occupied by the plasma and σ_r is the real component of conductivity of the plasma. In terms of the collision frequency the complex conductivity may be written

$$\sigma = \sigma_r + i\sigma_i = \frac{n_e e^2}{m} \frac{\nu - i\omega}{\nu^2 + \omega^2}, \tag{14.4}$$

where n_e is the electron concentration. If the disturbing field is of sufficiently high frequency $\nu \ll \omega$ and

$$\sigma_r = \frac{n_e e^2}{m} \frac{\nu}{\omega^2}. \tag{14.5}$$

On substitution in (14.3), assuming that n_e and ν do not vary appreciably throughout the plasma

$$P_e = \frac{1}{2} \frac{e^2}{m} \frac{\nu}{\omega^2} |E_0|^2. \tag{14.6}$$

To calculate the collision frequency ν as a function of t from (14.1) it remains only to consider the functional relation between ν and ε. The simplest assumption to make is that the collision cross section Q is independent of velocity. We then have

$$\nu = f Q n_0 (2\varepsilon/m)^{\frac{1}{2}} \tag{14.7}$$

where n_0 is the number of gas molecules per cm^3 and f is a factor arising from averaging over the electron velocity distribution. If this is Maxwellian as will be so if the electrons are in thermal equilibrium with the gas molecules in the absence of any disturbing field, it may be shown that $f = \frac{4}{3}$. We now have from (14.1) the equation for ν

$$\frac{d\nu}{dt} + \frac{1}{2} G\nu^2 = \frac{1}{2} G(\nu_0^2 + \nu_1^2) \tag{14.8}$$

where ν_0 is the collision frequency in the absence of the disturbing field and

$$\nu_1^2 = \frac{16 q n_0}{3\sqrt{3\pi}} \frac{e^2}{G m \omega^2} |E|^2. \tag{14.9}$$

When equilibrium has been reached in the presence of the field the new value ν' of the collision frequency is given by

$$\nu'^2 = \nu_0^2 + \nu_1^2. \tag{14.10}$$

If the field is suddenly removed after this equilibrium has been established, integration of (14.8) shows that ν changes with time t after cut-off according to the relation

$$\nu(t) = \nu_0 \left[\left(\frac{\nu' + \nu_0}{\nu' - \nu_0}\right) e^{G\nu_0 t} + 1\right] \Big/ \left[\left(\frac{\nu' + \nu_0}{\nu' - \nu_0}\right) e^{G\nu_0 t} - 1\right]. \tag{14.11}$$

If the difference between ν' and ν_0 is kept small this gives approximately

$$\frac{\nu - \nu_0}{\nu' - \nu_0} \sim e^{-G\nu_0 t}. \tag{14.12}$$

It may similarly be shown that, for sudden application of the field, $\nu - \nu'$ is also associated with a time constant $G\nu_0$. Hence, by observation of the time constant of the plasma when the disturbing field is suddenly established or cut-off, G may be obtained if ν_0 is known.

It is possible by further microwave observations to determine ν_0. If the plasma fills a section of wave guide several wave lengths long the propagation constant γ for guided waves of angular frequency ω passing through the plasma is given by

$$\gamma^2 = \left(\frac{2\pi}{\lambda_g}\right)^2 (\alpha + i\beta) \tag{14.13}$$

where

$$\left.\begin{aligned} \alpha &= -1 + \frac{A}{\nu^2 + \omega^2}\left(\frac{\lambda_g}{\lambda}\right)^2, \quad \beta = \left(\frac{\lambda_g}{\lambda}\right)^2 \frac{A\nu}{\omega(\nu^2 + \omega^2)}, \\ A &= 64\pi^4 n_e^2 e^4/m^2. \end{aligned}\right\} \tag{14.14}$$

λ_g is the wavelength in the air-filled guide and λ is the free space wavelength. It follows that, by measurement of the real and imaginary parts of the propagation constant, both n_e and ν may be determined. The real part may be obtained from measurement of the absorption of the waves within the plasma and the imaginary part from the phase shift produced by the plasma.

This technique was first applied by ANDERSON and GOLDSTEIN[1] to helium which provides a useful check of the method. In this case G is known to have the value appropriate to purely elastic collisions, so that measurement of the time constant $G\nu_0$ gives an independent value for ν_0. The plasma was produced in a thin-walled cylindrical pyrex tube 152 cm long and 1.85 cm inside diameter, located coaxially in a square brass wave guide with inside dimensions 2.07 cm of a side. Breakdown of the gas within the tube was achieved by application of a high voltage d.c. pulse of 2 microseconds duration to a hot cathode and anode external to the wave guide, the repetition frequency being 25 pulses per second. Microwave signals were propagated during each afterglow period at times when the electrons had reached thermal equilibrium with the gas molecules. The disturbing wave was a square pulse of microwave energy of frequency 8600 Mc/sec of 1 to 100 microseconds' duration, while the "wanted" wave was a continuous wave at 9600 Mc/sec with a power level lower by at least 20 db.

Fig. 20 illustrates the type of oscilloscope record made of the interaction between the waves. The sine wave 1 *a* is a 100 kc/sec timing wave. 1 *b* is the crystal-detected record of the disturbing wave pulse, showing that the disturbing wave pulse is present for about 20 microseconds. 2a is the record of the

[1] J. M. ANDERSON and L. GOLDSTEIN: Phys. Rev. **100**, 1037 (1955).

"wanted" wave transmitted through the plasma in the absence of any disturbance and 2b when this disturbance is present, the downward deflection indicating an increase in attenuation. From the time constants of the rise and fall of the interaction the collision frequency may be obtained and compared with that given via the formula (14.4) from phase shift and loss measurements.

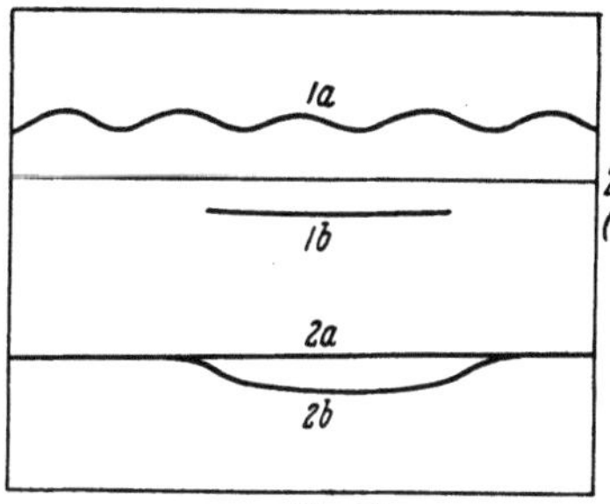

Fig. 20.

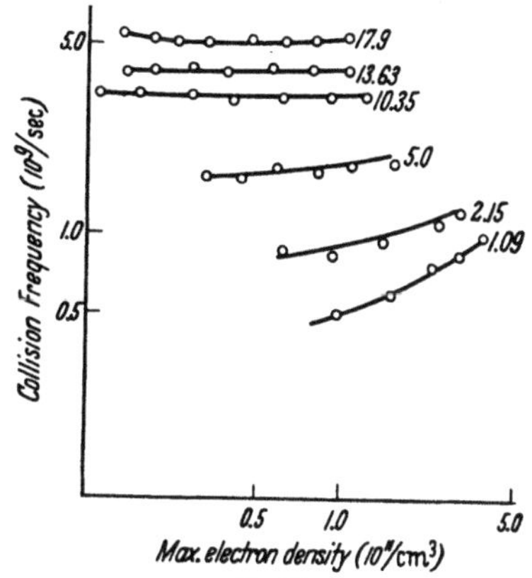

Fig. 21.

Fig. 20. Illustrating the forms of the oscillograph records obtained in microwave interaction experiments. *1a* 100 kc/sec timing waves; *1b* pulse due to disturbing wave pulse (8600 Mc/sec); *2a* transmitted wanted wave (9400 Mc/sec) in absence of disturbing wave; *2b* transmitted wanted wave in presence of disturbing pulse.

Fig. 21. Observed collision frequency of thermal electrons in helium as a function of electron concentration at different gas pressures (indicated in mm Hg by the numbers associated with each curve). ○ exptl. points.

Fig. 21 illustrates the observed collision frequency as a function of helium pressure. It will be seen that good agreement is obtained between values derived from the interaction time constant and from phase shift and loss measurements.

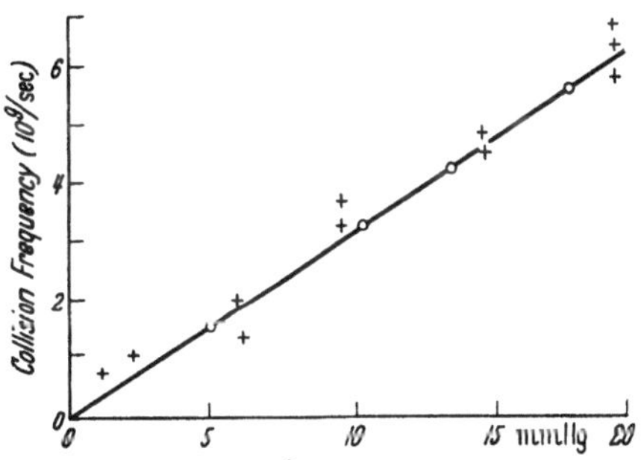

Fig. 22.

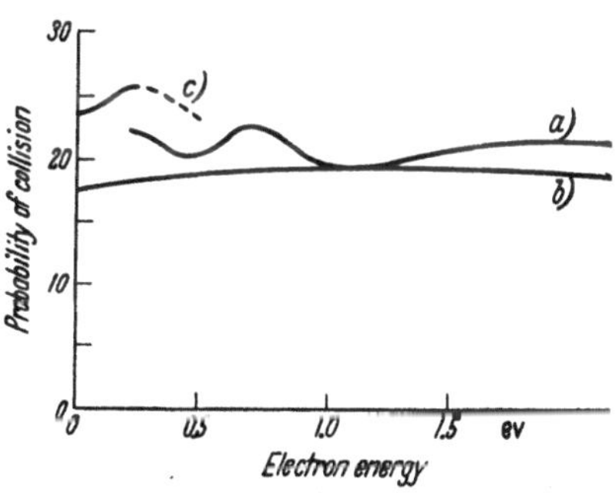

Fig. 23.

Fig. 22. Observed collision frequency of thermal electrons in helium as a function of gas pressure. ○ from measurements of the plasma microwave conductivity; + from wave interaction studies.

Fig. 23. Comparison of momentum loss cross sections for collisions of slow electrons in helium observed by different methods: (a) observed by NORMAND using the Ramsauer method; (b) observed by GOULD and BROWN; (c) observed by ANDERSON and GOLDSTEIN. In case (a) the cross sections are for electrons of homogeneous energy whereas in cases (b) and (c) the energy refers to the mean energy of the electrons which have an approximately Maxwellian distribution. For unit probability of collision the cross section $= \pi a_0^2$.

Furthermore, the observed results exhibit the required proportionality to pressure. At low pressures, however, the observed values show a variation with maximum electron density, as will be seen from Fig. 21. This arises through the contribution from electron-ion collisions which have been ignored in the preceding analysis. ANDERSON and GOLDSTEIN[1] have shown in considerable detail

[1] J.M. ANDERSON and L. GOLDSTEIN: Phys. Rev. **100**, 1037 (1955).

that this interpretation is valid. For present purposes, however, it is only necessary to confine attention to gas pressures high enough for electron-ion collisions to be neglected.

It is possible to use wave interaction to obtain information about the variation of mean collision frequency with mean electron energy, the disturbing wave being used to raise the electron temperature. Comparison of the variation of collision cross section with mean electron energy observed by the interaction method[1] with that observed by Gould and Brown[2], using a microwave method based on a somewhat different principle, is illustrated in Fig. 23. The agreement is reasonably good and both sets of results are not inconsistent with the observations made by the Ramsauer technique at high electron energies. This suggests that the interaction technique is sound in principle (see, however, the discussion below of the results for nitrogen).

15. Experimental results. *α) Hydrogen and deuterium.* Ramien[3], using the Hertz diffusion method, found an energy loss for electrons in hydrogen which could be interpreted as due to excitation of one vibrational quantum (0.54 ev), the chance per collision decreasing from 0.03 for 3.5 volt electrons to 0.02 for 7 volt electrons. There was no evidence of energy losses due to excitation of higher vibrational quanta. In particular, no loss corresponding to the dissociation energy (4.4 ev) of the molecule was observed.

The only other evidence concerning energy losses of very slow electrons in hydrogen comes by comparing the careful measurements of Crompton and Sutton[4] of the mean energies of electrons diffusing through the gas under the action of a uniform electric field, with the drift velocity measurements of Nielsen and Bradbury[5]. Table 6 gives the desired values of the mean fractional energy loss λ per collision as a function of mean electron energy. Two sets of values are given depending on whether the electron velocity distribution is taken to be Maxwellian or as given by Druyvestyn. The difference is not important for present purposes.

These data extend down to electron energies only some 20 to 30% greater than thermal. It will be seen that at the lowest mean energy λ is not much greater than for purely elastic collisions, but it soon rises to well above this

Table 6. *Observed mean fractional energy loss λ per collision of electrons in nitrogen.* M and D refer to values obtained using Maxwell and Druyvestyn distributions respectively.

Mean electron energy (ev)		$\lambda\times10^4$		Mean electron energy (ev)		$\lambda\times10^4$	
M	D	M	D	M	D	M	D
0.116	0.101	3.33	4.11	1.002	0.879	5.27	6.51
0.181	0.158	3.11	3.84	1.092	0.958	6.73	8.32
0.298	0.261	3.07	3.78	1.174	1.028	8.20	10.1
0.406	0.354	2.90	3.58	1.226	1.073	9.05	11.2
0.507	0.443	3.03	3.74	1.390	1.215	14.5	18.0
0.604	0.529	3.10	3.83	1.509	1.319	20.6	25.4
0.682	0.596	3.28	4.05	1.602	1.401	26.9	33.2
0.753	0.660	3.52	4.35	1.874	1.639	66.9	82.6
0.823	0.723	3.86	4.76	2.042	1.788	114	141
0.872	0.764	4.20	5.18	2.221	1.945	166	205
0.920	0.805	4.52	5.58				

[1] J. M. Anderson and L. Goldstein: Phys. Rev. **102**, 933 (1956).
[2] L. Gould and S. C. Brown: Phys. Rev. **95**, 897 (1954).
[3] H. Ramien: Z. Physik **70**, 353 (1931).
[4] R. W. Crompton and D. J. Sutton: Proc. Roy. Soc. Lond., Ser. A **215**, 467 (1952).
[5] R. A. Nielsen and N. Bradbury: Phys. Rev. **49**, 388 (1936).

value as the mean energy increases. There seems to be little doubt that this is due largely to rotational excitation.

An interesting experimental comparison of hydrogen and deuterium has been made by Miss HALL[1]. She found as expected that the cross sections of the two molecules towards electrons of a particular velocity are the same, but the mean fractional energy loss is twice as large for H_2 as for D_2. This would be expected for elastic losses and for rotational excitation.

β) Nitrogen. Evidence of single vibrational excitation in nitrogen has been obtained by HARRIES[2] using the Hertz diffusion method. He found that the chance of such excitation is about 1/79 per collision with 5.2 ev electrons.

The mean energy loss λ per collision has been measured by the swarm method down to mean electron energies about three times thermal by CROMPTON and

Table 7. *Observed mean fractional energy loss λ per collision for electrons in hydrogen.* M and D refer to values obtained assuming Maxwell and Druyvestyn distributions respectively.

Mean electron energy (ev)		$\lambda\times10^4$		Mean electron energy (ev)		$\lambda\times10^4$	
M	D	M	D	M	D	M	D
0.0503	0.0440	5.37	6.63	0.384	0.336	19.9	24.5
0.0645	0.0563	10.6	13.1	0.451	0.395	22.1	27.3
0.0950	0.0831	16.0	19.9	0.503	0.440	24.5	30.3
0.128	0.116	18.1	22.3	0.540	0.473	25.2	31.1
0.160	0.139	18.2	22.5	0.708	0.619	30.6	37.8
0.191	0.167	18.6	23.0	0.846	0.741	35.8	44.2
0.222	0.194	19.2	23.7	0.969	0.846	41.4	51.0
0.254	0.222	19.1	23.6	1.595	1.40	61.3	75.6
0.286	0.250	18.9	23.4	2.31	2.02	81.0	100
0.312	0.273	18.8	23.2	2.94	2.57	111	138
0.337	0.294	19.3	23.8				

SUTTON[3], and for electrons with energies only slightly in excess of thermal by ANDERSON and GOLDSTEIN[4] using the microwave interaction method. Table 7 summarizes the results of CROMPTON and SUTTON, as for hydrogen. Once again λ is much greater than for purely elastic collisions showing that rotational excitation is occurring.

ANDERSON and GOLDSTEIN found from the time constant of wave interaction effects $G\nu$ [in the notation of Eq. (14.12)] for thermal electrons $=6.4\times10^5$ sec^{-1}. Measurement of absorption and phase change due to the nitrogen plasma gave $\nu=6.6\times10^8$/sec/mm Hg and hence $G=9.8\times10^{-4}$. This is to be compared with the value 0.52×10^{-4} for purely elastic collisions. Extrapolation of the results of CROMPTON and SUTTON to thermal energies is a little uncertain, but suggests a value for G between 30 and 50×10^{-4}, substantially greater than that obtained by the microwave method. On the other hand the value of ν obtained by ANDERSON and GOLDSTEIN is about four times larger than that obtained by PHELPS, FUNDINGSLAND and BROWN[5] using a different microwave method. This latter value is not inconsistent with the extrapolation of CROMPTON and SUTTON'S measurements of mean free path and if it is assumed correct, the discrepancy

[1] B. HALL: Austral. J. Physics **8**, 468 (1955).
[2] W. HARRIES: Z. Physik **42**, 26 (1927).
[3] R.W. CROMPTON and D.J. SUTTON: Proc. Roy. Soc. Lond., Ser. A **215**, 467 (1952).
[4] J.M. ANDERSON and L. GOLDSTEIN: Phys. Rev. **102**, 388 (1956).
[5] A.V. PHELPS, A. FUNDINGSLAND and S.C. BROWN: Phys. Rev. **84**, 559 (1951).

between the estimates of G from microwave and swarm experiments largely disappears. It is clear that further experiment is needed to clear the matter up thoroughly. One point of importance which was checked by ANDERSON and GOLDSTEIN was the absence of any appreciable dissociation of the nitrogen in the plasma. This followed because no change of pressure was observed when the discharge was struck.

Table 8. *Observed mean fractional energy loss λ per collision of electrons in air.* Maxwell distribution of electron velocity assumed.

Mean electron energy (ev)	$\lambda \times 10^4$	Mean electron energy (ev)	$\lambda \times 10^4$	Mean electron energy (ev)	$\lambda \times 10^4$
0.149	6.80	0.484	15.7	1.230	23.0
0.220	8.84	0.533	16.0	1.394	29.9
0.268	11.2	0.581	16.0	1.502	38.8
0.309	13.1	0.674	16.0	1.777	91.6
0.350	14.1	0.809	16.2	1.975	154
0.391	15.1	0.920	17.0	2.217	218
0.440	15.5	0.987	17.7		

γ) Oxygen and air. Table 8 gives results obtained by CROMPTON and SUTTON for the mean fractional energy loss λ in air. It will be seen by comparison with Table 6 that λ must be substantially greater for oxygen than for nitrogen.

Because of its interest in the interpretation of radio wave interaction effects, considerable attention has been devoted, particularly by HUXLEY and his collaborators[1], to a study of the energy loss in air. The measurements, by the swarm method, cannot be taken down to electron mean energies below about four times thermal and interpolation to the thermal limit is difficult. This is apparent from Fig. 24 which illustrates λk as a function of k where $k = \varepsilon/\varepsilon_0$ is the ratio of the mean energy to thermal energy. Writing (14.2) in the form

$$\lambda k = G(k-1) \tag{15.1}$$

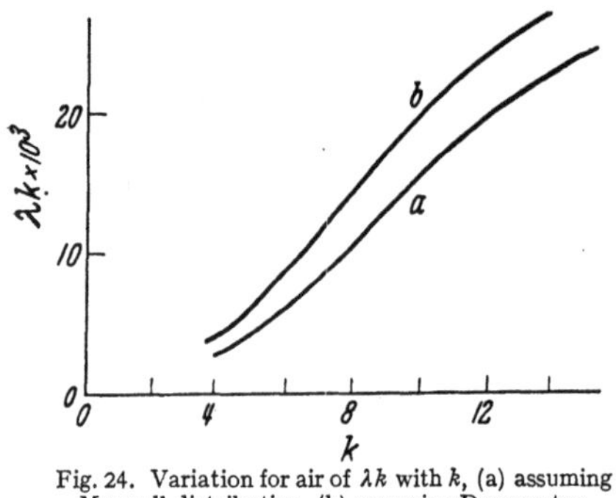

Fig. 24. Variation for air of λk with k, (a) assuming Maxwell distribution, (b) assuming Druyvestyn distribution.

it is seen that, if G is independent of k, the graph of λk should be a straight line passing through $k = 1$ with slope equal to G. Reference to Fig. 24 shows that this is not demanded by the observed data. The best constant value for G is given by CROMPTON, HUXLEY and SUTTON as 9.5×10^{-4}, but the same authors point out that the data are perhaps better fitted by taking

$$\lambda k = H \frac{(k-1)^2}{k} \tag{15.2}$$

where H is a constant.

δ) Other gases. Table 9 summarizes values of λ at different mean electron energies for several other gases. These have been obtained by earlier swarm experiments and are therefore somewhat less reliable than the results given in Tables 6 to 8. It is unlikely, however, that they are seriously in error. The very high values obtained for gases with heteronuclear molecules are noteworthy.

[1] R. W. CROMPTON, L. G. H. HUXLEY and D. J. SUTTON: Proc. Roy. Soc. Lond., Ser. A **218**, 507 (1953). — L. G. H. HUXLEY: Proc. Roy. Soc. Lond., Ser. A **218**, 520 (1953).

Table 9. *Observed mean fractional energy loss λ per collision of electrons with various gas molecules.*

Mean electron energy (ev)	$\lambda \times 10^4$					
	CO	NO	HCl	N_2O	CO_2	NH_3
0.1	45	—	250	600	360	285
0.2	70	110	600	1500	580	420
0.4	89	390	450	1300	650	370
0.6	70	450	350	1100	630	330
0.8	54	380	290	1000	600	305
1.0	62	320	—	900	575	280
1.2	75	275	—	880	560	255
1.4	110	275	—	860	550	240
1.6	150	280	—	800	540	220
1.8	195	320	—	740	535	215
2.0	250	—	—	690	535	230
2.5	360	—	—	600	515	—
3.0	470	—	—	540	—	—
$2m/M \times 10^4$	0.39	0.36	0.30	0.25	0.25	0.64

16. Theory of vibrational and rotational excitation. The theory of excitation of nuclear motion by electrons is still far from satisfactory and has been confined to the use of BORN's approximation.

We refer to the formula (9.4) of Sect. 9 which gives, according to BORN's first approximation, the differential cross section for an inelastic collision of an electron with a molecule in which the molecular quantum numbers change from $nvJM$ to $n'v'J'M'$. In that section attention was confined to transitions in which there was a change of the orbital electronic quantum numbers n. If there is no such change then, according to the approximation (9.8) there would be no vibrational excitation. This shows that the probability of such excitation, when there is no electronic excitation, will certainly be small provided BORN's first approximation is valid. To obtain an expression for it we proceed as follows.

In the notation of Sect. 9 we return to the formula (9.4) but write $M(R, \Theta, \Phi)$ in the form

$$M(R, \Theta, \Phi) = \frac{2\pi m}{h^2}\left(\frac{k'}{k}\right)^{\frac{1}{2}} \int V(\boldsymbol{r}, R)\, \mathrm{e}^{i(\boldsymbol{k}-\boldsymbol{k}')\cdot\boldsymbol{r}}\, d^3\boldsymbol{r} \tag{16.1}$$

where

$$V(\boldsymbol{r}, R) = e^2 \int \sum_m \frac{1}{|\boldsymbol{r}-\boldsymbol{r}_m|} |\psi(n\,|\,\boldsymbol{r}_m, R)|^2\, d^3\boldsymbol{r}_m \tag{16.2}$$

is the mean interaction energy of the incident electron with the molecule in its ground electron state, for fixed R. For the calculation of cross sections for vibrational excitation it is now usually assumed that $V(\boldsymbol{r}, R)$ can be represented as the sum of two terms depending respectively on the distance of the electron from each nucleus. Thus for a homonuclear molecule we write

$$V(\boldsymbol{r}, R) = U(|\boldsymbol{r} + \tfrac{1}{2}\boldsymbol{R}|) + U(|\boldsymbol{r} - \tfrac{1}{2}\boldsymbol{R}|). \tag{16.3}$$

With this approximation M takes the form

$$M(R, \Theta, \Phi) = \frac{4\pi m e^2}{h^2}\left(\frac{k'}{k}\right)^{\frac{1}{2}} \cos\frac{1}{2}(\boldsymbol{k}-\boldsymbol{k}')\cdot\boldsymbol{R} \int U(s)\, \mathrm{e}^{i\boldsymbol{k}(-\boldsymbol{k}')\cdot\boldsymbol{s}}\, d^3\boldsymbol{s}. \tag{16.4}$$

The relative probability of a particular $vJM \to v'J'M'$ transition is therefore proportional to

$$\left|\int \Phi(v\,|\,R)\,\chi(J, M\,|\,\Theta, \Phi) \cos\{\tfrac{1}{2}(\boldsymbol{k}-\boldsymbol{k}')\cdot\boldsymbol{R}\}\,\varphi(v'\,|\,R)\,\chi^*(J', M'\,|\,\Theta, \Phi)\, d^3\boldsymbol{R}\right|^2. \tag{16.5}$$

Calculations using this approximation have been carried out by MASSEY[1], WU[2] and MORSE[3]. The latter author points out that the rate of energy loss by nuclear excitation should not be much greater than in purely elastic collisions in contradiction to the experimental data. It is necessary therefore to examine the matter further.

Referring first to the problem of vibrational excitation it is to be noted that for low energies of impact the cosine in (16.5) is very close to unity for $R \approx R_0$, the equilibrium separation, so that the integral in (16.5) becomes very small due to the orthogonality of the nuclear wave functions. Under these conditions it is not possible to neglect the variation of U in (16.3) with R. If we write

$$U = U(R_0) + \left(\frac{\partial U}{\partial R}\right)_{R=R_0} (R - R_0) \tag{16.6}$$

we obtain in the expression for M an additional term

$$\frac{4\pi m e^2}{h^2} \left(\frac{k'}{k}\right)^{\frac{1}{2}} 2(R - R_0) \cos\left\{\frac{1}{2}(\boldsymbol{k} - \boldsymbol{k}') \cdot \boldsymbol{R}\right\} \int \left(\frac{\partial U}{\partial R}\right)_{R=R_0} e^{i(\boldsymbol{k}-\boldsymbol{k}')\cdot \boldsymbol{s}}\, d^3\boldsymbol{s}. \tag{16.7}$$

Allowance for this term was first made by MASSEY[1] in his calculations on the excitation of vibration in molecular hydrogen. His work has been extended by CARSON[4], who made the same assumptions concerning U. This was taken to be

$$U(s) = s^{-1}(1 + Z s)\, e^{-2Zs}, \tag{16.8}$$

with Z an effective nuclear charge depending on R so that

$$Z = Z_0 + (R - R_0) \left(\frac{\partial Z}{\partial R}\right)_{R=R_0}. \tag{16.9}$$

From variational calculations of nuclear wave functions $Z_0 = 1.166$ and $(\partial Z/\partial R)_{R=R_0} = -0.23/a_0$, R_0 being $1.40 a_0$.

Fig. 25 illustrates the results of CARSON's calculations. Although BORN's first approximation is of doubtful validity for the estimation of absolute cross sections for electrons with energies less than about 100 ev, the relative probabilities of different types of collision should be rather more reliable. The most significant part of CARSON's calculation is therefore the ratio of his inelastic to his elastic cross sections. Even allowing for the effect of the term in $\partial Z/\partial R$ which is very important at low energies, the chance of vibrational excitation rarely exceeds 0.1% which is an order of magnitude less than that found by RAMIEN[5] using the Hertz diffusion method. This is somewhat surprising, but it is not impossible that BORN's first approximation is unreliable even for the estimation of relative magnitudes of cross sections at these low energies.

Turning now to the excitation of rotation, we again find a large discrepancy, CARSON's calculations (Fig. 26) being in agreement with those of MORSE in predicting that rotational excitation will not increase the mean energy loss much above that given by purely elastic loss. The discrepancy is even more marked for nitrogen and far more so for the heteronuclear molecules.

The source of this discrepancy may be revealed as follows. In order to excite the molecular rotation an incident electron must change its angular momentum.

[1] H. S. W. MASSEY: Trans. Faraday Soc. **31**, 556 (1935).
[2] TA-YOU WU: Phys. Rev. **71**, 111 (1947).
[3] P. M. MORSE: Phys. Rev. **90**, 51 (1953).
[4] T. R. CARSON: Proc. Phys. Soc. Lond. A **67**, 908 (1954).
[5] H. RAMIEN: Z. Physik **70**, 353 (1931).

Thus an s incident electron must emerge as a p or d electron. At low energies, however, the probability of finding a p or d electron near either atom of the molecule is very small, so that a low probability of the transition is expected. If, however, the molecule has a permanent dipole moment, its interaction with the electron is of long range and will be appreciable even at the large distances from the molecule at which the p or d electron wave functions are close to their first maxima. In classical terms, if a long range interaction exists, electrons of sufficiently great angular momentum to excite nuclear rotation can interact with sufficient strength to produce this excitation. The assumption (16.3) with the functions U of the general form expected for interaction of an electron with an atom excludes the possibility of any multipolar interaction at large distances from the molecule and so gives too small a probability of rotational excitation by slow electrons.

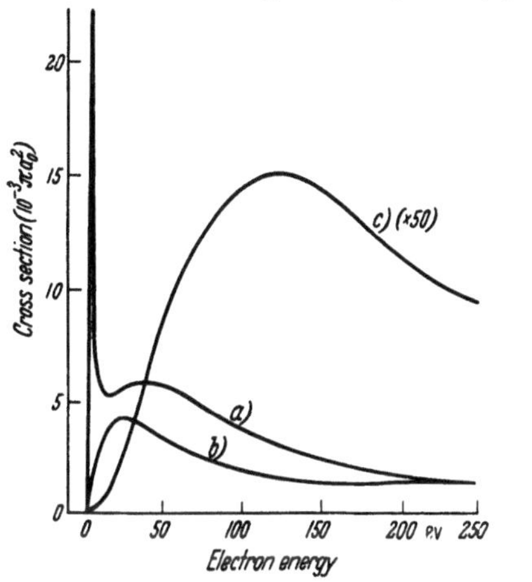

Fig. 25. Calculated cross sections for excitation of the first excited vibrational state from the ground state for H_2. (a) no associated rotational excitation; (b) rotational quantum number J changes from 0 to 2; (c) rotational quantum number J changes from 0 to 4.

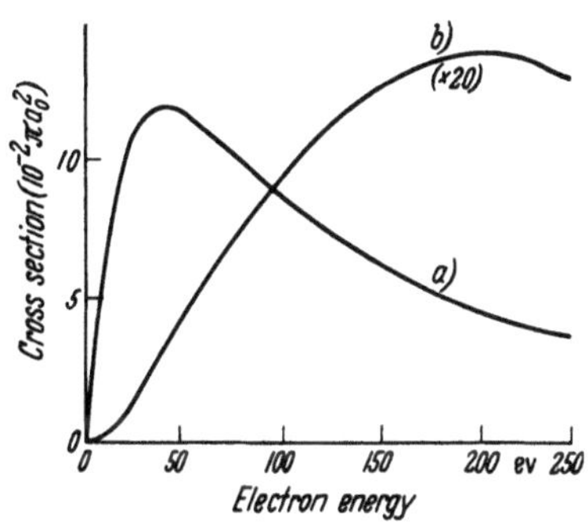

Fig. 26. Calculated cross section for excitation of rotation by electron impact in hydrogen (quadrupole effects ignored). (a) rotational quantum number J changes from 0 to 2; (b) rotational quantum number J changes from 0 to 4.

These possibilities were first pointed out by MASSEY[1], who showed that the cross section for excitation of rotation in a dipolar molecule can be quite large. This probably explains qualitatively the large values of λ found for heteronuclear molecules (Table 9). For homonuclear molecules there is naturally no permanent dipole, but GERJUOY and STEIN[2] have pointed out that in such cases the effect of the quadrupole moment is quite important. Assuming that, for the present purpose, the molecule may be treated as an electric quadrupole of moment qea_0^2, BORN's first approximation gives for the cross section for excitation of a rotational transition from $J \to J'$:

$$\left.\begin{aligned} Q &= 0, && J' \neq J \pm 2, \\ Q_{JJ'} &= \frac{8\pi q^2 a_0^2}{15}\,\frac{k'}{k}\,f(J), && J' = J \pm 2, \\ \text{with}\qquad f(J) &= \frac{(J+2)(J+1)}{(2J+3)(2J+1)}, && J' = J+2, \\ &= \frac{J(J-1)}{(2J-1)(2J+1)}, && J' = J-2. \end{aligned}\right\} \tag{16.10}$$

[1] H. S. W. MASSEY: Proc. Cambridge Phil. Soc. **28**, 99 (1932).
[2] E. GERJUOY and S. STEIN: Phys. Rev. **97**, 1671 (1955).

The rate at which an electron of velocity v and energy ε loses energy in rotational excitation is then given by

$$\frac{d\varepsilon}{dt} = v \sum_J n_J \left[Q_{J,J+2}(E_{J+2} - E_J) - Q_{J,J-2}(E_J - E_{J-2})\right], \tag{16.11}$$

where n_J is the number of molecules per cm³ with rotational quantum number J and rotational energy $E_J \approx BJ(J+1)$. If k'/k may be taken as unity, this gives

$$\frac{d\varepsilon}{dt} = \frac{32\pi q^2 a_0^2}{15} n B v \tag{16.12}$$

where n is the total number of molecules per cm³. The mean fractional energy loss λ per collision is now given by

$$\frac{1}{\varepsilon} \frac{d\varepsilon}{dt} \frac{1}{n Q v} + \frac{2m}{M}, \tag{16.13}$$

where Q is the total collision cross section. Hence

$$\lambda = \frac{2m}{M} + \frac{32\pi q^2 a_0^2}{15} \frac{B}{Q\bar{\varepsilon}}. \tag{16.14}$$

For N_2, $B = 0.249 \times 10^{-3}$ ev and $|q| = 0.96$. Using for Q the value $4.8\,\pi\, a_0^2$ obtained by PHELPS, FUNDINGSLAND and BROWN[1], we obtain for λ, in units of $2m/M$ as a function of $\bar{\varepsilon}$, the curve A of Fig. 27. In obtaining this curve allowance has been made for the difference between k' and k for very low energy collisions. It will be seen that the theoretical results are of the same order of magnitude as the observed and are adequate to explain the tenfold increase of λ above the value expected for purely elastic losses. Similar agreement is obtained with the microwave interaction measurements of G by ANDERSON and GOLDSTEIN[2], provided it is assumed that the collision frequency ν is as given from the constant cross section of $4.8\,\pi\, a_0^2$ and not from the much bigger cross section which they measured.

It remains to show that BORN's first approximation gives satisfactory results for the very low energy collisions considered. GERJUOY and STEIN[3] have discussed this in some detail and show that, if the molecular interaction is purely of quadrupole type, the approximation will be better the lower the electron energy. The effect of the inner field is not so clear, but as it alone would give rise to a very low cross section tor rotational excitation, it is not unreasonable to assume that it may be ignored.

Similar considerations should apply also to H_2, but in this case GERJUOY and STEIN[4] had less success. Fig. 28 illustrates a comparison of their calculated values for λ for H_2 with those observed by CROMPTON and SUTTON. The total cross section was taken to be $13.5\,\pi a_0^2$ independent of electron velocity, as measured by VARNERIN[5] using the microwave methods of PHELPS, FUNDINGSLAND and BROWN. It will be seen that the calculated values are definitely too small by more than can he ascribed to assumption of too large a value ıor the total cross section. The reason for this discrepancy is not clear though it may be that BORN's first approximation is not so reliable as the theoretical arguments suggest. It is unlikely that vibrational excitation is sufficiently important. GERJUOY

[1] A. V. PHELPS, A. FUNDINGSLAND and S. C. BROWN: Phys. Rev. **84**, 559 (1951).
[2] T. M. ANDERSON and L. GOLDSTEIN: Phys. Rev. **102**, 388 (1956).
[3] E. GERJUOY and S. STEIN: Phys. Rev. **97**, 1671 (1955).
[4] E. GERJUOY and S. STEIN: Phys. Rev. **98**, 1848 (1955).
[5] L. J. VARNERIN: Phys. Rev. **24**, 563 (1951).

and STEIN have pointed out that evidence concerning the importance of rotational excitation can be obtained by carrying out the swarm experiments in normal hydrogen, deuterium and parahydrogen at, say, liquid air temperatures. This is because at this temperature only a few rotational states are excited and the energy involved in excitation becomes important. No experiments on these lines have yet been carried out.

The theory of GERJUOY and STEIN in the simple form outlined above is only applicable to molecules such as N_2 and H_2 with $^1\Sigma$ electronic ground states. Otherwise there are complications due to coupling between electron spin and nuclear rotation and energy loss may also occur due to excitation of transitions between different spin states.

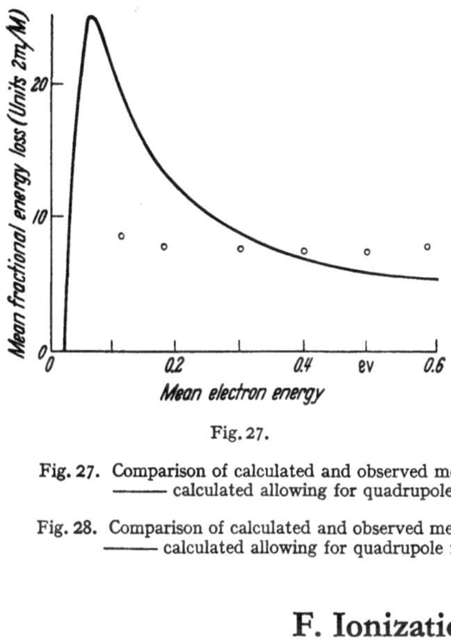

Fig. 27.

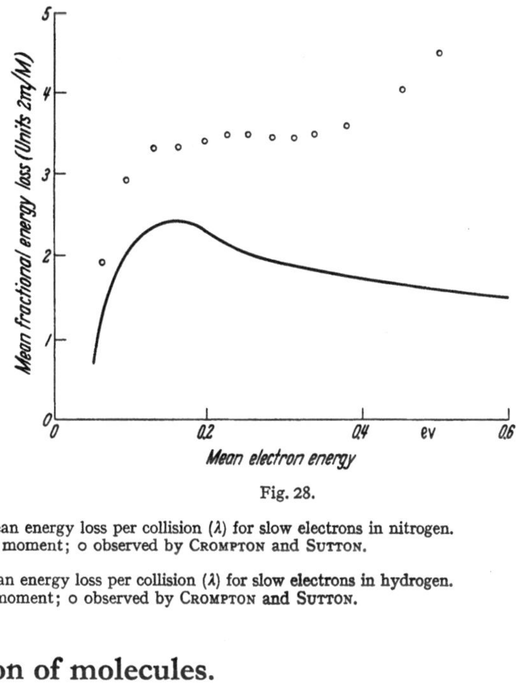

Fig. 28.

Fig. 27. Comparison of calculated and observed mean energy loss per collision (λ) for slow electrons in nitrogen. ——— calculated allowing for quadrupole moment; o observed by CROMPTON and SUTTON.

Fig. 28. Comparison of calculated and observed mean energy loss per collision (λ) for slow electrons in hydrogen. ——— calculated allowing for quadrupole moment; o observed by CROMPTON and SUTTON.

F. Ionization of molecules.

17. Introductory remarks—total ionization cross sections. Nearly all the experimental work on the ionization of molecules which has been carried out deals with the measurement of critical potentials and the application of the data to the study of bond dissociation energies and other molecular constants. It is unfortunate that little work has been carried out on the measurement of cross sections for excitation and ionization in molecular gases during, say, the last ten years. FOX and his collaborators, using their new pulse technique (see Vol. XXXVI, p. 313) are now obtaining cross sections for various gases and their work on SF_6 is discussed in Sect. 36α below. CRAGGS and his collaborators[1] are also making measurements of this kind, particularly for negative ion processes, and have obtained new data in oxygen. Apart from these exceptions, we only have the old, but carefully made, measurements of TATE and SMITH and others. The technique of these measurements is fully described in Vol. XXXVI pp. 309—313. Table 10 summarizes the ionization cross sections obtained for different diatomic molecules over a wide range of electron energies.

[1] J.D. CRAGGS, R. THORBURN and B.A. TOZER: Proc. Roy. Soc. Lond., Ser. A **240**, 473 (1957).

18. Appearance energies. In Sect. 1 the energy relations in dissociative collisions such as

$$\mathrm{AB} + e \to \mathrm{A}^+ + \mathrm{B} + 2e, \tag{18.1}$$

$$\mathrm{AB} + e \to \mathrm{A} + \mathrm{B}^- \tag{18.2}$$

were discussed in detail. It was shown that if $T(\mathrm{A}^+)$, $T(\mathrm{B}^-)$ are the kinetic energies of the ions produced in (18.1) and (18.2) respectively.

$$T(\mathrm{A}^+) = \frac{M_\mathrm{B}}{M_\mathrm{A} + M_\mathrm{B}} \{A - D(\mathrm{AB}) - X - I(\mathrm{A})\}, \tag{18.3}$$

$$T(\mathrm{B}^-) = \frac{M_\mathrm{A}}{M_\mathrm{A} + M_\mathrm{B}} \{A - D(\mathrm{AB}) + E\,A(\mathrm{B}) - X\}, \tag{18.4}$$

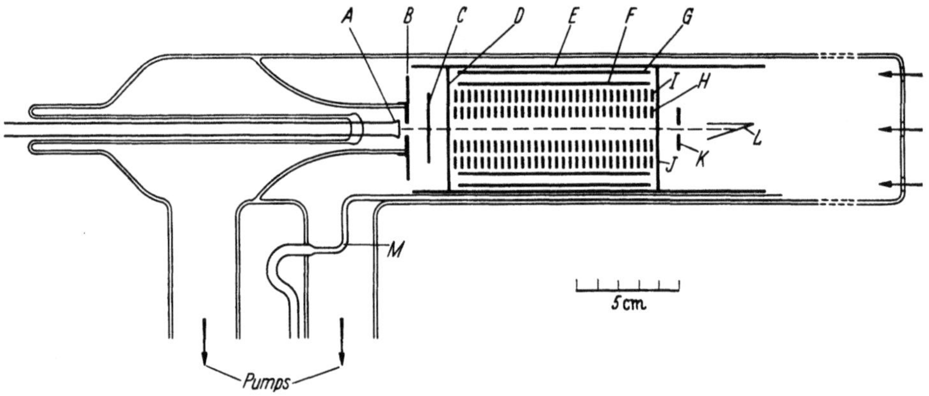

Fig. 29. Illustrating the design of an apparatus of the type introduced by LOZIER.

where A, usually known as the appearance energy, is the minimum energy of the bombarding electrons for which ions of the appropriate kinetic energy are produced. D is the dissociation energy of AB into normal atoms, X the sum of the excitation energies, if any, of the dissociation products, I the ionization potential of A and EA the electron affinity of B. M_A and M_B are the respective masses of A and B. Eqs. (18.3) and (18.4) follow from (1.7) and (1.9) respectively. It will be seen that the appearance potential is a linear function of the kinetic energy of the ion concerned.

Most of the experimental work we shall describe is concerned with the measurement of appearance energies for various dissociation processes in different molecules. The techniques involved in these measurements is described in a book by BARNARD[1], to which reference may be made. Important data concerning dissociation energies and electron affinities may be obtained from the measurements if information is also available about the kinetic energies of the ionized products.

The early ionization chambers used by many investigators for the measurement of ionization potentials and collision cross sections generally incorporated magnetic collimation of the electron beam, guard rings and various other refinements when necessary. A detailed account of these methods is given in Vol. XXXVI of this Encyclopedia. LOZIER[2] adapted and elaborated the normal chamber so that, essentially, a radial filtering system enabled the initial kinetic energies of

[1] G. P. BARNARD: Modern Mass Spectrometry. London: Institute of Physics 1953.
[2] W. W. LOZIER: Phys. Rev. **36**, 1285 (1930).

fragments to be measured. Since we shall often refer to this kind of measurement, the Lozier apparatus will first be described. A brief account will then be given of the use of the mass spectrograph in this work.

19. LOZIER's method. The principle of this method is illustrated in Fig. 29. An electron beam from the filament *A* passes through discs *B*, *C* and *D* and is collimated by a magnetic field of about 200 gauss, so as to pass to a collector *L* through two coaxial sets of circular discs *H* and *I*. Negative ions, for example, produced inside *H* (which is connected to *D*, *E* and *J*) may be collected at *F* if the discs *I* are made 5 to 10 volts positive with respect to *H*. Kinetic energy analysis may then be made by appropriate adjustment of the potential of *F*. The velocity component perpendicular to the electron beam is, essentially, measured. To avoid thermal dissociation effects, the region near the filament is differentially pumped—a procedure often difficult with mass spectrometer sources. The apparatus does not give mass discrimination and therefore can only be used in cases where a single positive or negative ion dominates the spectrum. Operating pressures generally lie between 10^{-4} and 10^{-6} mm Hg.

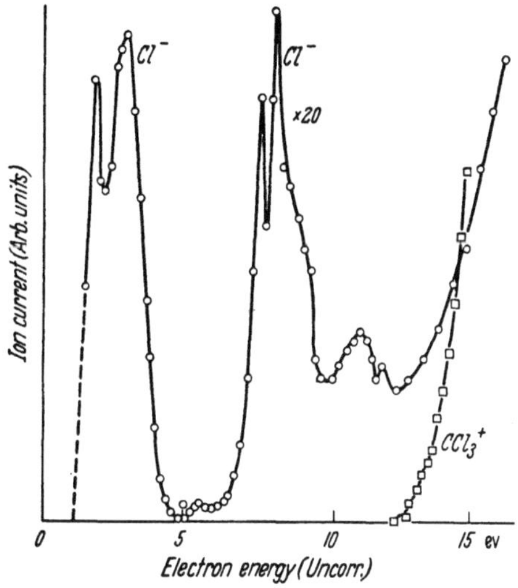

Fig. 30. Variation with electron energy of the principal ion currents, both negative and positive, produced by electron impact in CCl_4.

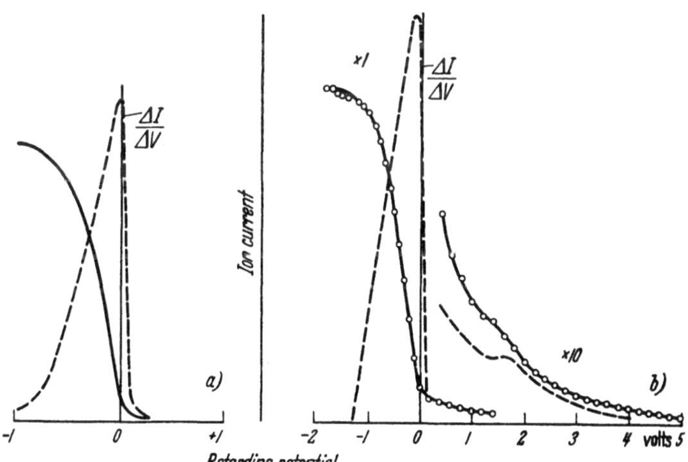

Fig. 31 a and b. Energy distribution of the positive ions produced by impact in oxygen of (a) 20 ev electrons, (b) 35 ev electrons.

Typical results are shown in Figs. 30 and 31. The former gives the appearance potential[1] curve for CCl_4^+ ions from CCl_4 and the latter shows the kinetic energy analysis[2] for O^+ from O_2. Further data are given below ($TiCl_4$, CCl_4, etc.). The

[1] J. MARRIOTT, R. THORBURN and J.D. CRAGGS: Proc. Phys. Soc. Lond. B **67**, 437 (1954).
[2] J. MARIOTT and J.D. CRAGGS: E.R.A. Report No. L/T 308 (1954).

apparatus may clearly be used also for the measurement of cross sections, as shown recently by CRAGGS, THORBURN and TOZER[1].

Some further data taken with oxygen are shown in Fig. 32. The reaction is $O_2 + e \rightarrow O + O^-$. Since in this case $M_A = M_B$ in (18.4) the slope of the line obtained by plotting the minimum kinetic energy against the appearance potential is $\frac{1}{2}$. Taking $I(O_2) = 12.19$ ev, $A(O^-)$ is found to be (4.9 ± 0.1) ev. On extrapolating to zero kinetic energy $A_0(O^-) = 3.2$ ev whilst LOZIER's[2] value, corrected to $I(O_2) = 12.2$ ev is (2.6 ± 0.2) ev, and HAGSTRUM's more recent value[3] is (2.9 ± 0.4) ev. Taking[4] $D(O_2) = 5.08$ ev, the new data, using $A_0(O^-) = 3.2$ ev give $EA(O) = 1.9$ ev. Further discussion of this important quantity is given in Sect. 22.

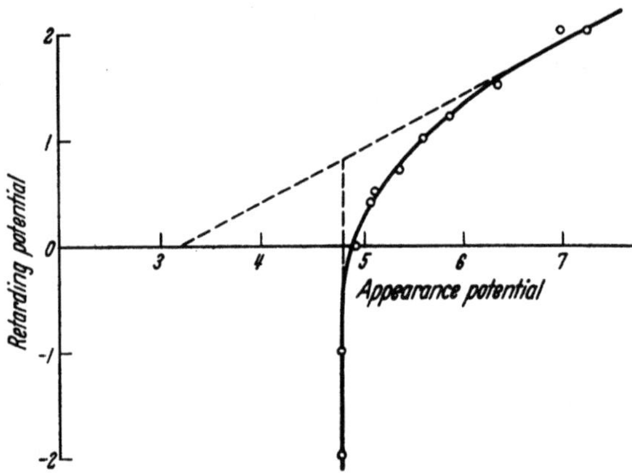

Fig. 32. Illustrating the retarding potential analysis of the kinetic energy of negative ions produced by electron impact in oxygen.

20. Use of the mass spectrometer. The most obvious method of examining the kinetic energies of the ions issuing from a mass spectrometer source is the retarding slit method used by HAGSTRUM[3] and others. Fig. 33 illustrates the electrode arrangement in HAGSTRUM's apparatus, which uses a 90° magnetic deflection analysing system with a retarding electrode at the collecting end of the tube to measure kinetic energies. The details of the ionization chamber will be clear from the figure. CD_1D_2E forms a nearly field-free enclosure at the zero of the potential, although some field penetration is to be expected from the region G_1G_2 in which the ions are produced. After passing through various focussing arrangements the ions impinge on a collector P whose potential, with respect to that of the source, is variable from -2 to $+10$ volts. The kinetic energy of the ions, at or near an appearance potential, can thus be analysed by plotting the collector current at P as a function of V_r, the retarding potential on P.

The instrument may be calibrated with ions formed in ionization processes of the type

$$\mathrm{Ne} + e \rightarrow \mathrm{Ne}^+ + 2e$$

[1] J. D. CRAGGS, R. THORBURN and B. A. TOZER: Proc. Roy. Soc. A **240** 473 (1957).
[2] I. W. W. LOZIER: Phys. Rev. **46**, 268 (1934).
[3] H. D. HAGSTRUM: Rev. Mod. Phys. **23**, 185 (1951).
[4] G. HERZBERG: Spectra of Diatomic Molecules. New York: Van Nostrand Co. Inc. 1950.

in which the parent atom (or molecule) loses only one or more electrons so that the ion possesses little more than thermal energy. Fig. 34 shows characteristic current—retarding potential curves for the collector. Reference should be made to HAGSTRUM's paper for a discussion of the factors responsible for the rounding-off of the curve of Fig. 34 above and below the zero potential.

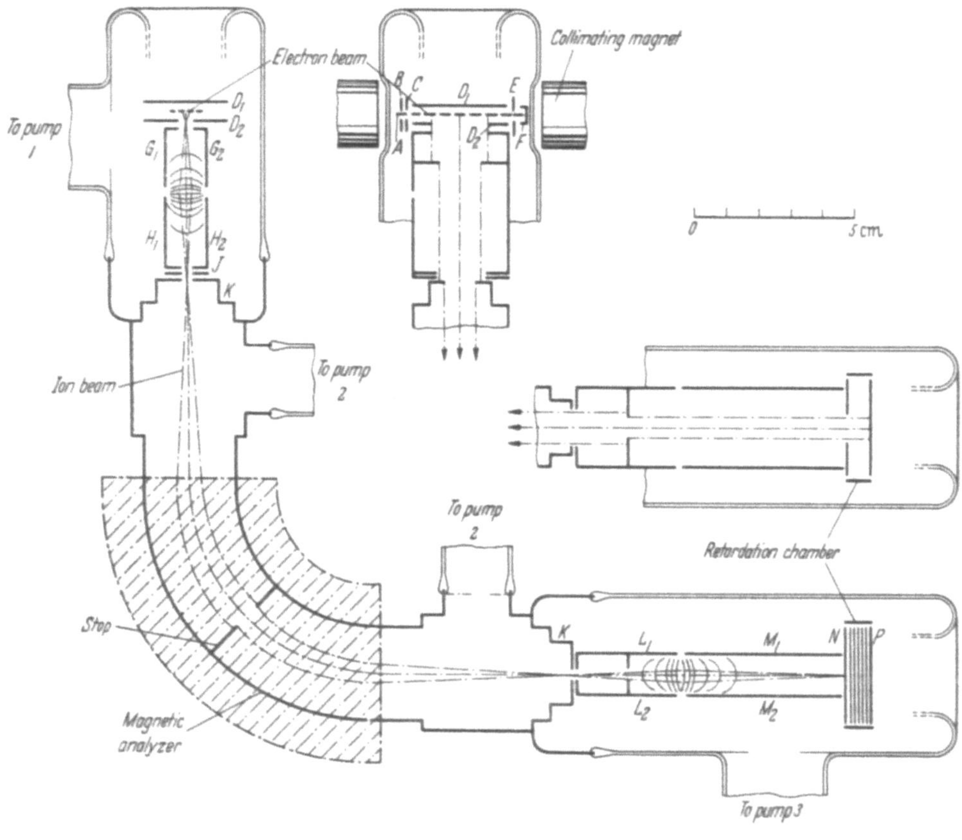

Fig. 33. Schematic view of HAGSTRUM's mass spectrometer. Views of the electron beam and retardation chamber ends of the apparatus at 90° from the principal view are shown to the right of these portions and above them, respectively. Equipotential lines are indicated in the G-H and L-M ion lenses and in the retardation chamber. E, 0.1 × 1.2 cm; D_2, 0.4 × 2.0 cm; G_1G_2, 0.4 cm wide; H_1H_2, 0.05 cm wide; J, 0.1 × 2.0 cm; K (entrance), 0.02 × 1.0 cm; K (exit), 0.03 × 1.0 cm; L_1L_2, 0.05 cm wide; stop in L_1L_2, 0.1 × 0.8 cm; N, 0.025 × 3.0 cm. Other dimensions may be estimated from the scale of the drawing. The shaded area indicates the extent of the analyser magnetic field. Typical electrode potentials for $E_2(e^-) =$ 100 ev follow: A, −100 volts; B, −80 volts; C, D_1, D_2, and E, zero; F, +45 volts; G_1-G_2, −15 volts; H_1-H_2, J, and K, −100 volts; $L-L_2$, −150 volts; M_1-M_2 and N, −30 volts; P, −2, +10 volts. [Reproduced from Rev. Mod. Phys. **23**, 185 (1951).]

HAGSTRUM's mass spectrometer has been used to measure critical (ionization or appearance) potentials, with typical results shown in Fig. 34. If the appearance potential for any chosen positive ion is plotted against the retarding potential, a curve similar to that of Fig. 35 (O^+ from CO) is obtained and from such a plot the kinetic energy term may be found. Similar plots (e.g. Fig. 36) may be obtained for negative ions, and we shall discuss the results of the measurements below.

Besides the retarding slit method various investigators have used line width methods. These are based on the fact that an initial spread of energies amongst the ions diffusing out from the mass spectrometer source will be apparent as a

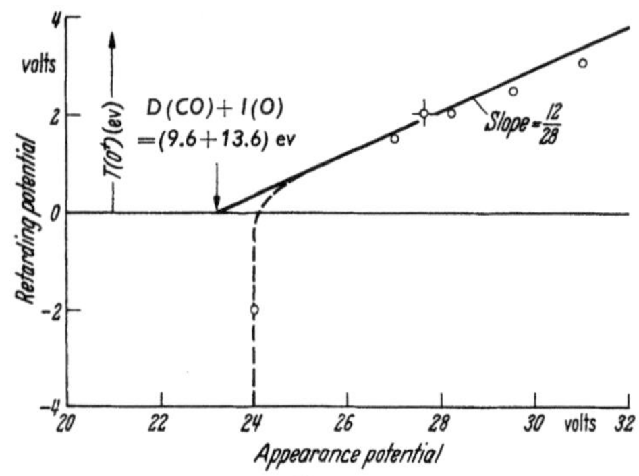

Fig. 35.

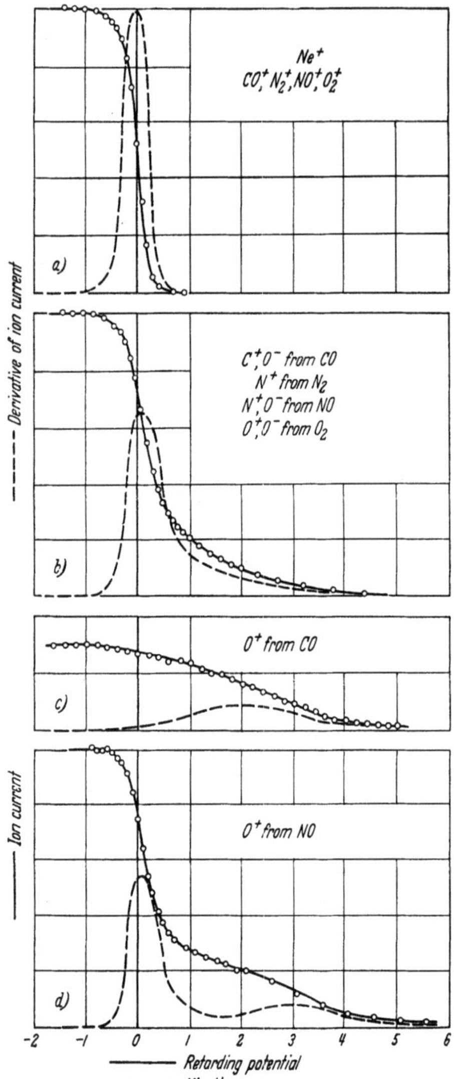

Fig. 34.

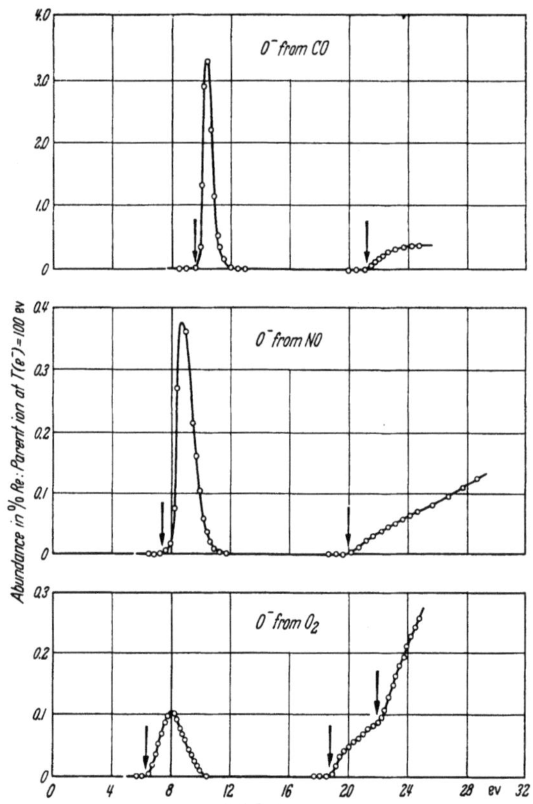

Fig. 36.

Fig. 34 a—d. Data obtained with the apparatus of Fig. 33, concerning ions formed by impact of 100 volt electrons in Ne, CO, N_2, NO and O_2. The full line curves give the variation of ion current with retarding potential and the dotted curves are the derived curves which give the energy distributions of the ions. (a) Ne^+ ions from Ne. Similar results are obtained for CO^+, N_2^+, NO^+ and O_2^+ ions. (b) C^+ and O^- ions from CO. Similar curves are obtained for N^+ from N_2, N^+ and O^- from NO, O^+ and O^- from O_2. (c) O^+ from CO. (d) O^+ from NO. In (a) the ions are formed with effectively zero kinetic energy, but in (b), (c) and (d) some ions are produced with a considerable kinetic energy.

Fig. 35. Further data obtained with the apparatus of Fig. 33. Plot of retarding potential against appearance potential for O^+ ions from CO. Open circles at $+2$ and -2 volts retarding potential show data from this work. Open circles represent data of HAGSTRUM and TATE, for which the retarding potential scale is replaced by the equivalent scale of displacement across the peak on the ion accelerating voltage scale. The line of slope 12/28 is drawn through the point $A(O^+) = D(CO) + I(O) = (9.6 + 13.6)$ ev $= 23.2$ ev; $T(O^+) = 0$ in the K, A portion of the graph. (Reproduced from Rev. Mod. Phys. **23**, 185.)

Fig. 36. Further data obtained with the apparatus of Fig. 33. Efficiency of production of O^- ions from CO, NO, and O_2 as a function of bombarding electron energy.

line width at the collector. This width may be used, in principle, to determine the energy spread[1]. Further details may be obtained by reference to review articles[2].

G. Experimental data for ionization of diatomic molecules.

21. Hydrogen. Some of the possibilities which arise in connection with the ionization of hydrogen will be apparent by reference to Fig. 37 which illustrates some potential energy curves of H_2^+ in addition to that for the ground state of H_2.

According to the Franck-Condon principle, final states of H_2^+ arising from ionization of H_2 must lie on potential energy curves within the shaded areas. Transitions to the ground $^2\Sigma_g$ state of H_2^+ should therefore give rise to stable H_2^+ ions, the appearance potential being about 15.5 volts. On the other hand, the first excited state of H_2^+, the $^2\Sigma_u$ state, is repulsive and transitions to this state should give rise to H atoms and protons with kinetic energies between 5 and 7 ev. The appearance potential for 5 volt protons should be 28 volts. Transitions to higher potential energy curves will occur as the incident electron energy increases until eventually, at 46 volts, a transition is possible to the potential curve for H_2^{++} which is of course repulsive. This transition at its onset potential should lead to two protons each with 7 ev kinetic energy.

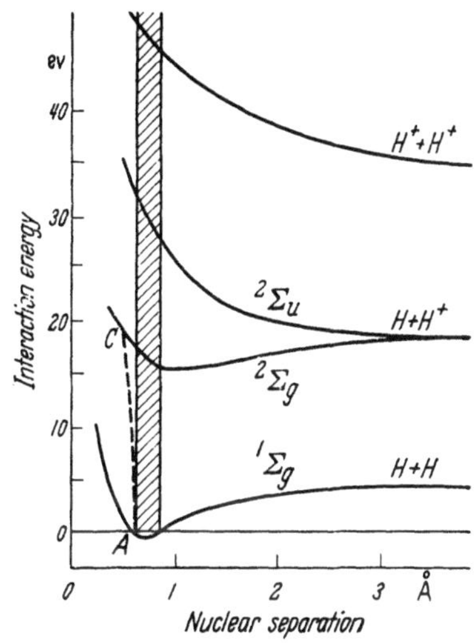

Fig. 37. Potential energy curves for certain states of H_2, H_2^+ and H_2^{++}.

All of these processes have been identified experimentally. TATE and SMITH measured the apparent cross section for ionization of H_2 using the apparatus described in Vol. XXXVI, p. 310. In fact this is effectively the cross section for the transition to the $^2\Sigma_g$ state. The observed values are given in Table 10.

In 1930 BLEAKNEY[3], using a mass spectrograph, detected not only the slow H_2^+ ions from

$$H_2 + e \rightarrow H_2^+ (^2\Sigma_g) + 2e \tag{21.1}$$

but also the energetic protons from

$$H_2 + e \rightarrow H_2^+ (^2\Sigma_u) + 2e \rightarrow H + H^+ + 2e. \tag{21.2}$$

In addition to these he also observed a group of slow protons with an appearance potential of 18 volts. Although he was unable to make quantitative measurements of the proton energies it was possible to tell from the width of the spectrograph lines whether or not they were appreciable.

---

[1] See especially the work of C. A. McDOWELL and J. W. WARREN, Faraday Soc. Disc. **10**, 53 (1951), for a discussion of a useful qualitative test which decides in some cases between zero and finite kinetic energy.

[2] H. S. W. MASSEY: Negative Ions. Cambridge: Cambridge University Press 1950. — H. S. W. MASSEY and E. H. S. BURHOP: Electronic and Ionic Impact Phenomena. Oxford: Oxford University Press 1952. — G. P. BARNARD: Modern Mass Spectrometry. London: Institute of Physics 1953.

[3] W. BLEAKNEY: Phys. Rev. **36**, 1303 (1930).

Table 10. *Cross sections for ionization of certain molecular gases by electron impact.*

Gas	H_2	O_2	N_2	CO	NO
Ionization potential (volts)	15.6	12.5	15.7	14.1	9.5
Electron energy (ev)	Cross section in 10^{-16} cm^2				
10	—	—	—	—	0.03
12	—	—	—	—	0.17
14	—	0.06	—	—	0.37
16	—	0.16	—	0.11	0.54
18	0.12	0.25	—	0.26	0.69
20	0.23	0.41	0.20	0.41	0.84
25	0.46	0.79	0.61	0.97	1.25
30	0.66	1.18	1.04	1.46	1.58
35	0.78	1.55	1.40	1.78	1.91
40	0.88	1.84	1.76	2.11	2.15
50	0.97	2.18	2.22	2.54	2.57
60	1.00	2.44	2.52	2.77	2.84
70	1.01	2.63	2.67	2.93	3.02
80	1.00	2.76	2.79	3.03	3.14
100	0.96	2.89	2.87	3.07	3.25
120	0.90	2.91	2.86	3.03	3.25
140	0.85	2.87	2.80	2.93	3.19
160	0.81	2.80	2.72	2.82	3.07
200	0.73	2.64	2.52	2.61	2.85
250	0.64	2.42	2.33	2.36	2.61
300	0.57	2.20	2.12	2.15	2.39
400	0.46	1.85	1.81	1.78	2.02
500	0.39	1.61	1.55	1.55	1.73
700	0.31	1.25	1.22	1.22	1.41

Reference to Fig. 37 shows that the only way in which the slow protons can arise is by an "inclined" transition, from A to C in that figure, which weakly violates the Franck-Condon principle. For the lightest molecule this is not too improbable, but even so one would expect the process to be very much less probable than (21.2). This did not appear to be so from BLEAKNEY's observations, but the collecting power of a spectrograph for ions with initial kinetic energy is much less than for initially very slow ions, so little weight could be given to his relative abundances. The matter was re-examined by NEWHALL[1].

He took advantage of the fact that an electron beam of radius ϱ and strength i amperes passing along the axis of a metal cylinder of radius R enclosing gas at a low pressure (5×10^{-5} mm Hg) is, through its space charge, a region with potential less than that of the walls by an amount V volts where

$$V = 1.52\times10^4\, i\,(1+2\log R/\varrho)/V_0^{\frac{1}{2}},$$

V_0 ev being the kinetic energy of the beam. This potential trough will prevent the escape of very slow ions produced in the gas by the beam. For thermal ions, this should be apparent for beam currents as low as 2 microamperes. Examination of the variation of ion yield with electron current at low currents should therefore reveal the production of slow ions.

In NEWHALL's apparatus a small fraction of the ions reaching the walls passed out through a slot into a mass analysing chamber. The yield of H_2^+ ions varied with current as expected for thermal ions, but no such effect was found

[1] H. F. NEWHALL: Phys. Rev. **62**, 11 (1942).

for protons. For electrons of energies greater than 28 ev departures from linearity in the yield-current relation were observed when the current exceeded 100 microamperes. From these departures the proton energy distribution was derived and found to be consistent with process (21.2). This shows that the production of slow protons occurs only very rarely. Even process (21.2) was found by NEWHALL to be very improbable, the cross section being only about 1/200 of that for process (21.1).

The double ionization process

$$H_2 + e \rightarrow H^+ + H^+ + 3e \tag{21.3}$$

was observed by LOZIER[1], who was able to measure the appearance potential for protons of different kinetic energies. His results are shown in Fig. 38. If $A(E_f)$ is the appearance potential for protons of kinetic energy T we have, according to the Franck-Condon principle

$$A(E_f) = 2T + D(H_2) + 2E_H \tag{21.4}$$

where $D(H_2)$ is the dissociation energy 4.4 ev of H_2 and E_H the energy (13.5 ev) of the ground state of H. It will be seen from Fig. 38 that LOZIER's results agree very closely with this relation.

It is of interest to notice also that LOZIER, in the course of these experiments, made similar observations for the process (21.2) for which (21.4) holds with $2E_H$ replaced by E_H. Reference to Fig. 38 shows that quite good agreement with expectation was also obtained in this case.

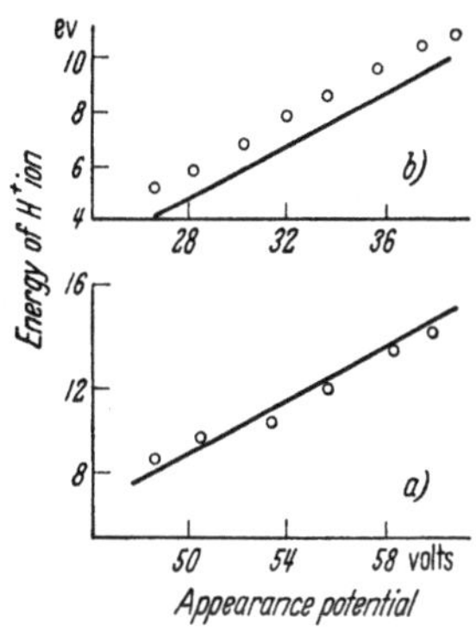

Fig. 38. Observed appearance potentials for H^+ ions of different kinetic energy produced (a) from $H_2 + e \rightarrow H^+ + H^+ + 2e$, (b) from $H_2 + e \rightarrow H^+ + H + 2e$. o observed points; ——— predicted curve.

22. Oxygen, carbon monoxide, nitric oxide and nitrogen. We shall consider this group of molecules together as the interpretation of the electron collision data obtained for them is closely linked. It provides evidence concerning the dissociation energies of CO and of N_2, the electron affinity of O and the possible existence of a stable excited state of O^-.

Whereas the dissociation energy of O_2 has been known accurately for a long time, the dissociation energies of CO and of N_2 have proved extraordinarily difficult to determine. Until recently two sets of values were extant, both of which could be supported by quite strong evidence. The higher pair of values[2] were 11.11 ev for CO and 9.76 ev for N_2 to be contrasted with the lower pair, 9.6 and 7.38 ev respectively. In the last year or so the evidence has piled up in favour of the higher values which are now universally accepted. We shall assume these values and show that they are consistent with the most recent electron impact data.

The dissociation energy of NO is linked to those of O_2 and N_2, through the reaction[3]

$$2\,NO \rightarrow N_2 + O_2 + 1.87\ ev.$$

[1] W. LOZIER: Phys. Rev. **36**, 1285, 1417 (1930).

[2] A. G. GAYDON: Dissociation Energies, Chap. IX. Chapman & Hall: London 1947. — A. E. DOUGLAS: Canad. J. Phys. **30**, 302 (1932).

[3] F. R. BICHOWSKY and F. D. ROSSINI: Thermochemistry of Chemical Substances. New York: Reinhold 1932.

If $D(O_2)$ is taken as 5.09 ev[1], D(NO) comes out to be either 6.49 or 5.29 ev, according as $D(N_2)$ is taken as 9.76 or 7.38 ev. In accordance with our assumptions concerning $D(N_2)$ we shall therefore assume $D(\mathrm{NO}) = 6.49$ ev.

The situation concerning the electron affinity of O has changed to a remarkable extent due to the new measurements made by BRANSCOMB and SMITH[1] of the threshold frequency for photodetachment of electrons from O^-. They find $EA(O^-) = (1.45 \pm 0.15)$ ev, whereas the previously accepted value was considerably higher—2.2 ev. This had been obtained from electron impact data in O_2 and was in agreement with that derived from METLAY and KIMBALL[2] from the measurement of the equilibrium constant for the reaction

$$O + e \rightarrow O^-$$

occurring in the neighbourhood of a hot filament. We shall assume the value of BRANSCOMB and SMITH in the analysis of the impact data and although it is not consistent with all the most recent results, the discrepancies are not very great.

The possible existence of an excited state of O^- was indicated from the earlier appearance potentials for O^- ions produced in CO and NO. On the whole, the evidence from these earlier measurements, combined with the high electron affinity of O derived from the O_2 impact data, supported the lower dissociation energies of CO and N_2. If, in turn, these were assumed, the appearance potentials for O^- in CO and NO required that the O^- should carry off nearly 1.9 ev excitation energy. The evidence of such a state would not be expected on theoretical grounds as shown by BATES and MASSEY[3]. Again, HASTED[4], in the course of an experimental study of the cross sections for detachment of electrons from O^- ions in collision with rare gas atoms, found no evidence that the ions, which according to the interpretation of the impact data, were in excited states, had detachment cross sections different from those for normal ions. As the excited state, if it existed, would be metastable with a lifetime sufficiently long to pass through the ion analyser in HASTED's equipment, the absence of any distinction aroused grave doubt about the reality of the excited state. Furthermore, no absorption is observed in the photodetachment experiments of SMITH and BRANSCOMB which would indicate the presence of metastable O^- ions in the beam. Since the beam is taken from a glow discharge source, it is reasonable to expect that if such a state existed it would be appreciably populated. It is therefore satisfactory that the most recent impact data may be interpreted without the assumption of any such state.

Summarizing the assumptions which we shall make in interpreting the impact data, we have:

$$D(O_2) = 5.09\text{ ev};\quad D(\mathrm{CO}) = 11.11\text{ ev},\quad D(N_2) = 9.76\text{ ev},\quad D(\mathrm{NO}) = 6.49\text{ ev},$$
$$I(\mathrm{O}) = 13.61_5\text{ ev};\quad I(\mathrm{C}) = 11.26_5\text{ ev},\quad I(\mathrm{N}) = 14.5_5\text{ ev},$$
$$EA(\mathrm{O}) = 1.45\text{ ev};\quad EA(\mathrm{C}) = 1.5\text{ ev}.$$

It is also of interest in these interpretations to note the excitation energies of the lowest excited states of the atoms and ions concerned. They are

$$\mathrm{O}(^1D)\ 1.96\text{ ev};\quad \mathrm{O}^+(^2D)\ 3.31\text{ ev};\quad \mathrm{C}(^1D)\ 1.25\text{ ev},\quad \mathrm{C}(^1S)\ 2.67\text{ ev};\quad \mathrm{N}(^2D)\ 2.37\text{ ev}.$$

---

[1] S. J. SMITH and L. M. BRANSCOMB: J. Res. Nat. Bur. Stand. **55**, 165 (1955).
[2] M. METLAY and G. E. KIMBALL: J. Chem. Phys. **16**, 774 (1948).
[3] D. R. BATES and H. S. W. MASSEY: Phil. Trans. Roy. Soc. Lond., Ser. A **239**, 269 (1943).
[4] J. B. HASTED: Proc. Roy. Soc. Lond., Ser. A **222**, 74 (1954).

We now consider the data for each gas in turn.

Oxygen. The ionization energy of the molecule according to the impact data agrees well with the spectroscopic value of 12.2 ev. Thus HAGSTRUM[1] finds (12.1 ± 0.2) ev and this agrees with most recent measurements by MARRIOTT[2] and by THORBURN[3] at Liverpool.

The production of the atomic ions O^+ and O^- has been studied by various workers during the last 20 years. The data published up to 1951 have been discussed and tabulated by HAGSTRUM[1]. An error in voltage calibration was corrected in a later paper[4]. The most recent work is that of THORBURN:, using the Lozier technique. We shall use this in discussing the interpretation of the processes involved.

In Figs. 39 and 40 (see also Figs. 34 and 36) the observed variation of the current of O^+ and O^- ions with energy of the colliding electrons is illustrated. A calibration curve for A^+ ions is also given in Fig. 39, I(A) being taken as 15.76 ev. It will be noticed that both the O^+ and O^- curves show inflexions and that the latter includes the well known low energy capture peak.

Table 11 summarizes the appearance potential data. Two sets of data are listed

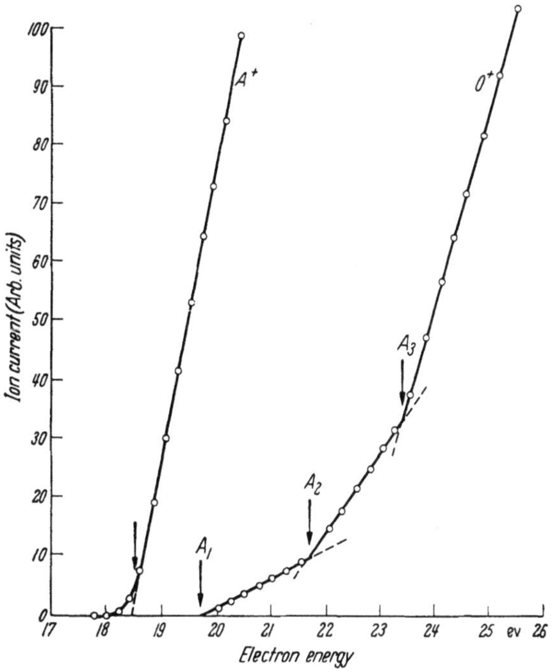

Fig. 39. Observed variation with electron energy of the O^+ current arising from electron bombardment of oxygen. The corresponding curve for A^+ ions from argon is used for voltage calibration.

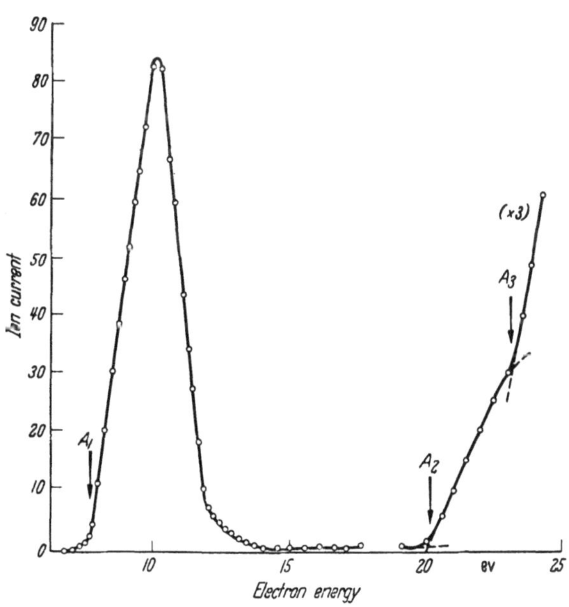

Fig. 40. Observed variation with electron energy of the O^- current arising from electron bombardment of oxygen.

[1] H. D. HAGSTRUM: Rev. Mod. Phys. **23**, 185 (1951).
[2] J. MARRIOTT: Ph. D. thesis, Liverpool University 1954.
[3] R. THORBURN: Applied Mass Spectrometry, p. 185. London: Institute of Petroleum 1954. — S. J. SMITH and L. M. BRANSCOMB: J. Res. Nat. Bur. Stand. **55**, 165 (1955).
[4] H. D. HAGSTRUM: J. Chem. Phys. **23**, 1178 (1955).

under "Earlier observations"—the values in brackets being those which included the voltage calibration error. It was on the basis of these incorrect values that the interpretations given in the last column were assigned. Thus the fact that the appearance potentials were taken to be nearly 2.0 ev too large seemed to require the existence of excited O^- ions.

Table 11. *Electron impact data in oxygen and its interpretation.*

THORBURN's observations			Earlier observations		
Observed appearance potential (volts)	Calculated appearance potential (volts)	Suggested process [$EA(O)=1.45$ ev]	Observed[2] appearance potential (volts)	Observed[3] appearance potential (volts)	Suggested process [$EA(O)=2.3$ ev]
4.9 ± 0.1 (3.2 ± 0.1 zero k.e.)	3.63	$O_2+e\rightarrow O(^3P)+$ $+O^-(^2P)$	[2.9 (zero k.e.)]	3.2	$O_2+e\rightarrow O(^3P)+$ $+O^-(^2P)$
16.9 ± 0.2 (O^+) 17.1 ± 0.2 (O^-)	17.25	$O_2+e\rightarrow O^+(^4S)+$ $+O^-(^2P)+e$	(18.6 ± 0.2) (18.8 ± 0.2)	17.2 ± 0.2 17.0 ± 0.2	$O_2+e\rightarrow O^+(^4S)+$ $+O^{-*}+e$
18.9 ± 0.2 (O^-)	18.69	$O_2+e\rightarrow O^+(^4S)+$ $+O(^3P)+2e$	(20.8 ± 0.2)	19.2 ± 0.2	$O_2+e\rightarrow O^+(^4S)+$ $+O(^1D)+2e$
20.5 ± 0.2 (O^+) 20.2 ± 0.2 (O^-)	20.57	$O_2+e\rightarrow O^+(^2D)+$ $+O^-(^2P)+e$	(22.0 ± 0.2) (22.0)	20.4 ± 0.2 (O^-)	$O_2+e\rightarrow O^+(^2D)+$ $+O^{-*}+e$

O^{-*} denotes the assumed excited negative ion with about 1.9 ev excitation energy.

It will be seen that the agreement between observed and calculated appearance potentials on the basis of the interpretations listed in the third column is quite good. The only exception is the first process for which there is a discrepancy of 0.4 ev. The reason for this discrepancy is not clear but recent measurements by RANDOLPH and GEBALLE[1], using the retarding potential difference method of FOX (see p. 357) and an energy scale calibrated from the appearance potential of SF_6^- (see p. 412), have given an appearance potential for O^- ions of zero kinetic energy as high as 3.5 ev which considerably reduces the discrepancy. The data from CO and NO do not show any inconsistency.

Table 12. *Electron impact data in carbon monoxide and its interpretation.*

Ion	Appearance potentials or ions of zero kinetic energy: Observed	Calculated	Present interpretation [Based on $D(CO)=11.1$ ev]	Earlier interpretation [Based on $D(CO)=9.6$ ev]
O^-	9.6 ± 0.2	9.66	$CO+e\rightarrow C(^3P)+O^-(^2P)$	$CO+e\rightarrow C(^3P)+O^{-*}(^2P)$
O^- C^+	21.2 ± 0.2 20.9 ± 0.2	20.92	$CO+e\rightarrow C^+(^2P)+O^-(^2P)+e$	$CO+e\rightarrow C^+(^2P)+O^{-*}(^2P)+2e$
C^+	22.8 ± 0.2	22.37	$CO+e\rightarrow C^+(^2P)+O(^3P)+2e$	$CO+e\rightarrow C^+(^2P)+O(^1D)+2e$
O^+	23.2 ± 0.2	23.22	$CO+e\rightarrow C^-(^4S)+O^+(^4S)+e$	$CO+e\rightarrow C(^4S)+O^+(^4S)+2e$

O^{-*} denotes the assumed excited negative ion with about 1.9 ev excitation energy.

[1] P. T. RANDOLPH and R. GEBALLE: Bull. Amer. Phys. Soc., Ser. II, 376 (1957).

[2] H. D. HAGSTRUM: Rev. Mod. Phys. 23, 185 (1951). — H. D. HAGSTRUM and J. T. TATE: Phys. Rev. 59, 354 (1941). — J. T. TATE and P. T. SMITH: Phys. Rev. 39, 270 (1932). — A. L. VAUGHAN: Phys. Rev. 38, 1687 (1931).

[3] H. D. HAGSTRUM: J. Chem. Phys. 23, 1178 (1955).

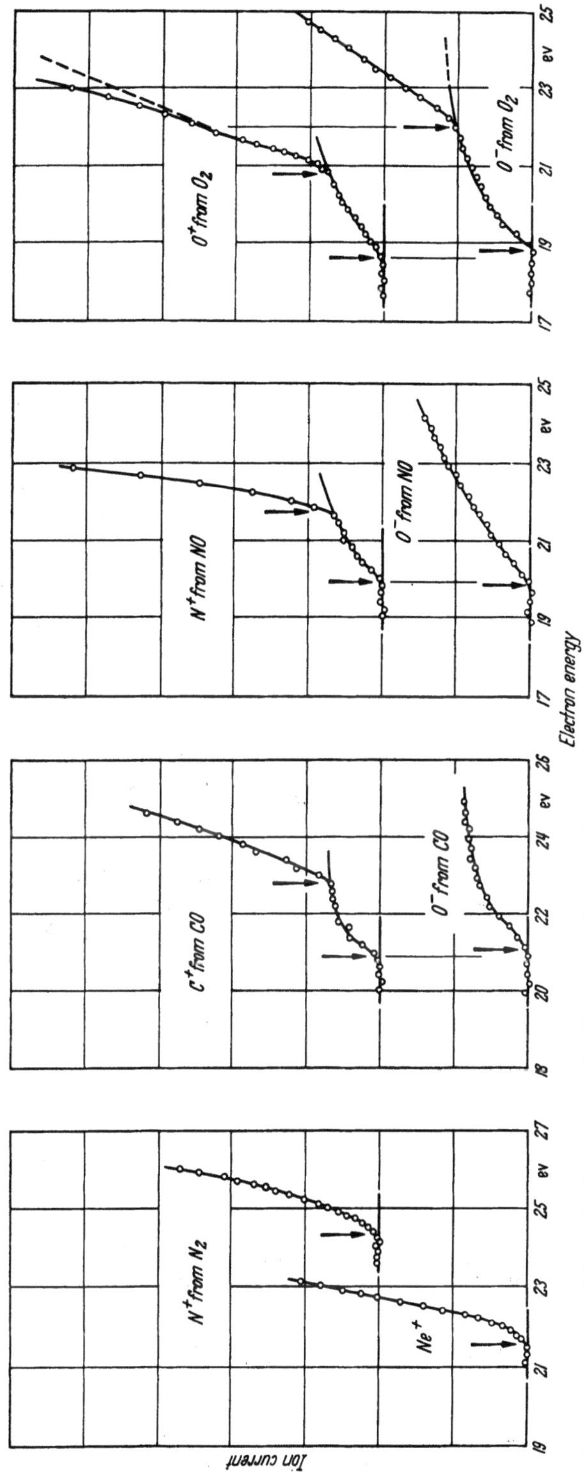

Fig. 41. Appearance potential data for ions of zero initial kinetic energy from dissociative ionization in N_2, CO, NO, and O_2. The electron energy scale is calibrated against $A(Ne^+) = I(Ne) = 21.56$ volts. The indicated unit of the ordinate scale is approximately 1% of the ion current at electron energy = 100 ev. The detection limit is approximately 5% of this or about one part in 2000 of this ion current.

Carbon monoxide. The appearance potentials and ion yield curves for this gas seem to be well established as the most recent work[1] gives results in substantial agreement with earlier measurements[2]. Figs. 34 and 36 illustrate the yield curves for the positive and negative-ions respectively. Appearance potential data and their interpretations are given in Table 12.

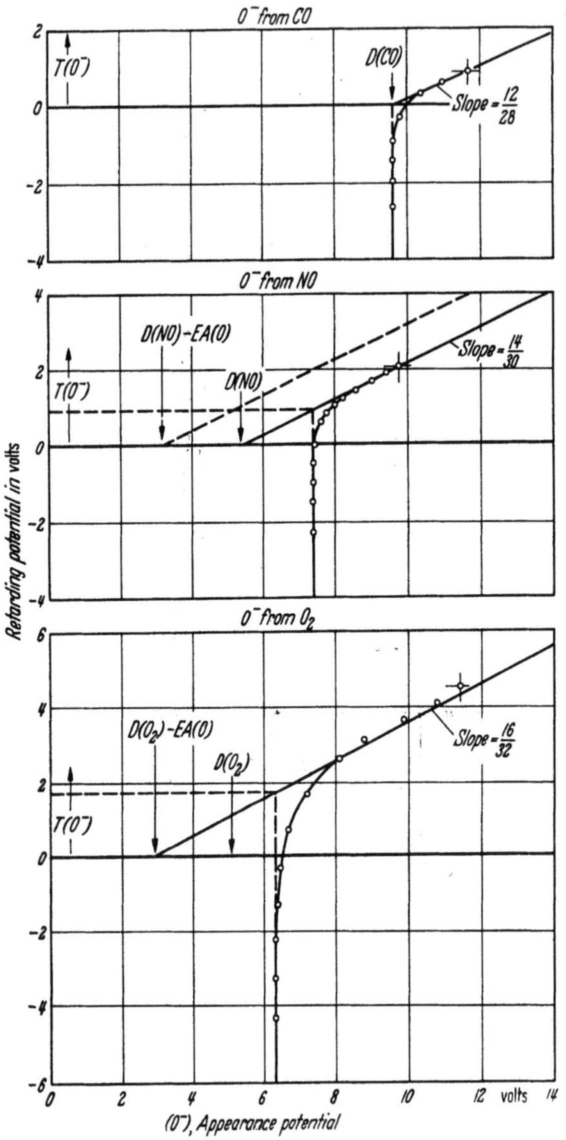

Fig. 42. Plots of retarding potential *vs.* appearance potential for O^- ions arising from the resonance electron capture processes in CO, NO and O_2. Possible experimental error is indicated at the datum point at highest appearance potential for each curve. The intercept on the *A* axis, locating the dissociation limit of the process, is interpreted in each case as indicated. The dashed *E*, *A* line of slope 14/30 for O^- (NO) is for the dissociative capture process observed by HANSON (see text).

The earlier interpretations were based on $EA(O^-) = 2.2$ ev and $D(CO) = 9.6$ ev and are also given in Table 12. Once again they involved the assumption of the stable excited O^- ion.

The only doubtful feature of the new interpretation is that of the process with appearance potential 23.22 ev. An electron affinity of 1.5 ev for carbon is quite reasonable[3]. On the other hand LAGERGREN[4], who has confirmed the formation of C^- ions, finds an electron affinity less than that of oxygen by (0.35 ± 0.05) ev i.e. 1.10 ev.

FOX and HICKAM[5] have applied their pulsed retarding potential difference method (see Vol. XXXVI, p. 313) to study the onset of ionization in CO. By carrying out an analysis on the same lines as for the ionization of rare gas atoms (see Fig. 43), they obtained evidence of the onset of two additional ionization processes occurring at potentials (2.4 ± 0.1) and (5.0 ± 0.1) ev above the ionization potential. These may probably be associated with production of the CO^+ ions

[1] W. A. CHUPKA and M. G. INGHRAM: J. Phys. Chem. **59**, 100 (1955). — H. D. HAGSTRUM: J. Chem. Phys. **23**, 1178 (1955).
[2] See footnote 2, p. 368.
[3] H. S. W. MASSEY: Negative Ions. Cambridge: Cambridge University Press 1950.
[4] C. R. LAGERGREN: Ph. D. Thesis, Univ. of Minnesota (1955).
[5] R. E. FOX and W. M. HICKAM: J. Chem. Phys. **22**, 2059 (1954).

in the excited $A\,^2\Pi$ and $B\,^2\Sigma$ states which according to the spectroscopic data lie respectively 2.52 and 5.66 ev above the ground state. The discrepancy for the $B\,^2\Sigma$ state is rather large, however, and Fox and Hickam suggest that the accuracy of determination of the onset potential in this case is reduced because the yield curve is modified by autoionization.

Nitric oxide. Nitrix oxide has been studied by various workers and reference should be made to HAGSTRUM's detailed review for references. Some of the ion yield curves found are illustrated in Fig. 34b, 36, 41 and 42. Appearance potential data, due to the most recent work of HAGSTRUM[1] are given in Table 13, together with their interpretation. Earlier data, with references, are cited by HAGSTRUM[2]. These data agree quite closely with those of Table 13, except for the dissociative attachment process. For this HANSON[3] obtained (3.2 ± 0.5) ev for O^- of zero kinetic energy. This was consistent with the values $EA(O) = 2.2$ ev and $D(NO) = 5.29$ ev.

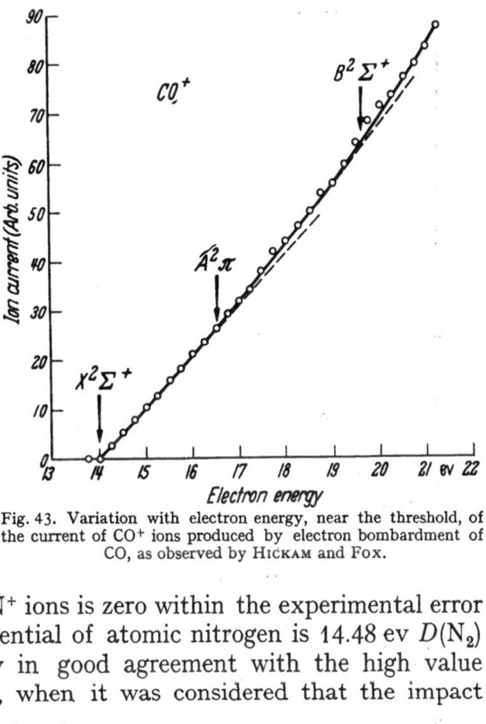

Fig. 43. Variation with electron energy, near the threshold, of the current of CO^+ ions produced by electron bombardment of CO, as observed by HICKAM and FOX.

Nitrogen. No negative ions are formed in this gas. The appearance potentials of the N_2^+ and N^+ ions are quite well established — the most recent measurements of BURNS[4], of THORBURN and CRAGGS[5] and of HAGSTRUM[1] give $A(N_2^+) = (15.7 \pm 0.2)$ ev and $A(N^+) = (24.3 \pm 0.3)$ ev in good agreement with earlier values. In the careful experiments of THORBURN and CRAGGS, a Lozier apparatus was used to show that the initial kinetic energy of the N^+ ions is zero within the experimental error of ± 0.1 ev. Since the ionization potential of atomic nitrogen is 14.48 ev $D(N_2)$ comes out to be $24.3 - 14.5 = 9.8$ ev in good agreement with the high value advocated by GAYDON[6]. Previously, when it was considered that the impact

Table 13. *Electron impact data in nitric oxide and its interpretation.*

Ion	Appearance potentials for ions of zero kinetic energy (volts)		Interpretation
	Observed	Calculated	
NO^+	9.4 ± 0.2	—	$NO + e \rightarrow NO^+ + 2e$
O^-	5.3 ± 0.4	5.04	$NO + e \rightarrow N(^4S) + O^-(^2P)$
O^-	19.8 ± 0.2	19.68 (for O^- and N^+)	$NO + e \rightarrow N^+(^3P) + O^-(^2P) + e$
N^+	19.9 ± 0.2		
N^+	21.7 ± 0.2	21.03	$NO + e \rightarrow N^+(^3P) + O(^3P) + 2e$
O^+	20.6 ± 0.2	20.10	$NO + e \rightarrow N(^4S) + O^+(^4S) + 2e$

[1] H. D. HAGSTRUM: J. Chem. Phys. **23**, 1178 (1955).
[2] H. D. HAGSTRUM: Rev. Mod. Phys. **23**, 185 (1951).
[3] E. E. HANSON: Phys. Rev. **51**, 86 (1937).
[4] J. F. BURNS: J. Chem. Phys. **23**, 1347 (1955).
[5] R. THORBURN and J. D. CRAGGS: Proc. Phys. Soc. Lond. B **69**, 682 (1956).
[6] A. G. GAYDON: Dissociation Energies, Chap. IX. London: Chapman & Hall. — A. E. DOUGLAS: Canad. J. Phys. **30**, 302 (1952).

data in NO favoured the lower value 7.4 ev of $D(N_2)$, the additional 2.4 ev was supposed to reside in excitation energy of the N atom. This offered no difficulty as the lowest excited state of the atom (^1D) lies 2.37 ev above the ground state. It is true that this possibility can still not be excluded, but consistency with all the impact data is only obtained if it is supposed that the N atoms are produced in their normal states.

FOX and HICKAM[1] have used the new pulsed retarding potential difference technique to study the onset of ionization in N_2. Using the ionization potential of Xe as standard they obtained (15.60 ± 0.01) ev for the ionization energy of N_2, which is to be compared with the spectroscopic value of 15.58 ev. Analysis of the ionization function (see Fig. 44) provides evidence of ionization producing N_2^+ ions in the $A\,^2\Pi$ and $B\,^2\Sigma$ states at threshold energies exceeding that for production of normal N_2^+ ions by (1.33 ± 0.04) and (3.24 ± 0.04) ev respectively. These values agree well with the spectroscopic values of 1.36 and 3.17 ev.

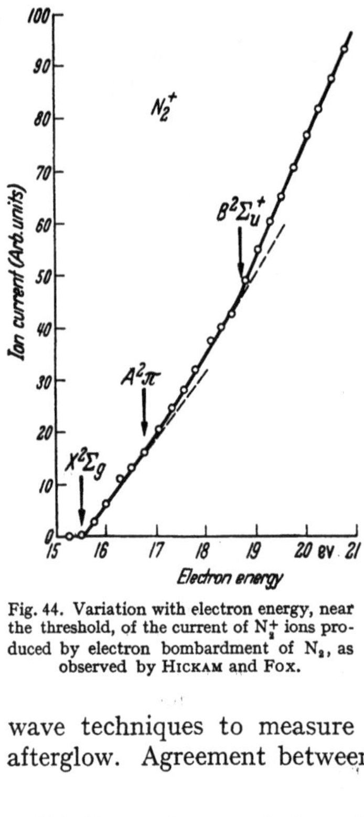

Fig. 44. Variation with electron energy, near the threshold, of the current of N_2^+ ions produced by electron bombardment of N_2, as observed by HICKAM and FOX.

23. The halogens. The study of electron impact processes in the halogens is rendered difficult by the reactivity of the vapours. Because of this, impurities are not easy to exclude, the hydrides being particularly hard to avoid.

As an indication of the unsatisfactory state of knowledge in this field, we may cite the results obtained by different investigators for impact phenomena in iodine. The formation of negative ions in this vapour has been studied by HEALEY[2] using the diffusion method of BAILEY (see Vol. XXI of this Encyclopedia), by BUCHDAHL[3] and by HOGNESS and HARKNESS[4] using homogeneous electron beams, by FOX[5] using the pulsed retarding potential difference method and by BIONDI[6] using microwave techniques to measure the rate of loss of free electrons in an iodine afterglow. Agreement between the results obtained is poor.

H. Experimental data for ionization of polyatomic molecules.

24. Introduction. Experimental work in this field has proceeded on the same lines as for diatomic molecules, most of the effort being concentrated on the determination of appearance potentials. Some attention has also been paid to measurement of the relative abundances of different kinds of ions produced by electrons of a particular energy, usually near 100 ev. These abundance measurements are difficult and it is doubtful whether the instrumental factors which affect the yield of negative and of positive ions have yet been sorted out. Mislead-

[1] R. E. FOX and W. M. HICKAM: J. Chem. Phys. **22**, 2059 (1954). See also Sect. 35α.
[2] R. H. HEALEY: Phil. Mag. **26**, 940 (1938).
[3] R. BUCHDAHL: J. Chem. Phys. **9**, 146 (1941).
[4] T. R. HOGNESS and R. W. HARKNESS: Phys. Rev. **32**, 784 (1928).
[5] R. E. FOX: Phys. Rev. **109**, 2008 (1958).
[6] M. BIONDI: Phys. Rev. **109**, 2005 (1958).

ing results may also be obtained because of the dependence of the collecting power on the initial kinetic energy of the ions.

An interesting phenomenon which has been observed is the occurrence of metastable positive and negative ions first observed by HIPPLE and CONDON[1]. Cases in which such ions are produced will be discussed below.

Interesting results can be expected from the application of the pulsed retarding-potential difference method of FOX and HICKAM. It has already been applied[2] to the study of the formation of negative ions by slow electrons in SF_6. The direct formation of SF_6^- by resonance electron capture has been observed and studied in detail. An account of this remarkable investigation will be given in Sect. 36.

Unambiguous interpretation of the data is naturally more difficult than for diatomic molecules and is rendered more difficult still by the fact that the initial kinetic energies of the observed ions are rarely measured. Information available about the excitation and dissociation energies of the fragments is very slight as these fragments are usually chemically unstable radicals or complex ions.

Some effort has been made towards interpreting known mass spectra ("cracking patterns" of molecules) in terms of the electronic structure of the molecules concerned. This has been considered by MCDOWELL[3] who discusses this question in relation to HCN, NH_3 and CH_4 in addition to the diatomic molecules H_2, N_2 and O_2. In this connection there is no theoretical background available which may be utilised. Some progress has been made, however, in the calculation of ionization potentials for complex molecules. Thus HALL[4] using an equivalent orbital theory has[5] calculated ionization potentials for saturated hydrocarbons which are in good agreement with experimental data due to HONIG. A further typical example is that of CO_2 for which theoretical calculations have been

Table 14. *Ionization potentials measured by* COX.

Molecule	Ionization potential volts	Molecule	Ionization potential volts
H_2O	12.58 ± 0.03	C_6H_6	9.58 ± 0.01
CH_3OH	10.86 ± 0.05	$C_6H_5CH_3$	9.20 ± 0.03
C_2H_5OH	10.39 ± 0.04	$C_6H_5C_2H_5$	9.17 ± 0.02
nC_3H_7OH	10.22 ± 0.06	$C_6H_5nC_3H_7$	9.14 ± 0.01
$(CH_3)_2O$	10.13 ± 0.03	$C_6H_5nC_4H_9$	9.13 ± 0.02
$CH_3OC_2H_5$	9.81 ± 0.02	$C_6H_5CH=CH_2$	8.96 ± 0.01
$(C_2H_5)_2O$	9.62 ± 0.06	$C_{10}H_8$	8.16 ± 0.02
CH_3F	12.84 ± 0.03	$C_6H_5C_6H_5$	8.92 ± 0.02
CH_3Cl	11.46 ± 0.03	$C_6H_5C\equiv CH$	9.39 ± 0.02
CH_3Br	10.70 ± 0.03	HCN	13.71 ± 0.01
CH_3I	9.55 ± 0.03	CH_3CN	12.48 ± 0.04
C_2H_5F	12.00 ± 0.04	HOD	12.58 ± 0.03
C_2H_5Cl	11.32 ± 0.03	D_2O	12.60 ± 0.01
C_2H_5Br	10.49 ± 0.03	CH_3OD	11.04 ± 0.04
C_2H_5I	9.47 ± 0.01	C_2H_5OD	10.45 ± 0.01
C_2H_6	11.83 ± 0.02	nC_4H_{10}	10.79 ± 0.04

[1] J. A. HIPPLE and E. A. CONDON: Phys. Rev. **68**, 54 (1945).

[2] R. E. FOX and W. M. HICKAM: J. Chem. Phys. **25**, 642 (1956). See also J. Chem. Phys. **22**, 2059 (1954) which reports some investigations by the same method in propylene and benzene.

[3] C. A. MCDOWELL: Applied Mass Spectroscopy, p. 129. London: Institute of Petroleum 1954.

[4] G. G. HALL: Proc. Roy. Soc. Lond., Ser. A **205**, 541 (1951). — Trans. Faraday Soc. **49**, 113 (1953).

[5] Sir J. E. LENNARD-JONES: Proc. Roy. Soc. Lond., Ser. A **198**, 1, 14 (1949).

carried out by MULLIKEN[1] and by MULLIGAN[2] and observations by MORRISON[3]. Systematic changes in the first ionization potentials for a molecular series may, in some cases, be interpreted in terms of equivalent orbital theory. For example, the effect of alkyl substitution in the ethylene and water molecules may be discussed in this way. Table 14 gives the ionization potentials of a number of series

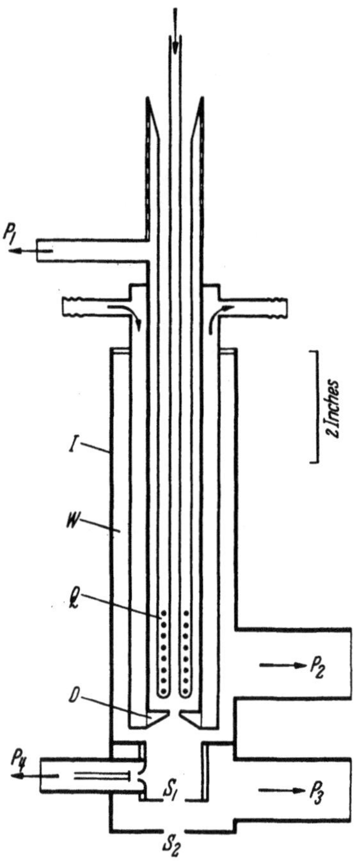

Fig. 45. A free-radical source due to ELTENTON. The low pressure quartz reactor is attached to ionization chamber of mass spectrometer. Ground joint J, water-cooled jacket W, quartz double-walled tube Q, diaphragm D, ion-collimating slits S_1 and S_2, pumping leads P_1, P_2, P_3, and P_4.

Table 15. *Ionization potentials of free radicals.*

Radical	Source	Ionization potential (volts)	Reference
methyl (CH_3)	lead tetramethyl	10.07 ± 0.1	a
	lead tetramethyl	9.85 ± 0.2	b
	mercury dimethyl	9.95 ± 0.03	c
	mercury dimethyl	10.11 ± 0.01 10.04 ± 0.12	d
deuteromethyl (CD_3)	mercury di-deuteromethyl	9.96 ± 0.03	c
methylene (CH_2)	diazomethane	11.9 ± 0.2	e
ethyl (C_2H_5)	lead tetraethyl	8.67 ± 0.1	a
	mercury diethyl	8.78 ± 0.05	f
alkyl	alkyl iodide	8.16 ± 0.03	c
($CH_2-C=CH_2$)	and 1-butene	8.15	
isopropyl ($(CH_3)_2CH$)	azo-isopropane	7.90 ± 0.05	f
propargyl ($CH_2-C\equiv H$)	propargyl iodide	8.25 ± 0.08	f
trichloromethyl (CCl_3)	chloropicrin	8.78 ± 0.05	g
benzyl ($C_6H_5CH_3$)	benzyl iodide	7.73 ± 0.08	c
	benzylamine	7.81 ± 0.08	h

(a) J. A. HIPPLE and D. P. STEVENSON: Phys. Rev. **63**, 121 (1943).
(b) P. J. LEGOFF: J. Chim. phys. **50**, 423 (1953).
(c) F. P. LOSSING, K. U. INGOLD and I. H. S. HENDERSON: J. Chem. Phys. **22**, 621 (1954); Applied Spectrometry, London: Institute of Petroleum **102** (1954).
(d) J. D. WALDRON: Trans. Faraday Soc. **50**, 102 (1954).
(e) A. LANGER and J. A. HIPPLE: Phys. Rev. **69**, 691 (1946).
(f) J. B. FARMER and F. P. LOSSING: Unpublished.
(g) F. P. LOSSING, J. B. FARMER and I. H. S. HENDERSON: Unpublished.
(h) J. B. FARMER, I. H. S. HENDERSON, C. A. MCDOWELL and F. P. LOSSING: J. Chem. Phys. **22**, 1948 (1954).

of the chemically related molecules which have been measured by COX[4] in some work as yet unpublished.

The results which have been obtained on the interpretation of ionization potentials have been reviewed by CRAGGS and MCDOWELL[5] and by MORRISON[3].

[1] R. S. MULLIKEN: J. Chem. Phys. **3**, 506, 514, 564 (1935).
[2] J. F. MULLIGAN: J. Chem. Phys. **19**, 347 (1951).
[3] J. D. MORRISON: Rev. Pure and Appl. Chem. **5**, 22 (1955).
[4] B. C. COX: Ph. D. Thesis, University of Liverpool 1953.
[5] J. D. CRAGGS and C. A. MCDOWELL: Rep. Progr. Phys. **18**, 374 (1955).

Useful information has been obtained about the ionization potentials of free radicals beginning with the pioneer work of ELTENTON[1] in 1942. These potentials are of special value in analysing energy balance equations for ionization processes in molecules so as to obtain bond dissociation energies or other quantities.

Fig. 45 illustrates ELTENTON's early source of radicals which were detected with a mass spectrometer. The first measurements of ionization potentials were carried out by HIPPLE and STEVENSON[2] using essentially a molecular beam source which provided CH_3 radicals from dissociation of $Pb(CH_3)_4$. Table 15 summarises some of the measurements which have been made[3].

25. H_2O, H_2S and H_2Se. MANN, HUSTRULID and TATE[4] studied the H_2O molecule, with the results and interpretations shown in Table 16. The critical potentials for O^+ (see Fig. 46) are interesting and it should be remembered that excited states of $O^+(^2D^0, {}^2P^0)$ at 3.3 and 5.0 ev respectively exist. H^- from the reaction

$$H_2O + e \rightarrow OH + H^-$$

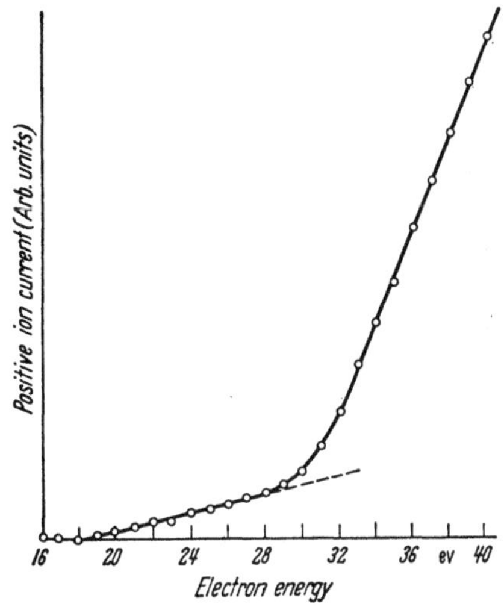

Fig. 46. Variation with electron energy of the current of O^+ ions produced by electron bombardment of H_2O.

should appear at 5.7 ev if LOZIER's kinetic energy data[5] are used, in good agreement with the observed critical potential of (5.6 ± 0.5) ev. It seems possible

Table 16. *Ions observed in water vapour. Electron affinity of oxygen taken as 1.5* ev, *of hydrogen as 0.7* ev.

	Relative intensity at 100 volts	Appearance potential (volts)	Probable process	Calculated minimum energy (ev)	Excess energy or deduced ioniz. pot. (ev)
H_2O^+	100	13.0 ± 0.2	$H_2O \rightarrow H_2O^+$		$I(H_2O) = 13.0$
OH^+	23.2	18.7 ± 0.2	$H_2O \rightarrow H + OH^+$	$I(OH) + 4.91$	$I(OH) = 13.8$
O^+	2.0	18.8 ± 0.5	$H_2O \rightarrow H_2 + O^+$	18.6	
		28.1 ± 1.0	$H_2O \rightarrow 2H + O^+$	23.1	$T(O^+) \leq 5.0$
H^+	5.0	19.5 ± 0.2	$H_2O \rightarrow OH + H^+$	18.4	$T(H^+) \leq 1.1$
H_2^+	0.07	23.0 ± 2.0	$H_2O \rightarrow O + H_2^+$	20.6	
H^-	0.6[6]	5.6 ± 0.5	$H_2O \rightarrow OH + H^-$	4.7	
			$H_2O \rightarrow O + H + H^-$	8.8	
O^-	0.15	7.5 ± 0.3	$H_2O \rightarrow 2H + O^-$	8.0	
		23.7 ± 0.5	$H_2O \rightarrow H + H^+ + O^-$	21.5	
		36.0 ± 3	$H_2O \rightarrow H^+ + H^+ + {}^-O$	35.0	

[1] G.C. ELTENTON: J. Chem. Phys. **10**, 403 (1942).
[2] J.A. HIPPLE and D.P. STEVENSON: Phys. Rev. **63**, 121 (1943).
[3] For a review of techniques employed in this field see F.P. LOSSING, K.U. INGOLD and I.H.S. HENDERSON: Applied Mass Spectrometry, p. 102. London: Institute of Petroleum 1954. See also F.P. LOSSING and A.W. TICKNER: J. Chem. Phys. **20**, 907 (1952); F.P. LOSSING, K.U. INGOLD and A.W. TICKNER: Disc. Faraday Soc. **14**, 34 (1953).
[4] M.M. MANN, J. HUSTRULID and J.T. TATE: Phys. Rev. **58**, 340 (1940).
[5] W.W. LOZIER: Phys. Rev. **52**, 843 (1930).
[6] At maximum.

that the subsidiary peaks in the O^- curve may be associated with excited states of H_2O^- and it would be of interest to check this with kinetic energy measurements.

The earlier work of Mann, Hustrulid and Tate has been repeated by Cox[1] and, as a result, a new interpretation of some of the data has been given. For

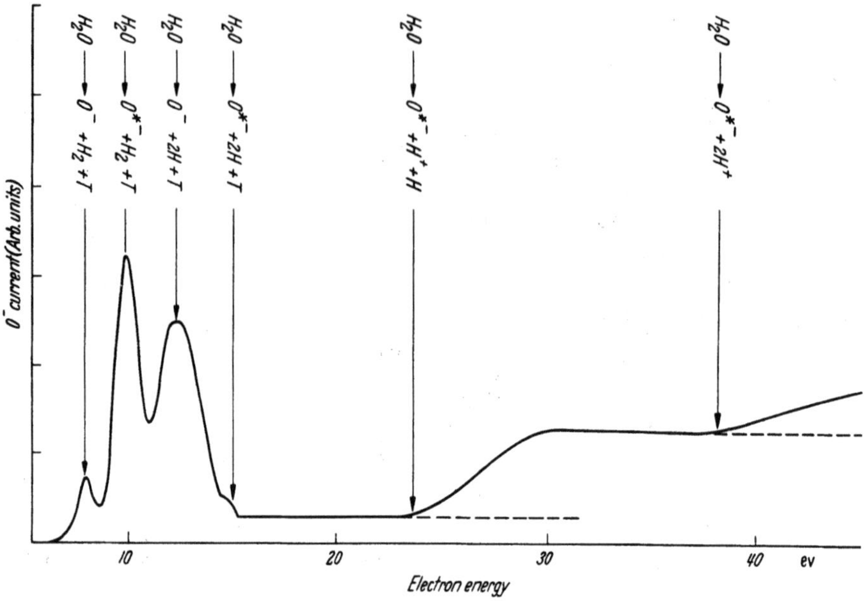

Fig. 47. Variation with electron energy of the current of O^- ions produced by electron bombardment of H_2O.

example, the production of O^- (Fig. 47) has been studied, and the first three appearance potentials are given in Table 17.

It can be shown that, ignoring kinetic energy and assuming $EA(O) = 1.5$ ev, the reactions

$$H_2O + e \rightarrow 2H + O^-,$$

and

$$H_2O + e \rightarrow H_2 + O^-,$$

require 8.1 and 3.8 ev respectively. Cox identifies the latter reaction with the process of appearance potential 6.55 volts (Table 17), with kinetic energy to resolve the balance.

Table 17. *Appearance potentials for O^- from H_2O.*

		Data of Mann et al.[2]	Data of Smyth and Mueller[3]
1st A. P.	6.55 ± 0.5	7.3	—
2nd A. P.	23.3 ± 0.5	23.7 ± 0.5	22 ± 3
3rd A. P.	36.4 ± 0.5	36 ± 3	36 ± 4

The above work on H_2O may be compared with a recent study of H_2S and H_2Se by Neuert[4], whose data on Se^- and S^- are of particular interest. The main collision processes may be put as

$$H_2X + e \rightarrow X^- + (2H \text{ or } H_2), \tag{25.1}$$

$$H_2X + e \rightarrow X^- + H_2^+ + e, \tag{25.2}$$

or

$$H_2X + e \rightarrow X^- + H^+ + H + e. \tag{25.3}$$

[1] B. C. Cox: Ph. D. thesis, University of Liverpool 1953.
[2] M. M. Mann, A. Hustrulid and J. T. Tate: Phys. Rev. **58**, 340 (1940).
[3] H. D. Smyth and D. W. Mueller: Phys. Rev. **43**, 116 (1933).
[4] H. Neuert: Z. Naturforsch. **8**a, 459 (1953).

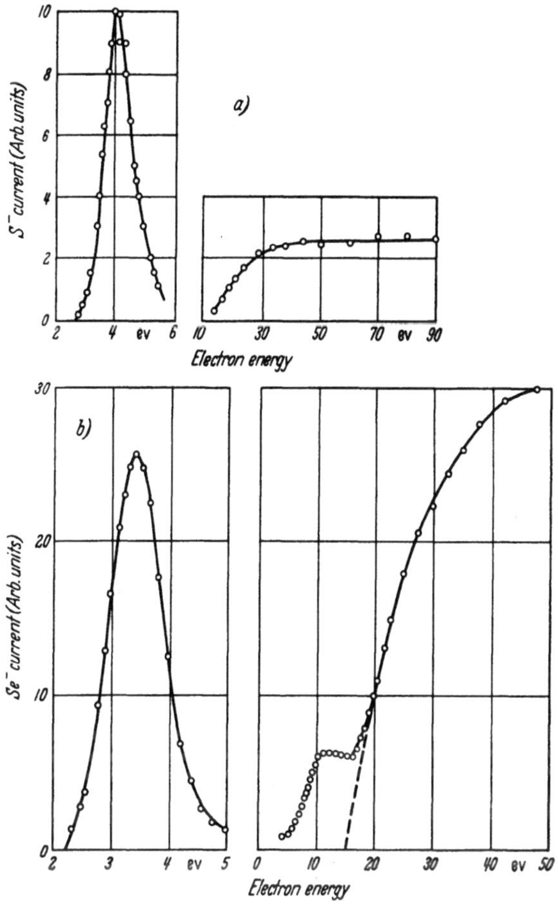

Fig. 48 a and b. Variation with electron energy of the current (a) of S^- ions produced by electron bombardment of H_2S, (b) Se^- ions produced by electron bombardment of H_2Se.

The negative ion yield curves for Se^- and S^- are shown in Figs. 48a and b. The results are summarised in Table 18 using the H_2O data given by MANN, HUSTRULID and TATE (Table 16).

Table 18.

	H_2O ev	H_2S ev	H_2Se ev
Dissociation energy for $H_2X \rightarrow X + 2H$	9.5	6.5	5.3
Dissociation energy for $H_2X \rightarrow X + H_2$	5.0	2.0	0.8
Appearance potential for process (25.1)	7.5	2.9 ± 0.3	2.3 ± 0.3
Electron affinity of X	1.5 to 1.9	about 3.5	about 3
Estimated appearance potential of process (25.2)	20.9	about 14.1	about 13.4
Estimated appearance potential of process (25.3)	18.3	about 16.6	about 15.9
Measured appearance potential process (25.2) or (25.3)	23.7	—	15.2 ± 1

26. NH_3. The ammonia molecule has been studied by MANN, HUSTRULID and TATE[1] with results which are summarised in Table 19.

Table 19. *Ions observed in ammonia.*

Ion	Relative intensity at 100 volts	Appearance potential (volts)	Probable process	Calculated minimum energy (ev)	Excess energy or deduced ioniz. pot. (ev)
NH_3^+	100	10.5 ± 0.1	$NH_3 \to NH_3^+$		$I(NH_3) = 10.5$
NH_2^+	78	15.7 ± 0.1	$NH_3 \to NH_2^+ + H$	$I(NH_2) + 3.8$ (?)	$I(NH_2) = 11.9$(?)
NH^+	4.5	19.4 ± 0.5	$NH_3 \to NH^+ + H_2$	$I(NH) + 3.0$	$I(NH) \leq 16.4$
		23.7 ± 0.5	$NH_3 \to NH^+ + 2H$	$I(NH) + 7.5$	
N^+	1.7	24.9 ± 0.5	$NH_3 \to N^+ + 3H$	25.4	
		28.0 ± 0.5(?)	$NH_3 \to N^{+*} + 3H$	27.3	$T \leq 0.7$
H^+	0.5	23.3 ± 0.5	$NH_3 \to NH + H + H^+$	21.0	
		26.9 ± 0.5	$NH_3 \to N^* + 2H + H^+$(?)	26.7	
H_2^+	0.02	15.5 ± 0.5	$H_2 \to H_2^+$	15.3	
			$NH_3 \to NH + H_2^+$(?)	18.3	
NH_3^{++}	0.02	42 ± 3	$NH_3 \to NH_3^{++}$		$I(NH_3^+) = 31.5$
H^-	0.15	5.8 ± 0.3	$NH_3 \to N + H_2 + H^-$	5.7	
		23.0 ± 1.0	$NH_3 \to NH^+ + H + H^-$	23.2	
NH_2^-	0.3	6.0 ± 0.5	$NH_3 \to H + NH_2^-$		

27. Halides of boron. α) *Boron trifluoride.* A study of collision processes occurring with this molecule is of interest because of its use in ionisation chamber work with neutrons (see ROSSI and STAUB's book[2] and other similar works). The only work so far available on the negative ions of this molecule is due to MARRIOTT[3] but OSBERGHAUS[4] independently carried out a brief study of the positive ions in the course of work on boron isotope abundances. The relative abundances and appearance potential data are given in Table 20.

MARRIOTT used the following data in his discussion of the above critical potential data: $D(BF_3) = 18.2$ ev[5]. $D(BF) = (4.5 \pm 1)$ ev[3]. $D(F_2) = (1.63 \pm 0.15)$ ev[4]. $EA(F) = (3.56 \pm 0.1)$ ev[6,7] and finally $I(B) = 8.296$ ev[8] and $I(F) = 17.422$ ev[8].

Table 20. *Electron collision data in* BF_3.

Ion	Relative abundance at 60 ev		Appearance potential (volts)	
	MARRIOTT	OSBERGHAUS	MARRIOTT	OSBERGHAUS
BF_3^+	6.1 (5.5)	3.0	15.7 ± 0.1	
BF_2^+	100 (100)	100	16.17 ± 0.05	17.0 ± 0.5
BF^+	0.82 (0.92)	1.1	(1) 11.2 ± 0.4	—
			(2) 25.2 ± 0.2	27.2 ± 1
B^+	1.3 (1.6)	3.0	30.6 ± 1	31.5 ± 1
BF_2^{++}	— (0.78)	—	—	—
F_2^+	— (0.12)	—	—	—
F^+	— (1.2)	—	31.5 ± 2	—
F^-	— (at peak 1.1)	—	11.4 ± 0.2	—

Bracketed abundances are those observed with 75 ev electrons.

[1] M. M. MANN, J. H. HUSTRULID and J. T. TATE: Phys. Rev. **58**, 340 (1940).
[2] B. ROSSI and H. H. STAUB: Ionization Chambers and Counters: Experimental Techniques. New York-Toronto-London: McGraw-Hill 1949.
[3] J. MARRIOTT: Ph. D. Thesis. Liverpool University 1954.
[4] O. OSBERGHAUS: Z. Physik **128**, 366 (1950).
[5] See reference 2, above.
[6] R. B. BERNSTEIN and M. METLAY: J. Chem. Phys. **19**, 1612 (1951).
[7] H. O. PRITCHARD: Chem. Rev. **52**, 529 (1953).
[8] G. HERZBERG: Atomic Structure and Spectra. New York: Dover Publications 1944.

BF_2^+. Since F^- is formed only in negligible quantities at electron energies higher than capture values, see Fig. 49, the relevant process must be

$$BF_3 + e \rightarrow BF_2^+ + F + 2e.$$

It follows then that $D(BF_2{-}F) + I(BF_2) \leq 16.2$ ev and taking a mean bond energy of 6 ev, $I(BF_2) \leqq 10$ ev.

BF^+. The ion current shows a tail, the onset of which is at 11.2 ev. If this is due to ionisation of thermally produced radicals at the filament then, as MARRIOTT pointed out, the energy required would be ≈ 12 ev, a very high value. The tail is believed to be due to an unknown impurity ($M/e = 30$). At 25.2 ev the BF^+ current is presumably due either to

$$BF_3 + e \rightarrow BF^+ + 2F + 2e,$$

or

$$BF_3 + e \rightarrow BF^+ + F_2 + 2e$$

giving $I(BF) \leq (11.5 \pm 1.2)$ ev and $\leq (13.1 \pm 1.3)$ ev respectively.

B^+. The critical potential of 30.6 volts is difficult to explain. The reactions

$$BF_3 + e \rightarrow B^+ + F_2 + F + 2e,$$

and

$$BF_3 + e \rightarrow B^+ + 3F + 2e,$$

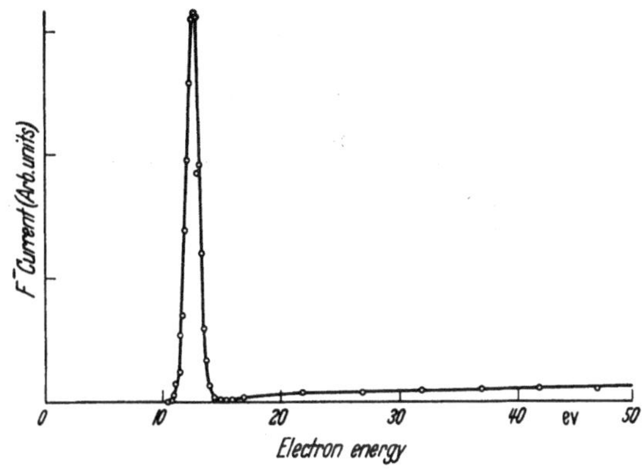

Fig. 49. Variation with electron energy of the current of F^- ions produced by electron bombardment of BF_3.

are possible, but they result in values of $T + X$ of (5.7 ± 1.2) ev and (4.1 ± 1.0) ev respectively, both unusually high.

F^+. The reaction at 31.5 ev may be

$$BF_3 + e \rightarrow BF_2 + F^+ + 2e \tag{27.1}$$

or

$$BF_3 + e \rightarrow BF + F^+ + 2e. \tag{27.2}$$

Then respectively, $D(BF_2{-}F) \leqq (14.1 \pm 2)$ ev and $T + X = (0.4 \pm 3.0)$ ev. Taking an average bond energy of 6 ev, the process (27.2) is more likely since (27.1) would involve an excess energy of ≈ 8 ev.

F^-. The molecule BF_3 is of unusual interest because the negative ion formation process is simple [as with $(CN)_2$] (see Fig. 49) but the critical energy is unusually high. This circumstance should make BF_3 a suitable molecule for resonance capture studies, since manipulation of the electron beam is less difficult than in many other cases (O_2, CBr_4, etc.) The following processes seem possible:

$$BF_3 + e \rightarrow BF_2 + F^-, \tag{27.3}$$

$$BF_3 + e \rightarrow BF^+ + F^- + F. \tag{27.4}$$

(27.3) seems relatively improbable since it would give $D(BF_2{-}F) \leqq 15$ ev and thus $(T + X) \approx 9$ ev to give a reasonable value of $D(BF_2{-}F)$. In (27.4) $(T + X)$ is calculated to be (1.4 ± 1.3) ev, a reasonable result. MARRIOTT[1] pointed out that (27.3) might be expected at a lower appearance potential, i.e. at, or slightly

[1] J. MARRIOTT: Ph. D. Thesis, Liverpool University 1954.

above, $D(BF_2-F)-EA(F)$ or ≈ 2.6 ev, but no F^- current could be found at this value of electron energy. The occurrence of the last process (27.4) without (27.3) is apparently rare and the reasons for the peculiar behaviour of BF_3 are, for the present, obscure.

β) Boron trichloride. The examination of BCl_3 by MARRIOTT[1] revealed the ease with which hydrolysis occurs—the resulting HCl may be readily identified with a mass spectrometer. Measurements on Cl^+, Cl^{++} and Cl^- are therefore valueless, and are not included in the data of Table 21.

The following data are used in the discussion of the collision processes in BCl_3: $D(BCl_3)=12.0$ ev[2], $D(BCl)=(3.5\pm0.5)$ ev[3], $I(Cl_2)=11.8$ ev[4]. Other requisite constants are given elsewhere in this chapter.

Table 21. *Electron collision data in* BCl_3.

Ion	Relative abundance at 60 ev		Appearance potential (volts)	
	MARRIOTT	OSBERGHAUS	MARRIOTT	OSBERGHAUS
BCl_3^+	33.4 (35)	27	12.03 ± 0.02	12.0 ± 0.5
BCl_2^+	100 (100)	100	13.01 ± 0.02	13.2 ± 0.5
BCl^+	6.6 (5.7)	8.2	18.54 ± 0.07	19.2 ± 0.5
B^+	2.4 (2.15)	6.2	13.6 ± 0.2	—
			18.4 ± 0.2	—
			22.35 ± 0.6	22.5 ± 0.5
BCl_2^{++}	— (2.6)	—	33.77 ± 0.07	—

Bracketed abundances are those observed with 75 ev electrons.

BCl_2^+. It appears that Cl^- is not formed in detectable amounts at 13 ev so that we expect

$$BCl_3 + e \rightarrow BCl_2^+ + Cl + 2e,$$

from which $D(BCl_2-Cl)+I(BCl_2)\lesssim 13.01$ ev. MARRIOTT thus shows, if an average B—Cl bond energy of 4 ev may be taken, that $I(BCl_2)\lesssim$ about 9 ev.

BCl^+. Again, there is no evidence of any accompanying Cl^- current, so either

$$BCl_3 + e \rightarrow BCl^+ + 2Cl + 2e$$

or

$$BCl_3 + e \rightarrow BCl^+ + Cl_2 + 2e$$

would appear to be operative. Then either $I(BCl)\lesssim(10.0\pm0.5)$ ev or $\lesssim(12.5\pm0.5)$ ev.

It was suspected that a low energy tail to the BCl^+ current curve might be caused by thermal decomposition at the filament. There is also evidence that B^+ (with an appearance potential of 13.6 ev) may be formed by ionisation of BCl radicals near the filament from the reaction:

$$BCl + e \rightarrow B^+ + Cl + 2e$$

which gives $T+X=(1.8\pm0.7)$ ev, a reasonable value.

[1] See footnote 1, p. 379.

[2] Selected Values of Chemical Thermodynamic Properties. Washington: Nat. Bur. Stand. Circ. No. 500 (1952).

[3] A. G. GAYDON: Dissociation Energies and Spectra of Diatomic Molecules, 2nd ed. London: Chapman & Hall 1952.

[4] J. D. MORRISON and A. C. J. NICHOLSON: J. Chem. Phys. **20**, 1021 (1952).

The appearance of B^+ at 18.4 ev may be due either to

$$BCl_3^+ + e \rightarrow B^+ + Cl_2 + Cl + 2e$$

or to

$$BCl_3 + e \rightarrow B^+ + 2\,Cl + Cl^- + e.$$

Substitution of the appropriate values in the energy balance equations give values of $(T+X)$ of respectively 0.6 ± 0.2 ev and 1.8 ± 0.2 ev both of which are reasonable, so that no certain decision may be made.

The B^+ current commencing at 22.35 ev may be due to

$$BCl_3 + e \rightarrow B^+ + 3\,Cl + 2e$$

from which $(T+X) = 2$ ev, a not unreasonable value.

28. Methane CH_4. We shall deal in detail with this molecule, since it is the simplest saturated hydrocarbon and provides an important basis of comparison with other similar molecules such as the fluoro-chloro-methanes and methyl halides. We do not propose to deal in detail with more complex hydrocarbons, for which BARNARD's book[1] or the literature should be consulted. Data for CH_4 are given in Tables 22 and 23.

Table 22. *Relative abundances of certain ions from* CH_4 *produced by impact with* 50 ev *electrons.*

Ion	Reference			
	a	b	c	d
CH_4^+	100 (standard)	100	100	100
CH_3^+	77.9	84.6	85	76.4
CH_2^+	8.3	15.9	15.7	15.8
CH^+	3.4	7.9	6.8	6.0
C^+	1.2	2.4	2.7	2.1

(a) L. SMITH: Phys. Rev. **51**, 553 (1937).
(b) V. B. DIBELER and F. L. MOHLER: J. Res. Nat. Bur. Stand. **40**, 25 (1948).
(c) J. GEERK and H. NEUERT: Z. Naturforsch. **5**, 502 (1950).
(d) C. A. McDOWELL and J. W. WARREN: Faraday Soc. Disc. **10**, 53 (1951).

Table 23. *Appearance potentials (volts) for certain ions from* CH_4.

Ion	Reference				
	a	b	c	d	e
CH_4^+	13.2 ± 0.4	13.0 ± 0.2	13.04 ± 0.02	13.3 ± 0.3	13.12 ± 0.02
CH_3^+	14.5 ± 0.4	14.4 ± 0.3	14.57 ± 0.05	14.5 ± 0.3	14.39 ± 0.02
CH_2^+	15.8 ± 0.5	20.0 ± 0.3	—	16.5 ± 0.3	15.30 ± 0.5
	23.0 ± 0.8	—	—	21.5 ± 1.0	20.1 ± 0.1
CH^+	23.4 ± 0.6	24.3 ± 0.5	—	22.5 ± 0.5	22.4 ± 0.1
	—	—	—	27.0 ± 1	—
C^+	26.8 ± 0.8	—	—	21.5 ± 1	26.2 (upper limit 27.3)
				29.5 ± 1	

(a) L. SMITH: Phys. Rev. **51**, 663 (1937).
(b) M. B. KOFFEL and R. A. LAD: J. Chem. Phys. **16**, 420 (1948).
(c) J. J. MITCHELL and F. F. COLEMAN: J. Chem. Phys. **17**, 44 (1949).
(d) J. GEERK and H. NEUERT: Z. Naturforsch. **5** A, 502 (1950).
(e) C. A. McDOWELL and J. W. WARREN: Faraday Soc. Disc. **10**, 53 (1951).

[1] G. P. BARNARD: Modern Mass Spectrometry. London: Institute of Physics 1953.

For details of the methods of measuring appearance potentials used in obtaining the data of Tables 22 and 23 the original papers should be consulted. The ion current-voltage curves[1] are of various shapes, as shown in Figs. 50, 51, 52 and 53, for CH_4^+, CH_3^+, CH_2^+ and C^+.

CH_3^+. This ion is produced in the reaction

$$CH_4 + e \rightarrow CH_3^+ + H + 2e,$$

from which $A(CH_3^+) \geq I(CH_3) + D(CH_3-H)$.

It was shown[1] that the CH_3^+ ion is produced with little or no kinetic energy. Also, the dissociation energy of the C—H bond or $D(CH_3-H)$ is to be taken[2] as

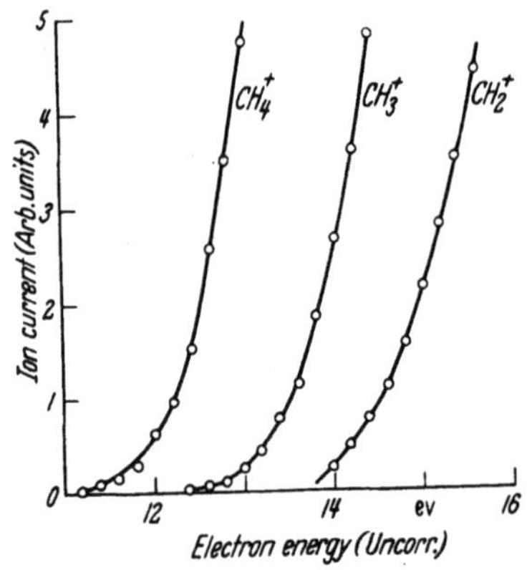

Fig. 50. Variation with electron energy, near the threshold, of the currents of CH_4^+, CH_3^+ and CH_2^+ ions produced by electron bombardment of CH_4.

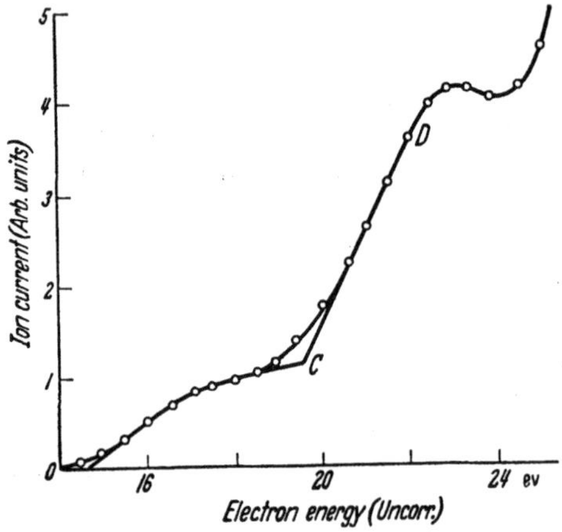

Fig. 51. Variation with electron energy of the current of CH_2^+ ions produced by electron bombardment of CH_4.

(4.43 ± 0.05) ev, and $A(CH_3^+) = (13.12 \pm 0.02)$ ev so that $I(CH_3^+) \leq (9.96 \pm 0.05)$ ev, in good agreement with the directly obtained value, due to HIPPLE and STEVENSON[3], of (10.08 ± 0.1) ev.

CH_2^+. There are two reactions to be considered, namely

$$CH_4 + e \rightarrow CH_2^+ + H_2 + 2e, \tag{28.1}$$

$$CH_4 + e \rightarrow CH_2^+ + 2H + 2e. \tag{28.2}$$

The difference in appearance potentials for ions of zero kinetic energy for these two processes will be $D(H_2)$ i.e. 4.48 ev. The value of $A(CH_2^+)$ of 15.3 ev may be taken for reaction (28.1). (28.2) will then begin at about 19.8 ev (see Fig. 51).

At 15.3 ev we have $A(CH_2^+) \geq I(CH_2) + D(CH_3-H) + D(CH_2-H) - D(H_2)$. There was found to be little or no kinetic energy of CH_2^+ involved, but in processes such as that being considered, where several fragments are produced, some of the lighter products of dissociation may carry away kinetic energy which would be difficult to detect or measure. LANGER and HIPPLE[4] give $I(CH_2) = 11.9 \pm 0.2$ ev so taking $D(CH_3-H)$ as (4.43 ± 0.05) ev, we obtain $D(CH_2-H) \lesssim 3.45 \pm 0.2$ ev.

[1] C. A. MCDOWELL and J. W. WARREN: Faraday Soc. Disc. **10**, 53 (1951).
[2] G. B. KISTIAKOWSKY and A. VAN ARTSDALEN: J. Chem. Phys. **12**, 462 (1944).
[3] J. A. HIPPLE and D. P. STEVENSON: Phys. Rev. **63**, 121 (1943).
[4] A. LANGER and J. A. HIPPLE: Phys. Rev. **59**, 691 (1946).

CH⁺. The reactions involved (see Fig. 52) are

$$CH_4 + e \rightarrow CH^+ + 3H + 2e,$$
$$CH_4 + e \rightarrow CH^+ + H_2 + H + 2e.$$

If the first of these is operative, then

$$A(CH^+) \geqq I(CH) + D(CH_3-H) + D(CH_2-H) + D(CH-H).$$

Using $I(CH) = (11.1 \pm 0.2)$ ev[1] and other data given above, we find $D(CH-H) \leqq (3.4 \pm 0.3)$ ev.

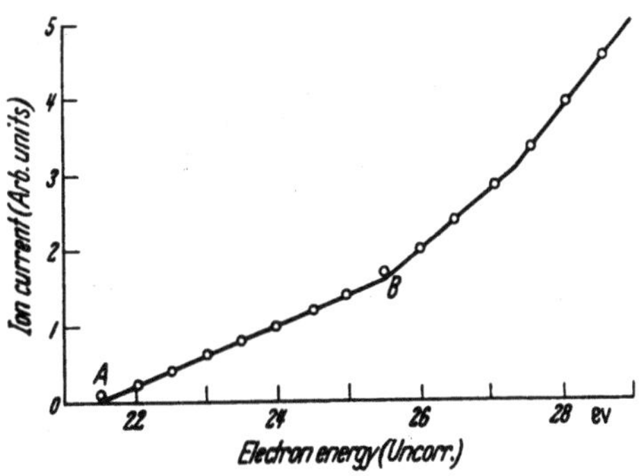

Fig. 52. Variation with electron energy of the current of CH^+ ions produced by electron bombardment of CH_4.

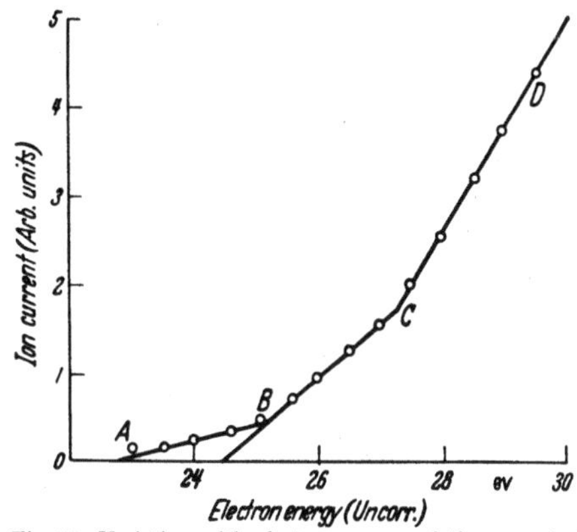

Fig. 53. Variation with electron energy of the current of C^+ ions produced by electron bombardment of CH_4.

C⁺ (see Fig. 53). The appropriate collision processes are:

$$CH_4 + e \rightarrow C^+ + 4H + 2e,$$
$$CH_4 + e \rightarrow C^+ + 2H_2 + 2e,$$
$$CH_4 + e \rightarrow C^+ + H_2 + 2H + 2e.$$

SMITH[2] chose the first of these and McDOWELL and WARREN[3] also favoured it, although it was admitted that no final proof could be given. Then

$$A(C^+) \geqq I(C) + D(CH_3-H) + D(CH_2-H) + D(CH-H) + D(C-H).$$

Since $I(C) = 11.27$ ev (BACHER and GOUDSMIT[4]) we find $D(CH) \leqq 3.6$ ev.

If a, b, c, and d denote the values of $D(CH_3-H)$, $D(CH_2-H)$, $D(CH-H)$ and $D(C-H)$ respectively, then[5] the latent heat of sublimation, L_1 of carbon into ground state atoms at 25° C is given by

$$L_1 = a + b + c + d - 226.1 \text{ kcal}.$$

We have seen that the electron impact value of $D(CH_3-H)$ is in good agreement with other values and, further, than $D(CH) \leqq 3.6$ ev agrees with HERZBERG's spectroscopic value[6] of 3.5 ev. KISTIAKOWSKY and ROSENBERG[7] have deduced

[1] A.E. DOUGLAS and G. HERZBERG: Canad. J. Res. A **20**, 71 (1942).
[2] L. SMITH: Phys. Rev. **51**, 663 (1937).
[3] C.A. McDOWELL and J.W. WARREN: Faraday Soc. Disc. **10**, 53 (1951).
[4] R.F. BACHER and S. GOUDSMIT: Atomic Energy States. New York-Toronto-London: McGraw-Hill 1932.
[5] L.A. LONG and R.G.W. NORRISH: Proc. Roy. Soc. Lond., Ser. A **187**, 337 (1946).
[6] G. HERZBERG: Molecular Spectra and Molecular Structure. I. Diatomic Molecules. New York: Prentice-Hall 1950.
[7] G.B. KISTIAKOWSKY and N.W. ROSENBERG: J. Amer. Chem. Soc. **72**, 321 (1950).

that $D(CH_2-H)$ must be 80 kcal/mole (3.5 ev) or even higher. The present electron impact value is $\leqq(3.45\pm0.2)$ ev, in fair agreement. The value of $D(CH-H)\lesssim(3.4\pm0.3)$ ev is even further from an accurate check, although[1]. recent theoretical studies suggest a value of about 80 kcal/mole or 3.5 ev. The value of $D(C-H)$ taken above i.e. $\lesssim 3.6$ ev, agrees reasonably with the spectroscopic value of 3.5 ev [2]. Using these values of a, b, c and d in the equation above, $L_1=125$ kcal/mole. This may be compared with values obtained by other methods and which vary from 125 up to about 170 kcal/mole (full references and discussion are given by McDOWELL and WARREN [3]). The latter value is in some disfavour, but it is not inconsistent with the process

$$CH_4+e\rightarrow CH^+ + H_2 + H + 2e$$

if the kinetic energy term were ≈ 2.4 ev, of which there appears to be little indication in the electron impact work.

Further, the appearance of the C^+ ion has been associated with $CH_4+e\rightarrow C^+ + 4H + 2e$ (at 26.2 ev) and putting $D(H_2)=4.5$ ev, $I(C)=11.27$ ev[4], we obtain

$$L_1\approx 5.2\text{ ev} = 120\text{ kcal/mole}.$$

However, if $A(C^+)=27.3$ ev, as shown above to be a possibility, then $L_1\approx$ 140 kcal/mole as a maximum.

CH^+ appears, it is suggested, from

$$CH_4+e\rightarrow CH^+ + 3H + 2e.$$

Putting $I(CH)=11.1$ ev[5] and $D(CH)=3.5$ ev[5] then the process

$$CH_4\rightarrow C + 4H$$

requires 14.8 ev. With a heat of formation of methane of 0.74 ev, $L_1\approx 120$ kcal per mole.

Further work with CH_3CN (Sect. 30) leads to another assessment of $D(CH-H)$.

29. CH_3OH and C_2H_5OH. *α) CH_3OH.* The positive ion spectra are illustrated in Fig. 54 and the appearance potentials are given in Table 24.

Table 24. *Appearance potentials of ions from* CH_3OH.

Ion	Appearance potentials (volts)		Ion	Appearance potentials (volts)	
	COX [6]	CUMMINGS and BLEAKNEY [7]		COX [6]	CUMMINGS and BLEAKNEY [7]
CH_3O^+	12.02 ± 0.03	11.8 ± 0.1	CH_3^+	14.48 ± 0.03	14.0 ± 0.5
CH_2O^+	11.00 ± 0.07	12.4 ± 0.1	CH_2^+	15.65 ± 0.15	15.4 ± 0.5
CHO^+	13.46 ± 0.11	14.2 ± 0.2	CH^+	23.4 ± 0.5	
CO^+	13.82 ± 0.1	—		27.5 ± 0.5	
OH^+	17.8 ± 0.5	—		29.9 ± 0.5	
O^+ and CH_4^+	13.9 ± 0.5	—	C^+	23.4 ± 0.5	
CH_3O^{++}	35.4 ± 2	—			

[1] G. B. KISTIAKOWSKY and A. VAN ARTSDALEN: J. Chem. Phys. **12**, 462 (1944). — L. A. LONG: Proc. Roy. Soc. Lond., Ser. A **198**, 62 (1949).

[2] G. HERZBERG: Molecular Spectra and Molecular Structure. I. Diatomic Molecules. New York: Prentice-Hall 1950.

[3] C. A. McDOWELL and J. W. WARREN: Faraday Soc. Disc. **10**, 53 (1951).

[4] R. F. BACHER and S. GOUDSMIT: Atomic Energy States. New York-Toronto-London: McGraw Hill 1932.

[5] A. E. DOUGLAS and G. HERZBERG: Canad. J. Res. A **20**, 71 (1942).

[6] B. C. COX: Ph. D. thesis, University of Liverpool 1953.

[7] C. S. CUMMINGS and W. BLEAKNEY: Phys. Rev. **58**, 787 (1940).

Cox has discussed the ion of mass 31, interpreted by CUMMINGS and BLEAKNEY[1] as $H_2C{=}O{-}H^+$ and not CH_3O^+. The CH_3^+ ion data lead to a value of $D(HO{-}H)$ of 131.4 kcal/mole (5.7 ev), whilst a recent reliable estimate[2] is 118 kcal/mole (5.1 ev), suggesting that about 0.5 ev of kinetic energy is produced at $A(CH_3^+)$ from CH_3OH. Cox interprets the ion of mass 29 as $HC{=}O^+$, produced in the reaction

$$CH_3OH + e \rightarrow HC{=}O^+ + H_2 + H + 2e,$$

from which $A(CH{=}O^+) \geq I(HC{=}O) + D(CH_3O{-}H) + D(OCH_2{-}H) + D(OCH{-}H) - D(C{=}O)$. The heat Q of the reaction $CH_3OH \rightarrow HC{=}O + H_2 + H$, i.e., $Q = Q_f(CHO) + Q_f(H) - Q_f(CH_3OH)$, gives the value of the sum of the dissociation energy terms. $Q_f(CHO)$ was calculated to be +0.3 kcal/mole (0.01 ev) giving $Q = 4.49$ ev, and so $I(CHO) \leqq 8.97$ ev.

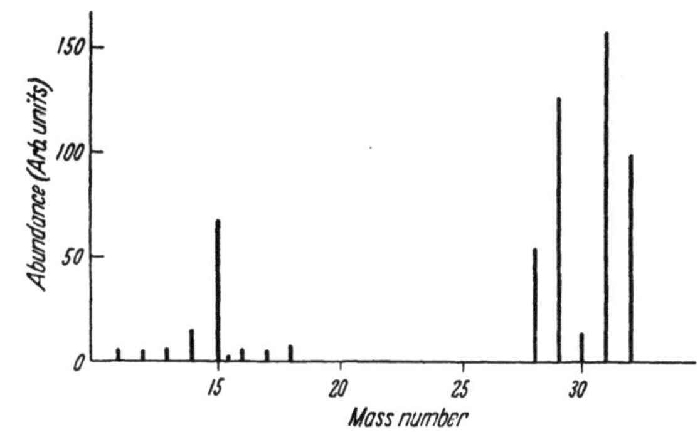

Fig. 54. Relative abundances of positive ions produced from CH_3OH by bombardment with 75 volt electrons.

β) C_2H_5OH. The ion spectrum is illustrated in Fig. 55 and appearance potential data are given in Table 25.

Table 25. *Appearance potentials of ions from* C_2H_5OH.

Ion	Appearance potentials (volts) Cox[3]	Appearance potentials (volts) CUMMINGS and BLEAKNEY[1]	Ion	Appearance potentials (volts) Cox[3]	Appearance potentials (volts) CUMMINGS and BLEAKNEY[1]
$C_2H_5O^+$	10.78 ± 0.05	11.0 ± 0.1	CH_2O^+	12.25 ± 0.05	12.8 ± 0.2
	12.67 ± 0.12		CHO^+ and $C_2H_5^+$	13.18 ± 0.11	
$C_2H_4O^+$	10.45 ± 0.12	11.4 ± 0.3	CO^+ and $C_2H_4^+$	11.28 ± 0.11	
$C_2H_3O^+$	11.57 ± 0.2			13.14 ± 0.1	
	12.92 ± 0.12			14.64 ± 0.07	
	13.44 ± 0.07		$C_2H_3^+$	14.51 ± 0.05	14.5 ± 0.2
	15.7 ± 0.2			17.47 ± 0.1	
$C_2H_2O^+$	13.7 ± 0.2		$C_2H_2^+$	12.87 ± 0.15	14.1 ± 0.2
	18.19 ± 0.2		C_2H^+	21 ± 1	
C_2HO^+	14.44 ± 0.07			27.4 ± 2	
	21.5 ± 0.3		C^+	31.5 ± 2	
CH_4O^+	11.78 ± 0.15		CH_3^+	14.99 ± 0.07	
	14.02 ± 0.2		CH_2^+	16.07 ± 0.02	16.5 ± 0.5
CH_3O^+	11.48 ± 0.08	11.3 ± 0.2		18.74 ± 0.25	
	13.9 ± 0.14				

Cox proposed the following interpretation of these data. The ion of mass 45 could have either the structure $C_2H_5O^+$ or $H_3CCH{=}OH^+$. The latter will require less energy and thus presumably occurs at 10.78 ev and the former at 12.67 ev. We have

$$A(C_2H_5O^+) \geq I(C_2H_5O) + D(C_2H_5O{-}H),$$

[1] C. S. CUMMINGS and W. BLEAKNEY: Phys. Rev. **58**, 787 (1940).
[2] R. J. DWYER and O. OLDENBERG: J. Chem. Phys. **12**, 351 (1944).
[3] B. C. Cox: Ph. D. thesis, University of Liverpool 1953.

and using REBBERT and LAIDLER's value[1] of 4.38 ev for $D(C_2H_5O-H)$, $I(C_2H_5O) \leq$ 8.29 ev.

The ion of mass 44 is probably CH_3CHO^+, with an ionisation potential of 10.42 volts[2] although the agreement between the calculated appearance potential and the experimental value of 10.45 ev is not good (a discrepancy of about 0.3 ev).

The ion of mass 31 is probably $H_2C{=}OH^+$, for which interpretation COX advanced detailed arguments as follows. The appearance potentials of this ion in C_2H_5OH and CH_3OH may be used, as a check on the above structure interpretation, to calculate $D(CH_3-H)$ as follows. We have

$$CH_3OH + CH_4 \rightarrow C_2H_5OH + 2H + Q$$

$$C_2H_5OH \rightarrow CH_2OH^+ + CH_3 + A_1,$$

$$CH_2OH^+ + H \rightarrow CH_3OH - A_2,$$

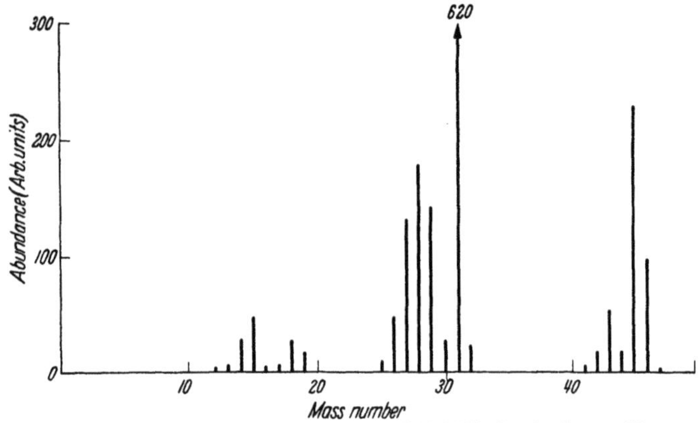

Fig. 55. Relative abundances of positive ions produced from C_2H_5OH by bombardment with 75 volt electrons.

so
$$D(CH_3-H) = Q - (A_2 - A_1) = (101 \pm 2) \text{ kcal/mole},$$

in excellent agreement with the accepted value.

The ion of mass 28 may be formed either by the process:

$$C_2H_5OH + e \rightarrow C_2H_4^+ + HOH + 2e, \text{ involving energy } Q_1,$$

or by

$$C_2H_5OH + e \rightarrow C_2H_4^+ + OH + H + 2e, \text{ involving energy } Q_2.$$

$$I(C_2H_4) = 10.62 \text{ ev}^3$$

so,

$$Q_1 = 10.62 + Q_f(C_2H_4) + Q_f(HOH) - Q_f(C_2H_5OH)$$

and

$$Q_2 = 10.62 + Q_f(C_2H_4) + Q_f(H) + Q_f(OH) - Q_f(C_2H_5OH)$$
$$= 16.3 \text{ ev}.$$

Q_1 corresponds to the measured appearance potential of 11.3 ev if 0.2 ev is allowed for initial kinetic energy.

[1] R. E. REBBERT and K. J. LAIDLER: J. Chem. Phys. **20**, 574 (1952).
[2] J. W. WARREN: Ph. D. thesis, University of London 1950.
[3] R. E. HONIG: J. Chem. Phys. **16**, 105 (1948).

The ion of mass 27 may similarly be interpreted as $H_2C{=}CH^+$ and, since the initial kinetic energy is apparently zero, within the limits of error, $I(C_2H_3) = (9.55 \pm 0.1)$ ev.

30. CH_3CN and $(CN)_2$. *α) CH_3CN.* McDowell and Warren[1] have investigated this molecule, studying both negative and positive ions. The appearance potentials were measured where possible by the method of extrapolated voltage differences but where the currents involved were too small for this technique to be used, linear intercepts were used with some loss of accuracy. Some of the data obtained are given in Tables 26 and 27.

Table 26. *Relative abundances of ions from* CH_3CN (50 ev *electrons*).

Ion	Relative abundance	Ion	Relative abundance
CH_3CN^+	100 (standard)	C_2^+	1.0
CH_2CN^+	50	CH_3^+	2.3
$CHCN^+$	14.7	CH_2^+, N^+	8.6
C_2N^+	8.7	CH^+	2.5
$C_2H_3^+$, HCN^+	4.8	C^+	3.7
$C_2H_2^+$, CN^+	4.3	CN^-	—
C_2H^+	2.9		

The appearance potential of CN^- from CH_3CN was obtained by using O^- from oxygen as a standard [$A(O^-) = 4.7$ ev from the work of Lozier (Sect. 22)].

We proceed to discuss the appearance potential data for the positive ions.

Table 27. *Appearance potentials of positive ions from CH_3CN.*

Ion	Appearance potential (volts)		Ion	Appearance potential (volts)	
	Extrapolated voltage differences	Linear intercepts		Extrapolated voltage differences	Linear intercepts
CH_3CN^+	12.52 ± 0.02	12.77 ± 0.03	CH_2^+	(i) 15.62 ± 0.04	16.0 ± 0.1
				(ii) —	20.0 ± 0.2
CH_2CN^+	14.30 ± 0.02	14.47 ± 0.06	CH^+	—	22.4 ± 0.2
					25.7 ± 0.2
$CHCN^+$	(i) 15.1 ± 0.1	15.7 ± 0.1	C^+	—	22.6 ± 0.2
	(ii) —	20.0 ± 0.2	CN^-	—	27.0 ± 0.3
CN^+	—	20.3 ± 0.1			5.6
CH_3^+	14.6 ± 0.1	14.8 ± 0.2			

CH_3CN^+. This arises from simple ionisation. The value $I(CH_3CN) = 12.52$ ev and is appreciably greater than Cutler's[2] spectroscopic value of 11.96 ev. This disagreement has not yet been satisfactorily resolved.

CH_2CN^+ is formed by

$$CH_3CN + e \rightarrow CH_2CN^+ + H + 2e,$$

so that

$$A(CH_2CN^+) \geq D(H{-}CH_2CN) + I(CH_2CN).$$

If we assume[3] that $D(H{-}CH_2CN)$ is the same as $D(CH_3{-}H)$ from methane then $I(CH_2CN) = 9.87$ ev.

[1] C. A. McDowell and J. W. Warren: Faraday Soc. Disc. **10**, 53 (1951). — Trans. Faraday Soc. **48**, 1084 (1952).

[2] J. A. Cutler: J. Chem. Phys. **16**, 136 (1948).

[3] G. B. Kistiakowsky and A. van Artsdalen: J. Chem. Phys. **12**, 462 (1944).

$CHCN^+$. This ion may be produced in the reactions

$$CH_3CN + e \rightarrow CHCN^+ + H_2 + 2e,$$

$$CH_3CN + e \rightarrow CHCN^+ + 2H + 2e,$$

and the appearance potentials, other quantities being equal, should differ by $D(H_2)$, i.e. 4.5 ev. This is indeed the case (see Table 27) and both processes therefore occur.

CH_3^+. There are two possible reactions, i.e.

$$CH_3CN + e \rightarrow CH_3^+ + CN + 2e, \tag{30.1}$$

$$CH_3CN + e \rightarrow CH_3^+ + CN^- + e. \tag{30.2}$$

Since the CN^- current is negligible at the appropriate electron energies we conclude that only reaction (30.1) occurs, with appreciable probability, so that

$$A(CH_3) = D(CH_3{-}CN) + I(CN_3) + (T + X).$$

It has been shown that CH_3^+ is produced without kinetic energy so, ignoring X,

$$A(CH_3^+) = D(CH_3CN) + I(CH_3).$$

Taking $I(CH_3) = (10.08 \pm 0.1)$ ev from the work of HIPPLE and STEVENSON[1], $D(CH_3{-}CN)$ is (4.52 ± 0.14) ev.

CN^+. The reaction

$$CH_3CN + e \rightarrow CH_3 + 2e$$

gives

$$A(CN^+) = D(CH_3CN) + I(CN) + T,$$

again ignoring X. The excess kinetic energy term T is probably not negligible in this case, so only an upper limit of 15.78 ev for $I(CN)$ may be found.

CH_2^+. There are two possible reactions, given by McDOWELL and WARREN[2] namely

$$CH_3CN + e \rightarrow CH_2^+ + HCN + 2e,$$

and

$$CH_3CN + e \rightarrow CH_2^+ + H + CN + 2e.$$

CN^- is not formed with CH_2^+ since the former is found only in a capture peak at lower energies [as in Fig. 56 for $(CN)_2$]. Beam width measurements show that the CH_2^+ has little or no kinetic energy[3] so we may write

$$A(CH_2^+) = D(CH_3CN) - D(HCH) + D(CH_2{-}H) + I(CH_2).$$

Using some of the methane data (Sect. 28) we have

$$D(CH_2{-}H) + I(CH_2) = (15.38 \pm 0.07)\ \text{ev},$$

and hence

$$D(CH_3{-}CN) - D(HCN) = (0.24 \pm 0.08)\ \text{ev}.$$

The difference between the two appearance potentials for the CH_2^+ ion i.e. ≈ 4.0 ev should give approximately $D(HCN)$ and so $D(CH_3{-}CN) = 4.2$ ev.

---

[1] J.A. HIPPLE and D.P. STEVENSON: Phys. Rev. **63**, 121 (1943).

[2] C.A. McDOWELL and J.W. WARREN: Faraday Soc. Disc. **10**, 53 (1951). — Trans. Faraday Soc. **48**, 1084 (1952).

CH^+. This ion appears at 22.4 ev with a break in the yield curve at 25.4 ev. We have

$$CH_3CN + e \rightarrow CH^+ + 2H + CN + 2e,$$

and

$$CH_3CN + e \rightarrow CH^+ + 2H + CN\,(B\,^2\Sigma^+) + 2e.$$

CH^+ also appears in methane at 22.4 ev. Thus

$$A(CH^+)\,(CH_3CN) = D(CH_3CN) + D(CH_2{-}H) + D(CH{-}H) + I(CH)$$

and

$$A(CH^+)\,(CH_4) = D(CH_3{-}H) + D(CH_2{-}H) + D(CH{-}H) + I(CH),$$

where $A(CH^+)\,(CH_3CN)$ denotes the appearance energy for CH^+ ions from CH_3CN. It was shown also by McDowell and Warren[1] in their work on CH_4 (Sect. 28) that $D(CH_2{-}H) \lesssim (3.45 \pm 0.2)$ ev; also $I(CH) = (11.13 \pm 0.2)$ ev[2], so, putting $D(CH_3CN) \approx 4.3$ ev, we obtain 3.5 ev for $D(CH{-}H)$. The latter value requires the latent heat of vaporization of carbon to be ≈ 125 kcal/mole, markedly different from the value of 170 kcal/mole obtained by the more direct and normal method (see also Sect. 28, p. 384).

C^+. This ion has critical potentials at (22.6 ± 0.2) and (27.0 ± 0.3) volts. The most likely reactions are:

$$CH_3CN + e \rightarrow C^+ + 3H + CN + 2e, \tag{30.3}$$

$$CH_3CN + e \rightarrow C^+ + 2H + HCN + 2e, \tag{30.4}$$

$$CH_3CN + e \rightarrow C^+ + H + H_2 + CN + 2e. \tag{30.5}$$

The first (30.3) is probably that occurring at 27 ev, and, since C^+ was shown[1] to be produced without appreciable kinetic energy, we have

$$A(C^+) = D(CH_3CN) + D(CH_2{-}H) + D(CH{-}H) + I(C).$$

Using the methane data (Sect. 28) we find

$$A(C^+)\,(CH_3CN) - A(C^+)\,(CH_4) = D(CH_3CN) - D(CH_3{-}H)$$

and since

$$A(C^+)\,(CH_3CN) \approx A(C^+)\,(CH_4),$$

then

$$D(CH_3CN) \approx D(CH_3{-}H).$$

It may be shown that the critical electron energies required for the processes (30.4) and (30.5) above are approximately equal. A balance set up for either reaction results in a value of $D(CH{-}H) \approx 4.0$ ev.

Table 28. *Appearance potentials of ions from C_2N_2.*

Ion	Appearance potentials (volts) according to reference			
	a	b	c	d
$C_2N_2^+$	13.5	14.1	—	13.57 ± 0.02
CN^+	18.0	21.3	20.31 ± 0.2	20.4 ± 0.3
CN^-				4.4 ± 0.2

(a) K. E. Dorsch and H. Kallmann: Z. Physik **60**, 376 (1930).
(b) J. T. Tate, P. T. Smith and A. L. Vaughan: Phys. Rev. **48**, 525 (1935).
(c) D. P. Stevenson: J. Chem. Phys. **18**, 1347 (1950).
(d) C. A. McDowell and J. W. Warren: Trans. Faraday Soc. **48**, 1084 (1952).

β) C_2N_2. Cyanogen has been studied in various laboratories. We use here the results of McDowell and Warren[3]. Appearance potentials are given in Table 28 and the yield curve for the CN^- ion in Fig. 56. The following comments may be made.

[1] C. A. McDowell and J. W. Warren: Faraday Soc. Disc. **10**, 53 (1951).
[2] A. E. Douglas and G. Herzberg: Canad. J. Res. A **20**, 71 (1942).
[3] C. A. McDowell and J. W. Warren: Trans. Faraday Soc. **48**, 1084 (1952).

$C_2N_2^+$. We have $C_2N_2 + e \rightarrow C_2N_2^+ + 2e$, whence $I(C_2N_2) = (13.57 \pm 0.02)$ ev. PRICE and WALSH[1] give a spectroscopic value of 13.8 ev.

CN^+. Since CN^- arises only from a capture process (see Fig. 56) the formation of CN^+ is probably by

$$C_2N_2 + e \rightarrow CN^+ + CN + 2e.$$

Qualitative measurements[2] indicated that CN^+ was produced with appreciable kinetic energy.

McDOWELL and WARREN[2] have worked out $D(\mathrm{CN{-}CN})$ and $D(\mathrm{HCN})$ from the above data and we shall quote their discussion in detail, as an example of the relevance of electron impact data to topical thermo-chemical matters.

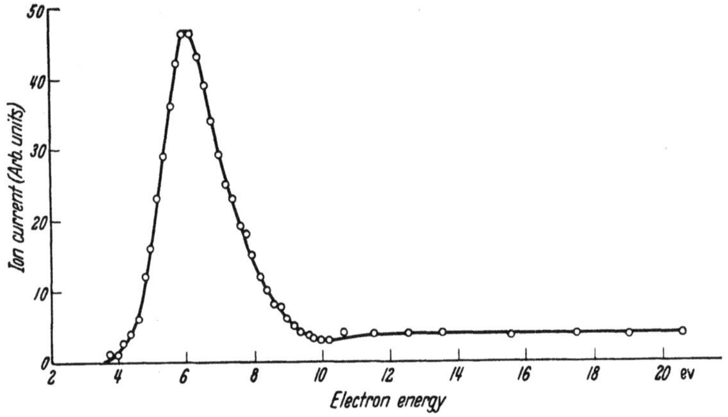

Fig. 56. Variation with electron energy of the current of CN^- ions produced by electron impact in C_2N_2.

$D(\mathrm{CH_3CN})$, which is 4.52 ev, equals the heat of reaction of the process

$$CH_3CN = CH_3 + CN,$$

which is $D(\mathrm{CH_3CN}) = \Delta H_f^0(\mathrm{CH_3}) + \Delta H_f^0(\mathrm{CN}) - \Delta H_f^0(\mathrm{CH_3CN})$. Here $\Delta H_f^0(\mathrm{X})$ refers to the heat of formation of X from its elements in their standard states at 0° K and the use of such quantities would result in $D(\mathrm{XY})$ at 0° K. It is more convenient and usual to work with heats of formation and dissociation energies determined at 25° C, so the latter will be used. $\Delta H_f^0(\mathrm{CH_3})$ at 25° C is 31 kcal/mole[3] and LONG[4] has given the heat of formation of gaseous CH_3CN as 19.8 kcal/mole at 18° C, which becomes 19.5 kcal/mole at 25° C. Using these data and $D(\mathrm{CH_3{-}CN}) = 104.2$ kcal/mole, in the last equation, $\Delta H_f^0(\mathrm{CN})$ is 92.7 kcal/mole at 25° C. The dissociation energy of the C—C bond in cyanogen follows from

$$D(\mathrm{CN{-}CN}) = 2\Delta H_f^0(\mathrm{CN}) - \Delta H_f^0((\mathrm{CN})_2).$$

Using KNOWLTON and PROSEN's[5] value of (73.84 ± 0.43) kcal/mole for $\Delta H_f^0((\mathrm{CN})_2)$, and the above estimate of $\Delta H_f^0(\mathrm{CN})$, $D(\mathrm{CN{-}CN})$ is found to be 111.6 kcal/mole (4.85 ev). The dissociation energy of the C—H bond in HCN may be found in a similar manner to be 114.9 kcal/mole (5.0 ev). Detailed considerations of the other determinations of $D(\mathrm{CN{-}CN})$ and related matters such

[1] W. C. PRICE and A. D. WALSH: Trans. Faraday Soc. **41**, 381 (1945).
[2] C. A. McDOWELL and J. W. WARREN: Faraday Soc. Disc. **10**, 53 (1951).
[3] J. D. MORRISON and A. C. J. NICHOLSON: J. Chem. Phys. **20**, 1021 (1952).
[4] L. A. LONG: Proc. Roy. Soc. Lond., Ser. A **198**, 62 (1949).
[5] J. W. KNOWLTON and E. J. PROSEN: J. Res. Nat. Bur. Stand. **46**, 489 (1951).

as $EA(\mathrm{CN})$, which McDowell and Warren estimate as approximately 80 kcal per mole (3.5 ev) are given in their paper[1].

31. Methyl halides. Electron impact studies with these molecules have been carried out by a number of investigators[2]. We shall merely choose the recent work of Dibeler and Reese and of Cox for detailed discussion.

Considerable interest attaches to the accurate determination and interpretation of appearance potentials and it is possible to determine from them C—X bond dissociation energies, X being the halogen. Thus if the ions are formed with zero kinetic and excitation energy

$$A(\mathrm{CH_3^+}) = D(\mathrm{CH_3{-}X}) + I(\mathrm{CH_3}), \tag{31.1}$$

$$A(\mathrm{X^+}) = D(\mathrm{CH_3{-}X}) + I(\mathrm{X}), \tag{31.2}$$

it being assumed that the respective reactions are

$$\mathrm{CH_3X} + e \rightarrow \mathrm{CH_3^+} + \mathrm{X} + 2e, \tag{31.3}$$

$$\mathrm{CH_3X} + e \rightarrow \mathrm{CH_3} + \mathrm{X^+} + 2e. \tag{31.4}$$

To apply the relations (31.1) and (31.2) in practice it is necessary not only to verify that the ions are produced with negligible kinetic energy, but also that the $\mathrm{CH_3^+}$ ions do not arise from the reaction

$$\mathrm{CH_3X} + e \rightarrow \mathrm{CH_3^+} + \mathrm{X^-} + e \tag{31.5}$$

in which case (31.1) is replaced by

$$A(\mathrm{CH_3^+}) = D(\mathrm{CH_3{-}X}) + I(\mathrm{CH_3}) - EA(\mathrm{X}).$$

If the use of (31.1) and (31.2) is thoroughly justified, it is possible to determine both $D(\mathrm{CH_3{-}X})$ and $I(\mathrm{CH_3})$. A useful check is provided by the existence of independent determinations not only of $D(\mathrm{CH_3{-}X})$, but also of $I(\mathrm{CH_3})$ for which Hipple and Stevenson[3] (see Table 15) give (10.08 ± 0.1) ev.

Fig. 57 illustrates some of the results obtained for $\mathrm{CH_3F}$ by Dibeler and Reese[4] and by Cox. A striking feature is that no capture process of formation of $\mathrm{F^-}$ seems to occur. Comparison of the $\mathrm{F^-}$ and $\mathrm{CH_3^+}$ production curves provides strong support for the assumption that the reaction responsible is of the form (31.5) with an appearance potential of 12.39 volts (Cox). A break in the $\mathrm{CH_3^+}$ curve at 15.41 volts (Cox) may be ascribed to a reaction of the form (31.3) since there is no corresponding break in the $\mathrm{F^-}$ curve.

Results for $\mathrm{CH_3Cl}$ are illustrated in Fig. 58. It is clear that, as for $\mathrm{CH_3F}$, $\mathrm{CH_3^+}$ arises from reactions of both types (31.3) and (31.5) with appearance potentials of (13.63 ± 0.3) and (10.85 ± 0.2) volts respectively. The $\mathrm{CH_3^+}$ ion formed from reaction (31.3) was not found to have appreciable kinetic energy. In contrast to $\mathrm{F^-}$, $\mathrm{Cl^-}$ ions are formed by a capture process at low electron energies.

For $\mathrm{CH_3Br}$ and $\mathrm{CH_3I}$, on the other hand (Figs. 59 and 60 respectively), the negative ions do not appear to be produced with appreciable probability at any electron energy by a process of type (31.5), but are certainly produced quite

---

[1] C.A. McDowell and J.W. Warren: Trans. Faraday Soc. **48**, 1084 (1952).

[2] S.H. Bauer and T.R. Hogness: J. Chem. Phys. **3**, 687 (1933). — J.D. Morrison and A.C.J. Nicholson: J. Chem. Phys. **20**, 1021 (1952). — C.A. McDowell and B.C. Cox: J. Chem. Phys. **20**, 1496 (1952). — H. Branson and C. Smith: J. Amer. Chem. Soc. **75**, 4133 (1953). — B.C. Cox: Ph. D. thesis, University of Liverpool 1953. — F.P. Lossing, K.V. Ingold and I.H.S. Henderson: J. Chem. Phys. **22**, 1489 (1954). — V.H. Dibeler and R.M. Reese: J. Res. Nat. Bur. Stand. **54**, 127 (1955).

[3] J.A. Hipple and D.P. Stevenson: Phys. Rev. **63**, 121 (1943).

[4] V.H. Dibeler and R.M. Reese: J. Res. Nat. Bur. Stand. **54**, 127 (1955).

strongly by a low energy process. It is true that some of the subsidiary maxima in the curves of Figs. 59 and 60 may be due to processes in which positive ions are also formed, but it will be noted that they correspond to very weak reactions.

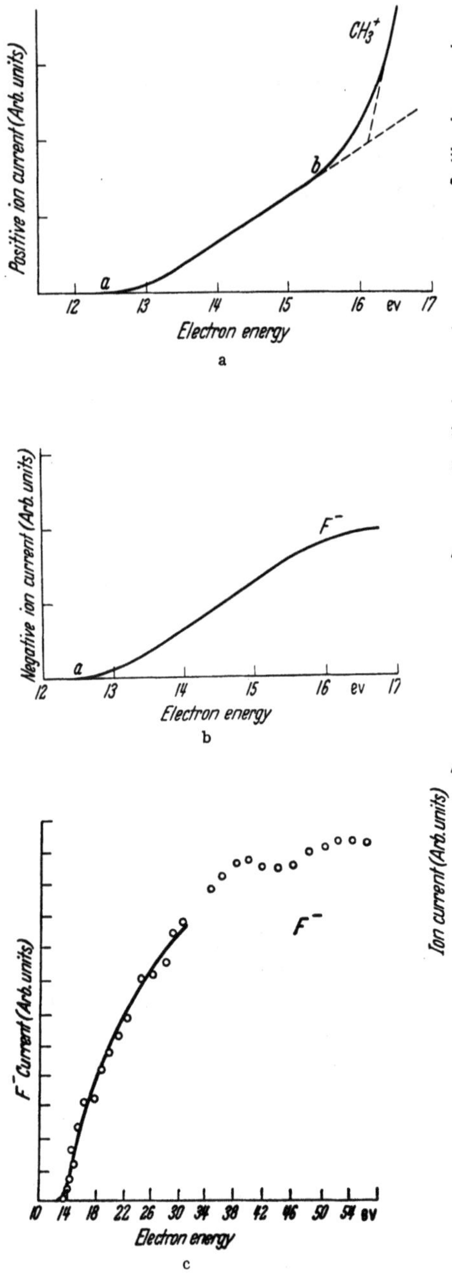

Fig. 57 a—c. Variation with electron energy of the currents of CH_3^+ and of F^- ions produced by electron impact in CH_3F. (a) and (b) observations by Cox. (c) observations by DIBELER and REESE.

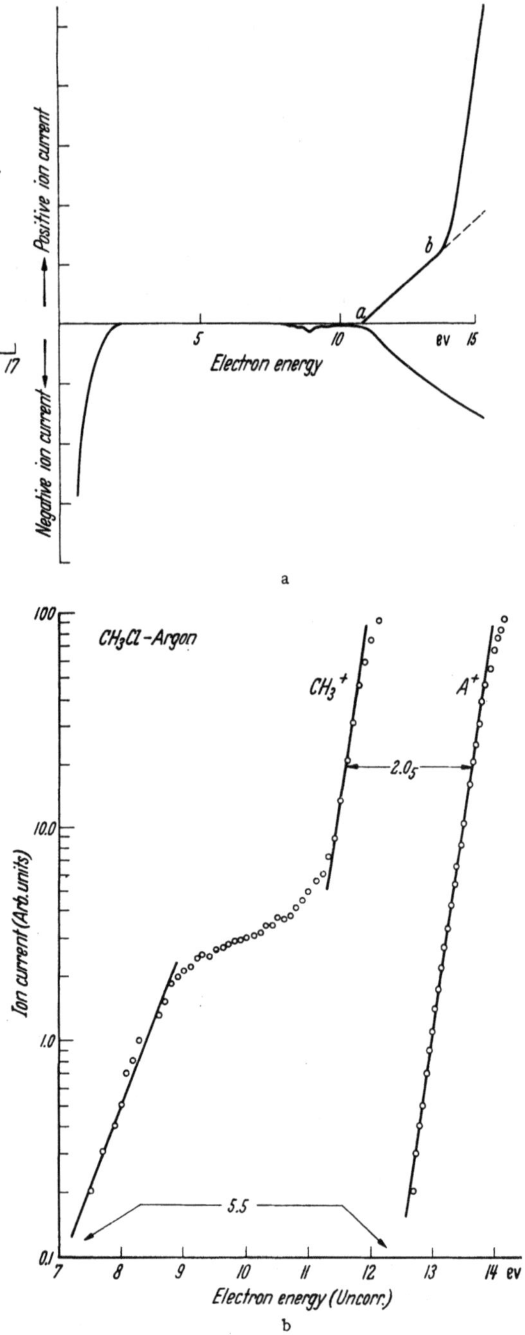

Fig. 58 a and b. (a) Variation with electron energy of the currents of CH_3^+ and of Cl^- ions produced by electron impact in CH_3Cl. (b) the initial part of the ionization function for CH_3^+ as observed by DIBELER and REESE.

Table 29 summarizes data obtained for $D(CH_3-X)$ from observed appearance potentials on the assumption of HIPPLE and STEVENSON's value (10.08 ± 0.1) ev for $I(CH_3)$.

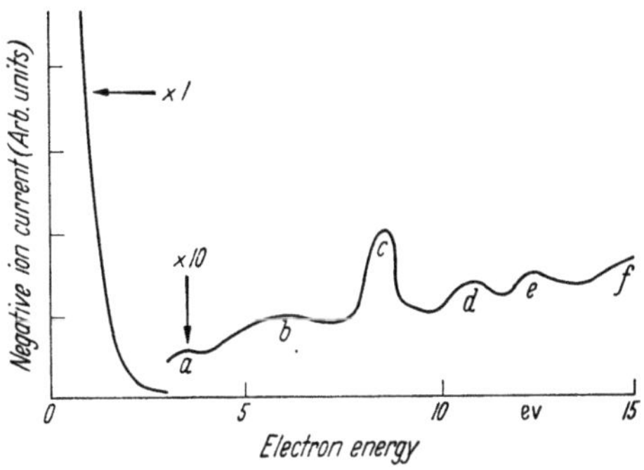

Fig. 59. Variation with electron energy of the current of Br^- ions produced by electron impact in CH_3Br.

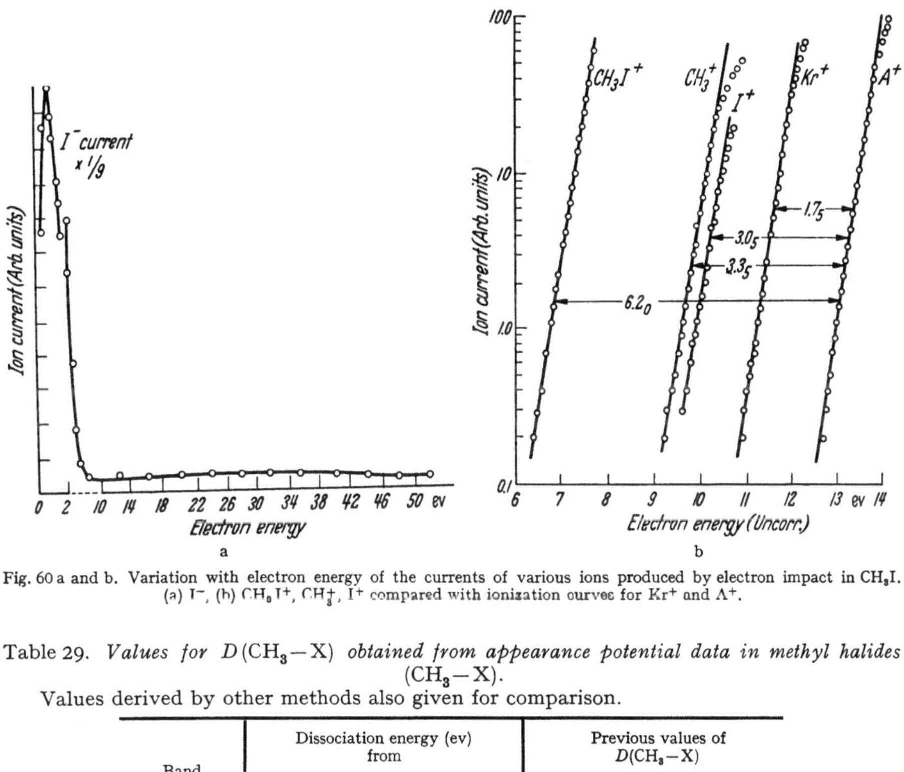

Fig. 60 a and b. Variation with electron energy of the currents of various ions produced by electron impact in CH_3I. (a) I^-, (b) CH_3I^+, CH_3^+, I^+ compared with ionization curves for Kr^+ and A^+.

Table 29. *Values for* $D(CH_3-X)$ *obtained from appearance potential data in methyl halides* (CH_3-X).
Values derived by other methods also given for comparison.

Band	Dissociation energy (ev) from		Previous values of $D(CH_3-X)$		
	$A(CH_3)$ a	$A(X^+)$ a	b	c	d
CH_3-F	4.8 ± 0.3	5.6 ± 0.5	4.6	—	—
CH_3-Cl	3.7 ± 0.2	3.5 ± 0.5	3.5	3.4	3.5
CH_3-Br	2.9 ± 0.1	3.0 ± 0.3	2.9	3.2	2.9
CH_3-I	2.4 ± 0.2	2.3 ± 0.2	2.2	2.3	2.4

(a) B.C. COX: Ph. D. thesis, University of Liverpool 1953.
(b) F.P. LOSSING: Unpublished.
(c) H. BRANSON and C. SMITH: J. Amer. Chem. Soc. **75**, 4133 (1953).
(d) J.S. ROBERTS and H.A. SKINNER: Trans. Faraday Soc. **45**, 229 (1949).

Fig. 61 illustrates the relative abundance of different ions formed from the methyl halides by impact of 50 volt electrons, the measurements being those of Cox[1]. It will be noted that the ion X^+ is formed in increasing relative abundance in proceeding from CH_3F to CH_3I. This is probably associated with a corresponding decrease in strength of the CH_3-X bonds (see Table 29). The relatively great abundance of CH_2F^+ is noteworthy, as also the behaviour of the CH_3^+ abundance.

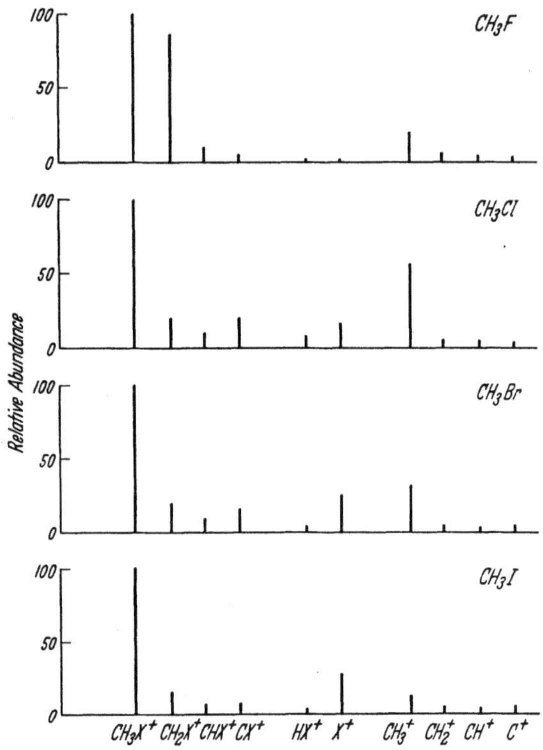

Fig. 61. Relative abundances of positive ions produced from the methyl halides by impact of 50 volt electrons.

32. Ethyl halides. Electron impact phenomena with these molecules have been studied by Cox[1].

The ionisation efficiency curves for $C_2H_5^+$ and F^- resemble those for CH_3^+ and F^- from CH_3F, no formation of F^- by capture being observed. The appearance potentials of the two reactions

$$C_2H_5F + e \rightarrow C_2H_5^+ + F^- + e,$$
$$C_2H_5F + e \rightarrow C_2H_5^+ + F + e,$$

were found to be 10.44 and 13.72 volts respectively. Assuming that in the second reaction the $C_2H_5^+$ ions are formed with zero kinetic energy and that $I(C_2H_5)$ is 8.67 ev[2] (see Table 15), we find $D(C_2H_5-F) = 5.02$ ev (115.9 kcal/mole). On the other hand, it is possible to compare $D(C_2H_5-F)$ with $D(C_2H_5-H)$ by comparing the appearance potentials for $C_2H_5^+$ ions produced by electron impact in C_2H_5F and C_2H_6 respectively. Assuming $D(C_2H_5-H) = 4.26$ ev (98.4 kcal/mole) we find $D(C_2H_5-F) = 5.26$ ev (121.5 kcal/mole). This is appreciably higher than the value obtained by the direct method and suggests that $I(C_2H_5)$ may be somewhat too high.

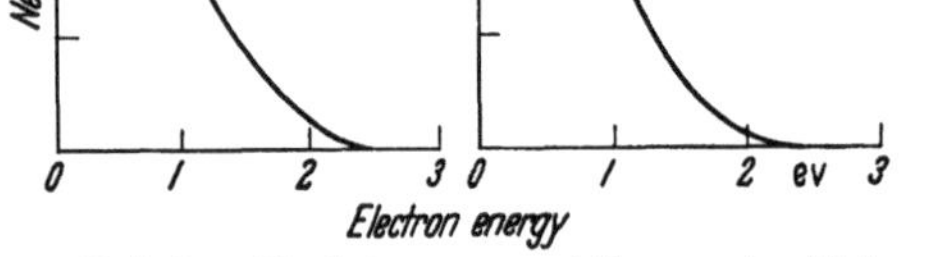

Fig. 62. Variation with electron energy of the currents of halogen negative ions produced by electron impact in ethyl halides.

It appears that for C_2H_5Cl, Cl^- ions are only produced by a resonance capture process at low electron energies (see Fig. 62) $C_2H_5^+$ ions appearing at a potential of 11.92 volts arising from

$$C_2H_5Cl + e \rightarrow C_2H_5^+ + Cl + 2e.$$

---

[1] B. C. Cox: Ph. D. thesis, University of Liverpool 1953.

[2] J. A. HIPPLE and D. P. STEVENSON: Phys. Rev. **63**, 121 (1943).

From the appearance potential $D(C_2H_5-Cl)$ is found to be (3.24 ± 0.10) ev $[(74.9 \pm 2.8)$ kcal/mole$]$ and (3.44 ± 0.10) ev $[(79.5 \pm 2)$ kcal/mole$]$ by the same direct and indirect methods respectively as used for CH_3Cl. The latter value agrees quite well with that $[(80.9 \pm 0.3)$ kcal/mole$]$ found by LANE, LINNETT and OSWIN[1] from the study of the equilibrium in the reaction

$$C_2H_4 + HCl \rightleftharpoons C_2H_5Cl.$$

This again suggests that $I(C_2H_5)$ is somewhat less than 8.67 ev.

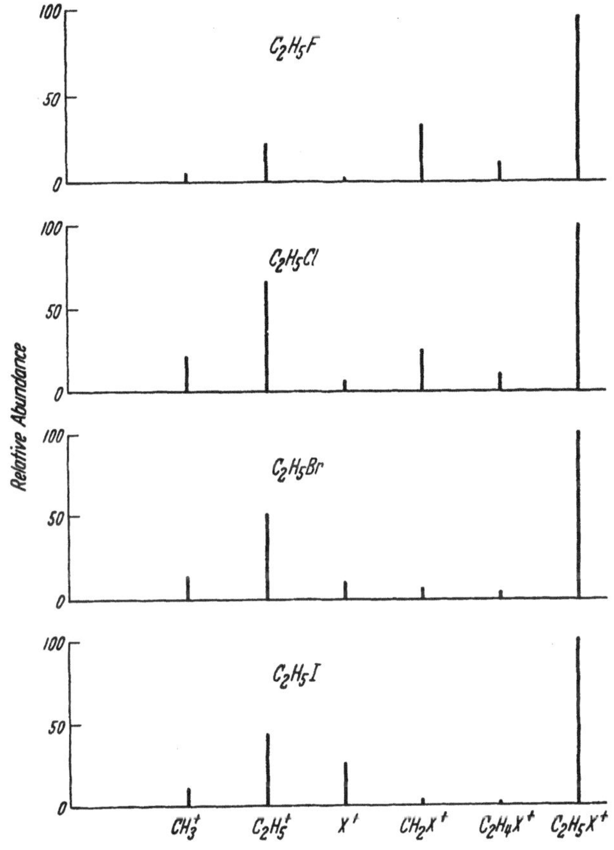

Fig. 63. Relative abundances of positive ions produced from the ethyl halides by impact of 50 volt electrons.

In both C_2H_5Br and C_2H_5I, the negative halogen ions are also formed only at low energies (see Fig. 62). The appearance potentials for $C_2H_5^+$ are found to be 11.25 and (10.67 ± 0.2) volts for the bromide and iodide respectively. From these $D(C_2H_5-Br)$ is found to be (62.2 ± 2.8) and (66.8 ± 2) kcal/mole by the direct and indirect methods respectively, the corresponding results for $D(C_2H_5-I)$ being (46.1 ± 2) and 50.7 kcal/mole.

The relative abundances of the different ions formed from the various molecules by impact of 50 volt electrons are illustrated in Fig. 63.

[1] M. R. LANE, J. W. LINNETT and H. G. O. OSWIN: Proc. Roy. Soc. Lond., Ser. A **216**, 361 (1953).

33. Carbon tetrahalides. The study of these molecules is of interest not only because of the information provided about the strengths of carbon-halogen bonds under different conditions, but also because of their high dielectric strengths.

Table 30 summarizes appearance potential and abundance data obtained in the most recent measurements for CCl_4, CCl_3F, CCl_2F_2, $CClF_3$, CF_4, CF_3Br and CF_3I. These present a sequence within which many significant comparisons may be made.

Table 30. *Appearance potentials and relative intensity of ions produced from carbon tetrahalides.*

CCl_4

Ion	CCl_3^+	CCl_2^+	CCl^+	Cl^+	C^+	Cl^-	CCl_2^{++}
Appearance potential (volts)	11.83 ±0.05	16.1 ±0.1	19.3 ±0.2	20.8 ±0.3	26.6 ±0.3		33.1 ±0.3
Relative intensity at 75 ev	100	20	22	14	3	40	4

CCl_3F

Ion	CCl_2F^+	CCl_3^+	$CClF^+$	CCl_2^+	CF^+	CCl^+
Appearance potential (volts)	12.27 ±0.05	13.8 ±0.2	17.0 ±0.1	18.9 ±0.2	19.1 ±0.2	21.2 ±0 2
Relative intensity at 75 ev	100	2	11	3.5	8	8

Ion	Cl^+	F^+	CCl_2F^{++}	Cl^-	F^-
Appearance potential (volts)	21,2 ±0.2	29.5 ±0.5	34.5 ±0.3		
Relative intensity at 75 ev	13	0.1	1.5	14	0.1

CCl_2F_2

Ion	CCl_2F^+	$CClF_2^+$	CCl_2F^+	CF_2^+	$CClF^+$	Cl^+
Appearance potential (volts)		12.55 ±0.05	15.8 ±0.2	19.2 ±0.5	20.2 ±0.2	20.6 ±0.2
Relative intensity at 75 ev	0.2	100	33	11	4	13

Ion	CF^+	CCl^+	F^+	$CClF_2^{++}$	CCl_2F^{++}	Cl^-	F^-
Appearance potential (volts)	21.7 ±0.3	25.7 ±0.2	31.9 ±0.5	36.3 ±0.5	36.5 ±0.5		
Relative intensity at 75 ev	12	5	3.0	—	—	4	0.2

$CClF_3$

Ion	$CClF_3^+$	CF_3^+	$CClF_2^+$	$CClF^+$	CF_2^+	Cl^+
Appearance potential (volts)	12.8 ±0.2	12.95 ±0.05	15.5 ±0.2	20.9 ±0.3	21.0 ±0.2	21.0 ±0.2
Relative intensity at 75 ev	0.7	100	30	0.6	10	10

Ion	CF^+	CCl^+	F^+	CF_2^{++}	$CClF^{++}$	Cl^-	F^-
Appearance potential (volts)	25.0 ±0.2	28.8 ±0.5	35.0 ±0.5	38.8 ±0.5	41.0 ±0.3	— —	— —
Relative intensity at 75 ev	4	1	0.4	2	1	0.04	0.1

Table 30 (Continued.)

CF_4

Ion	CF_3^+	CF_2^+	CF^+	F^+	C^+	CF^{++}	CF_3^{++}	F^-
Appearance potential (volts)	15.44 ± 0.05	22.4 ± 0.2	27.2 ± 0.5	39.8 ± 0.8		44.1 ± 0.3	45.5 ± 0.3	—
Relative intensity at 75 ev	100	7	3	1		1	1	0.1

CF_3Br

Ion	CF_3Br^+	CF_3^+	CF_2Br^+	CF_2^+	$CFBr^+$	CF^+	CBr^+
Appearance potential (volts)	11.82 ± 0.02	12.15 ± 0.02	15.0 ± 0.1	18.3 ± 0.1 21.4 ± 0.2	—	—	—
Relative intensity at 75 ev	36.9	100	31.5	6.5	0.34	0.07	0.96

Ion	Br^+	F^+	C^+	CF_2Br^{++}	Br^-	F^-
Appearance potential (volts)	(1) 12.2 ± 0.1 (2) 16.7 ± 0.1 (3) 21.0 ± 0.2	—	—	—	3.5 ± 0.2	(1) 0.3 ± 0.3 (2) 2.5 ± 0.3 (3) 4.4 ± 0.4 (4) 17.5 ± 0.5
Relative intensity at 75 ev	18.7	0.56	0.69	1.6	0.15	0.15

CF_3I

Ion	CF_3I^+	CF_3^+	CF_2I^+	CF_2^+	CFI^+	CF^+	CI^+
Appearance potential (volts)	10.64 ± 0.02	11.28 ± 0.02	14.58 ± 0.06	17.95 ± 0.1	—	19.9 ± 0.3	—
Relative intensity at 75 ev	100	47.6	18.5	1.72	0.18	1.85	0.35

Ion	I^+	F^+	C^+	CF_2I^{++}	FI^+	I^-	F^-
Appearance potential (volts)	(1) 10.6 ± 0.2 (2) 13.25 ± 0.07 (3) 16.4 ± 0.2	—	—	—	—	(1) 0 (2) 2.7 ± 0.2 (3) 14.1 ± 0.3	(1) 1.5 + 0.2 (2) 13.3 ± 0.2
Relative intensity et 75 ev	42.0	0.14	0.24	0.46	0.67	0.55	0.13

Notes to table 30.

The data given for CCl_4, $CFCl_3$, CF_2Cl_2, CF_3Cl and CF_4 are those of MARRIOTT and CRAGGS[1] and of CRAGGS and McDOWELL[2]. For CCl_4 the agreement with the work of BAKER and TATE is good with the exception of $A(C^+)$ for which the latter investigators give (23.5 ± 0.3) ev. This agreement applies even to abundance ratios which are difficult to measure accurately. The only appreciable differences are for Cl^+, C^+ and CCl^{++} for which BAKER and TATE[3] obtained the relative abundances 27.6%, 14.4% and 1.6% respectively. For CCl_2F_2 the agreement with BAKER and TATE's results is not quite so good. Thus the latter give $A(CCl^+)$ as (17 ± 0.5) ev less than $A(CClF^+)$ which they give as (18.1 ± 0.2) compared with (20.2 ± 0.2) ev

[1] J. MARRIOTT and J. D. CRAGGS: E.R.A. Report No. L/T 301 (1953).
[2] J. D. CRAGGS and C. A. McDOWELL: Rep. Progr. Phys. **18**, 374 (1955).
[3] R. F. BAKER and J. T. TATE: Phys. Rev. **53**, 683 (1938).

in Table 30. BAKER and TATE also find a much greater abundance of F^- so that they consider CCl_2F^+ ions to be formed primarily from

$$CCl_2F_2 + e \rightarrow CCl_2F^+ + F^- + e.$$

The disagreement extends also to the abundances of CCl_2F^+, Cl^+, C^+ and Cl^{++} for which BAKER and TATE find 8%, 6.4%, 18.5% and 9%.

Positive ion formation in CF_4 has also been investigated by DIBELER and MOHLER[1].

The data for CF_3Br and CF_3I are those of WARREN[2] and MARRIOTT[3].

For discussion of the data the following values (in ev) will be assumed:

$$I(\mathrm{F}) = 17.4,\ I(\mathrm{Cl}) = 12.96,\ I(\mathrm{Br}) = 11.84,\ I(\mathrm{I}) = 10.44,\ EA(\mathrm{F}) = 3.7,$$
$$EA(\mathrm{Cl}) = 3.7,\ EA(\mathrm{Br}) = 3.49 \pm 0.2,\ EA(\mathrm{I}) = 3.15 \pm 0.05,\ D(\mathrm{F}_2) = 1.55,$$
$$D(\mathrm{Cl}_2) = 2.47.$$

The recent measurements by FARMER et al.[4] (see Sect. 24) of the ionisation potentials of the radicals CF_3 and CCl_3 (10.10 ± 0.5 and 8.78 ± 0.05 volts respectively) are also very useful.

α) The yield of the undissociated ion. For all molecules CCl_nF_{4-n} the undissociated ion $CCl_nF^+_{4-n}$ is formed very rarely. It is particularly improbable from CCl_4 and CF_4. On the other hand for CF_3Br and CF_3I it is formed with comparatively high probability and for the latter case it is the ion formed most abundantly by 75 volt electrons.

β) The dissociation energies $D(CCl_3-X)$ and $D(CF_3-X)$. If it is possible to decide the nature of the reactions which give rise to CCl_3^+ or CF_3^+ ions, information may be obtained about the corresponding dissociation energies $D(CCl_3-X)$ and $D(CF_3-X)$.

It seems quite definite from a comparison of the abundances of the negative halogen ion and of the appropriate CX_3^+ ion that the appearance potential of CX_3^+ corresponds to

$$CX_3Y + e \rightarrow CX_3^+ + Y, \tag{33.1}$$

and not to

$$CX_3Y + e \rightarrow CX_3^+ + Y^-. \tag{33.2}$$

The only uncertain case is that of CCl_4 for which the abundance of Cl^- is comparable with that of CCl_3^+ at energies near the threshold. However, if one assumes that in all cases (33.1) applies and that the ions are formed with negligible kinetic energy, we obtain the values of $D(CCl_3-X)$ and $D(CF_3-X)$ given in Table 31. $I(CF_3)$ and $I(CCl_3)$ have been taken as given by FARMER et al[4].

Table 31. *Values of the dissociation energies $D(CCl_3-X)$, $D(CF_3-X)$.*

	ev	kcal/mole		ev	kcal/mole
$D(CCl_3-F)$	5.2 ± 0.3	120 ± 7	$D(CF_3-F)$	5.25 ± 0.2	121 ± 4
$D(CCl_3-Cl)$	3.0 ± 0.1	69.6 ± 3	$D(CF_3-Cl)$	2.9 ± 0.1	67.5 ± 3
$D(CCl_3-Br)$	2.1 ± 0.1	49.5 ± 3	$D(CF_3-Br)$	2.0 ± 0.2	46 ± 4
			$D(CF_3-I)$	1.2 ± 0.2	28 ± 4

[1] V. H. DIBELER and F. L. MOHLER: J. Res. Nat. Bur. Stand. **40**, 25 (1948).
[2] J. W. WARREN: Ph. D. thesis, University of London 1950.
[3] J. MARRIOTT: Ph. D. thesis, University of Liverpool 1954.
[4] J. B. FARMER, I. H. S. HENDERSON, F. P. LOSSING and D. G. H. MARSDEN: J. Chem. Phys. **24**, 348 (1956).

These values form quite reasonable sequences for both $CCl_3{-}X$ and $CF_3{-}X$ bonds. Furthermore, since the energies of dissociation of CF_4 and of CCl_4 into atoms are 19.9[1] and 12.2 ev[2] respectively, the average energy for the C—F and C—Cl bonds are 4.97 and 3.05 ev respectively. These are consistent with the above values of $D(CF_3{-}F)$ and $D(CCl_3{-}Cl)$.

γ) Appearance potentials of positive ions common to various molecules. For purposes of comparison, Table 32 gives the appearance potentials of positive ions common to the various molecules.

Table 32. *Appearance potentials of positive ions common to various molecules.*

Ion	Parent molecule	Appearance potential (volts)	Ion	Parent molecule	Appearance potential (volts)
CCl_3^+	CCl_4	11.83 ± 0.05	$CCl\,F^+$	CCl_3F	17.0 ± 0.1
CCl_3^+	CCl_3F	13.8 ± 0.2	$CCl\,F^+$	CCl_2F_2	20.1 ± 0.2
			$CCl\,F^+$	$CCl\,F_3$	20.9 ± 0.3
CCl_2^+	CCl_4	16.1 ± 0.1			
CCl_2^+	CCl_3F	18.9 ± 0.2	$CCl\,F_2^+$	CCl_2F_2	12.55 ± 0.05
CCl_2^+	CCl_2F_2		$CCl\,F_2^+$	$CCl\,F_3$	15.5 ± 0.05
CCl^+	CCl_4	19.3 ± 0.2	CF_3^+	CF_4	15.44 ± 0.05
CCl^+	CCl_3	21.2 ± 0.2	CF_3^+	$CCl\,F_3$	12.95 ± 0.05
CCl^+	CCl_2F_2	25.7 ± 0.2	CF_3^+	$CBr\,F_3$	12.15 ± 0.02
CCl^+	$CCl\,F_3$	28.8 ± 0.5	CF_3^+	$CI\,F_3$	11.28 ± 0.02
Cl^+	CCl_4	20.8 ± 0.3	CF_2^+	CF_3I	17.95 ± 0.1
Cl^+	CCl_3F	21.2 ± 0.2	CF_2^+	$CF_3\,Br$	18.3 ± 0.1
Cl^+	CCl_2F_2	20.6 ± 0.2	CF_2^+	CCl_2F_2	19.2 ± 0.5
Cl^+	$CCl\,F_3$	21.0 ± 0.2	CF_2^+	$CCl\,F_3$	21.0 ± 0.2
			CF_2^+	CF_4	22.3 ± 0.2
F^+	CCl_3F	29.5 ± 0.5			
F^+	CCl_2F_2	31.9 ± 0.5	CF^+	CCl_3F	19.1 ± 0.2
F^+	$CCl\,F_3$	35.0 ± 0.5	CF^+	CCl_2F_2	21.7 ± 0.3
F^+	CF_4	39.8 ± 0.8	CF^+	$CCl\,F_3$	25.0 ± 0.2
			CF^+	CF_4	27.2 ± 0.2
CCl_2F^+	CCl_3F	12.3 ± 0.05			
CCl_2F^+	CCl_2F_2	15.8 ± 0.2			

It is of interest to note that the appearance potential for Cl^+ is almost the same for all the molecules, implying that the removal and ionisation of a Cl atom does not involve the rest of the molecule. This is contrary to the case of F^+. For this ion, increasing amounts of energy are required for successive replacements of Cl by F in the molecules, a trend which appears to apply for all the other series of common ions.

The appearance potential of the most abundant ions show a gradual increase from CCl_4 to $CClF_3$ and then a superimposed sudden increase of about 2 ev in passing from $CClF_3$ to CF_4. This may be attributed to the difference in the appropriate C—Cl and C—F bond dissociation energies, provided the reaction involved is of the form (33.1) and not (33.2).

δ) Other regularities in the appearance potentials and abundances of positive ions. Many other regularities may be discussed. Thus if we consider the undissociated ions from CF_3X the appearance potentials increase regularly as X

[1] F. R. Bichowsky and F. D. Rossini: Thermochemistry of Chemical Substances. New York: Reinhold 1936.

[2] J. O. Craggs and J. W. Warren: Unpublished.

goes from Cl to I and the abundance ratio to CF_3^+ increases very rapidly from practically zero for CF_4 to 2.1 for CF_3Br.

Again, reference to Table 32 shows the increase in the appearance potential for CF_3^+ and for CF_2^+ ions in proceeding from CF_4 to CF_2I.

ε) C^+ ions from CCl_4 and CF_4. Since $D(CCl_4)=12.2$ ev and $I(C)=11.2$ ev, the appearance potential of normal C^+ ions of zero kinetic energy from the reaction

$$CCl_4 + e \rightarrow C^+ + 4\,Cl + e, \qquad (33.3)$$

should be 23.4 ev. This is not inconsistent with $A(C^+)$ in Table 30 if it is assumed that the kinetic energy released is 3.2 ev. It agrees even more closely with the value (26.6 ± 0.3) ev obtained for $A(C^+)$ by BAKER and TATE[1]. In either case it seems unlikely that the process responsible is, for example,

$$CCl_4 + e \rightarrow C^+ + 2\,Cl + Cl_2 + e, \qquad (33.4)$$

which would require only 20.9 ev, but this cannot be completely excluded.

For C^+ from CF_4 by the reaction analogous to (33.2) the minimum appearance potential is 31.1 ev whereas $A(C^+)$ according to DIBELER and MOHLER[2] is 27.5 ev. This would be consistent with

$$CF_4 + e \rightarrow C^+ + 3\,F + F^- \qquad (33.5)$$

or with

$$CF_4 + e \rightarrow C^+ + 2\,F_2 + 2e. \qquad (33.6)$$

The observations of MARRIOTT and CRAGGS[3] (see Fig. 65a) show that F^- is present at energies near the threshold for C^+ formation, but it is not possible to say whether it is sufficient to justify the interpretation (33.5). It is also noteworthy that the difference between $A(CF^+)$ and $A(C^+)$ is only 2.3 ev whereas the mean C—F bond energy in CF_4 is nearly 5 ev. This suggests that the CF^+ and C^+ ions are formed by essentially different processes, but the discrepancy might be ascribed to kinetic energy of the ions produced.

ζ) Cl^+ ions from CCl_4 and F^+ ions from CF_4. The minimum appearance energies of Cl^+ from the various reactions

$$CCl_4 + e \rightarrow C + Cl^+ + 3\,Cl + 2e, \quad 25.2 \text{ ev}, \qquad (33.7)$$
$$\rightarrow C + Cl^+ + Cl^- + 2\,Cl + e, \quad 21.5 \text{ ev}, \qquad (33.8)$$
$$\rightarrow C + Cl^+ + Cl_2 + Cl + 2e, \quad 22.7 \text{ ev}, \qquad (33.9)$$
$$\rightarrow C + Cl^+ + Cl^- + Cl_2 + e, \quad 19.0 \text{ ev}, \qquad (33.10)$$

are as indicated. It will be seen that the observed appearance potential is not consistent with (33.7) or (33.8), but may be with (33.9) or (33.10). The abundance of Cl^- ions is consistent with their formation together with Cl^+.

For F^+ from CF_4 the corresponding values are

$$CF_4 + e \rightarrow C + F^+ + 3\,F + 2e, \quad 37.3 \text{ ev}, \qquad (33.11)$$
$$\rightarrow C + F^+ + F^- + 2\,F + e, \quad 33.6 \text{ ev}, \qquad (33.12)$$
$$\rightarrow C + F^- + F_2 + F + 2e, \quad 35.75 \text{ ev}. \qquad (33.13)$$

The value of $A(F^+)$ given in Table 30 is as high as (39.8 ± 0.8) ev which suggests that (33.11) is the most probable reaction. On the other hand DIBELER and

[1] R. F. BAKER and J. T. TATE: Phys. Rev. **53**, 683 (1938).
[2] V. H. DIBELER and F. L. MOHLER: J. Res. Nat. Bur. Stand. **40**, 35 (1948).
[3] J. MARRIOTT and J. D. CRAGGS: E.R.A. Report No. L/T 301 (1953).

MOHLER find $A(F^+) = (35.5 \pm 1)$ ev which is not consistent with (33.11) and suggests either (33.12) or (33.13).

η) Br^+ from CF_3Br and I^+ from CF_3I. Fig. 64 illustrates the variation with electron energy of the current of Br^+ ions from CF_3Br. It will be seen that the curve shows two inflexions.

As the F^- current is very much less than that of the Br^+ at the energies concerned, the appearance potential at 16.7 volts may correspond to either of

$$CF_3Br + e \rightarrow CF_3 + Br^+ + 2e, \qquad (33.14)$$

$$CF_3Br + e \rightarrow CF_2 + F + Br^+ + 2e. \qquad (33.15)$$

If (33.14) is correct a kinetic energy of 2.1 ev must be involved, while if it is (33.15) $D(CF_2{-}F)$ must be ≤ 2.1 ev. As that is nearly 5 ev (33.14) must be taken as most probable.

The appearance potential at 21.0 volts may be ascribed either to

$$CF_3Br + e \rightarrow CF_2 + F + Br^+ + 2e. \qquad (33.16)$$

or to

$$CF_3Br + e \rightarrow CF_2 + F + Br^+ + 2e. \qquad (33.17)$$

With (33.16) no kinetic energy would be involved while with (33.17) it would be 2.4 ev.

A similar situation prevails with I^+ from CH_3I, but in this case the most probable reaction with the intermediate appearance potential of (13.25 ± 0.07) volts is

$$CF_3I + e \rightarrow CF_3 + I^+ + 2e. \qquad (33.18)$$

The highest appearance potential probably corresponds to

$$CF_3I + e \rightarrow CF_2 + F + I^+ + 2e. \qquad (33.19)$$

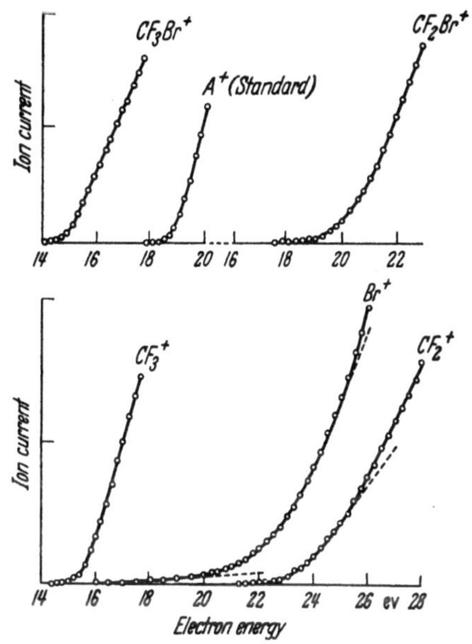

Fig. 64. Variation with electron energy, near the threshold, of the currents of various positive ions produced by electron impact in CF_3Br.

ϑ) Reactions involving negative ion formation. In Table 33 the ratio of the abundances of the different negative ions to the most abundant positive ion, for impact of 75 volt electrons, is given for the various molecules.

Table 33.

Molecule	Ion	Fraction of most abundant positive ion (indicated in brackets) for impact of 75 volt electrons		Molecule	Ion	Fraction of most abundant positive ion (indicated in brackets) for impact of 75 volt electrons	
CCl_4	Cl^-	0.4	(CCl_3^+)	CF_4	F^-	0.001	(CF_3^+)
CCl_3F	Cl^-	0.14	(CCl_2F^+)	$CBrF_3$	Br^-	0.0015	(CF_3^+)
	F^-	0.001			F^-	0.0015	
CCl_2F_2	Cl^-	0.04	$(CCl\,F_2^+)$	CIF_3	I^-	0.0055	(CF_3I^+)
	F^-	0.002			F^-	0.0013	
$CCl\,F_3$	Cl^-	0.0004	(CF_3^+)				
	F^-	0.001					

Fig. 65 illustrates the observed variation with electron energy of the yield of halogen negative ions from different tetrahalides.

The very low relative abundance of most of the negative ions for impact of 75 volt electrons is due to the fact that these ions are formed by resonance capture processes which involve slow electrons. On the other hand, the Cl^- ions formed with 75 volt electrons from CCl_4, CCl_3F and CCl_2F_2, probably result largely from ionic dissociation in which a positive ion is simultaneously produced. The identification of these processes is still far from certain (see p. 398). A further possible example is the reaction

$$CF_3Br + e \rightarrow CF_2^+ + F^- + Br + e \quad (33.20)$$

which may be associated with the rise in the F^- current from CF_3Br at (17.5 ± 0.5) ev (see Fig. 65).

Reference to Fig. 30 shows that Cl^- ions are also formed with high probability by impact of slow electrons in CCl_4. Fig. 65 illustrates the variation with electron energy of the production of the different negative halogen ions from CCl_3F, CF_4, CF_3Br and CF_3I.

The Br^- ions from CF_3Br are formed with unusually large probability at very low electron energy, presumably by

$$CF_3Br + e \rightarrow CF_3 + Br^- \quad (33.21)$$

for which the appearance energy is $T + X - 0.7$ ev. The second Br^- peak located at 3.5 ev is probably due to

$$\left.\begin{aligned} &CF_3Br + e \\ &\quad \rightarrow CF_2 + F + Br^-, \end{aligned}\right\} \quad (33.22)$$

which requires that

$$D(CF_2\text{—}F) \leq 4.2 \text{ ev}.$$

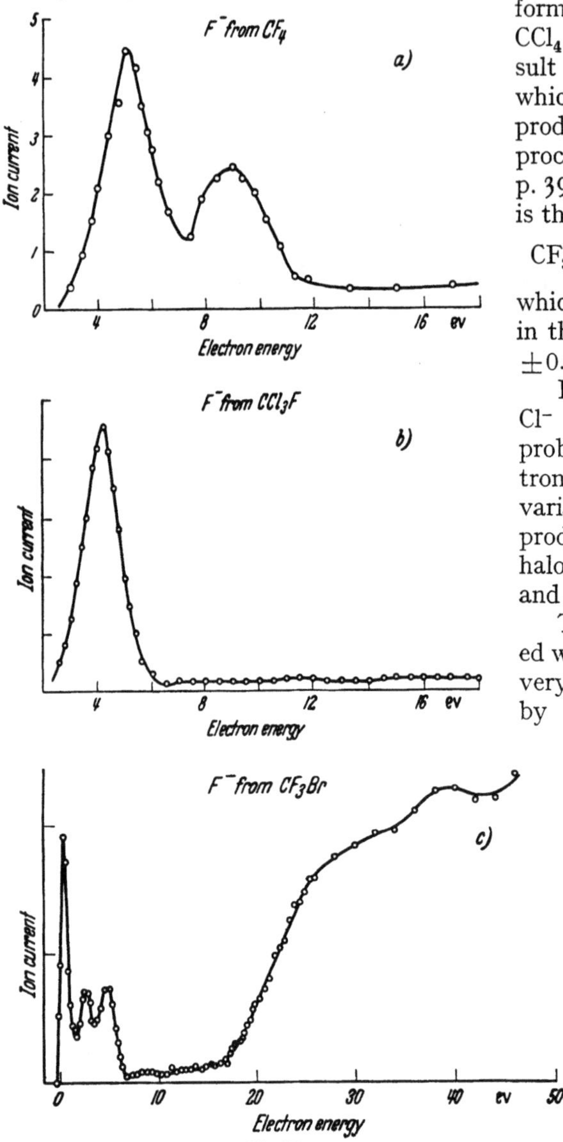

Fig. 65 a—c.

Fig. 65 a—f. Variation with electron energy of the currents of various halogen negative ions produced by electron impact in various carbon tetrahalides. (a) F^- from CF_4, (b) F^- from CCl_3F, (c) F^- from CF_3Br.

The production of Br^- ions at energies > 12 ev may well arise from secondary processes, due to slow electrons from reactions in which positive ions and neutral radicals are the primary products. Thus the rise in yield at 12 ev might be associated with the production of slow electrons from

$$CF_3Br + e \rightarrow CF_3Br^+ + 2e, \quad (33.23)$$

for which the appearance potentials is (11.82 ± 0.02) volts. These slow electrons could then produce Br^- through reaction (33.21).

Similar behaviour is found for I^- ions from CF_3I. Thus if the peak at 2.7 ev is ascribed to the reaction analogous to (33.22), $D(CF_2{-}F) \leqq 4.0$ ev, which is consistent with the value derived from Br^-.

The variation with electron energy of the F^- produced from CF_3Br is very complicated and the detailed shape is not well established. If the peak at the lowest electron energy is due to

$$\left.\begin{array}{l} CF_3Br + e \rightarrow \\ \qquad CF_2Br + F^- \end{array}\right\} \quad (33.24)$$

then $D(F{-}CF_2Br) \leq 3.9$ ev which does not seem unreasonable.

In contrast the F^- ions from CF_3I seem to result from a single capture process only with a peak at (1.5 ± 0.2) ev. If this is to be interpreted as

$$\left.\begin{array}{l} CF_3I + e \\ \qquad \rightarrow CF_2I + F^-, \end{array}\right\} \quad (33.25)$$

then

$$D(F{-}CF_2I) \leq 5.1 \text{ ev}.$$

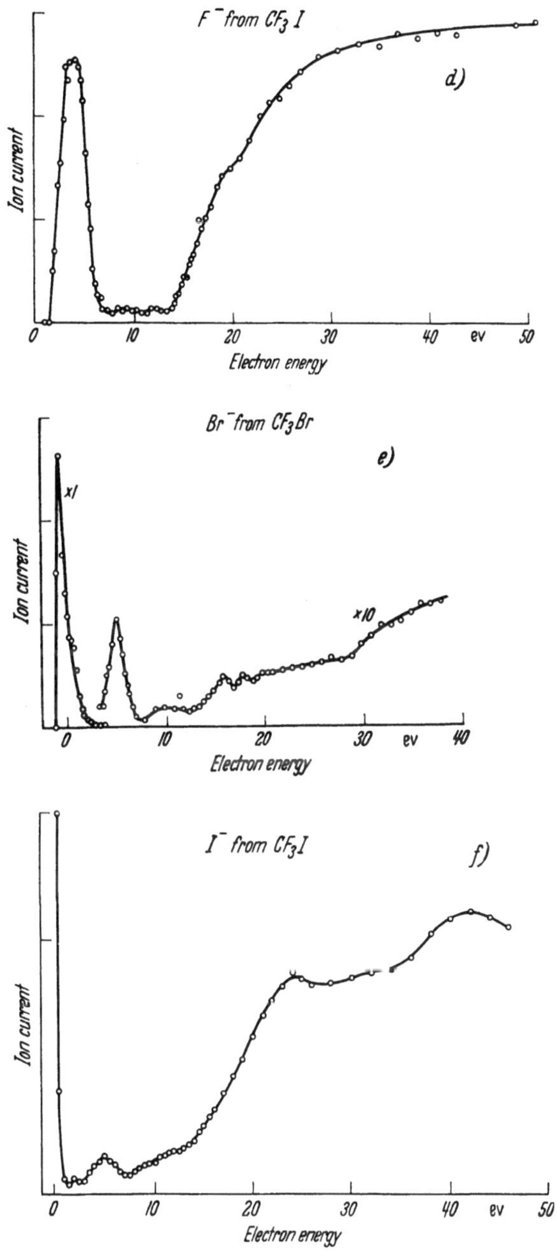

Fig. 65d—f. (d) F^- from CF_3I, (e) Br^- from CF_3Br, (f) I^- from CF_3I.

34. Silicon and germanium tetrahalides. The results obtained for these molecules are of interest for comparison with the corresponding results for the carbon tetrahalides, apart from new information which is provided about ionization potentials and dissociation energies of the various silicon and germanium halide radicals.

Results obtained for appearance potentials and relative intensity of ions produced by 75 ev electrons (50 ev for SiF_4) are given in Table 34 for SiF_4, $SiCl_4$ and $GeCl_4$.

SiF_4 has been investigated by Dibeler and Mohler[1] and they confined their attention to the positive ions only. Both positive and negative ion formation

[1] V. H. Dibeler and F. L. Mohler: J. Res. Nat. Bur. Stand. **40**, 25 (1948).

Table 34. *Appearance potentials and relative intensities of ions produced from SiF_4, $SiCl_4$ and $GeCl_4$.*

SiF_4

Ion	SiF_4^+	SiF_3^+	SiF_2^+	SiF^+	F^+	Si^+	SiF_3^{++}	SiF_2^{++}	SiF^{++}
Appearance potential (volts)	16.9 $\pm$0.5	16.1 $\pm$0.2	27.5 $\pm$0.5	—	—	—	—	—	—
Relative intensity at 50 ev	1.8	100	0.6	3.5	1.6	2.9	0.5	1.4	0.1

$SiCl_4$

Ion	$SiCl_4^+$	$SiCl_3^+$	$SiCl_2^+$	$SiCl^+$
Appearance potential (volts)	12.06 $\pm$ 0.02 (11.6 $\pm$ 0.2)	12.75 $\pm$ 0.05 (12.9 $\pm$ 0.2)	(1) 11.5 $\pm$ 0.1 (11.8 $\pm$ 0.3) (2) 18.15 $\pm$ 0.1 18.4 $\pm$ 0.3	(1) 14.3 $\pm$ 0.1 (2) 20.5 $\pm$ 0.1 (20.5 $\pm$ 0.3) (3) 21.5 $\pm$ 0.2
Relative intensity at 75 ev	63.5 (56)	100 (100)	4.8 (4.1)	11.7 (13)

Ion	Cl^+	Si^+	$SiCl_3^{++}$	$SiCl_2^{++}$
Appearence potential (volts)	(1) 14.7 $\pm$ 0.3 (2) 22.05 $\pm$ 0.1	25.5 $\pm$ 0.2 (27 $\pm$ 1)	32.6 $\pm$ 0.2 (33.8 $\pm$ 0.5)	36.2 $\pm$ 0.5 (37.3 $\pm$ 0.5)
Relative intensity at 75 ev	3.7 (5.5)	3.6 (14)	2.3 (4.9)	0.28 (0.65)

Ion	$SiCl^{++}$	Cl_2^+	$SiCl_2^-$	Cl^-
Appearance potential (volts)	(42 $\pm$ 1)		not observed (0.8 $\pm$ 0.03)	(1) 0 (1.4 $\pm$ 0.3) [1] (2) 2.1 $\pm$ 0.4 [1] (2.4 $\pm$ 0.3) [1] (3) 8.5 $\pm$ 0.4 [1] (8.2 $\pm$ 0.3) [1] (4) 10.1 $\pm$ 0.5 (5) 12 + 1 (6) 21.7 $\pm$ 0.2
Relative intensity at 75 ev	0.40 (1.2)	0.32	not observed (0.25)	0.26 (4 to 100)

$GeCl_4$

Ion	$GeCl_4^+$	$GeCl_3^+$	$GeCl_2^+$	$GeCl^+$	Ge^+	Cl^+
Appearance potential (volts)	11.90 $\pm$ 0.02	12,39 $\pm$ 0.02	(1) 11.0 $\pm$ 0.5 (2) 16.8 $\pm$ 0.3	(1) 12.1 $\pm$ 0.3 (2) 17.7 $\pm$ 0.2	(1) 16 $\pm$ 1 (2) 22.7 $\pm$ 0.2	(1) 18.9 $\pm$ 0.2 (2) 27.4 $\pm$ 0.3
Relative intensity at 75 ev	34.5	100	1.9	10.3	7.8	4.0

Ion	$GeCl_3^{++}$	Cl^-	$GeCl_2^-$	$GeCl_3^-$
Appearance potential (volts)	—	(1) 4.4 $\pm$ 0.1 (2) 11 $\pm$ 1	5.2 $\pm$ 0.2	—
Relative intensity at 75 ev	0.6	0.9 12 [2]	— 0 0.8 [2]	— 0 trace at low energies

[1] Potential at a peak in the production-energy relation.

[2] At peak of production.

in $SiCl_4$ has been studied thoroughly by VOUGHT[1] and by MARRIOTT[2]. They used similar apparatus except that the former used a permanent magnet in his mass spectrometer, so that voltage scanning was necessary. MARRIOTT, on the other hand, was able to use magnetic scanning. In Table 34 both sets of results are obtained, those of VOUGHT being enclosed in brackets. It will be seen that very good agreement is obtained for most ions. The most serious discrepancy is in the abundance of Cl^- ions and this is further discussed below. The results for $GeCl_4$ given in Table 34 are those obtained by MARRIOTT[2].

α) Comparison with results for CF_4 and CCl_4. First it will be noted by comparison of SiF_4 with CF_4 (Table 30) that, for both, the trifluoride ions SiF_3^+, CF_3^+ are by far the most abundant, while the ions SiF_4^+, CF_4^+ are only rarely produced. There seems to be appreciably more chance of forming SiF_4^+ than CF_4^+, however.

A similar comparison between SiF_4 and $SiCl_4$ shows that the trihalide ion is again the most abundant for both, but in $SiCl_4$ there is quite a high probability of producing $SiCl_4^+$. It is also of interest to notice that, for both, the ions $SiCl^+$, SiF^+ are more likely to be produced than $SiCl_2^+$, SiF_2^+ respectively.

Finally we may compare $GeCl_4$, $SiCl_4$ and CCl_4. For all these the trichloride positive ion is the most abundant. The relative abundance of the tetrachloride ion seems to behave in an anomalous fashion. In order of increasing probability at formation, we have $SiCl_4$, $GeCl_4$, CCl_4. On the other hand the ratio of the abundances of XCl_2^+ to XCl^+ (X=Cl, Si, Ge) decreases regularly on going from C to Ge, it being less than unity for all three.

A further point of interest is the fact that complex negative ions do not seem to be formed with CCl_4. There is some evidence for $SiCl_2^-$ while the production of $GeCl_2^-$ and possibly $GeCl_3^-$ seems to be quite definite.

β) Analysis of appearance potential data for $SiCl_4$ and $GeCl_4$. In the absence of any information about negative ion formation from SiF_4 we shall confine any analysis of appearance potential data to $SiCl_4$ and $GeCl_4$. The following will be assumed, in ev:

$$D(SiCl_4) = 15.1^{3}; \quad D(GeCl_4)^{3} = 13.7; \quad D(SiCl_2) = \tfrac{3}{5} D(SiCl_4)^{4} = 9.1;$$

$$D(GeCl_2) = \tfrac{3}{5} D(GeCl_4) = 8.2; \quad D(SiCl)^{5} = 3.3 \pm 0.5; \quad D(GeCl)^{6} = 3.1 \pm 0.5;$$

$$I(Si) = 8.15^{7}, \quad I(Ge) = 8.13^{7}; \quad D(Cl_2) = 2.47; \quad EA(Cl) = 3.72;$$

$$I(Cl)^{7} = 12.96.$$

We shall use X to denote either Si or Ge.

Cl_3^+. It appears from the low abundance of Cl^- ions at the appearance potentials of these ions that they are formed from

$$XCL_4 + e \rightarrow XCl_3^+ + Cl + 2e. \tag{34.1}$$

[1] R.M. VOUGHT: Phys. Rev. **71**, 93 (1947).

[2] J. MARRIOTT: Ph. D. thesis, University of Liverpool 1954.

[3] Selected Values of Chemical Thermodynamic Properties. Washington: Nat. Bur. Stand. Circ. No. 500 (1952).

[4] Y.P. PARTI and R. SAMUEL: Proc. Phys. Soc. Lond. **49**, 568 (1937).

[5] A.G. GAYDON: Dissociation Energies and Spectra of Diatomic Molecules. 2nd. ed. London: Chapman & Hall 1952.

[6] A.G. GAYDON and G. HERZBERG: Spectra of Diatomic Molecules. New York: Van Nostrand Co. Inc. 1950.

[7] G. HERZBERG: Atomic Spectra and Atomic Structure. New York: Dover Publications 1944.

It follows that

$$D(\mathrm{SiCl_3{-}Cl}) + I(\mathrm{SiCl_3}) \lesssim 12.75 \text{ ev},$$

$$D(\mathrm{GeCl_3{-}Cl}) + I(\mathrm{GeCl_3}) \lesssim 12.39 \text{ ev}.$$

Taking for $D(\mathrm{XCl_3{-}Cl})$ the mean bond energies 3.8 and 3.4 ev and assuming no excess energy $T+X$, $I(\mathrm{SiCl_3})$ and $I(\mathrm{GeCl_3})$ come out to be nearly 9 ev. As this is less than $I(\mathrm{Cl})$ STEVENSON's rule would suggest that $T+X$ is indeed small.

XCl_2^+. The first appearance potential may be due to ionisation of radicals $\mathrm{XCl_2}$ or $\mathrm{XCl_3}$ produced by thermal dissociation at the hot filament. On the other hand it could be interpreted without this assumption as arising from either of

$$\mathrm{XCl_4} + e \rightarrow \mathrm{XCl_2^+} + \mathrm{Cl_2} + 2e, \qquad (34.2)$$

$$\mathrm{XCl_4} + e \rightarrow \mathrm{XCl_2^+} + \mathrm{Cl} + \mathrm{Cl^-} + e. \qquad (34.3)$$

The evidence from the abundance of $\mathrm{Cl^-}$ near the appearance potential favours (34.2) but for $\mathrm{SiCl_4}$ this is far from definite. However, on this assumption $I(\mathrm{SiCl_2})$ and $I(\mathrm{GeCl_2})$ are both $\lesssim 8$ ev. Since $I(\mathrm{Cl_2}) = 11.8$ ev, STEVENSON's rule would suggest then that the equality sign applies.

At the higher appearance potential it is probable that the process occurring is

$$\mathrm{XCl_4} + e \rightarrow \mathrm{XCl_2^+} + 2\,\mathrm{Cl} + 2e. \qquad (34.4)$$

If this is so and $I(\mathrm{XCl_2}) \approx 8$ ev surplus energy $(T+X)$ of 3 ev or so would be involved.

XCl^+. Three processes may be involved here:

$$\mathrm{XCl_4} + e \rightarrow \mathrm{XCl^+} + \mathrm{Cl_2} + \mathrm{Cl^-} + e, \qquad (34.5)$$

$$\rightarrow \mathrm{XCl^+} + \mathrm{Cl_2} + \mathrm{Cl} + 2e, \qquad (34.6)$$

$$\rightarrow \mathrm{XCl^+} + 3\,\mathrm{Cl} + 2e. \qquad (34.7)$$

Identifying the appearance potentials for $\mathrm{SiCl^+}$ with these respective reactions we find $I(\mathrm{SiCl}) \lesssim 8.7$, 10.6 or 9.7 ev from each respectively. The abundance of $\mathrm{Cl^-}$ near the lowest appearance potential is much smaller than would be expected on these assumptions, but this may be instrumental. On the other hand, for $\mathrm{GeCl_4}$, the $\mathrm{Cl^-}$ abundance is comparable with that of $\mathrm{GeCl^+}$ at the lower appearance potential, from which $I(\mathrm{GeCl}) \lesssim (7.7 \pm 0.8)$ ev. The only other higher appearance potential observed must be ascribed to (34.6) or (34.7) giving $I(\mathrm{GeCl}) \lesssim$ 9.6 or 7.1 ev respectively.

X^+. The appearance potentials above 20 ev for each molecule presumably arise from similar reactions which may be either

$$\mathrm{XCl_4} + e \rightarrow \mathrm{X^+} + 2\,\mathrm{Cl_2} + 2e, \qquad (34.8)$$

$$\rightarrow \mathrm{X^+} + \mathrm{Cl_2} + 2\,\mathrm{Cl} + 2e, \qquad (34.9)$$

$$\rightarrow \mathrm{X^+} + 4\,\mathrm{Cl} + 2e. \qquad (34.10)$$

The third requires the least excess energy, $T+X$, of 2.3 ev for Si and 0.9 ev for Ge and is therefore the most likely.

For $\mathrm{Ge^+}$ there is a much lower appearance potential at 16 volts, which seems to have no analogue for $\mathrm{Si^+}$. From the apparent absence of $\mathrm{Cl^-}$ at 16 ev in $\mathrm{GeCl_4}$ it is probable that the process is of type (34.8), requiring $T+X = (-0.8 \pm 1)$ ev.

Cl^+. The identification of the processes giving rise to $\mathrm{Cl^+}$ is still very uncertain.

For $SiCl_4$ the lowest appearance potential may well correspond to

$$SiCl_4 + e \rightarrow SiCl_2 + Cl^+ + Cl^- + e \qquad (34.11)$$

which would require very little excess energy $T+X$. The higher potential (22.05 volts) might at first sight be attributed to production from HCl which results from hydrolysis of $SiCl_4$. Thus NIER and HANSON[1] find an appearance potential at (21.3 ± 0.5) volts for Cl^+ from HCl. However, there is no sign of the inflexion at 17.2 volts which would be expected on this interpretation. The most likely process arising from $SiCl_4$ is probably

$$SiCl_4 + e \rightarrow SiCl_2 + Cl^+ + Cl + 2e \qquad (34.12)$$

for which the excess energy would be about 1.8 ev.

Turning to $GeCl_4$, it seems that the most likely processes giving rise to appearance potentials at 18.9 and 27.4 volts respectively are

$$\left.\begin{aligned} GeCl_4 + e \rightarrow \\ GeCl_3 + Cl^+ + 2e, \end{aligned}\right\} \qquad (34.13)$$

and

$$\rightarrow Ge + 3\,Cl + Cl^+ + 2e. \qquad (34.14)$$

From the first it follows that $D(GeCl_3{-}Cl) \lesssim 5.9$ ev and for the second $T+X \approx 0.7$ ev, less than for other possible reactions.

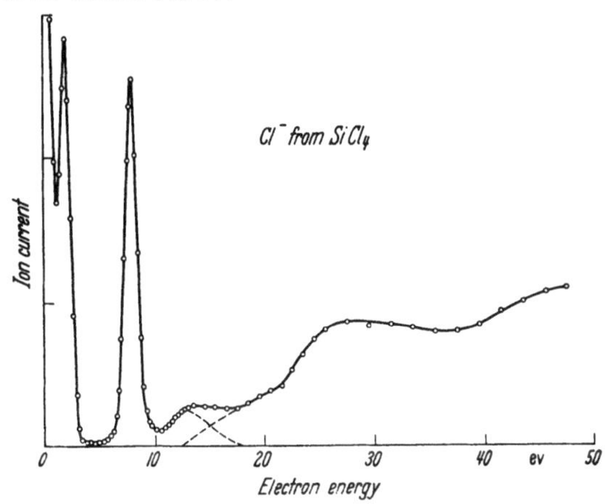

Fig. 66. Variation with electron energy of the currents of Cl^- ions produced by electron impact in $SiCl_4$ (as observed by MARRIOTT).

Cl^-. Marriott encountered difficulties in making his measurements on the Cl^- ion currents resulting from $SiCl_4$. It was found that, if negative ions were being collected, the electron beam would decrease with time, and, at low electron energies ($\leqq 10$ ev or so), would finally disappear. This was attributed to the formation of an electrically charged insulating layer on the slit facing the source filament. The effect could be temporarily removed by cleaning the slit. Despite these difficulties, some data were obtained and in Fig. 66 the ion yield as a function of electron energy is illustrated. VOUGHT[2] does not seem to have encountered the same difficulties and has made more extensive measurements illustrated in Fig. 67.

The yield of Cl^- from $SiCl_4$ at very low electron energies probably arises from

$$SiCl_4 + e \rightarrow SiCl_3 + Cl^- \qquad (34.15)$$

for which $T+X \approx 0$ if $D(SiCl_3{-}Cl)$ is taken as the mean bond energy, 3.8 ev.

At 2.1 ev the reaction which sets in may be

$$SiCl_4 + e \rightarrow SiCl_2 + Cl + Cl^- \qquad (34.16)$$

for which $T+X = (-0.2 \pm 0.4)$ ev, and at 6.4 ev

$$SiCl_4 + e \rightarrow SiCl + Cl_2 + Cl^- \qquad (34.17)$$

for which $T+X \approx 0.8$ ev.

[1] A.O. NIER and E.E. HANSON: Phys. Rev. **50**, 722 (1936).
[2] R.M. VOUGHT: Phys. Rev. **71**, 93 (1947).

A further appearance potential is suggested at 10.1 ev which may be ascribed to

$$SiCl_4 + e \rightarrow Si + Cl_2 + Cl + Cl^-, \tag{34.18}$$

with $T+X=1.2$ ev.

The steady production of Cl^- at electron energies beyond 10 ev presents some features of interest, in particular with respect to the apparent critical potentials at (12 ± 1) and (21.7 ± 0.2) volts. These may be due to secondary processes. Thus, just beyond 12 volts there is a considerable yield of very slow electrons from direct ionisation.

$$SiCl_4 + e \rightarrow SiCl_4^+ + 2e. \tag{34.19}$$

These slow electrons may then form Cl^- by the process (34.15). Similarly one could expect increased yields of very slow electrons just beyond the thresholds for $SiCl_3^+$ (12.76 ev) and $SiCl^+$ (21.5 ev) production. In the absence of data showing the variation of yield with electron current strength, this explanation can only be regarded as tentative.

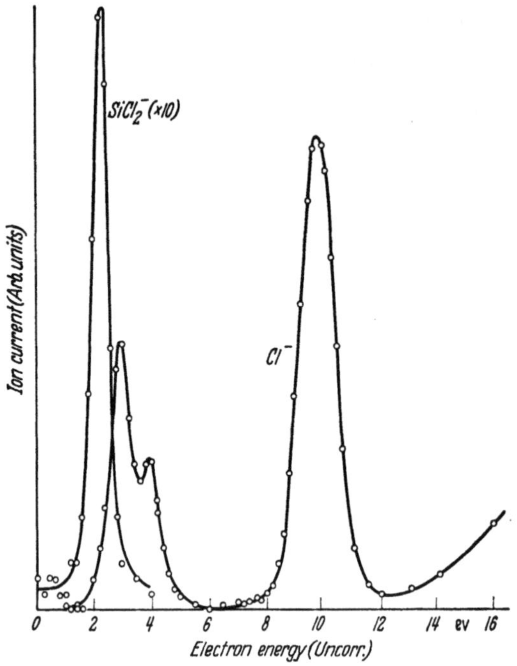

Fig. 67. Variation with electron energy of the currents of $SiCl_2^-$ and Cl^- ions produced by electron impact in $SiCl_4$ (as observed by VOUGHT).

Only one resonance capture peak was found in the production of Cl^- from $GeCl_4$. The process giving the least excess energy is

$$\left.\begin{array}{l} GeCl_4 + e \rightarrow \\ \qquad GeCl + Cl^- + Cl_2 \end{array}\right\} \tag{34.20}$$

for which $T+X \approx 0$. Steady formation of Cl^- at energies beyond 11 ev is observed, probably due to the reaction (see discussion of $GeCl^+$),

$$GeCl_4 + e \rightarrow GeCl^+ + Cl^- + Cl_2 + e. \tag{34.21}$$

XCl_2^-. Ions of this kind are certainly produced from $GeCl_4$ and probably also from $SiCl_4$. VOUGHT (see Fig. 67) observed the ions for very low electron energies and although they are not observed by MARRIOTT this may have been due to the difficulties he encountered when studying the negative ions.

The identification of the capture processes responsible for the production fo these ions is uncertain.

35. Titanium tetrachloride. The processes occurring in this molecule have been extensively studied both with a mass spectrometer and with the Lozier type of apparatus, so that rather more definite information is available about the nature of the reactions involved and the dissociation energies and ionisation potentials of the radicals concerned. The processes leading to negative ion formation were first studied by MARRIOTT, THORBURN and CRAGGS[1] and the further work carried

[1] J. MARRIOTT, R. THORBURN and J. D. CRAGGS: Proc. Phys. Soc. Lond. B 67, 437 (1954).

out by THORBURN[1]. Data on appearance potentials and relative abundance are given in Table 35.

Fig. 68 illustrates the variation of the negative and positive ion currents with electron energy as measured in the Lozier apparatus, the identification being obtained from complementary mass spectrograph measurements, which are illustrated in Fig. 69 for the positive ions.

Table 35. *Appearance potentials and relative abundances of ions produced from* $TiCl_4$.

Ion	Appearance potential (volts)	Relative abundance at 70 volts
$TiCl_4^+$	11.7 ± 0.2	58
$TiCl_3^+$	12.88 ± 0.5	100
$TiCl_2^+$	19.6 ± 0.5	5.6
Cl^+	26.0 ± 0.5	5
	27.2 ± 0.5	—
Cl^-	0 ± 0.4	—
	4.5 ± 0.4	
	13.2 ± 0.4	
$TiCl_3^-$	3.2 ± 0.3	—

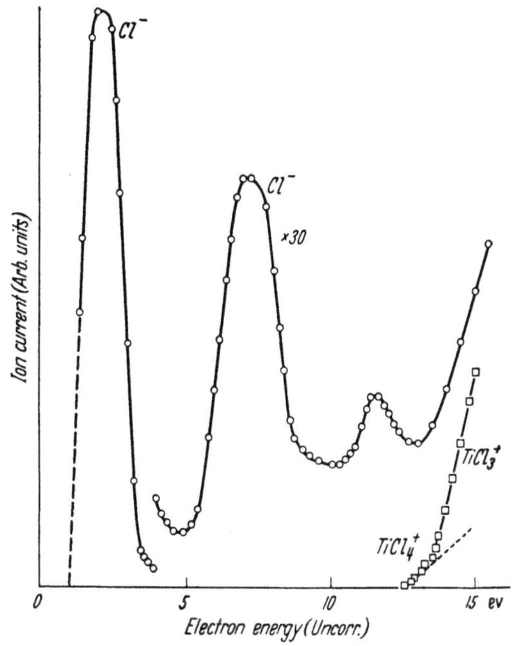

Fig. 68. Variation with electron energy of the predominant negative and positive ions produced by electron impact in $TiCl_4$, obtained with a Lozier type apparatus.

For the analysis of the data, the following will be assumed, in ev.:

$$D(TiCl_4) = 16.7^2,$$

$$D(TiCl_2) = \tfrac{3}{5} D(TiCl_4)^3 = 10,$$

$$EA(Cl) = 3.8,$$

$$I(Cl) = 13.0.$$

$TiCl_3^+$. Results obtained with the Lozier apparatus show that this ion results from the reaction

$$\left.\begin{array}{l} TiCl_4 + e \rightarrow \\ \qquad TiCl_3^+ + Cl^- + e \end{array}\right\} \quad (35.1)$$

as for CCl_4, but not $SiCl_4$ or $GeCl_4$. The kinetic energy of the $TiCl_3^+$ fragment was found to be $\lesssim 0.4$ ev and for $Cl^- < 0.6$ ev, so if $X=0$

$$12.5 \text{ ev} \leqq I(TiCl_4) \lesssim 12.9 \text{ ev}.$$

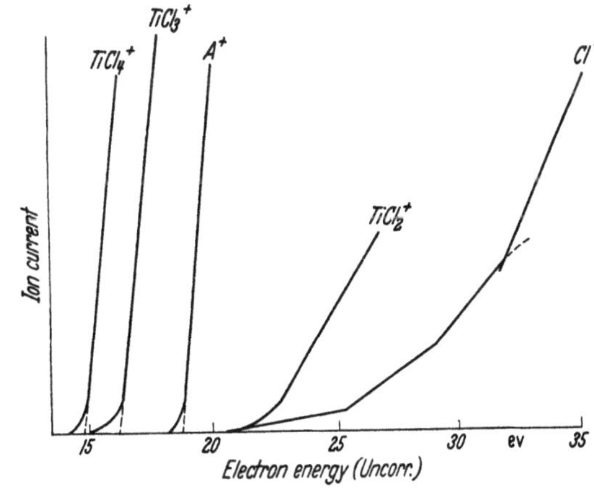

Fig. 69. Variation with electron energy, near the threshold, of the currents of various positive ions produced by electron mpact in $TiCl_4$.

[1] R. THORBURN and J.D. CRAGGS: To be published.
[2] F.R. BICHOWSKY and F.D. ROSSINI: Thermochemistry of Chemical Substances. New York: Reinhold 1936.
[3] Y.P. PARTI and R. SAMUEL: Proc. Phys. Soc. Lond. **49**, 568 (1937).

$TiCl_2^+$. This may be formed either by

$$TiCl_4 + e \to TiCl_2^+ + Cl_2 + 2e, \tag{35.2}$$

$$\to TiCl_2^+ + 2Cl + 2e. \tag{35.3}$$

From (35.2) $I(TiCl_2) \leq 15.3$ ev and from $(35.3) \leq 12.9$ ev.

$TiCl^+$. This ion seems to be only formed very rarely and no evidence of a corresponding appearance potential was obtained.

Cl^+. Of the four possible reactions likely to give rise to this ion, the most probable seem to be

$$TiCl_4 + e \to TiCl + Cl_2 + Cl^+ + 2e, \tag{35.4}$$

and

$$\to Ti + Cl_2 + Cl + Cl^+ + 2e \quad (35.5)$$

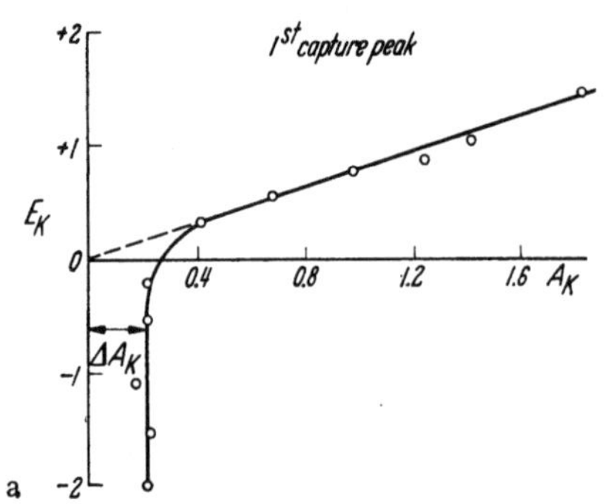

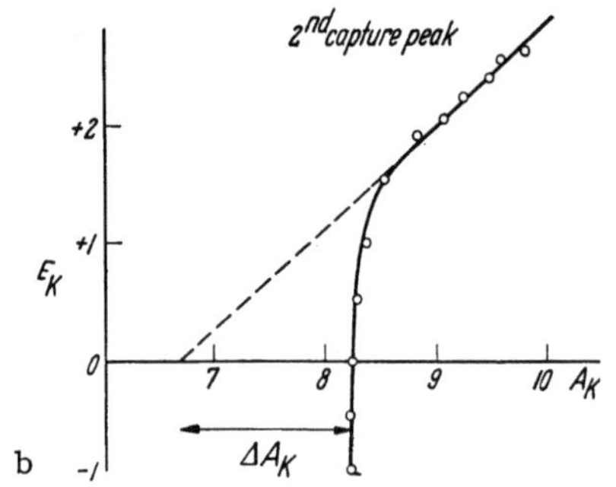

Fig. 70a and b. Illustrating the measurement of the appearance potentials for negative ions of zero kinetic energy resulting from electron capture by $TiCl_4$.

for which, if there is no excess energy, the appearance potentials would be 26.4 and 27.3 volts in good agreement with those observed.

Cl^-. The results obtained with the Lozier apparatus suggest that these ions are formed by three resonant capture processes and one of ionic dissociation. These may be identified tentatively as

$$TiCl_4 + e \to TiCl_3 + Cl^-, \tag{35.6}$$

$$\to TiCl_2 + Cl + Cl^-, \tag{35.7}$$

$$\left.\begin{array}{l} \to TiCl + Cl_2 + Cl^- \\ \text{or} \quad TiCl + 2Cl + Cl^-, \end{array}\right\} \tag{35.8}$$

$$\to TiCl_3^+ + Cl^- + e. \tag{35.9}$$

The appearance potentials for ions of zero kinetic energy have been measured for the first two of these processes (Fig. 70), identified with the two lowest energy capture peaks (see Fig. 68). The kinetic energies involved are found to be (0 ± 0.1) ev and (1.3 ± 0.3) ev respectively. It follows then from (35.6) that if the products are unexcited

$$D(TiCl_3 - Cl) = 3.8 \text{ ev}.$$

The appearance potential for (35.7) is given by

$$A(Cl^-) = D(TiCl_4) - D(TiCl_2) - EA(Cl) + (T + X).$$

Taking $A(Cl^-) = 4.6$ ev we find $T + X = (1.7 \pm 0.3)$ ev which is consistent with the measured value (1.3 ± 0.3) ev for T.

It is not easy to distinguish between the two possibilities (35.8) for the third peak. The first requires $D(TiCl_4) - D(TiCl) \leqq 13.6$ ev and the second that it is $\leqq 16$ ev.

For the ionic dissociation (34.9) we have

$$A(TiCl_3^+) = A(Cl^-) = D(TiCl_3{-}Cl) - EA(Cl) + I(TiCl_3) + T + X$$

giving

$$I(TiCl_3) \leqq 12.65 \text{ ev}.$$

This is consistent with the rather close limits imposed from the study of the $TiCl_3^+$ data.

36. Sulphur hexafluoride (SF_6). The study of the effects of electron collisions with this molecule is of some technical importance as sulphur hexafluoride has a dielectric strength appreciably higher than nitrogen and has been used in transformer insulation etc. Apart altogether from this special interest, the study of negative ion formation in SF_6 has led to some very remarkable results. Heavy negative ions, including in particular SF_6^- and SF_5^-, are formed by resonance capture of very slow electrons. HICKAM and FOX[1] have applied their retarding potential difference method to study these capture processes for electrons with energy less than 2 ev. With this method it is possible to use electrons with a very narrow energy distribution so that the sharpness of the resonance capture effects is not obscured. In particular, it was found that SF_6^- ions are formed by capture of electrons with energies lying within a range of less than 0.05 ev, about an energy less than 1.1 ev. The resonance peak is so sharp that FOX and HICKAM have suggested that it could be used to measure electron energy distributions.

The SF_6^- ions are formed in a single collision and must be metastable. Evidence in support of this has been obtained from observation of non-integral peaks in mass spectra of negative ions from SF_6 by AHEARN and HANNAY[2] and by MARRIOTT and CRAGGS[3].

These investigators have used conventional mass spectrometer methods to study appearance potentials and abundances for both negative and positive ions. The latter ions have also been studied in this way by DIBELER and MOHLER[4], while THORBURN and CRAGGS[5] have used a Lozier apparatus to study the yields of ions of both signs.

We shall begin by describing the experiments of HICKAM and FOX.

α) Resonance capture of very slow electrons to form SF_6^- and SF_5^-. The use of the retarding potential difference method for the study of the production of positive ions by electron impact at energies near the threshold has been described in Vol. XXXVI, p. 313. By a suitable electrode arrangement in the electron source the variation of the ionization probability with mean electron energy for electrons with a defined energy spread ΔE of about 0.1 ev may be obtained. Suppose that, between the source filament, at potential V_1 and the ionisation chamber at ground potential, there is interposed an electrode of potential V_2. Only those electrons which are emitted from the filament with energies greater than $V_2 - V_1$ can pass through to reach the ionisation chamber. Measurements are made of the ionisation produced for fixed V_1 and two values V_2 and $V_2 - \Delta V_2$ of V_2. The *difference* between these currents is that produced by electrons with

[1] W.M. HICKAM and R.E. FOX: J. Chem. Phys. **25**, 642 (1956).
[2] A.J. AHEARN and N.B. HANNAY: J. Chem. Phys. **21**, 119 (1953).
[3] J. MARRIOTT and J.D. CRAGGS: Brit. J. Electronics, Ser. IV **1**, 405 (1956).
[4] V.H. DIBELER and F.L. MOHLER: J. Res. Nat. Bur. Stand. **40**, 25 (1948).
[5] R. THORBURN and J.D. CRAGGS: To be published.

energies between eV_2 and $e(V_2 - \Delta V_2)$, which are emitted from the filament with energies between $e(V_1 - V_2)$ and $e(V_1 - V_2 + \Delta V_2)$. The variation of ionisation current with mean electron energy for electrons with an energy spread $e\Delta V_2$ can now be obtained by measuring the ion difference current as a function of V_2 with $V_1 - V_2$ and ΔV_2 fixed. ΔV_2 is usually about 0.1 volt.

An additional feature is usually introduced to eliminate the effect of penetration of the ion extraction field in producing energy inhomogeneity in the electron beam. The beam and the extraction voltage are pulsed so that electrons only reach the ionisation chamber when the extraction voltage is zero.

HICKAM and FOX applied this method to study negative ion formation in SF_6 by electrons with energies below 1.5 ev. In addition to a mass spectrometer

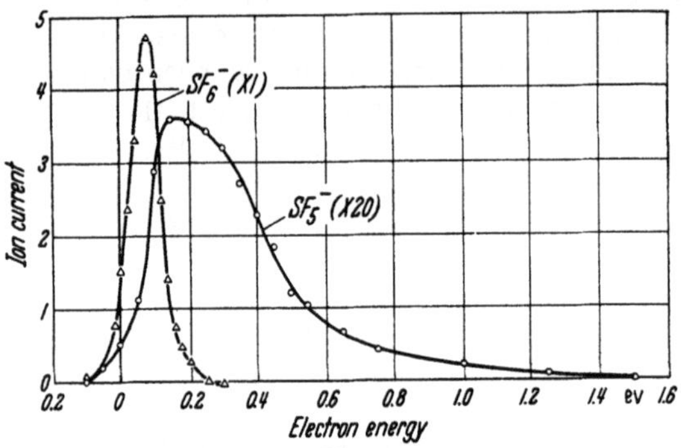

Fig. 71.

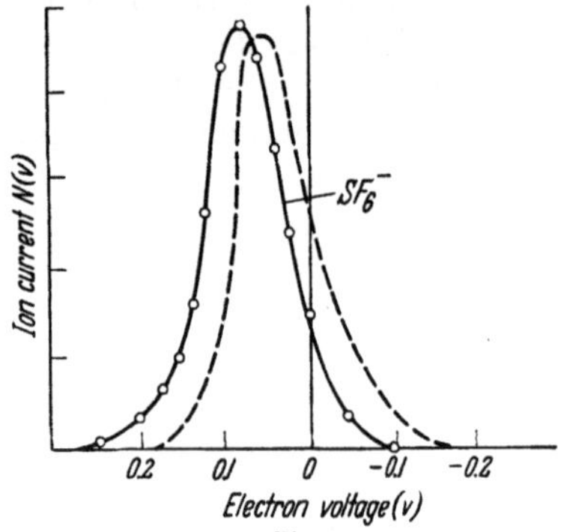

Fig. 72.

Fig. 71. Variation at low electron energies of the currents of SF_6^- and SF_5^- ions produced by electron impact in SF_6.

Fig. 72. Retarding potential analysis of the energy distribution of the electrons used in obtaining the data of Fig. 71 (dotted curve). The full line curve shows the observed variation of the current of SF_6^- ions (as in Fig. 71).

they also employed a Lozier type method. Although for both the difference method was employed, it was not necessary to pulse the beam with the Lozier type instrument.

Fig. 71 illustrates the variation with mean electron energy of the currents of SF_6^- and SF_5^- ions produced by electrons with an energy spread of 0.1 ev. In both the currents were proportional to electron beam intensity and SF_6 pressure.

The SF_6^- peak is very sharp. HICKAM and FOX compared it with the energy distribution function for the electrons producing the ionization. This they obtained by suitable retarding potential measurements. Fig. 72 shows the comparison. It will be seen that the two curves are very nearly of the same shape with the SF_6^- peak occurring at an energy about 0.03 ev above the peak of the energy distribution. This means that the capture process which gives rise to SF_6^- occurs only within an electron energy range considerably smaller than the energy spread of the electrons. Allowing for various experimental errors, it can be said that the production of SF_6^- occurs within an energy range less than 0.05 ev about a mean energy of less than 0.1 ev. Because of this, HICKAM and FOX suggested that the study of the apparent variation in the efficiency of production of SF_6^- by an electron beam could be used to measure the energy distribution of the electrons in the beam.

By using the Lozier type apparatus, HICKAM and FOX obtained evidence that the SF_5^- ions were formed with kinetic energy less than 0.1 ev. Good agreement was obtained between the variation with electron energy of the total ion current measured with the Lozier apparatus and that obtained by summing the SF_6^- and SF_5^- currents measured with the mass spectrometer.

If the SF_6^- capture process occurs over an energy range of 0.05 ev the mean cross section is at least 10^{-15} cm². This completely precludes the possibility that these ions result from radiative capture, for which the cross section would be about 10^{-5} times smaller and, in any case, would not be confined to such a narrow energy range. The process occurring must be one in which the electron is captured to a nearly stationary state of SF_6^-. This state has an energy slightly greater than that of normal SF_6, so that it will eventually break up again, ejecting the electron. In a complicated molecule such as SF_6 the surplus energy will be distributed over a great number of modes of motion. Breakup can only occur when this energy is concentrated in a suitable mode, before which a considerable time may elapse. It follows from the observation of SF_6^- in the experiments of HICKAM and FOX that the time must be greater than 10^{-6} sec. The situation is in some respects analogous to processes of capture of slow neutrons by nuclei.

β) Effects due to impacts of electrons of higher energy. Table 36 gives the appearance potentials and relative abundances (for impact of 70 ev electrons) of the different positive ions produced in SF_6. The bracketed values are those found by DIBELER and MOHLER[1], the others being those of MARRIOTT[2].

Table 36. *Appearance potentials and relative abundances of ions produced from* SF_6.

Ion	SF_5^+	SF_4^+	SF_3^+	SF_2^+	SF^+
Appearance potential (volts)	(1) 15.76 ± 0.05 (15.9 ± 0.2) (2) 17.4 ± 0.2	18.8 ± 0.15 (18.9 ± 0.2)	19.3 ± 0.2 (20.3 ± 0.3)	(1) 16 (2) 26.0 ± 0.5 (26.8 ± 0.5)	30.7 ± 0.5 (31.3 ± 0.3)
Relative abundance for 70 ev electrons	100 (100)	10 (9.6)	22 (29.6)	3.2 (5.6)	4.1 (9.8)

Ion	F^+	S^+	SF_5^{++}	SF_4^{++}	SF_3^{++}	SF_2^{++}	SF^{++}
Appearance potential (volts)	35 ± 2 (35.8 ± 1)	— (37.3 ± 1)	—	40 ± 1 (40.6 ± 0.5)	—	— (46.5 ± 0.5)	—
Relative abundance for 70 ev electrons	0.7 5.0	— (8.8)	2.4 —	5.8 9.8	1.3 1.1	2.4 6.0	— 0.7

The interpretation of these results is still very incomplete and uncertain. Thus the process giving rise to SF_5^+ might be either

$$SF_6 + e \to SF_5^+ + F + e, \tag{36.1}$$

or

$$\to SF_5^+ + F^-. \tag{36.2}$$

The second of these is favoured by AHEARN and HANNAY[3] and by LIU, MOE and DUNCAN[4] but MARRIOTT[2] has pointed out that near the threshold for SF_5^+ production the F^- current is only 1/100 of that due to SF_5^+. Even allowing for charge discrimination in collection this must be taken as more favourable to

[1] V.H. DIBELER and F.L. MOHLER: J. Res. Nat. Bur. Stand. **40**, 25 (1948).
[2] J. MARRIOTT: Ph. D. thesis, Liverpool University 1954.
[3] A.J. AHEARN and N.B. HANNAY: J. Chem. Phys. **21**, 119 (1953).
[4] T. LIU, G. MOE and A.B.F. DUNCAN: J. Chem. Phys. **19**, 71 (1951).

(36.1). If this is correct then

$$D(SF_5{-}F) + I(SF_5) \leqq 15.76 \text{ ev}.$$

The mean bond energy in SF_6 is 3.1 ev[1], so that $I(SF_5) \leqq 12.7$ ev.

The SF_5^+ ions which appear at 17.4 ev may possess an excitation energy of 1.6 ev.

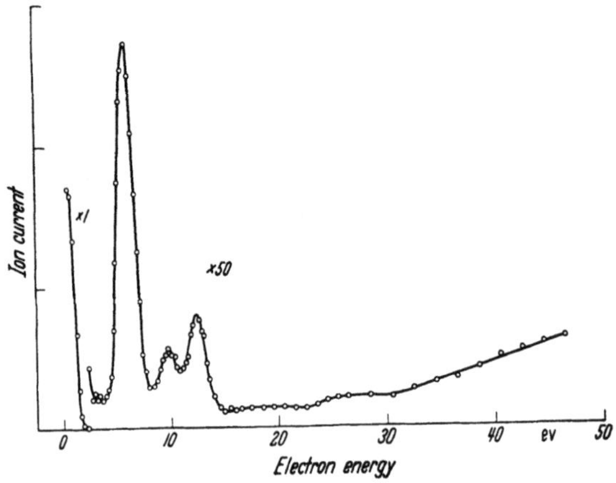

Fig. 73. Variation with electron energy of the current of F^- ions produced by electron impact in SF_6.

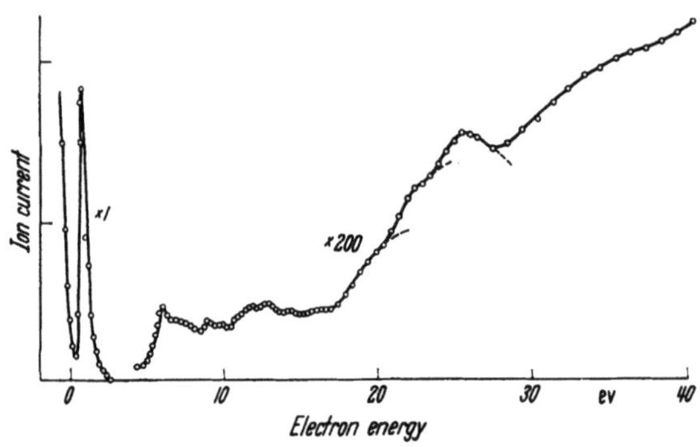

Fig. 74. Variation with electron energy of the current of SF_6^- ions produced by electron impact in SF_6.

The F^+ ions which appear at about 35 volts are probably associated with the F^- ions which appear at (32.5 ± 2) volts, the process of formation being

$$SF_6 + e \rightarrow S + 4F + F^+ + F^- + e, \quad (36.3)$$

with excess energy $T + X$ of about 26 ev.

The variation with electron energy of the F^- current is complex as will be seen reference to Fig 73. Four capture peaks occur. The voltage scale was obtained by comparison of the appearance potential for the second peak with that for the appearance of O^- from O_2 by electron capture. The latter was taken as 4.9 ev, but as discussed in Sect. 22, there is some uncertainty about this. However, assuming this to be correct, the appearance potentials for the four capture processes are near 9, 4.5 and 10.8 ev. The corresponding reactions may well be

$$\begin{aligned} SF_6 + e &\rightarrow SF_5 + F^-, \\ &\rightarrow SF_4 + F + F^-, \\ &\rightarrow SF_3 + 2F + F^-, \\ &\rightarrow SF_2 + 3F + F^-, \end{aligned}$$

which would give

$$D(SF_5{-}F) \leqq 3.6 \text{ ev} \quad (\text{the mean bond energy of } SF_6 = 3.1 \text{ ev}),$$

$$D(SF_5{-}F) + D(SF_4{-}F) \lesssim 8.1 \text{ ev},$$

$$D(SF_5{-}F) + D(SF_4{-}F) + D(SF_3{-}F) \leqq 11.7 \text{ ev},$$

$$D(SF_6) - D(SF_2) \lesssim 14.4 \text{ ev}, \quad \text{so}$$

$$D(SF_2) \lesssim 4.2 \text{ ev}.$$

At higher energies than about 15 ev the F^- production no longer exhibits resonance capture peaks and undoubtedly results from ionic dissociation of which (36.3) is one example.

[1] Selected Values of Chemical Thermodynamic Properties. Washington: Nat. Bur. Stand. Circ. No. 500 (1952).

Fig. 74 illustrates the variation of the SF_6^- current with electron energy up to 40 ev. It has a very complicated shape which of course cannot be interpreted in terms of ionic dissociation processes. At the lower energies it seems possible that direct capture into higher metastable states of the ion is occurring, but at higher energies there is little doubt that secondary effects are contributing. Thus AHEARN and HANNAY[1] found that the yield of SF_6^- ions by 35 ev electrons was proportional not to the SF_6 pressure p but to p^2 (Fig. 75). Presumably slow electrons resulting from some process in which positive ions are produced, are captured through the strong resonance which gives rise to the capture peak for very low energy electrons, investigated particularly by HICKAM and FOX (see Fig. 71).

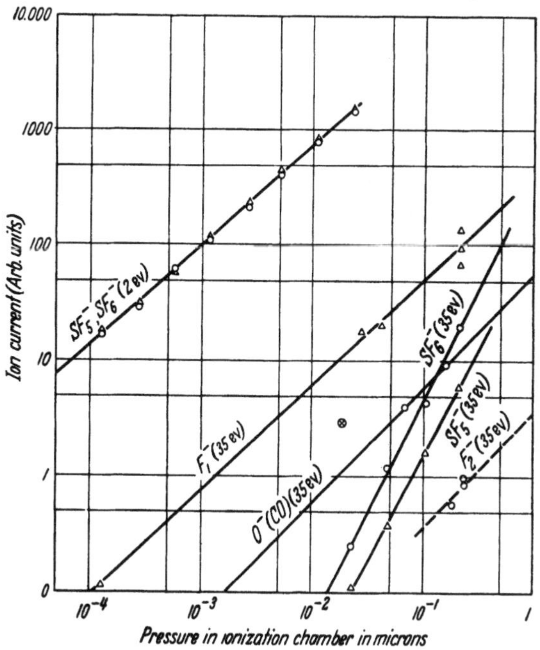

Fig. 75. Observed pressure dependence of negative ion currents produced by electron impact in SF_6 and in CO. Electron energies are indicated in brackets.

The variation of SF_5^- current with electron energy is also very complicated as may be seen by reference to Fig. 76. In this case, at the higher energies, ionic dissociation

$$SF_6 + e \rightarrow SF_5^- + F^+ + e$$

may contribute, but the existence of secondary processes at these energies has been shown in this case also by AHEARN and HANNAY (see Fig. 75). The data at the lower energies suggest that, just as for SF_6^-, capture into several relatively long-lived metastable states can occur.

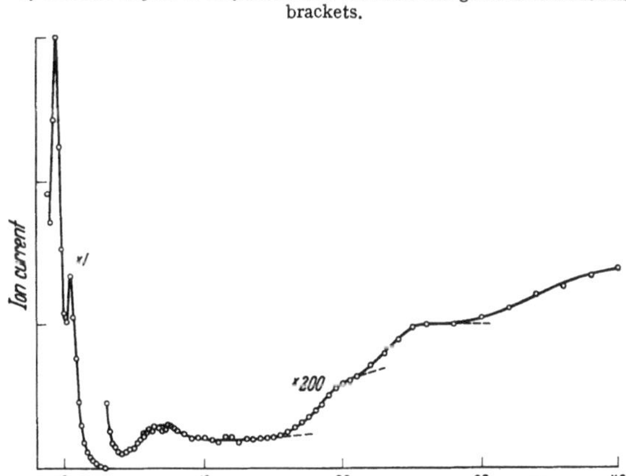

Fig. 76. Variation with electron energy of the currents of SF_5^- ions produced by electron impact in SF_6.

Acknowledgments.

We are indebted to Dr. L. BRANSCOMB for his assistance in reading the proofs and for the discussion of many points.

[1] A. J. AHEARN and N. B. HANNAY: J. Chem. Phys. **21**, 119 (1953).

Sachverzeichnis.

(Deutsch-Englisch.)

Bei gleicher Schreibweise in beiden Sprachen sind die Stichwörter nur einmal aufgeführt.

Subject Index.

(English-German.)

Where English and German spelling of a word is identical the German version is omitted.

MIX
Papier aus verantwortungsvollen Quellen
Paper from responsible sources
FSC® C105338

If you have any concerns about our products,
you can contact us on
ProductSafety@springernature.com

In case Publisher is established outside the EU,
the EU authorized representative is:
Springer Nature Customer Service Center GmbH
Europaplatz 3, 69115 Heidelberg, Germany

Printed by Libri Plureos GmbH
in Hamburg, Germany